Stress corrosion cracking

Related titles:

Gaseous hydrogen embrittlement of high performance metals in energy technologies
(ISBN 978-1-84569-677-1)
Hydrogen embrittlement is the process whereby various metals, particularly high strength steels, become brittle and crack following exposure to hydrogen at high temperatures. The process has increasingly been recognised as a key factor in fatigue and failure of components in the energy sector, including developing technologies using hydrogen as an alternative fuel source. This comprehensive reference summarises the wealth of recent research on how hydrogen embrittlement affects particular industries, its mechanisms and how it can be predicted to prevent component failure.

Thermal barrier coatings
(ISBN 978-1-84569-658-0)
Thermal barrier coatings are used to counteract the effects of high temperature corrosion and degradation of materials exposed to environments with high operating temperatures. The book covers both ceramic and metallic thermal barrier coatings as well as the latest advances in physical vapour deposition and plasma spraying techniques. Advances in nanostructured thermal barrier coatings are also discussed. The book reviews potential failure mechanisms in thermal barrier coatings as well as ways of testing performance and predicting service life. A final chapter reviews emerging materials, processes and technologies in the field.

Nanostructured metals and alloys
(ISBN 978-1-84569-670-2)
Nanostructured metals and alloys have enhanced tensile strength, fatigue strength and ductility and are suitable for use in applications where strength or strength-to-weight ratios are important. Part I of this important book reviews processing techniques for bulk nanostructured metals and alloys. Parts II and III discuss microstructure and mechanical properties, whilst Part IV outlines applications of this new class of material.

Details of these and other Woodhead Publishing materials books can be obtained by:

- visiting our web site at www.woodheadpublishing.com
- contacting Customer Services (e-mail: sales@woodheadpublishing.com; fax: +44 (0) 1223 832819; tel.: +44 (0) 1223 499140 ext. 130; address: Woodhead Publishing Limited, 80 High Street, Sawston, Cambridge CB22 3HJ, UK)
- contacting our US office (e-mail: usmarketing@woodheadpublishing.com; tel. (215) 928 9112; address: Woodhead Publishing, 1518 Walnut Street, Suite 1100, Philadelphia, PA 19102-3406, USA)

If you would like to receive information on forthcoming titles, please send your address details to: Francis Dodds (address, tel. and fax as above; e-mail: francis.dodds@woodheadpublishing.com). Please confirm which subject areas you are interested in.

Stress corrosion cracking

Theory and practice

Edited by
V. S. Raja and Tetsuo Shoji

WOODHEAD
PUBLISHING

Oxford   Cambridge   Philadelphia   New Delhi

Published by Woodhead Publishing Limited,
80 High Street, Sawston, Cambridge CB22 3HJ, UK
www.woodheadpublishing.com

Woodhead Publishing, 1518 Walnut Street, Suite 1100, Philadelphia, PA 19102-3406, USA

Woodhead Publishing India Private Limited, G-2, Vardaan House, 7/28 Ansari Road,
Daryaganj, New Delhi – 110002, India
www.woodheadpublishingindia.com

First published 2011, Woodhead Publishing Limited
© Woodhead Publishing Limited, 2011; Chapters 1, 2 and 18 © The Commonwealth of
Australia; Chapter 19 © Her Majesty the Queen in right of Canada, as represented by the
Minister of Natural Resources
The authors have asserted their moral rights.

British Library Cataloguing in Publication Data
A catalogue record for this book is available from the British Library.

Library of Congress Control Number: 2011934934

ISBN 978-1-84569-673-3 (print)
ISBN 978-0-85709-376-9 (online)

The publisher's policy is to use permanent paper from mills that operate a sustainable
forestry policy, and which has been manufactured from pulp which is processed using
acid-free and elemental chlorine-free practices. Furthermore, the publisher ensures that the
text paper and cover board used have met acceptable environmental accreditation standards.

Typeset by Replika Press Pvt Ltd, India
Printed by TJI Digital, Padstow, Cornwall, UK

Contents

S. P. LYNCH, Defence Science and Technology Organisation (DSTO), Australia

W. DIETZEL and P. BALA SRINIVASAN, Helmholtz-Zentrum Geesthacht, Germany and A. ATRENS, The University of Queensland, Australia

U. K. CHATTERJEE, Indian Institute of Technology Kharagpur, India and R. K. SINGH RAMAN, Monash University, Australia

M. Bobby Kannan, James Cook University, Australia and
P. K. Shukla, Southwest Research Institute, USA

H. Shaikh, T. Anita, A. Poonguzhali, R. K. Dayal and B. Raj,
Indira Gandhi Centre for Atomic Research, India

J. K. Lim, Chonbuk National University, South Korea

P. L. ANDRESEN and F. P. FORD (retired), GE Global Research, USA

S. P. LYNCH, Defence Science and Technology Organisation (DSTO), Australia

W. ZHENG, M. ELBOUJDAINI and R. W. REVIE, CANMET Materials Technology Laboratory, Canada

Contributor contact details

(* = main contact)

Editors

V. S. Raja
Department of Metallurgical
    Engineering and Materials
    Science
Indian Institute of Technology,
    Bombay
Mumbai 400 076
India

E-mail: vsraja@iitb.ac.in

T. Shoji
Fracture and Reliability Research
    Institute and New Industry
    Creation Hatchery Center
Tohoku University
6-6-10 Aramaki Aoba
Aoba-ku
Sendai 980-8579
Japan

E-mail: tshoji@rift.mech.tohoku.ac.jp

Chapters 1, 2 and 18

S. P. Lynch
Defence Science and Technology
    Organisation
PO Box 4331
Melbourne

Victoria 3001
Australia

E-mail: stan.lynch@dsto.defence.gov.au

Chapter 3

W. Dietzel* and P. Bala Srinivasan
Helmholtz-Zentrum Geesthacht
Institut für Werkstoffforschung
Max-Planck-Str. 1
D-21502 Geesthacht
Germany

E-mail: wolfgang.dietzel@hzg.de;
    pbalasrinivasan@gmail.com

A. Atrens
The University of Queensland
Division of Materials
Brisbane
Queensland 4072
Australia

E-mail: andrejs.atrens@uq.edu.au

Chapter 4

U. K. Chatterjee
Indian Institute of Technology
Kharagpur 721 302
India

E-mail: uday_chatterjee@yahoo.com

R. K. Singh Raman*
Department of Mechanical and
    Aerospace Engineering
Department of Chemical
    Engineering
Building 31
Monash University – Clayton
    Campus
Melbourne
Victoria 3800
Australia

E-mail: raman.singh@monash.edu

Chapter 5

V. Kain
Corrosion Science Section
Materials Science Division
Bhabha Atomic Research Centre
Mumbai 400085
India

E-mail: vivkain@barc.gov.in

Chapter 6

T. Shoji*, Z. Lu and Q. Peng
Fracture and Reliability Research
    Institute and New Industry
    Creation Hatchery Center
Tohoku University
6-6-10 Aramaki Aoba
Aoba-ku
Sendai 980-8579
Japan

E-mail: tshoji@rift.mech.tohoku.ac.jp;
    zhanpeng@rift.mech.tohoku.
    ac.jp;
    qpeng@rift.mech.tohoku.ac.jp

Chapter 7

R. B. Rebak
GE Global Research Center
1 Research Circle
Schenectady, NY 12309
USA

E-mail: rebak@ge.com

Chapter 8

M. Bobby Kannan*
Discipline of Chemical Engineering
School of Engineering and Physical
    Sciences
James Cook University
Townsville 4811
Australia

E-mail: bobby.mathan@jcu.edu.au

P. Bala Srinivasan
Helmholtz-Zentrum Geesthacht
Institut für Werkstoffforschung
Max-Planck-Str. 1
D-21502 Geesthacht
Germany

E-mail: pbalasrinivasan@gmail.com

V. S. Raja
Metallurgical Engineering and
    Materials Science
Indian Institute of Technology
Mumbai
India

E-mail: vsraja@iitb.ac.in

Chapter 9

A. Atrens*
The University of Queensland
Division of Materials
Brisbane
Queensland 4072
Australia

E-mail: andrejs.atrens@uq.edu.au

W. Dietzel and P. Bala Srinivasan
Helmholtz-Zentrum Geesthacht
Institut für Werkstoffforschung
Max-Planck-Str. 1
D-21502
Germany

E-mail: wolfgang.dietzel@hzg.de;
        pbalasrinivasan@gmail.com

N. Winzer
Fraunhofer Institute for Mechanics
   of Materials IWM
Woehlerstrasse 11
D-79198 Freiburg
Germany

E-mail: nickwinz@gmail.com

M. Bobby Kannan
Discipline of Chemical Engineering
School of Engineering and Physical
   Sciences
James Cook University
Townsville 4811
Australia

E-mail: bobby.mathan@jcu.edu.au

Chapter 10

I. Chattoraj
National Metallurgical Laboratory
Council of Scientific and Industrial
   Research (CSIR)
Jamshedpur 831007
India

E-mail: ichatt_62@yahoo.com

Chapter 11

M. Bobby Kannan
Discipline of Chemical Engineering
School of Engineering and Physical
   Sciences
James Cook University
Townsville 4811
Australia

E-mail: bobby.mathan@jcu.edu.au

P. K. Shukla*
Center for Nuclear Waste
   Regulatory Analyses
Southwest Research Institute
6220 Culebra Road
San Antonio, TX 78328
USA

E-mail: pshukla@cnwra.swri.edu

Chapter 12

H. Shaikh, T. Anita, A.
   Poonguzhali, R. K. Dayal and B.
   Raj*
Indira Gandhi Centre for Atomic
   Research
Kalpakkam 603102
Tamil Nadu
India

E-mail: dir@igcar.gov.in

Chapter 13

J. K. Lim
Department of Mechanical Design
Advanced Wind Power System
    Research Institute
Chonbuk National University
Jeonju 561-756
South Korea

E-mail: jklim@jbnu.ac.kr

Chapter 14

M. J. Esmacher
GE Water & Process Technologies
9669 Grogan's Mill Road
The Woodlands, TX 77380
USA

E-mail: mel.esmacher@ge.com

Chapter 15

M. Iannuzzi
GE Oil and Gas
Eyvind Lyches vei 10
PO Box 423
NO-1338 Sandvika
Norway

E-mail: mariano.iannuzzi@gmail.com

Chapter 16

R. J. H. Wanhill*
National Aerospace Laboratory
    NLR
P.O. Box 153
8300 AD Emmeloord
The Netherlands

E-mail: Russell.Wanhill@nlr.nl

R. T. Byrnes and C. L. Smith
Defence Science and Technology
    Organisation
PO Box 4331
Melbourne
Victoria 3001
Australia

E-mail: Rohan.Byrnes@dsto.defence.
    gov.au

Chapter 17

P. L Andresen*
GE Global Research Center
One Research Circle
Schenectady, NY 12309
USA

E-mail: andresen@crd.ge.com

F. P. Ford
19 Nott Road
Rexford, NY 12148
USA

E-mail: fpctford@aol.com

Chapter 19

W. Zheng
CANMET Materials Technology
    Laboratory
Natural Resources Canada
183 Longwood Road South
Hamilton, Ontario L8P 0A5
Canada

E-mail: Wenyue.Zheng@NRCan-RNCan.
    gc.ca

M. Elboujdaini
CANMET Materials Technology
    Laboratory
Natural Resources Canada
3303 – 33rd Street NW
Calgary, Alberta T2L 2A7
Canada

E-mail: Mimoun.Elboujdaini@NRCan-
    RNCan.gc.ca

R. W. Revie*
CANMET Materials Technology
    Laboratory
Natural Resources Canada
555 Booth Street, Room 126-E
Ottawa, Ontario K1A 0G1
Canada

E-mail: Winston.Revie@NRCan-RNCan.
    gc.ca

List of reviewers

The editors wish to acknowledge the help of a number of experts who kindly reviewed chapter drafts. Their comments and suggestions have added significantly to the chapters in this book. Their names are listed below.

Dr Vinod S. Agarwala, Naval Air Systems Command, USA

Dr Peter Andresen, GE Research and Development Center, USA

Prof. Jean-Yves Cavaille, University of Lyon, France

Mr Mel J. Esmacher, P.E., GE Water & Process Technologies, USA

Dr Vivekanand Kain, Bhabha Atomic Research Centre, India

Dr Russell D. Kane, iCorrosion LLC, USA

Dr Fraser King, Integrity Corrosion Consulting Ltd, Canada

Mr Mark Knop, Defence Science and Technology Organisation, Australia

Dr Stan Lynch, Defence Science and Technology Organisation, Australia

Prof. Dewidar Montasser, South Valley University, Egypt

Prof. V. S. Raja, Indian Institute of Technology Bombay, India

Dr Raul B. Rebak, GE Global Research, USA

Dr R. N. Singh, Bhabha Atomic Research Centre, India

Dr Narasi Sridhar, DNV Columbus, USA

Dr Zoran Sterjovski, Defence Science and Technology Organisation, Australia

Dr Alan Turnbull, National Physical Laboratory, UK

Dr R. J. H. Wanhill, National Aerospace Laboratory NLR, The Netherlands

Foreword

It is a pleasure to introduce this book, *Stress corrosion cracking: Theory and practice*. Stress corrosion cracking (SCC) is the dominating mechanical property in the reliability of most commercial equipment and applications. While the nominal mechanical properties seem to dominate design of equipment, in fact, in the longer term it is the interactions among the materials, their environments and their stresses that control the lifetime. Thus, to have such a text as *Stress corrosion cracking: Theory and practice* edited by Professors Raja and Shoji is an asset to those who design, manufacture, and utilize equipment regardless of the materials of construction.

The SCC phenomenon usually does not produce gross degradation such as rusting. In fact, little, if any, material is actually reacted. The exterior often retains its shiny metallic appearance. However, and unfortunately, the cracking phenomenon proceeds via narrow cracks with no local ductile deformation. Such cracks may proceed along grain boundaries or in the grains of materials – intergranular or transgranular.

Environments in which SCC occur vary with materials. Not all environments produce SCC in all materials, but all engineering materials sustain SCC in many common environments. Special environments are not required. For example, the quite tough Alloy 600, (78%Ni, 15%Cr, 7%Fe), sustains SCC in very pure deoxygenated water above about 250°C. Iodine gas from fissioning of uranium produces rapid SCC in zirconium alloys, which are used to clad nuclear fuels. Hydrogen gas produces rapid growth of SCC in high strength steels. The velocity of SCC in high strength steels increases by two orders of magnitude in HCl gas and three orders of magnitude in H_2S compared with the same pressure in dry hydrogen. The authors in this text describe such combinations of ordinary alloys in ordinary environments sustaining SCC.

SCC occurs broadly in high strength materials and in tough materials. It occurs in both ductile and brittle materials. SCC occurs in metals and ceramics. SCC occurs often at applied stresses as low as 10 percent of the annealed yield strength as in copper alloys. As a general matter it should be appreciated that the stresses that produce failure by SCC are almost always residual stresses and not the applied stresses as calculated by designers.

SCC initiates on surfaces which are absolutely smooth and requires no previous defects such as pits. SCC starts usually more readily at pre-existing defects such as pits, intergranular penetrations, and mechanical defects. When materials are cold worked as well as having high strength, the initiation and propagation of SCC is more rapid.

SCC occurs in pure metals such as pure copper and pure iron. SCC is often mistaken for fatigue although cyclic stressing accelerates SCC. SCC in a given type of machines is statistically distributed. In a given system, e.g. a steam generator in a pressurized water reactor, the occurrence of SCC may be a part of the statistical distribution but occurrences may be four orders of magnitude separated.

I have studied SCC for 55 years in a broad range of materials and environments. This book is the best available. This book, *Stress corrosion cracking: Theory and practice,* should be carefully studied by designers, operators, manufacturers, materials engineers and scientists.

Roger Washburne Staehle
North Oaks
Minnesota

Preface

Metals are known for their ductility and strength, making them the largest group of construction materials across all industries. However, ductility loss and cracking at low stresses on account of exposure to chemical environments is a great concern. This phenomenon was observed as early as 1873 by William H. Johnson in a simple laboratory experiment, when he exposed steel to acids. However, its direct impact on engineering components came as a surprise to engineers and public alike in early 19th century when several boiler explosions caused the loss of a large number of human lives. The phenomena occurred across several industrial components on a number of alloys. At that time, no general principles had been established so it was christened by several names such as caustic embrittlement, season cracking, chloride embrittlement, delayed failure, plating embrittlement – just to name a few. Due to its industrial importance, the problem received significant support for research and has been widely investigated. As the understanding of the underlying principles of the so-called brittle failures associated with chemical environment became better, such failures are now termed as stress corrosion cracking, caused by a host of chemical species (in aqueous and non-aqueous) under the influence of tensile stresses.

The subject of stress corrosion cracking (SCC) may be old, but it remains academically and industrially a very important topic. There is an ever growing need to operate plants with least failures, as these failures have adverse affect on, among other things, safety, environment, and global resources, which are of great concern to humanity. Among all the forms of failures affecting the integrity of plants, exposed to chemical environments, SCC is important in the sense that stressed components predominantly suffer by this failure. This is because of the fact that synergy exists between stress and environment to cause cracking of materials (components) well below their yield strength. So, to offer long-term reliable plant service not only do we need to predict and monitor SCC but we also need to develop materials and design components that can promote longevity.

In spite of its importance the subject of SCC has been dealt with only by a few conference proceedings and books. Therefore we feel the present publication will be very useful to the SCC community as well as industrial personnel. Further, the present publication is different from the previous publications in the sense that it attempts to cover all aspects of SCC – mechanisms, test methods, materials and industrial problems. Chapters have been written by academicians, researchers and engineers so as to give a holistic picture regarding SCC research and SCC problems. The book has 19 chapters, divided into four sections to deal with mechanisms, testing, materials and industrial problems written by 32 experts. It is obviously a difficult task to cover all aspects within a limited space, though the experts have covered most of the subject, and we hope this book helps academics, research and industry in their endeavor to solve SCC problems.

We received overwhelming support to bring out this book. The book had the privilege of over 40 experts who contributed as authors and or reviewers. Dr Stan Lynch who has contributed three chapters, took keen interest in general, and provided us with invaluable input all through the preparation of this book. His ideas were most useful. The authors and reviewers (listed on pages xiii-xix) shared valuable expertise and time. Professor Roger W. Staehle wrote a foreword to this book. We also received strong support from several staff of Woodhead Publishing. We appreciate all of them for their support in publication of this book.

V.S. Raja
Tetsuo Shoji

Part I

Fundamental aspects of stress corrosion cracking (SCC) and hydrogen embrittlement

Mechanistic and fractographic aspects of stress-corrosion cracking (SCC)

S. P. LYNCH, Defence Science and Technology Organisation
(DSTO), Australia

Abstract: Basic aspects of stress-corrosion cracking (SCC) in metallic materials are outlined, followed by a summary of the numerous mechanisms that have been proposed for SCC. The characteristics of transgranular and intergranular SCC in model systems, e.g. pure metal and single-phase alloy single crystals and bi-crystals under testing conditions that facilitate discrimination between mechanisms, are then described. The applicability of the various proposed mechanisms, such as those based on dissolution, hydrogen embrittlement, film-induced cleavage, and adsorption, are discussed in detail for these systems. Mechanisms of SCC in complex commercial alloys are then considered in the light of these studies on model systems.

Key words: stress-corrosion cracking, cleavage-like cracking, intergranular cracking, mechanisms, fractography, adsorption-induced cracking, dissolution, hydrogen embrittlement.

1.1 Introduction

Stress-corrosion cracking (SCC) is the generally accepted term for describing sub-critical cracking of materials under sustained loads (residual or applied) in most liquid and some gaseous environments. Sub-critical cracking of materials in gaseous hydrogen or hydrogen sulphide, and cracking due to internal hydrogen resulting from pre-exposure of materials to hydrogen-bearing environments, are considered to be forms of hydrogen embrittlement (HE) rather than SCC. However, SCC in some materials can involve generation and ingress of hydrogen at crack tips, and characteristics and mechanisms of SCC and HE have a lot in common. Sub-critical cracking in liquid-metal environments is also considered to be a separate phenomenon, usually called liquid-metal embrittlement (LME), but also has a number of similarities to SCC. An understanding of the mechanisms of HE and LME, which are not as complex as SCC, is therefore valuable in understanding SCC, and chapters on the fundamentals of HE (Chapter 2) and LME (Chapter 18) should be consulted in this regard.

SCC occurs in a wide range of materials/environments at rates varying from $\sim 10^{-2}$ m/s to $< 10^{-11}$ m/s (< 0.3 mm/yr) – with extremely low rates obviously significant in regard to the integrity of structures with projected lifetimes of 50 years or more. There are extensive databases regarding susceptible material:environment combinations [1–4], but failures involving SCC continue to occur, sometimes with catastrophic consequences. Many SCC failures occur because 'old' SCC-susceptible materials are present in ageing structures and components where it has not been economical to replace them with more recently developed SCC-resistant materials. In other cases, SCC failures occur because the detrimental environmental conditions have not been predicted, e.g. in crevices where impurities can concentrate due to evaporation/re-wetting cycles, or because testing conditions for determining SCC resistance have not been representative of service conditions.

Applied stress levels may also have been underestimated or residual stresses not considered. Transient conditions during start-up or shut-down of equipment, where environmental conditions, and stresses/strain-rates are often different from those during normal operation, are also not always taken into account during material selection and design. In other cases, the SCC resistance of welds (and associated heat-affected zones) may not have been fully considered, especially if welding of structures in practice has been carried out under different conditions than those used for the test specimens. Environmental conditions envisaged at the design stage can also be changed during service, e.g. to increase operational efficiencies or to slow down general corrosion, without fully considering the implications for SCC resistance.

A good example of the failure to take into account some of the above considerations led to SCC of an austenitic steel lever-arm-pin (worth about €10) in a military jet engine, leading to a series of events that caused the crash of the aircraft (worth about €10 million) (see Chapter 16). The particular alloy may not have been in any SCC databases, but it could (arguably) have been predicted that a (hot) concentrated aqueous chloride environment would have developed in crevices, that residual stresses would have been present, and that most austenitic alloys would be susceptible to SCC under such conditions. A better fundamental understanding of SCC by those who select materials and design structures and components would probably help prevent many failures.

In this chapter, various proposed processes and mechanisms of SCC are outlined, and then the applicability of these mechanisms for particular materials and environments are discussed. The present review of SCC takes a somewhat different approach from previous reviews on the topic by focusing on understanding SCC in model systems, e.g. pure metal and single-phase alloy single crystals and bi-crystals, in order to provide a sounder basis for understanding SCC in complex commercial alloys. Fractographic aspects of SCC are also emphasised more than in previous reviews. Before discussing SCC mechanisms, some basic aspects of SCC are summarised. More detailed coverage of these basic aspects and electrochemical/thermodynamical fundamentals can be found in previous reviews, conference proceedings, and books [3–23].

1.2   Quantitative measures of stress-corrosion cracking (SCC)

SCC can occur at remarkably low stresses in tensile specimens for some materials and environments, as illustrated by time-to-failure versus applied-stress data where threshold stresses for SCC can be as low as 5% of the yield stress. For pre-cracked specimens, threshold stress-intensity factors (K_{th}), can also often be only ~5% of the K value for fast fracture (K_{Ic}). Plots of crack velocity, v, versus K often show two regimes – region-I, just above K_{ISCC}, where the crack velocity increases rapidly with increasing K, and region-II where there is little or no dependence of crack velocity on K (termed the plateau velocity) (Fig. 1.1). In a few cases, several plateau velocities are observed, and there is sometimes a third region where crack velocity increases rapidly with increasing K just below K_{Ic}.

Data from slow-strain-rate testing of smooth or notched tensile specimens are often used as a measure of SCC susceptibility, and are popular since these tests generally do not take as long as other tests. The time-to-failure,

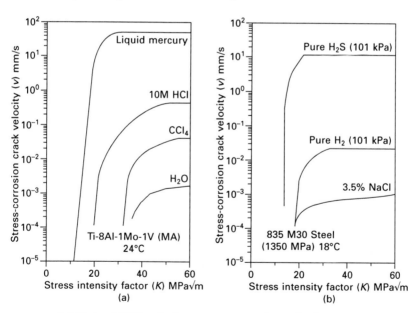

1.1 Plots of SCC velocity versus stress-intensity factor for (a) Ti alloy in various environments, (from data in [24]), and (b) high-strength steel in NaCl solution (from data in [25]). Data for LME (of a Ti alloy in mercury) and HE (of a steel in H$_2$S and H$_2$) are included for comparison.

Published by Woodhead Publishing Limited, 2011

reduction-of-area, or extent of SCC on fracture surfaces, are used to assess the degree of susceptibility. Slow, rising-load tests on pre-cracked fracture-mechanics specimens are also used to obtain SCC data. Threshold K values may be lower under rising-load conditions than for sustained-load conditions, and may sometimes be more applicable to practical situations.

Threshold stresses can also be lower and crack-growth rates can be higher when small cyclic loads are superimposed on sustained loads. These so-called 'ripple-load' tests are equivalent to corrosion-fatigue at high R-ratios (0.9–0.95). For corrosion fatigue at low R ratios, SCC processes can be superimposed on fatigue processes when there are hold times at maximum loads. Laboratory testing rarely simulates the precise conditions in service so that the relevance of data from specific tests to structural-integrity and remaining-life estimations needs to be assessed on a case-by-case basis. Further details of test methods and their relevance can be found in Chapter 3 and elsewhere [26].

1.3   Basic phenomenology of stress-corrosion cracking (SCC)

1.3.1   Crack-initiation

Initiation of cracks is sometimes not necessary since structures and components may contain pre-existing crack-like defects, e.g. porosity, hot tears. Welds are especially prone to such defects. Machining can also produce discontinuities such as deep grooves, laps, cracking of inclusions, and can produce microstructural changes, e.g. martensite transformations (and associated cracking) in steels. When such pre-existing cracks are present, initiation of SCC may require the development of specific environmental conditions within the cracks. SCC can occasionally initiate from fatigue cracks, e.g. when cyclic loading occurs in benign environments followed by sustained loads in aggressive environments, as can occur in aircraft where fatigue cracking occurs at altitude and then SCC occurs after flights due to condensation of water (containing Cl^- ions).

SCC initiation also often occurs from corrosion pits, which are usually associated with metallurgical inhomogeneities such as inclusions and grain boundaries, where oxide films are not as protective as in adjacent areas. Other forms of localised corrosion which can initiate SCC include crevice corrosion, de-alloying, and intergranular corrosion (associated with anodic grain-boundary precipitates or anodic solute-depleted regions adjacent to grain-boundary precipitates). Localised dissolution can also occur along slip bands in some materials, e.g. stainless steels, due to slip-induced rupture of oxide films, solute segregation to dislocations, and perhaps a greater intrinsic reactivity of deforming material compared with strain-free material.

A planar-slip mode, e.g. due to a high stacking-fault-energy, solute-ordering, or shearable precipitates, should promote such behaviour.

The transition from pits (and other localised corrosion sites) to SCC depends on the pit depth and shape, and local stresses/strains/stress-intensity factor, as discussed in detail elsewhere [27]. Growth of 'short' cracks can occur despite K values being below long-crack threshold K values, probably due to differences in crack-tip chemistry/potential between short and long cracks. When there are multiple, closely spaced crack-initiation sites, crack-coalescence usually occurs before long-crack behaviour is established, and modelling such behaviour is difficult. This difficulty is exacerbated if there are steep gradients in residual stresses near surfaces. A schematic summary of common crack-initiation sites, and micrographs showing crack initiation from pits and de-alloying, are shown in Figs 1.2 and 1.3.

1.3.2   Crack growth

On a macroscopic scale, fractures produced by SCC are characterised by a lack of macroscopic deformation and the absence of general corrosion (Fig. 1.4). There is, however, sometimes evidence of localised plasticity, and fracture surfaces are generally 'tarnished'. Corrosion products on fracture surfaces are often different from those formed on external surfaces due to differences between crack-tip and external environments, such as the lower oxygen concentrations in cracks. For example, external surfaces of stainless steels are shiny, but rust-coloured films are often observed on

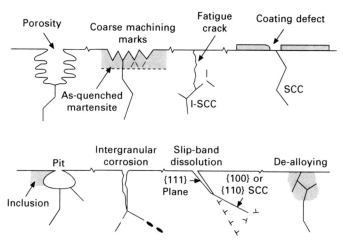

1.2 Schematic diagrams illustrating the most common crack-initiation sites for SCC. An appreciation of the 3-D geometry of crack-initiation sites is useful, and can be obtained by various techniques such as synchrotron X-radiation.

Published by Woodhead Publishing Limited, 2011

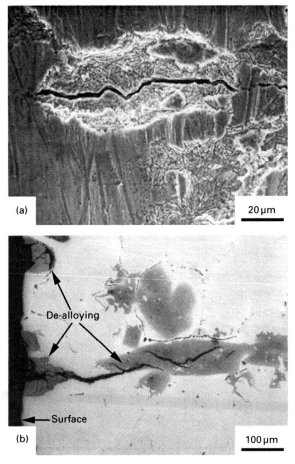

1.3 (a) SEM showing crack-initiation from a corrosion pit in a high-strength martensitic-steel aircraft component [28], and (b) optical micrograph of a polished section (not etched) showing crack-initiation from a de-alloyed Mn-rich phase in a Mn-Cu alloy (Sonoston) used for submarine propellers [28].

fracture surfaces (after they are exposed to air due to oxidation of ferrous ions that were present within cracks). For ferritic steels, black magnetite films can be present on fracture surfaces whereas external surfaces can be rusty. For materials that are susceptible to SCC in mild environments such as moist air, SCC fracture surfaces can be bright and shiny since only very thin protective oxide films are present.

On a microscopic scale, SCC in normally ductile materials produces brittle-intergranular or cleavage-like fracture surfaces parallel to low-index crystallographic planes, in contrast to dimpled fractures produced by fast fracture in inert environments (Fig. 1.5). Crack-branching and secondary

1.4 SEM (back-scattered mode) of macroscopic fracture-surface appearance of a 7075-T6 aluminium-alloy nut (see inset) (machined from an extruded bar), which failed by I-SCC due to exposure to moist air/condensation. Some areas are covered with corrosion product (exhibiting a mud-crack pattern) and other areas are relatively clean. Note also the 'lifted' areas associated with crack-branching along the elongated grain structure. (The underlying cause of failure was the use of the alloy in the T6 peak-aged condition rather than the more SCC-resistant T73 overaged condition that had been specified [28].)

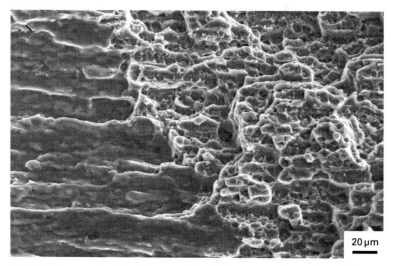

1.5 SEM (secondary-electron mode) of fracture surface of a similar nut as in Fig. 1.4, showing a transition from I-SCC to a dimpled 'overload' fracture. Details on intergranular facets are obscured by an oxide film [28].

Published by Woodhead Publishing Limited, 2011

cracking are also common for SCC (Fig. 1.6). Intergranular (I) fracture surfaces produced by SCC often appear to be fairly smooth and featureless, whereas transgranular (T) cleavage-like fracture surfaces exhibit numerous fine-scale features such as serrated steps, which form fan-like or herringbone patterns (Fig. 1.7). Both intergranular and cleavage-like facets also sometimes exhibit crack-arrest markings (CAMs) typically ~1 μm or so apart (Fig. 1.8).

SCC can usually be readily distinguished from fast (overload) fracture since the latter normally has a macroscopically fibrous fracture-surface appearance (except for shear lips adjacent to surfaces) and is microscopically dimpled on

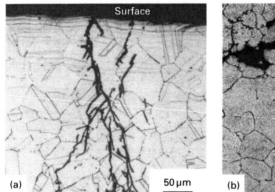 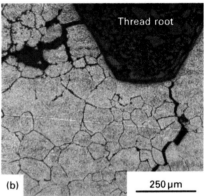

1.6 Optical micrographs of polished and etched sections of (a) 316 γ-stainless steel pipework that failed by T-SCC in aqueous chloride-containing environment (with the chloride emanating from insulation), and (b) intergranular SCC around a thread root in a sensitised AISI 431 martensitic stainless steel, showing extensive crack-branching in both cases [28].

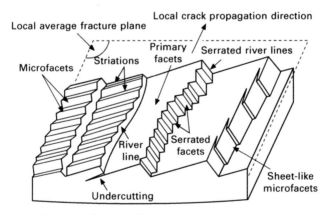

1.7 Schematic diagram illustrating some characteristics of cleavage-like SCC. Primary facets are often parallel to {100} or {110} planes in fcc materials. (Modified from Dickson et al. [29]).

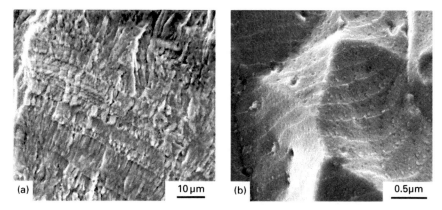

1.8 SEM of fracture surfaces produced by SCC in (a) an austenitic (Nitronic 80) alloy exposed to a hot aqueous chloride environment, showing cleavage-like appearance and crack-arrest markings [30], and (b) 7079 alloy tested in moist air showing crack-arrest markings and grain-boundary precipitates on intergranular facets [31].

a quite a coarse scale (Fig. 1.5). In some materials, however, fast fracture can result in brittle intergranular fracture surfaces (e.g. in temper-embrittled steels), or cleavage fractures (e.g. in ferritic steels at low temperatures). Fatigue-crack growth in inert environments can also sometimes produce intergranular and cleavage-like fracture surfaces, especially at low ΔK. Moreover, there are many similarities between SCC, HE and LME, as discussed later in this chapter and Chapter 2. Thus, diagnosing causes of failure in structures and components in service is not always straightforward [32].

1.4   Metallurgical variables affecting stress-corrosion cracking (SCC)

I-SCC and T-SCC can occur in many pure metals in specific environments – which is of interest in regard to understanding mechanisms of SCC – but alloys are much more susceptible and, obviously, of more practical interest. As would be expected, alloy composition, microstructure, and microchemistry, especially around grain boundaries, can have large effects on SCC resistance [3–23]. Alloy composition and microstructure also control strength, and there is a general trend of increasing SCC susceptibility with increasing strength for many materials, especially steels. However, atomic-scale and micro-scale structural and chemical features can often override the effects of strength, as illustrated in the following examples.

For I-SCC of 7xxx Al-Zn-Mg-Cu alloys in aqueous environments, overaging or retrogression and re-ageing (RRA) heat treatments decrease strength by only a small extent compared with the peak-aged condition, but can decrease SCC plateau velocities by about 3 orders of magnitude due to changes in grain-boundary microstructure and microchemistry (Fig. 1.9) [33]. In particular, an increase in the Cu content of grain-boundary precipitates is probably responsible for the beneficial effects of overaging [34]. For 5xxx Al-Mg alloys, precipitation of large area-fractions of the anodic β-phase can occur in structures and components subject to temperatures 50–80°C in tropical environments for long times, leading to marked increases in SCC susceptibility without significant effects on strength [35]. Incorrect heat treatments prior to service can also lead to 'sensitisation'. Similarly, stainless steels can be sensitised when chromium carbides are precipitated at grain boundaries, e.g. during welding and slow cooling, resulting in narrow, continuous, anodic crack paths in adjacent regions, thereby decreasing SCC resistance substantially [36].

For some solid-solution strengthened alloys such as Cu-Zn and Cu-Al, there is a strong correlation between alloy content, de-alloying rates, and SCC rates, with the de-alloying and SCC rates increasing markedly above a critical alloy content (15–18% in Cu-Zn and Cu-Al) (Fig. 1.10) [3, 11, 37]. Increasing Zn content of brasses, leading to a change from the fcc α-phase to the bcc β-phase, also increases SCC susceptibility, which may be associated with the structural change as well as the increase in zinc content. For T-SCC of γ-stainless steels, SCC resistance is also very dependent on the alloy composition, especially Ni content, with high sensitivities to SCC occurring for Ni contents in the range 2–25 at.% [36]. This compositional sensitivity may also be associated with the kinetics of film-formation/de-alloying, but is not well understood.

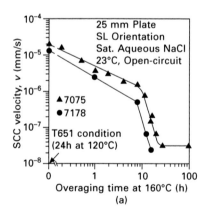

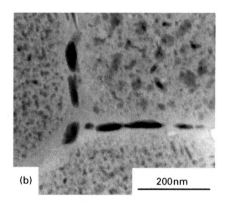

1.9 Plot of plateau crack-velocity versus overaging time for a 7075 Al-Zn-Mg-Cu alloy showing that overaging results in a marked decrease in the plateau crack velocity [33], and (b) TEM of typical grain-boundary microstructure for 7xxx alloys, showing matrix precipitates, grain-boundary precipitates, and a precipitate-free zone adjacent to grain boundaries [31].

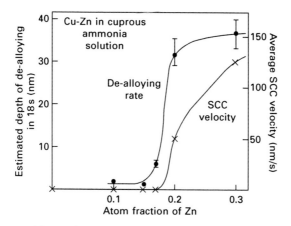

1.10 Rate of de-alloying and rate of SCC as a function of Zn content in Cu-Zn exposed to a cuprous-ammonia solution, showing a similar dependence of SCC and de-alloying on alloy composition. Similar correlations were found for Cu-Al [37].

Minor alloying elements or impurities can also have major effects on SCC resistance. For example, increasing levels of metalloid-impurity segregation at grain boundaries in high-strength steels can markedly increase SCC susceptibility by promoting hydrogen adsorption/absorption and by contributing to weakening of interatomic bonds in conjunction with hydrogen [3, 11]. For α-brass, small amounts of arsenic can substantially increase SCC resistance, possibly due to effects on surface-diffusion and de-alloying kinetics [11]. The effect of alloy compositional variations on SCC resistance can depend on the environment, e.g. for 7xxx Al alloys, increasing copper levels are usually beneficial for SCC resistance in aqueous environments, but low copper alloys are more resistant in humid air (<75% relative humidity), probably because oxide-film characteristics depend on both substrate composition and environment [31, 34].

When I-SCC occurs at much lower stresses than T-SCC, the grain morphology (and size) can have major effects on susceptibility. For example, many rolled/forged/extruded materials have elongated grain structures, and SCC resistance is low for material stressed in the short-transverse crack-plane orientations (S-L and S-T) since crack paths are fairly planar and normal to the applied stress. For T-L and L-T orientations, intergranular crack paths are tortuous and generally not normal to the applied stress so that SCC resistance is high. Grain-boundary-misorientation distributions (textures) are also important since certain special boundaries (low-angle and coincidence-site boundaries) are much more resistant to SCC than general high-angle grain boundaries. Increasing the proportion of resistant grain boundaries above a critical value by special processing and heat-treatment procedures (i.e. 'grain-boundary engineering') should substantially increase overall

SCC resistance [3, 11]. Crystallographic texture could also be important for T-SCC, e.g. crack velocities should be higher when textures are such that 'cleavage' planes are approximately normal to the applied stress direction. The degree of cold work is another important variable that can sometimes (but not always) affect SCC resistance. It is especially detrimental in some austenitic stainless steels where deformation-induced transformation to the less SCC-resistant martensite phase can occur.

1.5   Environmental variables affecting stress-corrosion cracking (SCC)

Solution composition (ionic species and their concentration, pH, dissolved oxygen content), temperature, and electrode potential usually have substantial effects on SCC resistance [3–23]. In some cases, only trace amounts of certain ions can have major effects, e.g. susceptibility of γ-stainless steels is increased markedly when chloride-ion concentration increases above a certain (ppm) level that depends on other variables. Other ions, e.g. nitrate ions for aluminium alloys, can completely inhibit SCC in aqueous environments. Increasing temperature generally decreases K_{ISCC} and increases crack-growth rates, although there are exceptions. The effect of temperature on SCC of γ-stainless steels is particularly noteworthy in that failure usually only occurs above 50–60°C unless environments are especially severe. Temperature variations can also lead to evaporation/drying cycles that increase ionic concentrations and thereby increase SCC rates.

For many materials, SCC occurs only in fairly narrow potential regimes, and these are often depicted on potential versus current-density (E-log i) plots. For example, for materials showing potential regimes where passivity occurs, SCC usually occurs in the regimes near (i) the transition from passivity to pitting, and (ii) the transition between general corrosion and passivity. SCC can also occur at negative potentials in some materials if copious amounts of hydrogen are generated and materials are susceptible to hydrogen embrittlement. Regimes of SCC for various environments are also depicted on Pourbaix (E-pH) diagrams, and whether or not the involvement of hydrogen is likely can be assessed from the potential-pH regimes for SCC and the line for water stability. However, differences in crack-tip chemistry/potential compared with bulk values, and differences in the composition of intergranular regions compared with the bulk, need to be considered. Details can be found in previous reviews [3–23].

1.6   Surface-science observations

Understanding the interactions between the environment and surfaces (especially crack-tip surfaces) is obviously critical to understanding SCC.

Crack-tip surfaces are inaccessible to experimental surface-science probes, but observations at plane surfaces may give an indication of what occurs on the atomic scale at crack tips during SCC. 'Plane' surfaces are made up of terraces, ledges, kink sites, surface vacancies, and sites were dislocations terminate (Fig. 1.11) [38], and adsorption, dissolution and other phenomena occur preferentially at these defects sites. In aqueous environments, specific adsorption weakens substrate interatomic bonds so that atoms at steps are remarkably mobile, as is evident from scanning tunnelling microscopy (STM) which shows 'frizzy' steps due to the rapid movement of atoms to and from steps during scans [39–41].

For 'clean' low-index crystallographic surfaces, STM and numerous other techniques such as low-energy electron diffraction (LEED) have shown that the lattice spacings up to 4–5 atomic layers beneath surfaces can be different from those in the bulk [42]. Contractions of the normal lattice spacings between the first and second layers of 5% or less for low-index crystallographic surfaces, but up to 30% for high-index surfaces, have been observed. Small expansions between the second and third layers, and small contractions between the third and fourth layers may also occur for a number of metal surfaces, although other sequences of contraction and expansion are sometimes present. The extent of these surface-lattice 'relaxations' depend on the material, crystal structure, temperature, as well as surface crystallographic plane.

For some metals, clean surfaces can be 'reconstructed', in that they have a different lattice structure or have periodic missing rows of atoms. Unlike relaxation, reconstruction involves lateral displacement of atoms as well as perpendicular displacements. For example, the (100) surfaces of some fcc metals (including Au, Pt and Ir) have a hexagonal crystal structure. The misfit between the reconstructed surface and the underlying bulk lattice often leads to regular surface corrugations, nano-scale faceting and other geometrical

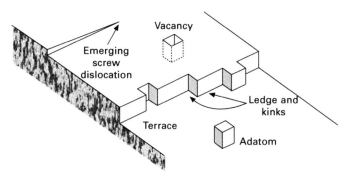

1.11 Schematic diagram showing terrace-ledge-kink model of a low-index crystallographic plane surface.

patterns. These clean-surface lattice relaxations and reconstructions occur essentially because atoms at the surface have fewer neighbours than those in the bulk and, hence, electron charge has to be redistributed. Thus, it is not surprising that adsorption of environmental atoms involving charge-transfer generally affects the surface-lattice perturbations.

Adsorption can (i) reduce the contractions between the first and second layers, (ii) produce an expansion between the first and second layers, (iii) result in reconstruction of previously unreconstructed clean surfaces, or (iv) change the nature of pre-existing reconstructions [42]. Some adsorbates (including metal atoms) can be incorporated into the topmost layers of substrate atoms, i.e. 'surface alloying' occurs, even in systems where there is no bulk miscibility [43]. The exact behaviour depends on the adsorbed species (and the type of adsorption/surface-coverage/degree of charge-transfer) as well as on temperature, electrode potential and surface-crystallography (Figs 1.12 and 1.13).

The effects of potential on surface-lattice perturbations of electrodes in aqueous environments are particularly interesting. For example, the hexagonal reconstruction of Au (100) (observed in vacuum) is stable in sulphuric acid, but only at negative potentials. Applying more positive potentials results in a change to the bulk fcc lattice structure at a critical potential, and the change can occur in millisecond to minute time-frames depending on the overpotential applied. Reverting back to negative potentials can result in the

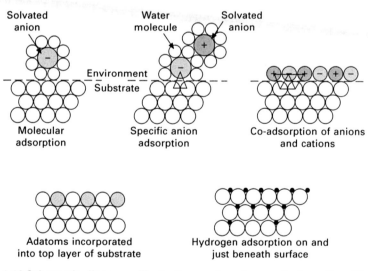

1.12 Schematic diagrams illustrating various types of adsorption. The type of adsorption and degree of charge transfer will determine the extent of substrate–substrate bond weakening. Adsorption sites (e.g. hollow, top, bridge sites) vary from system to system.

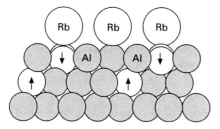

1.13 Schematic illustration showing the effect of Rb adsorption on Al (111) (side view). Arrows indicate the adsorption-induced displacements of some Al atoms, which results in atomic-scale surface rumpling [44]. (See also Fig. 2.12 for possible effects of adsorbed hydrogen on Ni surfaces.)

re-establishment of the hexagonal reconstruction, although this is a slower process than the lifting of the reconstruction, and involves nucleation and growth of reconstructed islands [39–41]. The potential-induced reconstructions are associated with enhanced surface-mobility due to anion adsorption, and the transition potential depends on ionic species. Essentially, the formation of a chemisorption bond between the anion and substrate atoms weakens substrate–substrate bonds. Specific adsorption of anions occurs in association with cations that may be surrounded by hydration sheaths in the outer Helmholtz plane or with cations that are stoichiometrically incorporated in the anion adlayer in order to ensure charge neutrality (Fig. 1.12) [41].

A reconstruction due to hydrogen adsorption (for a copper electrode in perchloric acid) has also been recently observed [45]. A transition from the unreconstructed (100) square lattice to a partially reconstructed surface with discrete stripes along [110] directions occurred within a few seconds after the potential was decreased into the hydrogen-evolution regime. Reconstructions due to hydrogen adsorption from the gas phase in many metals have also been observed [46–54]. In both cases, however, hydrogen 'adsorption' on the surface and in sites between the first and second atomic layers are probably involved. Sites between the first and second layers have deeper potential energy 'wells' than normal interstitial lattice sites and, hence, are also expected to have high hydrogen concentrations.[*]

Surface-relaxations of electrodes in aqueous environments (associated with anion and hydrogen adsorption) are also likely to be affected by the potential (and other variables) due to potential-induced changes to the electron distribution at the metal–solution interface. Unlike surface-reconstructions,

---

[*]There is no generally accepted term that defines both surface and just sub-surface sites for hydrogen. In the context of SCC (and HE) mechanisms, hydrogen on crack-tip surfaces and hydrogen within a few atomic distances of crack tips are referred to as adsorbed hydrogen for convenience.

breaking of interatomic bonds and lateral movement of atoms are not involved and, hence, the response of surface-relaxations to a potential change is probably extremely rapid (possibly on pico-second timeframes). However, detecting potential-induced relaxation changes is obviously more difficult than detecting reconstructions, and few experimental studies appear to have been made.

The surface-science observations summarised above have generally been made for planar, low-index surfaces in the absence of stress, and it would be expected (from theoretical considerations and from limited experimental data [55]) that stresses would influence adsorption and surface-lattice relaxations/reconstructions. For crack-tip surfaces, the curvature, varying crystallography, and high-density of defects, should also influence the surface-lattice perturbations. During crack growth, there should be time for relaxations to occur and be influenced by the environment, and reconstructions might occur if crack growth were sufficiently slow (although an equilibrium state may not be achieved).

Perturbations at clean crack-tip surfaces could well be greater than those observed at plane surfaces, so that reductions in the extent of these perturbations by adsorption could be particularly significant. If crack growth were intergranular, then grain-boundary structure and grain-boundary segregants could have significant effects on adsorption and crack-tip surface relaxations and reconstructions. The perturbations at crack tips, and related surface stresses [56–58], would be expected to affect decohesion (tensile separation of atoms) and dislocation emission from crack tips – processes that are central to determining whether ductile or brittle fracture behaviour occurs in inert and aqueous environments.

1.7   Proposed mechanisms of stress-corrosion cracking (SCC)

There are essentially only four atomistic processes that can produce crack growth. These are:

1. Removal of atoms at crack tips into solution.
2. Shear movement of atoms at crack tips – either emission of dislocations from crack tips or egress of dislocations (of opposite sign) exactly at crack tips.
3. Tensile separation of atoms at crack tips (decohesion) which could possibly involve incipient shear movement of atoms.
4. Surface-diffusion of atoms from crack tips to behind crack tips.

The last three processes do not require the presence of an environment, but can be aided by environmental effects. Mechanisms based on each of these processes (and combinations of them) have been proposed to account for

SCC, and those mechanisms that are considered to be viable (or still being debated) are outlined in the following. The applicability of these mechanisms, and how the effects of metallurgical and environmental variables on SCC could be explained by them, are discussed in subsequent sections.

1.7.1   Dissolution-based mechanisms

Mechanisms of SCC based on dissolution [59–63] require atoms at crack tips to be removed preferentially in a direction approximately normal to the applied stress, with only limited dissolution behind the crack tip. Directed dissolution could be promoted by either a chemically active path, such as anodic precipitates or segregants along grain boundaries, or an active path generated by stress/strain concentrated at crack tips. Limited dissolution behind crack tips relies on the formation of protective (or semi-protective) films on fracture surfaces, which can occur because (i) anodic microstructural/ microchemical features have been dissolved, (ii) stresses/strains are much lower than those at the crack tip, or (iii) environments behind crack tips are less aggressive than those at crack tips.

The slip-dissolution mechanism (also known as the film-rupture/anodic-dissolution mechanism) is the most widely cited dissolution-based SCC process. It involves the rupture of protective films by slip bands intersecting the crack tip, then dissolution along grain boundaries or low-index crystallographic planes, until repassivation occurs. Alternatively, crack tips may remain essentially film-free if the crack-tip strain-rate is high compared with the repassivation rate. In addition to rupturing protective films, slip processes increase the crack-tip-opening displacement, allowing solvent ions and solvated ions to diffuse to and from crack tips more rapidly so that dissolution does not become stifled by salt films or other corrosion products (Fig. 1.14).

1.7.2   Adsorption-based mechanisms

Adsorption mechanisms date back to 1928. Rebinder and co-workers [64] suggested that 'adsorption-active media' could affect deformation and fracture of solids due to decreases in the surface energy, and this proposal was subsequently taken up specifically for SCC by a number of workers [65, 66]. Benedicks (in 1945/48) [67] was perhaps the first to discuss the effects of adsorption on interatomic bonding, suggesting that the wetting effect of liquids facilitated cracking by decreasing the binding forces between surface atoms. He also mentioned the likelihood that surface bonding in the absence of wetting would be stronger than that in the bulk. Uhlig became an advocate of adsorption mechanisms for SCC in the late 1950s and 1960s [68–71], proposing that chemisorption of specific ions, e.g. complex Cu-NH$_3$ ions for SCC of α-brass, weakened the strained interatomic bonds at crack tips,

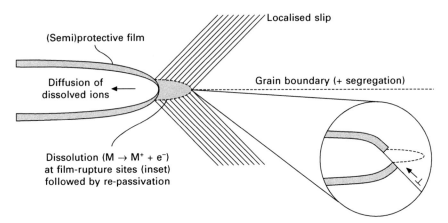

1.14 Schematic diagram illustrating slip-dissolution mechanism for SCC involving rupture of oxide films at crack tips due to localised slip and then repassivation behind crack tips.

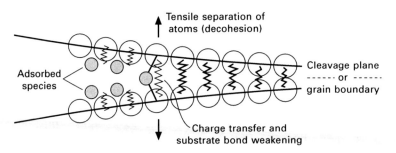

1.15 Schematic diagram illustrating the adsorption-induced decohesion mechanism.

thereby promoting crack growth by decohesion along a cleavage plane or a grain boundary (Fig. 1.15).

An adsorption-induced dislocation-emission (AIDE) mechanism for SCC, HE and LME (involving various adsorbed species, hydrogen, and metal atoms, respectively) was proposed by Lynch in 1976 and refined in subsequent papers [72–75]. In this model, adsorption weakens interatomic bonds at crack tips and thereby facilitates the nucleation of dislocations rather than decohesion. Dislocation nucleation involves the simultaneous formation of a dislocation core and a surface step by shear movement of atoms within a few atomic distances of crack tips – processes that are easier if interatomic bonds at crack tips are weakened by adsorption, providing that bonds between adsorbate and substrate are not too strong. Facilitating dislocation emission from crack tips promotes the coalescence of the crack tip with voids formed in the plastic zone ahead of cracks so that macroscopically brittle cleavage-like or intergranular fractures are produced.

Cleavage-like cracking along low-index crystallographic planes is explained in terms of the AIDE process on the basis that slip on planes on either side of cracks occurs alternately, so that the macroscopic fracture plane bisects the angle between the slip planes, resulting in a {100}⟨110⟩ or {110}⟨100⟩ fracture-surface crystallography for fcc materials. Approximately equal amounts of slip occur on either side of the crack because, if more dislocation activity occurred on one side, then this would increase the back-stress from emitted dislocations, thereby promoting dislocation emission on the other side. Voids formed in the plastic zone ahead of cracks contribute to crack growth (and re-sharpen crack tips), but crack growth occurs predominantly by AIDE. Voids can be nucleated at second-phase particles (which may be as small as ~50 nm), slip-band intersections, dislocation-cell boundaries, and vacancy clusters. Voids can be extremely small (on the nano-scale) such that the resultant dimples on fracture surfaces often may not be easily resolved (or may be obscured by corrosion products) (Fig. 1.16).

In inert environments where substrate bonds at surfaces are intrinsically strong (or in environments where adsorbates form strong bonds with the substrate), it is envisaged that dislocation nucleation at crack tips is difficult, and that link-up of the crack tip with voids occurs primarily by egress of dislocations around crack tips. Only a small proportion of dislocations activated from near-crack-tip sources exactly intersect crack tips to produce crack growth, so that extensive blunting occurs at crack tips. Large deep dimples (often containing small stretched dimples) are therefore produced on fracture surfaces (see Fig. 2.11 on page 109).

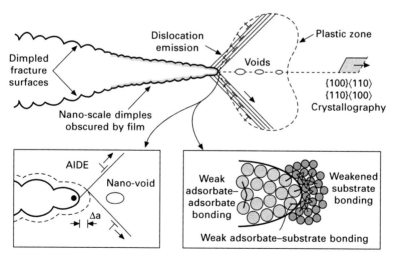

1.16 Schematic diagram illustrating the adsorption-induced dislocation-emission (AIDE) mechanism for transgranular SCC involving coalescence of cracks with nano-voids (or microvoids) in the plastic zone ahead of crack tips.

1.7.3   Hydrogen-based mechanisms

Mechanisms of SCC based on the effects of hydrogen on deformation and fracture are widely accepted for many materials, e.g. high-strength steels, nickel alloys, titanium alloys, aluminium alloys and magnesium alloys, since (i) hydrogen can be generated at crack tips by chemical dissociation of water molecules or by electrochemical reactions, (ii) these materials are known to be embrittled by hydrogen (after hydrogen-charging and testing in inert environments or during testing in gaseous hydrogen), and (iii) SCC in some of these materials can occur in moist air where dissolution and some other environmental interactions can be discounted.

The characteristics of fracture surfaces produced by SCC and HE in the materials mentioned above are often similar (except for differences in films or corrosion on fracture surfaces). Thus, mechanisms of SCC based on hydrogen are thought to be essentially the same as those proposed for HE, and involve either adsorbed hydrogen, dissolved hydrogen, or hydrides at crack tips, as described in detail in Chapter 2. Mechanisms of SCC based on hydrogen will therefore be only briefly summarised in the following, with subtle differences between HE and hydrogen-based SCC mechanisms highlighted. Chapter 2 should be consulted for more details of the effects of hydrogen on deformation and fracture, and a brief history regarding the development of various mechanisms. Figures illustrating hydrogen-based mechanisms can also be found in Chapter 2.

Mechanisms based on adsorbed hydrogen

These mechanisms involve weakening of substrate interatomic bonds leading to easier dislocation emission [72–76] or decohesion [76–82] at crack tips, as discussed for the adsorption of other species. However, unlike other adsorbed species, adsorbed hydrogen can readily diffuse ahead of crack tips, so that adsorption of hydrogen could occur not only at external crack tips but also at internal cracks (and voids just ahead of cracks). Another difference between hydrogen and other adsorbed species, related to the ease of hydrogen diffusion, is that high hydrogen concentrations are likely to be present within the first few atomic layers beneath crack tips as well as on the surface (see Fig. 2.1). Interatomic bonds involved in dislocation emission or decohesion at crack tips would probably be weakened to greater extents when hydrogen is present within the first few atomic distances as well as on the surface compared with when hydrogen is present only on the surface. For convenience, hydrogen on the surface and within a few atomic distances of surfaces is defined as 'adsorbed hydrogen' for the purposes of discussing mechanisms of HE and SCC, as already mentioned.

Mechanisms based on solute hydrogen

These mechanisms fall into two categories:

1. Where solute hydrogen weakens interatomic bonds and facilitates decohesion (as for adsorbed hydrogen but further ahead of crack tips) [76, 79–82].
2. Where solute hydrogen facilitates dislocation activity in the plastic zone ahead of cracks and promotes crack growth by a more localised-slip/microvoid-coalescence process than that which occurs in inert environments [83–85].

Hydrogen-enhanced decohesion (HEDE) mechanisms involve hydrogen concentrating in regions of high hydrostatic stress where the lattice is slightly dilated, at particle–matrix interfaces, or at grain boundaries. The hydrogen-enhanced localised-plasticity (HELP) mechanism also involves localised hydrogen concentrations ahead of crack tips due to high hydrostatic stresses (or due to hydrogen entry at crack tips), and it is based on hydrogen atmospheres around dislocations, or hydrogen at dislocation cores, facilitating dislocation movement in the localised regions where hydrogen is concentrated.

Mechanisms based on hydride formation

For hydride-forming materials, such as magnesium, zirconium and titanium alloys, hydrogen production and ingress at crack tips could potentially lead to the formation and brittle fracture of bulk hydrides around crack tips [86], as discussed for HE in such materials in Chapter 2. However, just because materials have a tendency to form hydrides does not necessarily indicate that that they will form under conditions that produce SCC, and adsorbed-hydrogen or solute-hydrogen-based mechanisms could be applicable in these materials. Even if hydrides are observed around cracks they could, of course, be formed after fracture had occurred by other hydrogen-based mechanisms.

Combinations of mechanisms

Combinations of AIDE, HELP and HEDE mechanisms could occur in many cases [76]. For example, dislocations nucleated from crack tips (due to AIDE) may move away from crack tips more readily due to HELP, thereby decreasing the back-stress on subsequent dislocation emission. For crack growth predominantly by AIDE, void-nucleation ahead of cracks could be promoted at slip-band intersections by HELP or by HEDE at particle–matrix interfaces. AIDE and HEDE could also occur sequentially, with AIDE occurring until the back-stress from emitted dislocations increased somewhat so that HEDE then occurred, followed by AIDE again when the crack tip had moved away from the stress-field of dislocations previously emitted. As well as AIDE and HELP possibly acting conjointly to localise plasticity, additional strain-localisation could occur due to effects of vacancies on deformation and fracture, with hydrogen reducing the vacancy-formation energy and stabilising vacancies produced by dislocation–dislocation interactions [87], as discussed in the next section.

1.7.4   Vacancy-based mechanisms

Mechanisms of SCC involving vacancies are based on (i) generation of vacancies at crack tips by dissolution, de-alloying or oxide-film formation, (ii) diffusion of vacancies (or di-vacancies) into the plastic zone ahead of cracks, and (iii) vacancy-induced changes to deformation and/or fracture behaviour at or ahead of crack tips (Fig. 1.17). If vacancy-induced embrittlement of a volume of material ahead of crack tips were involved, then crack growth could be discontinuous, involving repeated sequences of corrosion, vacancy injection, fracture, and crack-arrest.

For the selected-dissolution vacancy-creep (SDVC) model proposed by Hänninen and co-workers [12, 88], it has been proposed that vacancies ahead of cracks facilitate dislocation activity (such as climb) and can agglomerate to form voids, with both effects promoting fracture by localised shear. Another possibility, suggested by Jones [89, 90], is that di-vacancies accumulate along low-index crystallographic planes and promote cleavage. Yet another suggestion is that vacancies within the first few atomic layers at crack tips facilitate dislocation emission or decohesion by affecting surface/near-surface bonding [91, 92]. If dislocation emission were facilitated by vacancies, along with vacancy condensation to form nano-voids ahead of cracks, then the crack-growth process would be akin to the AIDE mechanism.

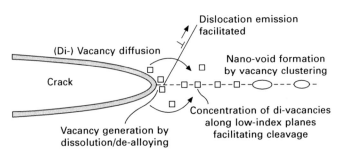

1.17 Schematic diagram illustrating various vacancy-based mechanisms for SCC.

1.7.5   Surface-mobility mechanism

The surface-mobility mechanism (SMM) for SCC, proposed by Galvele and coworkers [8, 93–96], is based on surface-diffusion of atoms from an elastically stressed, atomically sharp crack tip to an adjacent vacant lattice site behind the crack tip. Equivalently, in Galvele's words, 'crack growth occurs by the capture of vacancies by the stressed lattice at the crack tip' (Fig. 1.18). Vacancies are presumed to be supplied by, for example, selective dissolution of the crack surface just behind the crack tip, and crack growth is thought to be controlled by the rate of movement of vacancies along the crack surface to the crack tip (equivalent to the movement of adatoms away from the crack tip). Adatoms are subsequently dissolved or move into a vacant lattice site behind the crack tip. Surface-diffusion rates are assumed to be high due to the formation of low-melting-point 'surface contaminants', e.g. metal salts, formed by environmental interactions on fracture surfaces close to crack tips, while volume diffusion is assumed to be negligible (since SCC temperatures are less than half the homologous temperature of the material). The SMM mechanism has been proposed for a wide range of SCC systems (and for LME and HE) for both intergranular and transgranular crack paths.

1.7.6   Film-induced cleavage

The film-induced cleavage (FIC) model for SCC of normally ductile (fcc) materials, was proposed by Sieradzki and Newman in 1985 [97] and has been discussed in detail in subsequent reviews and papers [3, 98–103] by Newman and co-workers. The model involves repeated sequences of (i) the formation of an environmentally induced brittle film at crack tips, (ii) rapid brittle fracture of the film, (iii) continuation of brittle fracture into the underlying substrate for distances that are much greater (10–1000×) than the

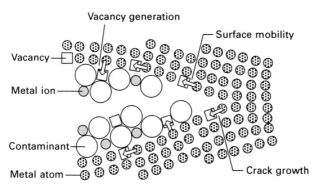

1.18 Schematic diagram illustrating surface-mobility mechanism of SCC, involving surface-diffusion of atoms away from an atomically sharp, elastic stressed crack tip [94].

film thickness, and (iv) crack-arrest and blunting (Fig. 1.19). Thus, the model predicts that crack growth is discontinuous and that crack-arrest markings should be present on fracture surfaces, as sometimes observed. FIC has been proposed for both transgranular (cleavage-like) and intergranular cracking, although the term should not, strictly speaking, be applied to the latter since cleavage is, by definition, transgranular.

A basic assumption of the model is that a crack in the brittle film propagates at such high velocities (hundreds of meters per second or more), and continues to propagate into the substrate without slowing down, such that only very limited dislocation activity occurs. It is assumed that crack velocities decrease as cracks propagate, due to occasional dislocation emission from crack tips or because cracks encounter obstacles such as slip bands, so that extensive dislocation activity (as usually occurs in fcc materials) then takes place, leading to crack-arrest and blunting.

For FIC to occur, the film must not only be brittle, but must also be strongly bonded to the substrate. Other film properties, such as the degree of coherency with the substrate, elastic modulus and film-thickness, may also be important. Nano-porous, de-alloyed films are considered to be particularly effective in promoting cleavage in the substrate, providing that the porosity is sufficiently fine. Coarsening of de-alloyed films is thought to somehow destroy their ability to inject a brittle crack into the substrate.

1.7.7　Corrosion-enhanced localised-plasticity mechanism

The corrosion-enhanced localised-plasticity (CELP) mechanism of transgranular SCC in fcc materials, involving dissolution, adsorbed hydrogen and absorbed hydrogen, was proposed by Magnin and co-workers in the 1980s [9, 104, 105]. SCC was thought to occur by the following sequence of events:

1. De-passivation/oxide-rupture.
2. Localised dissolution along a {111} slip plane.

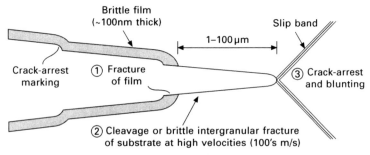

1.19 Schematic diagram illustrating the film-induced cleavage mechanism.

3. Localised shearing along this slip plane, promoted by the dissolution slot along the slip plane in conjunction with a lowering of the critical shear stress for dislocation activity due to adsorbed and absorbed hydrogen.
4. Pile-up of dislocations at obstacles, e.g. particles or other dislocations structures such as Lomer–Cottrell locks, which increases local stresses such that decohesion initiates at the head of the pile-up.
5. Propagation of a brittle crack back along a slip plane to link up with the dissolution slot.
6. Crack-arrest and possible re-passivation.

The above sequence of events then occurs along a different {111} slip plane, such that the crack plane macroscopically follows a {100} or {110} plane, but is composed of microscopic {111} corrugations (consistent with some observations) (Fig. 1.20). Crack advance primary occurs by decohesion, with only a small contribution from dissolution. However, anodic dissolution is considered essential not only for producing the necessary crack-tip geometry to promote localised shear, but also because the balancing cathodic reaction generates hydrogen that aids dislocation activity.

A somewhat similar 'corrosion-assisted cleavage' mechanism was proposed by Flanagan and Lichter [106–109]. Their model involved repeated sequences of dissolution along shear bands at crack tips followed by {110} cleavage, which is initiated from Lomer–Cottrell dislocation locks and triggered by a change in stress state.

1.7.8   Other hybrid mechanisms

Other combinations of dissolution/oxidation and mechanical fracture that have been proposed include:

- Dissolution of anodic precipitates and hydrogen embrittlement of regions between the precipitates [35].
- Dissolution, occurring irregularly along crack fronts (to produce corrosion

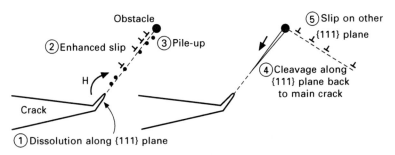

1.20 Schematic diagram illustrating the corrosion-enhanced localised-plasticity mechanism of SCC (adapted from [104]).

slots), plus ductile tearing of ligaments of material between the slots [110, 111].

- Internal oxidation along grain boundaries, involving (stress-assisted) oxygen diffusion promoted by vacancies resulting from oxidation, followed by 'brittle' fracture of the weakened boundaries [112].
- Dissolution or de-alloying followed by adsorption-induced fracture due to embrittling ions produced by corrosion [70, 113, 114].

Other considerations for dissolution-based mechanisms include the possibility that the dissolution rate is promoted by the presence of hydrogen (or vacancies) ahead of crack tips [115], and that hydrogen incorporated into films at crack tips affects their stability [116, 117] and could affect other properties, e.g. fracture strain. Hydrogen ahead of crack tips could also promote planar slip, and thereby facilitate the slip-dissolution process since slip concentrated in narrow bands is more likely to rupture oxide films than fine homogeneous slip. It has also been suggested that dissolution might facilitate dislocation emission from crack tips [118].

1.8   Determining the viability and applicability of stress-corrosion cracking (SCC) mechanisms

There is widespread agreement that there is a 'spectrum' of SCC mechanisms, ranging from those based primarily on dissolution for some materials to those based primarily on environmentally assisted mechanical fracture for other materials, as suggested by Parkins some time ago [119] and recently revisited by Newman [3]. However, there is presently little consensus on which specific mechanism (or combinations of mechanisms) applies to particular materials and environments. This is not really surprising given that such a variety of environment–material interactions often occurs, and that there are no techniques for directly observing crack-tip processes on the atomic scale in bulk material. Moreover, many fundamental studies of SCC have been carried out on fine-grained commercial materials with complex microstructures, resulting in fracture surfaces that are difficult to characterise. In addition, fracture surfaces produced by slow SCC are often somewhat corroded such that fine details produced by the fracture process, which would provide clues to the processes involved, are not apparent.

In the following, SCC in 'model' systems that do not suffer from some of the above problems will be reviewed to provide a sounder basis for elucidating mechanisms of SCC in more complex systems. T-SCC in pure metal and alloy single crystals (or very large-grained materials) will be described and discussed first since (i) complexities associated with SCC along grain boundaries are avoided, and (ii) fracture surfaces are macroscopically flat and fairly uniform in appearance, thereby facilitating the determination of crystallographic fracture planes and directions, active slip planes and other characteristics. I-SCC in some pure metal and alloy bi-crystal model systems is then considered. T-SCC and I-SCC in some commercial polycrystalline materials are then described and discussed.

Some of the model systems have been selected because, under certain conditions, they exhibit SCC at high velocities where there is no time for either crack-tip chemistry changes or time for some environmental interactions to occur. Moreover, corrosion after fracture is insignificant in such cases (providing specimens are quickly washed and dried) so that fine fractographic details are not dissolved or obscured by corrosion products. In these cases, stereographic pairs of SEM micrographs of mating areas of opposite fracture surfaces provide useful information. High-resolution TEM fractography of replicas, shadowed at low angles and examined under optimum conditions in order to resolve fine detail more clearly than by SEM, has also been carried out [73].

While very rapid SCC is not observed under normal conditions, it is instructive to compare the characteristics of SCC at high crack velocities with those produced at lower velocities. Comparing the characteristics of SCC with those produced by less complex environmentally assisted cracking phenomena, such as HE and LME, is also useful, and some of the model systems to be discussed have also been selected because such comparisons are available. The fractographic characteristics of SCC are described in particular since they are, arguably, the most important consideration in regard to the applicability and viability of the various SCC mechanisms. Mechanisms must, of course, be consistent with other observations, and the effects of variables on rates of SCC, but if mechanisms cannot account for the fractographic observations then they must be discounted or modified.

1.9   Transgranular stress-corrosion cracking (T-SCC) in model systems

1.9.1   T-SCC in pure magnesium

T-SCC in pure magnesium is considered first since complications associated with SCC in alloys are avoided, and most of the criteria for a good model system mentioned above are satisfied. In one study [120], magnesium specimens with large (as-cast) columnar grains ($3 \times 3 \times 20$ mm) were notched and fatigue pre-cracked such that subsequent cracking would occur within single grains for 10–15 mm. Tests were carried out in an aqueous environment (3.3%NaCl + 2%K$_2$CrO$_4$) at 20°C at open-circuit potential under dynamic loading in cantilever bending at various deflection rates. Some specimens were partially cracked in the aqueous environment and then cracked in dry air within the same grain. For rapid crack growth, crack velocities (v) were measured from high-speed cine films of the side surfaces of specimens. Specimens were also tested in liquid alkali metals, and detailed fractographic observations made after removing the alkali-metal deposits by dissolving them in alcohol.

Fractographic observations

Slow-strain-rate tests

SCC ($v \sim 10^{-4}$ mm/s) resulted in cleavage-like fracture surfaces on $\{0001\}$ planes (Fig. 1.21(a)) plus macroscopically brittle $\{10\bar{1}X\}$ facets (where X was usually between 1 and 2) (Fig. 1.21(b)). The $\{10\bar{1}X\}$ facets exhibited shallow flutes and dimples on a microscopic scale. In contrast, fracture in dry air produced relatively deep flutes/dimples on macroscopically flat $\{10\bar{1}X\}$ facets, with no signs of cleavage-like fractures.* No evidence of hydrides or crack-arrest markings (indicating that cracking was discontinuous) were evident on fracture surfaces. Other studies of slow SCC in pure magnesium (with a relatively small equi-axed grain size) have reported similar observations [122, 123].

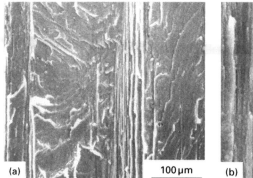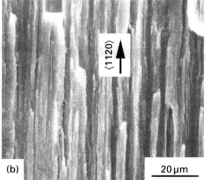

1.21 SEM of (a) cleavage-like fracture, and (b) fluted fracture produced by SCC ($v \sim 6 \times 10^{-4}$ mm/s) in pure magnesium specimens dynamically loaded in NaCl + K_2CrO_4 solution at open-circuit potential at 20°C [120].

---

*Unlike some other hcp metals, magnesium does not cleave in inert environments (even at low testing temperatures). Note also that fluted fracture surfaces have sometimes been mistaken for cleavage-like fracture surfaces. Flutes are elongated dimples, which have a concave profile on opposing fracture surfaces, and are formed when voids are nucleated along slip-band intersections [121].

High-strain-rate tests

Rates of SCC increased with increasing deflection rates and, at very high deflection rates (50°/s), SCC occurred at velocities ~50 mm/s in the coarse-grained material, as was evident from marked differences between fracture surfaces produced by rapid SCC and those produced in dry air (Fig. 1.22). The appearance of fracture surfaces produced by SCC at high velocities were essentially similar to those produced at lower crack velocities, namely cleavage-like {0001} and fluted {10$\bar{1}$X} fractures. Flutes were not as shallow as those produced at lower crack velocities, but were still much shallower than those produced by fracture in dry air. Cleavage-like fracture surfaces were relatively smooth except for many small twins (and occasional large twins), but there was very fine detail that may have been associated with localised plasticity occurring during fracture (Fig. 1.23).

Comparisons of SCC with LME and HE

Rapid cracking (10s mm/s) in liquid alkali metals produced fracture characteristics that were remarkably similar to those observed for cracking in aqueous environments, namely cleavage-like {0001} and fluted {10$\bar{1}$X} facets, although flutes were shallower than those produced by SCC (Fig. 1.24). Hydrogen-charging in aqueous environments (under conditions where hydrogen is generated on actively corroding surfaces), and then testing at slow strain rates in laboratory air after washing and drying, also produced cleavage-like and fluted fracture surfaces similar to those produced by SCC in aqueous environments [123, 124]. However, no embrittlement was observed for testing in aqueous environments under cathodic-charging conditions.

Discussion of T-SCC in pure Mg

For SCC at high crack velocities (~10s mm/s), mechanisms based on dissolution, films, vacancies, and solute hydrogen (e.g. HELP) can almost certainly be discounted since the various diffusion processes required would not be fast enough. For example, the hydrogen-diffusivity (D) for Mg is probably about 10^{-8}cm^2/s at 20°C [125] (See Fig. 9.15 on page 368 and related discussion). Using this value, the D/v ratio for a crack velocity (v) of 50 mm/s would be 2×10^{-9}cm, and analyses indicate that hydrogen should not diffuse more than a few atomic distances ahead of cracks for D/v ratios less than 10^{-8}cm [126]. Dislocation velocities around rapidly growing cracks are probably also too high for hydrogen to be transported by dislocations to regions directly ahead of cracks.

Most mechanisms can be discounted for rapid SCC on kinetic grounds and, since surface reactions can occur very rapidly for magnesium in aqueous

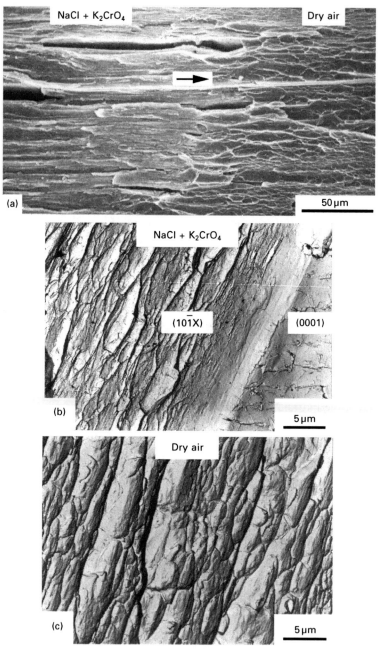

1.22 (a) SEM of fracture surface produced by rapid cracking
($v \sim$ 10 mm/s) first in NaCl + K$_2$CrO$_4$ solution and then in dry air,
showing that flutes/dimples are shallower after rapid cracking in the
salt solution than after cracking in air – shown more clearly in (b)
and (c) by TEM of replicas [120]. Note also secondary {0001} cracks
for SCC.

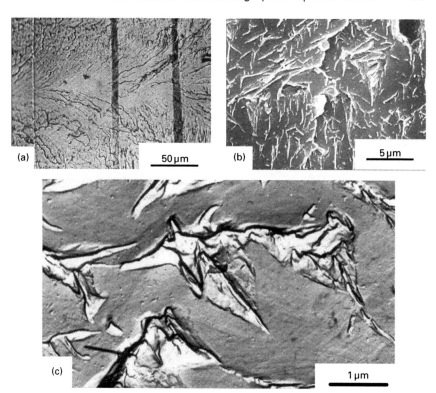

1.23 Cleavage-like {0001} fracture surface for pure Mg cracked rapidly in NaCl + K_2CrO_4 solution ($v \sim 50$ mm/s): (a) optical micrograph, (b) SEM and (c) TEM of replica, showing river lines and perturbations due to twinning [120].

environments, there is a strong case for an adsorption mechanism involving adsorption of hydrogen (produced by dissociation of water molecules at film-free crack-tip surfaces). The observations of small shallow dimples/flutes on {10$\bar{1}$X} SCC fracture surfaces, compared with deep dimples on fracture surfaces produced in dry air, are consistent with the adsorption-induced dislocation-emission (AIDE)/microvoid-coalescence (MVC) process. For the cleavage-like {0001} fractures, the adsorption-induced decohesion mechanism (or repeating sequences of decohesion and dislocation emission) could account for the observations, although an AIDE/nanovoid-coalescence process cannot be discounted since nano-dimples may not be clearly resolved.

The case for an adsorption mechanism for SCC of magnesium at high crack velocities is further supported by the close similarities between SCC and LME. It is generally accepted, as discussed in more detail in Chapter 18, that only adsorption can occur during rapid cracking for many liquid-metal:solid-metal couples, especially when mutual solubilities are low and there

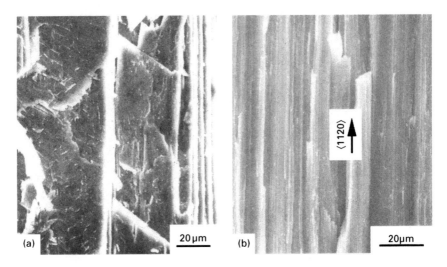

1.24 SEM of fracture surfaces produced by LME of pure Mg: (a) cleavage-like {0001} fracture surface produced by rapid cracking in liquid Cs at ~50°C, and (b) {10$\bar{1}$X} fracture surface with shallow flutes produced by rapid cracking in liquid Na at ~120°C [120].

is no tendency for compound formation [127], as is the case for magnesium in alkali metals. The LME observations also suggest that cleavage-like cracks are not atomically sharp since, if they were, capillary flow of liquid metal to crack tips, and adsorption of large atoms such as caesium at crack tips, would not be possible. (Atomic radii of Cs and Mg are 0.334 nm and 0.172 nm, respectively, and a crack-tip radius of at least several nanometers would probably be required to enable liquid to reach crack tips by capillary flow.)

For SCC at low velocities, there is time for other material-environmental interactions to occur in addition to hydrogen adsorption. However, adsorption mechanisms probably predominate at low as well as at high crack velocities for both cleavage-like and fluted fracture modes since fracture-surface characteristics are very similar at high and low velocities. When hydrogen does diffuse ahead of cracks during SCC, adsorption of hydrogen could occur at tips of voids ahead of cracks (and could promote void growth), and contributions from HEDE and HELP may occur as discussed in Section 1.7.3. Mechanisms based on bulk hydride formation can probably be discounted for both cleavage-like and fluted fractures since there was no evidence of such hydrides on fracture surfaces for the conditions studied (although a hydride mechanism could well occur in other circumstances since hydrides are known to form in magnesium).

1.9.2   T-SCC in various other pure metal and single-phase alloys

Metallographic and fractographic observations

SCC in pure iron (bcc) single crystals has been observed during slow-strain-rate testing while simultaneously cathodic charging in sulphuric acid (with a period of pre-charging) [128]. Cleavage-like fracture surfaces on {100} or {110} planes were observed depending on which plane was oriented normal to the stress axis. Crack fronts were a V-shape due to crack growth in two equivalent directions lying between ⟨110⟩ and ⟨112⟩ directions. Slip was observed on both {111} and {112} planes intersecting crack tips, and was much more extensive (with more lateral contraction of side surfaces and a greater crack-tip-opening angle) for specimens with the {110} orientation (Fig. 1.25).

In another study of hydrogen-charged pure iron single crystals, internal square-shaped cracks were observed, with cracks growing in four ⟨110⟩ directions on a {100} plane, and regularly spaced (1–2 µm) CAMs were

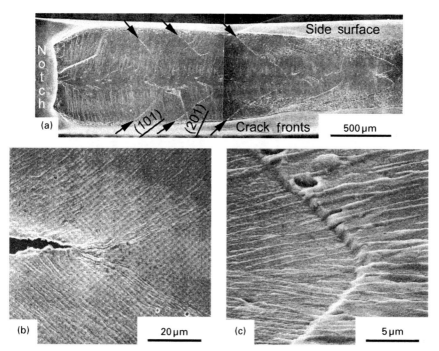

1.25 SEMs showing (a) macroscopic appearance of fracture surface of a pure iron single crystal ([110] orientation) after slow-strain-rate testing and simultaneously hydrogen-charging in sulphuric acid, (b) slip around crack tip on side surface, and (c) microscopic cleavage-like appearance of fracture surface [128].

sometimes observed [129]. TEM of replicas of mating halves of opposite fracture surfaces showed that they were rumpled on a fine scale, probably as a result of slip processes and perhaps nano-scale void formation ahead of cracks. Cleavage-like {100} and {110} fracture surfaces similar to those observed for SCC of pure iron have been observed for Fe-Si single crystals tested in hydrogen gas, and after hydrogen-charging and testing in air. Fe-Si single crystals cracked rapidly in liquid metals such as lithium also exhibit cleavage-like {100}⟨110⟩ fracture surfaces [73, 74].

SCC and HE of a number of other pure metal and single phase alloys also produces cleavage-like {100} fracture surfaces that are essentially identical to those produced by adsorption-induced LME. Examples include pure (fcc) nickel single crystals tested in hydrogen gas and in liquid mercury [73, 74], (bcc) β-Ti alloys tested in aqueous and liquid-mercury environments [130], and (bcc) β-brass tested in distilled water and liquid gallium [73, 74] (Fig. 1.26). For β-brass, CAMs have been observed on both SCC and LME fracture surfaces [131].

Copper is another pure (fcc) metal that exhibits SCC under specific (severe) conditions, e.g. slow-strain-rate testing of single crystals in a sodium-nitrite solution under anodic polarisation, and SCC is characterised by cleavage-like {110} fracture surfaces exhibiting regularly spaced (~1 μm) CAMs and considerable localised strains [132]. For SCC of pure copper in nitrite solutions, hydrogen-based mechanisms can be discounted, but it is not an ideal model system since fracture surfaces are somewhat corroded. However, the system is of interest because there are similarities between SCC of pure copper and copper alloys involving de-alloying.

Discussion

There is general agreement that T-SCC in Fe, β-Ti and Ni is usually associated with 'hydrogen-embrittlement' since hydrogen concentrations around SCC crack tips should be high. However, there is little agreement whether AIDE, HEDE or HELP mechanisms are involved. The fact that adsorption-induced LME can produce fracture surfaces that have the same appearance and crystallography as those produced by SCC indicates that a mechanism based on 'adsorbed' hydrogen (at internal or external crack tips) needs to be seriously considered. If an adsorption mechanism is operative, the AIDE mechanism seems most likely when there is substantial localised plasticity around crack tips with slip especially on planes intersecting crack tips (as occurs for pure iron {110} single crystals). When plasticity is very limited, adsorption-induced decohesion (or HEDE plus AIDE) is possibly applicable.

Whether hydrogen-induced decohesion due to solute hydrogen in normal interstitial lattice sites ahead of transgranular cracks occurs is questionable

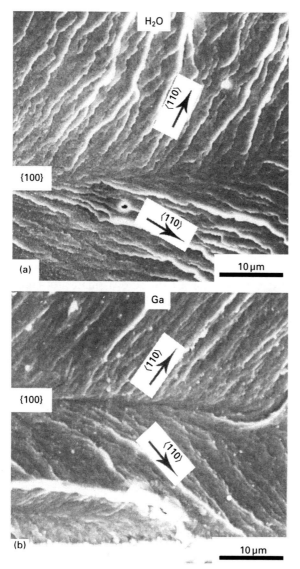

1.26 SEMs showing cleavage-like fracture surfaces produced by (a) SCC in distilled water ($v \sim 10^{-3}$ mm/s), and (b) LME in liquid gallium ($v \sim 10$ mm/s) in β-brass single crystals [73, 74].

since hydrogen concentrations in regions of hydrostatic stress ahead of transgranular cracks may not be sufficiently high. Very high elastic stresses near crack tips, possibly due to strain-gradient hardening, could possibly produce the requisite hydrogen concentrations in the lattice [10], but there is no unequivocal evidence that this occurs. High hydrogen concentrations are, however, undoubtedly present at particle–matrix interfaces (and other strong traps), and hydrogen-induced decohesion at these sites is more likely than in the lattice.

There is considerable evidence that solute hydrogen facilitates dislocation activity, as summarised in Chapter 2 and elsewhere [84, 85], but whether HELP makes a major contribution to cleavage-like cracking is debatable. In particular, it is difficult to envisage how a HELP mechanism, involving just facilitating and localising dislocation activity ahead of cracks, would produce a change from a non-crystallographic ductile fracture in inert environments to a cleavage-like fracture macroscopically along low-index planes in specific directions in hydrogen-bearing environments. However, HELP could make a minor contribution to cracking as discussed in Section 1.7.3.

For SCC of β-brass in distilled water, an adsorption mechanism probably predominates given the similarities between SCC and LME. The observation of CAMs for adsorption-induced LME of β-brass is interesting in that it indicates that CAMs are not necessarily produced by embrittlement of a volume of material ahead of cracks, e.g. due to solute hydrogen, as is sometimes assumed. CAMs can be produced by a number of processes, and different processes probably operate in different circumstances, as discussed further in Section 1.12.4.

1.9.3   T-SCC in precipitation-hardened Al-Zn-Mg single crystals

Transgranular SCC and corrosion-fatigue (CF) in precipitation-hardened Al-Zn-Mg alloys are considered next, and again most of the criteria for a good model system discussed previously are satisfied. Moreover, the size and spacing of age-hardening precipitates can be systematically varied, and differences in SCC characteristics between underaged, peak-aged, and overaged material are useful in elucidating mechanisms of cracking. The slip distributions in the interiors of peak-aged material can also be revealed by further ageing to 'decorate' dislocations with precipitates, then polishing and etching. In addition, SCC occurs not only in aqueous environments but also in moist-air environments where dissolution and other processes can be discounted, and where there is no significant corrosion of fracture surfaces even for slow crack growth. The effects of numerous variables, such as solution composition, electrode potential, pH, temperature, and strain rate, on SCC and CF have also been studied [73, 74, 133–135].

Metallographic and fractographic observations

Sustained-load and slow strain-rate tests

Slow SCC in Al-Zn-Mg single crystals in distilled water, aqueous NaCl solutions, and moist-air resulted in cleavage-like cracking. In contrast, ductile, dimpled fractures were produced in inert environments. Cleavage-like cracking on {100} planes (with up to 10° deviations from {100}) was most commonly observed, but {110} fracture planes have also been reported [136]. Fracture surfaces exhibited serrated steps on secondary {100} or {110} planes, tear ridges, slip lines and isolated pyramidal-shaped dimples centred on small second-phase impurity particles. For {100} fracture surfaces, the steps and ridges formed herringbone patterns as a result of crack growth occurring in two different ⟨110⟩ directions in adjacent regions (Fig. 1.27).

High-strain-rate tests

Rapid cleavage-like cracking at velocities as high as 10 mm/s were produced in notched, fatigue pre-cracked Al-Zn-Mg single crystals tested in cantilever bending at high deflection rates in distilled water and some salt solutions (Fig. 1.28). (Ductile dimpled fractures were produced by rapid cracking in a 2M sodium-nitrate solution, as for dry air environments [133].) The characteristics of rapid SCC were essentially the same as those described for slow SCC, notwithstanding that there was up to 8 orders of magnitude difference in crack velocities. There was, however, more extensive slip on {111} planes intersecting crack tips, and crack-opening angles were larger for rapid SCC compared with slow SCC.

Cleavage-like {100}⟨110⟩ cracking was observed for underaged, peak-aged, and overaged Al-Zn-Mg single crystals, providing that the strength was above a minimum value (~120 HV for rapid cracking), but there were subtle differences in fracture-surface appearance. Most importantly, high-resolution

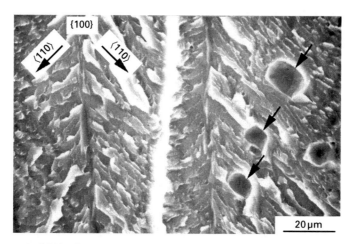

1.27 SEM of cleavage-like {100}⟨110⟩ fracture surfaces produced by SCC in peak-aged Al-Zn-Mg single crystals under sustained loading at 20°C in moist air (20% relative humidity) ($v \sim 10^{-7}$ mm/s). Isolated large dimples are arrowed. [73]

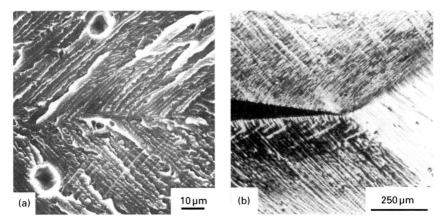

1.28 (a) SEM of cleavage-like {100}⟨110⟩ fracture surfaces, and (b) optical micrograph showing slip on specimen sides, after SCC of peak-aged Al-Zn-Mg single crystals in distilled water ($v \sim 10$ mm/s) under dynamic loading at high strain rates [73].

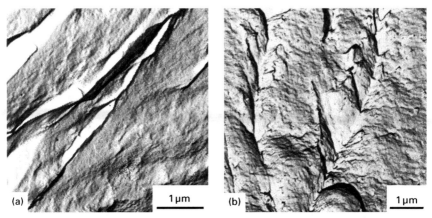

1.29 TEM of secondary carbon replicas of cleavage-like fracture surfaces of Al-Zn-Mg single crystals after SCC: (a) peak-aged condition in moist air (20% relative humidity) ($v \sim 10^{-7}$ mm/s), and (b) overaged condition in distilled water ($v \sim 10$ mm/s), showing tear ridges and nano-scale dimples [73, 74].

SEM and TEM fractography for material aged to different extents showed that cleavage-like fracture surfaces (produced under conditions where there was insignificant corrosion) were covered by nano-scale dimples (as well as the occasional large dimples already mentioned). The size and spacing of the small dimples corresponded to the spacing of the age-hardening precipitates, as would be expected if voids were nucleated from these precipitates. Dimples were clearly apparent (by SEM and TEM) for overaged conditions, but were only just resolved by TEM for peak-aged material (Fig. 1.29).

Published by Woodhead Publishing Limited, 2011

Cyclic loading

Cleavage-like {100} fractures were also produced under cyclic loading in Al-Zn-Mg single crystals in aqueous environments, and crack fronts (along ⟨110⟩ directions) were delineated by 'brittle' fatigue striations (due to the unloading part of the stress-cycle deforming part of the fracture surface produced during loading). Increasing cycle frequency in distilled water (or salt solutions) resulted in decreasing striation spacings, but brittle striations on {100} planes were observed even at high cycle frequencies (Fig. 1.30). (Striation spacings ~0.2 mm were observed at high ΔK at 24 Hz, indicating that crack growth had occurred at ~10 mm/s, substantiating the observations of rapid cracking under monotonic loading.)

Marked reductions (~20×) in crack-growth rates were observed when the environment was changed from distilled water to a 2 M sodium-nitrate solution, and ductile striation spacings were the same in the nitrate solution as in dry air [133]. A chromate solution (10%) was a much less effective inhibitor than the nitrate solution, and resulted in only a small decrease (~50%) in the spacings of brittle striations compared with those produced in distilled water (at 20°C) [133]. The effects of testing temperature depended on the environment, with increases in striation spacings (~80%) observed when increasing temperature from −5° to 65°C in 3.5% NaCl solution, and decreases in striation spacings (~40%) observed when increasing temperature from 5°C to 65°C in distilled water (for the cycle frequency and ΔK used) (Fig. 1.31).

1.30 SEM of fracture surface of Al-Zn-Mg single crystal tested at high ΔK in an aqueous KI solution showing effect of cycle frequency on striation spacings [135].

Published by Woodhead Publishing Limited, 2011

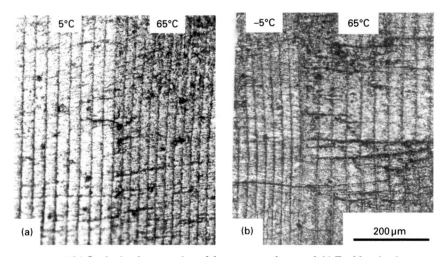

1.31 Optical micrographs of fracture surfaces of Al-Zn-Mg single crystals showing (a) decrease in striation spacing with increased temperature in distilled water, and (b) increase in striation spacing with increased temperature in 3.5% NaCl solution [133].

Comparisons of SCC with LME

Cleavage-like fractures in Al-Zn-Mg single crystals were not only produced in aqueous and moist-air environments but were also observed after rapid cracking in some liquid-metal environments. The characteristics of cleavage-like fractures produced by LME were essentially the same as those produced in aqueous and moist-air environments, except that the severity of embrittlement was much greater for LME. Thus, cleavage-like LME fractures occurred preferentially on {100} planes in ⟨110⟩ directions, and were associated with localised slip on {111} planes intersecting crack tips. Fractures also occurred along {110} planes (in ⟨100⟩ and ⟨112⟩ directions locally), but were much less common and occurred only for specimen orientations where {110} planes were highly stressed. Both {100} and {110} fractures exhibited serrated crystallographic steps like those observed after cracking in aqueous environments (Fig. 1.32).

Discussion of T-SCC in Al-Zn-Mg single crystals

SCC at high crack velocities (~mm/s) almost certainly occurs due to adsorbed hydrogen, produced by dissociation of water molecules at crack tips, since there is no time for other interactions to occur, as discussed for rapid T-SCC in magnesium. Hydrogen diffusivities quoted in the literature for aluminium alloys at 20°C vary from $\sim 10^{-10} cm^2/s$ to $\sim 10^{-7} cm^2/s$ [137]. Even if the higher value of D is used, D/v ratios for cracks growing at 10 mm/s are such that

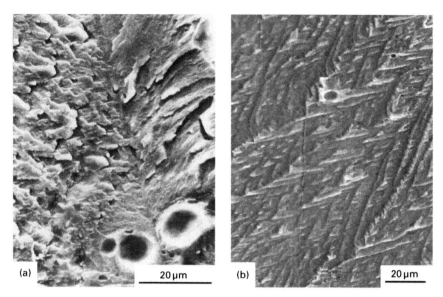

1.32 SEM of cleavage-like fracture surfaces after LME of Al-Zn-Mg single crystals: (a) {100} and (b) {110} facet [73, 125].

hydrogen should not diffuse more than a few atomic distances ahead of cracks. The presence of hydrogen ahead of cracks does not, in any case, appear to result in cleavage-like cracking given that ductile behaviour was observed in hydrogen-charged single crystals tested in dry air [73].

The case for an adsorption mechanism for SCC is also strengthened by close fractographic similarities between SCC and adsorption-induced LME, again paralleling the observations for Mg and other materials. The adsorption-induced dislocation-emission (AIDE) mechanism rather than the adsorption-induced decohesion mechanism best accounts for the observations since crack fronts lie along the line of intersection of {100} planes and {111} slip planes, localised slip was observed on planes intersecting crack tips, and nano-scale dimples were present on fracture surfaces. The isolated large dimples on fracture surfaces also indicate that strains just ahead of cleavage-like cracks must have been high.

The AIDE mechanism probably also applies to slow SCC (and CF) in aqueous and moist-air environments since the characteristics of cracking at low velocities were so similar to those at high velocities. Moreover, solute hydrogen does not appear to have significant effects on fracture (based on observations of hydrogen-charged single crystals [73]) and, in any case, cleavage-like cracking is difficult to explain by the HELP mechanism, as discussed in a previous section. In addition, dissolution-based mechanisms can be discounted for SCC in moist air (20% relative humidity) since a liquid electrolyte would not be present at crack tips at such a low relative humidity. The abrupt changes in striation spacings observed during CF due to abrupt changes in solution composition, temperature, or cycle frequency also support an embrittlement process that is at (or very close) to crack tips. The effects of these variables on corrosion fatigue (and SCC) could be explained in various ways, but are consistent with adsorption mechanisms, as discussed more generally in a subsequent section.

1.9.4   T-SCC of some binary single-phase alloys involving de-alloying

SCC in binary fcc metal alloys, such as Cu-Zn, Cu-Au, Ag-Au and Ag-Cd in aqueous environments is, in some ways, more complex than the other systems considered since de-alloying often occurs. De-alloying produces a macroscopically brittle, nano-porous film on surfaces and at crack tips (Fig. 1.33). It can also result in a highly concentrated ionic solution in the pores of the film, generation and diffusion of vacancies into the substrate, and stresses in the film and substrate [138–141]. On the other hand, hydrogen effects can be discounted in some systems, and some of the guidelines outlined previously for a good model system are satisfied, e.g. (i) the characteristics of SCC and the effects of variables have been studied for specifically oriented single crystals, (ii) high-resolution fractography has been carried out after SCC in environments where minimal corrosion had occurred, and (iii) critical experiments, involving fast fracture of superficially de-alloyed thin films and wires, have been performed.

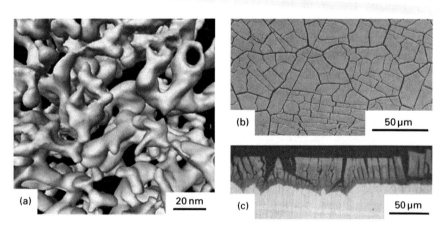

1.33 (a) High-angle annular dark-field TEM of nano-porous gold ligaments after de-alloying Ag20%Au in perchloric acid [139], and (b), (c) optical micrographs of plan view of surface and section normal to the surface, respectively, showing network of cracks in de-alloyed film of an Ag20%Au alloy after exposure (without stress) to perchloric acid [100].

Characteristics of slow T-SCC

T-SCC in aqueous environments under sustained loading or at slow strain rates produces cleavage-like fracture surfaces macroscopically parallel to {110} planes. High-resolution SEM observations have suggested that {110} primary facets were composed of {111} micro-facets, as illustrated schematically in Fig. 1.7. For notched Cu-Au single crystals, SCC on {110} planes occurred macroscopically in ⟨100⟩ directions for specimens notched to favour this direction, but crack growth occurred in two macroscopic ⟨112⟩ directions, so that crack fronts were a V-shape, when the notch favoured a ⟨110⟩ overall crack direction (Fig. 1.34(a)) [106–109]. CAMs indicative of discontinuous cracking were observed for the ⟨100⟩ crack-growth direction (where {111} slip planes intersect the crack front along its length) (Fig. 1.34(b)), but not for the ⟨112⟩ crack-growth directions (where {111} slip planes do not intersect the crack front along its length).

The spacings of CAMs varied between 20 and 100 μm, and depended on the applied potential in Cu-Au single crystals [106–109] (compared to the more typical 1-2 μm spacing in other materials). Significant slip on planes intersecting crack tips was observed on the side surfaces of specimens, with more marked slip bands coinciding with CAMs on fracture surfaces (Fig. 1.34(b)). Observations of load-drops, current-transients, and acoustic emission during crack growth were consistent with discontinuous cracking and CAMs. Crack velocities during crack jumps, estimated from these observations, were $\sim 10^{-1}$ mm/s, with some dependence on potential. Crack-arrest times also depended on the potential, and typically ranged from seconds to minutes so that average crack velocities were typically between 10^{-4} and 10^{-2} mm/s.

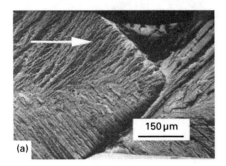
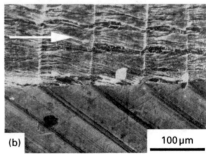

1.34 SEM of notched Cu25at.%Au single crystal tested at slow strain rates in a NaCl solution: (a) notched along ⟨110⟩ direction, showing cleavage-like {110} fracture surface with no CAMs except for the crack-front marking produced by changing the potential from 400 mV to 0 mV (where blunting occurred) and then back to 400 mV(SCE), and (b) specimen notched along ⟨100⟩ direction and tested at 400 mV (SCE), showing CAMs on cleavage-like {110} fracture surface and slip on specimen side surfaces [106–109].

For α-brass tested at slow strain rates in a de-aerated ammoniacal environment containing very high concentrations of copper-ion complexes [142], high-resolution TEM of fracture surfaces showed no evidence of CAMs or observable corrosion (de-alloying) at the nano-scale [74, 113]. These observations revealed serrated steps, undulations parallel to {111} slip directions, and a fine-scale topography that could possibly have arisen from a nano-scale void-coalescence process (Fig. 1.35). TEM studies of thin foils have shown that high dislocation densities are present just beneath fracture surfaces, consistent with a crack-growth process involving localised plasticity [91, 107, 143].

Characteristics of rapid T-SCC

Cleavage-like cracking at high velocities (at least several mm/s), with the same characteristics as those observed at low crack velocities (except for

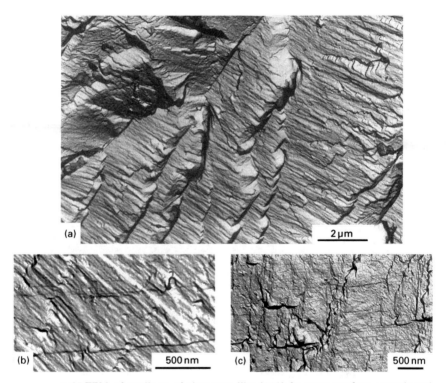

1.35 TEM of replicas of cleavage-like {110} fracture surfaces produced by SCC of α-brass in an ammoniacal environment containing a high concentration of Cu ions (so that insignificant corrosion occurred) showing fine serrated steps, slip lines, undulations and possibly nano-scale dimples [74, 113].

CAMs), has been reported for several fcc binary alloys, e.g. Cu-Zn, Cu-Au and Ag-Au, under special testing conditions [29, 98–100, 144]. In the simplest case, unstressed thin-foil or 'single-crystal' wire specimens (20–125 μm thickness/diameter) were pre-exposed to aqueous environments, e.g. perchloric acid under applied potentials for Ag-Au, to produce superficial de-alloying. Specimens were then fractured under the same environmental conditions at very high strain rates. The thickness of the de-alloyed films was usually less than 10% of the specimen thickness, and was sometimes only ~50 nm. Fracture initiated in the brittle, nano-porous de-alloyed films, and continued to propagate by 'cleavage' across the remainder of the specimen. In some cases, however, a transition to ductile fracture occurred after cleavage-like cracking had extended part of the way across the specimen.

The effects of a number of variations to the above procedure on fracture behaviour were investigated with the aim of establishing the mechanisms involved. For example, before testing partially de-alloyed (Ag-Au) specimens, the applied potential was changed to one where further de-alloying would not occur. This procedure often resulted in cleavage-like fractures provided that specimens were immediately tested after changing the potential (Fig. 1.36) [100]. If the specimens were held for more than a few seconds at the potential at which de-alloying was not possible, then ductile behaviour of the substrate was observed (although the de-alloyed film fractured in a brittle fashion). In other experiments, specimens were partially de-alloyed, removed from solution, plunged into liquid nitrogen (without washing or

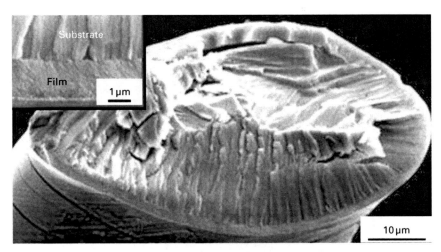

1.36 SEM of fracture surface of Ag20at%Au single crystal superficially de-alloyed at 975 mV(SCE) without applied stress, and then fractured at high strain rate at 0 V(SCE) at 20°C, showing cleavage-like appearance. Inset shows transition region from film to substrate fracture at a higher magnification (adapted from ref. [100]).

drying), removed from the liquid nitrogen, and then immediately fractured (at near −196°C) in air. This procedure also often resulted in cleavage-like fractures, whereas specimens that were not partially de-alloyed were ductile at −196°C (Fig. 1.37) [11, 98].

Discussion of T-SCC in de-alloying systems

Most of the mechanisms outlined in previous sections have been applied to SCC in these systems. However, mechanisms of SCC at high and low crack velocities are probably the same since (i) the fracture-surface characteristics produced by rapid fracture of superficially de-alloyed specimens are essentially the same as those produced by slow SCC, and (ii) 'normal' SCC occurs discontinuously by rapid crack jumps. For rapid cracking, mechanisms involving dissolution can be discounted since ionic diffusion rates are not sufficiently fast. Hydrogen-based mechanisms can also be discounted on thermodynamic/kinetic grounds. Thus, we are left with the surface-mobility mechanism, vacancy-based mechanisms, film-induced cleavage, and adsorption mechanisms.

Surface-mobility mechanism

The fundamental basis of the SMM has been questioned as discussed at length in the literature [145–151]. These references should be consulted for detailed arguments since they are often not easy to follow, and a summary

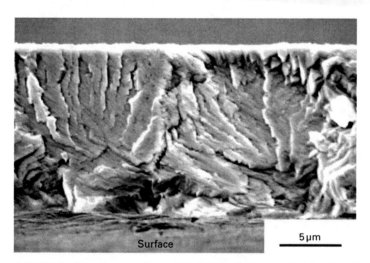

1.37 SEM of fracture surface of Cu30%Zn foil superficially de-alloyed without applied stress, and then fractured at high strain rate near −196°C, showing cleavage-like appearance [98, 125].

will not be attempted here. Rather, it is pertinent to the present review to point out that the SMM does not adequately consider the detailed metallographic and fractographic observations showing, for example, that SCC is often associated with locally high plastic strains, with slip occurring especially on planes intersecting crack tips. The fact that SCC is often discontinuous (producing CAMs) has also not been properly addressed by proponents of the SMM. In addition, SMM explanations for SCC on specific crystallographic planes in specific directions, based on differences in surface-diffusion rates for different planes/directions, are rather vague.

Vacancy-based mechanisms

High vacancy concentrations may be produced in the substrate due to de-alloying, and there have been a number of suggestions (summarised in Fig. 1.17) regarding how vacancies might produce embrittlement. However, there is no direct evidence that vacancies ahead of cracks can induce cleavage-like fracture. Moreover, vacancies would not diffuse sufficient distances to be able to account for brittle cleavage-like fracture over distances of ~80 μm after de-alloying for short times (<10 minutes) followed by immediate testing. The maximum diffusion distances for di-vacancies (which diffuse faster than single vacancies) in 10 minutes is estimated to be the order of 3 μm, even when stress-assisted di-vacancy diffusion is invoked [91]. Thus, if it is assumed that 'normal' SCC occurs by the same mechanism as for rapid SCC in superficially de-alloyed specimens, vacancy-based mechanisms are unlikely to play a major role.

Film-induced cleavage

The FIC mechanism is obviously consistent with the observations that prior de-alloying is necessary to produce fast fracture in the substrate, and that fractures in the substrate initiate from brittle cracks in the de-alloyed film. It is also conceivable (according to some modelling studies) that cracks in normally ductile fcc materials might be brittle if crack velocities were high enough, i.e. ~100s m/s, so that significant dislocation activity did not occur [152].

For the tests where thin foils/wires were cooled to near −196°C prior to cleavage-like fracture, it has been argued that the liquid in the nano-porous film would be frozen and, hence, that only the FIC mechanism could be applicable [11, 98]. However, the possibility that the frozen solution in the nano-porous film locally melts during deformation and fracture of the nano-scale ligaments of the film was not considered.

The ligaments of the nano-porous film probably deform and fracture by localised shear (as observed for nano-wires [153] and micrometer sized specimens [154]), and the frozen liquid in the nano-pores could also fracture by localised shear (as observed in ice) [155]. Localised shear fractures are known to produce local temperature increases of hundreds of degrees or more. For example, spectroscopic measurements of light emitted during fracture, and evidence of melting on fracture surfaces, for metallic glasses suggest that localised shear can result in temperature increases of up to 900°C [156, 157]. Significant increases in temperature should occur for any metallic material that possesses high strength, releases high stored elastic energies, and fractures by a localised shear mechanism [158]. Thus, it seems quite likely (but not proven) that local melting of frozen solution occurs during fracture of nano-porous films prior to fracture of the substrate and, hence, the low-temperature-fracture observations cannot be used to discount all other mechanisms.

While the FIC mechanism is consistent with some observations, a number of other observations are difficult to reconcile with this mechanism:

- The estimates of crack velocities in the foil/wire experiments, and during crack jumps for slow SCC, are only ~20 mm/s and ~100 μm/s, respectively, i.e. at least 1000 times slower than the 100s m/s that modelling studies suggest is necessary to suppress dislocation activity during crack growth in ductile fcc materials. Other theoretical calculations suggest that it is unlikely that fracture of nanometer-thick films would result in crack extension into the substrate for tens of micrometers [81].
- Model systems composed of contiguous ductile and brittle phases, e.g. Ni-Rh [159], Ag-Bi, Cd-Bi (Fig. 1.38) [74], give little or no support to the idea that rapidly propagating cleavage cracks in the brittle phase can

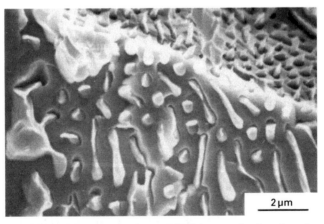

1.38 SEM of fracture surface of Bi-Cd eutectic showing cleavage of the Bi phase and ductile fracture of the Cd phase, even though a potential cleavage plane in the Cd was aligned with the cleavage plane of the Bi [74].

be injected a considerable distance (~10s μm) into normally ductile fcc (and hcp) materials.

- The detailed fracture-surface appearance, and observations of slip around cracks and beneath fracture surfaces for T-SCC, are not what would be expected for an atomically brittle cleavage process (accompanied by limited dislocation activity), but are consistent with crack growth occurring primarily by localised plasticity (which would be expected for the crack velocities observed).

- The fracture of nano-porous de-alloyed films probably involves sequential, localised shear of individual ligaments in the film, i.e. discrete events that would not be expected to affect substrate fracture behaviour since shear cracks in ligaments are not directly contiguous with the substrate ahead of the cracks (Fig. 1.39).

- The proposed explanation for the effect of hold times (after de-alloying prior to fracture), namely that coarsening of the porosity in the films 'somehow' precludes them from initiating brittle substrate fracture [11, 100, 160], is unconvincing, particularly as somewhat coarser films still fail in a macroscopically brittle manner.

- The reasons for large variations in crack-jump distances (from 1 μm to 100 μm), depending on material, potential, etc., are not adequately explained by the FIC model, e.g. proposed explanations for crack-arrest such as cracks encountering obstacles, or intermittently emitting dislocations, seem unlikely to be able to account for the variations. Also CAMs are not always observed, although it could be reasonably argued that crack growth is discontinuous without leaving a discernible indication in some cases.

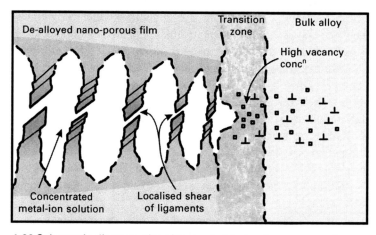

1.39 Schematic diagram showing probable fracture process in nano-porous de-alloyed films involving localised shear of discrete ligaments that are not contiguous with the substrate.

In view of the above issues/analogies and, if it is assumed that liquid was present in all cases where cleavage-like fracture was observed, an adsorption mechanism needs to be considered.

Adsorption mechanisms

The possibility that the AIDE mechanism could account for SCC in systems involving de-alloying is attractive since environmental transport and adsorption processes are sufficiently fast to account for the high crack velocities observed in the superficially de-alloyed specimens subjected to high strain rates. Moreover, unlike other mechanisms, adsorption mechanisms can account for the metallographic and fractographic observations, showing that cleavage-like cracking occurs on {110} planes in ⟨100⟩ and ⟨112⟩ directions and is associated with localised slip. As mentioned previously, cleavage-like cracks with these characteristics are sometimes observed after rapid adsorption-induced fracture of aluminium alloys in liquid-metal environments.*

The observations of fast fracture in superficially de-alloyed specimens could be explained if an embrittling solution were produced by de-alloying. The solution in the nano-pores produced by de-alloying would be highly concentrated so that adsorption of metal-ions co-adsorbed with anions (producing salt-like adlayers) could perhaps be responsible for embrittlement. There is no *a priori* reason why adsorption of ions of an element of an alloy should not embrittle the alloy, providing that charge-transfer (or partial charge-transfer) between adsorbate ions and substrate atoms occurs, and substrate bonds are thereby weakened [114]. Moreover, significant effects of specific adsorption of ions on substrate bonding (surface-reconstructions and relaxations) have been observed, as summarised in Section 1.6.

For the adsorption-induced-fracture hypothesis, the role of de-alloying is not only to produce an embrittling solution, but also to produce a nano-porous film that acts as a reservoir for the solution and inhibits diffusion of ions away from the film-substrate interface. In addition, when specimens are not pre-cracked, fracture of the film exposes the substrate, and results in a stress concentration at the film-substrate interface, so that adsorption-induced crack initiation and growth into the substrate is facilitated. Thus, this mechanism could then be considered to be a combination of adsorption-induced cleavage and film-induced cleavage.

The effect of delay times (prior to testing) in preventing rapid cleavage-like cracking of superficially de-alloyed specimens could be explained by

---

*There are no reports of liquid-metal-induced cleavage-like cracking in the binary alloys being considered. For copper alloys, T-LME in single crystals does not occur in liquid metals such as Hg, Ga, even though these environments produce intergranular LME. Thus, direct comparisons of T-SCC with T-LME in Cu alloys are not possible.

diffusion of concentrated solution away from the film-substrate interface during the delay time, so that the solution adjacent to the interface became diluted and was no longer embrittling. As the solution becomes diluted, the type of adsorption and the degree of charge transfer would change, e.g. the surface-coverage of salt-like adlayers would decrease to be replaced with adsorbed anions with cations surrounded by hydration sheaths in the outer Helmholtz plane. Dilution times of several seconds would probably be expected based on effective diffusivities of ions in viscous solutions within nano-porous films, which is consistent with the delay times observed under conditions where no further de-alloying could occur.

Washing and drying of specimens after superficially de-alloying also generally prevented brittle substrate behaviour [29, 98–100], as would be expected if embrittlement were due to a concentrated solution produced by de-alloying. Exceptions could occur if liquid were not completely removed from nano-porous films or if subsequent condensation occurred due to deliquescent salt residues and high relative humidity. Premature film fracture (due to film-induced stresses) prior to applying external stress also inhibited fast cleavage-like fracture of superficially de-alloyed specimens [100], as would also be expected because dilution of concentrated solution would be promoted in these circumstances.

For 'slow' discontinuous SCC, the following repeating sequence of events (involving adsorption) could explain the observations:

1. De-alloying and production of embrittling solution during crack-arrest periods.
2. Fracture of the film when it reached a critical thickness/critical strain.
3. Adsorption-induced cleavage-like fracture of the substrate.
4. Crack-arrest and blunting, leading to CAMs on fracture surfaces, when the supply of embrittling solution was exhausted.

The increment of cleavage-like cracking would depend on the volume of embrittling solution contained in the nano-porous film, which would depend on the rate of de-alloying (e.g. controlled by potential) and time of de-alloying prior to film fracture (e.g. determined by applied stresses plus film-induced stresses – with film-induced stresses also depending on de-alloying conditions). Continuous SCC could occur in some circumstances, e.g. if high concentrations of metal-ion complexes were present in the solution prior to immersion of specimens.

Conclusion

The above considerations for cleavage-like cracking in binary metal alloys suggest that the AIDE mechanism, involving adsorption of metal-ion complexes produced by de-alloying, more convincingly accounts for the experimental observations than other mechanisms. However, a number of issues regarding adsorption mechanisms require further clarification, as discussed in a subsequent section on adsorption mechanisms in general.

1.10   Intergranular stress-corrosion cracking (I-SCC) in model systems

1.10.1   I-SCC in single-phase binary alloys

For polycrystalline Ag-Au, Cu-Au, Ag-Cd and Ag-Pd alloys, rapid I-SCC (1–20 mm/s) occurred in thin foils/wires tested under similar conditions as those that produced rapid cleavage-like fractures in single crystals, e.g. when immediately tested at high strain rates after superficially de-alloying (Fig. 1.40) [96, 100, 144]. These conditions included those where the potential was dropped to one where further de-alloying would not occur, and where the temperature was decreased to near –196°C before testing. In some cases, transitions from brittle intergranular fracture to cleavage-like fracture occurred as cracks progressed across specimens. Transitions to ductile fracture also sometimes occurred, especially if de-alloying times were short, as observed for cleavage-like fractures.

Delay times before fracture (after de-alloying and dropping the potential) resulted in ductile behaviour rather than brittle intergranular fracture, as also observed for single crystals [100]. However, compared with cleavage-like fractures, longer delay times (10–15 s for Cu-Au in NaCl solutions and up to 60 s for Ag-Au in perchloric acid) were required to eliminate the occurrence of brittle intergranular fracture. When polycrystalline Cu-Au specimens de-alloyed in NaCl solutions were left for times greater than 15 s

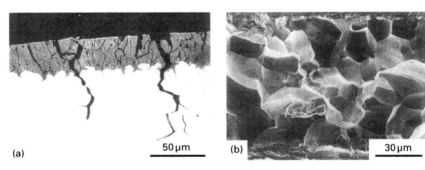

(a)   50 µm   (b)   30 µm

1.40 (a) Optical micrograph of polished section of a Ag20%Au alloy strained rapidly (after de-alloying to a depth of ~20 µm), showing brittle intergranular cracks extending to ~60 µm beneath the de-alloyed layer [160], and (b) SEM of brittle intergranular fracture surface for Cu30%Au de-alloyed without applied stress and then impact loaded in bending [144].

(such that ductile behaviour would have occurred during subsequent rapid fracture), but were then re-immersed so that further de-alloying occurred, brittle intergranular fracture was again observed [144].

The above observations suggest that rapid brittle intergranular fracture and cleavage-like fracture in superficially de-alloyed specimens occur by a similar mechanism. The arguments for discounting (or doubting) other mechanisms, such as those based on hydrogen-assisted cracking, film-induced cleavage, and vacancy-induced embrittlement, discussed for cleavage-like fractures in these materials, also apply to brittle intergranular fractures. Thus, adsorption-induced fracture, with embrittling ions produced by de-alloying and retained in the nano-porous film, best accounts for the observations. Brittle intergranular fractures were relatively featureless and there was much less evidence of plasticity compared with cleavage-like fractures, so that the adsorption-induced decohesion mechanism rather than the AIDE mechanism could be applicable.

Rapid brittle intergranular fracture occurs in preference to cleavage-like fracture in many superficially de-alloyed polycrystalline specimens probably because (i) de-alloyed films are thicker (by up to a factor of two) at grain boundaries than in adjacent regions (producing a greater stress concentration when cracks in the film occur at this location), and (ii) adsorption occurs preferentially and/or weakens interatomic bonding more at grain boundaries than within grains (as discussed in Chapter 18 for LME where adsorption-induced fracture often occurs preferentially along grain boundaries).

1.10.2   I-SCC in Fe-Si bi-crystals

Notched bi-crystals with defined grain-boundary orientations were tested (with grain boundaries generally normal to the applied stress) in ammonium-carbonate solutions as 70°C. The SCC susceptibility was determined as a function of (i) grain-boundary orientations with respect to slip systems in adjacent grains, (ii) applied potential, (iii) crack-tip-strain rate, and (iv) degree of phosphorus segregation at grain boundaries [161, 162]. For the anodic potentials used, hydrogen effects were not likely to be significant, and potential-drops/pH changes within cracks were probably small so that electrochemical measurements on bulk surfaces may give a reasonable indication of the dissolution/repassivation rates at crack tips. Fractographic observations enabled crack-growth rates and local directions of crack growth to be measured from markings produced by intermittent unloading. Crack-tip-opening angles were also used as a measure of SCC susceptibility.

The main observations for this model system were:

- SCC occurred only in the active/passive regime at anodic potentials, where crack-growth rates increased with increasing anodic polarisation.

- Rates of SCC were higher for grain-boundary misorientations where slip planes intersecting crack tips were favourably inclined to the grain-boundary plane, i.e. subject to high shear stresses.
- Higher levels of phosphorus segregation increased SCC susceptibility.
- SCC susceptibility decreased with increasing crack-tip-opening rates, with ductile behaviour occurring at high rates.

All the above observations were consistent with a slip-dissolution mechanism involving sequences of alternate-slip on planes intersecting crack tips, directed dissolution along grain boundaries, and passivation (or partial repassivation) between slip events. Moreover, there was good agreement between the measured crack-growth rates (and crack-tip-opening angles) and the predictions of a slip-dissolution model using dissolution rates predicted from electrochemical tests on bulk surfaces (Fig. 1.41). The occurrence of directed dissolution along grain boundaries (due to phosphorus and possibly other segregants or even the grain-boundary structure itself) was supported by observations of SCC in single crystals (under the same conditions as in bi-crystals) where the susceptibility was much less than that for bi-crystals and was associated

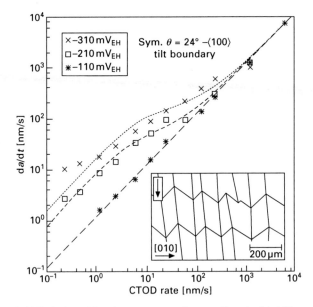

1.41 Relationships between crack-growth rate (da/dt) and crack-tip-opening displacement (CTOD) rate for Fe-Si bi-crystals with a 45°-⟨100⟩ tilt grain boundary (P concentration of 0.11%) at various potentials. Dotted lines are the predictions of a slip-dissolution mechanism, which show good fits to the experimental data. Inset shows schematic of intergranular fracture surface with a zig-zag crack front (adapted from ref. [161]).

with the formation of a thick oxide film [161, 162]. Mechanisms based on hydrogen effects for I-SCC were discounted, not only because hydrogen concentrations would be low under the conditions where SCC occurred, but also because changing from aqueous to hydrogen-gas environments during testing resulted in a change from I-SCC to transgranular {100} cleavage-like hydrogen-assisted cracking.

The critical role of slip was particularly evident from fracture surfaces of bi-crystals oriented so that slip on planes subject to high shear stresses would not intersect crack fronts if crack fronts across the specimen were straight (and normal to the crack tip sides). In such cases, crack fronts were a zig-zag shape (Fig. 1.41 inset), and each segment of the fracture surface corresponding to a different direction of cracking was tilted with respect to the adjacent segment as a result of slip on planes with high shear stresses intersecting crack tips along each segment of the crack front.

1.10.3   I-SCC in Mg and Al-Zn-Mg bi-crystals

I-SCC of Mg bi-crystals in aqueous environments was observed at very high crack velocities (10s mm/s) at high strain rates (i.e. under similar conditions as described for single crystals) suggesting that 'adsorbed' hydrogen on the crack-tip surface and within a few atomic distances of the crack tip may be responsible, since there may not be time for hydrogen diffusion further ahead of cracks.* The fracture surfaces produced at high velocities for pure Mg in aqueous environments were dimpled/fluted on a very fine scale, whereas deep dimples were observed after intergranular fracture in inert environments (Fig. 1.42) [73, 120]. SCC at high velocities in Al-Zn-Mg bi-crystals also exhibited shallower dimples than those produced by cracking in dry air, but only for material with a wide PFZ and widely spaced grain-boundary precipitates [163]. In such cases, the AIDE mechanism best accounts for the observations.

1.10.4   I-SCC of various other materials in hydrogen-bearing environments

There are a number of systems, e.g. steels, Ni alloys, Al alloys, where (i) hydrogen is known to be generated at tips of intergranular cracks in aqueous environments, and (ii) it has been established – based on tests in inert environments for hydrogen-charged specimens – that hydrogen segregated at grain boundaries can produce brittle intergranular fracture in normally ductile

---

*Assuming that D for lattice diffusion is $\sim 10^{-8}\,cm^2/s$ and that grain-boundary-diffusion rates are not significantly faster – which is, of course, debatable since there are no data.

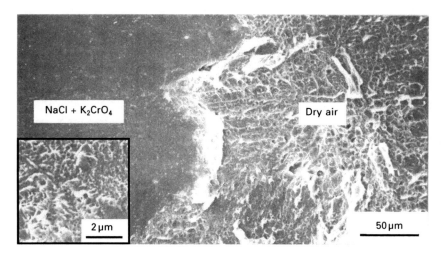

1.42 SEM of pure Mg bi-crystal tested at high deflection rates (crack velocity ~50 mm/s) in NaCl + K$_2$CrO$_4$ solution and then in dry air, showing transition from 'brittle' intergranular fracture to dimpled intergranular fracture. Inset (at a higher magnification) shows shallow dimples/flutes on the macroscopically brittle area [73, 120].

materials [10, 79, 81, 82, 164]. In many cases, little plasticity is evident on brittle intergranular facets produced by either SCC or by fracture of hydrogen-charged specimens, suggesting that cracking occurred by decohesion due to weakening of interatomic bonds by hydrogen segregated at grain-boundary sites at and ahead of crack tips. In addition, quantum-mechanical modelling supports the concept of weakening of interatomic bonds by hydrogen (and other segregants) across grain boundaries resulting in decohesion [165, 166]. Thus, there are good grounds for assuming that SCC in these materials involves hydrogen generation at crack tips, diffusion of hydrogen ahead of cracks, and then an increment of crack growth when a critical hydrogen concentration is achieved over a critical distance.

1.11 Stress-corrosion cracking (SCC) in some commercial alloys

SCC in commercial alloys is only briefly described here since details for specific materials can be found in other chapters and in previous reviews. Fractographic observations in some commercial alloys are discussed in the following, and the most likely mechanisms operating in commercial alloys are considered in the light of the insights obtained from studies in the model systems. Materials considered are Mg, Zr, Ti, Al, Cu, Ni alloys, and γ-stainless and ferritic steels.

1.11.1 Mg, Zr and Ti alloys

The characteristics of SCC in Mg alloys [86, 124, 167, 168] are generally similar to those in pure Mg, in that cleavage-like and fluted fracture surfaces are often observed (Fig. 1.43). Intergranular cracking is also observed in some circumstances. Cleavage-like {0001} and fluted fractures are also observed for SCC of Ti and Zr alloys in a variety of environments (acids, water ($+Cl^-$), chlorinated hydrocarbons, hot molten salts, and dry halogen vapours) [86, 169–171].* For some titanium alloys in aggressive environments, SCC plateau velocities at high K can be as high as ~0.1 mm/s (Fig. 1.1) so that the D/v ratios are ~10^{-8} cm. Thus, hydrogen generated at crack tips should not diffuse more than a few atomic distances ahead of cracks. Characteristics of cracking are similar at high and low velocities, and are also essentially the same as those produced by rapid cracking in liquid metals such as mercury (Fig. 1.44). Cleavage-like facets exhibiting numerous steps and tear ridges, and (possibly) nanoscale dimples are also present (Fig. 1.45).

The above observations are all consistent with an adsorption mechanism for SCC, involving adsorbed hydrogen for acid and aqueous environments, adsorbed metal atoms for liquid (and solid) metal environments, and adsorbed

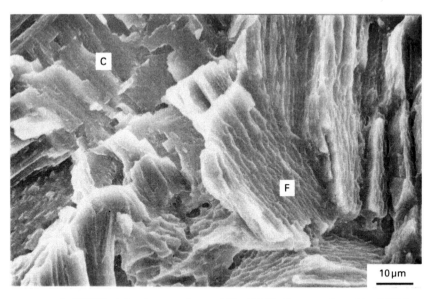

1.43 SEM of fracture surface of Mg-7%Al alloy produced by SCC in an aqueous chloride-chromate solution showing cleavage-like (C) and fluted (F) areas [125].

---

*Cleavage-like cracking on planes inclined 9–17° from basal planes have also been reported for SCC and LME in titanium alloys.

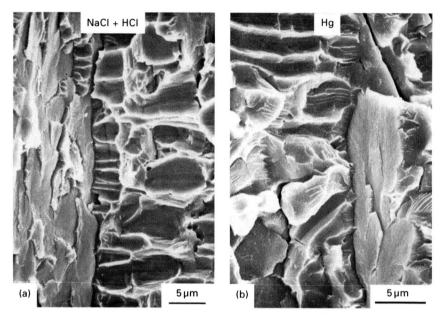

1.44 SEM of cleavage-like and fluted fracture surfaces for a Ti6Al4V alloy after sub-critical cracking in (a) NaCl + HCl solution, and (b) liquid mercury, at 20°C [73].

1.45 TEM of SCC fracture surface of a Ti6Al4V alloy, showing nanoscale tear ridges on a cleavage-like facet [125].

halogens and halides for dry halogen vapours and organic halogen solutions. For dry halogen vapours, reactions other than adsorption, e.g. dissolution, hydrogen effects, can be discounted, as they can for LME and SMIE. The

AIDE process or a combination of AIDE and adsorption-induced decohesion are more consistent with the observations than a decohesion process alone. Other mechanisms could be applicable in some circumstances, e.g. I-SCC predominates in Zr alloys in alcohol solutions containing halogens during the crack-initiation stage. Intergranular facets are featureless and crack growth is thought to occur by a localised dissolution process. There is generally no evidence for hydride formation during SCC, although these alloys can form hydrides in some circumstances, e.g. during cracking in hydrogen gas. Thus, a hydride mechanism cannot be discounted for SCC under unusual conditions.

1.11.2 Aluminium alloys

SCC of high-strength precipitation-hardened 7xxx Al-Zn-Mg-Cu alloys in aqueous and moist-air environments is intergranular in most circumstances, and cleavage-like fracture surfaces like those observed for the model Al-Zn-Mg single crystals occur only occasionally in service (Fig. 1.46) [164]. Such cases involve severe environments, with high stresses applied parallel to a pancake-shaped grain structure in plate/extrusions/forgings (when cracking along grain boundaries is difficult due to a tortuous intergranular crack path). Cleavage-like fracture surfaces on {100} and {110} planes (as in Al-Zn-Mg single crystals) do, however, commonly occur during corrosion fatigue of 7xxx alloys, especially when cycle frequencies are low. In such cases, there

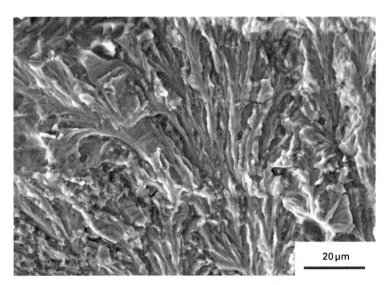

20 μm

1.46 SEM of cleavage-like fracture surface of 7075-T651 component exposed to an acidic environment during service resulting in T-SCC [125].

is no reason why the mechanisms of cracking should be different from those in the model single crystals – where it was argued that the AIDE mechanism best accounted for the observations. Suggestions that solute hydrogen ahead of cracks could be involved for SCC and CF in 7xxx alloys have been based on observations that cleavage-like cracking has been observed in hydrogen-charged specimens. However, such behaviour probably occurs because subsequent tests were carried out in moist air or because there was residual liquid or moisture present in pits or oxide films – such that adsorbed hydrogen would be present at crack tips during testing.

For I-SCC of 7xxx alloys in aqueous environments, fracture surfaces are often corroded, or fracture-surface details are obscured by corrosion products (thick oxide films exhibiting mud-crack patterns) (Figs 1.4 and 1.5) so that elucidating mechanisms of SCC from fractographic (and other) observations is difficult. However, it is reasonable to assume that mechanisms of cracking are the same in aqueous and in moist-air environments (where dissolution cannot occur). For example, there is no discontinuity in crack velocities as a function of relative humidity, and condensation at crack tips would be expected at very high relative humidity. Brittle intergranular fractures produced by SCC sometimes exhibit CAMs [31, 172] but no signs of dimples (from void-nucleation around grain-boundary precipitates), consistent with discontinuous cracking by hydrogen-induced decohesion (Fig. 1.47).

SCC of thin foils examined by TEM also showed that cracking occurred exactly along a grain boundary and around grain-boundary precipitates with

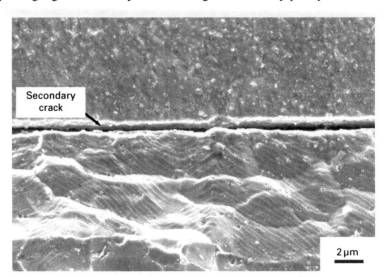

1.47 SEM of fracture surface produced by SCC in moist air for a 7079-T651 Al-Zn-Mg-Cu alloy showing CAMs on partially recrystallised grain facets but not on the flat grain facet in the upper half of the micrograph [31].

no evidence of plasticity [173]. Thus, in addition to the similarities between brittle intergranular fracture in hydrogen-charged specimens and in SCC specimens, the evidence for a hydrogen-induced decohesion mechanism is reasonably compelling. For sensitised 5xxx alloys with large area-fractions of anodic β-phase precipitates at grain boundaries, dissolution of these precipitates should make a significant contribution to crack growth, but hydrogen embrittlement of material between the precipitates is probably required [35].

1.11.3   Austenitic stainless steels

SCC in austenitic stainless steels in hot aqueous chloride or high-temperature pressurised-water environments can produce intergranular or transgranular cleavage-like cracking depending on (i) the degree of sensitisation, (ii) the extent of cold work, and (iii) environmental factors, e.g. temperature, oxidising versus reducing conditions. For temperatures <200°C in material that has not been sensitised, cleavage-like fracture surfaces parallel to {100} and {110} planes, and occasionally other low-index crystallographic planes, are usually observed [136, 174–176]. For {100} fractures, crack growth occurred in ⟨110⟩ directions, with slip on {111} planes intersecting cracks. CAMs (with a spacing ~1 μm), are also sometimes observed (Fig. 1.8(a)). Similar fracture surfaces have also been observed after SCC in high-temperature water (280°C) under some environmental conditions (Fig. 1.48) [177].

The characteristics of cleavage-like cracking in γ-stainless steels are remarkably similar to those observed for SCC (and LME) of Al-Zn-Mg

1.48 SEM of fracture surface produced by SCC in a concentrated neutral solution of chlorides and sulphates at 280°C for a 321 stainless steel, showing cleavage-like fracture surface with serrated steps and faint CAMs [177].

Published by Woodhead Publishing Limited, 2011

single crystals and SCC of Cu-Zn, Cu-Au and Ag-Au alloys, suggesting that similar mechanisms could be involved. For these model alloys, it was concluded that the AIDE mechanism occurred, involving adsorbed hydrogen (for Al-Zn-Mg alloys) and adsorbed ions produced by de-alloying (for Cu and Ag alloys). For stainless steels, hydrogen generation is known to occur in some environments, and there is evidence from hydrogen-charging experiments that hydrogen can be embrittling (at least for temperatures less than ~200°C). There is also some evidence that de-alloying may occur in some environments (or a porous oxide film might be produced at crack tips) [178, 179] so that a mechanism based on adsorption of metal-ion/anion complexes is also a possibility. Slip-dissolution mechanisms and film-induced-cleavage mechanisms have been proposed for T-SCC of austenitic stainless steels, but the former is generally considered to be inconsistent with a cleavage-like fracture-surface morphology, and the latter seems unlikely owing to the issues discussed for SCC in the model single-phase alloys. Other proposed mechanisms such as HELP also do not satisfactorily account for cleavage-like cracking, as discussed previously.

For sensitised material (and sometimes for the non-sensitised condition), I-SCC usually occurs, and crack tips have been characterised by cross-sectional analytical TEM [180–182]. These observations (for high-temperature I-SCC) have shown that cracks are remarkably narrow near the tips (typically ~5 nm wide extending up to ~100 nm behind crack tips). Various types of oxide were present in cracks, there was evidence of de-alloying, and nano-voids and high dislocation densities were observed ahead of cracks in some cases. One possible explanation for the observation of 'tight' cracks is that environmentally induced 'brittle' fracture occurs, perhaps involving adsorbed species or vacancy effects (or both) [182].

1.11.4   Copper alloys

SCC in commercial Cu alloys can produce both I-SCC (exhibiting essentially featureless facets) and T-SCC (exhibiting cleavage-like facets), depending on the material and environmental conditions [183, 184]. For fcc Cu alloys, the characteristics of SCC are the same as those observed for the model alloys exhibiting de-alloying. For α-β brasses, SCC occurs predominantly through the β-phase (producing cleavage-like fractures similar to those observed for SCC and LME of β-brass single crystals) and along α-β interfaces, with fracture surfaces often exhibiting CAMs [185, 186]. The mechanism of SCC involved for these commercial alloys is most probably the same as for the model copper-based alloys, where it was concluded that adsorption-induced SCC occurred. Indeed, a number of workers have concluded that adsorption-induced mechanisms are responsible for SCC in some of these commercial alloys [185–187].

Published by Woodhead Publishing Limited, 2011

1.11.5   Nickel alloys

For Ni alloys, most attention has been given to SCC in Ni-Cr-Fe alloys (e.g. alloy 600) in high-temperature water due to major SCC problems in this system in the nuclear industry [188, 189]. Predominantly I-SCC occurs but T-SCC has been observed occasionally. Cross-sectional TEM has been carried out and 'tight' cracks with similar dimensions as those described for stainless steels have been observed [180–182], suggesting that environmentally induced 'brittle' fracture might be involved. Hydrogen embrittlement seems unlikely at high temperatures for various reasons [112], and cracking is more likely to be associated with oxidation/de-alloying at crack tips, with crack advance occurring by adsorption-induced dislocation emission/decohesion or by fracture of grain boundaries weakened by internal oxidation/vacancy injection [112].

1.11.6   Ferritic steels

For high-strength martensitic steels, I-SCC along prior-austenite grain boundaries is much more common than T-SCC, but cleavage-like cracking does occur to some extent, depending on steel purity and other variables [15, 190] (Fig. 1.49). For precipitation-hardening martensitic stainless steels, SCC can be completely transgranular and cleavage-like in appearance (Fig. 1.50). For low strength steels, such as those used in pipelines, the crack path

1.49 SEM of fracture surface produced by SCC of a 300 M high-strength martensitic steel aircraft component in an aqueous environment, showing predominantly intergranular facets with small areas of cleavage-like cracking [28].

Published by Woodhead Publishing Limited, 2011

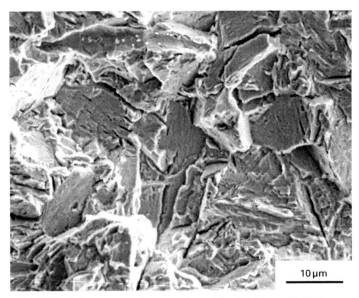

1.50 SEM of fracture surface produced by SCC in a 17-4PH martensitic stainless steel component from a yacht, showing cleavage-like facets [192].

depends on the local environment (especially pH), with I-SCC predominating for high pH conditions in concentrated carbonate/bi-carbonate solutions, and T-SCC occurring for low or near neutral pH solutions [5, 191].

I-SCC in high-strength steels is usually attributed to an effect of hydrogen (probably HEDE with contributions from AIDE and perhaps HELP), while I-SCC in low-strength steels is usually attributed to localised dissolution [5, 15, 190, 191]. Such hypotheses are reasonable based on the observations for the model systems, and comparisons of SCC with HEE, IHE and LME. For cleavage-like cracking, the AIDE mechanism probably predominates, possibly with minor contributions from HEDE and HELP, as discussed previously for other systems exhibiting cleavage-like cracking. When intergranular and cleavage-like facets are present in adjacent regions on fracture surfaces, it is tempting to assume that the same mechanism is operating for both fracture paths. While this may often be the case, it should be remembered that the local composition of the material and environment may be different for I-SCC and T-SCC so that different mechanisms could well occur.

1.12   General discussion of stress-corrosion cracking (SCC) mechanisms

Metallographic and fractographic observations, and other considerations, especially in model systems under specific environmental and testing conditions, suggest that most cases of SCC involve predominantly one of three environmental interactions:

1. Adsorption at tips of cracks/voids, especially for T-SCC.
2. Hydrogen generation, diffusion and segregation at grain boundaries ahead of crack tips for slow I-SCC.
3. Dissolution at intergranular crack tips.

Both the first and the second interactions involve weakening of substrate interatomic bonds at or ahead of crack tips, which facilitate decohesion or dislocation emission. Mechanisms based predominantly on dissolution at crack tips appear to be applicable to only a few materials/environments where chemically active paths along grain boundaries are present (due to segregation of solute/impurities or to precipitation/solute-depletion). If active paths are not continuous, contributions from decohesion or dislocation emission due to adsorbed/absorbed hydrogen are probably then necessary.

I-SCC and T-SCC due to hydride formation around crack tips appear to be applicable in only limited circumstances, even in hydride-forming materials. For the film-induced cleavage mechanism there are doubts whether it occurs in any circumstances, for reasons discussed on pages 50–51. The surface-mobility mechanism is considered to be unviable by a number of workers [3, 11, 146, 148], and fails to properly account for the metallographic and fractographic observations, as mentioned previously. The HELP mechanism is unlikely to make a major contribution to SCC in most circumstances since (i) there is no obvious reason why HELP should produce cleavage-like fracture surfaces, and (ii) if high hydrogen concentrations were present around grain boundaries, the highest concentration would be exactly at boundaries, tending to favour HEDE or AIDE.

These views will undoubtedly be controversial, but an objective assessment of the evidence has been attempted. Moreover, this assessment is based on the approach using model systems described in detail earlier, which is more discriminatory in regard to elucidating mechanisms than other approaches in the author's opinion. The basis for the conclusion that mechanisms based on adsorption, grain-boundary hydrogen segregation, and dissolution appear to cover most SCC systems are considered further in the following section, and unresolved issues concerning these mechanisms are discussed.

1.12.1 Adsorption-based mechanisms

Summary of evidence

Adsorption-based mechanisms for T-SCC in many materials and I-SCC in some materials are supported by (or are consistent with):

• An ability to account for cleavage-like and brittle intergranular fractures that are either associated with little or no plasticity if adsorption-induced decohesion occurs, or associated with considerable localised plasticity and dimpled on a nano/microscopic scale if the AIDE mechanism occurs.

- Observations of SCC at high velocities, where processes other than adsorption can probably be discounted, and a similar fracture-surface appearance, crystallography, and slip distribution around cracks grown at high and low crack velocities.
- Remarkable similarities in fracture-surface appearance and crystallography (and slip distributions around cracks) between SCC in aqueous environments and cracking in liquid-metal environments (in many materials) and in halogen vapours (for Zr and Ti alloys), where interactions other than adsorption are not likely to occur.
- Observations of abrupt transitions from brittle to ductile behaviour (and vice versa), or abrupt changes in the degree of embrittlement, when environmental conditions are changed.
- Surface-science observations showing that adsorption can have significant (and sometimes dramatic) effects on bonding at plane surfaces, which would also be expected to occur at crack-tip surfaces.
- Quantum-mechanical calculations indicating that hydrogen adsorption can weaken substrate interatomic bonds, and thereby facilitate dislocation emission or decohesion (as discussed in more detail in Chapter 2).
- An ability to account for the effects of variables on SCC, as discussed further in Section 1.12.5.

Historical objections to adsorption mechanisms

Various objections have been raised regarding adsorption mechanisms for SCC in the past [5, 111, 193], but these objections can be discounted for some systems now that adsorption and fracture processes are better understood. The main objections to adsorption mechanisms, and rebuttals to them, are as follows:

- Early adsorption mechanisms for SCC were based on surface-energy reductions, and it was argued that the overall fracture energy (dominated by a plastic-work term) would not be significantly reduced by decreasing the surface energy. This objection is not applicable to the AIDE mechanism, where the plastic strain required for crack growth is smaller (more localised) than that for cracking in inert environments.
- An adsorption mechanism has also been discounted because not all adsorbates produce embrittlement. It is now known that different adsorbates can have quite different effects on surface bonding, and that embrittlement would not be expected when strong adsorbate-substrate bonds are formed, as discussed in more detail in Section 1.12.4.

Published by Woodhead Publishing Limited, 2011

- An adsorption mechanism based on decohesion was considered questionable on the grounds that significant dislocation activity was observed and, hence, cracks were unlikely to be atomically sharp during SCC in ductile materials. This argument is reasonable for many cases, but obviously does not apply to the AIDE mechanism where various degrees of crack-tip blunting can occur depending on the stresses required for AIDE.

- Observations of discontinuous SCC were considered inconsistent with adsorption mechanisms. However, there are a number of reasons why adsorption-induced SCC could be discontinuous, such as the intermittent generation and exhaustion of embrittling adsorbates generated by de-alloying, as discussed in Section 1.9.4. Other reasons for discontinuous cracking, which could apply to adsorption and other mechanisms, are considered in Section 1.12.4.

- It has been suggested that significant rates of dissolution observed in straining-electrode tests are not consistent with an adsorption mechanism. However, dissolution might be required to produce embrittling adsorbates (as proposed for de-alloying systems) and dissolution could occur just behind the crack tip after crack advance by adsorption-induced decohesion or dislocation emission in some systems. (In other systems, I-SCC could well occur predominantly by dissolution, as indicated previously.)

- Surprise has sometimes been expressed that void nucleation and growth ahead of cracks (which are an essential part of the AIDE model for sustained-load cracking) can be affected by adsorption at crack tips, given that the range of influence of adsorption is only a few atomic distances. It seems not to have been appreciated that facilitating crack growth at external crack tips (by adsorption or other processes) would reduce the extent of crack-tip blunting and plasticity ahead of cracks, thereby affecting void nucleation and growth in the plastic zone ahead of cracks. For SCC involving hydrogen, diffusion of hydrogen to, and adsorption at, tips of voids could sometimes occur so that void growth could be directly affected.

- It has been suggested that rates of SCC are slower than would be expected if an adsorption mechanism was responsible, possibly on the basis that adsorption-induced LME usually (but not always) occurs very rapidly. However, rate-controlling steps for SCC and LME are probably different, and slow steps for SCC (which are not applicable to LME) could include: (i) chemical dissociation of water to produce hydrogen, (ii) anodic and cathodic reactions, (iii) generation of embrittling adsorbates by de-alloying, (iv) hydrogen diffusion from surface sites to lattice sites between the first few atomic layers, and (v) film-rupture to enable adsorption to occur.

1.12.2   Summary of evidence for hydrogen-induced decohesion mechanisms

SCC mechanisms based on decohesion due to solute hydrogen segregated at grain boundaries ahead of cracks are based on:

- Experimental evidence that high concentrations of hydrogen are present at grain boundaries, e.g. detection of hydrogen (using mass spectroscopy) released when hydrogen-charged specimens are fractured (in-situ) along grain boundaries.
- Brittle intergranular fracture with no obvious signs of localised plasticity or dimples on fracture surfaces for hydrogen-charged specimens tested in inert environments, as well as for specimens fractured in aqueous and moist-air environments under conditions where hydrogen diffusion can occur.
- Quantum-mechanical calculations showing that hydrogen atoms (as well some impurities) can weaken interatomic bonds across grain boundaries.
- CAMs on fracture surfaces, which could be explained in terms of crack jumps and arrests corresponding to the development of a critical hydrogen concentration over a critical distance – although other explanations are also possible as discussed in Section 1.12.4.

As already mentioned, mechanisms based on adsorption and on hydrogen segregated at grain boundaries both involve weakening of interatomic bonds and, in some ways, there are only subtle differences between these mechanisms. For example, crack nuclei could form ahead of cracks by hydrogen-induced decohesion and propagate back towards the main crack tip. However, once such cracks are nucleated, their growth would be facilitated essentially by 'adsorbed' hydrogen, i.e. hydrogen atoms at and within a few atomic distances of the crack tip, providing that hydrogen was present at most grain-boundary sites. Hydrogen-induced decohesion could occur not only at grain-boundary sites but also at grain-boundary-precipitate/matrix interfaces. Once nucleated, cracks could propagate by AIDE, or AIDE plus HEDE, rather than by HEDE alone.

1.12.3   Summary of evidence for dissolution-based mechanisms

A dissolution mechanism for SCC in some materials is supported by observations that dissolution occurs preferentially along grain boundaries in unstressed specimens, especially when anodically active grain-boundary microstructural features are present. A slip-dissolution process is also supported by the observations for the model Fe-Si bi-crystals (Section

1.10.2). Correlations between maximum crack-growth rates and current densities expected at crack tips (from scratching electrode tests) for some materials are also consistent with a dissolution mechanism [12], although such correlations cannot be considered to be definitive evidence. Fractographic observations, showing featureless or corroded intergranular facets, are also consistent with a dissolution mechanism, but again should not be taken as evidence in support of such mechanisms. Dissolution could occur behind crack tips growing by other mechanisms and thereby obliterate any features that might have been produced by another crack growth mechanism.

1.12.4 Some unresolved issues

Adsorbate specificity

The reason why some adsorbates, but not others, are embrittling is not fully understood, although there are general guidelines for LME [127, 194, 195]. For example, embrittlement is not generally observed when there is a tendency for intermetallic compound formation between the metals, i.e. strong bonds form between adsorbate and substrate atoms. Embrittling adsorbates are probably those that (i) weaken substrate interatomic bonds, and (ii) form only weak bonds with the substrate and between themselves, because all these bonds have to be sheared during the nucleation of a dislocation at a crack tip. The evidence outlined in previous sections suggests that embrittling adsorbates include (i) hydrogen atoms (on the surface and just sub-surface) for HE and SCC in many materials, (ii) halogens (for SCC in a few materials), (iii) some complex ions, such as those produced by de-alloying (for SCC in some binary metal alloys), and (iv) some metal atoms (for LME and SMIE).

Some liquid-metal environments do not produce embrittlement even though weak bonding would be expected between adsorbate and substrate atoms. For example, nickel is severely embrittled by lithium, moderately embrittled by sodium (and mercury), but not by potassium and rubidium [125, 194, 195]. Magnesium, on the other hand, is embrittled by all the alkali metals, but not by mercury [125, 194]. For copper, mercury results in severe intergranular LME, but transgranular LME does not occur [125], whereas LME in most other materials can occur along both intergranular and transgranular crack paths. This specificity has been a long-standing, unresolved issue for LME, and needs to be addressed using state-of-the-art quantum-mechanical modelling techniques.

Adsorption of ions on metals in aqueous environments is more complex than adsorption of metal atoms, but there is no reason why embrittlement should not occur in some cases providing charge-transfer (or partial charge-transfer) occurs. Surface-science observations show that specific adsorption of ions can result in changes in bonding at surfaces, and that small changes in environmental conditions (e.g., solution concentration, temperature and potential) can have large effects. Many of these effects at plane, unstressed surfaces are not well understood, and the effects of adsorption at crack-tip surfaces are even less well understood, and are likely to be significantly different from those at plane surfaces due to high crack-tip stresses and crack-tip curvature.

Crack growth by decohesion versus localised plasticity

Distinguishing between crack growth by an atomically brittle decohesion process along cleavage planes or grain boundaries and crack growth by a localised slip/nanovoid coalescence process is difficult when plasticity is extremely localised. Dimples produced by a nanovoid (or microvoid) coalescence process can be extremely shallow, with cusps as small as several nanometres, and may not be resolved by conventional fractographic techniques. Thus, fracture surfaces produced by localised plasticity can appear to be flat (between steps) like those that would be produced by decohesion. Determining whether crack growth occurs by a combination of decohesion and dislocation emission, rather than by just one of these processes, is similarly difficult owing to resolution issues. For SCC, even minor corrosion/film-formation on fracture surfaces could obscure fine details, thereby compounding the problem. Decohesion almost certainly occurs in some circumstances, but concluding that decohesion has occurred based on fractographic observations, e.g. the presence of featureless areas in SEM micrographs (at ~10,000×), is not justified (but is common).

Fracture planes and directions

Why T-SCC in some fcc metals occurs on {110} planes in some cases and on {100} planes in other cases is not understood. Fracture on {100} planes in ⟨110⟩ directions would be expected if the AIDE alternate-slip mechanism operated since (i) crack fronts lie along the line of intersection of {111} slip planes and crack planes, and (ii) there are two ⟨110⟩ directions available, enabling crack fronts to have a zig-zag shape on a microscopic scale so that cracks can grow in directions favoured by the applied stress on a macroscopic scale. An alternate-slip mechanism can also account for SCC on {110} planes if crack growth occurs in ⟨100⟩ directions. However, crack growth sometimes occurs in other directions that cannot be explained in terms of a simple alternate-slip process. Crack growth could still occur by a localised-slip mechanism, but perhaps one involving alternate-shear, with complex slip within shear bands. Such a process has been proposed for fatigue crack growth where macroscopically planar {110} fractures in directions other than ⟨100⟩ have been observed [196].

The presence of fine-scale {111} serrations which are sometimes observed on macroscopic {110} planes (Fig. 1.7) is also not understood. Mechanisms such CELP (Fig. 1.20), involving localised dissolution along a {111} plane followed by shear plus fracture on this plane and then fracture on an alternate slip plane, have been suggested to try to explain such features. However, such processes are not consistent with some observations, e.g. the presence of the above features on fracture surfaces produced by rapid cracking after superficial de-alloying where dissolution would not have time to occur. It is also not clear why stresses due to dislocation pile-ups against obstacles on one slip plane would be relieved by fracture along the slip plane rather than by slip on another slip plane.

Continuous versus discontinuous cracking

Why SCC is discontinuous in some cases, but appears to be continuous in other cases, is not well understood. There are, however, numerous reasons why SCC could be discontinuous, thereby producing CAMs on fracture surfaces, although which explanation is applicable in particular cases is often not clear [197]. Possible explanations for discontinuous cracking include:

- Formation and intermittent fracture of ligaments of material lagging behind the main crack tip, so that variations in the effective stress-intensity factor occurs.
- Variations in the dislocation sub-structure ahead of cracks that could result in intermittent dislocation avalanches that blunt crack tips.
- Intermittent generation of embrittling adsorbates during crack growth, as discussed in Section 1.9.4 for SCC systems involving de-alloying.
- Crack-growth increments when a critical hydrogen concentration is achieved over a critical distance, followed by crack-arrest when hydrogen concentrations fall below the critical value.
- Development of nano-voids ahead of crack tips leading to a sudden crack jump when a critical number of voids have formed over a critical distance.

1.12.5   Predicting SCC rates and the effects of variables

The effects of many variables on SCC can be rationalised by a number of mechanisms. For example, surface-science observations have shown that electrode-potential, pH, solution-composition, temperature, and stress can all influence adsorption characteristics, with the effects depending on the material. Thus, the stresses required for dislocation emission or decohesion at crack tips could be influenced by the effect of these variables on bonding and, hence, their effect on SCC rationalised in terms of adsorption mechanisms.

Published by Woodhead Publishing Limited, 2011

Other variables, such as crack-tip strain rate, could change the degree of adsorption-induced embrittlement due to variations in surface-coverage of adsorbed hydrogen (on the surface and between the first few atomic layers). However, a deeper level of understanding (at the electronic level) regarding why certain adsorbed species but not others are embrittling is currently not available, even for the relatively simple case of adsorption-induced LME (as mentioned in Section 1.12.4).

Many of the above variables could also affect the kinetics of film formation at crack tips, as well as film characteristics such as fracture strain, thereby determining the time-intervals when crack tips are film-free such that adsorption could occur on the substrate surface. Environmental changes that promote more rapid repassivation could not only inhibit adsorption-induced SCC but could also inhibit dissolution and ingress of hydrogen at crack tips. Thus, it is difficult to discriminate between mechanisms based on these various processes on the basis of the effects of variables on SCC susceptibility (which is one reason why metallographic and fractographic observations of model systems reviewed in this chapter are the best way of discriminating between mechanisms).

The effects of temperature and electrode-potential on SCC/CF of aluminium alloys, where the evidence that adsorption is responsible for T-SCC/CF is compelling, illustrate the difficulties in definitively accounting for the effects of variables. As shown in Fig. 1.31, increasing temperature increases crack-growth rates for Al-Zn-Mg single crystals in NaCl solutions but decreases crack-growth rates in distilled water. Similar effects of temperature on crack-growth rates have been observed for commercial Al-Zn-Mg-Cu alloys [198]. On the other hand, the effects of potential on crack-growth rates are different for Al-Zn-Mg single crystals compared with Al-Zn-Mg-Cu commercial alloys. For the latter, changing from anodic to cathodic potentials resulted in changes from brittle striations to ductile striations (and a significant decrease in striation spacing) [199], whereas there was no effect of changing potential on brittle striation spacing for Al-Zn-Mg single crystals [133]. The presence of copper (~1.5 wt%) in the commercial alloys is probably responsible for the difference in behaviour. Thus, there are clearly subtle effects involved, and considerable further work is required to provide detailed explanations.

The effects of microstructure and strength on SCC resistance of materials in general can also be rationalised in various ways, and it is also difficult to be definitive regarding precise explanations. For example, increasing strength is likely to inhibit general dislocation activity in the plastic zone ahead of cracks to a greater extent than inhibiting short-range dislocation emission from crack tips, so that embrittlement due to AIDE would be increased. Hydrostatic stresses ahead of crack tips should also be greater for high-strength materials so that higher solute hydrogen concentrations would occur, thereby leading to greater embrittlement due to HEDE. Changing strength (via modifications to heat-treatment or composition) could also lead to changes in grain-boundary microstructure and microchemistry, thereby affecting SCC involving dissolution/repassivation at grain boundaries. Strength would also influence crack-tip strain rates so that oxide-rupture rates and dependent phenomena could be affected. In addition, microstructural features will affect hydrogen trapping/diffusion and nucleation of voids ahead of cracks, and thereby impact on SCC by AIDE and HEDE processes.

Making quantitative predictions based on most proposed mechanisms is generally not possible, and only the slip-dissolution mechanism has been developed to any extent in this regard. For this mechanism, the crack-growth rate depends on the anodic current density via Faraday's laws and on the rate of repassivation at crack tips controlled by the crack-tip strain rate [62, 200]. For cleavage-like T-SCC, an additional term incorporating an increment of crack growth due to mechanical fracture, e.g. film-induced cleavage, has been introduced into the equation [62]. However, predictions cannot be treated with any confidence given that there are questions concerning the viability of the film-induced cleavage mechanism, and the likelihood that adsorption-induced cracking and hydrogen-enhanced decohesion (which are difficult to quantify) probably make some contribution to SCC even when dissolution predominates.

A quantitative expression for SCC rates has also been suggested for Galvele's surface-mobility mechanism in terms of environmentally assisted surface-diffusion coefficients and other parameters [8, 93–96], but also cannot be used with confidence given the doubts expressed [3, 11, 146, 148, 150, 151] regarding the fundamental basis of this mechanism. On the other hand, Galvele and co-workers have drawn attention to correlations between crack-growth rates and surface phenomena, which need to be explained. Interestingly, the concept of environmentally enhanced surface-mobility is, in some respects, also the basis of the AIDE mechanism. Thus, dislocation nucleation at crack tips (i.e. the formation of a dislocation core and surface step by shear movement of atoms) involves mobility of surface and near-surface atoms, which is facilitated by adsorption-induced weakening of substrate interatomic bonds. Some of the predictions of the SMM, and correlations between rates of SCC and measures of surface mobility, e.g. exchange-current density, might therefore arise because of a link between surface mobility and dislocation nucleation at crack tips – a suggestion made by the present author and taken up by Newman in a recent review [3].

Attempting to quantify the AIDE mechanism is not possible given that so little is known regarding the effects of adsorption on substrate bonding and the consequent effects on dislocation emission. Realistically modelling the effects of adsorption on dislocation emission is challenging because the process depends not only on the structure and bonding at crack tips, but also on a number of other parameters [201] such as: (i) stress-mode and the extent of 'shear-softening' due to tensile stresses across slip planes, (ii) whether emission occurs on inclined planes or on oblique planes at ledges along the crack front, and the angle of these planes with respect to the crack plane, (iii) the crack-tip radius and core-width of nucleating dislocations, (iv) the shape of the emitted dislocation loop, and whether full or partial dislocations are involved, and (v) the extent of shielding/back-stresses from previously emitted dislocations, which will depend on the mobility of emitted dislocations and their ability to cross-slip, which may depend on the HELP phenomenon.

Quantifying the HEDE mechanism for I-SCC is also not possible since, amongst other issues, rates of hydrogen diffusion along grain boundaries are not well established in many materials, and probably depend on grain-boundary misorientation and degree of grain-boundary segregation. Whether hydrogen transport by dislocations from crack tips to sites ahead of crack tips is significant is also not known. The degree of bond-weakening due to hydrogen (and other segregants) is also difficult to quantify, although attempts have been made using quantum-mechanical-based modelling techniques.

While the major aim of mechanistic studies, i.e., predicting rates of crack growth so that structures and components can be inspected/repaired/replaced before failures occur, may be unrealistic, there are some benefits from an appreciation of mechanisms of SCC (and other fracture modes). Thus, mechanistic insights have had some input into (i) designing more fracture-resistant materials, e.g. by compositional and microstructural changes, (ii) modifying environmental conditions to inhibit cracking, (iii) enabling better judgements to be made regarding whether data from laboratory testing are appropriate to structural-integrity assessments of components in service, and (iv) diagnosing modes and causes of failure, e.g. by having a better idea of what processes can produce various fracture-surface features. The mechanistic understanding obtained to date is also useful in indicating directions for future studies, as indicated in Section 1.12.4, in order to advance understanding of mechanisms of SCC.

1.13   Conclusions

1. Detailed metallographic and fractographic studies of single crystals and bi-crystals of selected materials, using a range of environments and testing conditions, are the best way of determining the viability and applicability of SCC mechanisms.
2. The model-system observations suggest that only three mechanisms have wide applicability: (i) adsorption-induced dislocation-emission (AIDE) or decohesion at crack tips, (ii) hydrogen-enhanced decohesion (HEDE) ahead of crack tips, and (iii) localised strain-assisted dissolution.
3. Combinations of the above processes may occur in some materials

and environments, with the predominant mechanism depending on the fracture mode, which in turn depends on the material/environment/ testing conditions. Other processes, e.g. HELP due to solute hydrogen, and vacancy effects ahead of cracks, possibly play a minor role in some cases.

4. Adsorption-induced SCC can be associated with (i) hydrogen atoms on and within a few atomic distances of crack tips, (ii) some ionic species (including those produced by dissolution/de-alloying in some systems), and (iii) halogens (for Zr and Ti alloys), all of which can weaken substrate interatomic bonds at crack tips. Adsorption of metal atoms can also facilitate cracking (i.e., produce liquid-metal embrittlement (LME)), and there are remarkable similarities between the metallographic and fractographic characteristics of SCC/HE and LME.

5. The AIDE mechanism probably predominates for cleavage-like transgranular SCC when substantial localised strains are involved, as appears to be the case in most systems. The AIDE mechanism probably also contributes to intergranular SCC in some systems when localised strains are high.

6. The hydrogen-enhanced decohesion process at or ahead of crack tips probably predominates for intergranular SCC where little plasticity is evident, with hydrogen generated at crack tips accumulating at grain-boundary sites and weakening interatomic bonds across grain boundaries (in conjunction with other segregants in some cases).

7. Dissolution-based mechanisms are applicable for intergranular SCC in materials with anodically active grain-boundary paths, resulting from some types of grain-boundary precipitation or segregation. If such paths are not continuous, HEDE or AIDE processes may also be necessary in addition to dissolution.

8. The film-induced-cleavage hypothesis for SCC, in systems where de-alloying occurs, is at odds with a number of observations, e.g. (i) the substantial localised plasticity often associated with SCC, (ii) crack velocities (during crack jumps) much less than that theoretically required for cleavage in fcc materials, and (iii) the fact that cracks in the ligaments of de-alloyed films are not contiguous with the substrate ahead of cracks.

9. Fractographic aspects of SCC in most materials are well characterised, but there are a number of aspects that are not well understood. For example, crack-arrest markings on fracture surfaces, produced by discontinuous crack growth, could occur for a number of reasons, and which explanation is applicable in specific circumstances is not always clear. The factors determining the crystallographic planes for cleavage-like SCC (besides the tensile stress normal to the fracture plane) are also not well established.

Published by Woodhead Publishing Limited, 2011

10. The specificity of adsorbates (for SCC and LME), (i.e., why some adsorbates facilitate cracking while others do not) is also not well understood except that, in general, adsorbate–substrate and adsorbate–adsorbate bonds should not be too strong, and should weaken substrate–substrate bonds.

11. The effects of variables, such as electrode potential, solution composition, and temperature on SCC can be rationalised in terms of some of the proposed mechanisms in terms of their effects on adsorption, film formation and crack-tip hydrogen concentration, but it is not always clear which explanation is applicable.

12. Quantitative, mechanistically based, predictions of crack-growth rates cannot be made with any confidence due to the uncertainties discussed in the previous section. However, current mechanistic understanding of SCC (and other fracture modes) is useful, notwithstanding the limitations regarding quantitative predictions, since it provides general guidance in regard to aspects of SCC such as testing procedures, interpretation of data, and analysing failures.

1.14   Acknowledgements

Thanks are due to Alan Turnbull (National Physical Laboratory, UK), Geoff Scamans (Innoval Technology, UK), Anna Hojna (Nuclear Research Institute, Czech Republic), José Galvele (CNEA, Argentina), and Mark Knop (DSTO, Australia) for their critical reading of the chapter and suggested modifications. Discussions with Roger Newman (University of Toronto) and Karl Sieradzki (Arizona State University) regarding the roles of de-alloying in SCC have also been valuable. Rohan Byrnes and other colleagues at DSTO have also kindly provided some of their unpublished micrographs.

1.15   References

1. R.H. Jones, Stress corrosion cracking, pp. 145–163 in *ASM Metals Handbook*, Vol. 13, *Corrosion*, 9th edn, ASM, Metals Park, OH, 1987.
2. *Stress Corrosion Cracking*, NACE data survey, Houston, TX, March 1974.
3. R.C. Newman, Stress corrosion cracking, pp. 864–901, in *Shreir's Corrosion*, 4th edn, Vol. 2, J.A. Richardson *et al.* (eds), Elsevier, Amsterdam, 2010.
4. R.H. Jones (ed.), *Stress-Corrosion Cracking: Materials Performance and Evaluation*, ASM, Metals Park, OH, 1992.
5. R.N. Parkins, Current understanding of stress-corrosion cracking, *J. of Metals*, 12–19, Dec. 1992.
6. A. Turnbull, Modelling of environment assisted cracking, *Corros. Sci.*, 34, 921–960, 1993.
7. P. Neumann, Embrittlement due to gaseous and liquid environments, *Mater. Sci. and Engng*, A176, 9–18, 1994.
8. J.R. Galvele, Electrochemical aspects of stress corrosion cracking, pp. 233–358

in *Modern aspects of electrochemistry 27*, R.E. White *et al.* (eds), Plenum Press, New York, 1995.

9. T. Magnin, Environment sensitive fracture, pp. 207–263 in *Materials Science and Technology: A Comprehensive Treatment*, R.W. Cahn, P. Haasen, and E.J. Kramer (eds), Corrosion and Environmental Degradation, Vol. 1, Volume Editor: M. Schütze, Wiley, New York, 2000.

10. R.P. Gangloff, Hydrogen assisted cracking of high strength alloys, pp. 31–101 in *Comprehensive Structural Integrity*, I. Milne, R.O. Ritchie, and B. Karihaloo (eds), Vol. 6, Environmentally Assisted Fracture, Elsevier, Amsterdam, 2003.

11. R.C. Newman, Stress-corrosion cracking mechanisms, pp. 399–450 in *Corrosion Mechanisms in Theory and Practice*, 2nd edition, P. Marcus and J. Oudar (eds), Marcel Dekker, New York, 2002.

12. H. Hänninen, Stress corrosion cracking, pp. 1–29 in *Comprehensive Structural Integrity*, Vol. 6, I. Milne, R.O. Ritchie, and B. Karihaloo (eds), vol. 6, Environmentally Assisted Fracture, Elsevier, Amsterdam, 2003.

13. R.W. Staehle, A.J. Forty, and D. van Rooyen (eds), *Fundamental Aspects of Stress Corrosion Cracking*, NACE, Houston, TX, 1969.

14. J.C. Scully (ed.), *The Theory of Stress Corrosion Cracking*, NATO, Brussels, 1971.

15. R.W. Staehle, J. Hochmann, R.D. McCright and J.E. Slater (eds), *Stress Corrosion Cracking and Hydrogen Embrittlement of Iron Base Alloys*, NACE, Houston, TX, 1977.

16. P.R. Swann, F.P. Ford and A.R.C. Westwood (eds), *Mechanisms of Environment Sensitive Cracking of Materials*, The Metals Society, London, UK, 1977.

17. R.H. Jones and W.W. Gerberich (eds), *Modeling Environmental Effects on Crack Growth Processes*, TMS, Warrendale, PA, 1986.

18. R.P. Gangloff and M.B. Ives (eds), *Environment-induced Cracking of Metals*, NACE-10, NACE, Houston, TX, 1990.

19. S.M. Bruemmer, E.I. Meletis, R.H. Jones, W.W. Gerberich, F.P. Ford and R.W. Staehle (eds), *Parkins Symposium on Fundamental Aspects of Stress Corrosion Cracking*, TMS, Warrendale, PA, 1992.

20. T. Magnin and J.-M. Gras (eds), *Corrosion-Deformation Interactions*, Les Editions de Physique, Les Ulis, 1993.

21. T. Magnin (ed.), *Corrosion-Deformation Interactions*, The Institute of Materials, London, 1997.

22. N.R. Moody, A.W. Thompson, R.E. Ricker, G.W. Was and R.H. Jones (eds), *Hydrogen Effects on Materials Behavior and Corrosion Deformation Interactions*, TMS, Warrendale, PA, 2003.

23. S.A. Shipilov, R.H. Jones, J.-M. Olive and R.B. Rebak (eds), *Environment-Induced Cracking of Metals*, Elsevier, Amsterdam, 2008.

24. J.A. Feeney and M.J. Blackburn, The status of stress corrosion cracking of titanium alloys in aqueous solutions, pp. 355–398, in *The Theory of Stress Corrosion Cracking*, J.C. Scully (ed.), NATO, Brussels, 1971.

25. P. McIntyre, The relationships between stress corrosion cracking and sub-critical flaw growth in hydrogen and hydrogen sulphide gases, pp. 788–815 in *Stress Corrosion Cracking and Hydrogen Embrittlement of Iron Base Alloys*, R.W. Staehle *et al.* (eds), NACE, Houston, TX, 1977.

26. W. Dietzel and A. Turnbull, Stress corrosion cracking, pp. 43–74 in *Comprehensive Structural Integrity, Addendum 2007: Mechanical Characterization of Materials*, K.-H. Schwalbe (Ed.), Elsevier, Oxford, 2007.

Published by Woodhead Publishing Limited, 2011

27. A. Turnbull, D.A. Horner and B.J. Connolly, Challenges in modelling the evolution of stress corrosion cracks from pits, *Engng Fract. Mech.*, 76, 633–640, 2009.

28. Defence Science and Technology Organisation, Melbourne, Australia, unpublished work.

29. J.I. Dickson, Li Shiqiong, J.-P. Bailon and D. Tromans, The fractography of transgranular SCC in F.C.C. metals: mechanistic implications, pp. 303–322 in *Parkins Symposium on Fundamental Aspects of Stress Corrosion Cracking*, S.M. Bruemmer *et al.* (eds), TMS, Warrendale, PA, 1992.

30. R.J.H. Wanhill, National Aerospace Laboratory, The Netherlands, unpublished work.

31. S.P. Knight, Stress corrosion cracking of Al-Zn-Mg-Cu alloys, Ph.D thesis, Monash University, 2008.

32. *Metals Handbook*, 9th edn, Vol. 11, Failure Analysis and Prevention, ASM, Metals Park, OH, 1986.

33. M.O. Speidel, Current understanding of stress corrosion crack growth in aluminum alloys, pp. 289–344 in *The Theory of Stress Corrosion Cracking*, J.C. Scully (ed.), NATO, Brussels, 1971.

34. S.P. Lynch, S.P. Knight, N. Birbilis and B.C. Muddle, Stress corrosion cracking of Al-Zn-Mg-Cu alloys: effects of composition and heat treatment, pp. 243–250 in *Effects of Hydrogen on Materials, Proc. of the 2008 Int. Hydrogen Conf., Wyoming, USA*, B. Somerday, P. Sofronis, and R. Jones (eds), ASM, Metals Park, OH, 2009.

35. R.H. Jones, D.R. Baer, M.J. Danielson and J.S. Vetrano, Role of magnesium in the stress corrosion cracking of an Al-Mg alloy, *Metall. and Mater. Trans. A*, 32A, 1699–1711, 2001.

36. A.J. Sedriks, Stress corrosion cracking of stainless steels, pp. 91–130 in *Stress-Corrosion Cracking: Materials Performance and Evaluation*, R.H. Jones (ed.), ASM, Metals Park, OH, 1992.

37. K. Sieradzki, J.S. Kim and R.C. Newman, Relationship between dealloying and transgranular stress-corrosion cracking of Cu-Zn and Cu-Al alloys, *J. Electochem. Soc.*, 134, 1635–1639, 1987.

38. R.M. Latanision, Surface effects in crystal plasticity: general overview, pp. 3–47 in *Surface Effects in Crystal Plasticity*, R.M. Latanision and J.T. Fourie (eds), NATO Advanced Study Inst. Series E: Applied Science No. 17, Noordhoff Int. Publ., 1977.

39. D.M. Kolb, An atomistic view of electrochemistry, *Surface Science*, 500, 722–740, 2002.

40. D.M. Kolb, Reconstruction phenomena at metal-electrolyte interfaces, *Progress in Surf. Sci.*, 51, 109–173, 1996.

41. O.M. Magnussen, Ordered anion adlayers on metal electrode surfaces, *Chem. Rev.*, 102, 679–725, 2002.

42. M.A. Van Hove, Crystal surfaces, in *Materials Science and Technology: A Comprehensive Treatment*, R.W. Cahn, P. Haasen and E.J. Kramer (eds), 1, 485–531, 1991, and references therein.

43. D.P. Woodruff (ed.), *The Chemical Physics of Solid Surfaces, Vol. 10, Surface Alloys and Alloy Surfaces*, Elsevier, Amsterdam, 2002.

44. D.L. Adams and J.N. Andersen, Alkali-aluminum surface alloys, pp. 225–276 in *The Chemical Physics of Solid Surfaces, Vol. 10, Surface Alloys and Alloy Surfaces*, D.P. Woodruff (ed.), Elsevier, Amsterdam, 2002.

45. H. Matsushima, A. Taranovskyy, C. Haak, Y. Gründer and O.M. Magnussen, Reconstruction of Cu(100) electrode surfaces during hydrogen evolution, *J. Am. Chem. Soc.*, 131, 10362–10363, 2009.

46. J.W. Davenport and P.J. Estrup, Hydrogen on metals, *Rev. of Mod. Phys.*, pp. 1–37, 1992, and references therein.

47. T.E. Fischer, Hydrogen on metal surfaces, pp. 135–148 in *Advanced Techniques for Characterizing Hydrogen in Metals*, N.F. Fiore and B.J. Berkowitz (eds), Met. Soc. AIME, 1982.

48. K. Christmann, Some general aspects of hydrogen chemisorption on metal surfaces, *Progress in Surf. Sci.*, 48, 15–26, 1995.

49. E. Protopopoff and P. Marcus, Surface effects on hydrogen entry into metals, pp. 53–96, in *Corrosion Mechanisms in Theory and Practice*, 2nd edn, P. Marcus (ed.), Marcel Dekker, New York, 2002.

50. J. Oudar, Introduction to surface reactions: adsorption from the gas phase, pp. 19–51 in *Corrosion Mechanisms in Theory and Practice*, 2nd edn, P. Marcus (ed.), Marcel Dekker, New York, 2002.

51. W. Moritz, R.J. Behm, G. Ertl, G. Kleinle, V. Penka, W. Reimer and M. Skottke, Relaxation and reconstruction on Ni(110) and Pb(110) induced by adsorbed hydrogen, pp. 207–213 in *The Structure of Surfaces II*, Springer-Verlag, Berlin, 1988.

52. Y. Kuk, P.J. Silverman and H.Q. Nguyen, Adsorbate-induced reconstruction in the Ni (111)-H system, *Phys. Rev. Lett.*, 59 (13), 1452–1455, 1987.

53. R. Stumpf, H-Induced reconstruction and faceting of Al surfaces, *Phys. Rev. Lett.*, 78 (23), 4454–4457, 1997.

54. A. Pundt and R. Kirchheim, Hydrogen in metals: Microstructural aspects, *Annual Rev. Mater. Res.*, 36, 555–608, 2006.

55. D.R. Baer, M.T. Thomas and R.H. Jones, Influence of stress on H_2S adsorption on iron, *Metall. Trans. A*, 15A, 853–860, 1984.

56. R.A. Oriani, On the possible role of surface stress in environmentally induced embrittlement and pitting, *Scripta Metall.*, 18, 265–268, 1984.

57. R. Thomson, T.-J. Chuang and I.-H. Lin, The role of surface stress in fracture, *Acta Mater.*, 34, 1133–1143, 1986.

58. W. Haiss, Surface stress of clean and adsorbate covered solids, *Rep. Prog. Phys.*, 64, 591–648, 2001.

59. D.A. Vermilyea, A film rupture model for stress corrosion cracking, pp. 208–217 in *Stress Corrosion Cracking and Hydrogen Embrittlement of Iron Base Alloys*, R.W. Staehle *et al.* (eds), NACE, Houston, TX, 1977.

60. J.C. Scully, The mechanism of dissolution controlled cracking, pp. 1–18 in *Mechanisms of Environment Sensitive Cracking of Materials*, P.R. Swann *et al.* (eds), The Metals Society, London, UK, 1977.

61. R.N. Parkins, Stress corrosion cracking, pp. 1–19, in *Parkins Symposium on Fundamental Aspects of Stress Corrosion, Cracking*, S.M. Bruemmer *et al.* (eds), TMS, Warrendale, PA, 1992.

62. F.P. Ford and P.L. Andresen, Corrosion in nuclear systems: environmentally assisted cracking in light water reactors, pp. 605–642 in *Corrosion Mechanisms in Theory and Practice*, 2nd edn, P. Marcus and J. Oudar (eds), Marcel Dekker, New York, 2002.

63. M.M. Hall, Film rupture model for aqueous stress corrosion cracking under constant and variable stress intensity factor, *Corrosion Sci.*, 51, 225–233, 2009.

Published by Woodhead Publishing Limited, 2011

64. P. A. Rebinder and E. D. Shchukin, Surface phenomena in solids during deformation and fracture processes, *Prog. in Surf. Sci.* 3, 97–180, 1972.

65. E.G. Coleman, D. Weinstein and W. Rostoker, On a surface energy mechanism for stress-corrosion cracking, *Acta Metall.*, 9, 491–496, 1961.

66. H. Nichols and W. Rostoker, Analogies between stress-corrosion cracking and embrittlement by liquid metals, *Trans. ASM*, 56, 494–507, 1963.

67. C. Benedicks, The 'wetting effect' strongly affecting the tensile strength of solids; 'Liquo-striction', A new effect resulting, pp. 196–201 in *Pittsburgh International Conference on Surface Reactions*, The Corrosion Publ. Co., Pittsburgh, PA, 1948.

68. H. H. Uhlig, New perspectives in the stress corrosion problem, pp. 1–17 in *Physical Metallurgy of Stress Corrosion Fracture*, T. Rhodin (ed.), Interscience Publ., New York, 1959.

69. H. Uhlig, K. Gupta and W. Liang, Critical potentials for stress corrosion cracking of brass in ammoniacal and tartrate solutions, *J. Electrochem. Soc*, 122, 343–350, 1975.

70. H.H. Uhlig, An evaluation of stress corrosion cracking mechanisms, pp. 86–97 (and comment on p. 710) in *Fundamental Aspects of Stress Corrosion Cracking*, R.W. Staehle *et al.* (eds), NACE, Houston, TX, 1969.

71. H.H. Uhlig, Stress sorption cracking and the critical potential, pp. 174–179 in *Stress Corrosion Cracking and Hydrogen Embrittlement of Iron Base Alloys*, R.W. Staehle *et al.* (eds), NACE, Houston, TX, 1977.

72. S.P. Lynch, Mechanisms of liquid metal embrittlement and stress corrosion cracking in high-strength aluminium alloys and other materials, pp. 201–212 in *Mechanisms of Environment Sensitive Cracking of Materials*, P.R. Swann *et al.* (eds), The Metals Society, London, UK, 1977.

73. S.P. Lynch, Environmentally assisted cracking: Overview of evidence for an adsorption-induced localised-slip process, *Acta Metall.*, 20, 2639–2661, 1988, and references therein.

74. S.P. Lynch, Metallographic contributions to understanding mechanisms of environmentally assisted cracking, *Metallography*, 23, 147–171, 1989, and references therein.

75. S.P. Lynch, Comments on 'A unified model of environment-assisted cracking', *Scripta Mater.*, 61, 331–334, 2009.

76. S.P. Lynch, Mechanisms of hydrogen assisted cracking – a review, pp. 449–466 in *Hydrogen Effects on Materials Behavior and Corrosion Deformation Interactions*, N.R. Moody *et al.* (eds), TMS, Warrendale, PA, 2003.

77. R.A. Oriani, A decohesion theory for hydrogen-induced crack propagation, pp. 351–358 in *Stress Corrosion Cracking and Hydrogen Embrittlement of Iron Base Alloys*, R.W. Staehle *et al.* (eds), NACE, Houston, TX, 1977.

78. A.S. Tetelman and S. Kunz, A unified model for hydrogen embrittlement, liquid metal embrittlement, and temper embrittlement, pp. 359–375 in *Stress Corrosion Cracking and Hydrogen Embrittlement of Iron Base Alloys*, R.W. Staehle *et al.* (eds), NACE, Houston, TX, 1977.

79. R.H. Jones, Analysis of hydrogen-induced subcritical crack growth of iron and nickel, *Acta Metall. Mater.*, 38, 1703–1718, 1990.

80. C. J. McMahon, Jr., Hydrogen-induced intergranular fracture of steels, *Engng Fract. Mech.*, 68, 773–788, 2001.

81. W.W. Gerberich and S. Chen, Environment-induced cracking of metals: fundamental

processes: micromechanisms, pp. 167–192 in *Environment-induced Cracking of Metals*, R.P. Gangloff and M.B. Ives (eds), NACE, Houston, TX, 1990.

82. W.W. Gerberich, P. Marsh, J. Hoehn, S. Venkataraman and H. Huang, Hydrogen/plasticity interactions in stress corrosion cracking, pp. 325–353 in *Corrosion-Deformation Interactions*, T. Magnin and J.-M. Gras (eds), Les Editions de Physique, Les Ulis, 1993.

83. C.D. Beachem, A new model for hydrogen assisted cracking (hydrogen embrittlement), *Metall. Trans.*, 3, 437–451, 1972.

84. H.K. Birnbaum and P. Sofronis, Hydrogen-enhanced localised plasticity – a mechanism for hydrogen-related fracture, *Mater. Sci. Eng.*, A176, 191–202, 1994.

85. H.K. Birnbaum, I.M. Robertson, P. Sofronis and D. Teter, Mechanisms of hydrogen related fracture – a review, pp. 172–195 in *Corrosion-Deformation Interactions*, T. Magnin (ed.), The Institute of Materials, London, 1997.

86. D. Hardie, The environment-induced cracking of hexagonal materials: magnesium, titanium, and zirconium, pp. 347–361 in *Environment-induced Cracking of Metals*, R.P. Gangloff and M.B. Ives (eds), NACE, Houston, TX, 1990.

87. M. Nagumo, Hydrogen related failure of steels – a new aspect, *Mater. Sci. Tech.*, 20, 940–950, 2004.

88. P. Aaltonen, T. Saario, U. Ehrnstén, M. Itäaho and H. Hänninen, Selective dissolution-vacancy-creep model for EAC of brass, pp. 35–44 in *Corrosion-Deformation Interactions*, T. Magnin (ed.), The Institute of Materials, London, 1997.

89. D.A. Jones, A unified mechanism of stress corrosion and corrosion fatigue cracking, *Metall. Trans. A*, 16 A, 1133–1141, 1985.

90. D.A. Jones, Localized surface plasticity during stress corrosion cracking, *Corrosion*, 52, 356–362, 1996.

91. E.I. Meletis, K. Lian and W. Huang, Vacancy-dislocation interactions and transgranular stress corrosion cracking, pp. 69–81 in *Corrosion-Deformation Interactions*, T. Magnin and J.-M. Gras (eds), Les Editions de Physique, Les Ulis, 1993.

92. K. Lian and E.I. Meletis, Environment-induced deformation localization during transgranular stress corrosion cracking, *Corrosion*, 52, 347–355, 1996.

93. J.R. Galvele, A stress corrosion cracking mechanism based on surface mobility, *Corrosion Sci.*, 27, 1–33, 1987.

94. J.R. Galvele, Past, present, and future of stress corrosion cracking (W.R. Whitney award lecture), *Corrosion*, 55, 723–731, 1999.

95. J.R. Galvele, Recent developments in the surface-mobility stress-corrosion-cracking mechanism, *Electrochimica Acta*, 45, 3537–3542, 2000.

96. S.A. Serebrinsky and J.R. Galvele, Effect of the strain rate on stress corrosion crack velocities in face-centred cubic alloys. A mechanistic interpretation, *Corrosion Sci.*, 46, 591–612, 2004.

97. K. Sieradzki and R.C. Newman, Brittle behaviour of ductile metals during stress corrosion cracking, *Philosophical Magazine A*, 51, 95–132, 1985.

98. R.C. Newman, T. Shahrabi and K. Sieradzki, Film-induced cleavage of alpha-brass, *Scripta Metall.*, 23, 71–74, 1989.

99. M. Saito, G.S. Smith and R.C. Newman, Testing the film-induced cleavage model of stress-corrosion cracking, *Corrosion Sci.*, 35, 411–413, 1993.

100. A. Barnes, N.A. Senior and R.C. Newman, Film-induced cleavage of Ag-Au alloys, *Metall. Mater. Trans. A*, 40A, 58–68, 2009.

Published by Woodhead Publishing Limited, 2011

101. K. Sieradzki and R.C. Newman, Stress corrosion cracking, *J. Phys. Chem. Solids*, 48, 1101–1113, 1987.
102. R.C. Newman and R.P.M. Proctor, Stress corrosion cracking: 1965–1990, *Br. Corros. J.*, 25, 259–269, 1990.
103. R.C. Newman and N.A. Senior, A revised interpretation of ultra-fast stress corrosion cracking experiments by Serebrinsky and Galvele, *Corrosion Sci.*, 52,1541–1544, 2010.
104. T. Magnin, A. Chambreuil and B. Bayle, The corrosion-enhanced plasticity model for stress corrosion cracking in ductile fcc alloys, *Acta Mater.*, 44, 1457–1470, 1996.
105. T. Magnin and J. Lépinoux, Metallurgical aspects of the brittle S.C.C. in austenitic stainless steels, pp. 323–339 in *Parkins Symposium on Fundamental Aspects of Stress Corrosion Cracking*, S.M. Bruemmer et al. (eds), TMS, Warrendale, PA, 1992.
106. W.F. Flanagan and B.D. Lichter, A mechanism for stress-corrosion cracking, *Int. J. Fracture*, 79, 121–135, 1996.
107. L. Zhong, W.F. Flanagan and B.D. Lichter, The competitive role of slip and dissolution in transgranular stress corrosion cracking, pp. 309–322 in *Corrosion-Deformation Interactions*, T. Magnin and J.-M. Gras (eds), Les Editions de Physique, Les Ulis, 1993.
108. W.F. Flanagan, M. Wang, M. Zhu and B.D. Lichter, A fully plastic microcracking model for transgranular stress-corrosion cracking in planar-slip materials, *Metall. and Mater. Trans. A*, 25A, 1391–1401, 1994.
109. B.D. Lichter, R.M. Bhatkal and W.F. Flanagan, Corrosion-assisted cleavage in copper-gold single crystals, pp. 279–302 in *Parkins Symposium on Fundamental Aspects of Stress Corrosion Cracking*, S.M. Bruemmer et al. (eds), TMS, Warrendale, PA, 1992.
110. H.W. Pickering and P.R. Swann, Electron metallography of chemical attack upon some alloys susceptible to stress corrosion cracking, *Corrosion*, 19, 373t–389t, 1963.
111. R.W. Staehle, Stress corrosion cracking of the Fe-Cr-Ni system, pp. 223–288 in *The Theory of Stress Corrosion Cracking*, J.C. Scully (ed.), NATO, Brussels, 1971.
112. P.M. Scott and P. Combrade, On the mechanism of stress corrosion crack initiation and growth in Alloy 600 exposed to PWR primary water, pp. 29–38 in *11th Int. Conf. on Environmental Degradation of Materials in Nuclear Systems, Stevenson, WA*, 10–14 Aug. 2003.
113. S.P. Lynch, Concerning the mechanism of transcrystalline stress corrosion cracking of brass, *Scripta Metall.*, 18, 321–326, 1984.
114. M.B. Hintz, Stress corrosion cracking and crack-tip adsorption: considerations regarding the role of alloy dissolution products, *Scripta Metall.*, 19, 1445–1450, 1985.
115. L.J. Qiao and L.L. Luo, Hydrogen-facilitated dissolution of austenitic stainless steels, *Corrosion*, 54, 281–288, 1998.
116. J.G. Yu, J.L. Luo and P.R. Norton, Electrochemical investigation of the effects of hydrogen on the stability of the passive oxide film on iron, *Electrochemica Acta*, 47, 1527–1536, 2002.
117. J. Hou, Q.J. Peng, K. Sakaguchi, Y. Takeda, J. Kuniya and T. Shoji, Effects of hydrogen on Inconel alloy 600 on corrosion in high temperature oxygenated water, *Corrosion Sci.*, 52, 1098–1101, 2010.

Published by Woodhead Publishing Limited, 2011

118. K.W. Gao, W.Y. Chu, B. Gu, T.C. Zhang and L.J. Qiao, *In-situ* transmission electron microscopic observation of corrosion-enhanced dislocation emission and crack initiation of stress corrosion, *Corrosion*, 56, 515–522, 2000.

119. R.N. Parkins, Stress corrosion spectrum, *Br. Corros. J.*, 7, 15–28, 1972.

120. S.P. Lynch and P. Trevena, Stress corrosion cracking and liquid metal embrittlement in pure magnesium, *Corrosion*, 44, 113–124, 1988.

121. S.P. Lynch, Ductile and brittle crack growth: fractography, mechanisms, and criteria, *Materials Forum*, 11, 268–283, 1988.

122. E.I. Meletis and R.F. Hochman, Crystallography of stress corrosion cracking in pure magnesium, *Corrosion*, 40, 39–45, 1984.

123. R.S. Stampella, R.P.M. Proctor and V. Ashworth, Environmentally-induced cracking of magnesium, *Corrosion Sci.*, 24, 325–341, 1984.

124. N. Winzer, A. Atrens, G. Song, E. Ghali, W. Dietzel, K.U. Kainer, N. Hort and C. Blawert, A critical review of the stress corrosion cracking (SCC) of magnesium alloys, *Adv. Engng Mater.*, 7, 659–693, 2005.

125. S.P. Lynch, unpublished work.

126. H.H. Johnson, Hydrogen gas embrittlement, pp. 35–49 in *Hydrogen in Metals*, I.M. Bernstein and A.W. Thompson (eds), ASM, Metals Park, OH, 1974.

127. S.P. Lynch, Metal-induced embrittlement of materials, *Materials Characterization*, 28, 279–289, 1992.

128. S. Hinotani, Y. Ohmori and F. Terasaki, Hydrogen crack initiation and propagation in pure iron single crystal, *Mater. Sci. Tech.*, 10, 141–148, 1994.

129. N. Takano, K. Kidani, Y. Hattori and F. Terasaki, Fracture surface of hydrogen embrittlement in iron single crystals, *Scripta Metall. Mater.*, 29, 75–80, 1993.

130. R.W. Lycett and J.C. Scully, Stress corrosion crack propagation in Ti–13V-11Cr-3Al alloy in methanolic solutions, *Corrosion Sci.*, 19, 799–817, 1979.

131. J.T. Lukowski, D.B. Kasul, L.A. Heldt and C.L. White, Discontinuous crack propagation in Ga induced liquid metal embrittlement of β-brass, *Scripta Metall. Mater.*, 24, 1959–1964, 1990.

132. K. Sieradzki, R.L. Sabatini and R.C. Newman, Stress-corrosion cracking of copper single crystals, *Metall. Trans. A*, 15A, 1941–1946, 1984.

133. S.P. Lynch, Environmentally induced cleavage-like cracking in aluminium alloys, pp. 401–413 in *Corrosion-Deformation Interactions*, T. Magnin and J.-M. Gras (eds), Les Editions de Physique, Les Ulis, 1993.

134. S.P. Lynch, Mechanisms of environmentally asissted cracking in Al-Zn-Mg single crystals, *Corrosion Sci.*, 22, 925–937, 1982, and Further observations of environmentally assisted cracking in Al-Zn-Mg single crystals, *Corrosion Sci.*, 24, 375–378, 1984.

135. S.P. Lynch, Mechanisms of fatigue and environmentally assisted fatigue, pp. 174–213 in *Fatigue Mechanisms*, J.T. Fong (ed.), ASTM STP 675, ASTM, Philadelphia, PA, 1979.

136. E.I. Meletis and R.F. Hochman, A review of the crystallography of stress corrosion cracking, *Corrosion Sci.*, 26, 63–90, 1986.

137. G.A. Young and J.R. Scully, Hydrogen production, absorption, and transport during environmentally assisted cracking of an Al-Zn-Mg-(Cu) alloy in humid air, pp. 893–907 in *Hydrogen Effects on Materials Behavior and Corrosion Deformation Interactions*, N.R. Moody *et al.* (eds), TMS, Warrendale, PA, 2003.

138. J. Erlebacher, An Atomic description of dealloying: porosity evolution, the critical

potential, and rate-limiting behavior, *J. Electrochem. Soc.*, 151, C614–C626, 2004.

139. H. Rosner, S. Parida, D. Kramer, C.A. Volkert and J. Weissmüller, Reconstructing a nanoporous metal in three dimensions: an electron tomography study of dealloyed gold leaf, *Adv. Engng Mater.*, 9(7), 535–541, 2007.

140. A.C. Van Orden, Dealloying, pp. 229–239, in *Corrosion Tests and Standards: Applications and Interpretation*, R. Baboian (ed.), ASTM Manual Series: MNL 20, ASTM, Philadelphia, PA, 1995.

141. J. Snyder, K. Livi and J. Erlebacher, Dealloying silver/gold alloys in neutral silver nitrate solution: porosity evolution, surface composition, and surface oxides, *J. Electrochem. Soc.*, 155, C464–C473, 2008.

142. U. Bertocci, F.I. Thomas and E.N. Pugh, Stress corrosion cracking of brass in aqueous ammonia in the absence of detectable anodic dissolution, *Corrosion*, 40, 439–440, 1984.

143. M.J. Kaufman and J.L. Fink, Evidence for localised ductile fracture in the 'brittle' transgranular stress corrosion cracking in ductile F.C.C alloys, *Acta Metall.*, 36, 2213–2228, 1988.

144. J.S. Chen, M. Salmeron and T.M. Devine, Intergranular and transgranular stress corrosion cracking of Cu-30Au, *Corrosion Sci.*, 34, 2071–2097, 1993.

145. G.S. Duffo and J.R. Galvele, Surface mobility stress corrosion cracking mechanism in silver alloys, pp. 261–263, and discussion by R.A. Oriani, pp. 263–264, in *Environment-induced Cracking of Metals*, R.P. Gangloff and M.B. Ives (eds), NACE, Houston, TX, 1990.

146. K. Sieradzki and F.J. Friedersdorf, Notes on the surface mobility mechanism of stress corrosion cracking, *Corrosion Sci.*, 36, 669–675, 1994.

147. J.R. Galvele, Comments on 'Notes on the surface mobility mechanism of stress-corrosion cracking' by K. Sieradzki and F.J. Friedersdorf, *Corrosion Sci.*, 36, 901–910, 1994.

148. E.M. Gutman, Notes on the discussion concerning the 'surface mobility mechanism' of stress corrosion cracking, *Corrosion Sci.*, 45, 2105–2117, 2003.

149. J.R. Galvele, Reply to 'Notes on the discussion concerning "surface mobility mechanism" of stress corrosion cracking', by E.M. Gutman, *Corrosion Sci.*, 45, 2119–2128, 2003.

150. (a) E.M. Gutman, Comments on the 'Stress corrosion cracking of zirconium and zircaloy-4 in halide aqueous solutions' by S.B. Farina, G.S. Duffo, J.R. Galvele, *Corrosion Sci.*, 46, 1801–1806, 2004, and (b) J.R. Galvele, Reply to E.M. Gutman's comments on the 'Stress corrosion cracking of zirconium and zircaloy-4 in halide aqueous solutions', *Corrosion Sci.*, 46, 1807–1812. (See also E.M. Gutman, An inconsistency in 'surface mobility mechanism' of stress corrosion cracking, *Corrosion*, 61, 197–200, 2005.)

151. R.C. Newman, Stress corrosion cracking of noble metals and their alloys in solutions containing cations of the noble metal: review of observations relevant to competing models of SCC, *Corrosion Sci.*, 50, 1807–1810, 2008.

152. I.-H. Lin and R.M. Thomson, Dynamic cleavage in ductile materials, *J. Mater. Res.*, 1, 73–80, 1986.

153. R. Dou and B. Derby, The strengths of gold nanowire forests, *Scripta Mater.*, 59, 151–154, 2008.

154. D. Kiener, W. Grosinger, G. Dehm and R. Pippan, A further step towards an understanding of size-dependent crystal plasticity: *in situ* tension experiments of miniaturized single-crystal copper samples, *Acta Mater.*, 56, 580–592, 2008.

155. M.A. Rist, High-stress ice fracture and friction, *J. Phys. Chem. B*, 101, 6263–6266, 1997.
156. J.J. Lewandowski and A.L. Greer, Temperature rise at shear bands in metallic glasses, *Nature Materials*, 5, 15–18, 2006.
157. C.J. Gilbert, J.W. Ager III, V. Schroeder, R.O. Ritchie, J.P. Lloyd and J.R. Graham, Light emission during fracture of a Zr-Ti-Ni-Cu-Be bulk metallic glass, *Appl. Phys. Lett.*, 74, 3809–3811, 1999.
158. J. Eckert, G. He, Z.F. Zhang and W. Löser, Fracture-induced melting in glassy and nanostructured composite materials, *J Metastable Nanocrystalline Materials*, 20–21, 357–365, 2004.
159. R.E. Ricker, J.L. Fink, J.S. Harris and A.J. Shapiro, Evidence for film-induced cleavage in rhodium plated nickel, *Scripta Metall. Mater.*, 26, 1019–1023, 1992.
160. R.G. Kelly, A.J. Young and R.C. Newman, The characterization of the coarsening of dealloyed layers by EIS and its correlation with stress corrosion cracking, pp. 99–112 in *Electrochemical Impedance: Analysis and Interpretation*, ASTM STP 1188, J.R. Scully, D.C. Silverman and M.W. Kendig (eds), ASTM, Philadelphia, PA, 1993.
161. H. Vehoff, H. Stenzel and P. Neumann, Experiments on bicrystals concerning the influence of localized slip on the nucleation and growth of intergranular stress corrosion cracks, *Z. Metallkde*, 78, 550–556, 1987.
162. H. Stenzel, H. Vehoff and P. Neumann, Intergranular stress corrosion cracking in bicrystals, pp. 225–241 in *Modeling Environmental Effects on Crack Growth Processes*, R.H. Jones and W.W. Gerberich (eds), TMS, Warrendale, PA, 1986.
163. S.P. Lynch, Mechanisms of stress-corrosion cracking and liquid-metal embrittlement in Al-Zn-Mg bicrystals, *J. Mater. Sci.*, 20, 3329–3338, 1985.
164. N.J.H. Holroyd, Environment-induced cracking of high-strength aluminum alloys, pp. 311–345 in *Environment-induced Cracking of Metals*, R.P. Gangloff and M.B. Ives (eds), NACE, Houston, TX, 1990.
165. K. Yoshino and C.J. McMahon, Jr., The cooperative relation between temper embrittlement and hydrogen embrittlement in a high strength steel, *Metall. Trans. A*, 5, 363–370, 1974.
166. R.P. Messmer and C.L. Briant, The role of chemical bonding in grain boundary embrittlement, *Acta Mater.*, 30, 457–467, 1982.
167. N. Winzer, A. Atrens, W. Dietzel, G. Song and K.U. Kainer, Fractography of stress corrosion cracking of Mg–Al alloys, *Metall. Mater. Trans.*, 39A, 1157–1173, 2008.
168. W.K. Miller, Stress corrosion cracking of magnesium alloys, pp. 251–263 in *Stress-Corrosion Cracking: Materials Performance and Evaluation*, R.H. Jones (ed.), ASM, Metals Park, OH, 1992.
169. B. Cox, Environmentally-induced cracking of zirconium alloys – a review, *J. Nuclear Materials*, 170, 1–23, 1990.
170. R.W. Schutz, Stress corrosion cracking of titanium alloys, pp. 265–297 in *Stress-Corrosion Cracking: Materials Performance and Evaluation*, R.H. Jones (ed.), ASM, Metals Park, OH, 1992.
171. Te-Lin Yau, Stress corrosion cracking of zirconium alloys, pp. 299–311 in *Stress-Corrosion Cracking: Materials Performance and Evaluation*, R.H. Jones (ed.), ASM, Metals Park, OH, 1992.
172. G.M. Scamans, Evidence for crack-arrest markings on intergranular stress corrosion fracture surfaces in Al-Zn-Mg alloys, *Metall. Trans. A*, 11A, 846–850, 1980.

Published by Woodhead Publishing Limited, 2011

173. M.O. Speidel, Hydrogen embrittlement of aluminum alloys?, pp. 249–273 in *Hydrogen in Metals*, I.M. Bernstein and A.W. Thompson (eds), ASM, Metals Park, OH, 1974.

174. M. Marek and R.F. Hochman, Crystallography and kinetics of stress corrosion cracking in type 316 stainless steel single crystals, *Corrosion*, 27, 361–370, 1971.

175. J.M. Silcock, Orientations of transgranular stress corrosion cracking in austenitic steels tested in $MgCl_2$ solutions, *Br. Corros. J.*, 16, 78–93, 1981.

176. R. Liu, N. Narita, C. Altstetter, H. Birnbaum and E.N. Pugh, Studies of the orientations of fracture surfaces produced in austenitic stainless steels by stress corrosion cracking and hydrogen embrittlement, *Metall. Trans. A*, 11A, 1563–1574, 1980.

177. A. Brozova and S. Lynch, Transgranular stress-corrosion cracking in austenitic stainless steels at high temperatures, pp. 149–161 in *Corrosion Issues in Light Water Reactors: Stress Corrosion Cracking*, D. Féron and J.-M.Olive (eds), Woodhead, Cambridge, 2007.

178. R.C. Newman, R.R. Corderman and K. Sieradzki, Evidence for dealloying of austenitic stainless steels in simulated stress corrosion crack environments, *Br. Corros. J.*, 24, 143–148, 1989.

179. R.C. Newman and A. Mehta, Stress corrosion cracking of stainless steels, pp. 489–509 in *Environment-induced Cracking of Metals*, R.P. Gangloff and M.B. Ives (eds), NACE, Houston, TX, 1990.

180. S.M. Bruemmer and L.E. Thomas, High-resolution analytical electron microscopy characterization of corrosion and cracking at buried interfaces, *Surf. Interface Anal.*, 31, 571–581, 2001.

181. L.E. Thomas and S.M. Bruemmer, High resolution characterization of intergranular attack and stress corrosion cracking of alloy 600 in high-temperature primary water, *Corrosion*, 56, 572–587, 2000.

182. R.W. Staehle, Critical analysis of 'tight cracks', *Corrosion Reviews*, 28, 1–103, 2010.

183. U. Bertocci, E.N. Pugh, and R.E. Ricker, Environment-induced cracking of copper alloys, pp. 273–286 in *Environment-induced Cracking of Metals*, R.P. Gangloff and M.B. Ives (eds), NACE, Houston, TX, 1990.

184. J.A. Beavers, Stress corrosion cracking of copper alloys, pp. 211–231 in *Stress-Corrosion Cracking: Materials Performance and Evaluation*, R.H. Jones (ed.), ASM, Metals Park, OH, 1992.

185. M.B. Hintz, L.J. Nettleton and L.A. Heldt, Stress corrosion cracking of alpha-beta brass in distilled water and sodium sulphate solutions, *Metall. Trans. A*, 16A, 971–978, 1985.

186. M.B. Hintz, W.K. Blanchard, P.K. Brindley and L.A. Hintz, Further observations of SCC in alpha-beta brass: considerations regarding the appearance of crack-arrest markings during SCC, *Metall. Trans. A*, 17A, 1081–1086, 1986.

187. U.K. Chatterjee and S.C. Sircar, An analysis of evidence to support an adsorption model for stress corrosion cracking of alpha-brass, pp. 397–400 in *Environment-induced Cracking of Metals*, R.P. Gangloff and M.B. Ives (eds), NACE, Houston, TX, 1990.

188. R.H. Jones, Environment-induced crack growth processes in nickel-based alloys, pp. 287–310 in *Environment-induced Cracking of Metals*, R.P. Gangloff and M.B. Ives (eds), NACE, Houston, TX, 1990.

189. N. Sridhar and G. Cragnolino, Stress corrosion cracking of nickel-base alloys, pp. 131–179 in *Stress-corrosion Cracking: Materials Performance and Evaluation*, R.H. Jones (ed.), ASM, Metals Park, OH, 1992.

190. P.G. Marsh and W.W. Gerberich, Stress corrosion cracking of high-strength steels (Yield strengths greater than 1240 MPa), pp. 63–90 in *Stress-corrosion Cracking: Materials Performance and Evaluation*, R.H. Jones (ed.), ASM, Metals Park, OH, 1992.

191. S.W. Ciaraldi, Stress corrosion cracking of carbon and low-alloy steels (yield strengths less than 1241 MPa), pp. 41–61 in *Stress-corrosion Cracking: Materials Performance and Evaluation*, R.H. Jones (ed.), ASM, Metals Park, OH, 1992.

192. R.J.H. Wanhill, Failure of backstay rod connectors on a luxury yacht, *Practical Failure Analysis*, 3, 33–39, 2003.

193. J. Kruger, Failure by stress corrosion cracking – current approaches towards failure prediction, pp. 5–36 in *Stress Corrosion Cracking*, J. Yahalom and A. Aladjem (eds), Freund Pub, Tel Aviv, Israel, 1980.

194. N.S. Stoloff, Recent developments in liquid-metal embrittlement, pp. 486–518 in *Environment-Sensitive Fracture of Engineering Materials*, Z.A. Foroulis (ed.), Met. Soc. AIME, New York, 1979.

195. N.S. Stoloff, Metal induced embrittlement – a historical perspective, pp. 3–26 in *Embrittlement by Liquid and Solid Metals*, M.H. Kamdar (ed.), Met. Soc. AIME, New York, 1984.

196. P. Rieux, J. Driver and J. Rieu, Fatigue crack propagation in austenitic and ferritic stainless steel single crystals, *Acta Metall.*, 27, 145–153, 1979.

197. S.P. Lynch, Progression markings, striations, and crack-arrest markings on fracture surfaces, *Mater. Sci. Eng. A.*, 468–470, 74–80, 2007, and references therein.

198. L.B. Vogelesang and J. Schijve, Environmental effects on fatigue fracture mode transitions observed in aluminium alloys, *Fat. Eng. Mat. Struct.*, 3, 85–98, 1980.

199. R.M.N Pelloux, Corrosion fatigue crack propagation, pp. 731–744 in *Fracture 1969, Proc. 2nd Int. Conf. on Fracture, Brighton*, Chapman and Hall, London, 1969.

200. Q.J. Peng, J. Kwon and T. Shoji, Development of a fundamental crack-tip strain rate equation and its application to quantitative prediction of stress corrosion cracking in high temperature oxygenated water, *J. Nucl. Mater.*, 324, 52–61, 2004.

201. Papers in *Scripta Metall.*, Viewpoint set 10 on Dislocation Emission from Cracks, Vol. 20, 1986.

2

Hydrogen embrittlement (HE) phenomena and mechanisms

S.P. LYNCH, Defence Science and Technology Organisation, Australia

Abstract: Mechanisms of hydrogen embrittlement in steels and other materials are described, and the evidence supporting various hypotheses, such as those based on hydride-formation, hydrogen-enhanced decohesion, hydrogen-enhanced localised plasticity, adsorption-induced dislocation-emission, and hydrogen-vacancy interactions, are summarised. The relative importance of these mechanisms for different fracture modes and materials are discussed based on detailed fractographic observations and critical experiments.

Key words: hydrogen-embrittlement, mechanisms, fractography, hydride embrittlement, hydrogen-enhanced decohesion, hydrogen-enhanced localised-plasticity, adsorption-induced dislocation-emission.

2.1   Introduction

The damaging effects of hydrogen in iron and steel were first recognised in the early 1870s, and the first tentative explanation, namely 'hydrogen in interspaces impeding the movement of iron molecules', was proposed at this time [1]. Since then, there have been intensive efforts to characterise and understand the phenomenon of hydrogen-embrittlement (HE) of materials in general, especially over the last 50 years as more advanced characterisation techniques became available. The present chapter describes the mechanisms of HE that are now considered to be viable, and summarises the evidence supporting them. The relative importance of these mechanisms for different fracture modes, which depends on the material, microstructure/strength, and testing conditions, is then discussed.

Mechanisms of HE are relevant to stress-corrosion cracking (SCC) (and corrosion-fatigue) in many materials since hydrogen is generated at crack tips in aqueous or moist-air environments. Mechanisms of SCC (with and without the involvment of hydrogen) are discussed in detail in Chapter 1, with reference to this chapter for more details regarding hydrogen effects. Before describing and discussing the mechanisms of HE, some background is provided regarding: (i) types of HE, (ii) the susceptibility of various materials to HE, (iii) sources of hydrogen, (iv) sites and traps of hydrogen

90

in materials, (v) hydrogen diffusion, (vi) measures of HE susceptibility, (vii) the main variables affecting HE, and (viii) the general characteristics of HE. Further details regarding these aspects, and original references, can be found in previous reviews, conference proceedings, and other compilations of papers on HE [2–14]. A chapter on 'Metallographic and fractographic techniques for characterising and understanding hydrogen-assisted cracking' [15] by the present author also provides additional information, albeit with a significant overlap with the present chapter in some areas.

2.1.1   Types of hydrogen embrittlement and terminology

There are a number of manifestations of HE, and various terminologies are used to describe different phenomena depending on the source of hydrogen and the type of damage produced. 'Internal hydrogen embrittlement' (IHE) involves concentration of pre-existing hydrogen in regions of high hydrostatic stress (applied or residual), resulting in cracking under sustained stresses well below the yield stress. 'Hydrogen-environment embrittlement' (HEE), on the other hand, involves sub-critical cracking of materials under sustained loads in hydrogen or hydrogen-sulphide gases. In hydride-forming materials containing high concentrations of hydrogen, embrittlement often involves formation and fracture of brittle hydrides at and ahead of cracks, and is termed 'hydride-embrittlement'.

IHE, HEE, and hydride embrittlement are sometimes referred to simply as 'hydrogen-assisted cracking' (HAC) (with 'internal' or 'environmental' sometimes appropriately inserted). For HAC under cyclic loads, the term 'hydrogen-assisted fatigue' is sometimes used. Other terminologies for IHE are used in specific industries. For example, the term 'hydrogen-assisted cold cracking' (HACC) is often used in the welding industry (when welds crack after cooling to room temperature). In the oil and gas industry, the term 'stress-oriented hydrogen-induced cracking' is used when cracking has a ladder-like morphology due to cracking from elongated inclusions distributed along the rolling direction of plate. The term 'environmental hydrogen embrittlement' (EHE) is sometimes used when cracking occurs in cathodically protected structures.

Other forms of hydrogen damage are 'blistering' and 'hydrogen-attack'. Blistering occurs when solute hydrogen re-combines to form high-pressure hydrogen gas at cracked inclusion/matrix interfaces near surfaces. The high pressure causes voids to expand by plastic deformation and cracking, causing surface swelling. Hydrogen-attack, also called 'hydrogen-reaction embrittlement', occurs when some materials such as steels and copper are exposed to hot hydrogen-bearing gases, leading to hydrogen diffusion to, and reaction with, carbides and oxides: High-pressure methane and steam, respectively, can then be produced, leading to internal voids and cracks.

Neither of these forms of hydrogen damage is considered further in this chapter.

2.1.2   Susceptibility of materials to HE

High-strength (and ultra-high-strength) martensitic steels with yield strengths greater than about 1400 MPa (Hardness > 38 on Rockwell C scale) are extremely susceptible to IHE, with solute hydrogen concentrations as little as 0.5–1 ppm (wt.) causing cracking. For ferritic steels with strengths less than about 750 MPa, relatively high hydrogen concentrations (~10 ppm) are often necessary for IHE to be significant. Titanium alloys with α-β microstructures are susceptible to IHE when hydrogen concentrations are greater than 100–200 ppm. Nickel alloys, aluminium alloys, (stable) austenitic steels, and copper alloys exhibit little (if any) susceptibility to IHE, and industrial problems due to IHE in these materials are rare. The susceptibilities of materials to HEE are generally similar to those for IHE, but high-strength nickel alloys seem to be an exception in that they can be very susceptible to HEE, but only slightly embrittled by internal hydrogen unless concentrations are high.

2.1.3   Sources of hydrogen

The main sources of hydrogen leading to IHE of high-strength steels are the solutions used to clean and apply (or re-apply) protective coatings. Electroplating solutions, especially if out-of-specification, are particularly prone to produce IHE. Other solutions include those used for pickling, phosphating, paint-stripping, and cathodic-cleaning. Welding (or melting/casting the original material) in moist atmospheres or with moisture in rust, fluxes, etc., has also led to a significant number of failures by IHE. Heat-treatment in hydrogen-bearing atmospheres can also lead to hydrogen uptake, as can applied cathodic-protection with impressed current devices or sacrificial anodes. Similarly, corrosion of sacrificial coatings during service can generate hydrogen, especially if coatings are porous.

High-pressure gaseous (and liquid) hydrogen have been used as a fuel source in the space industry for some time, and failures of steel tanks due to HEE first occurred in the 1960s [16]. Hydrogen gas and combustion/reaction products containing hydrogen are also encountered in the nuclear, chemical and other industries. The use of hydrogen as an energy source to replace fossil-fuels is projected to increase dramatically in the future, as climate-change and pollution become even more of a concern. Many more structures and components used for hydrogen-production, hydrogen-storage and hydrogen-transportation (e.g. pipelines), will therefore be exposed to hydrogen gas.

2.1.4   Sites and traps for hydrogen

Solute hydrogen occupies and diffuses between interstitial lattice sites in metals, and can be trapped to various degrees at other sites (i.e. occupy lower potential-energy sites relative to normal interstitial sites). These other sites, listed roughly in order of trapping strength, include: (i) some solute atoms, (ii) free surfaces and sites between the first few atomic layers beneath surfaces, (iii) mono-vacancies and vacancy clusters (which are present in concentrations well in excess of thermal equilibrium values because hydrogen reduces the vacancy formation energy), (iv) dislocation cores and strain fields, (v) grain boundaries (including prior-austenite grain boundaries in martensitic steels), (vi) precipitate/matrix interfaces and strain-fields around precipitates, (vii) inclusion/matrix interfaces, and (viii) voids and internal cracks (Fig. 2.1).

The high concentrations of hydrogen on or just beneath surfaces ('chemisorbed hydrogen')* will also be present at (non-atomically sharp)

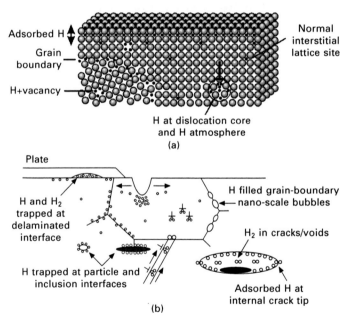

2.1 Schematic illustrations of sites and traps for hydrogen in materials (a) on the atomic scale (adapted from [12]), and (b) on a microscopic scale.

---

*There is no generally accepted term that defines both surface and just sub-surface sites for hydrogen, although some papers do discuss 'chemisorption into sites below the surface' as well as hydrogen on the surface. In the context of HE mechanisms, hydrogen on crack-tip surfaces and hydrogen within a few atomic distances of crack tips are referred to as adsorbed hydrogen for convenience.

crack-tip surfaces. For HEE, hydrogen gas will dissociate and adsorb at crack-tips, while for IHE, solute hydrogen will diffuse to, and adsorb at, internal crack-tip surfaces. Solute hydrogen can also precipitate as hydrogen gas (sometimes resulting in high pressures) in voids and at delaminated non-porous plating/substrate interfaces, and can then be re-absorbed and diffuse under applied stresses. Hydrogen in normal interstitial lattice sites will concentrate ahead of notches and cracks where high hydrostatic stresses result in a slightly expanded lattice.

2.1.5   Hydrogen diffusion

Rates of hydrogen diffusion in pure metals at ambient temperatures depend especially on the crystal structure, with hydrogen-diffusion coefficients, D, generally 4 to 5 orders of magnitude higher for body-centred cubic (bcc) metals compared with face-centred cubic (fcc) and hexagonal close-packed (hcp) metals at 20°C (Fig. 2.2) [17, 18]. However, there are exceptions such as Pd (fcc) and Co (hcp) which have D values several orders of magnitude greater than most other fcc and hcp metals, presumably because quantum-effects override lattice-packing effects. For some metals, such as aluminium, D values quoted in the literature span 4 to 5 orders of magnitude at 20°C,

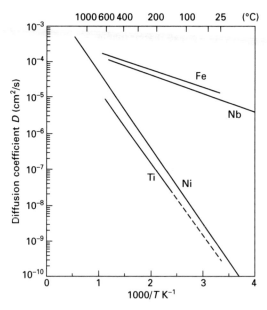

2.2 Hydrogen diffusion coefficients for Fe and Nb (bcc), Ni (fcc) and Ti (hcp) as a function of temperature (plotted as inverse absolute temperature) (data from [17, 18]).

possibly due to experimental difficulties related to controlling surface conditions (oxide films) that affect hydrogen ingress and egress [19].

The hydrogen-diffusion distance $\sim(2Dt)^{1/2}$ in bcc iron with a D value $\sim10^{-5}\,cm^2/s$ at 20°C is about 50 µm in a time, t, of one second, whereas for nickel with a D value of $\sim10^{-10}\,cm^2/s$, the corresponding distance is about 0.1 µm. In alloys with complex microstructures, effective D values at ambient temperatures decrease with increasing number and strength of traps. For ferritic steels, effective D values vary by 3 to 4 orders of magnitude at 20°C for different microstructures. Higher-strength steels tend to have lower effective D values owing to a larger volume fraction of carbide strengthening phases and a higher dislocation density.

Hydrogen-diffusion rates increase with increasing temperature, with a stronger dependence on temperature for most close-packed metals compared with bcc metals (Fig. 2.2). As the temperature is increased, hydrogen is released from deeper traps, and can diffuse to, and egress at, surfaces more readily. Thus, it is usual to minimise hydrogen contents in steels by 'baking' at elevated temperatures after processing and plating. For example, porous cadmium-plated high-strength steels are generally baked at 190°C for 23 h as soon as possible after plating. Optimum baking times and temperatures depend on the plating material, porosity, thickness, steel strength, and other factors that affect hydrogen egress.

Hydrogen diffuses through the lattice in response to gradients in (i) hydrogen concentration, (ii) temperature, and (iii) hydrostatic-stress fields (with the last factor being the most important in regard to IHE). Hydrogen can be transported more rapidly by mobile dislocations than by lattice diffusion when hydrogen is present at dislocation cores or as atmospheres around dislocations. Dislocation-transport of hydrogen may be important in moving hydrogen from grain interiors to grain boundaries, thereby promoting intergranular fracture. Hydrogen may also diffuse more rapidly along grain boundaries than through the lattice in some cases, but if there is a higher trap density at boundaries than elsewhere, then grain-boundary diffusion of hydrogen could be slower than through the lattice.

2.1.6   Testing procedures for assessing the degree of HE

A wide variety of parameters has been used to assess susceptibilities of materials to HE, and to determine the effects of variables on HE. Measures include (i) reductions in the strain to failure (elongation, reduction-of-area) and in the fracture stress of tensile specimens subjected to increasing loads, especially at low strain rates, (ii) delayed-failure times for tensile and C-ring specimens under sustained loads, and (iii) for hydride-containing material, reductions in fracture energies in impact tests (Fig. 2.3(a) and (b)). Fracture-

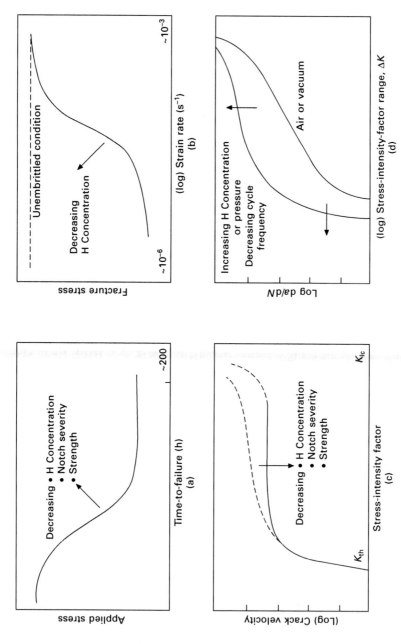

2.3 Schematic plots of measures of HE susceptibility: (a) time-to-failure (versus applied stress), (b) fracture stress (at low strain rates), (c) threshold stress-intensity factors and plateau velocities, and (d) relative rates of fatigue-crack growth (versus ΔK).

mechanics parameters, such as threshold stress-intensity factors (K_{Ith}) and crack velocities at high K values for statically loaded specimens, are other common measures of susceptibility (Fig. 2.3(c)). For cyclic loading, ΔK threshold values and crack-growth rates per cycle compared with those in air (or vacuum) for hydrogen-free material provide measures of HE susceptibility (Fig. 2.3(d)).

Testing procedures for checking whether or not batches of steel components have been embrittled by plating processes are detailed in various standards [8]. For example, notched tensile specimens of a susceptible high-strength steel are considered to be not significantly embrittled if specimens survive 200 h at a sustained load of 75% of the notched tensile strength. Components subject to the same process conditions are then deemed to be safe to use. (It is worth noting that specimens are not usually tested for longer than 200 h and, when they are, failures do occasionally occur at longer times.) For notched tensile specimens subjected to slow strain rates ($2 \times 10^{-4} s^{-1}$), achieving a fracture stress of greater than ~75% of the notched tensile strength indicates that similarly processed components are not significantly embrittled by hydrogen.

2.1.7   Main variables affecting HE

Important variables include stress/stress-intensity factors, hydrogen content or pressure, temperature, strain-rate (or rise-time/cycle-frequency for fatigue), and alloy-strength/microstructure/impurity level (Fig. 2.3). As would be expected, increasing hydrogen contents (for IHE) and increasing hydrogen-gas pressures (for HEE) result in increasing susceptibility, although there can be a threshold hydrogen concentration for embrittlement and there is sometimes little change in the degree of embrittlement with increasing hydrogen pressure once a certain critical 'saturation' level is reached. Susceptibility to HE is often greatest around ambient temperature and gradually decreases with decreasing temperature in Ni alloys and steels, with susceptibility decreasing more steeply for IHE than for HEE for high-strength steels (Fig. 2.4). Susceptibility decreases precipitously above a critical temperature (often less than 100°C) for both IHE and HEE in some steels. Strain rate (during tensile tests) is another important variable, with the extent of HE decreasing with increasing strain rate. HE is usually not observed under impact loading unless high hydrogen (and impurity) concentrations are present at grain boundaries, or unless hydrides are present.

The susceptibility of most materials to HE increases with increasing strength providing that differences in microstructure are not too large, e.g. for tempered martensitic steels with strengths greater than about 900 MPa. However, microstructure can sometimes override the effects of strength, and HE susceptibility can sometimes increase with decreasing strength

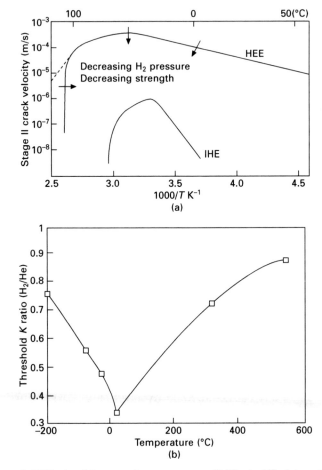

2.4 Effects of temperature on susceptibility to HE: (a) sustained-load crack-growth rates versus $1/T$ for IHE and HEE of high-strength steels [13, 20, 21], and (b) ratio of threshold stress-intensity factor for Inconel 718 in high-pressure hydrogen (34.5 MPa) compared with inert (helium) environment as a function of temperature [16].

below ~800 MPa due to coarsening of spheroidised carbide microstructures. Other important microstructural features that influence HE susceptibility of steels include grain size, amount of retained austenite, and the type of martensite (with twinned martensite being much more susceptible than lath martensite).

Materials with higher impurity contents can be more susceptible (for a given strength) owing to impurity segregation to grain boundaries or increase in size or number of inclusions. Thus, plots of HE susceptibility versus strength, without considering microstructure, can exhibit considerable scatter, especially for lower strength steels. In principle, microstructures with strong traps (such as inclusions) that are easily fractured should be particularly susceptible to IHE since these traps are not only associated with high hydrogen concentrations but also act as crack-initiation sites. On the other hand, numerous moderate-to-strong traps, such as some small precipitates that do not easily initiate cracks, should inhibit hydrogen diffusion to crack-initiation sites, thereby increasing resistance to IHE.

2.1.8   Characteristics of HE

Cracking produced by HE is associated with little or no macroscopic deformation, but does generally exhibit signs of localised plasticity. Fracture surfaces are generally bright (unless atmospheric corrosion occurs after cracking), and are usually faceted due to cracking along grain boundaries or along low-index crystallographic cleavage planes. For high-strength steels, cracking usually occurs along prior-austenite grain boundaries for normal levels of metalloid-impurity segregation at boundaries (Fig. 2.5), but can occur along martensite–lath interfaces in high-purity steels.

IHE initiates at or very close to surfaces (or at electroplate-steel interfaces) for high-strength steels, whereas IHE for low-strength steels can initiate well below the surface. In the latter case, 'fish-eyes' or 'flakes' with a bright, faceted appearance, surrounded by either ductile overload fractures or fatigue fractures are observed (Fig. 2.6). Further details of the metallographic and fractographic characteristics of IHE and HEE are discussed in subsequent sections of this chapter, especially with regard to whether or not the various proposed mechanisms of HE are consistent with the observations.

 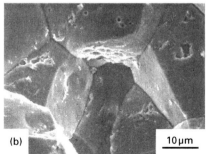

(a)   50 µm   (b)   10 µm

2.5 (a) Metallographic section through fracture surface showing cracking along prior-austenite grain boundaries of the tempered-martensite structure, and (b) SEM of fracture surface of a high-strength steel showing brittle intergranular facets.

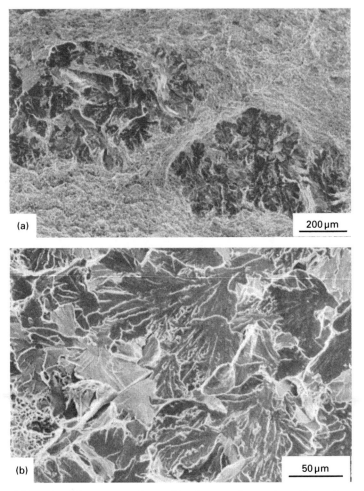

2.6 SEMs of fracture surface of hydrogen-charged medium-strength steel showing (a) fisheyes – bright, faceted, cleavage-like areas surrounded by dimpled areas, and (b) cleavage-like areas at a higher magnification.

2.2   Proposed mechanisms of hydrogen embrittlement (HE) and supporting evidence

2.2.1   Hydride formation and fracture

Description

A mechanism based on the formation and fracture of hydrides at crack tips was first proposed by Westlake in 1969 [22]. The basic mechanism is thought to involve repeated sequences of: (i) hydrogen diffusion to regions of high hydrostatic stress ahead of cracks, (ii) nucleation and growth of a hydride phase, (iii) cleavage of the hydride when it reaches a critical size, and (iv) crack-arrest at the hydride–matrix interface (Fig. 2.7).

A hydride mechanism only occurs in the temperature and strain-rate regime where hydrogen has time to diffuse to regions ahead of crack tips, and only at temperatures where the hydride phase is stable (and brittle) [23, 24]. When hydrogen is present as solute, embrittlement is not observed, and tensile specimens sometimes undergo considerable plastic strain before hydrides are nucleated and brittle fracture occurs. Variations to the hydride mechanism illustrated in Fig. 2.7 may occur in some circumstances, e.g. (i) pre-existing hydrides may be present, and may dissolve and reorient so that hydride plates are perpendicular to the tensile stress axis, (ii) hydrides may dissolve after fracture owing to relaxations of hydrostatic stresses, and (iii) ductile tearing may occur between discrete, closely spaced hydrides. HE in some hydride-forming materials can, however, also occur in the absence of hydrides by other mechanisms described in subsequent sections.

Supporting evidence

A hydride mechanism is generally accepted for HE of certain materials such as V, Zr, Nb, Ta, and Ti [23, 24] since (i) there is a strong thermodynamic driving force for hydride formation, (ii) hydrides are known to be brittle, (iii) hydrides have been observed on both halves of cleavage-like fracture surfaces, (iv) the process has been directly observed in hydrogen-charged thin foils by high-voltage transmission-electron microscopy (TEM), and (v) crack-arrest markings (CAMs) have been observed on fracture surfaces of these materials, consistent with the discontinuous cracking process illustrated in Fig. 2.7.

2.2.2   Hydrogen-enhanced decohesion (HEDE)

Description

A decohesion theory was first proposed in 1926 by Pfeil [25], who proposed that 'hydrogen decreased the cohesion across cubic cleavage planes' (and grain

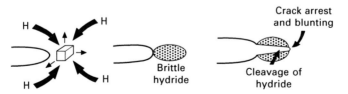

2.7 Schematic diagram illustrating sub-critical crack growth involving hydrogen diffusion to hydrostatically stressed regions, then formation and fracture of a brittle hydride at a crack tip.

boundaries). In 1959 Troiano [26] suggested that weakening of interatomic bonds (in iron) was due to donation of the hydrogen 1s electron to the unfilled 3d shell of the iron atoms. The decohesion hypothesis – involving charge-transfer and weakening of interatomic bonds so that tensile separation of atoms (decohesion) occurred in preference to slip – was subsequently quantitatively developed by Oriani [27] and others [13, 28–30]. Decohesion is usually envisaged as a simple, sequential tensile separation of atoms when a critical crack-tip-opening displacement (CTOD) (approximately half the interatomic spacing) is reached. However, separation of atoms at crack tips is constrained by surrounding atoms and, hence, the separation process could be more complex and involve incipient shear movement of atoms ('atomic shuffles') to enable a critical CTOD to be achieved [31].

Some dislocation activity may accompany decohesion, and may locally increase stresses at decohesion sites, but should be fairly limited so that atomically sharp crack tips are not continually blunted. High concentrations of hydrogen and the decohesion event could occur at a variety of locations, namely: (i) at atomically sharp crack tips due to adsorbed hydrogen, (ii) several tens of nanometres ahead of cracks where dislocation shielding effects result in a tensile-stress maximum, (iii) positions of maximum hydrostatic stress (several micrometres in high-strength steels), and (iv) particle–matrix interfaces ahead of cracks (Fig. 2.8).

Very high elastic stresses are probably required to produce sufficiently high concentrations of hydrogen in interstitial lattice sites ahead of crack tips to produce decohesion, and whether such stresses can be achieved, e.g. due to strain-gradient hardening [13], is debatable. Decohesion at grain boundaries

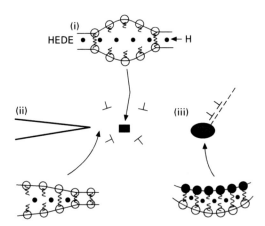

2.8 Schematic diagrams illustrating the HEDE mechanism, involving tensile separation of atoms owing to weakening of interatomic bonds by (i) hydrogen in the lattice (ii) adsorbed hydrogen and (iii) hydrogen at particle–matrix interfaces [32].

Published by Woodhead Publishing Limited, 2011

could occur at or ahead of crack tips due to hydrogen trapped at specific sites, and may occur as a result of bond weakening by both hydrogen and metalloid-impurity segregation. Hydrogen diffusion to the aforementioned sites is required for both IHE and HEE, except for cases where hydrogen adsorption at tips of external cracks is solely responsible for HEE.

Supporting evidence

High hydrogen concentrations have been observed at grain boundaries and particle–matrix interfaces by various techniques, and quantum-mechanical calculations support the concept of weakening of interatomic bonds by hydrogen leading to HEDE, particularly if slip planes around crack tips are not favourably oriented for slip [33, 34]. Direct experimental evidence of HEDE is, however, difficult to obtain since there are no techniques for directly observing events on the atomic scale at crack tips in bulk material. Perhaps the most direct experimental evidence that hydrogen can weaken interatomic bonds is the easier field-evaporation of surface atoms observed during field-ion microscopy when hydrogen is used as the imaging gas [35].

A featureless fracture surface sometimes observed by SEM at high magnifications is often taken as evidence that decohesion has occurred [13], but it is debatable whether SEM is able to resolve sufficiently small, shallow dimples that crack growth by very localised plasticity can be ruled out. However, TEM of fractured thin foils shows that atomically brittle intergranular fracture can occur in materials with segregated impurities at grain boundaries [36], e.g. S in Ni, and atomistic calculations indicate that such impurities can, like hydrogen, weaken interatomic bonds [37, 38]. Thus, it would not be surprising if hydrogen segregated at grain boundaries and other interfaces also resulted in decohesion, especially if hydrogen and impurity elements were both present.

2.2.3   Hydrogen-enhanced localised plasticity (HELP)

Description

In 1972 Beachem [39] was the first to suggest, partly on the basis of fractographic observations, that HAC occurred because solute hydrogen facilitated the movement of dislocations. This idea was subsequently promoted by Birnbaum, Sofronis, Robertson, and co-workers from the 1980s onwards [23, 24, 40, 41]. It was proposed that, since hydrogen concentrations were localised near crack tips due to hydrostatic stresses or entry of hydrogen at crack tips, deformation was localised near crack tips as a result of solute hydrogen facilitating dislocation activity. It was argued that sub-critical crack growth should then occur by a more localised microvoid-coalescence (MVC) process than that which would occur in inert environments (Fig. 2.9).

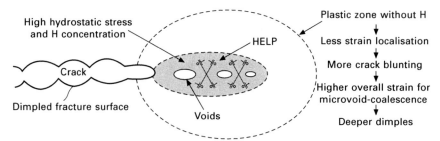

2.9 Schematic diagram illustrating the HELP mechanism, involving a microvoid-coalescence process, with plasticity localised and facilitated in regions of high hydrogen concentrations.

Hydrogen diffusion to local regions ahead of crack tips is obviously required for operation of the HELP mechanism for both IHE and HEE. Crack paths could be transgranular or intergranular depending on whether locally high hydrogen concentrations were present within grain interiors or adjacent to grain boundaries.

Supporting evidence

The idea that solute hydrogen facilitates dislocation activity is supported by: (i) elasticity theory, (ii) atomistic calculations, (iii) *in-situ* transmission-electron microscopy observations of dislocations in thin foils exposed to hydrogen gas, (iv) softening and strain localisation in bulk specimens under some conditions, (v) comparisons of slip-line characteristics in hydrogen-charged compared with hydrogen-free specimens, and (vi) nano-indentation tests.

Elasticity theory

Linear-elastic and finite-element calculations indicate that dislocations are 'shielded' from the full force of repulsive interactions between dislocations and obstacles (including other dislocations) when hydrogen is present [23, 24, 40, 41]. This is thought to arise due to 're-configuration' of hydrogen atmospheres around dislocations as they approach and pass obstacles during deformation providing, of course, that hydrogen atmospheres can keep up with dislocations. Edge dislocations should be more effectively shielded than screw dislocations since the dilatational stress-field around edge dislocations promotes hydrogen atmospheres around them.

Atomistic calculations

Atomistic simulations indicate that hydrogen lowers the dislocation-core energy, and thereby reduces the Peierls stress for dislocation movement (in Al) by more than an order of magnitude [42]. This effect occurred for edge, screw and mixed dislocations, although the effect of hydrogen on the binding energy depended on the dislocation character such that cross-slip was inhibited and slip-planarity was promoted. These effects were thought likely to be applicable to other metals besides aluminium, and that the effect of hydrogen on the core energies of dislocations could act in conjunction with hydrogen effects on elastic interactions between dislocations [42].

In-situ TEM observations

Direct observations of hydrogen-induced increases in dislocation activity have been made using high-voltage TEM, where hydrogen gas is introduced into an environmental cell around thin foils subjected to stress [43]. On introducing hydrogen, which rapidly dissociates under the influence of the electron beam and then diffuses into specimens, dislocations that were stationary start to move, those that were moving increase their velocity (by typically 10 to 100 times), and the rate of dislocation generation from sources is increased. These effects have been observed for edge, screw, and mixed dislocations, which may be entirely within foils or intersect the surface, in a wide range of materials. Decreases in the spacings of dislocations in pile-ups and decreases in the extent of cross-slip have also been observed upon the introduction of hydrogen.

Softening in bulk specimens

Lower flow stresses have sometimes been observed during tensile tests for hydrogen-charged specimens (or specimens tested at very slow strain rates in hydrogen gas) compared with hydrogen-free specimens tested in air. However, the influence of hydrogen on stress–strain curves depends on the material, its purity, strain rate, temperature, and other variables, and hardening is observed in some circumstances. The degree of softening, when it occurs, is usually quite small, e.g. ~10% in hydrogen-charged aluminium and nickel, but is sometimes large, e.g. in hydrogen-charged pure iron single crystals at low temperatures and low strain rates [3, 23, 24, 40, 41, 44].

Observations of slip lines and dislocation arrangements

Increases and decreases in the spacing and height of slip steps, and in the number of active slip systems (produced by tensile or indentation tests) have been reported for hydrogen-charged specimens compared with hydrogen-free specimens. The exact behaviour depends on the material, hydrogen content, microstructure, degree of strain, and other variables, and HELP has been invoked to explain all these effects [23, 24, 44–51]. An ability to accommodate higher shear strains in slip bands (owing to HELP) could result in fewer active slip systems and coarser slip steps in some circumstances, while decreased elastic interactions between adjacent slip bands (owing to HELP) could lead to more closely spaced slip bands and smaller slip steps in other circumstances. Preferential activity of edge-dislocation components over screw components (owing to HELP) could lead to inhibited cross-slip and strain localisation, resulting in coarser slip bands, but could also lead to enhanced hardening in slip bands leading to other slip systems being activated.

Nano-indentation experiments

Besides observations of the effects of hydrogen on slip characteristics around indentations, load-depth and load-time data obtained from nano-indentation tests on hydrogen-free and hydrogen-charged material can also provide information on hydrogen effects during deformation [52]. For example, for nickel single crystals (with a {111} surface) introducing hydrogen into specimens decreased the pop-in load (under increasing loads) and decreased the pop-in time (under constant load). Hydrogen could be introduced and removed electrochemically, and the effects on pop-in were reversible, suggesting that they were intrinsic ones associated with solute hydrogen. Oxide-film effects could be discounted, and the results were explained on the basis that solute hydrogen reduced the shear modulus and facilitated the homogeneous nucleation of dislocations in a volume of material underneath the indenter.

Fractographic observations and direct TEM observations of HAC

Fractographic observations, showing smaller, shallow dimples on fracture surfaces after HAC than after fast fracture in air, are often taken as evidence for HELP, and were what prompted Beachem to first propose this mechanism, as already mentioned. Direct observations of HAC in thicker regions of thin foils by high-voltage TEM also show that cracking occurs by a localised nano-void coalescence process [53]. However, such observations could also be explained by an adsorption-induced dislocation-emission mechanism (AIDE) (or a combination of AIDE and HELP), as discussed in the next section.

2.2.4   Adsorption-induced dislocation emission (AIDE)

Description

The AIDE mechanism was first proposed by Lynch in 1976 and further developed in subsequent papers [32, 54–56]. Theories of HE based on adsorbed hydrogen had been suggested earlier by others, such as Petch in 1956 [57], but were couched in thermodynamic terms, e.g. surface-energy reductions, rather than in mechanistic terms. In 1975 Clum [58] was the first to suggest, based on field-ion microscopy observations, that adsorption (of hydrogen) might facilitate dislocation nucleation at surfaces. However, he did not indicate how such a process could lead to embrittlement.

The AIDE mechanism is, in some respects, more complex than the HEDE or HELP mechanisms and, hence, requires a lengthier explanation. For the AIDE model, the term 'dislocation-emission' encompasses both nucleation and subsequent movement of dislocations away from crack tip, and it is important to note that it is the nucleation stage that is critical and facilitated by adsorption. Once nucleated, dislocations can readily move away from the crack tip under the applied stress. The nucleation stage involves the simultaneous formation of a dislocation core and surface step by co-operative shearing of atoms (breaking and re-forming of interatomic bonds) over several atomic distances. Thus, weakening of interatomic bonds over several atomic distances by 'adsorbed' hydrogen can facilitate the process.

In the AIDE model, crack growth under sustained or monotonically increasing stresses occurs not only by dislocation emission from crack tips, but also involves nucleation and growth of microvoids (or nano-voids) ahead of crack tips. Nucleation and growth of voids at second-phase particles, slip-band intersections, or other sites in the plastic zone ahead of cracks occurs because stresses required for dislocation emission are sufficiently high that some general dislocation activity occurs ahead of cracks. Void formation contributes to crack growth, and also serves to re-sharpen crack tips and results in small crack-tip-opening angles. However, crack growth primarily occurs by dislocation emission from crack tips (Fig. 2.10).

For IHE, hydrogen diffusion to, and adsorption at, internal crack tips or voids is necessary for AIDE, while diffusion is not essential for HEE – just dissociation of hydrogen molecules and adsorption at external crack tips is required. However, hydrogen diffusion to voids ahead of cracks may occur during HEE, resulting in AIDE at tips of voids as well at the external crack tip. Dislocation transport as well as lattice diffusion could be involved in some circumstances. For IHE, if internal crack tips break through to the external surface, hydrogen adsorption would be inhibited by preferential adsorption of oxygen (from air environments), but could still occur at tips of voids just ahead of crack tips.

To understand why facilitating dislocation emission from crack tips results in embrittlement, it is necessary to consider how crack growth occurs in inert (non-embrittling) environments for ductile materials. Ductile crack growth appears to occur predominantly by egress of dislocations, nucleated from sources in the plastic zone ahead of cracks, with little or no emission of dislocations occurring from crack tips. Dislocation emission from crack

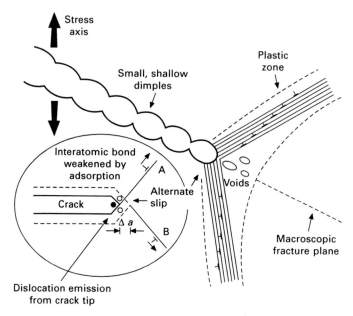

2.10 Schematic diagram illustrating the adsorption-induced dislocation emission (AIDE) mechanism for HEE, which involves crack growth by alternate-slip (for transgranular paths) from crack tips, facilitating coalescence of cracks with voids formed in the plastic zone ahead of cracks [55].

tips is probably difficult in inert or air environments because interatomic bonding at crack tips is intrinsically strong or because strong bonds are formed between oxygen and substrate bonds, respectively.

When dislocation egress around crack tips predominates, only a small proportion of dislocations emanating from near-crack-tip sources exactly intersect crack tips to produce crack advance – most produce only blunting or contribute to the strain ahead of cracks. Large strains ahead of cracks are therefore needed to produce crack growth by microvoid-coalescence, and deep dimples with smaller dimples within them are produced on fracture surfaces. The small dimples within large dimples arise because coalescence of large voids (nucleated from large particles) involves nucleation and growth of small voids (nucleated from smaller particles or other sites at higher strains) between large voids (Fig. 2.11).

When hydrogen adsorption weakens interatomic bonds and thereby promotes dislocation emission from crack tips, a greater proportion of dislocation activity results in crack growth since dislocation emission on suitably inclined slip planes produces crack-advance as well as crack-opening. Thus, coalescence of cracks with voids occurs at lower strains and shallower dimples are produced on fracture surfaces when AIDE occurs. The dimples

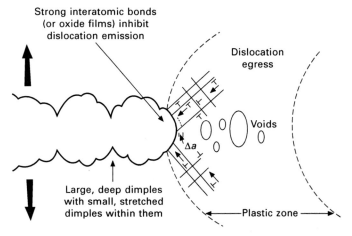

2.11 Schematic diagram illustrating ductile crack growth involving coalescence of cracks with voids by egress of dislocations nucleated from near-crack-tip sources [55].

resulting from an AIDE/MVC process also appear to be smaller (as well as shallower) than those produced by ductile fracture since, for the latter, small dimples within large dimples are often stretched and difficult to resolve.

Crack paths produced as a result of the AIDE mechanism could be intergranular or transgranular depending on where dislocation emission and void formation occurred most easily. For transgranular cracking, alternate-slip on planes on either side of cracks would tend to occur in order to minimise the back-stress from previously emitted dislocations. Macroscopic planes for transgranular cracking would therefore bisect the angle between the slip planes, and crack fronts would lie along the line of intersection of crack planes and slip planes, e.g. along {100} planes in ⟨110⟩ directions when {111} or {112} slip planes were active in fcc or bcc materials. However, deviations from low-index planes and directions would occur if unequal amounts of slip occurred on either side of cracks owing to large differences in shear stresses on the different slip planes. Deviations of fracture planes away from low-index planes could also occur depending on the location of void nuclei ahead of cracks.

Background and supporting evidence

As detailed in the following, an AIDE mechanism for IHE and HEE in some materials is supported by: (i) the presence of high concentrations of hydrogen adsorbed on surfaces (and within a few atomic distances of surfaces), (ii) surface-science observations, (iii) atomistic modelling, (iv) observations of HE at 'high' crack velocities relative to hydrogen diffusivities, (v) metallographic and fractographic observations, including remarkable similarities between HE and LME, and (vi) other miscellaneous observations.

Hydrogen adsorbed at surfaces

For clean metal surfaces of many metals, e.g. Fe, Ni, Ti, exposed to hydrogen gas, it is well established that hydrogen molecules dissociate, and that hydrogen atoms chemisorb at specific surface sites with a deep potential-energy trough relative to normal interstitial lattice sites. The interstitial sites between the first and second atomic layers (and perhaps second and third layers) are also a stronger trap site for hydrogen than normal interstitial sites in some materials [12, 59–62]. Thus, much higher hydrogen concentrations would be expected at (crack-tip) surface sites and just sub-surface sites than at normal interstitial lattice sites (Fig. 2.1). The adsorption process can be complex, involving physical adsorption of hydrogen molecules and other precursor steps, and is known to be strongly influenced by the substrate metal, strain, temperature, surface-crystallography, surface-coverage and roughness. Hydrogen diffusion to internal voids/cracks would produce similarly high concentrations of 'adsorbed' hydrogen at these internal surfaces. (As mentioned in the introduction, surface and just sub-surface sites are referred to as 'adsorbed hydrogen' in regard to the AIDE and HEDE mechanisms.)

Surface-science observations

For clean, stress-free metal surfaces (without adsorbates or films), surface-lattice 'relaxations' or 'reconstructions' are observed, i.e. the lattice spacings or lattice structure within several atomic distances of surfaces are different from those in the bulk [63–67]. These surface-lattice perturbations occur essentially because atoms at the surface have fewer neighbours than those in the bulk and, hence, electron-charge is redistributed. Thus, it is not surprising that electron-charge transfer owing to adsorption of hydrogen and other environmental atoms often affects the extent and nature of the perturbations. These changes can sometimes produce buckling or rumpling of the surface, involving lateral or vertical shear movements of atoms (Fig. 2.12). The exact effect of adsorption depends on variables such as the adsorbed species, surface-crystallography, surface-coverage of adsorbed species and temperature. Generalisations are difficult to make because such a diversity of effects has been observed, and an understanding of all these effects is far from complete.

When crack tips are not atomically sharp but have a finite radius, surface-lattice perturbations (and modifications thereof by adsorption) should occur at crack tips during sub-critical crack growth. Indeed, the phenomenon of adsorption-induced LME, involving sub-critical cracking at velocities of the

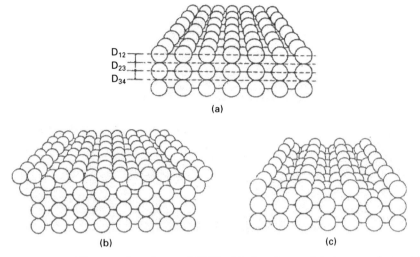

2.12 Perspective views of {110} nickel surfaces: (a) clean surface with oscillatory contractions and expansions in the first few layers, and (b), (c) possible reconstructions resulting from adsorbed hydrogen (not shown) [65].

order of 100 mm/s, shows that adsorption-induced changes to bonding at crack tips must have time to occur even at high crack velocities. However, the crack-tip surface-lattice perturbations and effects of adsorption are probably significantly different from those at plane surfaces. For example, there may be insufficient time to achieve an equilibrium state during crack growth, and high elastic (and plastic) strains and crack-tip curvature are likely to have significant effects.

Crack-tip surface-lattice perturbations/interatomic-bond strengths would be expected to influence dislocation nucleation from crack tips because the nucleation process, i.e. the simultaneous formation of a surface step and a dislocation core, involves the co-operative shear movement of atoms over several atomic distances at crack tips. In general, small crack-tip perturbations (and weak adsorbate–substrate and weak adsorbate–adsorbate bonding) should favour emission of dislocations. (Adsorbate–substrate and adsorbate–adsorbate bonds as well as metal–metal bonds have to be sheared.) Conversely, large crack-tip lattice perturbations (or strong adsorbate–substrate bonding) would be expected to inhibit dislocation emission.

Atomistic modelling

Atomistic calculations of crack growth in nickel using the embedded-atom method indicate that adsorbed hydrogen facilitates dislocation emission from crack tips, providing that slip planes are favourably oriented with respect

Published by Woodhead Publishing Limited, 2011

to the crack plane [33, 34]. Another study showed that dislocations (in Al) should be emitted more easily when hydrogen is present due to lowering of the unstable stacking-fault energy [42]. Theoretical calculations also suggest that dislocation emission from crack tips could be affected by adsorption-induced changes to surface stresses (which are linked to the surface lattice perturbations being restrained by the sub-surface lattice) [68, 69]. In one study, cluster-calculations for adsorbed hydrogen on beryllium suggested that high surface tensile stresses could be induced by adsorption, thereby facilitating dislocation emission [70].

Observations of HEE at low D/v ratios

Analyses have shown that hydrogen should not diffuse ahead of cracks (for more than a few atomic distances) for D/v ratios of less than $\sim 10^{-8}$ cm, were D is the hydrogen diffusivity and v is the crack velocity [19, 71]. HEE in nickel can occur at crack velocities such that D/v ratios are less than 10^{-8} cm [54, 55], which suggests that adsorption is responsible and, since considerable localised plasticity occurs, that AIDE rather than HEDE predominates. Dislocation-transport of hydrogen directly ahead of crack tips was considered to be unlikely for the particular circumstances involved. More details are provided in a subsequent section.

Metallographic and fractographic observations

Remarkably similar (dimpled) fracture surfaces produced by HE and LME in some materials suggest that the AIDE mechanism is responsible. For LME, only adsorption occurs at crack tips in many systems because crack growth occurs so rapidly (up to 200 mm/s) that there is no time (or tendency) for other interactions to occur [54, 55]. Details of the close similarities between HE and LME are shown in subsequent sections for specific materials, where the pros and cons of AIDE, HELP and HEDE mechanisms are discussed in detail. Other observations supporting an adsorption mechanism for some fracture modes are also discussed in more detail in subsequent sections.

2.2.5   Hybrid mechanisms and possible role of vacancies

Combinations of AIDE, HELP, and HEDE mechanisms could, of course, occur in many cases, with the relative importance depending on the fracture mode, which depends on the material and other variables [32]. For example, dislocations nucleated from crack tips (due to AIDE) may move away from crack tips more readily due to HELP, thereby decreasing the back-stress on subsequent dislocation emission. For crack growth predominantly by AIDE, void-nucleation ahead of cracks could be promoted at slip-band intersections

by HELP or by HEDE at particle–matrix interfaces (Fig. 2.13(a)). AIDE
and HEDE could also occur sequentially, with AIDE occurring until the
back-stress from emitted dislocations increased somewhat so that HEDE
then occurred, followed by AIDE again when the crack tip had moved
away from the stress-field of dislocations previously emitted (Fig. 2.13(b)).
Such a process could produce small crack-tip-opening angles without void
formation ahead of cracks [10, 32].

As well as AIDE and HELP possibly acting conjointly to localise
plasticity, additional strain-localisation could occur due to hydrogen–vacancy
interactions. Vacancy effects probably play only a secondary role, although
there has been a somewhat radical suggestion that HE may be primarily the
result of a high concentration of vacancies ahead of cracks, due to hydrogen-
induced reductions in the vacancy-formation energy, rather than hydrogen
effects *per se* [72]. A high concentration of vacancies ahead of cracks could
occur not only due to high hydrogen concentrations at regions of high
triaxial stress but also because hydrogen stabilises vacancies produced by
dislocation–dislocation interactions [73]. Vacancy agglomeration could lead
to nano-void formation [74], and autocatalytic void-growth and coalescence
might occur if the number or size of nano-voids reached a critical level.
Vacancies could also facilitate dislocation-climb and cross-slip, and reduce
strain-hardening, thereby promoting strain localisation, possibly in conjunction
with hydrogen-induced slip-localisation due to AIDE or HELP.

2.3   Relative contributions of various mechanisms for different fracture modes

The likelihood that different mechanisms predominate for different fracture
modes (in non-hydride forming materials) is discussed in the following. The
characteristics of various fracture modes are described in detail, followed by

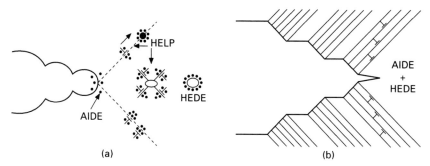

2.13 Schematic diagrams illustrating hybrid mechanisms of
hydrogen-assisted cracking: (a) AIDE with contributions from HELP
and HEDE, and (b) AIDE alternating with HEDE [32].

Published by Woodhead Publishing Limited, 2011

a discussion of how well different mechanisms account for the observations. Intergranular fracture produced by HE in high-strength steels and other materials is considered first since this is the most common fracture mode. Transgranular fractures in high-strength steels are then discussed, followed by consideration of cleavage-like fractures, and slip-band fractures.

2.3.1   Intergranular fracture in high-strength steels and other materials

Intergranular fracture along prior-austenite grain boundaries is the most common fracture mode for IHE, HEE and LME of high-strength steels. For normal metalloid-impurity contents, especially in steels tempered in the range where impurity segregation is prone to occur, SEM indicates that intergranular fracture surfaces are featureless in areas between ridges and isolated dimples (Fig. 2.14). However, in some cases, the relatively smooth areas appear to be dimpled on a very fine scale when they are examined by TEM of replicas (Fig. 2.15). TEM resolves small, shallow features better than SEM, providing that replicas are shadowed at low angles (10–15°), and examined at low kV at high tilt angles (~30°) [54].

In other cases, e.g. for high-strength steels tempered at high temperatures, SEM (and TEM) show that intergranular facets produced by HEE and LME are completely covered by well-defined dimples (Fig. 2.16). Such observations lend credibility to the possibility that apparently smooth areas observed by SEM for other heat-treatments could be dimpled on a scale not resolved by SEM. The dimples on fracture surfaces produced by HE appeared to be smaller and shallower than those produced by fast fracture in air (where transgranular fracture generally occurred). Comparison of dimple sizes is,

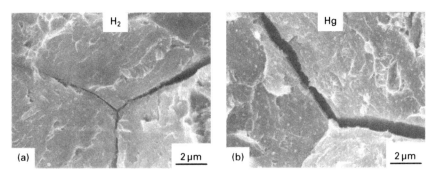

2.14 SEM of fracture surfaces produced by sub-critical cracking ($v \sim 10^{-5}$ m/s) in (a) gaseous hydrogen (101 kPa), and (b) liquid mercury, at 20°C for a high-strength martensitic steel (tempered at 550°C), showing 'brittle' intergranular facets exhibiting tear ridges and isolated dimples [54].

2.15 TEM of replica of fracture surface produced by slow crack growth in gaseous hydrogen (101 kPa) for a high-strength martensitic steel (tempered at 550°C), showing fine details between tear ridges [54].

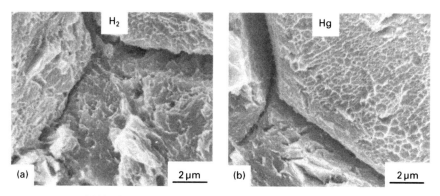

2.16 SEM of fracture surface produced by sub-critical cracking ($v \sim 10^{-5}$ m/s) in (a) gaseous hydrogen (101 kPa), and (b) in liquid mercury, at 20°C in a high-strength martensitic steel (tempered at 650°C), showing dimpled intergranular facets [54].

however, difficult since the large, deep dimples produced by fracture in the absence of hydrogen contain small, stretched dimples that are often difficult to resolve.

For completely dimpled fractures, AIDE or HELP mechanisms rather than HEDE are most likely to be primarily responsible for facilitating and localising microvoid-coalescence processes. The AIDE mechanism probably predominates because higher hydrogen concentrations are likely to be present exactly at grain boundaries than adjacent to grain boundaries. The essentially identical, dimpled intergranular fracture mode produced by adsorption-induced LME (Fig. 2.16) as produced for HEE (for the same steel and the same heat-treatment condition) also supports an AIDE mechanism.

For intergranular fractures where it is debatable whether or not dimples are present, either an AIDE/MVC process operates on such a fine scale that dimples are difficult to resolve or a HEDE process occurs. It is also possible that both processes occur, with some grain-boundary structures and metalloid-impurity contents locally favouring HEDE and other grain-boundary structures and compositions favouring AIDE. A combined process is reasonable since both HEDE and AIDE are based on hydrogen-induced weakening of interatomic bonds. The same considerations regarding AIDE, HELP and HEDE mechanisms, as discussed for the high-strength steels, also apply to intergranular fracture in other materials.

2.3.2   Transgranular dimpled fracture in high-strength steels and other materials

For high-strength martensitic steels with low levels of metalloid impurities, especially at tempering temperatures that do not promote metalloid-impurity segregation to prior-austenite grain boundaries, HEE can produce completely dimpled transgranular fractures [54]. (Crack-paths were transgranular with respect to prior-austenite grain boundaries, but sometimes followed martensite-site lath boundaries.) Dimples appeared to be smaller and were shallower than those produced by fast fracture in air, as discussed for dimpled intergranular fractures. Transgranular dimpled fracture surfaces essentially identical to those produced by HEE were observed for adsorption-induced LME (for the same steel and tempering temperature) (Fig. 2.17) [54]. Lower strains-to-fracture and differences in dimple size (generally smaller and with a smaller depth-to-width ratio) have also been observed in other (lower strength) hydrogen-charged steels compared with hydrogen-free steels after tensile testing [75]. The HEE and LME comparisons suggest that an AIDE mechanism most likely predominates, with possible contributions from HELP and HEDE (and vacancy effects), as discussed in previous sections.

2.3.3   Cleavage-like fractures

Hydrogen-induced cleavage-like fractures (involving slow, stable crack growth) are less common than intergranular fractures, but have been extensively studied, especially in single crystals (e.g. Ni, Fe, Fe-Si), on the basis that it should be easier to elucidate the mechanisms of HE in such materials than in materials with more complex microstructures [10, 54, 55, 76–80].

For Ni single crystals, cleavage-like fractures have been observed after

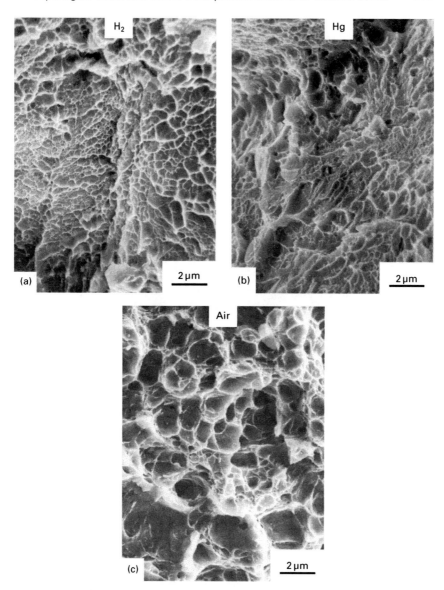

2.17 SEM of transgranular fracture surfaces of martensitic steel (290°C temper) resulting from slow crack growth in (a) gaseous hydrogen (101 kPa), (b) liquid mercury, and (c) overload in air, at 20°C, showing shallower dimples after LME and HEE compared with overload fracture [54].

crack growth in notched and fatigue pre-cracked specimens tested in tension or cantilever-bending in gaseous hydrogen [10, 54, 55, 76]. Increasing deflection rates in cantilever bending increased rates of crack growth, and cleavage-like fracture surfaces were observed even at crack velocities of up to 0.5 mm/s [54, 55]. Fracture planes were generally near {100}, although significant deviations occurred for specimen orientations where {100} planes were at large angles to the tensile-stress axis, especially at high deflection rates. Crack-growth directions on {100} planes were near $\langle 110 \rangle$, and Y-shaped tear ridges, isolated large dimples, and coarse slip lines were observed on fracture surfaces (Fig. 2.18(a)). TEM of replicas showed that the regions between the tear ridges and large dimples had a rumpled appearance due to fine slip markings and (possibly) nano-scale dimples (Fig. 2.18(b)). Extensive slip on {111} planes intersecting crack tips was observed on the side surfaces of specimens (Fig. 2.19).

For iron and iron-silicon single crystals, cleavage-like fractures, generally along {100} planes, have been observed after HEE and IHE of Fe and Fe-Si single crystals under conditions where ductile fracture would occur in the absence of hydrogen [77–79]. Cracking often occurred in two different crystallographic directions ($\langle 112 \rangle$ in Fe and $\langle 110 \rangle$ in Fe-Si) in adjacent regions, resulting in a herringbone pattern of steps and tear ridges. Fracture surfaces also sometimes exhibited regularly spaced crack-arrest markings (CAMs) (~ 1 μm apart) (Fig. 2.20).

Slip on {111} or {112} planes intersecting crack fronts was observed on the side surfaces of specimens (and in the interior for Fe-Si after sectioning and etching). Other evidence that cleavage-like fracture surfaces are associated with locally high strains includes diffuse electron-channelling patterns (and electron back-scattered diffraction patterns), and high dislocation densities

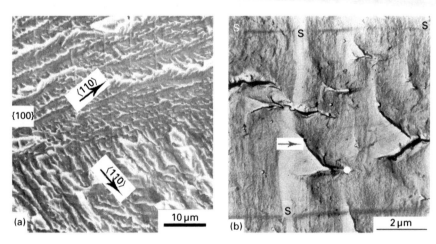

2.18 Cleavage-like fracture surfaces in a nickel single crystal cracked in gaseous hydrogen (101 kPa) at 20°C ($v \sim 10^{-3}$ mm/s): (a) SEM showing tear ridges and steps forming a herringbone pattern, and (b) TEM of secondary-carbon replica showing fine details between the tear ridges and coarse slip lines (S-S) [54].

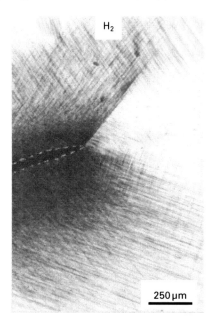

2.19 Optical micrograph showing slip around crack tip after fatigue crack growth of nickel single crystal in hydrogen gas (101 kPa) at 20°C. The slip distribution was similar, but more extensive, after cracking under monotonically increasing displacements [54].

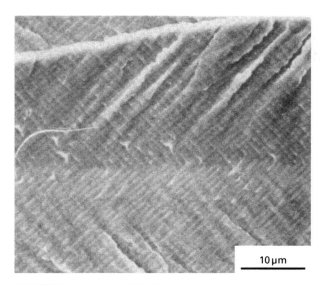

2.20 SEM of cleavage-like fracture surface produced in an Fe-Si single crystal in gaseous hydrogen, showing crack-arrest markings [77].

observed just beneath fracture surfaces by TEM [28, 77]. Fracture surfaces were also rumpled on a fine scale, which was particularly apparent when replicas were examined by TEM [54]. Hydrogen-induced cleavage-like cracking has been observed in Fe at temperatures ranging from −100°C to 20°C, and occurred regardless of whether hydrogen resulted in hardening (at lower temperatures) or softening (at higher temperatures) [80].

The fact that cleavage-like fracture surfaces could be produced in nickel single crystals at velocities as high as 0.5 mm/s at 20°C suggests that 'adsorbed' hydrogen, by itself, can produce such fractures since D/v ratios are $\sim 10^{-8}$ cm. As already mentioned, this is the value below which hydrogen should not have time to diffuse more than one or two atomic distances ahead of cracks. Detailed analyses of the effects of hydrogen-gas pressure and temperature on slow crack growth in nickel single crystals also suggest that the embrittlement site is close (within about eight atomic distances) to crack tips [10, 76]. Assuming some uncertainties in the analyses, this is not inconsistent with an adsorption mechanism, where adsorption is considered to include not only hydrogen on the surface but also hydrogen trapped within a few atomic distances of the crack tip. Observations that cleavage-like cracking in iron-silicon single crystals in gaseous hydrogen can be arrested almost immediately (within 1s) of adding oxygen to the environment also suggest that cracking is due to a surface or very near-surface effect rather than HELP in the plastic zone ahead of cracks [10, 76].

Similarities between fractures produced by HEE and adsorption-induced LME in Fe-Si and Ni single crystals (Figs 2.21 and 2.22) also support an adsorption mechanism for cleavage-like cracking. These similarities are even more striking than those described in the previous section for high-strength steels. For Ni single crystals, partially cracking specimens in liquid mercury, evaporating the mercury, then continuing crack growth in hydrogen gas (at $v \sim 0.1$–0.5 mm/s in both mercury and hydrogen), showed that there were essentially no differences in the detailed fracture-surface appearance and crystallography (Fig. 2.21). Moreover, the detailed appearance and crystallography depended somewhat on the crystal orientation with respect to the stress-axis, but was the same in mercury and hydrogen environments for each orientation. The extent and distribution of slip lines on the side surfaces of specimens were also the same for cracking in mercury and hydrogen [54, 55].

Since plastic strains are locally high, with extensive slip on planes intersecting crack fronts, and there is evidence of void formation ahead of cracks, at least in some cases, it appears that an AIDE/MVC mechanism rather than an adsorption-induced decohesion mechanism predominates. However, it has been suggested that crack growth involves sequences of alternate-slip and decohesion (as shown in Fig. 2.13(b)) on the basis that dimples were not apparent on fracture surfaces [10, 76]. Comments made

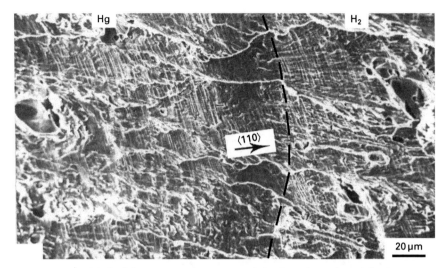

2.21 SEM of fracture surface around the junction (dotted line) between regions produced in mercury and gaseous hydrogen (101 kPa) at 20°C, showing steps, tear ridges, slip lines and isolated dimples for both environments. Note that the crack-tip-opening displacement was decreased slightly after evaporating mercury from cracks prior to cracking in hydrogen in order to produce a crack-front marking [54, 55]. (Note also that tests on separate specimens resulted in the same fracture characteristics in mercury and hydrogen.)

regarding the difficulty of resolving very small, shallow dimples on fracture surfaces made previously are, however, relevant here, and further work is required to resolve this issue.

Aside from evidence discussed above showing that an adsorption mechanism can account for cleavage-like fractures, it is difficult to envisage how a HELP mechanism could explain the fractographic observations. Thus, just facilitating and localising dislocation activity ahead of cracks would not be expected to produce a change from a non-crystallographic ductile fracture in inert environments to a cleavage-like fracture macroscopically along low-index planes in specific directions in hydrogen environments. On the other hand, the fracture characteristics are exactly what one would expect for an AIDE (or AIDE plus HEDE) mechanism. However, minor contributions from HELP (and vacancy effects) cannot be discounted at crack velocities where hydrogen can diffuse ahead of cracks, as discussed in Section 2.2.5.

A characteristic of cleavage-like cracking which at first sight appears to be more consistent with HEDE (or HELP) than AIDE are the crack-arrest markings sometimes observed on fracture surfaces, which suggest that cracking occurs discontinuously with ~1 μm crack jumps. Discontinuous cracking can be explained by a HEDE or HELP mechanism in terms of an

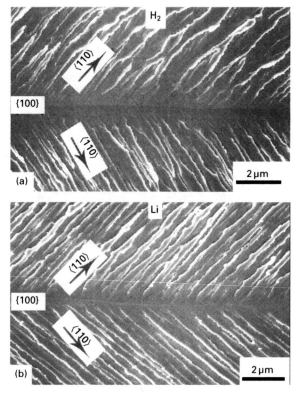

2.22 SEM of cleavage-like {100} fracture surfaces produced in Fe2.6%Si single crystals by (a) slow crack growth (~10^{-3} mm/s) in gaseous hydrogen (101 kPa) at 25°C, and (b) rapid crack growth (>1 mm/s) in liquid lithium at 210°C, showing herringbone pattern of steps owing to crack growth in two ⟨110⟩ directions in adjacent regions [54, 55].

increment of cracking occurring when a critical concentration of hydrogen is built up over a critical distance ahead of cracks, followed by crack-arrest and blunting when cracks run into a region of lower hydrogen concentration. However, discontinuous cracking could occur for other reasons [81], and does not preclude an AIDE mechanism. For example, crack-arrest markings could be produced by intermittent bursts of dislocation egress around cracks associated with a dislocation substructure ahead of cracks.

2.3.4   Slip-band fractures

Hydrogen-assisted cracking along slip bands, while less common than other fracture modes, is sometimes observed in materials where there is an inherent tendency for planar slip, e.g. in stainless steels and nickel-based superalloys

[82–84]. In some cases, slip-band fractures occur only in hydrogen-charged specimens, while in other cases, slip-band fractures occur in hydrogen-free specimens as well as hydrogen-charged specimens, with the latter exhibiting lower strains to failure. Small, shallow dimples are observed on fracture surfaces indicating that cracking occurs by a localised microvoid-coalescence process along slip bands.

Slip bands appear to be more sharply defined (less diffuse) in hydrogen-charged specimens than in hydrogen-free specimens in some materials, and the HELP mechanism is probably significant in promoting more planar slip as a result of a greater effect of hydrogen on edge-dislocation activity than screw-dislocation activity, so that the tendency for cross-slip is reduced [43]. Reductions in stacking-fault energy due to solute hydrogen could also reduce the tendency for cross-slip, but such reductions are thought to be small. Greater slip planarity would increase the strains within slip bands leading to void initiation at lower overall strains.

Void-initiation appears to occur at slip-band intersections and at precipitates, and contributions from HEDE (at precipitate–matrix interfaces) could be involved. Once a significant number of voids were initiated along a slip band, shear-localisation would intensify, resulting in more voids, and the localisation process would become autocatalytic. Void-coalescence would then occur rapidly and contributions from hydrogen effects would probably not be significant.

2.4   General comments

The conclusion that IHE and HEE involving localised plasticity predominately occur by AIDE (for cleavage-like and some intergranular fractures) is at odds with the more general popularity of HELP. The widespread acceptance of HELP seems to stem from (i) the diverse and convincing evidence that the phenomenon does occur, and (ii) the fact that hydrogen can diffuse rapidly in many materials so that solute hydrogen is present ahead of cracks in most circumstances. However, whether the magnitude of the HELP effect in bulk material* is sufficiently large to promote sub-critical cracking, and whether HELP is consistent with other observations is not usually critically considered.

Mechanisms of HE must be able to account, in particular, for (i) the detailed fracture characteristics and crystallography, and (ii) the effects of temperature,

---

*Hydrogen effects on dislocations in thin foils observed by TEM do appear to be large, but these effects may only be partly due to HELP, with contributions to increased dislocation activity on introducing hydrogen arising from non-uniform hydrogen concentrations that induce stresses, and from surface effects, e.g. hydrogen-induced reductions of oxide films.

crack-tip strain rate, hydrogen-pressure and other variables. A mechanism based predominantly on HELP cannot account for these observations for many fracture modes and, considering the evidence overall, it appears that HELP only plays a dominant role in initiating slip-band fractures. In most circumstances, HELP is probably too small to have major effect on sub-critical crack growth. The absence of any embrittlement when only solute hydrogen is present ahead of crack tips (without adsorbed hydrogen at internal cracks/ voids) in some materials supports such a view [15, 54, 55].

The HEDE mechanism also has many adherents because essentially featureless fracture surfaces are often observed using scanning-electron microscopy (SEM). However, as already mentioned, it is debatable whether SEM (even when instruments with field-emission guns are used) is capable of resolving sufficiently shallow dimples to conclude that atomically brittle decohesion, rather than an extremely localised plasticity process, has occurred when fracture surfaces are featureless. The fact that dimples could be as small as ~25 nm diameter, with the cusps only ~4 nm high, needs to be borne in mind. The additional secondary-electron emission from such cusps may not be sufficiently above background levels for them to be apparent. Thus, HEDE could be less common than generally supposed, although probably does dominate for brittle intergranular fractures when high concentrations of other embrittling impurities are present at grain boundaries.

The popularity of solute-hydrogen-based mechanisms possibly also arises because adsorption mechanisms have often been discounted in the past for various reasons. For example, early adsorption mechanisms were based on surface-energy reductions, and it was argued that, since significant localised plasticity usually occurred, the overall fracture energy (dominated by a plastic-work term) would not be significantly reduced by decreasing the surface energy. Such an argument is clearly not applicable to the AIDE mechanism since this process reduces the plastic strains required for crack growth by increasing the proportion of dislocations that produce crack growth. Since AIDE facilitates the coalescence of cracks with voids ahead of them, voids do not grow to the same extent as they would in an inert environment – a point that has eluded some workers who have objected to an AIDE/MVC mechanism on the grounds that the range of influence of adsorption is only a few atomic distances and, hence, should not affect growth of voids ahead of crack tips.

The suggestion that embrittlement results from *easier* dislocation emission from crack tips is also contrary to the predictions of most criteria for ductile versus brittle behaviour. For example, the widely-cited Rice–Thomson (R-T) criterion [85], based on the competition between decohesion and dislocation-emission, assumes that, if dislocation emission occurs at a lower stress than decohesion, then ductile fracture will ensue. However, the R-T criterion, and variants of it, do not address whether macroscopically ductile behaviour will occur, but simply whether crack growth occurs by decohesion or dislocation activity. Such criteria are therefore not relevant to transitions from macroscopically brittle (but microscopically ductile) fracture to a more macroscopically ductile fracture, which is often the case when changing from a hydrogen environment to an inert environment.

An adsorption mechanism has also sometimes been discounted because not all adsorbed species produce embrittlement. However, only weakly adsorbed species such as hydrogen and some metal atoms, e.g. mercury, on particular metals, are likely to result in weakened metal–metal bonds *and* weak adsorbate–adsorbate bonds plus weak metal–adsorbate bonds, and thereby facilitate dislocation emission, as mentioned previously. However, a good understanding of why only some weakly adsorbing metal species produce embrittlement is lacking, and further quantum-mechanical modelling of the effects of various adsorbates on bonding would be worthwhile.

Realistically modelling the effects of adsorption on dislocation emission is challenging because the process depends not only on the structure and bonding at crack tips, but also on a number of other parameters such as: (i) stress-mode, and the extent of 'shear-softening' due to tensile stresses across slip planes, (ii) whether emission occurs on inclined planes or on oblique planes at ledges along the crack front, and the angle of these planes with respect to the crack plane, (iii) the crack-tip radius and core-width of nucleating dislocations, (iv) the shape of the emitted dislocation loop, and whether full or partial dislocations are involved, and (v) the extent of shielding/back-stresses from previously emitted dislocations, which will depend on the mobility of emitted dislocations, and their ability to cross-slip, etc. Further quantum-mechanical-based atomistic and multi-scale modelling of HE (and LME) to get a better understanding of these aspects, and HE in general, are clearly merited.

2.5   Conclusions

Several mechanisms of hydrogen-assisted cracking have significant experimental and theoretical support, namely those based on: (i) repeated formation and fracture of brittle hydrides at crack tips, (ii) adsorption-induced dislocation-emission (AIDE) (involving hydrogen at crack-tip surfaces and within a few atomic distances of crack/void tips), (iii) hydrogen-enhanced localised-plasticity (HELP) in the plastic zone ahead of cracks, and (iv) hydrogen-enhanced decohesion (HEDE) (involving hydrogen at crack tips or hydrogen segregated at grain boundaries and at particle–matrix interfaces ahead of cracks). There is also emerging evidence that hydrogen-vacancy complexes and vacancy clusters may play some role. In other words, as suggested by Oriani some time ago [86], hydrogen can be considered as a 'versatile embrittler'.

Published by Woodhead Publishing Limited, 2011

The mechanisms of cracking mentioned above may occur conjointly in some circumstances, and the dominant mechanism probably depends on the fracture-path and fracture-mode which, in turn, depend on the material, microstructure, environment, temperature, stress-intensity factor, and other variables. In the absence of hydrides, the AIDE mechanism probably predominates for cleavage-like fractures and for dimpled intergranular/transgranular fractures in steels. The HEDE mechanism probably predominates for brittle intergranular fractures (where SEM shows essentially flat facets) when high hydrogen concentrations, in conjunction with segregated embrittling impurities, are present at grain boundaries. The HELP mechanism contributes to the initiation of slip-band fractures by localising slip in some materials, e.g. Ni-base superalloys and stainless steels, but probably plays only a minor role for other fracture modes.

2.6   References

1. W.H. Johnson, On the remarkable changes produced in iron and steel by the action of hydrogen and acids, *Proc. Royal Society of London*, 23, 168–180, 1874–75.
2. R.W. Staehle, J. Hochmann, R.D. McCright, and J.E. Slater (eds), *Stress Corrosion Cracking and Hydrogen Embrittlement of Iron Base Alloys*, NACE, Houston, TX, 1977.
3. J.P. Hirth, Effects of hydrogen on the properties of iron and steel, *Metall. Trans. A*, 11, 861–890, 1980.
4. C.G. Interrante and G.M. Pressouyre (eds), *Current Solutions to Hydrogen Problems in Steels*, ASM, Metals Park, OH, 1982.
5. N.F. Fiore and B.J. Berkowitz (eds), *Advanced Techniques for Characterizing Hydrogen in Metals*, Met. Soc. AIME, New York, 1982.
6. R.A. Oriani, J.P. Hirth and M. Smialowski (eds), *Hydrogen Degradation of Ferrous Alloys*, Noyes Publ., Norwich, NY, 1985.
7. M.R. Louthan, Jr., The effect of hydrogen on metals, pp. 329–365 in *Corrosion Mechanisms*, F. Mansfeld (ed.), Marcel Dekker, New York, 1987.
8. L. Raymond (ed.), *Hydrogen Embrittlement: Prevention and Control*, ASTM STP 962, ASTM, Philadelphia, PA, 1988.
9. R.P. Gangloff and M.B. Ives (eds), *Environment–Induced Cracking of Metals*, NACE-10, NACE, Houston, TX, 1990.
10. H. Vehoff, Hydrogen related material properties, *Topics in Appl. Phys.*, 73, 215–278, 1997.
11. P. Sofronis (guest ed.), Special Issue on Recent Advances in Engineering Aspects of Hydrogen Embrittlement, *Engineering Fracture Mechanics*, 68, 2001.
12. A. Pundt and R. Kirchheim, Hydrogen in metals: microstructural aspects, *Ann. Rev. Mater. Res.*, 36, 555–608, 2006.
13. R.P. Gangloff, Hydrogen assisted cracking of high strength alloys, pp. 31–101 in *Comprehensive Structural Integrity*, I. Milne, R.O. Ritchie, and B. Karihaloo (eds), Vol. 6, Environmentally Assisted Fracture, Elsevier, Amsterdam, 2003.
14. B. Somerday, P. Sofronis, and R. Jones (eds), *Effects of Hydrogen on Materials, Proc. of the 2008 Int. Hydrogen Conf., Wyoming, USA*, ASM, Metals Park, OH,

2009. (See also proceedings of previous Wyoming conferences and other meetings listed in reference 13.)

15. S.P. Lynch, Metallographic and fractographic techniques for characterising and understanding hydrogen-assisted cracking in metals, in *Gaseous Hydrogen Embrittlement of High Performance Metals in Energy Systems*, R.P. Gangloff and B. Somerday (eds), Woodhead, Cambridge, in press.

16. L.G. Fritzmeier and W.T. Chandler, Hydrogen-embrittlement – rocket motor applications, pp. 491–524 in *Superalloys, Supercomposites and Superceramics*, Academic Press, London, 1989.

17. J Völkl and G. Alefeld, Hydrogen diffusion in metals, pp. 231–302 in *Diffusion in Solids*, A.S. Nowick and J.J. Burton (eds), Academic Press, New York, 1975.

18. H.-J. Christ, M. Decker and S. Zeitler, Hydrogen diffusion coefficients in the titanium alloys IMI 834, Ti 10–2–3, Ti 21 S, and alloy C, *Metall. Mater. Trans. A*, 31A, 1507–1517, 2000.

19. G.A. Young and J.R. Scully, Hydrogen production, absorption, and transport during environmentally assisted cracking of an Al-Zn-Mg-(Cu) alloy in humid air, pp. 893–907 in *Hydrogen Effects on Materials Behavior and Corrosion Deformation Interactions*, N. R. Moody, A.W. Thompson, R.E. Ricker, G.W. Was and R.H. Jones (eds), TMS, Warrendale, PA, 2003.

20. T. Livne, X. Chen, and W.W. Gerberich, Temperature effects on hydrogen assisted crack growth in internally charged AISI 4340 steel, *Scripta Metall.*, 20, 659–662, 1986.

21. R.P. Gangloff and R.P. Wei, Gaseous hydrogen embrittlement of high strength steels, *Metall. Trans. A*, 8A, 1043–1053, 1977.

22. D.G. Westlake, A generalized model for hydrogen embrittlement, *Trans. ASM*, 62, 1000–1006, 1969.

23. H.K. Birnbaum, Mechanisms of hydrogen related fracture of metals, pp. 639–658 in *Hydrogen Effects on Materials Behavior*, N.R. Moody and A.W Thompson (eds), TMS, Warrendale, PA, 1990.

24. H.K. Birnbaum, I.M. Robertson, P. Sofronis and D. Teter, Mechanisms of hydrogen related fracture – a review, pp. 172–195 in *Corrosion-Deformation Interactions*, T. Magnin (ed.), Institute of Materials, London, 1997, and references therein.

25. L.B. Pfeil, The effect of occluded hydrogen on the tensile strength of iron, *Proc. Roy. Soc. (London)*, A112, 128–195, 1926.

26. A.R. Troiano, The role of hydrogen and other interstitials in the mechanical behavior of metals, *Trans. ASM*, 52, 54–80, 1960.

27. R.A. Oriani, A mechanistic theory of hydrogen embrittlement of steels, *Ber. Bunsenges. Phys. Chem.*, 76, 848–857, 1972.

28. W.W. Gerberich and X. Chen, Environment-induced cracking of metals – fundamental processes: micromechanisms, pp. 167–186 in *Environment-induced Cracking of Metals*, R.P. Gangloff and M.B. Ives (eds), NACE-10, NACE, Houston, TX, 1990.

29. W.W. Gerberich, P.G. Marsh and J.W. Hoehn, Hydrogen induced cracking mechanisms – are there critical experiments?, pp. 539–553 in *Hydrogen Effects in Materials*, A.W. Thompson and N.R. Moody (eds), TMS, Warrendale, PA, 1996.

30. C.J. McMahon, Jr., Hydrogen-induced intergranular fracture of steels, *Engng Fract. Mech.*, 68, 773–788, 2001.

31. J.F. Knott, Fracture toughness and hydrogen-assisted crack growth in engineering alloys, pp. 387–408 in *Hydrogen Effects in Materials*, A.W. Thompson and N.R. Moody (eds), TMS, Warrendale, PA, 1996.

Published by Woodhead Publishing Limited, 2011

32. S.P. Lynch, Mechanisms of hydrogen assisted cracking – a Review, pp. 449–466 in *Hydrogen Effects on Materials Behavior and Corrosion Deformation Interactions*, N.R. Moody, A.W. Thompson, R.E. Ricker, G.W. Was, and R.H. Jones (eds), TMS, Warrendale, PA, 2003.

33. M.S. Daw and M.I. Baskes, Application of embedded atom method to hydrogen embrittlement, pp. 196–218 in *Chemistry and Physics of Fracture*, R.H. Jones and R.M. Latanision (eds), Martinus Nijhoff, Leiden, The Netherlands, 1987.

34. R.G. Hoagland and H.L. Heinisch, An atomic simulation of the influence of hydrogen on the fracture behaviour of nickel, *J. Mater. Res.*, 7, 2080–2088, 1992.

35. M. Wada, N. Akaiwa and T. Mori, Field evaporation of iron in neon and hydrogen and its rate-controlling processes, *Phil. Mag. A*, 55, 389–403, 1987.

36. T.C. Lee, I.M. Robertson and H.K. Birnbaum, An HVEM *in situ* deformation study of nickel doped with sulfur, *Acta Mater.*, 37, 407–415, 1989.

37. K. Yoshino and C.J. McMahon, Jr., The cooperative relation between temper embrittlement and hydrogen embrittlement in a high strength steel, *Metall. Trans. A*, 5, 363–370, 1974.

38. R.P. Messmer and C.L. Briant, The role of chemical bonding in grain boundary embrittlement, *Acta Mater.*, 30, 457–467, 1982.

39. C.D. Beachem, A new model for hydrogen assisted cracking (hydrogen embrittlement), *Metall. Trans.*, 3, 437–451, 1972.

40. H.K. Birnbaum and P. Sofronis, Hydrogen-enhanced localised plasticity – a mechanism for hydrogen–related fracture, *Mater. Sci. Eng.*, A176, 191–202, 1994.

41. H.K. Birnbaum, I.M. Robertson and P. Sofronis, Hydrogen effects on plasticity, pp. 367–381 in *Multiscale Phenomena in Plasticity*, J. Lépinoux *et al.* (eds), Kluwer Academic, Norwell, MA, 2000.

42. G. Lu, Q. Zhang, N. Kioussis and E. Kaxiras, Hydrogen enhanced local plasticity in aluminum: an *ab-initio* study, *Phys. Rev. Lett.*, 87, 095501, 2001.

43. I.M. Robertson, The effect of hydrogen on dislocation dynamics, *Engng Fracture Mech.*, 68, 671–692, 2001.

44. H.K. Birnbaum, Hydrogen effects on deformation – relation between dislocation behavior and the macroscopic stress strain behavior, *Scripta Metall. Mater.*, 31, 149–153, 1994.

45. D. Abraham and C.J. Altstetter, Hydrogen-enhanced localization of plasticity in an austenitic stainless steel, *Metall. Mater. Trans.*, 26A, 2859–2871, 1995.

46. A.M. Brass and J. Chene, Influence of deformation on the hydrogen behavior in iron and nickel base alloy: a review of experimental data, *Mater. Sci. Engng*, A242, 210–221, 1998.

47. E. Lunarska, V. Novsak, N. Zarubova and S. Kadeckova, Effect of electrolytic hydrogen charging on flow stress and slip line pattern in iron single crystals, *Scripta Metall.*, 17, 705–710, 1983.

48. W.A. McInteer, A.W. Thompson and I.M. Bernstein, The effect of hydrogen on the slip character of nickel, *Acta Metall.*, 28, 887–894, 1980.

49. C. Wang and I.M. Bernstein, The effect of strain on hydrogen-induced dislocation morphologies in single crystal iron, *Acta Metall.*, 34, 1011–1020, 1986.

50. I.M. Robertson and H.K. Birnbaum, Effect of hydrogen on the dislocation structure of deformed nickel, *Scripta Metall.*, 18, 269–274, 1984.

51. K.A. Nibur, D.F. Bahr and B.P. Somerday, Hydrogen effects on dislocation activity in austenitic stainless steel, *Scripta Metall.*, 54, 2677–2684, 2006.

52. A. Barnoush and H. Vehoff, Electrochemical nanoindentation: a new approach to probe hydrogen/deformation interaction, *Scripta Metall.*, 55, 185–198, 2006.

Published by Woodhead Publishing Limited, 2011

53. H.E. Hanninen, T.C. Lee, I.M. Robertson and H.K. Birnbaum, *In situ* observations on effects of hydrogen on deformation and fracture of A533B pressure vessel steel, *J. Mater. Engng Perf.*, 2, 807–818, 1993.

54. S.P. Lynch, Environmentally assisted cracking: overview of evidence for an adsorption-induced localised-slip process, *Acta Metall.*, 20, Overview No. 74, 2639–2661, 1988, and references therein.

55. S.P. Lynch, Metallographic contributions to understanding mechanisms of environmentally assisted cracking, *Metallography*, 23, 147–171, 1989, and references therein.

56. S.P. Lynch, Comments on 'A unified model of environment-assisted cracking', *Scripta Mater.*, 61, 331–334, 2009.

57. N.J. Petch, The lowering of fracture stress due to surface adsorption, *Phil. Mag.*, 1, 331–337, 1956.

58. J.A. Clum, The role of hydrogen in dislocation generation in iron alloys, *Scripta Metall.*, 9, 51–58, 1975.

59. J.W. Davenport and P.J. Estrup, Hydrogen on metals, pp. 1–37 in *The Chemical Physics of Solid Surfaces and Heterogeneous Catalysis*, D.A. King and D.P. Woodruff (eds), Vol 3, Elsevier, Amersterdam, 1990.

60. T.E. Fischer, Hydrogen on metal surfaces, pp. 135–148 in *Advanced Techniques for Characterizing Hydrogen in Metals*, N.F. Fiore and B.J. Berkowitz (eds), Met. Soc. AIME, New York, 1982.

61. K. Christmann, Some general aspects of hydrogen chemisorption on metal surfaces, *Progr. Surf. Sci.*, 48, 15–26, 1995.

62. E. Protopopoff and P. Marcus, Surface effects on hydrogen entry into metals, pp. 53–96, in *Corrosion Mechanisms in Theory and Practice*, 2nd edn, P. Marcus (ed.), Marcel Dekker, New York, 2002.

63. M.A. Van Hove, Crystal surfaces, pp. 485–531 in *Materials Science and Technology: A Comprehensive Treatment*, R.W. Cahn, P. Haasen and E.J. Kramer (eds), John Wiley, New York, 1991, and references therein.

64. D.P. Woodruff (ed.), *The Chemical Physics of Solid Surfaces, Vol. 10, Surface Alloys and Alloy Surfaces*, Elsevier, Amsterdam, 2002.

65. W. Moritz, R.J. Behm, G. Ertl, G. Kleinle, V. Penka, W. Reimer and M. Skottke, Relaxation and reconstruction on Ni(110) and Pb(110) induced by adsorbed hydrogen, pp. 207–213 in *The Structure of Surfaces II*, J.F. van der Veen and M.A. van Hove (eds), Springer-Verlag, Berlin, 1988.

66. Y. Kuk, P.J. Silverman and H.Q. Nguyen, Adsorbate-induced reconstruction in the Ni (111)-H system, *Phys. Rev. Lett.*, 59 (13), 1452–1455, 1987.

67. R. Stumpf, H-induced reconstruction and faceting of Al surfaces, *Phys. Rev. Lett.*, 78 (23), 4454–4457, 1997.

68. R.A. Oriani, On the possible role of surface stress in environmentally induced embrittlement and pitting, *Scripta Metall.*, 18, 265–268, 1984.

69. R. Thomson, T.-J. Chuang and I.-H. Lin, The role of surface stress in fracture, *Acta Mater.*, 34, 1133–1143, 1986.

70. B.N. Cox and C.W. Bauschlicher Jr., Surface relaxation and induced stress accompanying the adsorption of H on Be (0001), *Surf. Sci.*, 102, 295–311, 1981.

71. H.H. Johnson, Hydrogen gas embrittlement, pp. 35–49 in *Hydrogen in Metals*, I.M. Bernstein and A.W. Thompson (eds), ASM, Metals Park, OH, 1974.

72. M. Nagumo, Hydrogen related failure of steels – a new aspect, *Mater. Sci. Tech.*, 20, 940–950, 2004.

73. R.B. McLellen and Z.R. Xu, Hydrogen-induced vacancies in the iron lattice, *Scripta Mater.*, 36, 1201–1205, 1997.

74. A.M. Cuitino and M. Ortiz, Ductile fracture by vacancy condensation in F.C.C. single crystals, *Acta Mater.*, 44, 427–436, 1996.

75. A.W. Thompson, The mechanism of hydrogen participation in ductile fracture, pp. 467–477 in *Effect of Hydrogen on Behavior of Materials*, A.W. Thompson and I.M. Bernstein (eds), Met. Soc. AIME, New York, 1976.

76. H. Vehoff and W. Rothe, Gaseous hydrogen embrittlement in FeS- and Ni-single crystals, *Acta Metall.*, 31, Overview No. 30, 1781–1793, 1983.

77. X. Chen and W.W. Gerberich, The kinetics and micromechanics of hydrogen-assisted cracking in Fe-3 Pct Si single crystals, *Metall. Trans. A*, 22A, 59–70, 1991.

78. T.J. Marrow, M. Aindow, P. Prangnell, M. Strangwood and J.F. Knott, Hydrogen-assisted stable crack growth in iron-3wt% silicon steel, *Acta Metall.*, 44, 3125–3140, 1996.

79. X. Chen and W.W. Gerberich, Evidence for 1 μm sized instabilities in hydrogen assisted cracking of Fe–3%Si crystals, *Scripta Metall.*, 22, 1499–1502, 1988.

80. N. Takano, K. Kidani, Y. Hattori and F. Terasaki, Fracture surface of hydrogen embrittlement in iron single crystals, *Scripta Metall. Mater.*, 29, 75–80, 1993.

81. S.P. Lynch, Progression markings, striations, and crack-arrest markings on fracture surfaces, *Mater. Sci. Eng. A.*, 468–470, 74–80, 2007.

82. P.D. Hicks and C.J. Altstetter, Hydrogen-enhanced cracking of superalloys, *Metall. Trans. A*, 23A, 237–249, 1992.

83. N.R. Moody and F.A. Greulich, Hydrogen-induced slip band fracture in an Fe-Ni-Co alloy, *Scripta Metall.*, 19, 1107–1111, 1985.

84. N.R. Moody, R.E. Stoltz and M.W. Perra, The effect of hydrogen on fracture toughness of the Fe-Ni-Co superalloy IN903, *Metall. Trans. A*, 18A, 1469–1482, 1987.

85. J.R. Rice and R. Thomson, Ductile versus brittle behaviour of crystals, *Phil. Mag.*, 29, 73–97, 1974.

86. R.A Oriani, Hydrogen – the versatile embrittler, *Corrosion*, 43, 390–397, 1987.

Part II

Test methods for determining stress corrosion cracking (SCC) susceptibilities

3

Testing and evaluation methods for stress corrosion cracking (SCC) in metals

W. DIETZEL and P. BALA SRINIVASAN,
Helmholtz-Zentrum Geesthacht, Germany and A. ATRENS,
The University of Queensland, Australia

Abstract: The development of standardised methods of testing is a key element in the prevention of stress corrosion cracking (SCC) in order to provide a unified reference framework. A range of test methods is available in the form of international and national standards and procedures, comprising a wide variety of test methods for assessing SCC and also a variety of specimen configurations. The chapter describes the various test methods, comments on their advantages and disadvantages and discusses the interrelationships between results obtained from the various methods. In this way, a framework for drawing intelligent comparisons of SCC data and conclusions from these data is provided. At the end of the chapter potential future developments of SCC test methods are given.

Key words: accelerated stress corrosion cracking test, fracture mechanics approach to stress corrosion cracking, hydrogen embrittlement, sub-critical crack extension, threshold stress.

3.1   Introduction

It is important to prevent stress corrosion cracking (SCC) in service and to reduce the likelihood of consequent unexpected failures without being unnecessarily conservative. Laboratory testing plays an important role. Since there is no generalized analytical approach to allow prediction of combinations of material and environment that result in SCC, the avoidance of SCC has to be based either on past experience and/or on testing in the laboratory. The current design approaches presume that short-term laboratory-scale experimental data can be extrapolated to predict long-term structural performance. In principle, these tests consist of a stressed piece of the material, i.e. the specimen, exposed to a presumably aggressive environment. The aim of the test is to determine if and how SCC occurs in this specimen during the test period and to screen out materials which are susceptible to SCC or to predict long-term service behaviour.

Various methods of conducting SCC tests are specified in great detail in appropriate test standards issued by the American Society for Testing and Materials (ASTM), the International Organization for Standardization

133

(ISO) and the National Association of Corrosion Engineers (NACE), in documents and procedures issued by other organizations (see Section 3.7), and in various handbooks (McEvily, 1990; Sedriks, 1990; Sprowls, 1992, 1996; Turnbull, 1992; Wanhill, 1991). This chapter deals with the basic principles of laboratory SCC testing, providing an overview of the general characteristics of the various SCC test methods and discusses the applicability of the various test standards.

3.2   General aspects of stress corrosion cracking (SCC) testing

SCC tests determine a number of parameters that are of varying usefulness and relevance for service operation. To reliably predict the service behaviour of a structure or component under particular loading conditions, SCC testing has to take into account the materials and structures used in service, the environmental conditions and the actual service load history as realistically as possible.

The objectives of SCC testing may comprise (Turnbull, 1992):

- alloy development,
- materials selection,
- assessment of relative environment aggressiveness,
- evaluation of protective schemes,
- production of design data,
- quality assurance,
- investigation of mechanisms.

Since SCC typically develops and extends over long periods of time, laboratory testing often needs to be accelerated or, alternatively, the testing needs to be terminated after a duration that is short compared to service lifetimes. The acceleration of testing may involve applying more severe environments and/or higher mechanical stresses, raising the temperature, and applying anodic or cathodic polarization. In all cases, the SCC mechanism in the accelerated test must be essentially the same as that occurring in service, and it is wise to ensure that the test is not so severe as to exclude otherwise acceptable and less costly alloys. On the other hand, testing has to be sufficiently discriminating to eliminate any unsafe application of susceptible materials.

3.2.1   SCC test philosophies

SCC test methodologies follow the two basic philosophies underlying structural design, i.e. (Wanhill, 1991):

- the safe-life approach, and
- the damage-tolerance approach.

The older safe-life approach undertakes to design the structure for a finite service life during which there is no significant damage, i.e., there is no critical crack. Consequently, this approach does not include in-service inspection. The damage-tolerance approach, which evolved later, includes designing for an adequate service life without significant damage, but also enables operation beyond the actual life at which such damage starts to occur. It must be shown that the damage, and in particular cracks, is detected by routine inspection before they propagate to the extent that they decrease the residual strength of the structure below a safe level.

The damage-tolerance philosophy goes beyond the safe-life approach by including the possibility that cracks or flaws already exist in new structures. Either in-service inspection has to ensure structural integrity during the service life, or it has to be ensured that any initial cracks grow sufficiently slowly during the design service life, so that they do not reach a critical crack size, which would cause a premature failure.

3.2.2   Specimen configuration and loading mode

To reflect the different approaches in laboratory SCC testing and to obtain the relevant SCC data, testing can be based on either smooth or pre-cracked specimens, depending on the purpose of the test and the specific nature of the application (Sedriks, 1990). Testing using a smooth (i.e., non-pre-cracked) specimen represents a situation in which there are no initial flaws, whereas the use of pre-cracked specimens accounts for the existence of flaws or cracks according to the damage-tolerant approach. Figure 3.1 shows an example of specimens used for SCC testing (Dietzel, 1991).

3.1 Specimen types typically used for SCC testing (Dietzel, 1991): smooth (non-pre-cracked) tensile specimen; bolt-loaded double cantilever beam specimen, DCB; pre-cracked compact specimen, C(T); surface-cracked panel, SC(T).

Selecting a specimen configuration and a loading method for SCC testing requires first a definition of the test objective and the type of information needed, whether qualitative or quantitative. There are four basic kinds of SCC tests using combinations of test specimens and testing conditions:

1. Smooth specimens – static load tests.
2. Smooth specimens – dynamic (monotonic increasing) load tests.
3. Pre-cracked specimens – static load tests.
4. Pre-cracked specimens – dynamic (monotonic increasing) load tests.

Each of these tests has its merits. For example, a material may be immune to SCC under static loading but may be highly susceptible under continuous slow plastic straining (Parkins, 1972). The use of pre-cracked specimens acknowledges the difficulty of ensuring the total absence of crack-like defects that may be introduced during manufacture, fabrication or subsequent service. The presence of such defects can cause susceptibility to stress corrosion cracking, and pre-cracked specimens may be necessary to reveal this susceptibility, which in some material/environment systems like titanium alloys in aqueous media is not evident from tests under constant load using smooth specimens (Wanhill, 1975). The use of smooth specimens would be based on the assumption that structures are free from initial defects and may be operated over more than 90 per cent of their total service life before a flaw develops. Since in practice no structure is really free from such defects, SCC tests on smooth specimens are essentially used as screening criteria. Due to their characteristics, the different testing combinations may be regarded as complementing each another. In addition, testing in a corrosive environment may be carried out with unloaded samples, either smooth or pre-cracked, to evaluate, for example, the existence of residual stresses and their effect on crack initiation and propagation.

In this chapter, testing of smooth and of pre-cracked specimens is considered separately because of their distinctive geometric characteristics and the kind and amount of information they provide. A detailed overview of the various specimen types and a discussion of the many aspects and problems associated with SCC tests using these various specimen types and loading modes can be found in Turnbull (1992).

3.3   Smooth specimens

In conducting a test programme to select a material for engineering service using screening tests on smooth specimens, usually one or two types of tests are adopted, often based on an industry standard. A wide variety of smooth or mildly notched specimens are in use for SCC testing, and these are described in detail in the various test standards. Apart from the generally used tensile specimens, many of these specimen types are associated with the corresponding method of mechanical loading, as is described below.

3.3.1   Constant load tests

The most important characteristic of these tests is that the load is maintained constant throughout the period of testing. The specimens in constant load tests can be cylindrical or flat, sometimes notched, and are usually loaded in tension. In principle, there are two kinds of constant load tests, i.e. specimens in a self-loading frame and specimens that are loaded by an external load frame.

The obvious advantages of a self-loading frame are the reduced equipment costs and much less space required for testing. This is especially important for tests in environmental chambers and autoclaves. A potential disadvantage is often that the specimens must be loaded before environmental exposure. Pre-loading can lead to non-conservative results, although this problem may be more relevant to fracture mechanics based tests (see Section 3.4.4).

Testing with an external load frame is carried out with dead weight loads or with springs. The number of load frames required can be minimized by testing chains of specimens, which may be connected by loading links suitable to prevent unloading upon fracture of one of the specimens.

Constant load SCC tests usually produce a plot of the initial applied stress against the total time to failure, t_f, such as illustrated in Fig. 3.2. This plot is usually used to estimate the threshold stress, σ_{th}, below which SCC failure does not occur. For pragmatic reasons it is usual to terminate the tests after an arbitrary chosen time if failure has not occurred. In this case, the number of cracks per unit length of material is sometimes used for a comparative assessment of susceptibility, especially with respect to the initiation of stress corrosion cracks.

An approximate estimate of the crack velocity can be derived from these tests by dividing the crack length by the total testing time; the crack length is measured as the depth of the largest crack on the surface of the failed specimen, or on sections through a specimen that has not proceeded to total failure. This measure of crack velocity assumes that crack initiation occurs at the start of the test and that the deepest crack also initiated immediately at that time. These assumptions are not necessarily correct, and the approach is moreover non-conservative in that the SCC crack growth velocities are typically underestimated. A more accurate measurement of the average crack growth velocity is feasible if crack initiation can be detected using, for example, an electrical resistance measurement method as discussed in Section 3.4.7.

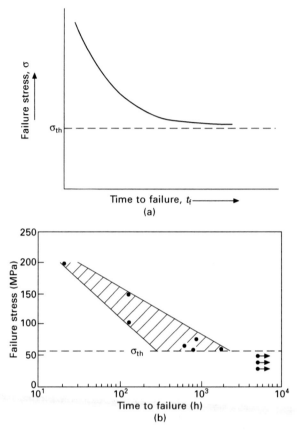

3.2 Representation of SCC test data produced by constant load tests: (a) schematic influence of the initial stress, σ_{init}, on the time to failure, t_f (b) experimental data measured in constant load tests on tensile specimens of the aluminum alloy AA 2024 T351 in chromated 3.5% sodium chloride solution.

3.3.2   Constant displacement tests

Also widely used are the constant displacement tests. These are sometimes called constant strain tests, which is not entirely correct. The specimens are either self-loaded or stressed in a self-loading frame. Although SCC testing under constant displacement is straightforward and inexpensive, there are some disadvantages. The specimen must be loaded before environmental exposure, and again this pre-loading can lead to non-conservative results. Further, the determination of the threshold stress, σ_{th}, and the time to failure, t_f, may involve subjective judgment resulting from the problem of terminating the tests after an arbitrarily chosen period of time.

The most common types of smooth constant displacement specimens are bent beam, C-ring and U-bend specimens. The preparation and use of these specimens are described in ASTM, ISO, and other standards and procedures (see Section 3.7.1).

3.3.3   Slow strain rate tests

The slow strain rate test, also designated CERT test (CERT = constant extension rate tensile) is widely used and is considered to be a severe SCC test. In this test, specimens are subjected to constantly increasing elongations, typically causing an increase in tensile strain, while being exposed to a specified environment (Parkins, 1993). Tests may be conducted either in tension or in bending, on initially plain or notched specimens. In particular, notched specimens can be used in order to restrict cracking to a location of specific interest, for example, when sampling a weld region including the heat-affected zone, and they may further be used to limit the load requirements.

An important characteristic of the slow strain rate test is the relatively low applied strain rate. Nevertheless, a principal advantage is the rapidity with which SCC susceptibility can be assessed. Figure 3.3 is an example of the results of slow strain rate tests on the austenitic steel AISI 316H in a 100 ppm NaCl solution at 90°C and in laboratory air (Dietzel, 1999). The tests were performed at strain rates between 10^{-8} and 5×10^{-6} s^{-1} by two different laboratories. Even at the highest applied strain rate a significant difference existed between the reduction in area measured in air (~65%) and in the

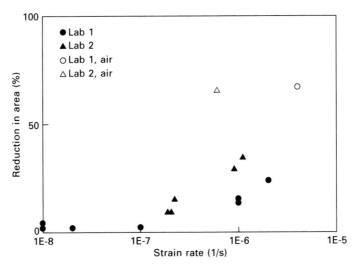

3.3 Results of constant extension rate tests (CERT) on the austenitic steel AISI 316H in 100 ppm NaCl solution at 90°C and in laboratory air, performed at strain rates between 10^{-8} and 5×10^{-6} s^{-1} in two different laboratories (Dietzel, 1999).

corrosive environment (≤35%). At the lowest strain rate used, 10^{-8} s^{-1}, there was almost no reduction in area (<5%). The figure also demonstrates the scatter which is often seen between test data obtained in different laboratories, even though in this case the specimens were machined from the same piece of material and the environmental conditions were nominally identical.

In many material/environment systems, tensile strain rates of 10^{-5} or 10^{-6} s^{-1} are used to study SCC, but the absence of SCC in tests conducted at these rates should not be construed as indicative of immunity to SCC until tests have been performed at slower strain rates. This is of some importance, since for certain combinations of material and environment, the susceptibility to SCC appears to decrease, or even vanish, at low strain rates, a phenomenon which is mainly attributed to repassivation effects (Fig. 3.4; Parkins, 1993). Of more importance is that in some material/environment systems the SCC susceptibility only becomes apparent at strain rates of 10^{-8} or 10^{-9} s^{-1} (Speidel and Atrens, 1980).

For initially smooth specimens the strain rate at the onset of the test is clearly defined, but once cracks have initiated, the effective strain rate cannot be measured and only approximate estimates may be derived. When necking starts in a ductile material stressed in tension, the effective strain rate in the necked region may increase by more than an order of magnitude, possibly causing the strain rate to move in or out of the critical region.

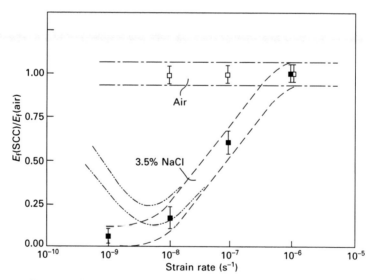

3.4 Results of slow strain rate tensile tests performed at various strain rates; the Y-axis shows the ratio of fracture energy (E_{f_r}) measured in environment, to fracture energy measured in air, $E_f(SCC)/E_f(air)$. The recovery of this ratio at very low strain rates, indicated by the dot-and-dash lines, was not observed in this case but was reported in the literature (Parkins, 1993).

Comparison between identical specimens exposed to the test environment and to an inert environment may be used for assessing the susceptibility to SCC. The parameters commonly measured are:

- time to failure,
- ductility assessed by, e.g., reduction in area or elongation to fracture,
- maximum load achieved,
- area bounded by nominal stress/elongation curve representing the fracture energy,
- percentage of stress corrosion cracking on the fracture surface.

A more suitable SCC parameter is the threshold stress for stress corrosion cracking, which can be relatively easily measured using a potential drop technique (Strieder *et al.*, 1994; Winzer *et al.*, 2008).

The choice of the failure criterion depends on the characteristics of the failure process and it is recommended that more than one failure criterion is used in appropriate circumstances. The ratio of the parameter measured in the environment relative to the value measured in an inert environment yields an index of the SCC susceptibility. According to ISO 7539-Part 7 increasing susceptibility to cracking is indicated by increasing departure from unity of the ratio

$$\frac{\text{results from specimen in test environment}}{\text{result from specimen in inert environment}}$$

applied to one or more of the above parameters at the same initial strain rate.

Figure. 3.5 is an example of data generated in a slow strain rate test. Figure 3.5(a) presents the results of two slow strain rate tensile tests on flat tensile specimens of the magnesium alloy AZ31 in air and in a mildly aggressive solution according to ASTM D1384 as corrosive environment, in the form of stress–strain curves. The strain rate in both tests was 10^{-8} s^{-1}. Figure 3.5(b) shows photographs of the specimens after failure in air and in the corrosive environment. The susceptibility index was in this case below 0.1.

An important aspect of testing, when failure is associated with absorption of hydrogen atoms, is the total time of testing relative to the time for hydrogen atom ingress, and comparison of results from different strain rates can be meaningless unless there is some degree of uniformity of hydrogen charging of the specimens. Also, account must be taken of possible specimen thinning due to corrosion in longer exposure tests at low strain rates.

The average stress corrosion crack velocity, for initially plain specimens, may be estimated from the total testing time and the size of the deepest crack, measured on the fracture surfaces of a specimen that has failed or from sections through a specimen for which the test had been interrupted prior to failure. But, as in the case of constant load experiments, caution should

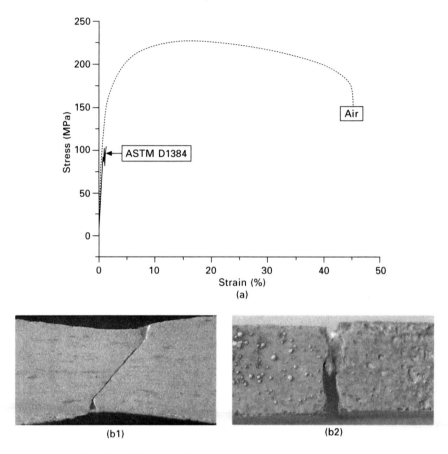

3.5 Results of slow strain rate tensile tests performed on flat tensile specimens of the magnesium alloy AZ31 in air and in corrosive environment (solution according to ASTM D1384), at a strain rate of 10^{-8} s^{-1}: (a) stress-strain curves, (b) photographs of the specimens after failure in air (left) and in corrosive solution (right).

be taken since SCC crack growth velocities are typically underestimated by this method due to the fact that the time of crack initiation is not known but is typically after the start of the test. Nevertheless, these crack velocity measurements can provide useful insights into SCC susceptibility.

Although the strain rate is a key parameter in determining SCC, it is not a parameter which is readily measured in service and for this reason the results cannot sensibly be used for design purposes. However, these tests provide a basis for assessing relative SCC properties of materials by using the index of the SCC susceptibility. The test method is also well suited for mechanistic studies.

3.3.4   Breaking load and linearly increasing stress tests

These SCC tests are not standardized. They represent interesting alternatives or innovations to the above methods, to which they are, to some extent, complementary. In the breaking load test, smooth specimens are self-loaded or stressed in a self-loaded frame before environmental exposure (Sprowls *et al.*, 1984). Following the environmental exposure, the specimens are subjected to residual static strength tests, which quantify the damage due to SCC. A further development of this test method is the ASCOR (automated stress corrosion ring) test (Schra and Groep, 1993). Smooth cylindrical or sheet specimens are stressed in strain-gauged loading rings and are exposed to the corrosive environment. Continuous monitoring of the load enables the detection of crack initiation and the construction of stress–lifetime curves.

The linearly increasing stress test (LIST) introduced by Atrens *et al.* is essentially similar to the slow strain rate test (Atrens *et al.*, 1993). The primary difference is that the slow strain rate test is displacement controlled whereas the LIST test is under load control. The load applied to the specimen, typically in tension, is continuously increased, independent of the increasing compliance of the specimen after cracking has initiated and leading to increasing strain rates. This has an accelerating effect as compared to a slow strain rate test, and the method is particularly suited for determining threshold values, σ_{th}, by monitoring, for example, the increase in speed of the test machine occurring upon crack initiation. The LIST allows a relatively straightforward measurement of the threshold stress for SCC, σ_{th}, using an electrical potential drop technique. Winzer *et al.* have subsequently shown that CERT can also be used to measure σ_{th} when employing the electrical potential drop technique (Dietzel and Schwalbe, 1986; Winzer *et al.*, 2008). At the threshold stress, the measured potential drop begins to increase faster than linear when the potential drop is plotted against the engineering stress in the specimen (Fig. 3.6). The LIST test and the CERT test are identical up to crack initiation. Thereafter, the LIST test ends when the crack grows to a critical size. The CERT test can take much longer because there can be significant specimen extension as the stress corrosion cracks gape open.

3.4   Pre-cracked specimens – the fracture mechanics approach to stress corrosion cracking (SCC)

The damage-tolerant design practice takes into consideration that initial flaws may already exist in a structure or component, and fracture mechanics is applied to characterize the initiation and growth of cracks from these flaws. Usually, stress corrosion cracks are considered to be brittle, i.e. they occur

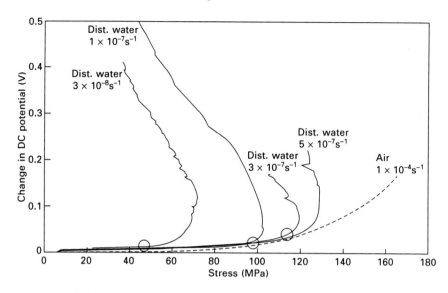

3.6 Potential drop vs. stress curves determined in CERT tests on cylindrical tensile specimens in laboratory air and in distilled water; the circles indicate crack initiation as determined by DC potential drop measurent (material: Mg alloy AZ91).

at stresses below general yield and propagate in an essentially elastic body, even though local plasticity may be necessary for the cracking process. Hence, linear elastic fracture mechanics (LEFM) concepts are applied. The plane strain stress intensity factor in the opening mode, K_I, is used to quantify the stress field at the crack tip in a pre-cracked specimen or structure. The stress intensity factor controls crack initiation and its extension. Elastic-plastic fracture mechanics (EPFM) is needed for ductile alloys and thinner sections for which the principle of small scale yielding becomes invalid. EPFM parameters include the crack tip opening angle, crack tip opening displacement and the J integral.

3.4.1   Linear elastic fracture mechanics

Experimental evidence has shown that, for a given material/environment combination, a unique relation exists between K_I and the growth rate of a stress corrosion crack. The usual way of representing this relationship is to plot the crack growth rate, da/dt (or v), as a function of K_I (Speidel, 1971). Figure 3.7 presents a typical diagram, the v-K plot, which in principle consists of three regions, two in which the crack growth rate is K dependent, separated by a region in which it is almost independent of K. A low value of K_I, the stress corrosion threshold stress intensity factor, K_{ISCC}, characterizes the

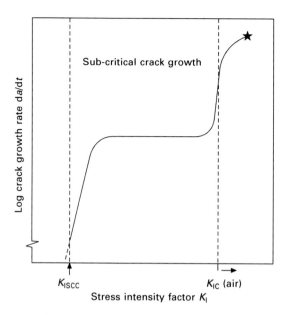

3.7 Influence of the stress intensity factor, K_I, on the stress corrosion crack growth rate, da/dt in the region of sub-critical crack growth (after Speidel, 1971).

stress intensity factor at which the first measurable crack extension occurs. This does not necessarily imply that there exists, under all circumstances, a threshold in the sense of a real cut-off. Contrariwise, for many material/ environment combinations, the existence of a true threshold appears rather doubtful. Yet, a practical meaning of K_{ISCC} lies in the fact that, below this value of the stress intensity factor, the growth rates of stress corrosion cracks fall below a certain lower limit, which is often set at 10^{-10} m/s, corresponding to a crack increment of roughly 3 mm per year.

Once the threshold, K_{ISCC}, is exceeded, the crack velocity becomes strongly dependent on K in region I. For larger values of K, the v-K curve may have a distinct plateau for which the crack growth rate can become more or less independent of K. In this region II, transport and electrochemical processes are the rate-limiting parameters, controlling the crack growth kinetics and, if a real plateau of the crack velocity exists, determining its absolute level. At still higher loads, in region III, the crack velocity again increases with the applied stress intensity factor. Finally, the applied stress intensity factor approaches the fracture toughness of the material, K_{Ic}, and pure mechanical rupture dominates over sub-critical crack extension caused by SCC. Some material/environment combinations may not show regions I and III, and in other cases there may be no plateau.

3.4.2   Requirements of fracture mechanics SCC testing

SCC tests based on fracture mechanics are performed with the primary aim of determining the threshold, K_{ISCC}, and the rate of crack growth, da/dt. The specimens contain initial cracks, so that these tests preclude study of crack initiation from an initially smooth surface, and these tests avoid the problem of separating the environmental influence on both crack initiation and growth.

It is not clear to what extent the presence of the corrosive environment affects the plastic deformation required for the initiation of stress corrosion cracking. Therefore, standard documents pertaining to fracture mechanics based SCC tests demand that the specimen dimensions must be sufficient to maintain predominantly plane strain conditions, in which plastic deformation is limited to the vicinity of the crack tip. The minimum requirements with respect to specimen thickness, B, pre-crack length, a, and ligament length, $(W - a)$ where W is the specimen width, are thus in accordance with those applied to plane strain fracture toughness tests, i.e.

$$a, B, (W - a) \geq 2.5 \left(\frac{K_I}{\sigma_y} \right)^2 \qquad\qquad 3.1$$

where σ_y is the yield strength of the material.

Because of the time dependence of the SCC process, SCC tests must allow sufficient time for SCC growth, according to the kinetics of the mechanics/metallurgy/chemistry interactions. For some materials, notably steels, crack growth incubation times can be long at low stress intensity factors. Too short a test duration would therefore lead to an over-estimation of K_{ISCC} and to non-conservative predictions. As a consequence, ASTM and ISO standards for determining K_{ISCC} from tests under constant load or constant displacement recommend test durations of up to 10 000 hours. A recent ISO test standard on rising load or rising displacement SCC testing on pre-cracked samples could reduce the test duration in a similar way as do CERT and LIST in the case of smooth specimen SCC testing (see also Section 3.4.6).

3.4.3   Loading mode

K_{ISCC} and da/dt data can be determined using:

- crack initiation tests under constant load,
- crack arrest tests on constant displacement (self-loaded) specimens,
- rising load or rising displacement tests.

The essential advantage of constant load, and even more of constant displacement tests, is the moderate requirements with respect to the equipment needed for performing these tests. Nevertheless, a number of shortcomings are inherent to both these modes of loading. Although the necessary test times, as recommended in the respective test standards, can be quite long, there often remains uncertainty as to whether the K values measured in these tests really represent the threshold K_{ISCC} value of the material/environment combination. Inter-laboratory test programmes in which these techniques had been applied to investigate nominally identical material–environment combinations in various laboratories have revealed a high degree of scatter between the results (Wei and Novak, 1987; Yokobori *et al.*, 1988; Dietzel, 2000). Another potential problem arises from the fact that testing is performed under static loading conditions whereas in certain material–environment combinations, dynamic loading (i.e., increasing plastic deformation) appears necessary to initiate the type of SCC observed in some practical applications (Ford, 1988).

3.4.4   Constant load tests

Constant load tests on pre-cracked specimens can be used for the determination of K_{ISCC} by the initiation of a stress corrosion crack from the pre-existing fatigue crack using a series of specimens loaded to different levels of the stress intensity factor, and to some extent also for measuring crack growth rates, da/dt. The most commonly used specimen types are compact specimens, C(T) (see Fig. 3.1), and notched bend specimens SE(B).

Constant load specimens can be loaded during exposure to the test environment in order to avoid the risk of unnecessary incubation periods associated with change in crack tip strain rate. In constant load tests, once the crack has initiated, the crack extension results in increasing crack opening so that there is less likelihood that oxide films either block the crack or wedge it open. On the other hand, since a growing crack causes a rapidly increasing stress intensity factor at the crack tip, the failure mode eventually shifts from environmental cracking to ductile rupture, so that measured crack growth rates do not necessarily represent those associated only with SCC. Subsequent inspection of the fracture surface is necessary to ensure reliable $(da/dt)_{SCC}$ data. The tests require an external loading system. This can, in the case of bend specimens, be a relatively simple cantilever beam arrangement, whereas specimens subjected to tension loading require dead-weight loaded testing machines similar to those used in creep rupture experiments. Chains of specimens connected by loading links that prevent unloading if one of the specimens should fail, can be used in order to minimize the equipment needed. The size of the loading systems makes it difficult to test constant load specimens under operating conditions, although they can be tested in environments that are bled off from operating systems.

Constant load specimens can be of two distinct types (Turnbull, 1992):

- Specimens in which the stress intensity factor increases with increasing crack length. These are suitable for the determination of K_{ISCC} and, to some extent, for measuring crack growth rates, $(da/dt)_{\mathrm{SCC}}$, as a function of K_{I}.
- Specimens, in which the stress intensity factor is effectively independent of crack length, can be used for fundamental studies of stress corrosion mechanisms.

Constant load specimens, in which K increases as the crack grows, can be subjected to either tension or bending load. Depending on the design, tension loaded specimens can experience stresses at the crack tip which are predominantly tensile (as in remote tension types such as the centre-crack plate) or contain a significant bend component (as in crack line loaded types such as compact specimens). The presence of significant bending stress at the crack tip can adversely affect the crack path stability during stress corrosion testing and can facilitate crack branching in certain materials. Bend specimens can be loaded in three-point, four-point or cantilever bend fixtures.

Constant load specimens in which K remains constant can be subjected to either torsion loading, or to tension loading, as in the case of contoured double cantilever beam specimens. In this latter case, although loaded in tension, the design of the contoured DCB specimens may produce crack line bending with an associated tendency for crack growth out of plane. This can be reduced by the use of side grooves.

Another approach for the determination of K_{Iscc} has been developed in recent years making use of small-diameter, circumferential-notched tensile (CNT) specimens (Stark and Ibrahim, 1986). The CNT specimens are claimed to be the smallest specimen type that can produce valid plane strain crack loading conditions and thus enable testing of thin section components, as well as testing at reduced costs because of lower material requirements for the specimens and the testing rigs.

3.4.5   Constant displacement tests

Constant displacement specimens are usually self-loaded and hence require no external stressing equipment. Their compact dimensions also facilitate exposure to operating service environments. Like constant load specimens, they can be used for the determination of K_{ISCC} by the initiation of stress corrosion cracks from the fatigue pre-crack, in which case a series of specimens must be used to pin-point the threshold value. More common is the determination of K_{ISCC} by the arrest of a propagating crack, since under constant displacement testing conditions, the stress intensity decreases progressively as crack propagation occurs. This means that, unlike in constant load or in rising load/rising displacement SCC tests, the v-K curve is traversed in the direction of decreasing stress intensity factor. In principle a single specimen suffices, but in practice the use of several specimens, not less than three, appears necessary.

The procedure for determining K_{ISCC} by constant displacement tests on self-loaded double cantilever beam (DCB) specimens is illustrated in Fig. 3.8. Three specimens made from the aluminium alloy AA7010 T651 were stressed to various initial K values, immersed in the corrosion environment, and the crack extension was recorded over a period of about 1.5 years. From the crack increments measured at discrete time intervals, crack growth rates, $\Delta a/\Delta t$, were calculated and plotted against the actual values of K_I, leading to a unique value of K_{ISCC} determined from crack arrest (Fig. 3.8(b)).

Figure 3.8(a) indicates that intermediate periods of apparent crack arrest existed which were followed by subsequent crack growth. If an apparent 'K_{ISCC}' were derived from such a first, apparent crack arrest, it would have significantly overestimated K_{ISCC}. Figure 3.8 further indicates that crack extension was also observed for a reference specimen tested in laboratory air. This specimen had been loaded to an initial K value of 13 MPa√m with the crack coming to arrest at about 10 MPa√m, whereas the K_{ISCC} value determined from the specimens which were tested in the corrosive environment corresponded to about 6 MPa√m. In the case of this aluminium alloy it was speculated that under long-term test conditions even laboratory air could be considered an aggressive environment because of the fairly high relative humidity: fracture toughness tests of this material performed in air yielded a K_{IC} value of 20 MPa√m. Hence, in such cases the reference tests in air would better be performed using a strong desiccant, e.g. phosphorous pentoxide, or in air dried by silica gel (Rieck *et al.*,1989).

Other problems arising in constant displacement tests are:

- The applied loads can only be measured indirectly by the displacement changes.
- Oxide formation or corrosion products can either wedge open the crack surfaces, thus changing the applied displacement and load, or can block the crack mouth, thus preventing the ingress of corrodent. Oxides can impair the accuracy of crack length measurements by electrical resistance methods.
- Crack branching, blunting or growth out of plane can invalidate crack arrest data (Fig. 3.9).
- Crack arrest must be defined by crack growth below some arbitrary rate, which can be difficult to measure accurately (though, arguably, the same concerns apply to the constant load method and are intrinsically associated with the definition of K_{ISCC}).
- Elastic relaxation of the loading system during crack growth can cause increased displacement and higher loads than expected.

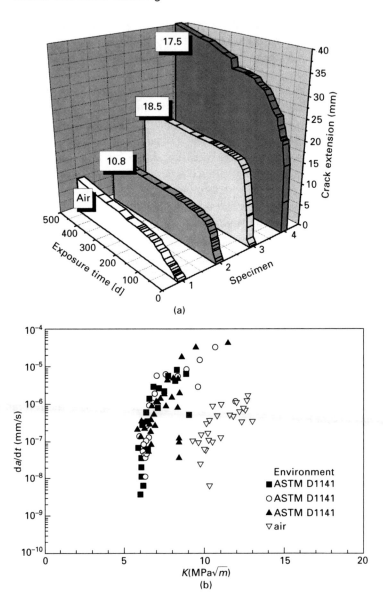

3.8 Determination of the threshold value, K_{ISCC}, from tests using bolt-loaded double cantilever beam specimens: (a) crack extension measured using four specimens (one reference test in air) for a total exposure time of 460 days, (b) curves of crack growth rates vs. stress intensity factor, K_I, as calculated from the data shown in (a) (material: aluminium AA 7010, environment: ASTM D1141 substitute ocean water).

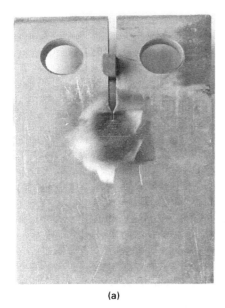
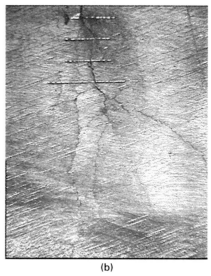

(a)   (b)

3.9 Severe crack branching in a wedge-loaded DCB specimen (material: austenitic steel AISI 316H, environment: 100 ppm NaCl solution, 90°C): (a) overview of wedge-loaded DCB specimen after testing, (b) detailed view of area indicated in (a).

- Plastic relaxation due to time-dependent processes within the specimen can cause lower loads than expected.
- It is sometimes impossible to introduce the test environment prior to application of the load. This can retard crack initiation during subsequent testing.

On the other hand, self-loaded constant displacement specimens offer significant economical benefits, so their use is often a practical solution for SCC investigations.

3.4.6   Rising load and rising displacement tests

In dynamic SCC testing, the stress intensity factor at the crack tip is increased whilst the specimen under test is immersed in the test environment. The tests can reveal cases of SCC susceptibility which would remain undetected in tests under static loading.

The rising load K_{ISCC} test was the first of these accelerated fracture mechanics based SCC tests. It applies a constant loading rate (McIntyre and Priest, 1972; Clark and Landes, 1976). A modified version of this test technique is the step-loading test, in which the load is increased in discrete increments, if there is no indication of crack initiation in a predetermined time interval (Raymond and Crumly, 1982). Another approach consists of a displacement-controlled rising load K_{ISCC} test, in which a constant displacement, or constant extension rate is applied (Dietzel *et al.*, 1989). The loading mode of the rising displacement test is similar to that of the CERT or LIST on smooth or notched specimens. The rising load K_{ISCC} test, like the LIST test on smooth specimens, is particularly suited for the determination of the SCC threshold K_{ISCC}, whereas the rising displacement test is favourable for the study of failure mechanisms. As in the case of smooth specimen testing, the magnitude of the loading or displacement rate applied is the key parameter for this test technique (Clark and Landes, 1976; Dietzel *et al.*, 1989). This rate needs to be sufficiently low to allow for the environmental cracking to develop without being overridden by pure mechanical rupture.

Typical load-line displacement rates in rising displacement tests are of the order of 1 to 10 μm/h. Despite these low rates, such dynamic tests have an accelerating nature and yield information about the SCC susceptibility of a material–environment system within reasonable test times. Figure 3.10 shows data from a rising displacement SCC test on a compact specimen where the crack length was continuously monitored using an electrical potential drop technique (Dietzel and Schwalbe, 1986; Dietzel, 1991). Although the displacement rate applied in the test was low, 3.6×10^{-11} m/s (0.2 μm/h), there was crack initiation after about 16 days, and the subsequent measurement of crack extension yielded the velocity of sub-critical cracking, i.e. region II in

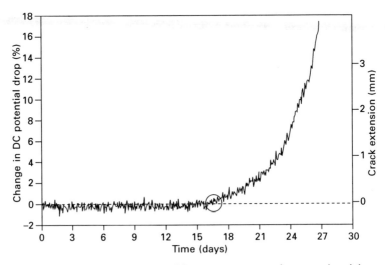

3.10 Crack extension measured in a compact specimen under rising displacement at a rate of 5.6×10^{-11} m/s (0.2 μm/h) in a chromated 3.5% sodium chloride solution (material: aluminum AA 2024 T351). The figure shows the signal obtained using the electrical resistance measurement technique as described by Dietzel and Schwalbe (1989); the circle indicates crack initiation.

the da/dt vs. K curve. The total test time was 27 days. Figure 3.11 compares a v-K curve determined with this technique (the full curve) with the data generated using self-loaded DCB specimens of the same material (see Fig. 3.8(b)), for which the total test time was more than 400 days.

To evaluate K_{ISCC} from rising displacement tests, a number of identical specimens are tested at different displacement rates. In each of these tests, the onset of cracking is identified, usually by an indirect crack length measuring technique as described in Section 3.5.1. A curve as shown in Fig. 3.12 is established. The threshold value, K_{ISCC}, is inferred as that corresponding to the lower values of the sigmoidally shaped curve.

In order to maintain the accelerating nature of dynamic SCC testing, it is necessary to limit the number of tests which have to be performed for evaluating K_{ISCC}. This requires a simple estimation of the loading rate (or displacement rate), which corresponds to the lower shelf regime of the curve in Fig. 3.12, and which yields the value of K_{ISCC}. ISO standard 7539-9 gives recommendations for this, based on the assumption that the displacement rate, $(dq/dt)_{SCC}$, for the rising displacement test in a corrosive environment, can be derived from the ratio of the measured crack growth velocity in a rising displacement test in an inert environment, $(da/dt)_{inert}$, and the crack growth velocity in the plateau region, $(da/dt)_{SCC}$.

3.4.7   Crack growth measurement

The measurement of crack initiation and growth is an important aspect of SCC testing. For a fracture mechanics based investigation of SCC, the crack

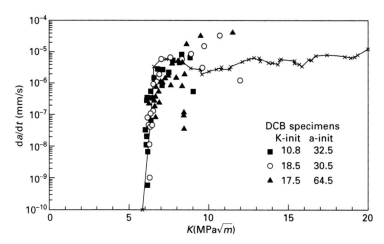

3.11 Comparison of the da/dt vs. K data of Fig. 3.8(b) and the v-K curve determined in a rising displacement test using a compact specimen with the same combination of material and environment.

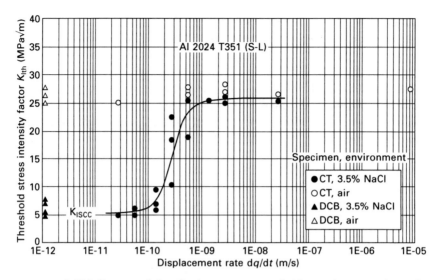

3.12 Influence of the displacement rate, dq/dt, on the stress intensity factor at crack initiation, K_{Ith}, measured in rising displacement tests using compact specimens (circled symbols). For comparison the results of long-term constant displacement tests with a duration of 10 000 hours using double cantilever beam specimens (triangles) are shown.

length is, together with the applied load, the key parameter for calculating the stress intensity factor. *In-situ* measurement of the crack length provides useful information about the onset of cracking, about the actual crack length, and about the prevailing crack growth velocity. The electrical resistance measurement technique, e.g. in the form of a comparator technique, is also a useful tool in tests on smooth specimens (Dietzel and Schwalbe, 1986; Strieder *et al.*, 1994; Winzer *et al.*, 2008). It provides a more precise identification of crack initiation and hence allows a better quantitative evaluation of these tests without the danger of underestimating crack growth velocities, as discussed in Section 3.3.4.

Optical methods of measurement are often precluded by the environment and the test chamber and, in any case, apply only to the surface length of a crack. Enhancement of crack visibility by removal of corrosion products may perturb the local electrochemistry and is thus problematic. Methods that measure the average crack length across the thickness of the specimen, such as electrical resistance or compliance techniques, are generally preferred. AC and DC potential drop measurements are suitable, but care has to be taken to ensure that the application of a current has no detectable influence on the electrochemistry and on the rate of crack extension and also to eliminate galvanic effects. Compliance methods based on measurement of displacement across the notch or of strain in the back face of the specimen opposite the notch are used only for SCC testing of pre-cracked specimens. A detailed description of indirect methods of crack length measurement can be found in Annex C of ISO standard 7539-9.

3.5   The elastic-plastic fracture mechanics approach to stress corrosion cracking (SCC)

The applicability of linear elastic fracture mechanics to SCC and the use of K_I as driving force parameter builds on the assumption of limited plasticity and of predominant plane strain conditions. In cases of SCC in which there is a higher amount of plasticity, neither plane strain nor linear elastic conditions are satisfied, so that elastic-plastic fracture parameters should be used instead of K.

For LEFM to be applicable, the specimens must satisfy the minimum size requirements imposed by the linear elastic fracture mechanics concept, as presented in Eq. 3.1. For lower strength and/or more ductile materials this may result in large specimen dimensions, particularly with respect to thickness.

3.5.1   The J integral

For these reasons, the elastic-plastic J integral concept is also applied in fracture mechanics treatment of SCC (Anderson and Gudas, 1984; Abramson et al.,1985; Dietzel et al., 1989). Figure 3.13 shows the effect of hydrogen embrittlement on J-R curves, measured in rising displacement tests on C(T) specimens of a high strength steel (Dietzel and Schwalbe, 1989). The difference between two J-R curves, one obtained in air and the second one under cathodic protection in a corrosive environment (substitute ocean water according to ASTM D1141) reflects the significant reduction of the fracture resistance caused by the ingress of atomic hydrogen leading to hydrogen embrittlement. Figure 3.13 shows that the value of the J integral at crack initiation, $J_{0.2}$, was significantly decreased due to the environmental degradation caused by hydrogen. Also, the slope of the crack growth resistance curve measured in the corrosive environment was almost negligible, indicating that only a little mechanical energy was needed for crack extension.

3.5.2   Crack tip opening angle/crack tip opening displacement

There is increasing interest in other elastic-plastic fracture parameters, in particular the crack tip opening angle, CTOA, and the crack tip opening displacement, CTOD, which are especially well suited for fracture toughness assessments of thin-walled structures and for investigating the initiation and

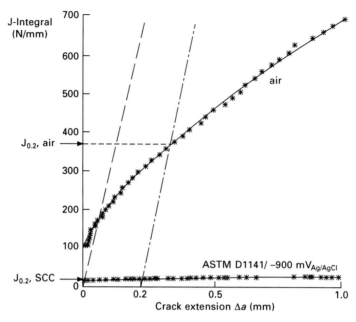

3.13 J-R curves determined using pre-cracked compact specimens of a low-alloyed structural steel (FeE 690T) in air and in ASTM D1141 substitute ocean water under cathodic protection, both tests performed at a load line displacement rate of 1 μm/h (Dietzel and Schwalbe, 1989).

growth of stress corrosion cracks in such structures. These parameters can also successfully be applied to study SCC.

The crack tip opening displacement, CTOD, often designated as δ, is specified as the relative displacement of the crack surfaces normal to the original, un-deformed crack plane at the tip of the fatigue pre-crack. In practice, this parameter can be determined at the surface of a specimen or a component, e.g. by using a specially designed clip-on gauge which measures the CTOD over a gauge length of 5 mm (δ_5) as demonstrated in Fig. 3.14 (Hellmann and Schwalbe, 1986). This figure also illustrates the position of the current leads ('I') and the measuring location for the potential drop across the crack when using the DCPD (see Section 3.4.7). In SCC experiments performed in a corrosive environment, a direct measurement of the variable δ_5 is difficult, since the clip-on gauge should be immersed in the corrosive environment during the test, and would hence need careful protection against corrosion. Calibration experiments performed using a number of materials and under various environmental conditions have confirmed that excellent agreement exists between directly measured δ_5 data and values which were calculated from the load and the crack mouth opening displacement, CMOD. The expression correlating δ_5 and the CMOD is derived from the British

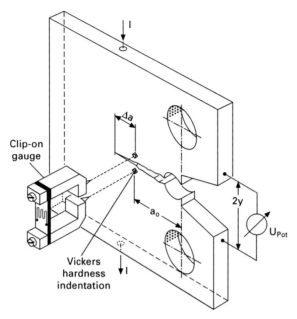

3.14 Clip-on gauge for measuring the CTOD over a gauge length of 5 mm ('δ_5'); the figure also illustrates the position of the current leads ('I')and the measuring location for the potential drop across the crack when using the DCPD (Hellmann and Schwalbe, 1986).

Standard 5762 and modified with respect to crack extension (Schwalbe, 1995):

$$\delta^M_{BS} = \frac{K^2}{2\sigma_y E'} + \frac{0.6\Delta a + 0.4(VW - a_0)}{0.6(a_0 + \Delta a) - 0.4W + z} \cdot v_{pl} \qquad 3.2$$

In this modified CTOD, the motion of the rotation centre of the specimen due to crack extension is taken into account. In Eq. 3.2, v_{pl} is the plastic portion of the crack mouth opening displacement (CMOD), E' the Young's modulus for plane strain, and z is the distance between the load line and the actual measuring position for v.

The crack tip opening angle (CTOA), ψ, is the angle included by the flanks of an extending crack. Experimental evidence has shown that, after an initial transition period, the crack extends in such a way that the crack tip opening angle remains almost constant, independent of the amount of crack extension (Newman *et al.*, 2003). In experimental studies, a measure of the CTOA, ψ, can be obtained from measurement of the change of the crack tip opening displacement, $\Delta\delta_5$, with increasing crack extension, Δa, by (Dietzel and Schwalbe, 1989):

$$\tan \Psi \approx \frac{\Delta \delta_5}{\Delta a} \hspace{3cm} 3.3$$

It has been shown that when the CTOA concept is used in SCC, the magnitude of ψ, which is governed by the fracture process in the process zone, becomes dependent on the deformation rate and decreases in the presence of an aggressive environment. Figure 3.15 presents a measure of the CTOA, $d\delta_5$/ da, corresponding to tan ψ, determined using pre-cracked compact specimens of a high strength structural steel at various constant displacement rates. The displacement rates have been converted into rates of change of the CTOD, $d\delta_5$/dt, to provide a measure of the applied deformation rate. There was a significant decrease of the CTOA due to the increasing influence of SCC at low deformation rates, $d\delta_5$/dt. The reference tests in air show no such change in CTOA with decreasing deformation rate. In this example, the drop in CTOA is caused by hydrogen embrittlement, resulting from the ingress of atomic hydrogen into the steel. A similar effect of the displacement rate on the CTOA was observed for material/environment combinations in which the failure mechanism was intergranular SCC.

3.6   The use of stress corrosion cracking (SCC) data

When employing the fracture mechanics approach, the results obtained from the different types of SCC tests and on various specimen configurations,

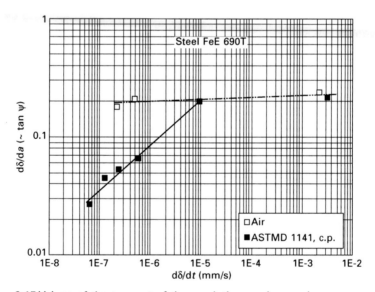

3.15 Values of the tangent of the crack tip opening angle, ψ, determined in rising displacement tests using compact specimens of a high strength steel (FeE 690T) subjected to hydrogen embrittlement due to cathodic protection.

are identical if all specific requirements are carefully met. Data obtained from through-cracked specimens can be used to predict crack initiation and growth in plates containing small semi-elliptical surface cracks which in turn simulate defects in real components and structures.

The advantage of using a threshold parameter like K_{ISCC} lies in the ability to predict the combinations of stress level, flaw size and shape which lead to SCC. K_{ISCC} may be used as a design criterion for ensuring no SCC growth in service, provided that the stress levels, minimum detectable flaw sizes, and environmental conditions are well defined, and that the service loads are essentially constant, i.e. that cyclic loading is not significant. Figure 3.16 illustrates in a schematic way how K_{ISCC} and σ_{th} values can be included in an assessment of structural integrity based on the failure assessment diagram (FAD) (Dietzel, 2001). In the case of SCC, the fracture toughness of the material, K_{mat}, is replaced by K_{ISCC}, and the threshold stress for SCC, σ_{th}, determined on smooth tensile specimens of the same material and in the same environment, yields a new cut-off line on the stress axis, L_r. Thus, an acceptable region with respect to immunity against SCC is defined. Depending on the susceptibility of the material–environment combination under investigation, this region can be significantly smaller than the original acceptable region as derived from tests in air.

As a screening parameter for susceptibility classifications of materials, K_{ISCC} is widely accepted for providing guidance to develop and/or select materials that exhibit sufficiently high threshold values for applications in which SCC must be prevented. Together with crack growth data, it yields information about the severity of environments, which can promote SCC, and about the effectiveness of countermeasures and means of protection. The crack growth rate, da/dt, vs. K_I data can be used to establish sub-critical crack growth allowable for both new designs and existing structures, i.e., to decide whether a period of safe crack extension exists, and if so, to specify inspection intervals for parts assumed or known to have flaws. SCC growth life predictions could in principle be performed by integrating da/dt vs. K_I data, thus yielding the time when the critical crack size is reached, as is demonstrated schematically in Fig. 3.17 (Wanhill, 1991). This in turn requires that the K values of components can be calculated and that residual and/or assembly stresses are either known or are negligible. A rough estimate of the significance of sub-critical crack growth may furthermore be obtained from the value of da/dt in the plateau region of the v-K curve.

3.7   Standards and procedures for stress corrosion cracking (SCC) testing

A number of standards and procedure documents provide detailed guidelines for SCC testing which are essentially aimed at determining σ_{th}, K_{ISCC} and

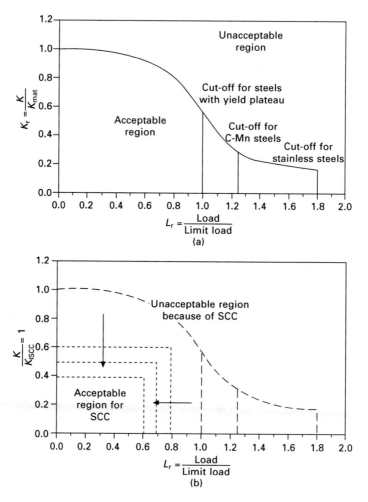

3.16 Schematic presentation illustrating the inclusion of K_{ISCC} data into a failure assessment diagram (FAD). By substituting the fracture toughness of the material, K_{mat}, for K_{ISCC}, and the cut-off line on the stress axis, L_r, for the SCC threshold stress, σ_{th}, an acceptable region with respect to non-growing SC cracks is defined; this region can be significantly smaller than the original acceptable region as determined from tests in air.

da/dt vs. K_I, in which testing of smooth and pre-cracked specimens is treated separately. The following list is not complete but should provide assistance in selecting an appropriate document for SCC testing. A more detailed guide to the selection of an appropriate mechanical SCC test method for a particular problem can be found in Dietzel and Turnbull (2007).

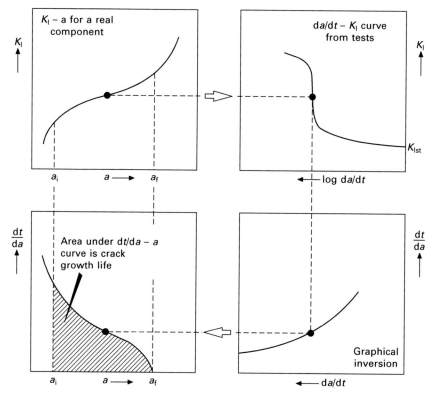

3.17 Schematic illustration of SCC crack growth life prediction performed by integrating da/dt vs. K_I data yielding the time remaining until the critical crack size, a_f, is reached (Wanhill, 1991).

3.7.1   Standards

Smooth specimens

Constant load

- ASTM C47 85 (1990)
- ISO 7539-4:1989
- NACE TM 0177-90

Constant displacement

Bend specimens

- ASTM G39 79 (1990)
- ISO 7539-2:1990
- NACE TM 0177-90

C ring specimens

- ASTM G38 73 (1990)
- ISO 7539-5:1989
- NACE TM 0177-90

Uniaxially strained tension specimens

- ASTM G49 85 (1990)
- ISO 7539-4:1989
- NACE TM 0177-90

Constant total strain/plastic strain specimens

- ASTM G30 90
- ISO 7539 3:1989

Dynamic straining/Slow strain rate test

- ASTM G129-95
- ISO 7539-7:1989
- NACE T-1F-9

SCC testing of specific materials

- ISO 9591:2003 (aluminium alloys)

Pre-cracked specimens

Constant load

- ASTM E 1681-03
- BS 6980
- ISO 7539-6:2003

Constant displacement

- BS 6980
- ISO 7539 6:1989
- NACE TM 0177-90

Rising load/rising displacement

- BS EN ISO 7539-9:2008

- EN ISO 7539-9:2008
- ISO 7539-9:2003

Test environments

- ASTM D 1141 - 90 (substitute ocean water)
- ASTM D 1384 - 96 (corrosive water/engine coolant)

3.8   Future trends

Concerning the application of test data, SCC fracture control can be a particular problem and may require verification by extensive small-scale laboratory testing as well as full-scale service qualification tests. The present chapter has defined, categorized and discussed the requirements for SCC testing. It has also described the ways in which fracture mechanics and non-fracture mechanics test data and parameters can be used for such SCC fracture control. Difficulties and limitations in practical application can occur especially with respect to actually controlling SCC, i.e. to establish alowable sub-critical crack growth.

Such control of SCC may not always be feasible, and hence SCC fracture control will in these cases rather have to be directed to preventing SCC in service. Guidance on the appropriate steps to be taken when a stress corrosion crack has been detected in service and an assessment of the implications for structural integrity is required (Wanhill, 1991). Documents that aim at providing such guidance are in preparation by ISO and CEN Committees.

Although there is good agreement between results obtained using different SCC test methods in terms of relative performance, differences can occur due to, for example, differences in environmental variables, e.g. the electrode potential, pH, flow rate, solution composition, and in the time of testing. The specification, monitoring and, where appropriate, control of environmental variables is a necessary requirement for improved reliability of data. But, despite appropriate indications in the various standards, the increased complexity and costs of testing can influence the experimental rigour, and economic pressure represents an additional burden.

In recent years, increased computer power has made modelling a useful tool for simulating mechanisms of SCC and hydrogen embrittlement. This has not only improved the understanding of the micro-mechanisms involved in environment-assisted failure but can also be complementary to laboratory SCC testing. Multi-scale modelling describing the influence of a corrosive environment on crack susceptibility of components and structures may allow prediction of deformation behaviour and crack initiation in different material configurations under different external load and environmental conditions. It can give predictions of critical flaws and fracture toughness of existing components and structures and should be able to suggest the optimal microstructural configuration of new constructions with respect to withstanding environmentally assisted cracking. It is believed that in the future it can replace, at least to some extent, expensive and time-consuming tests. In addition, new nano-mechanical test methods are being developed that can validate and provide information obtained from such models and can thus provide a safe basis for avoiding cases of SCC and to assess the criticality of existing components and structures with respect to SCC.

3.9   References

Abramson G, Evans J T and Parkins R N (1985), 'Investigation of stress corrosion crack growth in Mg alloys using J-integral estimations', *Met Trans*, 16 A, 101–108.

Anderson D R and Gudas J P (1984), 'Stress corrosion evaluation of titanium alloys using ductile fracture mechanics technology', in: *Environment Sensitive Fracture, Evaluation and Comparison of Tests Methods*, ASTM STP 821, S W Dean, E N Pugh and G M Ugiansky (eds), American Society for Testing and Materials, West Conshohocken, PA, 98–113.

Atrens A, Brosnan C C, Ramamurthy S, Oehlert A and Smith I O (1993), 'Linearly increasing stress test (LIST) for SCC research', *Meas Sc Technol* 4, 1281–1292.

Clark W G Jr and Landes J D (1976), 'An evaluation of rising load K_{ISCC} testing', in: *Stress Corrosion – New Approaches*, ASTM STP 610, H L Craig Jr (ed.), American Society for Testing and Materials, West Conshohocken, PA, 108–127.

Dietzel W (1991), 'Zur Anwendung bruchmechanischer Methoden bei der Untersuchung des Umgebungseinflusses auf die Rißausbreitung bei zügiger Beanspruchung' (in German), Report GKSS 91/E/27, Geesthacht.

Dietzel W (1999) 'Characterization of susceptibility of metallic materials to environmentally assisted cracking', Report GKSS 99/E/24, GKSS-Forschungszentrum Geesthacht GmbH, Geesthacht.

Dietzel W (2000), 'Standardization of rising load/rising displacement SCC testing', in: *Environmentally Assisted Cracking: Predictive Methods for Risk Assessment and Evaluation of Materials, Equipment and Structures*, ASTM STP 1401, R D Kane (ed.), American Society for Testing and Materials, West Conshohocken, PA, 317–326.

Dietzel W (2001), ' Stress corrosion cracking in metals: mechanics', in: *Encyclopedia of Materials: Science and Technology*, K H J Buschow, R W Cahn, M C Flemings, B Ilschner, E J Kramer and S Mahajan (eds), Elsevier Science, Amsterdam, 8883.

Dietzel W and Schwalbe K-H (1986), Monitoring stable crack growth using a combined AC/DC potential drop technique, *Z. Materialprüfung/Mats. Testing*, 28, 11, 368–372.

Dietzel W and Schwalbe K-H (1989), 'A study of the hydrogen induced stress corrosion cracking of a low alloy steel using fracture mechanics techniques'; in: *Hydrogen Effects on Material Behaviour*, N R Moody and A W Thompson (eds), The Minerals, Metals & Materials Society, 975–983.

Dietzel W and Schwalbe K-H (1993), 'Application of the rising displacement test to SCC investigations', in: *Slow Strain Rate Testing for the Evaluation of Environmentally Induced Cracking: Research and Engineering Applications*, ASTM STP 1210, R D Kane (ed.), American Society for Testing and Materials, Philadelphia, pp. 134–148.

Dietzel W and Turnbull A (2007), 'Stress corrosion cracking', in: *Comprehensive Structural*

Integrity, New Online Volume 11: Mechanical Characterization of Materials, K-H Schwalbe (ed.), Elsevier, Oxford, 43–74. (See also Dietzel W and Turnbull A (2007), 'Stress corrosion cracking', Report GKSS 2007/157, GKSS-Forschungszentrum Geesthacht GmbH, Geesthacht.)

Dietzel W, Schwalbe K-H and Wu D (1989), 'Application of fracture mechanics techniques to the environmentally assisted cracking of aluminium 2024', *Fat Fract Engng Mats Struct*, 12, 6, 495–510.

Ford F P (1988), 'The crack-tip system and its relevance to the prediction of cracking in aqueous environments', NACE-10, National Association of Corrosion Engineers, 139–166.

Hellmann D and Schwalbe K-H (1986), 'On the experimental determination of CTOD based R-curves', in: *The Crack Tip Opening Displacement in Elastic-Plastic Fracture Mechanics*, Workshop on CTOD Metodology, K-H Schwalbe (ed.), Springer-Verlag, Berlin, 115–132.

McEvily A J Jr (1990), *Atlas of Stress-Corrosion and Corrosion Fatigue Curves*, Materials Park, OH, ASM International.

McIntyre P and Priest A H (1972), 'Accelerated test technique for the determination of K_{ISCC} in steels', British Steel Corporation Report MG/31/71, London.

Newman J C Jr, James M A and Zerbst U (2003), 'A review of the CTOA/CTOD fracture criterion', *Engng Fract Mechan*, 70, 371–385.

Parkins R N (1972), 'Stress corrosion test methods', *Brit Corr J*, 7, 154–167.

Parkins R N (1993), 'Slow strain rate testing – 25 years experience'; in: *Slow Strain Rate Testing for the Evaluation of Environmentally Induced Cracking*, Research and Engineering Applications; ASTM STP 1210, R D Kane (ed.), American Society for Testing and Materials, Philadelphia, pp. 7–21.

Raymond L and Crumly W R (1982), 'Accelerated lowcost test method for measuring the susceptibility of HY-steels to hydrogen embrittlement', in: *Proc First Hydrogen Problems in Steel, Washington DC*, ASME, American Society for Metals, Metals Park, OH, 477–480.

Rieck R M, Atrens A and Smith I O (1989), 'The role of crack tip strain rate in the stress corrosion cracking of high strength steels in water', *Met Trans*, 20A, 889–895.

Schra L and Groep F F (1993), 'The ASCOR test: a simple automated method for stress corrosion testing of aluminum alloys', *J Test Eval*, 21, 1, 44–50.

Schwalbe K-H (1995), 'Introduction of δ_5 as an operational definition of the CTOD and its practical use', in: *Fracture Mechanics: 26th Volume*, ASTM STP 1256, W G Reuter, J H Underwood and J C Newman (eds), American Society for Testing and Materials, Philadelphia, PA, 763–778.

Sedriks A J (1990), 'Stress corrosion cracking – test methods', in: *Corrosion Testing Made Easy*, NACE, Houston, TX.

Speidel M O (1971), 'SC crack growth in aluminium alloys', in: *The Theory of Stress Corrosion Cracking*, Scully J C (ed.), NATO, Brussels, 289–344.

Speidel M O and Atrens A (1980), 'Immunity of austenitic retaining ring steel Fe-18%Mn-4.5%Cr-0.5%C to gaseous hydrogen embrittlement', in: *Hydrogen Effects in Metals*, I M Bernstein and A W Thompson (eds), TMS-AIME, 951–959.

Sprowls D O (1992), 'Evaluation of stress-corrosion cracking', in: *Stress-Corrosion Cracking – Materials, Performance and Evaluation*, R H Jones (ed.), ASM International, Materials Park, OH, 363–415.

Sprowls D O (1996), Evaluation of stress-corrosion cracking, in: *ASM Handbook Volume 13 Corrosion*, ASM International, Metals Park, OH, 245–282.

Sprowls D O, Bucci R J, Ponchel B M, Brazill R L and Bretz P E (1984), 'A study of environmental characterization of conventional and advanced aluminium alloys for selection and design. Phase II – the breaking load test method', NASA Contractor Report CR-172387, National Aeronautics and Space Administration, Washington, DC.

Stark H L and Ibrahim R N (1986), 'Estimating fracture toughness from small specimens', *Engng Fract Mechan*, 25, 4, 395–401.

Strieder K, Daum K-H, Dietzel W and Müller-Roos J (1994), 'The use of slow strain rate tests for measuring the velocity of environmentally assisted cracking', in: *Structural Integrity: Experiments – Models – Applications, Proceedings of the 10th European Conference on Fracture, ECF 10, Berlin, 20-23 September*, K-H Schwalbe and C Berger (eds), 715–720.

Turnbull A (1992), 'Test methods for environment-assisted cracking', NPL Report DMM(A)66, National Physical Laboratory, Teddington, Middlesex.

Wanhill R J H (1975), 'Aqueous stress corrosion in titanium alloys', *Brit Corr J*, 10, 69–78.

Wanhill R J H (1991), 'Fracture control guidelines for stress corrosion cracking of high strength alloys', NLR Technical Publication TP 91006 L, Amsterdam, The Netherlands.

Wei R P and Novak S R (1987), 'Interlaboratory evaluation of K_{ISCC} and da/dt determination procedures for high-strength steels', *J Test Eval*, 15, 38-75.

Winzer N, Atrens A, Dietzel W, Song G and Kainer K U (2008) 'Comparison of the linearly increasing stress test and the constant extension rate test in the evaluation of transgranular stress corrosion cracking of magnesium alloys', *Mater Sc Engng A*, 472, 1–2, 97–106.

Yokobori T, Watanabe J, Aoki T and Iwadate T (1988), 'Evaluation of the K_{ISCC} testing procedure by round robin tests on steel', in: *Fracture Mechanics: Eighteenth Symposium, ASTM STP 945*, D T Read and R P Reed (eds), American Society for Testing and Materials, Philadelphia, PA, 843–866.

Part III

Stress corrosion cracking (SCC) in specific materials

4
Stress corrosion cracking (SCC) in low and medium strength carbon steels

U. K. CHATTERJEE, Indian Institute of Technology
Kharagpur, India, and R. K. SINGH RAMAN,
Monash University, Australia

Abstract: Stress corrosion cracking of plain carbon steels having yield strengths of less than 1240 MPa is reviewed. Apart from the classically known hydroxide and nitrate media, stress corrosion cracking of such steels has been reported in a number of other environments, namely aqueous chloride, carbonate, phosphate, ammonia, CO-CO_2-H_2O, high purity water containing oxygen, fuel ethanol and several others. Cathodic stress corrosion cracking, i.e. hydrogen embrittlement, is encountered principally in hydrogen sulphide, hydrogen gas and in solutions containing hydrogen compounds. Salient observations of investigations carried out in the past and in recent times are highlighted in this review.

Key words: stress corrosion cracking, hydrogen embrittlement, low strength steels, medium strength steels, high strength low alloy steels, caustic cracking.

4.1 Introduction

The chapter reviews the stress corrosion cracking (SCC) of steels containing less than approximately 5% total alloying elements and having yield strengths of less than about 1240 MPa (180 ksi) of basic microstructural types of ferritic-pearlitic or tempered martensite (YS > 620 MPa/90 ksi). Several publications provide useful information on the phenomenon and mechanism of SCC of carbon steels [1–5].

Originally, a distinction used to be made between stress corrosion cracking and hydrogen embrittlement on the basis of their response to electrochemical conditions, the former getting enhanced with anodic polarization and the latter with cathodic polarization. However, they are referred as anodic SCC and cathodic SCC as well [6] and it has been suggested that SCC under both anodic and cathodic conditions is fundamentally the same process [7]. In view of this, the review covers the conventional SCC behaviour as well as hydrogen embrittlement behaviour of the steels under consideration.

For conventional SCC, the specificity of environment is a distinct characteristic feature. Not all alloys produce SCC in all environments, and the specificity is related to either adsorption or the ability to produce a borderline

169

passivity condition for SCC to occur. On the other hand, any environment providing atomic hydrogen can lead to hydrogen embrittlement. Atomic hydrogen produced as a result of anodic dissolution or corrosion, so to say, is the cause of hydrogen embrittlement in many cases, but quite a few other sources of hydrogen are readily available in environmental and industrial exposures. Thus the list of agents producing SCC in steels gets longer.

Among the identified and reported agents for anodic SCC, the most common are hydroxides (caustic), nitrates, carbonates, phosphates, CO-CO_2-H_2O and aqueous chlorides, and among the less common are ammonia, high purity water containing oxygen, phosphorus trifluoride, sulphuric acid, hydrogen-antimony/aluminium chloride and $MgCl_2$-NaF. Some of them (e.g. aqueous chlorides, sulphuric acid, hydrogen-antimony/aluminium chloride) have been reported to produce cathodic SCC as well. The environments in which cathodic SCC or hydrogen embrittlement are encountered are hydrogen sulphide, hydrogen gas, hydrogen fluoride and hydrogen cyanide.

The environmental and material variables that affect the SCC behaviour of alloys are material chemistry, microstructure, cold work, stress level, concentration of corrosive species, additives to the environment, pH, temperature and electrochemical conditions, and the vast literature on SCC of low and medium strength steels are associated with the study of one or more of these variables.

4.2   Dissolution-dominated stress corrosion cracking (SCC)

4.2.1   SCC in hydroxide solutions

Stress corrosion cracking of steels due to hydroxide solutions (generally, sodium hydroxide) is commonly known as 'caustic cracking' or 'caustic embrittlement'. The first documented cases of caustic cracking were reported in riveted locomotive boilers [8], where dilute alkali in the feed waters would build up in concentration at rivet holes as steam escaped, and the combination of moist hot caustic and stresses (due to riveting) caused SCC. Caustic cracking of boilers had been mitigated by employing welding (instead of riveting) and/or by adding inhibitors (such as sodium sulphate, trisodium phosphate, sodium nitrate, tannin, lignosulphates, etc.) into the boiler water [9]. However, caustic cracking continues to be a problem wherever stressed steel comes in contact with hot alkaline solutions.

Pulp and paper production is one industry that makes extensive use of caustic solutions of varying formulations in the Kraft process. Cracking of continuous digesters used in the Kraft process have been attributed to caustic embrittlement [10]. Depending on material, temperature and exposure conditions, the cracking susceptibility regime is expanded to caustic concentrations as low as 4% to values as high as 75% [9]. The minimum caustic concentration required to produce caustic cracking in boiling solutions is 4–5% [9, 11]. Corrosion rates of mild steels are reported [12] to be independent of caustic concentration in the range of 0.1–4% NaOH. Though the cracking is reported in the caustic concentration range of 4–75%, the intermediate range is believed to be most conducive for caustic cracking.

Caustic cracking can also occur over a wide range of temperatures (100–350°C). Figure 4.1 shows a typical caustic cracking susceptibility (caustic concentration temperature) map [13]. This map suggests that caustic cracking can occur at temperatures as low as 40–50°C in highly concentrated (30–50%) caustic solutions. The cracking is generally intergranular. Transgranular SCC of carbon steel has also been reported in caustic at elevated temperatures and pressures under freely corroding conditions [14], hydrogen-induced mechanism being prevalent in the latter case.

It is generally believed [15–18] that caustic cracking occurs under conditions for the formation of a passive oxide film and its dissolution i.e., in the regime of active–passive transition for a given potential–current diagram. The active-passive transition region for a given nominal caustic solution can be shifted considerably due to the presence of certain impurities or additives [9, 16, 17]. For pipeline steel X-70, it has been reported that dissolution-controlled SCC prevails at relatively less negative potential whereas hydrogen-based mechanism prevails at higher negative potentials [19]. The role of hydrogen in enhancing SCC susceptibility has also been reported for X-80 steel [20]. Sarioglu [21] has attributed the SCC of insufficiently tempered 4140 steel due to hydrogen embrittlement of martensitic band structure. Lukito and Szklarska-Smialowska [22] have shown that the steels with higher values of hydrogen trapping constants and hydrogen entry fluxes are highly susceptible to hydrogen-induced cracking in cathodically charged specimens in caustic solutions.

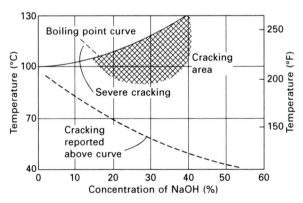

4.1 Caustic cracking susceptibility diagram [13].

Consistent with the passivation-dissolution mechanism, caustic cracking of steels has been shown to be highly facilitated in a narrow electrochemical potential range near the active current peak on the potential-current diagrams [9, 15, 16] (Fig. 4.2). Sriram and Tromans [24] have reported the current-density at pre-passivation peak (active current peak) to increase (from 40 to $100 \, Am^{-2}$) with caustic (OH^-) concentartion (2–5 M). The potential range is believed to correspond to that region of the potential-current curves where the steel surface is partially covered with a protective oxide (namely, magnetite, Fe_3O_4) [15, 25]. Bohnenkamp [26] has found that under the condition of an active current peak, the time-to-failure (of tensile specimens in constant load tests) decreases systematically with increase in temperature of a caustic solution (33% NaOH) in the range of 70–120°C. In plain caustic solutions, the active current peak on the potential-current diagrams is located at ~ $-76 \, mV_{SHE}$ in 12% NaOH [16] which increases to $-70 \, mV_{SHE}$ in 35% NaOH [26–28]. The cracking mechanism was consistent with electrochemical dissolution. In fact, caustic cracking was found to cease at a more positive potential at which the steel surface was completely covered with magnetite [15]. Flis and Ziomek-Moroz [29] have reported that carbon at low contents deteriorates the passivation of iron, whereas at high contents it promotes the formation of magnetite, and the high resistance to SCC of high carbon steels is explained by an intense formation of magnetite on these steels.

A caustic cracking insusceptible condition can turn into a susceptible

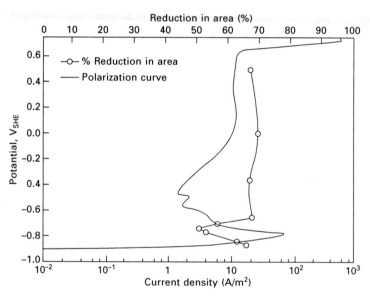

4.2 Effect of potential on reduction in area at strain rate 3.33 × $10^{-4} s^{-1}$, and superimposed anodic polarization diagram, for carbon steel in aqueous 3.35M NaOH at 92°C [16].

condition by the imposition of an electrochemical potential in the active-passive regime [30]. In SSRT tests, specimens of mild steel were not susceptible to caustic cracking in a 300 g/l caustic solution at 100°C. However, when potential of −800 mV was imposed, the steel showed a considerable degree of caustic cracking (Fig. 4.3). In order to establish active-passive regime of the mild steel in 30% NaOH at 100°C, anodic polarization curve was developed that suggested a regime of transition from active dissolution to passivity in the potential range of −850 mV to −775 mV.

The results of the studies by Singbeil and Tromans [16, 31, 32] have shown that the crack propagation rate was 25 times higher in the caustic solution containing sulphide than in the solution free from sulphide. The addition of sulphide ions was shown to have raised the active-passive transition potential to a more noble value.

Caustic cracking in both the active-passive transition and passive regions of the polarization curve generated for the solutions containing AlO_2 was reported by Sriram and Tromans [33] who attributed this behaviour to the nature of the passive film developed in the presence of AlO_2^-. This film is believed to contain an amorphous component of the form $Fe_{3-x}Al_xO_4$, as opposed to Fe_3O_4 which forms in absence of AlO_2^-. The deleterious influence of incorporation of AlO_2^- in the passive layer is also reported by Lee and Ghali [17] and Reinoehl and Berry [9].

There are conflicting reports [13] on the influence of silicon compounds. Some researchers suggested only solutions containing Na_2SiO_3 promote caustic cracking, whereas others have found that SiO_2 was not necessary to produce/ stimulate cracking. In an early report on the topic [34], it is suggested that it is not necessary to have SiO_2 in the caustic solution for cracking; however, there are indications that Na_2SiO_3 may have some inhibitive effect when its concentration exceeds that of NaOH. It is also suggested [9] that when silica

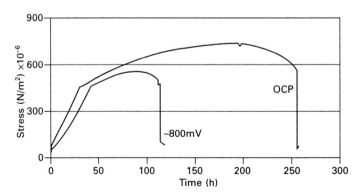

4.3 Engineering stress vs time curves generated by SSRT tests of a carbon steel in a 300 g/l caustic solution at 100°C, under open circuit potential (OCP) and imposed potential of −800 mV [30].

(and several other compounds) is present within critical concentration ranges they promote caustic cracking by only slightly moving the active–passive potential in the positive direction. However, at relatively high concentrations, these substances act as inhibitors, and decrease cracking susceptibility by considerably moving the potential outside the range of the active–passive transition region.

Cracking is promoted by small amounts of dissolved oxygen, but stopped at higher concentrations [35]. Also, bubbling of oxygen has been found [36] to prevent cracking of mild steel in boiling NaOH.

Chloride is believed [35] to delay cracking. There are also reports [37] suggesting that Na_2SO_4, which is believed to prevent caustic cracking, is effective only in the presence of considerable concentrations of Cl^-. There are reports [9] suggesting that small amounts of Cl^- accelerate cracking whereas large amounts tend to reduce this effect.

The oxidizing agents that promote cracking when present in low concentrations include: lead oxide [38], silicate [39], sulphate [37], nitrate [39], permanganate [39, 40] and chromate [37]. These agents also retard cracking when present at high concentrations. Other agents that retard cracking include phosphates, carbonates and some organic compounds (namely, tannins and butyric acid) [9, 17, 37].

4.2.2   SCC in nitrate solutions

Nitrates were first identified as cracking agents of intergranular failure of evaporation equipment containing NH_4NO_3 [41]. Stress corrosion cracking of low and medium carbon steels has been observed in the fertilizer industry [42] and in equipment used in the production of sodium nitrate [43]. Calcium nitrate, ammonium nitrate and their mixtures are most corrosive. Low carbon steel cracks in a 60% sodium nitrate solution, but the time to failure is much longer [44].

The susceptibility of steel to stress corrosion cracking in nitrate solutions is influenced by the concentration and temperature [27, 45]. Increased nitrate concentrations reduce SCC failure times and threshold stress levels [46, 47]. An increase in temperature decreases the time to failure, presumably due to increased crack growth rates [27, 48, 49]. A typical Arrhenius behaviour is exhibited. However, SCC can occur at quite low nitrate concentrations in boiling solutions, and an increase in concentration above a certain level does not affect SCC susceptibility in such solutions [50]. On the other hand, initiation and slow crack growth can occur over a wide temperature range given sufficient time [51].

If iron is immersed in a nitrate solution, its pH increases with time because of the reaction [52]:

$$Fe + NO_3^- + H_2O \rightarrow Fe^{2+} + NO_2 + 2OH^-$$

The rate of stress corrosion cracking is not affected within the pH range 3–7, but is greatly decreased if the alkalinity of the solution is increased above pH 7 [53].

The SCC of steels in nitrate solutions occurs within a well-defined potential range [23, 27]. This potential decreases as temperature is increased. The range of potential is considerably wider in nitrates than in caustics (Fig. 4.4) and nitrates are more potent cracking agents. The range corresponds to the transition between active corrosion and strong passivity. Anodic polarization leads to an enhancement of cracking and cathodic polarization results in a decreased failure [50, 51, 54], suggesting a dissolution-controlled mechanism of SCC. The association of a surface film of oxide/hydroxide in the susceptible potential-pH range supports a strain-induced film rupture mechanism [55] (Fig. 4.5). The inhibitive action of NaOH addition to nitrate

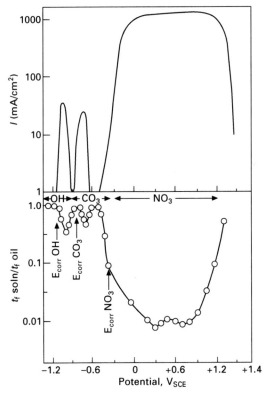

4.4 Current density differences between fast and slow sweep rate polarization curves and stress corrosion cracking susceptibility as a function of potential for C-Mn steel in NO_3^-, OH^-, and carbonate-bicarbonate environments. The free corrosion potential is indicated for each environment [4].

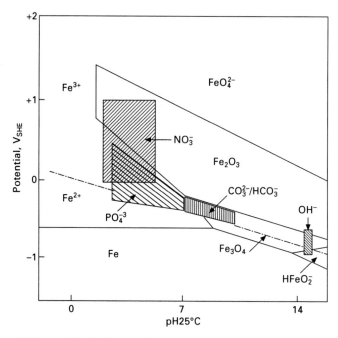

4.5 Potential-pH diagram for iron with cracking domains for some of the potent cracking agents [55].

solution is viewed to be due to the shift of the potential from the susceptible potential range.

Certain additions to nitrate solutions have been shown [46] to significantly affect SCC of mild steel. Oxidizing agents greatly accelerate cracking, while reducing agents or species that promote formation of insoluble iron products retard cracking (Table 4.1).

Studies on low carbon steels prepared by melting of pure iron have shown that steels containing less than 0.02% carbon are, in general, not susceptible to stress corrosion cracking in nitrate solutions [56], while steels containing about 0.0001% carbon and 0.003% nitrogen were susceptible [57]. An increase in the carbon content >0.2% decreased the susceptibility of steels to SCC [56, 58]. Radekar and Mishra [59] found that steel specimens containing more than 0.25% carbon and 0.001–0.008% nitrogen were not susceptible to SCC in nitrate solutions (Fig. 4.6). However, such steels with higher nitrogen content have been reported to be susceptible [60, 61]. According to Parkins [56], the increased resistance is associated with a decrease in cementite globules along grain boundaries and an increase in the amount of pearlite.

The effect of alloying additions of Ni, Mn and Cr in low carbon iron (C+N < 0.001–0.003%) on SCC in nitrate solutions was investigated by Long and Lockington [62], and Ni and Mn additions were reported to increase

Table 4.1 Effects of additions to boiling 4N NaNO$_3$ on cracking severity of a mild steel (initial stress 197.5 MPa)

Addition to 4 N NaNO$_3$	Average cracking time (h)	Average no. of cracks
None	26	2
1% K$_2$CRO$_4$	19	2
1% K$_2$CR$_2$O$^-$	22	18
0.1% KMnO$_4$	19	3
1% KMnO$_4$	22	19
1% NaHCO$_3$	500 n.f.[a]	0
1% Na$_2$CO$_3$	500 n.f.	0
1% Na$_2$HPO$_4$	500 n.f.	0
1% NaNO$_2$	19	4
0.5% CH$_4$N$_2$S	650	6
0.5% C$_6$H$_5$-COONa	260	2
1% C$_4$H$_4$O$_6$	27	Numerous
1% C$_6$H$_6$O$_7$	17	9

[a]n.f. = no failure in time indicated.

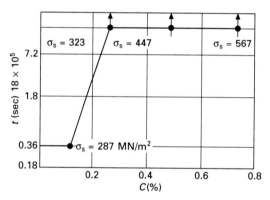

4.6 Dependence of the susceptibility of steel to stress corrosion cracking in nitrate solutions on its carbon content [59].

susceptibility. No such study in commercial steels has been reported. However, deoxidation and purity of steels are reported to influence the resistance to SCC in nitrate solutions [63]. The addition of uranium up to 0.15% has been reported to reduce SCC susceptibility of steels of strength levels within 135 000 to 155 000 psi in boiling nitrate solutions [64].

Chatterjee *et al.* [65] studied the SCC behaviour of plain carbon steels containing 0.19, 0.39 and 0.68% carbon (in pearlitic and spheroidized conditions) in boiling sodium nitrate ammonium nitrate solution. From the effect of applied stress, applied current and inhibitor addition, it was observed that the susceptibility to SCC decreased with increase in carbon content, the difference being more significant in pearlitic steels.

Prior plastic deformation has, in general, been reported to decrease the susceptibility of steels to SCC in nitrate solutions. Uhlig and Sava [66] have attributed the increase in the resistance to SCC to a decrease in the excess amount of interstitial atoms in the ferrite matrix and along grain boundaries, due to precipitation of carbides and nitrides during plastic deformation.

4.2.3   SCC in carbonate solutions

The first documented incident of stress corrosion cracking in carbonate solutions is that of a natural gas transmission pipeline that took place in 1965 [67]. A number of similar failures have occurred since then, and caustic accumulation due to cathodic protection was initially thought to have promoted such crackings. But detailed studies of the problem [68, 69] have shown that Na_2CO_3-$NaHCO_3$ is the agent responsible for failures. Caustic build-up at the pipe surface enhances adsorption of CO_2 from the soil to form the caustic environment.

For low strength steels, the SCC in carbonate solutions occurs only within a narrow electrochemical potential range [70–72]. This range corresponds to the transition between active corrosion and strong passivity, and is therefore similar to steel behaviour observed in nitrates and hydroxides. The cracking susceptibility, however, is considerably lower than that in nitrate solutions (Fig. 4.4). Parkins and Zhou [73] have reported that the potential range for intergranular (IG) cracking of a C-Mn steel in the pH regime of 6.7–11 to be ~100 mV wide, i.e., ~5 V (SCE) to ~ 6 V (SCE) (Fig. 4.7). However, at lower pH, a different morphology of quasi-cleavage cracking was observed over a considerable potential regime.

Might and Duquette [74] have carried out SCC tests under constant load of 1013 and 1002 steels in 170 g/l $(NH_4)_2CO_3$ solution at 70°C and have reported a ~100 mV wide critical potential range centered around 525 mV (SCE). The dynamic polarization scan exhibited two major anodic peaks located at ca –700 mV and ca –400 mV. A differential passivity between the bulk alloy and the grain boundary has been held responsible for promoting IG SCC.

The susceptible pH range for SCC of mild steel in carbonate solutions has been reported to be 8–10.5 [75]. The increasing pH has a narrowing effect on the susceptible potential range (Fig. 4.8, Parkins). The susceptible temperature range has been reported to be 22–90°C [76]. The threshold concentration of carbonate-bicarbonate for promoting SCC has been reported to be 0.25 N [75]. Alonso and Andrade [77] have reported the critical concentrations of 0.1 M for bicarbonate and 0.01 M for carbonate.

Berry et al. [78] found that phosphates, silicates, aluminates and chromates were effective inhibitors under potentiostated conditions, but only chromates

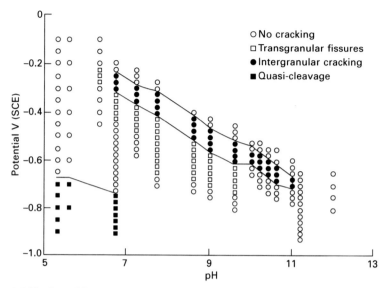

4.7 Modes of fracture resulting from the application of different potentials in SSRT on low-carbon steel in carbonate solutions of various pH values at ambient temperature [73].

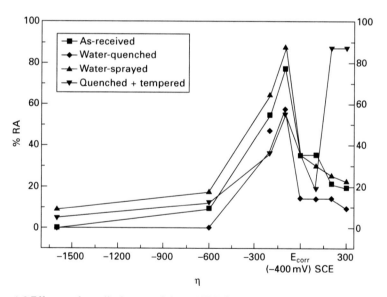

4.8 Effects of applied potential on %RA for steels in various heat treated conditions in 0.005 M Na_2HCO_3 at 50°C [89].

were effective over the entire potential range for cracking at carbonate concentrations below 0.1 wt%.

Abedi *et al.* [79] have reported the SCC of oil products API 5L X52 transmission pipeline due to the formation of carbonate-bicarbonate as a result of loosening and disbanding of the applied polyethylene tape coating. Sulphate reducing bacteria (SRB) activities have intensified pitting corrosion and related cracking processes.

A number of recent investigations [80–92] have been devoted to the study of SCC of pipeline steels in carbonate-bicarbonate media and in solutions simulating soil environment conditions. In general, the cracking has been reported to be transgranular and the SCC has been ascribed to anodic dissolution, while hydrogen plays a role in the cracking process, particularly at higher negative potentials [80–83]. The susceptibility increases with temperature [84, 85] and at higher stress levels [86]. In regard to microstructural dependence, steels in quenched condition are reported to be most susceptible [87, 88]. Torres-Islas *et al.* [89] reported that X-70 steel was susceptible in as-received, water quenched, quenched and tempered and water sprayed conditions; a minimum concentration of 0.005 м $NaHCO_3$ was necessary for cracking, the SCC susceptibility decreased in microstructural conditions at cathodic potential values close to their rest potential values, whereas higher cathodic potentials promoted SCC in all conditions. The cracking mechanism was dominated by anodic dissolution in as-received and water quenched conditions, but by hydrogen embrittlement for quenched and quenched and tempered steels (Fig. 4.8).

Atrens *et al.* [90] have shown that the grain boundary carbon concentration for the X70 steel was *ca* 10% or less and that gave its resistance to IG SCC, while X52 was prone in a 1N carbonate + 1N bicarbonate solution at 70°C. The difference in the initiation sites of SCC cracks in zinc coated, mechanically polished and electropolished X65 steel has been highlighted by Wang and Atrens [91]. A fundamentally-based mathematical model was developed to predict the crack growth rate of high pH SCC of buried steel pipelines [92].

4.2.4   SCC in CO-CO_2-H_2O environments

Stress corrosion cracking of mild steels in moist gases containing CO and CO_2 was first reported in the 1960s [93, 94], and cracks were transgranular. SCC occurs within a narrow electrochemical potential range, and the maximum susceptibility to SCC is observed in the potential range where the passivating effect is most pronounced.

Oxygen reportedly increases crack growth rates under potentiostatic conditions [95]. Kowaka and Nagata [96] observed cracking of U-bend specimens in liquid and vapour within one week at 40–70°C, but specimens tested under similar conditions at 100 and 150°C did not crack.

Laboratory simulation with SSRT of blower components in a coal gas refinery system and field reproduction of IG SCC through two-year exposure test in coal gas transport lines using U-bend specimens have shown that IG SCC of mild steel always initiates at corrosion pits with a critical size of *ca* 70 μm in depth [97]. A double-layer coating system consisting of an etching primer and a coal-tar epoxy urethane coating for prevention of SCC in CO-CO_2-H_2O environments has been developed [98].

4.2.5   SCC in aqueous chlorides

Though SCC of carbon and low-alloy steels in chloride solutions is quite common due to the prevalence of chloride in various environments, controversy remains regarding its exact mechanism. While in high strength steels the cracking is exclusively due to hydrogen embrittlement [99], in low strength steels failure is caused by anodic dissolution, though at considerable current densities, the failure is aggravated by hydrogen embrittlement [100]. In neutral chloride solutions, the corrosive medium is accumulated in pits and is acidified by an autocatalytic process; as a result the metal absorbs hydrogen. Therefore, initiation of cracks in steel tested in a 3% NaCl solution occurs after a time sufficient for pit formation and for acidification of the electrolyte [101].

For steels with yield strengths of less than 1240 MPa (180 ksi), increasing chloride content has been found to reduce failure times [102, 103]. In general, pH values below 1–3 enhance SCC, and pH values above 9 retard SCC [104, 105], both indicative of hydrogen embrittlement. The facts that very often the cracking is pronounced near room temperature and that cracking can be prevented by an increase of temperature to some threshold value [106, 107], demonstrate a case of hydrogen embrittlement. However, SCC of mild steel has been reported at high temperature of 316°C in water containing ferric chloride or chloride-bearing slurries of ferric oxides and hydrated oxides [108].

The additives known to promote hydrogen entry into steel, such as phosphorus, arsenic, selenium and antimony have been reported to accelerate SCC [109–111], whereas the presence of high levels of oxygen has been reported to have inhibited SCC [112, 113].

Cathodic polarization has little effect on the failure time of low strength steels [109, 114, 115], but anodic polarization has been reported to promote SCC [116, 117], apparently due to the occurrence of pitting, resulting in reduced SCC initiation times. SSRT test of cathodically charged HSLA-80 and HSLA-100 steels in synthetic seawater showed a distinct regime of hydrogen induced SCC at and below –900 mV (SCE) [118] (Fig. 4.9). Based on the SCC tests conducted on welded HY-130 steel in a non-flowing 3.5% NaCl solution with single notched bend specimens freely corroding or coupled to zinc anodes at 24°C, Fujii [119] has reported the K_{1scc} values to have increased with increase in YS value.

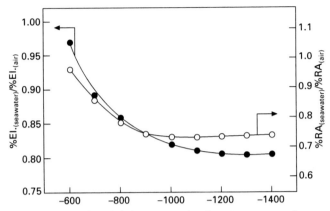

4.9 Variation of embrittlement ratios for stress corrosion cracking of a HSLA-80 steel in synthetic seawater under cathodic charging conditions [119].

In regard to the microstructural dependence of SCC, while limited improvements due to the presence of retained austenite in steel were observed in both threshold and crack growth rates, these were neither consistent nor more effective than other heterogeneities [120]. Banerjee and Chatterjee [121] have observed that in HSLA-80 and HSLA-100 steels, lath martensite or acicular ferrite structure was found to offer a greater resistance to hydrogen induced SCC, whereas a mixed bainitic-martensitic or polygonal ferrite structure showed higher susceptibility.

Zucchi *et al.* [122] have reported the occurrence of transgranular cracking of a 0.058% C steel in 5% NaCl + 0.5% acetic acid solution in the presence of 1 mM thiosulphate at 25°C. The SCC susceptibility increased under cathodically charged conditions. A few organic compounds, e.g. n-dodecyclquinolinium bromide, n-dodecylpyridinium chloride and 2-n-tridecyl-3-imidazoline at 10^{-4} concentration, completely inhibited the effect of thiosulphate.

SCC of a 0.5% carbon steel in a NaCl + CH_3COOH solution bubbled with H_2S was reported to have occurred at stress levels >75% of yield stress [123]. SCC of AISI 1018 and 4340 steels in polluted seawater, natural seawater and drinking tapwater, where the role of chloride is apparent, has been reported [124]. SSRT of a 0.25% carbon steel in 3.5% chloride solution containing sulphate ions and sulphate reducing bacteria (SRB) has revealed [125] that the embrittlement process is enhanced by the slowing down of the recombination kinetics of hydrogen atoms by SRB activity (Fig. 4.10).

4.2.6   SCC in Phosphates

The first report on SCC of steel in sodium phosphate solution dates back to 1936 [126]. Subsequently, cracking in various orthophosphates has been reported [43].

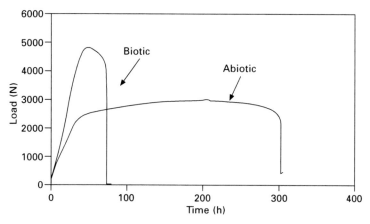

4.10 Comparison of SSRT of mild steel in different conditions: abiotic, with enhanced hydrogen intake (abiotic potentiostatic) and mixed culture biotic environments; solution used is NaCl + CH$_3$ COOH bubbled with H$_2$ [125].

Parkins *et al.* [127] performed systematic potential-controlled slow strain rate tests of the cracking of mild steel in phosphate solutions of various pHs. The potential-pH range involving Fe$_3$(PO$_4$)$_2$ and Fe$_3$O$_4$ film formation and inducing SCC was determined (Fig. 4.11). At higher negative potentials at pH 4, hydrogen-induced failures were indicated. A decrease in susceptibility with increasing temperature was reported.

4.2.7   SCC in ammonia

Intergranular stress corrosion cracking of cold-formed or welded steel tanks in which liquid or gaseous ammonia was stored under a pressure of 1.75 MN/m^2 has been reported [128]. Oxygen or the combined presence of oxygen and carbon dioxide is required for SCC [129, 130]. Cracking has also been reported to be either transgranular or mixed depending on testing conditions, and addition of more than 0.01% water stopped cracking [131]. Phelps [132] also reported inhibition of cracking in ammonia in the presence of 0.2% water, but vapour phase cracking was reported to be predominant in ammonia containing 20–300 ppm water in addition to oxygen [130].

Farrow *et al.* [133] have shown that 1–5 ppm oxygen promoted SCC of steel in anhydrous ammonia and water, hydrazine, ammonium carbonate, bicarbonate and carbamate were effective inhibitors of SCC in anhydrous ammonia. The cracking was transgranular, and passivation film rupture was the suggested mechanism of SCC, a view supported in the work of Deegan *et al.* [134] which showed the passivation to take place at a potential near +400 mV. SCC of NiCrMo low-alloy steel in liquid ammonia has been

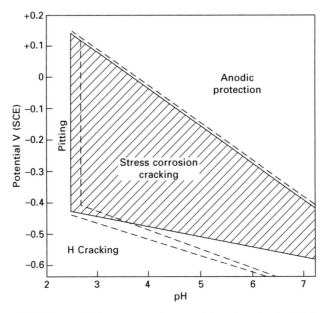

4.11 Observed stress corrosion cracking domain for C-Mn steel in phosphate solutions and domains (indicated by broken lines) calculated from equations involving $Fe_3(PO_4)_2$ and Fe_2O_3 as the filming substances in the cracking domain [127].

reported [135]. SCC has also been reported in ammonium carbonate [136] as well as in ammonium acetate [137] solutions.

4.2.8   SCC in water

Slow strain rate tests performed on carbon and low-alloy steels have shown susceptibility to SCC in high-temperature high purity water containing oxygen [138–141]. Although susceptibility appears greatest at about 200–250°C, SCC at temperatures up to 320°C has been reported [142]. Susceptibilty increased with increasing temperature. The cracking was transgranular and occurred at stress levels close to the ultimate tensile strength of the material. The addition of Ni or Cr to Fe-C-Mn steel retarded cracking, whereas Mo enhanced the susceptibility [143].

Oxygen and dynamic strain rate are considered important factors controlling crack-growth rates [144], and SCC is considered to occur through an anodic dissolution mechanism [143]. Chloride addition to high purity water reduces the time to failure [139, 145]. McDonald et al. [139] have found that small additions of chloride decreased the failure times of a 0.2%C steel in slow strain rate tests in high purity water containing 8 ppm oxygen at 250°C (Fig. 4.12).

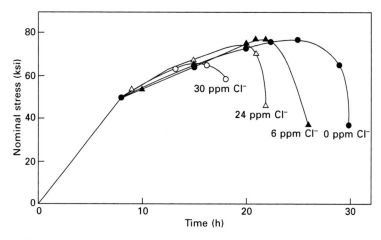

4.12 Load-time curves for a low-alloy steel (A508-C12) tested at strain rate of $10^{-6}\,sec^{-1}$ at 250°C in high purity water containing 8 ppm O_2 and various oncentrations of Cl^- [139].

Hwang *et al.* [146] have reported that in slow strain rate testing in deionized deaerated supercritical water on T91 ferritic-martensitic steel, no SCC was observed at 500, 550 and 600°C. Ferritic-martensitic steel HT-9 exhibited intergranular stress corrosion cracking when subjected to CERT tests in an environment of supercritical water at 400°C and 500°C, and also in an inert environment of argon at 500°C [147]. The susceptibility to cracking increased in going from an argon environment to the deaerated water to the 300 ppb oxygen containing water at the same temperature. Evidence of IG cracking in HT-9 steel in CERT tests in oxygenated water at 100–230°C is also reported.

Linearly increasing stress tests (LIST) conducted on as-quenched 4340 and 3.5 NiCrMoV turbine rotor steels in aerated distilled water at 30°C have shown that SCC occurred at all applied stress rates for 4340 steel, whilst only at applied rates less than or equal to 0.002 MPa/s for the turbine rotor steel [148].

4.2.9   SCC in other environments

Stress corrosion cracking of steels has been observed in methanol [149, 150]. In aerated solutions at 20°C, the critical amount of water that produces SCC is shown to be less than 0.05% to just below 1% by volume, with the highest probability at 0.2% water in methanol. SCC of steels has also been reported in ethanol in the 1980s [135, 150]. The recent surge in the use of ethanol as a fuel has triggered interest in the study of SCC of carbon steel in fuel grade ethanol [151–155]. Maldonado and Kane [156] have presented a number of industry experience and laboratory studies with SCC in fuel ethanol. Highly stressed or dynamically stressed welds and components are conducive to SCC. The cracking in ethanol appeared qualitatively less severe than in methanol. The cracking in methanolic and ethanolic environments was comparable in many ways to SCC of steel in liquid ammonia where susceptibility can be affected by minor additions of water.

Sridhar *et al.* [155] have shown corrosion potential, as influenced by oxygen, to have the greatest effect on causing SCC in fuel ethanol. The lower critical potential for SCC ranges from 25 mV (SCE) to 300 mV (SCE) depending on the presence of chloride and methanol as impurities (Fig. 4.13 and 4.14). SCC was reported to have been observed in SSRT tests with aerated ethanol-gasoline blends having ethanol concentrations of 20% or higher [153]. For mitigation of SCC in fuel grade ethanol, the use of inhibitors in existing pipelines and for new pipelines, the use of internal surface treatments that induce compressive residual stresses have been suggested [151]. SCC of NiCrMo low-alloy steel has been observed in methanol, ethanol, propanol, dimethylformaldehyde and acetonitrile [135].

Stress corrosion cracking of low-strength steels has been reported in dilute hydrogen cyanide solution [43, 157], and also in phosphorus trifluoride, hydrogen-antimony/aluminium chlorides, $MgCl_2$-NaF and hydrogen fluoride [158, 159] and in methanol [148].

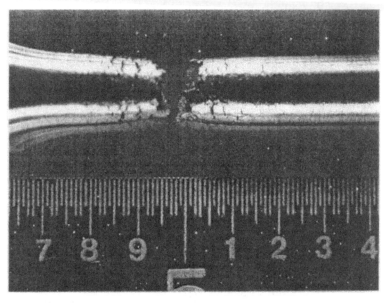

4.13 SCC produced on welded carbon steel in fuel ethanol [151].

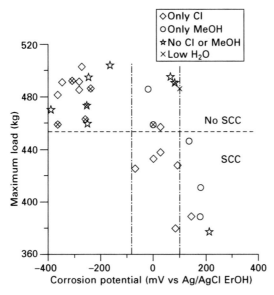

4.14 Effect of corrosion potential on SCC susceptibility as measured by maximum load to failure of notched specimens [155].

4.3   Hydrogen embrittlement-dominated stress corrosion cracking (SCC)

4.3.1   SCC in acid solutions

In non-oxidizing acids, the processes of dissolution of metals at anodic regions and absorption of hydrogen at cathodic regions take place simultaneously. Under the influence of tensile stresses each of these processes can lead to stress corrosion cracking of steel. However, cathodic processes (hydrogen embrittlement) are the main factor responsible for SCC of high strength steels in acid solution. Studies on low and medium strength steels are scarce.

The earlier works of Azhogin and co-workers [160] indicated higher susceptibility of high strength steels to dissolution-controlled SCC in HCl solutions or chloride containing H_2SO_4 solutions than in pure sulphuric acid solutions; in the latter, cracking due to hydrogen absorption was prevalent. The high susceptibility of steels to SCC in solutions containing chloride ions is attributed to the weak adsorption of chloride ions at sites where tensile stresses accumulate. For hardened and low-tempered steels, the reduction in short-time strength was considerably higher in hydrochloric acid, sulphuric acid, H_2SO_4 + NaCl and boiling NH_4NO_3 solutions than in boiling NaOH or boiling NaCl solutions.

In sulphuric acid, a wide temperature range for SCC, from room temperature to 240°C, has been reported [161]. Enhancement of susceptibility to SCC by both anodic and cathodic polarization has been reported [109, 162]. Siddiqui and Abdulla [163] carried out an investigation to assess the effect of hydrogenation time on SCC of a cathodically charged 0.31% carbon steel in a 0.4 M H_2SO_4 solution and found that a minimum of 6 h charging was necessary, apparently to attain a critical concentration of hydrogen.

In nitric acid, SCC is caused only by anodic processes, since there is no hydrogen absorption. However, inhibition of the solutions with thiourea slowed down corrosion, but considerably increased hydrogen absorption and subsequent rapid failure due to hydrogen embrittlement [164].

The SCC of rock bolts was evidenced in a number of mines in Australia [165], and the conditions (apparently acidic) leading to SCC were associated with abundant hydrogen evolution. Villalba and Atrens [166] carried out linearly increasing stress tests on commercial steels 1008, X65, X70, 4140 and 4145 H in a dilute pH 2.1 sulphate solution to simulate the mine conditions. The steel 4140 failed at −700 mV influenced by hydrogen, but there was no evidence of SCC either at their free corrosion potential or at negatively applied potential values up to −1500 mV vs Ag/AgCl.

Yamaguchi *et al.* [167] have used wedge open load (WOL) specimens to determine threshold stress intensity factor, K_{1H}, in a buffer solution of pH 3.6 for hydrogen-assisted cracking of carbon steels of various heat treatments and strength values, and have reported a critical tensile strength for HAC to be between 998 and 1396 MPa.

4.3.2   SCC in hydrogen sulphide

Corrosion is aqueous solutions of hydrogen sulphide can be represented by:

$$Fe + H_2S \rightarrow FeS + 2H$$

The recombination of atomic hydrogen is inhibited by the formation of iron sulphides, which facilitate the penetration of hydrogen into the metal and intensify embrittlement. Failures have been widely encountered in gas/oil well tubulars, well head equipment, pipelines, process piping and pressure vessels in the production, transmission and refining of oil and gas. The term 'sulphide stress cracking' is often used to describe such failures.

Stress corrosion cracking in hydrogen sulphide requires the presence of moisture [168]. Although hydrogen blistering can be formed at very low concentrations [ppm level) of hydrogen sulphide in the solution [169, 170], SCC aggravates with increased hydrogen sulphide concentrations [70, 171]. Higher ductility loss of steel with higher concentration and longer immersion time has been reported [172]. With increase in ageing time and temperature, the ductility is restored, providing evidence for a cathodic process controlled (hydrogen embrittlement) SCC. Increased temperatures have been observed to decrease the propensity for SCC [173–176], the maximum susceptibility being observed near room temperature, again indicating hydrogen embrittlement. On the other hand, Vermilyea [177, 178] has supported a film-rupture mechanism for SCC of steels in hydrogen sulphide at 289°C.

Cathodic polarization has, in general, been observed to promote SCC and anodic polarization has a retarding effect [50, 179], again supporting the hydrogen absorption mechanism. However, an accelerating effect of anodic polarization, probably caused by increased electrochemical reaction kinetics, has also been reported [179, 180].

Cracking in sulphide solutions has been shown to depend on many metallurgical factors associated with composition, heat treatment, strength, hardness, etc. For low alloy steels, susceptibility to SCC was reduced by heat treatment that eliminated the martensite [181]. Cold work showed no effect on the critical stress for cracking [175]. In welded samples, cracking usually initiated in the heat affected zone [182]. While cracking was observed in steel specimens with a hardness value of RC 34 in solutions containing a very low concentration of hydrogen sulphide [170], steels with a hardness value of less than RC 22 are generally considered to be immune. Ductility versus hydrogen content for quenched and tempered steels at various strength levels are shown in Fig. 4.15.

In aqueous solutions of hydrogen sulphide, the susceptibility of steels to SCC is intense at low pH levels, cracking usually not observed above pH 9

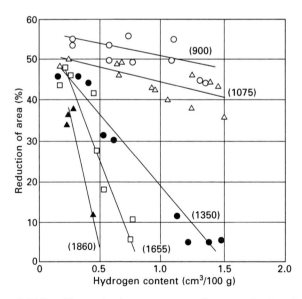

4.15 Ductility vs hydrogen content for quenched and tempered steels at various strength levels. Figures in parentheses indicate the ultimate tensile stength in MPa.

in low and medium strength steels [70, 183, 184]. Lower pH solutions have been found to decrease SCC threshold stress levels [180, 181, 185].

The cracking was observed in low strength sour crude oil pipelines in the regions of stress concentration and was termed 'stress-oriented hydrogen-induced cracking' or SOHIC [186]. The cracking was believed to have resulted from the linking together of small hydrogen fissures.

Quenched and tempered Cr-Mo steels have a significantly high susceptibility to SCC at yield strength over 830 MPa [187, 188] (Fig. 4.16). In a steel with yield stress of 753 MPa, a decrease in ductility was observed with decrease strain rate and increase of temperature in a 0.5% CH_3COOH-5%

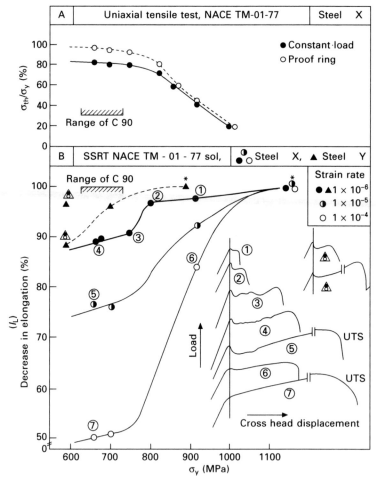

4.16 Relationship between SCC susceptibility and yield strength [181].

NaCl solution saturated with 1 atm H_2S [189]. Dvoracek [190], however, observed that quenched and tempered steels were more resistant to cracking than the normalized and tempered steels and recommended the use of steels having YS < 90 000 psi, though the presence of surface notches or flaws would be detrimental even in such steels. Solution pH, H_2S concentration, temperature and cold work had pronounced effects on SCC. Jimfeng et al. [191] have reported that the SCC sensitivity of 1Cr13 steel in H_2S solution containing CO_2 was influenced by temperature, solution pH and chloride ion concentration (in that order). A FEP-bonded MoS_2 coating has been reported [192] to be effective in inhibiting sulphide stress cracking in steels.

4.3.3   SCC in hydrogen gas

A number of moderate strength steel pressure vessel failures by high-pressure hydrogen gas have been reported [193, 194]. In general, increased pressure promote SCC by gaseous hydrogen and notched specimens rather than smooth specimens were found to be affected [195, 196]. The susceptibility is maximum at or near room temperature [197], but cracking has been reported over a wide temperature range (–46 to 112°C) [195, 198]. SCC is reduced under high strain-rate conditions [199, 200]. The dependence of crack growth rate in AISI 4340 steel on stress intensity at various hydrogen pressures at 24°C is shown in Fig. 4.17.

4.4   Conclusions

The vast literature produced since the first reporting of the occurrence of caustic cracking about a century back shows that the environments causing stress corrosion cracking of steels are not limited to only a few 'specific' ion-containing solutions. The application of slow strain rate testing and fracture mechanics has contributed to a great extent to the understanding of the phenomenon and mechanism of the process. The electrochemical conditions favourable for film formation and hydrogen entry have, in general, been identified. Addition of inhibitors has been shown in some studies to be effective in the retardation or prevention of stress corrosion cracking. Though the results have not been conclusive, the effects of alloying additions and microstructural variations are reported in some investigations.

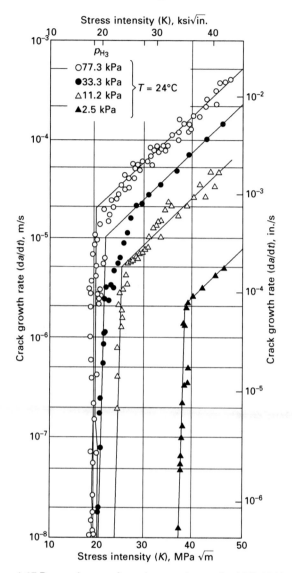

4.17 Dependence of crack growth rate in AISI 4340 steel on stress intensity at various hydrogen pressures at 24°C.

4.5   References

1. H. L. Logan, *The Stress Corrosion of Metals*, John Wiley, New York, 1966
2. L. L. Shreir (ed.), *Corrosion, Vol. 1*, 2nd edn, Newnes-Butterworths, London, 1976
3. G. V. Karpenko and I. I. Vasilenko, *Stress Corrosion Cracking of Steels*, Freund Publishing House, Tel Aviv, 1979
4. S. W. Ciaraldi, in *Stress Corrosion Cracking* (ed. R. H. Jones), pp. 41–61

5. J. A. Beavers, N. G. Thompson and R. N. Parkins, *Nucl. Chem. Waste Mgmnt.*, 5, 1985, pp. 279–296
6. B. F. Brown, *Problems with the Load Carrying Capacity of High Strength Steels*, DMIC Report 210, 1964, p. 91
7. G. E. Kearns, M. T. Wang and R.W. Staehle, in *Stress Corrosion Cracking and Hydrogen Embrittlement in Iron Base Alloys, Vol. 6*, National Association of Corrosion Engineers, 1977, pp. 700–735
8. Stomeyer Report to the Manchester Steam Users Association, *J. Iron Steel Inst.*, Vol. 1, 1909
9. J. E. Reinoehl and W. E. Berry, *Corrosion*, 28 (4), 1972, pp. 151–160
10. R. A. Yeske, Corrosion by Kraft Pulping Liquors, in *Metals Handbook*, 9th Edn, ASM International, 1987, p. 1210
11. A.A. Berk and W.F. Waldeck, *Chem. Engng.*, 57, 1950, p. 235
12. R.S. Thornhill, Centre Belge Etude et Document, CEBEDEAU, 1961, p. 213
13. *Metals Handbook*, 9th edn, Vol. 13, 1987, ASM International, Metals Park, OH, p. 332
14. F. Adcock and A.J. Cook, *J. Iron Steel Inst.*, 143, 1941, p. 117
15. H. Grafen, *Corros. Sci.*, 17, 1967, p. 177
16. D. Singbeil and D. Tromans, *J. Electrochem. Soc.*, 128, 1981, pp. 2065–2070
17. H. H. Lee and E. Ghali, *J. Appl. Electrochem.*, 22, 1992, p. 396
18. J. F. Newana, *Corros. Sci.*, 21, 1981, pp. 487–503
19. Z. Liang, L. Xiaogang, D. Cuiwei and H. Yizhong, *Mater and Design*, 30, 2009, pp. 2259–2263
20. P. Liang, C. Du, X. Li, Y. Chen and Z. Liang, *Int. J. Metals, Metallurgy and Materials*, 16, 2009, pp. 407–413
21. F. Sarioglu, *Mat. Sci. Eng A*, 315, 2001, pp. 98–102
22. H. Lukito and Z, Szklarska-Smialowska, *Corros. Sci.*, 39, 1997, pp. 2151–2169
23. R. N. Parkins, *Corros. Sci.*, 20, 1980, p. 147
24. R. Sriram and D. Tromans, *Corros. Sci.*, 25, 1985, p. 79
25. H. Grafen, *Werkst. Korros.*, 20, 1969, p. 305
26. K. Bohnenkamp, in *Stress Corrosion Cracking and Hydrogen Embrittlement of Iron Base Alloys* (ed. R. W. Staehle, J. Hochmann, R. D. McCright and J. E. Slater), NACE, Houston, 1977, p. 374
27. H. Mazille and H. H. Uhlig, *Corrosion*, 28, 1972, p. 427
28. R. N. Parkins, W. K. Blanchard and B. S. Delanty, *Corrosion*, 50, 1994, p. 394
29. J. Flis and M. Ziomek-Moroz, *Corros. Sci.*, 50, 2008, pp. 1726–1733
30. R. K. Singh Raman, *Mat. Sci. Eng. A*, 441, 2006, pp. 342–348
31. D. Singbeil and D. Tromans, *Met. Trans. A*, 13A, 1982, p. 1091
32. D. Singbeil and D. Tromans, *J. Electrochem. Soc.*, 129, 1982, p. 2669
33. R. Sriram and D. Tromans, *Corrosion*, 41, 1985, p. 381
34. E. P. Partridge, C. E. Kaufmann and R. E. Hill, *Trans. Am. Soc. Mech. Engns.*, 65, 1942, p. 417
35. H. J. Engell and A. Baumel, in *Physical Metallurgy of Stress Corrosion Fracture* (ed. T. N. Rhodin), 1959, p. 341
36. W. Radekar and H. Grafen, *Stahl u Eisen*, 76, 1956, p. 1716
37. M. J. Humphries and R. N. Parkins, *Corros. Sci.*, 7, 1967, p. 761
38. H. Grafen, *Werkst. Korros.*, 20, 1969, p. 305
39. F. P. A. Robinson and L. G. Nel, *Proc. 2nd Int. Congr. Metallic Corrosion*, NACE, New York, 1966, p. 172
40. A. A. Berk and W. F. Waldeck, *Chem. Engng.*, 57, 1950, p. 235

41. J. A. Jones, *Trans. Faraday Soc.*, 17, 1921, p. 102
42. W. Shearn and W. Dunwoody, *Ind. Eng. Chem.*, 45, 1953, p. 496
43. G. L. Schwartz and M. M. Kristal, *Corrosion of Chemical Apparatus*, Consultants Bureau Inc., New York, 1959, p. 87
44. E. Herzog, *Corrosion Anticorr.*, 2, 1954, p. 3
45. U. Nurnberger, *Werkst. Korros.*, 28, 1977, p. 312
46. R. N. Parkins and R. Usher, *Proc. 1st Int. Congr. Metallic Corrosion*, London, 1961, p. 289
47. G. T. Spare, *Wire and Wire Prod.*, 29, 1954, p. 1421
48. W. D. Everling, *Wire and Wire Prod.*, 30, 1955, p. 316
49. J. A. Donovan, *Corrosion/77*, Paper No. 84, NACE, Houston, 1977
50. J. D. Gilchrist and R. Narayan, *Corros. Sci.*, 11, 1971, p. 281
51. H. L. Logan, *The Stress Corrosion of Metals*, John Wiley, New York, p. 43
52. M. Smialowski and S. Szklarska-Smialowska, *Corrosion*, 18, 1962, p. 1
53. S. Szklarska-Smialowska, *2nd Int. Congr. Metallic Corrosion*, New York, 1963, p. 205
54. A. Baumell and H. G. Engell, *Arch. Eisenhutt.*, 32, 1961, p. 379
55. F. P. Ford, in *Corrosion Processes* (ed. R. N. Parkins), Applied Science Publishers, 1982
56. R. N. Parkins, *J. Iron Steel Inst.*, 172, 1953, p. 149
57. M. McIlree and H. Michels, *Cor. Res. Conf.*, Chicago, 1974
58. R. Munster and H. Grafen, *Arch. Eisenhutt.*, 36, 1965, p. 277
59. W. Radekar and B. Mishra, *Werkst. Korros.*, 17, 1966, p. 193
60. E. de Garmo and I. Cornet, *Welding J.*, 34, 1955, p. 472
61. C. McKinsey, *Welding J.*, 33, 1954, p. 161
62. L. Long and N. Lockington, *Corros. Sci.*, 7, 1967, p. 447
63. G. Pocheptsova and S. Timchenko, *Khim. Neft. Machin.*, 10, 1966
64. R. D. McDonald, *Corrosion*, 24, 1968, pp. 214–217
65. U. K. Chatterjee, S. Bhattacharya and S. C. Sircar, *Proc. 10th Int. Congr. Metallic Corrosion*, Madras, Oxford-IBH Publishing, New Delhi, 1987, pp. 2121–2132
66. H. H. Uhlig and J. Sava, *Trans. ASM*, 56, 1963, p. 361
67. Final Staff Report on Investigation of Tennessee Gas Transmission Company Pipeline, No. 100-1, Bureau of Natural Gas, Washington DC, 1965
68. R. R. Fessler, *Oil Gas*, 74, 1976, p. 81
69. G. J. Ogundele and R. D. Venter, *Corrosion/89*, Paper No. 572, NACE, 1989
70. C. M. Hudgins, R. L. McGlassan, P. Mehdizadeh and W. M. Rosborough, *Corrosion*, 1966, p. 238
71. J. H. Kmetz and D. J. Traux, *Corrosion/90*, Paper No. 206, NACE, 1990
72. R. N. Parkins, in *Theory of Stress Corrosion Cracking in Alloys* (ed. J. C. Scully), NATO, 1971, p. 167
73. R. N. Parkins and S. Zhou, *Corros. Sci.*, 39, 1997, pp. 159–173
74. J. A. Might and D. J. Duquette, Environmental considerations in the SCC of mildsteel in carbonate solutions, *Critical Issues in Reducing the Corrosion of Steels*, NACE, TX, 1985, p. 3
75. R. N. Parkins, in *Fifth Symposium on Line Pipe Research*, Houston, 1974
76. J. M. Sutcliffe, R. R. Fessler, W. K. Boyd and R. N. Parkins, *Corrosion*, 28, 1972, p. 313
77. M. C. Alonso and M. C. Andrade, *Corros. Sci.*, 29, 1989, pp. 1129–1139
78. W. E. Berry, T. J. Barlo, J. H. Payer and R. R. Fessler, *Corrosion/78*, Paper No. 64, NACE, 1978

79. S. Sh. Abedi, A. Abdolmaleki and N. Abidi, *Eng. Failure Analysis*, 14, 2007, pp. 250–261
80. Y. F. Cheng, *Int. J. Hydrogen Energy*, 32, 2007, pp. 1269–1276
81. M. C. Li and Y. F. Cheng, *Electrochim. Acta*, 52, 2007, pp. 8111–8117
82. X. Liu and X. Mao, *Scripta Metal. et Materialia*, 33, 1995, pp. 145–150
83. B. W. Pan, X. Peng, W. Y. Chu, Y. J. Su and L. J. Qiao, *Mat. Sci. Eng. A*, 434, 2006, pp. 76–81
84. I. Cerny and V. Linhart, *Eng. Fracture Mech.*, 71, 2004, pp. 913–921
85. P. G. Fazzini and J. L. Otegui, *Int. J. Pressure Vessels and Piping*, 84, 2007, pp. 739–748
86. X. Tang and Y. F. Cheng, *Electrochim. Acta*, 54, 2009, pp. 1499–1505
87. Z. Y. Liu, X. G. Li, C. W. Du, G. L. Zhai and Y. F. Cheng, *Corros. Sci.*, 50, 2008, pp. 2251–2257.
88. A. Torres-Islas, V. M. Salinas-Bravo, J. L. Albarran and J. G. Gonzalez-Rodriguez, *Int. J. Hydrogen Energy*, 30, 2005, pp. 1317–1322
89. A. Torres-Islas, J. G. Gonzalez-Rodriguez, J. Uruchurtu and S. Serna, *Corros. Sci.*, 50, 2008, pp. 2831–2839
90. A. Atrens, J. Q. Wang, K. Stiller and H. O. Andren, *Corros. Sci.*, 48, 2006, pp. 79–92
91. J. Q. Wang and A. Atrens, *Corros. Sci.*, 45, 2003, pp. 2199–2217
92. F. M. Song, *Corros. Sci.*, 51, 2009, pp. 2657–2674
93. M. Kowaka and S. Nagata, *Corrosion*, 24, 1968, p. 427
94. P. J. Ras, *Nat. Gas*, 4, 1964, p. 81
95. A. Brown, J. T. Harrison and R. Wilkins, *Corros. Sci.*, 10, 1970, pp 547–548
96. M. Kowaka and S. Nagata, *Corrosion*, 32, 1976, p. 395
97. E. Sato and T. Murata, *Corrosion*, 44, 1988, pp. 770–775
98. T. Fuga, S. Osuga, A. Murao and T. Takeda, *Corros. Prev. Control*, 33, 1986, pp. 120–124
99. H. L. Logan and J. Wehrung, *Corrosion*, 22, 1966, p. 265
100. R. Kripyakevich, Yu Babei, A. Litvin and B. Kachmer, in *Vliyanie Rabochikh Sred na Svoistva Mater.*, No. 3, Naukova Dumka, Kiev 1964
101. J. Truman, R. Perry and G. Chapman, *J. Iron Steel Inst.*, 202, 1964, p. 745
102. S. Yamamoto and T. Futija, in *Fracture 1969* (ed. P. L. Pratt), Chapman and Hall, 1969, p. 424
103. S. Fukui and A. Asada, *Trans. Iron Steel Inst. Japan*, 9, 1969, p. 448
104. G. Sandoz, in *Stress Corrosion Cracking in High Strength Steels and in Titanium and Aluminium Alloys* (ed. B. F. Brown), Naval Research Laboratory, 1969
105. M. J. May and A. H. Priest, in *The Influence of Solution pH on Stress Corrosion Cracking Resistance of a Low Alloy High Strength Steel*, MG/A/45/68, BISRA, British Steel Corp., 1968
106. H. H. Johnston and A. M. Willner, *Appl. Mater. Res.*, 4, 1965, p. 33
107. J. B. Greer, E. L. Va Rasenberg and J. Martinez, *Corrosion*, 28, 1972, pp. 378–384
108. M. B. Strauss and M. C. Bloom, *Corrosion*, 16, 1960, p. 109
109. C. G. Interrante, *Weld. Res. Coun. Bull*, 145, 1969
110. A. Tirman, E. G. Haney and P. Fugassi, *Corrosion*, 25, 1969, p. 342
111. W. McCleod and R. R. Rodgers, *Corrosion*, 22, 1966, p. 143
112. C. G. Hancock and H. H. Johnson, *Trans. Met. Soc. AIME*, 236, 1966, p. 513
113. C. F. Barth and A. P. Troiano, *Corrosion*, 28, 1972, p. 259
114. R. J. Scmitt and E. H. Phelps, *J. Met.*, 1970, p. 47

115. NACE Unit Committee T-1H, *Mater. Protect.*, 6, 1967, p. 65

116. B. F. Brown, in *Theory of Stress Corrosion Cracking in Alloys* (ed. J. C. Scully), NATO, Brussels, 1971

117. C. F. Barth, E. A. Steigenwald and A. R. Troiano, *Corrosion*, 25, 1969, pp. 353–358

118. K. Banerjee and U. K. Chatterjee, *ISIJ Int*, 39, 1999, pp. 47–55, and *Mater. Sci. Technol.*, 16, 2000, pp. 517–523

119. C. T. Fujii, *Metal. Trans A*, 12, 1981, pp. 1099–1105.

120. F. Salona, T. Takamadate, I. M. Bernstein and A. W. Thompson, *Metal. Trans. A*, 18, 1987, pp. 1023–1028

121. K. Banerjee and U. K. Chatterjee, *Metal. Trans. A*, 34A, 2003, pp. 1297–1309

122. F. Zucchi, G. Trabanelli, C. Monticelli and V. Grassi, *Corros. Sci.*, 42, 2001, pp. 505–515

123. T. L. Prakash and A. U. Malik, *Desalination*, 123, 1999, pp. 215–221

124. A. Alwai, A. Ragab and M. Saban, *Eng. Fracture Mech.*, 22, 1989, pp. 29–37

125. R. K. Singh Raman, R. Javaherdashti, C. Panter and E.V. Pereloma, *Mat. Sci. Technol.*, 21 (2005) p. 1094

126. W. C. Shroeder and E. P. Partridge, *Trans. ASME*, 58, 1936, p. 223

127. R. N. Parkins, N. J. H. Holroyd and R. R. Fessler, *Corrosion*, 34, 1978, p. 253

128. T. J. Dawson, *Weld J.*, 35, 1956, p. 568

129. J. M. Sutcliffe, R. R. Fesseler, W. K. Boyd and R. N. Parkins, *Corrosion*, 28, 1972, pp. 313-320

130. L. Lunde, *Ammonia Plant Safety & Related Facilities*, 24, 1984, pp. 154–166

131. L. Long and N. Lockington, *Corros. Sci.*, 7, 1967, p. 447

132. E. H. Phelps, *Ammonia Plant Safety & Related Facilities*, 16, 1974, pp. 32–38

133. K. Farrow, J. Hutchings and G. Sanderson, *Corros. Sci.*, 25, 1985, pp. 395–414

134. D. C. Deegan, B. E. Wilde and R. W. Staehle, *Corros. Sci.*, 32, 1976, pp. 139–143

135. C. A. Farina and U. Gressini, *Electrochim. Acta*, 32, 1987, pp. 977–980

136. H. Hixon and H. H. Uhlig, *Corrosion*, 32, 1976, p. 56

137. J. Parker and W. Pearce, *Corrosion*, 30, 1974, p. 18

138. S. P. Pednekar, T. Mizuno, S. Szklarska-Smialowska and D. D. McDonald, *Corrosion/82*, Paper No. 244, NACE, 1982

139. D. D. McDonald, S. Szklarska-Smialowska and S. P. Pednekar, *EPRI NP 2853*, 1983

140. P. M. Scott, *AERE Report 12783*, 1987

141. H. Hanninen *et al.*, *CSNI Report 141*, 1987

142. M. O. Speidel and R. M. Magdowski, *Corrosion/88*, Paper No. 283, NACE, 1988

143. J. Clogleton and R. N. Parkins, *Corrosion*, 44, 1988, pp. 290–298

144. P. M. Scott and D. R. Rice, *Nucl. Eng. Des.*, 119, 1990, p. 399

145. M. B. Strauss and M. C. Bloom, *Corrosion*, 16, 1960, p. 109

146. S. S. Hwang, B. H. Lee, J. G. Kim and J. Jang, *J. Nucl. Mater.*, 372, 2008, pp. 177–181

147. G. Gupta, P. Ampornrat, X. Ren, K. Sridharan, T. R. Allen and G. S. Was, *J. Nucl. Mater.*, 361, 2007, pp. 160–173

148. K. Matsukura and K. Sato, *J. Japan Iron Steel Inst.*, 63, 1977, p. 1016

149. K. Matsukura and K. Sato, *Trans. ISIJ*, 18, 1978, pp. 554–563

150. G. Capobianco, G. Farina, G. Faita and C. A. Farina, *9th Int. Congr. Metallic Corrosion*, Ottawa, 1984, pp. 532–537

151. J. Beavers, F. Giu and N. Sridhar, *Corrosion/2010*, Paper No. 10072, NACE, 2010

152. F. Gui, N. Sridhar, E. Trillo and M. Sing, *Corrosion/2010*, Paper No. 10075, NACE, 2010

153. J. Beavers, N. Sridhar and C. Zamarin, *Corrosion/2010*, Paper No. 09532, NACE, 2010

154. J. G. Moldonado and N. Sridhar, *Corrosion/2007*, Paper No. 07574, NACE, 2007

155. N. Sridhar, K. Price, J. Buckingham and J. Dante, *Corrosion*, 62, 2006, pp. 687–702

156. J. G. Maldonado and R. D. Kane, in *Environment-induced Cracking of Materials, Vol 2*, (eds. S. A. Shipilov *et al.*), Elsevier, 2008, pp. 337–347

157. H. Buckholtz and R. Pusch, *Stahl Eisen*, 62, 1942, p. 445

158. R. S. Treseder and A. Wachter, *Corrosion*, 5, 1949, p. 383

159. R. L. Schuyler, *Mater. Perform.*, 8, 1979, p. 9

160. F. Azhogin, Zash. Metallov, 1966; Korroziya i Zash. Metallov, 1966, *Fiz-Khim. Mech. Mater.*, 1967

161. H. L. Logan, in *Physical Metallurgy of Stress Corrosion Fracture* (ed. T. N. Rhodin), Interscience, 1959, p. 295

162. H. T. Effinger *et al.*, *Oil Gas J.*, 2, 1951, p. 99

163. R. A. Siddiqui and H. A. Abdulla, *J. Mater. Process. Tech.*, 170, 2005, pp. 430–435

164. S. Balezin, in *Ingibitory Korrozii*, *Vol. 2*, Profizdat, Moscow, 1957

165. A. Crosky, B. Smith and B. Hebblewhite, *Prac. Failure Anal.*, 14, 2007, pp. 1351–1393

166. E. Villalba and A. Atrens, *Eng. Failure Anal.*, 16, 2009, pp. 164–175

167. Y. Yamaguchi, H. Nonaka and K. Yamakawa, *Corrosion*, 53, 197, pp. 147–155.

168. H. Kihara *et al.*, *Proc. World Petroleum Congr.*, Elsevier, 1967

169. C. Edward and W. Wright, *Corrosion*, 18, 1962, p. 119

170. C. M. Hudgins, R. McGlasson and W. Rosborough, *Proc. 2nd Int. Congr. Metallic Corrosion*, New York, 1963, p. 364

171. H. Ishizuki and K. Onishi, *J. Japan Inst. Met.*, 30, 1966, p. 846

172. G. V. Karpenko and I. I. Vasilenko, Stress Cracking of Steels, Freund Publishing, Tel Aviv, 1979, p. 69

173. J. P. Frazer and R. S. Treseder, *Corrosion*, 8, 1952, p. 342

174. H. Ishizuki and K. Onishi, *Japan Chem. Qrly.*, 3, 1967, p. 30

175. L. M. Dvorecek, *Corrosion*, 26, 1970, p. 1977

176. H. E. Townsend, *Corrosion*, 28, 1972, p. 39

177. D. Vermilyea, *Corrosion*, 29, 1973, p. 442

178. D. Vermilyea, *J. Electrochem. Soc.*, 119, 1972, p. 405

179. I. Class, *Proc. 2nd Int. Congr. Metallic Corrosion*, New York, 1963, p. 342

180. G. B. Kohut and W. J. McGuire, *Mater. Protect.*, 7, 1968, p. 17

181. E. Shape, F. Schaller and R. Farbes-Jones, *Corrosion*, 25, 1969, p. 381

182. C. Biefer, *Corrosion*, 32, 1976, p. 378

183. R. N. Tuttle, *Mater. Protect.*, 9, 1970, p. 11

184. D. R. Johnston and T. G. McCord, *Proc. NACE 26th Conf.*, 1970

185. R. S. Treseder and T. M. Swanson, *Corrosion*, 24, 1968, p. 31

186. R. D. Merrick, *Mater. Perform.*, 28, 1989, pp. 53–55

187. A. Ikeda, S. Nagata, T. Tsumura, Y. Nara and M. Kwaka, API Paper No. SS-5:1, *Symp. On Line Pipe and Tubular Goods*, Florida, 1977

188. J. A. Straatmann, P. J. Grobner and D. L. Sponseller, ASME Preprint 77-Pet-48, *Enery Tech. Conf.*, *ASME*, Houston, 1977

189. A. Ikeda, T. Kaneko and T. Moroishi, The dependence of the sulphide stress cracking of low alloyed steels on environmental factors and materials, *Critical Issues in Reducing the Corrosion of Steels*, NACE, TX, p. 51

190. L. M. Dvoracek, *Corrosion*, 26, 1970, pp. 177–188

191. P. Jimfeng, Z. Guixin, Y. Shunshi and D. Xuefeng, *Adv. Mater. Res.*, 79–28, 2009, pp. 1005–1008

192. T. Sriskandarajah, H. C. Chan and A. C. Tseung, *Corros. Sci.*, 25, 1985, pp 395–414

193. W. T. Chandler and R. T. Walter, *Effects of High Pressure Hydrogen on Steels*, paper presented to ASM and AWS, American Society for Metals, 1967

194. J. S. Laws *et al.*, *NASA CR-1305*, 1969

195. W. Haufman and W. Rouls, *Weld J.*, 44, 1965, p. 225s

196. R. T. Walter and W. T. Chandler, *Mater. Sci. Eng*, 8, 1971, p. 90

197. D. P. Williams and H. G. Nelson, *Metall. Trans*, 1, 1970, p. 63

198. R. H. Cavet and H. C. Van Ness, *Weld J.*, 42, 1963, p. 3165

199. H. G. Nelson, D. P. Williams and A. S. Tetelman, *Metall. Trans*, 2, 1971, pp. 953–959

200. G. Sandoz, *Metall. Trans*, 3, 1972, p. 1169

5

Stress corrosion cracking (SCC) in stainless steels

V. KAIN, Bhabha Atomic Research Centre, India

Abstract: The five categories of stainless steels and key features essential for stress corrosion cracking are introduced. The most damaging forms of stress corrosion cracking in austenitic stainless steels are chloride and caustic stress corrosion cracking and cracking in high temperature, high pressure aqueous environments. The influence of chemical composition, stacking fault energy, microstructure and environmental variables on stress corrosion cracking are summarized. The developments related to grain boundary engineering of austenitic stainless steels and the beneficial role of a high fraction of low energy grain boundaries against sensitization and sensitization induced stress corrosion cracking is described. Various fabrication parameters including cold/warm working, welding and residual stresses and surface finish on stress corrosion cracking are shown to influence stress corrosion cracking. Mitigation measures against stress corrosion cracking are briefly described.

Key words: stainless steels, stress corrosion cracking, chloride, sulfide, caustic, ambient temperature, stacking fault, grain boundary nature, thermomechanical processing, residual stress, surface finish, mitigation measures.

5.1   Introduction to stainless steels

Stainless steels are alloys of iron and carbon containing more than 10.5 wt% chromium. Stainless steels may contain other alloying elements. These alloys do not rust or 'stain' in unpolluted atmospheres (D A Jones 1992, Sedriks 1996, Grubb *et al*. 2005) hence the name 'stainless'. The 'stainless' property comes from the chromium oxide (Cr_2O_3) film that instantaneously forms over the surfaces. This film protects stainless steel from rusting/corrosion due to atmospheric exposure and is commonly referred to as 'passive' film. This film forms on exposure to air and is a thin (25–50 Å), impervious and self-healing film. The passive film formed over type 304 stainless steel, after exposure to 20% nitric acid at 50°C for 30 minutes, is shown to be only 2.4 nm thick (Hannani *et al*. 1997). On damage (e.g., by scratching), the film repairs itself (repassivates) and continues to protect the stainless steel (Sedriks 1986, 1996, D A Jones 1992, Lo *et al*. 2009).

5.1.1   Role of alloying additions

While about a minimum of 11 wt% chromium addition facilitates formation of a stable passive film that spontaneously forms on its surfaces and protects it against general corrosion, higher amounts of Cr strengthen the passive film and its ease of self-repair (self-healing). Cr is known as a strong ferrite phase former and it suppresses the ferrite–austenite phase transformation. In the phase diagram for low carbon alloys (C up to 0.10%), steels with over 13% Cr remain in the ferrite region regardless of the temperature to which these are heated. Straight chromium stainless steels fall into two groups. The first group does not become austenitic on heating and consequently cannot be hardened on cooling. Others transform to austenite on heating and can be hardened on cooling (due to phase transformation to martensite). The Fe–Cr alloys are especially prone to transformation to sigma phase, especially when heated between 500 and 900°C. The intermetallic sigma phase is rich in Cr (but may develop Cr depletion around it), is hard and brittle and may also lower the corrosion resistance of the alloy. It typically has low toughness. Nickel has a strong favourable effect on the rate of formation of the passive film. When added in sufficient quantity, it stabilizes the austenite phase. It imparts toughness, ductility and ease of weldability. Molybdenum acts to stabilize the passive film and improves the resistance to localized corrosion like pitting, crevice corrosion and stress corrosion cracking. Molybdenum imparts high temperature strength but increases the probability of formation of deleterious sigma phase. Carbon additions are known to increase strength and hardness and it used to be considered as an unavoidable impurity in stainless steels. With the advent of modern steel making and refining practices, lowering the carbon content has become a common practice.

The main problem with high carbon stainless steels is sensitization during exposure to a temperature range of 450–850°C. This exposure may occur during cooling in the welding process, during heat treatments and fabrication steps and also during use at high operating temperatures. During exposure to this temperature range, chromium-rich carbides (e.g. $Cr_{23}C_6$) precipitate at grain boundaries with its concomitant depletion of Cr around the precipitates. Once the Cr content in the depletion regions is below typically 11–12%, the depleted region does not form a strong passive film and this film is liable to be broken during use in an aggressive environment (e.g. by chloride ions, strong acids) and cause corrosion damage. Sensitization is also the basic cause of inter-granular stress corrosion cracking. As welding of austenitic stainless steels is a common practice, low carbon stainless steels (e.g. type 304L) containing carbon levels typically < 0.03% are used to avoid sensitization related problems in the heat affected zones (HAZ) of these stainless steels. However, it is not only the control of carbon, but a combination of carbon, chromium and nickel that needs to be controlled (Kain *et al.* 1995) to avoid sensitization and intergranular corrosion (IGC).

Carbon also has a deleterious effect on toughness, especially in the ferritic grades. Except for the martensitic grades, it is desirable to reduce carbon content in stainless steels. Though addition of nitrogen cannot be avoided during melting in air, it is sought to be deliberately added to some grades to increase strength, hardness and corrosion resistance. Nitrogen is an austenite stabilizer and the low carbon high nitrogen (LN) austenitic grades have become popular due to a good combination of high strength and high corrosion resistance. Other elements may be used in stainless steels for specific purposes. For example, to improve machinability, sulfur and selenium are added, to improve oxidation resistance, aluminum and silicon are added, and to improve mechanical or processing characteristics, vanadium, zirconium, boron and rare earths are added. Copper is added to improve corrosion resistance in sulfuric acid. Manganese is added as a partial substitute for nickel in some austenitic stainless steels (in AISI 200 series).

5.1.2   Classification of stainless steels

As the stabilizing effect of each element for a given phase is known, it is possible to make use of simple diagrams like the Schaeffler diagram to predict which phase is the stable phase for a given composition of stainless steel (Pickering 1976, Schneider 1960). The Schaeffler diagram (see Fig. 5.1) is prepared by establishing the stable phase (at room temperature) for

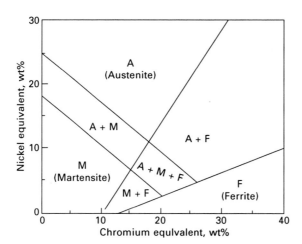

5.1 Schaeffler diagram showing stable phases for a given alloy composition after heating to 1050°C for 30 minutes and quenching in water. Here % Ni equivalent = %Ni + %Co + 30(%C) + 25(%N) + 0.5(% Mn) + 0.3(%Cu) and %Cr equivalent = %Cr + 2%Si + 1.5(%Mo) + 5(%V) + 5.5(%Al) + 1.75(%Nb) + 1.5(%Ti) + 0.75 (%W). From Pickering 1976 and Schneider 1960.

different compositions of stainless steels after heating an alloy to 1050°C for 30 minutes and then quenching it in water.

Figure 5.1 shows that four main categories of stainless steels (austenitic, ferritic, martensitic and duplex) can be produced by varying the chemical composition. In addition a fifth category – precipitation hardenable stainless steels – is also a popular category.

5.2   Introduction to stress corrosion cracking (SCC) of stainless steels

Stress corrosion cracking (SCC) is the brittle failure at relatively low constant tensile stress of an alloy exposed to an environment. A synergistic action of corrosive environment and tensile stress on the material is required to cause SCC. Figure 5.2 shows the three factors required to cause SCC.

When stainless steels are in contact with aqueous environment, the passive film consists of chromium oxide that is hydrated (Sedriks 1986). The uniform corrosion rates of stainless steels remain very low due to the protective nature of this passive film. However, the passive film may be broken locally at select locations due to (a) the weakening of the passive film at heterogeneities in the material, e.g. inclusions, precipitates, phases or grain boundaries, or (b) the segregation of aggressive species in the environment at certain locations on the material surface, e.g. accumulation of chloride

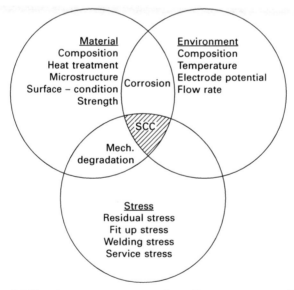

5.2 Simultaneous presence of tensile stress, susceptible metallurgical condition and critical corrosive solution required for stress corrosion cracking (adapted from Speidel 1984).

ions at surface roughness. This local breakage of the passive film leads to different forms of localized corrosion, e.g. pitting corrosion and SCC. Due to differences in chemical composition of the stainless steels, the passive film composition also changes for different types of stainless steels. However, chromium remains the dominant constituent of the passive film providing protectiveness of the surface film.

The service conditions over a period of time may produce conditions conducive for causing SCC. Therefore, a view is now emerging (Andresen *et al.* 2001) that a specific material-environment combination is not necessary for SCC. The tensile stress required for SCC can be below the macroscopic yield stress. However, residual stresses due to welding/thermal stresses, fit up stresses, etc., that are lower than the yield stress can result in SCC provided that, locally at defects in materials, the resultant stress exceeds the yield strength of the alloy. The SCC can be intergranular stress corrosion cracking (IGSCC), transgranular stress corrosion cracking (TGSCC) or be exhibited in a mixed mode of cracking. There have been a number of reviews (Logan 1966, Hanninen 1979, Scott 1985, R H Jones 1992, Parkins 1979, 1992, Jones 2003) covering stress corrosion cracking of stainless steels. There have been significant developments in the field of stress corrosion cracking of stainless steels in the last two decades, especially the understanding of grain boundary engineering and working (fabrication) affecting stress corrosion cracking. This review brings together the knowledge and understanding available so far on stress corrosion cracking of stainless steels. However, it is not aimed to cover the mechanisms of stress corrosion cracking and these are covered elsewhere (Sridhar 2011).

5.3   Environments causing stress corrosion cracking (SCC)

Certain alloys are known to undergo SCC in specific environments, e.g. steels in hot caustics and brass in ammoniacal environments. Austenitic stainless steels are known to undergo SCC (Fontana 1987, R H Jones 1992, Sedriks 1996) in hot concentrated chloride solutions, chloride contaminated steam, oxidizing high temperature high purity water, hot caustics, polythionic acid and sulfide environments. Stress corrosion cracking of austenitic stainless steels in ambient temperature atmospheric environment is now an accepted fact and a subject of many research projects. Salient features of SCC in these known environments are detailed below.

5.3.1   SCC induced by halides

The most common environmental species that is known to cause SCC of austenitic stainless steels are the halides. Chloride ions are the most potent ions that cause SCC. Chloride induced SCC is known to be typically TGSCC and exhibits branching. Bromide and iodide are heavier ions, hence not well known for SCC. However, cold worked type 304 and 316 stainless steels tested at high applied load of 426 MPa in deaerated 55% lithium bromide at 120 and 140°C showed SCC in 388 h tests at pH values of 6 and 8 (Itzhak and Elias 1994) but at a pH of 11.6 better passivation prevailed and hence there was lesser susceptibility to SCC. Fluoride ions are smaller in size than chloride ions but tend to form strong metal fluoride bonds tending to cause more of a heavy uniform dissolution than localized attack. However, clear stress corrosion cracking of sensitized austenitic stainless steels has been reported in the presence of fluoride (Zucchi *et al.* 1988). The minimum concentration of fluoride that produces cracking is 5×10^{-5} M up to temperatures of 50°C and 1×10^{-5} M at 80°C. Fluoride aggressivity was shown to be greater than that of chloride at all temperatures with the differences clearly visible at room temperature. The fluoride induced SCC of sensitized type 304 stainless steel was shown to be controlled by an electrochemical process (anodic dissolution of Cr depleted regions at grain boundaries).

Typically SS is not known to undergo SCC at low temperatures (i.e. below 60°C) in chloride solutions. An exhaustive study was carried out by Truman (1977) on the effect of chloride ion concentration and temperature of exposure on pitting and SCC behaviour, and a summary of the results is shown in Fig. 5.3 (Sedriks 1996, after Truman 1977). This study included annealed, heat treated, welded and bent stainless steel samples for 13 500 h in solutions with varying chloride ion concentration, pH and temperature. This study showed that at pH 7, the SCC of austenitic stainless steels did not take place until the chloride concentration increased beyond 10^5 parts per million (ppm). At pH ~ 2, it took only 10^4 ppm chloride ions to cause SCC. Even pitting corrosion took place at chloride concentrations above 500 ppm at 60°C in water with pH ~ 7. Using a fracture mechanics approach (measuring SCC growth rate at different temperatures in 22% sodium chloride solution), it has been shown (Speidel 1981) that stress corrosion cracks do grow at even 50°C in the case of type 304 (solution annealed condition) and even at lower (ambient) temperatures for type 304 (sensitized condition), although the rate of growth was very low (of the order of 10^{-11} m/s).

Boiling solutions of chloride have long been used to test the susceptibility to SCC. Magnesium chloride is known to be a particularly aggressive solution that produces SCC for stainless steels. The boiling point of magnesium chloride changes with the concentration of the salt. A standard procedure (G36 ASTM, 2006) is to test in 45% solution of magnesium chloride that boils at 155°C. The magnesium and calcium chlorides hydrolyse to produce acidic solutions and therefore are more aggressive than sodium chloride in producing SCC of stainless steels. The pH of the solution at the tip of a crack in type 304 stainless steel undergoing SCC in boiling magnesium chloride test is shown

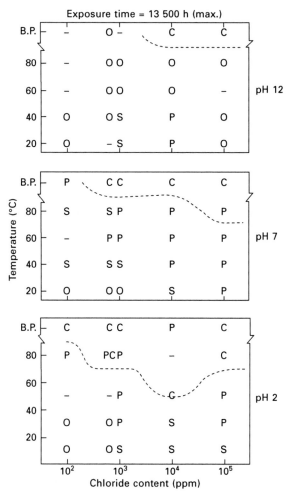

5.3 The effect of temperature, pH and chloride concentration in producing SCC of type 304 stainless steel in sodium chloride solutions (from Sedriks 1996, after Truman 1977). C: SCC, P: pits, S: stains and O: no effect.

(Baker *et al.* 1970) to be between 1.2 and 2.0. Long-term boiling leads to reduction in acidity of the solution and it is known (Maier *et al.* 1985) to reduce the aggressiveness (by making it more alkaline) of the solution to cause SCC. It is also reported (Thomas *et al.* 1964) that with lowering of pH (by acid addition) in magnesium chloride solution, cracking propensity increases while there is no change in cracking propensity when pH is lowered for calcium chloride solution. The boiling magnesium chloride solutions are known to be the most aggressive chloride solution to cause SCC, and austenitic stainless steels are known to undergo SCC in this solution even when only residual stresses are present (i.e. without application of external stress) in a few hours. Calcium chloride solution that boils at high temperature (155°C) and also forms acidic solution upon dissolution is also known to cause severe SCC (Truman 1971). Sodium chloride solution (22%) acidified to a pH of 1.5 is used in other standard test procedures (G123 ASTM, 2005) to evaluate the susceptibility to SCC.

The crack growth rates in boiling 22% chloride solution have been measured and for type 304/304L stainless steels in sensitized condition were of the order of 10^{-8} m/s (Speidel 1981). These were three orders of magnitude higher than that measured at ambient temperatures. Lowering the chloride concentration leads to a very high resistance to SCC. At a boiling temperature of 100°C, type 304 stainless steel showed extensive cracking (in a few days) in 1500 ppm chloride solution compared to over 6 months that it took to obtain the same extent of cracking in a 10 ppm chloride solution (Warren 1960). At 250°C in an autoclave test, type 347 stainless steel was shown to crack readily in a few hours in 3% chloride solution but it took over 1000 hours to crack it when the chloride concentration was reduced to 0.01% (Truman 1969). This reinforces the view that the laboratory tests carried out to obtain SCC in a short testing time (by accelerating the test either by increasing the aggressiveness of the solution, temperature or stress) are difficult to correlate to the long-term behaviour in field applications. However, a relative ranking of the resistance to SCC of different alloys is readily obtained in laboratory tests.

An exhaustive investigation (Edeleanu 1953) established the corrosion behaviour of low carbon 18Cr-10Ni stainless steel (type 304L) in various (boiling) chloride solutions and listed time to failures in each case. It clearly shows accelerated cracking in chloride solutions in which ferric chloride or sodium dichromate are added but no cracking (heavy uniform corrosion) in mercuric or chromic chloride solutions. Therefore, the cation present in the solution also has a role in determining if passivity prevails over most of the metallic surface and the chloride ion is allowed to cause a local breakage of the passive film to initiate SCC.

The influence of alloying additions (0.006–0.165 wt% nitrogen, 0.004–0.033 wt% sulfur, 0.003–0.05 wt% phosphorus and 14–35 wt% nickel) to austenitic stainless alloys containing 17–21 wt% chromium was evaluated (Cihal 1985) using experimental alloys in three different chloride solutions of 42% magnesium chloride at 154°C, 25% sodium chloride with 0.5% potassium dichromate at 200°C, 1.57 MPa and an aqueous solution with 200 g/l chloride ions at 350°C, 16.8 MPa. It was shown that increasing nitrogen content to 0.16 wt% in alloys with around 35% nickel (and also lower phosphorus) produced significant effect on increase in susceptibility to SCC. It was also shown that increasing sulfur to 0.033% had little effect on the susceptibility to SCC. Reducing the nickel content to 14%, even at low phosphorus levels had significantly reduced the SCC resistance in chloride solutions.

5.3.2   SCC at ambient temperature environments

Now it is commonly accepted that iron contamination on components made of stainless steels (and in solution annealed condition) results in transgranular SCC even when stored in coastal environments at ambient temperature (Gnanamoorthy 1990, Dillon 1990). The corrosion of the embedded iron leads to (a) development of stresses due to a wedging action and (b) concentration of chloride ions. The SCC of components (partially) fabricated from austenitic stainless steel and stored long term in a coastal environment has been reported (Kain *et al.* 2002) even when there was no iron contamination on the surfaces. The SCC of stainless steels used as roofing in covered swimming pools in Europe in the 1990s is another example of SCC occurring at ambient temperature (Oldfield and Todd, 1990). Presence of various salts of chlorides (Al, Zn, Ca chlorides) in the corrosion products were related (Torchio 1980, Oldfield and Todd 1991, Prosek *et al.* 2008) to formation of a high chloride content and low pH on SS surfaces of the roof and the support structure above the pools. Chloramines (from body sweat) helped to carry the chlorides to the roof and pools with fountains and surf makers further helped in the carryover of chlorides to the roof and the support structures and also caused high relative humidity.

In industries, the presence of highly oxidizing ions (Fe^{3+}, Cr^{6+}, etc.) has been identified in cases where SS had undergone SCC at room or near ambient temperatures. Laboratory studies have also reportedly demonstrated (Dillon 1990) that the presence of ferric ions in chloride solutions leads to faster SCC of austenitic stainless steel at room temperature. Type 304L stainless steel is reported (Dillon 1990) to have undergone SCC in an aqueous solution of 500 ppm chloride and hydrogen sulfide at room temperature in overnight exposures. Austenitic stainless steels are also reported (Dillon 1990) to have cracked by SCC in hydrochloric acid vapours at $-4°C$. Austenitic stainless steels are also reported (Kain 1990) to have shown SCC at ambient temperature in marine environments. It is estimated that 95% of the industrial and service occurrences of chloride SCC are due to sodium chloride (Dillon 1990). However, all this cracking is predominantly transgranular in nature.

The crack growth rate in 22% chloride solution (from ambient to boiling temperatures) has been measured (Speidel 1981) for type 304/304L stainless steels in annealed and sensitized conditions. At ambient temperatures, the crack growth rates were found to be of the order of 10^{-11} m/s which is of consequence for design life of components (10^{-11} m/s leads to a crack growth of 0.3 mm in one year). This test does not assess the SCC initiation process but suggests that once initiated, the stress corrosion cracks would certainly grow at these slow rates in components fabricated from stainless steels.

The chloride can come in contact with hot stainless steel surfaces from the thermal insulation itself. This is possible if the mineral insulation material used over hot stainless steel piping contains leachable impurities. It has been shown (USNRC Regulatory Guide 1973) that if the leached sodium silicate, chloride and fluoride are in a given range of concentration, transgranular SCC can occur from the outer surfaces of stainless steel in sensitized or non-sensitized conditions.

5.3.3   SCC in high temperature high pressure aqueous environments

Stainless steels are known to undergo SCC in high temperature aqueous environments. The most common example is the high purity demineralized water (specific conductivity of $0.055\,\mu S/cm$ at $25°C$) at the operating temperature of boiling water reactors (BWRs), typically at $288°C$. The presence of dissolved oxygen in this high purity water makes it more oxidizing, hence makes austenitic stainless steels more prone to SCC. It is to be noted that the chloride levels in these reactors are controlled to a few parts per billion (ppb) levels and the SCC is not caused by the presence of chloride ions. The oxidizing species get produced in the reactor by radiolysis (Turnbull and Psaila-Dombrowski 1992, Kain 2008). These species include various radicals of oxygen and hydrogen peroxide and are expressed commonly in equivalent dissolved oxygen content. The presence of dissolved oxygen in high purity water raises the electrochemical potential of stainless steel in a sigmoidal manner (Scott 1994) as shown in Fig. 5.4(a). The crack growth rate of stainless steels has been shown to increase in a sigmoidal manner with increasing potential (Andresen and Morra 2008) and is shown in Fig. 5.4(b). The SCC does not take place (R L Jones 1991) when the electrochemical potential is below a threshold value of $-235\,mV_{SHE}$ (standard hydrogen electrode: SHE). This is achieved in the aqueous phase in the reactor by injecting hydrogen gas to scavenge the dissolved oxygen (oxidizing species formed due to radiolysis) to a value lower than 10 ppb (typically) and the treatment (Cowan *et al.* 1986, R L Jones 1991, Cowan and Kiss 1993, Tipping 1996) is referred to as hydrogen water chemistry (HWC) as against the normal practice of not injecting hydrogen in the system (normal water chemistry: NWC). However, the hydrogen gas is not able to scavenge the oxidizing species in the steam phase and the equivalent dissolved oxygen levels remain high in steam (R L Jones 1991). This makes the top of the core components (operating in steam) prone to IGSCC even when operating with HWC.

Other types of water chemistries, e.g. noble metal chemical addition (NMCA), are in use in many BWRs and offer more efficient removal of oxidizing species at the metallic surfaces (Andresen 1995, Hettiarachchi *et al.* 1995, 2000). This is done by incorporation of fine particles of noble metals, e.g. platinum and rhodium, on the surfaces of stainless steels. These noble metals (typically $0.2\ \mu g/cm^2$) get incorporated in the surface oxide and act as a catalyst for oxygen reduction reaction. Therefore, the bulk of the oxygen reduction reaction now takes place on the metal surface, reducing the availability of oxidizing species on the metal surface. In the bulk of the water, the concentration of the oxidizing species may not come down. NMCA reduces the amount of hydrogen that is required to be injected as well as delaying SCC (Andresen 1993) by reducing the oxidizing species concentration at metal surfaces where it is required to be reduced.

In the BWR and BWR simulated environment, austenitic stainless steels in sensitized condition have been known (R L Jones 1991) to crack by IGSCC since the 1970s. The most common cases of SCC are for the welded austenitic stainless steels that have sensitized region at the heat affected zone (HAZ) of weldments. At the time of fabrication by welding, the degree of sensitization may not be high enough to make it prone to IGSCC. However, during long-term operation at the reactor operating temperature (typically after 10 years at around 288°C), the low temperature sensitization (LTS)

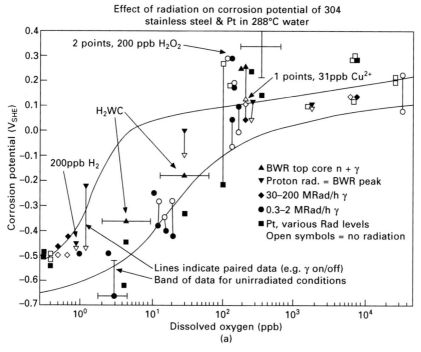

5.4 (a) Effect of irradiation (dissolved oxygen) on corrosion potential for stainless steel in high temperature BWR simulated water (from Scott 1994) and (b) variation in the crack growth rate in high temperature water for sensitized and un-sensitized stainless steels with change in corrosion potential (from Andresen and Morra 2008).

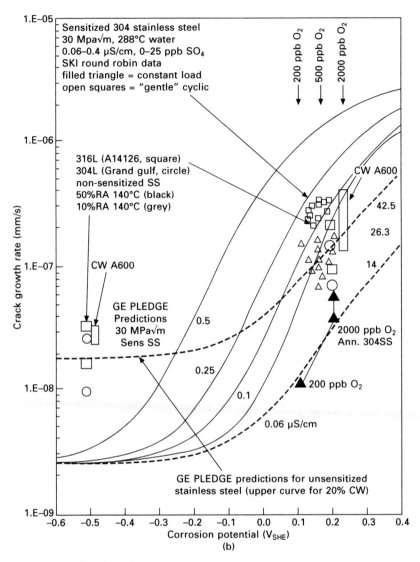

5.4 Continued.

causes increase in the degree of sensitization to a level that makes it prone to IGSCC (Povich 1978, Povich and Rao 1978, Kekkonen *et al.* 1985, Aaltonen *et al.* 1988). The pre-existing carbides grow by diffusion of chromium from the matrix at the reactor operating temperature of around 300°C but there is no new nucleation of chromium carbides. The threshold level of sensitization above which stainless steels become prone to IGSCC in the BWR simulated environment (Clarke *et al.* 1978) is the Pa value of $2\,C/cm^2$ in the single loop electrochemical potentiokinetic reactivation (EPR) (G108, ASTM 2010) or, DL – EPR value of 1 (Kain *et al.* 2004) from the double loop EPR test. Therefore, IGSCC takes place at the weakest location, i.e. the chromium depletion regions around the carbides at the HAZ of the weldments. Alloying additions of nitrogen (Bali *et al.* 2009, 2011) and cerium (Watanabe *et al.* 1999, 2000) have been shown to increase the resistance of austenitic stainless steels to sensitization, LTS and IGSCC in BWR simulated environments. Type 316 nuclear grade (316NG) was developed and is currently used in BWRs as a stainless steel that remains resistant to sensitization during welding (R L Jones 1991) and hence to LTS during long-term operation. Therefore, sensitization related IGSCC is not reported when components are fabricated from type 316NG and used in BWRs.

It has also been shown by laboratory tests that even a small extent of sensitization (chromium depletion of a few per cent at grain boundaries) is sufficient to cause IGSCC of austenitic stainless steels in the BWR simulated environment (Bruemmer *et al.* 1993). Therefore, the threshold level of sensitization needs to be considered from this context of even low levels of chromium depletion (much above 12–13 wt% chromium in the depletion regions) leading to IGSCC in BWR simulated environment in laboratory studies.

However, it has been reported since around 2000 (Andresen *et al.* 2000, Gott 2001) that IGSCC also occurs in BWRs in non-sensitized (no chromium depletion at grain boundaries) stainless steels. The core shroud cracking of austenitic stainless steel is attributed to development of plastic deformation next to the interface of the fusion zone and the parent material in the constrained weldments (Horn *et al.* 1997, Angeliu *et al.* 1998, Gott 2001). In such weldments, there may be up to 20% plastic deformation developed during the cooling stage after welding. This results in an increase in the strength of the deformed material due to strain aging. These stronger (higher yield strength) regions of the austenitic stainless steel weldment are prone to IGSCC in oxidizing conditions of the BWR environment (Andresen *et al.* 2000). These regions are also prone to IGSCC at lower values of dissolved oxygen. At the normal levels of dissolved oxygen or higher, the crack growth rates for the non-sensitized (high yield strength) stainless steels are higher than those for the sensitized stainless steels in 288°C high purity water ($< 0.06\,\mu S/cm$ at 25°C) and is shown in Fig. 5.5, and reported (Andresen and Morra 2008) to be of the same order in 288°C water ($< 0.10\,\mu S/cm$ at 25°C) but still high enough to cause IGSCC failures in the lifetime of the reactor (Andresen and Morra 2007). The addition of nitrogen has been shown (Roychowdhury *et al.* 2011a) to be beneficial in simulated BWR environment for the case of sensitized type 304L stainless steel by crack growth measurement studies but in the non-sensitized condition, nitrogen addition is shown (Roychowdhury *et al.* 2011b) to increase the crack growth rates for the 20% warm worked stainless steel. It has long been recognized that higher strength alloys lead

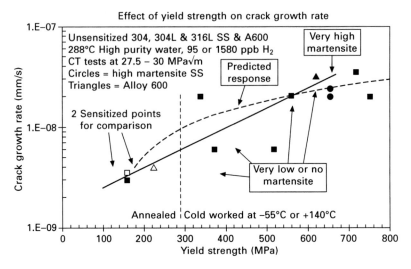

Effect of yield strength on crack growth rate

Unsensitized 304, 304L & 316L SS & A600
288°C High purity water, 95 or 1580 ppb H₂
CT tests at 27.5 – 30 MPa√m
Circles = high martensite SS
Triangles = Alloy 600

Very high martensite

Predicted response

2 Sensitized points for comparison

Very low or no martensite

Annealed | Cold worked at –55°C or +140°C

5.5 Crack growth rate as a function of strength of non-sensitized stainless alloys (from Andresen and Morra 2008).

to enhanced susceptibility to SCC as recorded in a review paper (Truman 1981): 'it is certainly the case that the increasing complexity of alloying and the development of enhanced strength do tend to increase susceptibility'.

The localized enhanced oxidation of crack tips is the main feature in SCC in high temperature aqueous environment (Thomas and Bruemmer 2002, Bruemmer and Thomas 2005, Shoji *et al.* 2010, Arnoux 2010). It was shown that oxidation at the crack tip is selective with much higher oxidation of chromium resulting in the formation of a nickel-rich spinel along the crack walls, while the crack centres were filled with an iron-rich spinel. The chromium-rich inner oxide films in cold worked stainless steel had thicknesses comparable to those found on the service samples from BWR, while the sensitized type 304 stainless steel showed lower chromium and thicker films indicating less passive film formation over sensitized regions (Bruemmer and Thomas 2005). The nickel-rich de-alloyed regions were found ahead of crack tips in 26 years' service samples from BWR but such de-alloyed regions were not found ahead of crack tips in short-term (1.5 months) tested laboratory samples. This is argued to suggest that nickel de-alloying ahead of crack tips occurs when crack growth slows down or stops, as in case of long-term service samples. This selective oxidation at the crack tip and in regions ahead of the crack tip (Fig. 5.6) is in contrast to the oxidation of stainless steel surfaces exposed to 288°C water forming chromium spinel oxide (Kim 1996).

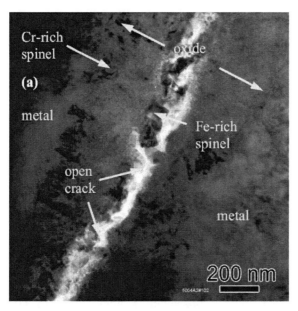

5.6 Oxides filling the crack tip and the boundaries ahead of the crack tip in type 304 stainless steel in high temperature oxidizing water (from Bruemmer and Thomas 2005).

5.3.4   SCC induced by sulfides

Polythionic acids ($H_2S_xO_6$, x = 3, 4, or 5) are known to attack chromium depletion regions in sensitized stainless steels. The damage has been shown to follow the kinetics observed for sensitization (Samans 1964). In addition to polythionic acids, SCC has also been reported in thiosulfate solutions at ambient temperatures (Isaacs *et al.* 1982). Very low concentrations of thiosulfate (0.1 ppm) are required for SCC at ambient temperature (Watanabe and Kondo 2000). It has therefore been argued (Samans 1964, Sedriks 1996) that polythionic acid induced SCC is actually stress assisted intergranular corrosion of sensitized stainless steels. The role of the stress is to open up the intergranularly corroded regions (salt layer), exposing fresh chromium depletion regions at grain boundaries to intergranular corrosion (dissolution) by polythionic acid.

Other sulfur containing solutions known to cause SCC for austenitic stainless steels are (Perillo and Duffo 1990) thiocyanate solutions (at concentrations greater than 10^{-4} mol/L) but it is less aggressive than SCC in thiosulfate or tetrathionate solutions. In addition, type 403 stainless steel in tempered martensitic microstructural condition has been shown (Bavarian *et al.* 1982) to undergo pitting and SCC in 0.01 M Na_2SO_4 at 75–100°C. Such cracking was shown not to occur at lower temperatures of 25 and 50°C. This was attributed to pits acting as sites for SCC cracks and the pits had preferentially nucleated at manganese sulfide inclusions. Even for ferritic stainless steel (low interstitial 26Cr-1Mo), SCC has been suspected (Hoxie 1977) at 132°C in water containing chloride, hydrogen sulfide, ammonia and traces of oil, thiocyanate and organic acids.

5.3.5   SCC induced by caustics

It is well known that certain combinations of caustic concentration and temperature result in SCC of stainless steels (Truman and Perry 1965). Caustics are encountered in the production of caustic soda and steam generators operating (until the 1990s) with caustic buffer solutions tended to concentrate caustics in the crevices on the secondary side (between tubes and tube sheets). The chemical (e.g. sodium hydroxide in the production of caustic soda), petrochemical and pulp and paper industries too have concerns about caustic SCC. A summary (Speidel 1977) of caustic SCC data for stainless steels suggests that there is an inherent danger of caustic SCC in strong caustic solutions at temperatures close to 100°C. At lower caustic concentrations, sensitization is detrimental. At higher operating temperatures (e.g. steam generators in nuclear power plants, which were previously operated with caustic buffer solutions at around 300°C), rapid SCC has been known to occur even in dilute solutions. Deaerated solutions of caustics produce a lesser extent of SCC compared to oxygen containing caustics. While austenitic stainless steels are also highly prone in deaerated solutions, increasing nickel content in alloys improves their resistance to caustic SCC. Aerated solutions require (McIlree and Michels 1977) more nickel and chromium in the alloys for improved resistance to caustic SCC. While removing sensitization and minimizing stresses by solution annealing has been found to be ineffective in preventing caustic SCC for austenitic stainless steels, control measures have been focused on use of phosphates (Wheeler and Howells 1960) that prevent formation of free caustics. Also additions of chromates (Staehle and Agarwal 1976) and chlorides (Ryabchenkov *et al.* 1966) in the environment have been reported to inhibit caustic SCC.

A detailed investigation of stainless steels other than austenitic (Wilson *et al.* 1977) reports that at 90% of yield strength (a condition that does not result in austenitic stainless steels to undergo caustic SCC), heat treatments that cause 475°C embrittlement are detrimental (to caustic SCC) for high chromium ferritic stainless steels. The lower chromium ferritic stainless steel (e.g. type 405) showed heavy uniform corrosion in such an environment. Similarly, duplex and martensitic stainless steels too showed more proneness to caustic SCC after heat treatments that cause 475°C embrittlement. In as-annealed conditions, duplex stainless steels show (Kowaka and Kudo 1977, Sedriks 1996) higher resistance to caustic SCC than ferritic and martensitic stainless steels.

5.4   Effect of chemical composition on stress corrosion cracking (SCC)

5.4.1   Stacking fault energy

The presence of stacking faults is known (Latanision and Ruff 1971) to influence the movement of dislocation. A low stacking fault energy (SFE) indicates presence of a wider gap between the dislocation partials. The wider the gap between dislocation partials in a stacking fault, the more difficult it will be for a dislocation to cross slip. Such a situation would cause more dislocations to move along the slip plane promoting planar movement of dislocation. This causes emergence of slip steps on the surface of the material where the slip plane meets the outer surface of the material. The repeated emergence of slip steps leads to disturbance in the passive film causing it to become prone to breakage at that spot. The same process of passive film breakage occurs at the crack tip during the growth of the stress corrosion crack (Ford 1982, Ford *et al.* 2006). Therefore, materials with low stacking fault energy are more prone to SCC. Austenitic stainless steel type 304/304L has low stacking fault energy (typically 20 mJ/m^2) and is particularly prone to chloride SCC. Nickel and nickel-based alloys have much higher stacking fault energy (typically 75 mJ/m^2) and are therefore more resistant to chloride SCC.

The SFE is determined by the chemical composition of the material. A typical correlation is shown in Eqs 5.1 (Rhodes and Thompson 1977) and 5.2 (Dai *et al.* 2002):

$$SFE = -25.7 + 2\%Ni + 0.6\%Cr + 7.7\%Mn - 44.7\%Si \qquad 5.1$$

$$\begin{aligned}
SFE = \gamma^0 &+ 1.59\%Ni - 1.34\%Mn + 0.06Mn^2 - 1.75\%Cr \\
&+ 0.01\%Cr^2 + 15.12\%Mo - 5.59\%Si - 60.69(\%C \\
&+ 1.2\%N)^{1/2} + 26.27(\%C + 1.2\%N)(\%Cr + \%Mn \\
&+ \%Mo)^{1/2} + 0.61[\%Ni(\%Cr + \%Mn)]^{1/2} \qquad 5.2
\end{aligned}$$

The role of nitrogen has been variously described by different researchers. Most of the researchers agree that at levels prevalent in commonly used austenitic stainless steels, nitrogen slightly reduces the SFE (Schramm and Reed 1975, Lo *et al.* 2009).

The SFE is reported (Fujita *et al.* 1994) to reduce with increase in temperature. It is therefore important to compare the SFE values at the temperatures at which the SCC resistance is aimed to be compared for different materials. At high temperatures of 300–400°C, there are other modes of dislocation movement and its obstruction that become operative, e.g. easier cross slip and dynamic strain aging, that may determine if dislocations would move in a planar manner. However, at ambient temperatures and at boiling temperatures of solutions (at ambient pressure), the SFE may be used to compare SCC resistance of various materials including stainless steels.

5.4.2   Effect of alloying elements

The effect of alloying additions on SCC susceptibility is also obvious from its influence on SFE (Eqs 5.1 and 5.2). Increasing nickel content increases the SFE hence increases the resistance of austenitic stainless steel to chloride SCC. When nickel is present in small quantities in ferritic stainless steels, it has a detrimental effect (Steigerwald *et al.* 1977) on chloride SCC. The famous 'Copson' curve (Copson 1959) indicates that stainless steels become resistant to SCC when nickel content is raised to a very high (50%) level. The results are derived from boiling magnesium chloride tests. Magnesium chloride is a very severe test. Tests in 22% sodium chloride solution at 105°C, using a fracture mechanics approach establishing (Speidel 1981) threshold stress intensity factors, showed that stainless steels become resistant to chloride SCC when the nickel levels are raised to about 30–40%. This is in line with the general industrial experience that stainless steels with nickel at or above 42% do not undergo chloride SCC. Alloying with silicon has been shown (Sedriks 1996) to be beneficial in increasing the resistance to chloride SCC using the boiling magnesium chloride test. However, such an effect is reportedly not clearly visible when using boiling sodium chloride tests. In high temperature aqueous environment, it has been reported (Andresen *et al.* 2003, Andresen and Morra 2005) that silicon addition (also segregation of silicon at grain boundary, e.g. due to neutron irradiation) enhances the crack growth rates, possibly because of formation of silica and its dissolution tendency in such environments. Similarly the effect of molybdenum alloying appears to depend on the type of test used. The magnesium chloride tests show (Hines and Jones, 1961) that molybdenum first decreases and then increases the resistance to SCC with a minima in resistance occurring at 1.5% molybdenum. Sodium chloride tests, on the other hand, show (Speidel 1981) SCC resistance (threshold stress intensity for SCC) increasing with increasing molybdenum content in austenitic stainless steels. It is possible that in the mentioned study the sodium chloride tests assessed the SCC propagation as compared to initiation as well as propagation in the tests with magnesium chloride solutions.

The effect of nitrogen on SFE has been reported to be variable (Lo *et al.* 2009). However, there is a general opinion that nitrogen in type 304/316 stainless steels tends to reduce SFE (Schramm and Reed 1975). Moreover, nitrogen is known to improve the resistance to sensitization (Kain *et al.* 2007, Bali *et al.* 2009), makes passive film more stable and helps to destroy the incipient chemistry (Palit *et al.* 1993) in a pit/crack, hence SCC resistance of the LN stainless steels after a given sensitization heat treatment is higher than that of its plain carbon variants.

High purity alloys (Fe-16Cr-20Ni and Fe-18Cr-14Ni) prepared by plasma furnace melting and containing total metallic impurities of only 1 ppm and non-metallic impurities of only 10 ppm showed (Sedriks 1996) very high resistance to SCC (tested by magnesium chloride tests). Similarly, it was shown (Kowaka and Fujikawa 1972) that lowering the phosphorus content in Fe-18Cr-10Ni stainless steel to 0.003 wt% raised its resistance to SCC in boiling magnesium chloride tests. The low phosphorus alloys did not show SCC even after 400 h of exposure while the commercial alloys (type 304 and 316) cracked in a few hours in the boiling magnesium chloride tests.

5.4.3   SCC susceptibility of type 304 vs 316 stainless steel

Type 304 stainless steel is highly susceptible to SCC in chloride solution as demonstrated by the results of exposure of constant strain samples (Kain 1997, Kain *et al.* 2002) in boiling acidified 25% sodium chloride solution (G123, ASTM 2005). Type 304 stainless steel showed extensive TGSCC in solution annealed condition in 144 h exposure tests. The cracking mode changed to IG in the sensitized condition. Type 316 stainless steel did not show any SCC in the same test even after exposure for 200 h. Another study reported (Speidel 1981) increased values of threshold stress intensity with increasing molybdenum content in the austenitic stainless steels in 22% boiling sodium chloride solution indicating higher resistance of molybdenum added stainless steel to SCC in boiling sodium chloride solution. Such differences in susceptibility to SCC are not recorded (Edeleanu 1953, Kowaka and Fujikawa 1972) for these two grades of stainless steels in the highly aggressive boiling magnesium chloride solutions. In fact almost the same time to failure has been reported (Edeleanu 1953) in boiling 42% magnesium chloride test for types 304, 304L, 316 and 321 stainless steels. The main difference between type 304/304L and type 316/316L is the addition of 2–3 wt% molybdenum in the latter. In addition to this, the nickel content in type 316 stainless steel is higher than that in type 304 stainless steel. The chemical composition of these two types of stainless steels is given in Table 5.1.

The higher nickel (and molybdenum) content makes type 316/316L more

Table 5.1 Chemical composition of type 304 and type 316 stainless steels (as per ASTM A240)

Element	C	Cr	Ni	Mn	Si	S	P	Mo
304	0.07	17.5–19.5	8–10.5	<2	<0.75	<0.03	<0.045	<0.50
304L	0.03	17.5–19.5	8–12	<2	<0.75	<0.03	<0.045	<0.50
316	0.08	16–18	10–14	<2	<1	<0.03	<0.045	2–3
316L	0.03	16–18	10–14	<2	<1	<0.03	<0.045	2–3

resistant to SCC by increasing its SFE (Schramm and Reed 1975) while chromium does not affect the SFE much. Lowering the carbon content reduces the SFE but not to a large extent. Therefore, the SFE of 304 and 304L is not quite different and also the SFE of 316 and 316L is quite similar. Hence the tendency for stress corrosion cracking of the plain 304 or 316 stainless steels (in solution annealed condition) is not considered to be different from its low carbon varieties. The presence of 2–3 wt% molybdenum in the alloy makes the passive film stronger (Kain *et al.* 2002, Newman and Shahrabi 1987, Sugimoto and Sawada 1976). This happens by enrichment of the passive film with more chromium making its breakage by an aggressive ion more difficult. However, molybdenum itself does not get incorporated in the passive film. This together with slightly higher stacking fault energy (Schramm and Reed 1975) are the main reasons for better resistance of type 316/316L stainless steel than type 304/304L to stress corrosion cracking in sodium chloride solutions.

5.5   Microstructure and stress corrosion cracking (SCC)

Microstructure of the stainless steel affects its susceptibility to SCC. Stainless steels may have austenitic, ferritic or martensitic crystal structure. Among these the duplex structure (austenite and ferrite) is known to be the most resistant to SCC.

The microstructure consists of the matrix and other features present in the material. In addition, there may be inclusions, precipitates, grain boundaries, chromium deletion regions around chromium carbides, etc., present in the material. The non-metallic inclusions and other precipitates exposed to the surface do not allow a good passive film to form over its surfaces as chromium is not available to form the chromium-rich film over these inclusions precipitates. This leads (Sedriks 1986) to easy breakage of the weak passive film at the sites of inclusions/precipitates, hence easy initiation of SCC. Similarly chromium depletion regions around chromium-rich carbides (sensitization) are well known to develop surface film that is not protective. This is also clear from (Kain *et al.* 1989) the lower values of pitting potential in chloride solution for the sensitized type 304 stainless steel compared to that for the de-sensitized and solution annealed stainless steel. These cause easy breakage of the surface film over the chromium depleted regions and also easy propagation as these chromium depletion regions provide a weak path for crack propagation.

5.5.1   Ferritic stainless steels

It is a general perception that ferritic stainless steels are more resistant to SCC than the austenitic stainless steels. While this is true, it does not imply that all the ferritic stainless steels are resistant to chloride SCC. Various studies have shown that types 405, 430, 444, 448, 409, 439 and 444 are resistant to chloride SCC. However, types 434, 430 (Bednar 1979) and Mo containing Fe-18Cr stainless steels have been reported (Bond and Dundas 1968) to undergo chloride SCC. Sensitization, cold working, high temperature embrittlement, etc., have been reported (Sedriks 1996) to increase the susceptibility of ferritic stainless steels to chloride SCC. It is to be noted that addition of austenite stabilizer elements, e.g. nickel, would make formation of austenite at high temperatures easier. However, nickel reduces the solubility of carbon in the austenitic phase. Hence formation of sensitization would become easier, making these stainless steels more prone to chloride SCC.

In ferritic stainless steels, presence of microstructural features that reduce its ductility (carbo nitrides, cold working, alpha prime phase formation due to 475°C embrittlement) tend to make it more prone to chloride SCC and are revealed in boiling magnesium chloride tests. Increasing the carbon content from 20 to 171 ppm is shown (Streicher 1975) to lead to SCC in magnesium chloride tests for a 28.5Cr–4Mo ferritic stainless steel. Sulfur too is known (Brown 1977) to be detrimental against chloride SCC as type 430F with 0.15% S was shown to undergo SCC in marine environment. Similarly copper and nickel contents have been shown (Steigerwald et al. 1977) to introduce SCC in boiling magnesium chloride tests in welded Fe-18Cr-2Mo alloy. However, the detrimental effects of nickel on SCC observed in boiling magnesium chloride tests are not clearly visible in boiling sodium chloride tests. In the presence of nickel, molybdenum addition was also shown (Bond and Dundas 1968) to be detrimental in a magnesium chloride solution boiling at 140°C.

5.5.2   Duplex stainless steels

The duplex (austenite and ferrite) stainless steels are known to be highly resistant to chloride SCC. It again needs to be pointed out that not all varieties of duplex SS are resistant but certain varieties are resistant and some others do undergo SCC but have a much higher (Speidel 1981) threshold stress for SCC (three times higher than that for austenitic stainless steels). In a particular study it was shown (Truman 1981) that as the ferrite content of the stainless steel is increased, its resistance to SCC increases. However, above 40% ferrite, the SCC resistance starts decreasing. The microstructural features that deteriorate SCC resistance in ferritic stainless steels (carbo nitrides, alpha prime phase, cold working and high temperature sensitization/ formation of other precipitates) also deteriorate SCC resistance in duplex stainless steels. The presence of sigma phase has also been shown (Spaehn 1990) to be detrimental to SCC resistance for duplex stainless steels. It has been shown (Adhe et al. 1996, Wilms et al. 1994) that it is the depletion of chromium and molybdenum around the sigma phase that is responsible for deterioration in its susceptibility to localized corrosion rather than the sigma phase itself. Therefore, welds of duplex stainless steels are much more prone to chloride SCC than the solution annealed duplex stainless steels.

The presence of ferrite phase together with austenite phase would increase the grain boundary length that the moving stress corrosion crack has to navigate. This is because of the higher resistance of the ferrite phase to SCC. Therefore, the presence of smaller grains of ferrite would help to increase the resistance to SCC. Coarsening of the ferrite phase due to welding of duplex stainless steel would reduce its resistance to SCC. With the advent of super duplex stainless steels, these are now claimed to resist SCC in aggressive environments as well as at higher temperatures. It was shown (Singh Raman and Siew 2010) that a super duplex stainless steel suffered IGSCC at a strain rate of 4×10^{-7} s^{-1} in 30% magnesium chloride solution at 180°C but no SCC occurred at higher strain rates. Addition of up to 2800 ppm nitrite to this solution effectively suppressed SCC but higher amounts of nitrite addition accelerated SCC.

It was shown (Tsai and Chen 2000) that for type 2205 duplex stainless steel, it was immune to SCC in near neutral sodium chloride solutions at concentrations up to 26% and temperatures up to 90°C. A critical potential existed above which there was a dramatic reduction in SCC susceptibility and this potential was the same as the pitting potential. Pitting was shown to assist in crack initiation and dissolution of alpha (ferrite) phase was shown to assist in crack propagation.

5.5.3   Role of delta ferrite

Delta ferrite in austenitic stainless steel welds increases its resistance to SCC. The effect is similar to the beneficial effect of increasing ferrite phase content in duplex stainless steels. In the welds of austenitic stainless steels (Kain and De, 2003), presence of fine, isolated pools of ferrite force the propagating crack to take a tortuous path, increasing its resistance to SCC. The effect of low temperature thermal aging (simulating the long-term exposure of the welds to the operating temperature of the nuclear reactors) on SCC in high temperature aqueous environment has suggested (Abe and Watanabe 2008) that the threshold stress level for SCC initiation was lower for the austenite to ferrite mode of solidification weld as compared to that for the ferrite to austenite solidification mode weld (for type 316L stainless steel). Since the ferrite phase solidified in the austenite to ferrite mode had a higher chromium content, it was suggested to have a higher resistance to SCC than the weld that solidified in the ferrite to austenite mode.

5.5.4 Martensitic and precipitation hardenable stainless steels

These stainless steels are used in heat treatment conditions that impart high strength and toughness to these alloys. Under these conditions, these are prone to hydrogen embrittlement, and stress corrosion cracking is not exactly a major issue (Logan 1966). It was shown (Logan 1966, Trozzo and McCartney 1960) that type 410 stainless steel (tempered at 345°C) showed cracking in nuclear reactor simulated environment and the cracking followed prior austenitic grain boundaries. The austenite at the boundaries had transformed to un-tempered martensite providing an easy path for crack propagation. Similarly type 410 stainless steel cracked in 2–8 weeks under stress of 267 MPa in air saturated water at ~150°C in an autoclave test but did not fail under similar conditions of testing when ammoniated water (pH 8.5–9.1) was used (Suss 1962). Another study reported (Lillys and Nehrenberg 1956) cracking for types 410, 420, 422 and 436 stainless steels in salt spray cabinet (5% sodium chloride) and in test conditions that cause hydrogen embrittlement. No cracking was observed after 75 days of exposure to the sodium chloride atmosphere when the specimens were tempered at ~370°C (and were also found to be resistant to hydrogen embrittlement). However, after tempering at 480°C, these speciemen were most prone to cracking and the crack path followed prior austenite boundaries.

5.6 Nature of the grain boundary and stress corrosion cracking (SCC)

The grain boundary energy in type 304 stainless steel is known (Murr 1975) to vary from 20 to 835 mJ/m^2. This has been attributed to the nature of the grain boundary or the misorientation angle (proximity to low coincident site lattice – CSL). The CSL approach (Brandon 1966, Smith and Pond 1976) defines sigma (Σ) as the inverse of the coincident sites at grain boundaries. For example $\Sigma3$ indicates one out of three sites at grain boundaries is coincident. Lower values of Σ denote lower energy of grain boundaries. Generally Σ values up to 29 are taken (Wasnik *et al.* 2002) as 'special' or low energy boundaries, and above it are taken as 'random' or high energy boundaries. Low energy boundaries are known to be resistant to initiation and growth of precipitates. The last two decades have seen numerous studies (Wasnik *et al.* 2002, Shimada *et al.* 2002, Jones and Randle 2010) on thermomechanical processing to tailor (engineer) the grain boundary nature of stainless steels.

5.6.1   Thermomechanical processing for grain boundary engineering

It has been shown that a small level of cold deformation followed by low temperature annealing helps in achieving (Shimada *et al.* 2002) a high fraction of 'special' boundaries. Increased levels of cold working (followed by high temperature solution annealing) tend to assist formation of random boundaries (Wasnik *et al.* 2002). This happens due to the increased tendency for recrystallization and it is commonly accepted that the new strain free grains favour formation of 'random' boundaries. To obtain a high fraction of 'special' or especially the twin ($\Sigma 3$) boundaries, cold rolling to obtain 5% reduction in thickness and annealing at 927°C for 24–72 h has been shown (Shimada *et al.* 2002) to be effective for type 304 stainless steel. Using thermomechanical processing with single and multiple cycles of 5% strain and annealing 1050°C for 30 minutes, it was shown (Jones and Randle 2010) that the fraction of $\Sigma 3^n$ boundaries increased for type 304 stainless steel. It was shown in this study that 97% of the $\Sigma 3$ boundaries were immune to sensitization (sensitization heat treatment at 650°C for 4 h) while 80% of the $\Sigma 9$ boundaries were immune to sensitization. Other 'special' boundaries ($\Sigma 27$ or others with Σ value less than 27) were not found to be resistant to sensitization. For type 316 stainless steel, 3% reduction in thickness by cold deformation and annealing at 967°C for 72 h has been shown (Michiuchi *et al.*, 2006) to significantly improve the fraction of 'special' boundaries.

5.6.2   Effect of grain boundary nature on SCC

Therefore, a stainless steel with a high fraction of 'special' boundaries is taken to be resistant to sensitization and therefore to IGSCC. This has been clearly shown on stainless steels in which the fraction of 'special' boundaries has been improved by thermomechanical processing. The nucleation and growth of chromium-rich carbides is dictated by the average grain boundary energy in a given material. This has been modified (Wasnik *et al.* 2002) to include the distribution (fraction) of various grain boundaries with different energies and is denoted by EGBE (effective grain boundary energy). The EGBE = $((\Sigma \gamma_i f_i) 4/d)\gamma_{max}$ where γ_i is the energy of the boundaries having fraction f_i, d is the grain diameter and γ_{max} is the energy of the boundary. It was shown (Wasnik *et al.* 2002) that the degree of sensitization increases with increasing EGBE but at and above a threshold value, it drastically drops to low values. The susceptibility to sensitization was shown to improve when the fraction of random boundaries increased above 75%. Such a high fraction also ensures grain to grain connectivity through 'random or special' boundaries, so that the characteristic feature of a given type of boundary is made use of throughout the thickness of the material. It has been reported (bond percolation theory) that in a one-dimensional path consideration, the fraction of boundaries required for grain to grain connectivity (Gaudett and Scully 1993, 1994, Engelberg *et al.* 2004) is 23% and in a two-dimensional consideration it is 89%. This implies improvement in IGSCC of stainless steel is obtained after thermomechanical processing when the fraction of 'special' or 'random' boundaries is increased above a threshold value mentioned above.

The improved resistance of grain boundary engineered stainless steels to IGSCC has been demonstrated in boiling chloride solutions (Wasnik *et al.* 2003, Jin *et al.* 2007). It should be noted that grain boundary engineering has been applied to alter the fraction of 'special' or 'random' boundaries in stainless steel and it affects the tendency to sensitization, hence susceptibility to IGSCC. It was shown (Jin *et al.* 2007) that the crack prefers to grow along the random boundaries and avoided the segments of special boundaries. Alteration of grain boundary nature does not change its SFE and the susceptibility of the alloy to SCC (TGSCC) in the as received condition (without sensitization) has not yet been shown to improve.

5.7   Residual stress and stress corrosion cracking (SCC)

SCC is known to occur under tensile stresses that are mostly static. These stresses are usually below the macroscopic plastic yield stress but sufficient to cause local (microscopic) yielding. This is based on the understanding of the need for the passive film breakage being assisted by dislocation movement towards the point of stress concentration (e.g. crack tip). Therefore, dislocation movement, though at localized regions in the material, is the essential feature for SCC. Hence, residual stresses in the material are sufficient to cause SCC if the level of residual stress is high enough to cause dislocation movement (strain or yielding) at localized (defect) regions. The maximum value of macroscopic stress is the yield stress of the material. However, the minimum value of residual stress that can cause SCC will be that level of residual stress that causes microscopic yielding at the defects present in the material.

The minimum level of residual stress in stainless steels that would ensure absence of SCC for the entire service life of components has been a matter of debate. For type 304/304L stainless steel, the threshold stress for feedwater heaters below which SCC would not occur in chloride environment at temperatures (also typically for condenser tubing applications) is taken (EPRI 1990) as 15% of the yield strength (of 230 MPa or 2350 kgf/cm^2). However, demonstration of such a correlation is difficult as attainment of such levels of residual stresses and SCC testing in the laboratory is not easy. A plausible approach is to measure residual stress in fabricated condenser tubes and select the tubes with residual stress levels varying from low (5% of yield strength) to high, up to 30% of the yield strength. Such a fabricated tube when tested in chloride solutions (either boiling magnesium chloride solution as per G36, ASTM, or boiling acidified 25% sodium chloride solution as per G123, ASTM) would provide an indication about the threshold level of residual stress that would not cause SCC. Such a study showed (Kain *et al.* 2010, Ghosh *et al.* 2011) that the stress corrosion cracks always appeared perpendicular to the direction of tensile stress and the extent of cracking was also indicative of the level of tensile stress. It is to be noted that such a threshold stress would change if the nature of defects present in stainless steel is different.

The SCC requires conjoint action of stress and environment. One effect of increased stress at defect (crack tip) location could be enhanced interaction with the material. Sieradzki and Freidersdorf (1994) showed that the synergistic effect of stress on dissolution rate is negligible by the concept of stored elastic energy reducing the equilibrium metal/metal ion reversible potential. However, recent work (Thomas and Bruemmer 2002, Bruemmer and Thomas 2005, Shoji *et al.* 2010, Arnoux 2010) has shown that increased stress at the SCC crack tip leads to enhanced oxidation in high temperature aqueous environment. This results in a voluminous oxide formation at the crack tips (that are shown to be of atomic dimensions) resulting in separation of crack faces exposing bare metal at the crack tip to accelerated oxidation. If the stresses at the crack tip are not able to break open the excessive oxidized regions, repassivation or stoppage of crack growth would occur. It is not yet reported if this oxidation (or formation of corrosion products in crack tips) mechanism is operative for cases of SCC in boiling chloride solutions.

5.7.1   Welding stresses

Welding is known to lead to residual stresses in stainless steels. The high coefficient of thermal expansion and low heat transfer rate of stainless steels is the main reason for distortion or high level of residual stress in weldments. To avoid distortion of weldments, fixtures are provided during welding of stainless steel components. Additionally, the sections that are welded last, develop a high level of residual strain during the process of cooling after the welding stage. The weld pool being at the highest temperature expands the most, while the low heat transfer through stainless steel causes much less expansion of the base material being welded. During the cooling stage, the weld pool contracts the most causing a tensile loading on the material immediately next to the weld pool. Therefore, during the last stages of cooling, at around 200–300°C, when the region immediately next to the weld pool is under maximum stress, it plastically deforms. The extent of plastic strain in this region has been measured (Angeliu *et al.* 2000) to be 20%. This plastic deformation occurs at temperatures around 200–300°C without any phase transformation in austenitic stainless steels. There is, therefore, resultant increase in the yield strength of the material in this zone. Hence the maximum level of residual stress in this region increases (maximum being up to the enhanced yield strength). This has been shown to cause IGSCC of even the non-sensitized stainless steels in BWR environment/BWR simulated environment (Andresen and Morra 2008). This phenomenon now occurs in the region of weldment that is immediately next to the weld fusion zone (width of a few grains to a few tens of grains). This is distinctly different from the location of IGSCC that usually occurs in weldment of austenitic stainless steels at the heat affected zone (HAZ) which is a few mm away from the weld fusion zone. The IGSCC in HAZ of weldment is due to sensitization occurring in that zone (Aaltonen *et al.* 1988), while IGSCC of the high strength welds occurs without any trace of sensitization (i.e. in a non-sensitized condition).

Apart from weld thermal stresses, even fit-up stresses can cause SCC. In a chloride environment the mode of SCC would be transgranular. Typically stainless steel products that are subjected to cold/warm working during fabrication stages are finally given a solution annealing heat treatment. However, distortion during solution annealing necessitates some of these components to be straightened in the final stages. The mechanical straightening, e.g. roller or stretch straightening of stainless steel tubes, results in development of residual stresses (EPRI 1990, Kain *et al.* 2010, Ghosh *et al.* 2011). Roller straightening, for example, results in circumferential residual stresses and stretch straightening in longitudinal stresses. The SCC always takes place in a direction perpendicular to the direction of residual/applied stresses. It has been clearly demonstrated in boiling magnesium chloride tests that the SCC takes place in a longitudinal direction for roller straightened tubes and in a circumferential direction for the stretch straightened tubes as shown in Fig. 5.7.

5.7.2   Cold working and warm working

Cold/warm worked stainless steels have higher yield and ultimate tensile strength than a solution annealed stainless steel. This implies that the level of residual stress that a cold/warm worked stainless steel can retain is higher than that in a solution annealed stainless steel. This is another reason for increased susceptibility of cold/warm worked stainless steels to stress corrosion cracking.

5.8   Surface finishing and stress corrosion cracking (SCC)

Surface finishing is an indispensable stage in the industrial fabrication process. Surface finishing is also dictated by the last of the fabrication operations to be carried out, e.g. machining, grinding, wire brushing, shot

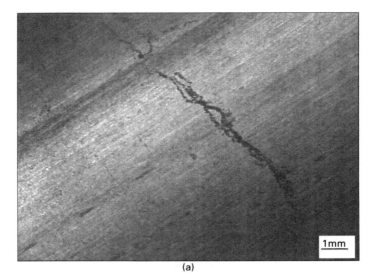

(a)

(b)

5.7 SCC in as fabricated type 304 stainless steel tubes in boiling magnesium chloride test for (a) stretch straightened tubes with circumferential cracks and (b) roller straightened tubes with longitudinal cracks.

peening, etc., of components. The surface states can compromise corrosion resistance (pitting corrosion and SCC) of stainless steels. These processes affect the electrochemical and mechanical stabilities of passive film and that of the near-surface layers, by changing the surface reactivity and altering the near-surface residual stress/strain state. Surface preparation operations are shown to alter the susceptibility of the steel to SCC and its resistance to the initiation and propagation of pitting (Ben Rhouma *et al.* 2001, Braham *et al.* 2005). Surface finishing operations affect predominantly the surface layers of a component and stress corrosion cracks initiate from the surfaces. Therefore, these operations are likely to affect the phenomenon of crack initiation. The types of damage that are introduced by machining or grinding and the effects on SCC are given below.

5.8.1   Increase in surface roughness

Roughness has a significant effect on stress-corrosion crack initiation. The greater the roughness, the deeper are the grooves wherein the aggressive species would concentrate. These grooves also act as stress raisers thus reducing the incubation time to cause SCC. Figure 5.8 shows (Cochran and Staehle 1968) that surface preparation affects the SCC behavior of type 304 stainless steel in boiling magnesium chloride solution. Especially in case of chloride stress corrosion cracking, it has been shown that accumulation of chloride ions and consequent destruction of the passive film would be greater in the presence of deep grooves on the surface (Shoji 2003).

5.8.2   Generation of tensile residual stresses on the surface

These stresses arise as a synergistic effect of thermal and transformation changes taking place in the material during surface machining/grinding operations. During machining, the surface layers become hot due to frictional heating, expand and exert compressive stresses on the bulk owing to the restraining effect of the cold bulk of the component and this is referred to as the thermal effect. On subsequent cooling, residual tensile stresses are generated in the surface layers due to shrinkage of the surface layer. Phase changes often accompany volume expansion which leads to tensile residual stresses in the surface layers and are referred to as transformation effects. Detailed recent studies (Ghosh and Kain 2010a,b) have shown that type 304L stainless steel when subjected to heavy surface machining resulted in the formation of a highly work hardened layer near the surface having sub-micron grain size, high density of deformation bands and a high volume fraction of martensite. This layer, having high tensile stresses, extended up to a depth of ~150 μm and made the material highly susceptible to SCC (Ghosh and Kain 2010a,b, Kain *et al.* 2010).

5.8.3   Formation of less protective surface oxide

Surface machining results in an increase in the dislocation density at the metal surface by orders of magnitude. As the metal on the surface layers is plastically deformed during machining/grinding, slip bands and deformation twins exist throughout the layer adjacent to the surface. Planar dislocation

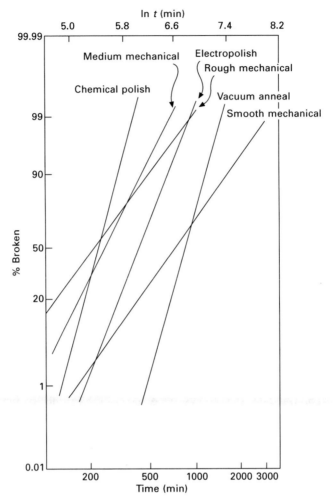

5.8 Effect of surface preparation on SCC of type 304 stainless steel in boiling magnesium chloride solution (from Cochran and Staehle 1968).

arrays are high stress raisers and cause easy rupture of the oxide film under the presence of stress and environment. These deformed surface layers are highly (electrochemically) active compared to the base, undeformed material (Ghosh and Kain 2010a,b).

In view of the above factors, it is important that careful control be kept over the machining parameters so as to improve appreciably the durability of these materials by reducing the surface electrochemical reactivity and their susceptibility to SCC. Development of final fabrication techniques that leave much lower strain (and the effects of strain) on the material is now being accorded recognition as a major factor to control SCC, especially initiation of SCC. Since surface finishing affects the microstructure and/or the state of stress up to a limited depth from the surface, it affects initiation of SCC rather than growth of the stress corrosion cracks beyond the depth of the affected surface layers.

5.9   Other fabrication techniques and stress corrosion cracking (SCC)

5.9.1   Cold and warm working

Cold working leads to two major changes in the microstructure of materials. The first (Byun *et al.* 2004, Karlsen *et al.* 2010) is increased dislocation density and dislocation entanglement. For the case of warm working (Almeida *et al.* 1998), the tendency for formation of dislocation cellular structure increases. Twinning in austenitic stainless steels has been reported (Mullner *et al.* 1994, Byun *et al.* 2004, Karlsen *et al.* 2010) to occur at higher strain levels or at lower deformation temperatures while dislocation motion (a thermally activated process) is predominant at lower strain levels and higher deformation temperatures. Lower strain rates favour dislocation motion and higher strain rates favour twinning (Karlsen *et al.* 2010). In the laboratory tested sample at 288°C for crack growth rate measurement, it was reported (Bruemmer and Thomas 2005) that deformation twinning at the crack tip was much more and shear transformation products appeared at several crack tips for type 304 stainless steel.

The second effect is the formation of strain induced (epsilon) martensite (Karlsen *et al.* 2010) or stress induced (alpha) martensite (Lee and Lin 2001). Higher strain rates and lower temperatures are shown to favour formation of alpha prime martensite. The Md30 has been defined (Kuniya *et al.* 1988, Angel 1954) as the temperature at which a true strain of 30% produces 50% martensite in a solution annealed stainless steel. The cold working involved in fabrication typically results in a much lower percentage of martensite. Cold working also results in increase in dislocation density and formation of dislocation entanglement. All of these result in increased strength of the stainless steel accompanied by reduced ductility. A study (Cochran and Staehle 1968) using type 310 stainless steel (that has little tendency for martensite formation) showed the minimum time to failure at a cold work level of 10% in a boiling magnesium chloride solution. Type 316 stainless steel has low tendency for martensite transformation upon cold working. For type 316 stainless steel, it has been shown (Kowaka and Fujikawa 1972) that in boiling magnesium chloride solution, there is a sharp decrease in time to failure when the extent of cold working approaches 5%, but there is little difference in time of cracking when the extent of cold working is further increased up to 40%. Type 304 stainless steel (which shows both the effects of cold working), on the other hand, shows a drastic reduction in time to failure upon initial low levels of cold working but again after about 20% cold working, the time to failure starts increasing, indicating improved resistance to SCC. In general, above about 35% cold working, austenitic stainless steels have increased resistance to SCC.

Using a constant load test, the effect of temperature on stress corrosion cracking of various austenitic stainless steels in magnesium chloride test was evaluated (Alousif and Nishimura 2006). The cracking mechanism for type 304 SS was transgranular between 140 and 155°C and intergranular between 139 and 130°C. For type 316 SS, it was transgranular between 151 and 155°C and a mixed (trans- and intergranular) between 151 and 135°C. For type 310 SS, it was transgranular at all temperatures above 135°C. The intergranular cracking for types 304 and 316 was attributed to formation of strain-induced martensite in these alloys at the grain boundaries. Type 310 did not form any martensite due to loading, hence did not show any intergranular SCC at any of the temperatures used in the study.

Warm working, on the other hand, is carried out at temperatures above which martensite does not form in stainless steels upon working. Typically above 150°C, martensite phase does not form in stainless steels upon working. However, multiplication of dislocations and to a slightly lesser degree, dislocation entanglement and formation of cellular structures do form in stainless steels (Frechard *et al.* 2006) at this temperature and result in increased strength and reduced ductility. The effect of warm working of austenitic stainless steel would be the same as for cold working (increased dislocation density and entanglement), which does not result in formation of martensite. However, dislocation cellular structure formation would take place (Roychowdhury *et al.* 2011c) during warm working that does not take place during cold working. Therefore, it would be expected that warm working would reduce the resistance against SCC and the minima would occur at a higher level (about 10–20% of warm working) as against 10% in the case of cold working. Detailed and systematic studies to show these differences between the effects of warm and cold working on SCC are not yet reported.

Increased strength of stainless steel has been shown to result in IGSCC of even non-sensitized stainless steels (constraint geometry welds) in BWR simulated environment (Gott 2001). Cold worked or the warm worked stainless steels have been shown to follow the same dependence (Andresen and Morra 2008). This dependence has been shown in various laboratory studies since 2000 as well as from the plant experience.

5.9.2   Cold working and sensitization

Increasing amounts of cold working in the stainless steel is known to increase the susceptibility of materials to sensitization and to intergranular corrosion.

There exists (Bose and De 1987) a peak proneness to sensitization at cold working levels between 15 and 20% and beyond this level of cold working, the precipitation of chromium carbides starts to take place inside the grain matrix itself. These effects have been shown by EPR studies as well as by IGC tests on cold worked stainless steels. It was shown (Solomon 1985) that the continuous cooling rates required to produce sensitization increased by a factor of 7 with around 11% prior to cold working. This implies that a stainless steel with 15% cold working is highly prone to sensitization during welding or during exposure to high temperatures during fabrication stages. This in turn makes these stainless steels more prone to IGSCC due to increased levels of sensitization.

It has been shown (Garcia *et al.* 2001) that lower levels of cold working (less than 10% for type 304 stainless steel) followed by sensitization heat treatment increased the IGSCC susceptibility and for higher deformations (over 30% for type 304 stainless steel) followed by sensitization heat treatment, the susceptibility to TGSCC increased.

5.10   Controlling stress corrosion cracking (SCC)

As shown in Fig. 5.2, three factors are required for SCC to take place: material, environment and stress/strain. Control of any one or multiple factors would result in avoidance/delay of SCC in stainless steels. The major factors that can be controlled are discussed below in brief.

5.10.1   Remedial measures related to materials

As has been described in earlier sections, addition of elements that increase stacking fault energy would result in increased resistance to SCC. Therefore increased nickel, molybdenum and nitrogen (above a threshold value) would result in higher resistance to SCC. It is to be noted that nitrogen addition up to a limit to austenitic stainless steels increases resistance to sensitization (Kain *et al.* 2004) and IGSCC (Bali *et al.* 2009, 2011). Beyond this limit, formation of chromium nitride upon sensitization heat treatment/welding occurs and, therefore, nitrogen acts as detrimental to SCC susceptibility. A highly resistant material, type 316NG (nuclear grade) has been used since the early 1990s in nuclear reactors against SCC (R L Jones 1991). Cerium addition to type 316 stainless steel has also been shown to be beneficial in reducing its susceptibility to sensitization, LTS (Watanabe *et al.* 2000) and sensitization induced IGSCC in BWR simulated environment (Watanabe *et al.* 1999).

5.10.2   Inhibitive coatings

Weldments are known to be the main locations of SCC failures. Previously the main reason for this was sensitization in the heat affected zones of weldments. Now, with the use of sensitization resistant stainless steels, the formation of high strain (strength) in the regions immediately adjacent to the weld fusion zone is recognized to be the main reason for SCC. In either case, covering the weldment with an inhibitive coating would avoid the contact of the specific environment with the susceptible material. This approach is being addressed by many researchers (Kim and Andresen 1998). Coatings like yttria stabilized zirconia (YSZ) have been tried. The main factors to be controlled are the adhesion with the base stainless steel and porosities in the YSZ coating. The adhesion issue is tackled by having an intermediate 100–300 μm thick layer of NiCrAlY and an equally thick layer of YSZ (Kim and Andresen 1998). It has been shown (Kim and Andresen 2003) that this coating mitigates SCC of stainless steels in high temperature aqueous environment. In addition to this, incorporation of zirconia fine particles in the surface passive layer of stainless steel was shown to be effective in catalysing the reactions to protect against IGSCC in BWR simulated environment (Yeh *et al.* 1997, 2002, 2004, Zhou *et al.* 2007).

5.10.3   Stress–strain-related remedial measures

Reducing residual stresses in the material helps to increase resistance against SCC. There exists a threshold limit below which SCC is not observed in laboratory tests. The SCC starts from the surfaces, therefore SCC initiation can be controlled by control of surface stresses. Changing the nature of surface stresses from tensile to compressive, e.g. by shot peening, has been shown (Sedriks 1996) to be beneficial against SCC initiation. It is also known that the stresses induced by shot peening gradually relax (Prevey and Hornbach 2008) during operation of components (especially at high temperatures). It has been shown (Angeliu *et al.* 2001) that shot peening does not affect the crack growth rates of austenitic stainless steels in high temperature oxygenated water when the crack depth is beyond the depth affected by peening. Therefore, shot peening must be repeated after a certain interval to maintain compressive stresses to avoid initiation of SCC. If the cold working introduced by surface treatment/peening is less than 5%, the introduced compressive stresses are expected to be stable (Prevey and Hornbach 2008).

Welding-induced stresses are a major factor leading to SCC. Reducing heat input during welding is a major approach to reduce the residual stresses. Newer welding techniques like narrow gap welding lead to lower heat input during the welding operation and also result in less distortion/strain in regions adjacent to the weld fusion zone. This results in increased resistance against SCC.

5.10.4   Environment-related remedial measures

Reducing the concentration of the species responsible for SCC in the environment (Ljungberg *et al.* 1988, Ruther *et al.* 1988, Kain 2008) is another approach to avoid/delay onset of SCC. Reduction in the chloride concentration and dissolved oxygen levels in the environment helps in reducing the extent of SCC. However, there are no certain levels of these species below which SCC would never occur for stainless steels in long-term operation of components (Warren 1960).

5.10.5   Cathodic protection

SCC essentially occurs in anodic regime for stainless steels. Therefore it is possible to control SCC by applying cathodic potentials (Logan 1966, Spaehn 1990). All grades of stainless steels can be protected by application of cathodic potentials. It has to be kept in mind that while applying cathodic potentials, the cracking susceptibility due to hydrogen should not be increased (especially for martensitic, ferritic and precipitation hardenable stainless steels). However, practical constraints, e.g. electrical isolation of components, size of the components, hydrogen production in the system, restrict application of cathodic protection in industry.

5.11   Sources of further information

Proceedings of a number of regular conferences are a rich source of detailed information on SCC of stainless steels and SCC in general. The NACE (National Association of Corrosion Engineers) annual technical conferences have a session devoted to SCC/environmentally assisted cracking (EAC). The conference on environmental degradation of materials in nuclear power systems – water reactors also has a number of papers on SCC. In addition several handbooks and research journals cover the SCC of stainless steels.

5.12   Conclusions

Austenitic stainless steels are known to be the workhorse of industry and are the most commonly employed stainless steels due to the combination of corrosion resistance, mechanical properties and ease of fabrication. Other categories of stainless steels, namely ferritic, martensitic, duplex and precipitation herdenable, are used in applications requiring specific properties like high mechanical strength or high resistance to corrosion. It is recognized that stress corrosion cracking of stainless steels occurs commonly by chloride solutions (95% of chloride SCC cases being by sodium chloride). It has been shown in this review that chloride SCC does occur even at ambient temperature and more commonly at temperatures up to the boiling temperature of chloride solutions. Magnesium chloride is an aggressive solution and is commonly used in accelerated tests to assess the comparative susceptibility of stainless steels to chloride SCC. The high temperature, high pressure aqueous environment, e.g. in nuclear power plants, is also known to cause SCC of stainless steels. It is an accepted fact that sensitization is not an essential feature required for SCC of stainless steels. Hot caustics and sulfides are the other known environments that cause SCC of stainless steels. Alloying elements affect the susceptibility to SCC, mainly by influencing the stacking fault energy of the alloy. Low stacking fault energy stainless steels are especially prone to TGSCC in chloride environments. Even non-metallic impurities and elements like phosphorus have a measurable influence on SCC. The ferritic, martensitic, duplex and precipitation hardenable stainless steels also show susceptibility to SCC under certain conditions of heat treatments and range of alloying elements. Grain boundary engineering to introduce a very high fraction of special (or random) grain boundaries has been shown in the last decade to be effective against sensitization and sensitization-induced SCC. Residual stresses and strains including those from welding, fit-up or due to various fabrication processes like cold or warm working, surface machining/grinding, etc., are now known to be responsible for SCC. Even surface finishing of the component has a clear influence on the initiation of SCC. The understanding developed so far has been summarized in this chapter and leads to mitigation measures against SCC by control over material composition and microstructure, control over environmental variables and control over stress–strain in the components.

5.13   References

Aaltonen P, Hanninen H, Nenonen P, Aho-Mantila I and Hakala J (1988), 'Aging related degradation of AISI 304 steel piping welds in BWR conditions', *Proceedings of the 3rd Environmental Degradation of Materials in Nuclear Power Systems – Water Reactors*, eds G.J. Theus, J.R. Weeks, The Materials Society, Warrendale, PA, 351–358.

Abe H and Watanabe Y (2008), 'Low temperature aging characteristics and SCC behavior of type 316L welds', *Proceedings of the 13th International Conference on Environmental Degradation of Materials in Nuclear Power Systems – Water Reactors*, The Canadian Nuclear Society, 1–9.

Adhe K N, Kain V, Madangopal K and Gadiyar H S (1996), 'Influence of sigma phase formation on the localized corrosion behavior of a duplex stainless steel', *Journal of Materials Engineering and Performance*, 5, 500–506.

Almeida L H, Le May I and Emygdio P R O (1998), 'Mechanistic modeling of dynamic strain aging in austenitic stainless steels', *Materials Characterization*, 41, 137–150.

Alousif O M and Nishimura R (2006), 'The effect of test temperature on SCC behavior of austenitic stainless steels in boiling saturated magnesium chloride solution', *Corrosion Science*, 48, 4283–4293.

Andresen P L (1993), 'Effect of noble metal coating and alloying on the stress corrosion crack growth rate of stainless steel in 288°C water', *Proceedings of the 6th International Symposium on Environmental Degradation of Materials in Nuclear Power Systems – Water Reactors*, eds Gold R E and Simonen E P, The Minerals, Metals and Materials Society, 245–253.

Andresen P L (1995), 'Application of noble metal technology for mitigation of stress corrosion cracking in BWRs', *Proceedings of the 7th International Symposium on Environmental Degradation of Materials in Nuclear Power Systems*, pp. 563–578.

Andresen P and Morra M M (2005), 'Effects of silicon on SCC of irradiated and unirradiated stainless steels and nickel alloys', *Proceedings of the 12th International Conference on Environmental Degradation of Materials in Nuclear Power Systems – Water Reactors*, eds Allen T R, King P J and Nelson L, The Minerals, Metals and Materials Society, 87–108.

Andresen P and Morra M (2007), 'Emerging issues in environmental cracking in hot water', *Proceedings of the 13th International Conference on Environmental Degradation of Materials in Nuclear Power Systems – Water Reactors*, Canadian Nuclear Society, The Minerals, Metals and Materials Society, 1–25.

Andresen P and Morra M M (2008), 'IGSCC of non-sensitized stainless steels in high temperature water', *Journal of Nuclear Materials*, 383, 97–111.

Andresen P L, Angeliu T M, Catlin W R, Young L M and Horn R M (2000), 'Effect of deformation on SCC of un-sensitized SS', *CORROSION 2000*, Paper no. 203, NACE, Houston, TX.

Andresen P L, Angeliu T M and Young L M (2001), 'Immunity, thresholds, and other SCC fiction', *Proc. Staehle Symp. on Chemistry and Electrochemistry of Corrosion and SCC*, The Materials Society, Feb. 2001.

Andresen P, Amigh P, Morra M and Horn R (2003), 'Effect of yield strength, corrosion potential, stress intensity factor, silicon and grain boundary character on SCC of stainless steels', *Proceedings of the 11th Conference on Environmental Degradation of Materials in Water Cooled Nuclear Power Plants*, The Minerals, Metals and Materials Society, American Nuclear Society and NACE International, Washington, DC, 816–833.

Angel T (1954), 'Formation of martensite in austentic stainless steels – effect of deformation, temperature and composition', *J. Iron and Steel Inst.*, 177, 165–174.

Angeliu T, Young L M, Andresen P L, Dunning G, Horn R, Willies E and van Diemen P (2001), 'The IGSCC behavior of shot peened austenitic alloys in BWR environments', *Proceedings of the 10th International Symposium on Environmental Degradation of Materials in Nuclear Power Systems – Water Reactors*, The Minerals, Metals and Materials Society, 1–9.

Angeliu T M, Andresen P L, Hall E, Sutliff J A and Sitzman S (2000), 'Strain and microstructural characterization of austenitic stainless steel weld HAZs', *CORROSION 2000*, NACE paper no. 186, NACE, Houston, TX.

Angeliu T M, Andresen P L and Pollack M L (1998), 'The deformation response of L-grade stainless steels relative to IGSCC in 288°C water, *Corrosion 1998*, Paper no 98135, NACE Houston.

Arnoux P (2010), 'Atomistic simulations of stress corrosion cracking', *Corrosion Science*, 52, 1247–1257.

Baker H R, Bloom M C, Bolster R N and Singleterry C R (1970), 'Film and pH effects in the stress corrosion cracking of type 304 stainless steel', *Corrosion*, 26, 420–426.

Bali S C, Kain V and Raja V S (2009), 'Effect of low temperature sensitization on IGSCC behavior of austenitic stainless steels in simulated boiling water reactor environment', *Corrosion*, 65, 726–740.

Bali S C, Takeda Y, Kain V, Raja V S and Shoji T (2011), 'Effect of nitrogen addition in type 304L stainless steel on low temperature sensitization and IGSCC behavior in simulated BWR environment', *Corrosion Science*, forthcoming.

Bavarian B, Szklarska-Smialowska Z and Macdonald D D (1982), 'Effect of temperature on the stress corrosion cracking of tempered type 403 martensitic stainless steel in sodium sulfate solution', *Corrosion*, 38(12), 604–608.

Bednar L (1979), 'Chloride stress corrosion cracking in non-sensitized ferritic stainless steels', *Corrosion*, 35, 96–100.

Ben Rhouma A, Braham C, Fitzpatrick M E, Lédion, J and Sidhom H (2001), 'Effects of surface preparation on pitting resistance, residual stress and stress corrosion cracking in austenitic stainless steels', *Journal of Materials Engineering and Performance*, 10, 507–514.

Bond A P and Dundas H J (1968), 'Effects of composition on the stress corrosion cracking of ferritic stainless steels', *Corrosion*, 24, 344–352.

Bose A and De P K (1987), 'An EPR study on the influence of prior cold work on the degree of sensitization of AISI 304 stainless steel', *Corrosion*, 43, 624–631.

Braham C, Ben Rhouma A, Lédion J and Sidhom H (2005), 'Effect of machining conditions on residual stress corrosion cracking of 316L SS', *Materials Science Forum*, 490–491, 305–310.

Brandon D G (1966), 'The structure of high-angle grain boundaries', *Acta Metall*, 14, 1479–1484.

Brown B F (1977), Stress corrosion cracking control measures, NBS monograph 156.

Bruemmer S M and Thomas L E (2005), 'High resolution characterization of stress corrosion cracks in austenitic stainless steels from crack growth rate tests in BWR simulated environment', *Proceedings of the 12th International Conference on Environmental Degradation of Materials in Nuclear Power Systems – Water Reactors*, eds Allen T R, King P J and Nelson L, The Minerals, Metals and Materials Society, 189–198.

Bruemmer S M, Arey B W and Charlot L A (1993), 'Grain boundary chromium concentration effects on IGSCC and IASCC of austenitic stainless steels', *6th International Symposium on Environmental Degradation of Materials in Nuclear Power Systems – Water Reactors*, eds R E Gold and E P Simonen, The Minerals, Metals and Materials Society.

Byun T S, Hashimoto N and Farell K (2004), 'Temperature dependence of strain hardening and plastic instability behaviours in austenitic stainless steels', *Acta Materialia*, 52, 3889–3899.

Cihal V (1985), 'The influence of nitrogen, phosphorus, sulfur and nickel on the stress corrosion cracking of austenitic Fe-Ni-Cr alloys', *Corrosion Science*, 25(8/9), 815–819.

Clarke W L, Cowan R L and Walker W L (1978), 'Comparative methods for measuring degree of sensitization in stainless steel', ASTM STP 656 in: Stegerwald R F (eds), *Intergranular Corrosion of Stainless Alloys*, ASTM, West Conshohocken, PA.

Cochran R W and Staehle R W (1968), 'Effects of surface preparation on stress corrosion cracking of type 304 stainless steel in boiling 42% magnesium chloride', *Corrosion*, 11, 369–378.

Copson H R (1959), *Physical Metallurgy of Stress Corrosion Fracture*, Interscience.

Cowan R L and Kiss E (1993), 'Optimizing BWR water chemistry', *Proceedings of the 6th International Symposium on Environmental Degradation of Materials in Nuclear Power Systems*, eds Gold R E and Simonen E P, The Minerals, Metals and Materials Society, 889–895.

Cowan R L, Indig M E, Kass J N, Law R J and Sundberg L L (1986), 'Experience with hydrogen water chemistry in boiling water reactors', *Proceedings of the International Conference on Water Chemistry of Nuclear Reactor Systems 4*, Bournemouth, 13–17 October, British Nuclear Energy Society, London, 29-36.

Dai Q-X, Wang A-D, Cheng X-N and Luo X-M (2002), 'Stacking fault energy of cryogenic austenitic steels', *Chinese Physics*, 11, 596–600.

Dillon C P (1990), 'Imponderables in chloride stress corrosion cracking of stainless steels', *Materials Performance*, 29(12), 66–67.

Edeleanu C (1953), 'Transgranular stress corrosion in Cr-Ni stainless steels', *J. Iron Steel Inst.*, 173, 140–146.

Engelberg D L, Marrow T J, Newman R C and Babout L (2004), 'Grain boundary engineering for crack bridging: a new model for IGSCC propagation', in *Environment Induced Cracking of Material*, Shipilov S A, Jones R H, Olive J M and Rebak R B (eds), Elsevier, 2007, 69–79. Proceedings of the International conference on environment induced cracking of materials, EICM-2, held at Banff Centre, Alberta, Canada, 19–23 September, 2004.

EPRI GS 6913 (1990), Feedwater heaters: replacement specification guidelines, Part 1.4, Tubing selection and preparation, EPRI final report GS 6913, project 2504-5, August.

Fontana M G (1987), *Corrosion Engineering*, 3rd edn, McGraw-Hill.

Ford F P (1982), 'Mechanisms of environmental cracking peculiar to the power generation industry', Report NP2589, EPRI, Palo Alto, September.

Ford F P, Gordon B M and Horn R M (2006), 'Corrosion in boiling water reactors', in *ASM Metals Handbook, Vol. 13C, Corrosion-Environments and Industries*, ASM, 341–361.

Frechard S, Redjaimia A, Lach E and Lichtenberger A (2006), 'Mechanical behavior of nitrogen-alloyed austenitic stainless steel hardened by warm rolling', *Materials Science and Engineering A*, 415, 219–224.

Fujita M, Kaneko Y, Nohara A, Saka H, Zauter R and Mughrabi H (1994), 'Temperature dependence of the dissociation width of dislocations in a commercial 304L stainless steel', *ISIJ International*, 34, 697–703.

G108–94(2010), ASTM, Standard test method for electrochemical reactivation for detecting sensitization of AISI 304 and 304L stainless steels, 2010.

G123–00(2005), ASTM, Standard test method for evaluating stress corrosion cracking resistance of alloys with different nickel content in boiling acidified sodium chloride solution, 2005.

G36–94(2006), ASTM, Standard practice for evaluating stress corrosion resistance of metals and alloys in a boiling magnesium chloride solution, 2006.

Garcia C, Martin F, de Tiedera P, Heredero J A and Aparicio M L (2001), 'Effects of cold work and sensitization heat treatment on chloride stress corrosion cracking in type 304 stainless steels', *Corrosion Science*, 43, 1519–1539.

Gaudett M A and Scully J R (1993), 'Distributions of Cr depletion levels in sensitized AISI 304 stainless steel and its implications concerning intergranular corrosion phenomena', *Journal of the Electrochemical Society*, 140, 3425–3435.

Gaudett M A and Scully J R (1994), 'Application of bond percolation theory to intergranular stress corrosion cracking of sensitized AISI 304 stainless steel', *Metall. Mater. Trans.* 25A, 775–787.

Ghosh S and Kain V (2010a), 'Effect of surface machining and cold working on the ambient temperature chloride stress corrosion cracking susceptibility of AISI 304L', *Materials Science and Engineering A*, 527, 679–693.

Ghosh S and Kain V (2010b), 'Microstructural changes in AISI 304L stainless steel due to surface machining: effect on its susceptibility to chloride stress corrosion cracking', *Journal of Nuclear Materials*, 403, 62–67.

Ghosh S, Rana V P S, Kain V, Mittal V and Baveja S K (2011), 'Role of residual stresses induced by industrial fabrication on stress corrosion cracking susceptibility of austenitic stainless steels', *Materials and Design*, 32(7), 3823–3831.

Gnanamoorthy J B (1990), 'Stress corrosion cracking of un-sensitized stainless steels in ambient temperature coastal atmosphere', *Materials Performance*, 29(12), 63–65.

Gott K (2001), 'Cracking data base as a basis for risk informed inspection', *Proceedings of the 10th International Symposium on Environmental Degradation of Materials in Nuclear Power Systems – Water Reactors*, The Minerals, Metals and Materials Society, 5–9 August, 2001.

Grubb J F, DeBold T and Fritz J D (2005), 'Corrosion of wrought stainless steels', in *ASM Handbook 13B*, ASM International, 54–77.

Hannani A, Kermiche F, Pourbaix A and Belmorke K (1997), 'Characterization of passive film on AISI 304 stainless steel', *Trans IMF*, 75(1), 7–9.

Hanninen H E (1979), 'Influence of metallurgical variables on environment sensitive cracking of austenitic alloys', *International Metals Reviews*, 24(3), 85–135.

Hettiarachchi S, Wozaldo G P and Diaz T P (1995), 'A novel approach for nobel metal deposition on surfaces for intergranular stress corrosion cracking mitigation in boiling water reactor internals', *CORROSION 95*, paper no. 413, NACE International, Orlando, FL.

Hettiarachchi S, Cowan R L, Law R J, Miller W D and Diaz T P (2000), 'Noble metal chemical addition for IGSCC mitigation of BWRs – field successes', *CORROSION 2000*, Paper no. 184, NACE, Houston, TX.

Hines J G and Jones E R W (1961), 'Some effects of alloy composition on the stress corrosion behaviour of austenitic Cr-Ni steels', *Corrosion Science*, 1, 88–107.

Horn R M, Gordon G M, Ford F P and Cowan R L (1997), 'Experience and assessment of stress corrosion cracking in L grade stainless steels BWR internals', *Nuclear Engineering and Design*, 174, 313–325.

Hoxie E C (1977), 'Some corrosion considerations in the selection of stainless steels for pressure vessels and piping', in *Pressure Vessels and Piping: A Decade of Progress, Vol. 3*, American Society of Mechanical Engineers.

Isaacs H S, Vyas B and Kendig M W (1982), 'The stress corrosion cracking of sensitized stainless steels used in thiosulfate solutions', *Corrosion*, 38 (3), 130–136.

Itzhak D and Elias O (1994), 'Behaviour of type 304 and type 316 austenitic stainless steels in 55% lithium bromide heavy brine environment', *Corrosion*, 50, 131–137.

Jin W, Yang S, Kokawa H, Wang Z and Sato Y S (2007), 'Improvement of intergranular stress corrosion crack susceptibility of austenite stainless steel through grain boundary engineering', *Journal of Materials Science and Technology*, 23(6), 785–789.

Jones D A (1992), *Principles and Prevention of Corrosion*, 2nd edn, Prentice-Hall.

Jones R and Randle V (2010), 'Sensitization behavior of austenitic stainless steel', *Materials Science and Engineering A*, 527, 4275–4280.

Jones R H (1992), *Stress Corrosion Cracking – Materials Performance and Evaluation*, ASM International, Materials Park, OH.

Jones R H (2003), 'Stress corrosion cracking', in *ASM Handbook 13A*, ASM International, 346–366.

Jones R L (1991), 'Prevention of stress corrosion cracking in boiling water reactors', *Materials Performance*, 30(2), 70–73.

Kain R M (1990), 'Marine atmospheric stress corrosion cracking of austenitic stainless steels', *Materials Performance*, 29(12), 60–62.

Kain V (1997), 'Sensitization and susceptibility of austenitic stainless steels to intergranular corrosion and intergranular stress corrosion cracking', PhD thesis, Indian Institute of Technology, Bombay.

Kain V (2008), 'Water chemistry: cause and control of corrosion degradation in nuclear power plants', *Proceedings of the National Symposium on Operational and Environmental Issues concerning use of Water as Coolant in Power Plants and Industries*, Kalpakkam, 15–16 December 2008, BRNS and INS, Kalpakkam, pp. 19–28.

Kain V and De P K (2003), 'Transformation of delta ferrite during high heat input welding of austenitic stainless steels', *Materials Performance*, 42(3), 50–54.

Kain V, Palit G C, Chouthai S S and Gadiyar H S (1989), 'Effect of heat treatment on the critical pitting potential of SS 304 in 0.01 N NaCl', *Journal of the Electrochemical Society of India*, 38(1), 505–509.

Kain V, Prasad R C, De P K and Gadiyar H S (1995), 'Corrosion assessment of AISI 304L stainless steel in nitric acid environments – an alternate approach', *ASTM Journal of Testing and Evaluation*, 23(1), 50–54.

Kain V, Prasad R C and De P K (2002), 'Testing sensitization and predicting susceptibility to intergranular corrosion and intergranular stress corrosion cracking', *Corrosion*, 58, 15–38.

Kain V, Chandra K, Adhe K N and De P K (2004), 'Effect of cold work on low temperature sensitization behaviour of austenitic stainless steels', *Journal of Nuclear Materials*, 334(2–3), 115–132.

Kain V, Samataray R, Acharya S, De P K and Raja V S (2007), 'Influence of low temperature sensitization on stress corrosion cracking of 304LN stainless steels, in *Environment-Induced Cracking of Materials: Prediction, Industrial Developments and Evaluation*, Shipilov S A, Jones R H, Olive J M, and Rebak R B (eds), Elsevier, Amsterdam, 163–172.

Kain V, Ghosh S, Mittal V and Baveja S K (2010), 'Effect of residual stress and strain generated during manufacturing process on the stress corrosion cracking susceptibility of austenitic stainless steel', *CORROSION 2010*, Paper No. 10304 NACE International, Houston, TX.

Karlsen W, Diego G and Devrient B (2010), 'Localised deformation as a key precursor to initiation of intergranular stress corrosion cracking of austenitic stainless steels employed in nuclear power plants', *Journal of Nuclear Materials*, 406(1), 138–151.

Kekkonen T, Aaltonen P and Hanninen H (1985), 'Metallurgical effects on the corrosion resistance of a low temperature sensitized welded AISI type 304 stainless steel', *Corrosion Science*, 25, 821–836.

Kim Y J (1996), 'Investigation of oxide film formed on 304 SS in high temperature water containing oxygen, hydrogen and hydrogen peroxide', *Corrosion 1996*, Paper no. 102, NACE International, 1–22.

Kim Y J and Andresen P L (1998), 'Application of insulated protective coatings for reduction of corrosion potentials of type 304 stainless steel in high temperature water', *Corrosion*, 54, 1012–1017.

Kim Y J and Andresen P L (2003), 'Application of insulated protective coating for 304 SS SCC mitigation in 288°C water', *Proceedings of the 11th International Conference on Environmental Degradation of Materials in Nuclear Power Systems – Water Reactors*, The Minerals, Metals and Materials Society, Stevenson, Washington, 526-537.

Kowaka M and Fujikawa H (1972), *Sumitomo Search*, No. 7, p. 10.

Kowaka M and Kudo T (1977), 'Stress corrosion cracking behavior of ferritic and duplex stainless steels in a caustic solution', Japan Society of Corrosion Engineers, Tokyo, May.

Kuniya, J, Masaoka I and Sasaki R (1988), 'Effect of cold work on the stress corrosion cracking of nonsensitized AISI 304 stainless steel in high temperature oxygenated water', *Corrosion*, 44, 21–28.

Latanision R M and Ruff A W (1971), 'The temperature dependence of stacking fault energy in Fe-Cr-Ni alloys', *Metallurgical Transactions*, 2, 505–509.

Lee W-S and Lin C-F (2001), 'Impact properties and microstructure evolution of 304L stainless steels', *Materials Science and Engineering A*, 308, 124–135.

Lillys P and Nehrenberg A E (1956), 'Effect of tempering temperature on stress corrosion cracking and hydrogen embrittlement of martensitic stainless steels', *Trans ASM*, 48, 327–355.

Ljungberg L G, Cubicciotti D and Trolle M (1988), 'Effects of impurities on the IGSCC of stainless steel in high temperature water', *Corrosion*, 44, 66–72.

Lo K H, Shek C H and Lai J K L (2009), 'Recent developments in stainless steels', *Materials Science and Engineering R*, 65, 39–104.

Logan H L (1966), 'Stress corrosion cracking of stainless steels', in *The Stress Corrosion of Metals*, John Wiley and Sons, New York, 100–155.

Maier I A, Manfredi C and Galvele J R (1985), 'The stress corrosion cracking of an austenitic stainless steel in HCl + NaCl solutions at room temperature', *Corrosion Science*, 25, 15–34.

McIlree A R and Michels H T (1977), 'Stress corrosion behavior of Fe-Cr-Ni and other alloys in high temperature caustic solutions', *Corrosion*, 33, 60–67.

Michiuchi M, Kokawa H, Wang Z J, Sato Y S and Sakai K (2006), 'Twin-induced grain boundary engineering for 316 austenitic stainless steel', *Acta Materialia*, 54, 5179–5184.

Mullner P, Solenthaler C and Speidel M O (1994), 'Second order twinning in austenitic steel', *Acta Metallurgica et Materialia*, 42, 1727–1732.

Murr L E (1975), *Interfacial Phenomena in Metals and Alloys*, Reading, MA, Addison Wesley (Reprinted 1991 by Tech Books, Fairfax, VA).

Newman R C and Shahrabi T (1987), 'The effect of alloyed nitrogen or dissolved nitrate on the anodic behavior of austenitic stainless steels', *Corrosion Science*, 27(8), 827–838.

Oldfield J W and Todd B (1990), 'Ambient temperature stress corrosion cracking of austenitic stainless steels in swimming pools', *Materials Performance*, 29(12), 57–58.

Oldfield J W and Todd B (1991), 'Room temperature stress corrosion cracking of stainless steels in indoor swimming pool atmospheres', *British Corrosion Journal*, 26(3), 173–182.

Palit G C, Kain V and Gadiyar H S (1993), 'Electrochemical investigations of pitting corrosion in nitrogen bearing 316LN stainless steels', *Corrosion*, 49, 977–991.

Parkins R N (1979), 'Environment-sensitive fracture-controlling parameters', in *Proceedings of 3rd International Conference on Mechanical Behavior of Materials*, ed. K J Miller and R F Smith, Cambridge, 20–24 August 1979, Vol. 1, Pergamon, pp. 139–164.

Parkins R N (1992), 'Environment sensitive fracture of metals', *Canadian Metallurgical Quarterly*, 31(2), 79–94.

Perillo P M and Duffo G S (1990), 'Stress corrosion cracking of sensitized type 304 stainless steel in thiocyanate solutions', *Corrosion*, 46(7), 545–546.

Pickering F B (1976), 'Physical metallurgy of stainless steel developments', *International Metals Reviews*, 21, 227–268.

Povich M J (1978), 'Low temperature sensitization of type 304 stainless steel', *Corrosion*, 34, 60–65.

Povich M J and Rao P (1978), 'Low temperature sensitization of welded type 304 stainless steel', *Corrosion*, 34, 269–275.

Prevey P and Hornbach D (2008), 'SCC mitigation in nuclear components using low plasticity burnishing (The importance of cold work in thermal relaxation as it affects SCC initiation)', Workshop on Detection, Avoidance, Mechanisms, Modeling, and Prediction of SCC Initiation in Water-Cooled Nuclear Reactor Plants, 7–12 September, Beaune, France.

Prosek T, Iversen A and Taxen C (2008), 'Low temperature stress corrosion cracking of stainless steels in the atmosphere in presence of chloride deposits', *CORROSION 2008*, Paper no. 8484, NACE International.

Rhodes C G and Thompson A W (1977), 'The composition dependence of stacking fault energy in austenitic stainless steels', *Metallurgical Transactions A*, 8, 1901–1906.

Roychowdhury S, Kain V and Prasad R C (2011a), 'Effect of nitrogen content in sensitized austenitic stainless steel on crack growth rate in simulated BWR environment', *Journal of Nuclear Materials*, 410(1–3), 59–68.

Roychowdhury S, Kain V, Gupta M and Prasad R C (2011b), 'IGSCC crack growth in simulated BWR environment – effect of nitrogen content in non-sensitized and warm rolled austenitic stainless steel', *Corrosion Science*, 53(3), 1120–1129.

Roychowdhury S, Neogy S, Gupta M, Kain V, Srivastava D, Dey G K and Prasad R C (2011c), 'Effect of test temperature and prior straining on the deformation mode of austenitic stainless steel during tensile testing', Paper No. 119, pp. 119–126 in the proceeding of the EPD Congress 2011, *Characterization of Minerals, Metal and Materials*, 27 February–3 March 2011, TMS, San Diego, CA, USA.

Ruther W E, Soppet W K and Kassner T F (1988), 'Effect of temperature and ionic impurities at very low concentrations on stress corrosion cracking of AISI 304 stainless steel', *Corrosion*, 44, 791–799.

Ryabchenkov A V, Gerasimov V I and Sidorov V P (1966), *Prot. Met. (USSR)*, 2, p. 217.

Samans C (1964), 'Stress corrosion cracking susceptibility of stainless steels and nickel based alloys in polythionic acid and copper sulfate solutions', *Corrosion*, 20(8), 256t–262t.

Schneider H (1960), 'Investment casting of high hot strength 12-per-cent chrome steel', *Foundary Trade Journal*, 108, 562–563.

Schramm R E and Reed R P (1975), 'Stacking fault energies of seven commercial stainless steels', *Metallurgical Transactions A*, 6A, 1345–1351.

Scott, P M (1985), 'A review of environment sensitive fracture in water reactor materials', *Corrosion Science*, 25(8/9), 583–606.

Scott P (1994), 'A review of irradiation assisted stress corrosion cracking', *Journal of Nuclear Materials*, 211, 101–122.

Sedriks A J (1986), 'Effect of alloy composition and microstructure on passivity of stainless steels', *Corrosion*, 42(7), 376–389.

Sedriks A J (1996), *Corrosion of stainless steels*, 2nd edn, John Wiley and Sons, New York.

Shimada M, Kokawa H, Wang Z J, Sato Y S and Karibe I (2002), 'Optimization of grain boundary character distribution for intergranular corrosion resistant 304 stainless steel by twin-induced grain boundary engineering', *Acta Metallurgica*, 50(9), 2331–2341.

Shoji T (2003), 'Progress in the mechanistic understanding of BWR SCC and its implication to the prediction of SCC growth behavior in plants', *Proceedings of the 11th International Conference on Environmental Degradation of Materials in Nuclear Systems*, Stevenson, WA, 10–14 August.

Shoji T, Lu Z and Murakami H (2010), 'Formulating stress corrosion cracking growth rates by combination of crack tip mechanics and crack tip oxidation kinetics', *Corrosion Science*, 52, 769–779.

Sieradzki K and Freidersdorf F J (1994), 'Notes on the surface mobility mechanism of stress corrosion cracking', *Corrosion Science*, 36(4), 669–675.

Singh Raman, R K and Siew W H (2010), 'Role of nitrite addition in chloride stress corrosion cracking of a super duplex stainless steel', *Corrosion Science*, 52, 113–117.

Smith D A and Pond R C (1976), 'Bollmann's O-lattice theory; a geometrical approach to interface structure', *International Metals Reviews*, 21, 61–74.

Solomon H (1985), 'Influence of prior deformation and composition on continuous cooling sensitization of AISI 304 stainless steel', *Corrosion*, 41, 512–517.

Spaehn H (1990), in *Environmental-induced Cracking of Materials*, R.P. Gangloff and M.B. Ives (eds), NACE, Houston, TX, 449–487.

Speidel M O (1977), Stress corrosion cracking of austenitic stainless steels, Ohio State University report to the Advanced Research Projects Agency, ARPA order no. 2616, Contract no. N00014-75-C-0703.

Speidel M O (1981), 'Stress corrosion cracking of stainless steels in NaCl solutions', *Metallurgical Transactions A*, 12, 779–789.

Speidel M O (1984), 'Stress corrosion cracking and corrosion fatigue–fracture mechanics', in *Corrosion in Power Generating Equipment*, Speidel M O and Atrens A (eds), Plenum Press, New York, 331–357.

Sridhar N (2011), 'Mechanisms of stress corrosion cracking', in *Stress Corrosion Cracking*, Raja V S and Shoji T (eds) Woodhead Publishing, Cambridge.

Staehle R W and Agarwal A K (1976), Corrosion, stress corrosion cracking and electrochemistry of the Fe and No based alloys in caustic environments, Ohio State University Report to ERDA, Contract E(11-1)-2421.

Steigerwald R F, Bond A P, Dundas H J and Lizlovs E A (1977), 'The new Fe-Cr-Mo ferritic stainless steels', *Corrosion*, 33, 279–295.

Streicher M A (1975), 'Stress corrosion of ferritic stainless steels', at *CORROSION 1975*, Paper no. 68, NACE International, Ohio.

Sugimoto K and Sawada Y (1976), 'The role of alloyed molybdenum in austenitic stainless steel in the inhibition of pitting in neutral halide solutions', *Corrosion*, 32(9), 347–352.

Suss H (1962), 'Untempered martensite affects corrosion of type 410 stainless', *Metal Progress*, 82(5) 89–91.

Thomas K C, Ferrari H M and Allio R J (1964), 'Stress corrosion of type 304 stainless steel in chloride environments', *Corrosion*, 20, 89t–92t.

Thomas L E and Bruemmer S M (2002), 'Analytical transmission electron microscopy characterization of stress corrosion cracks in irradiated type 316 stainless steel core component', *Proc. Fontevraud*, 5, 23–27 September.

Tipping P (1996), 'Lifetime and aging management of nuclear power plants: a brief overview of some light water reactor component aging degradation problems and ways of mitigation', *International Journal of Pressure Vessel and Piping*, 66, 17–25.

Torchio S (1980), 'Stress corrosion cracking of type 304 stainless steel at room temperature: influence of chloride concentration and acidity', *Corrosion Science*, 20, 555–561.

Trozzo P S and McCartney R F (1960), 'Relationship of microstructure and stress corrosion cracking of type 410 stainless steel', *Corrosion*, 16(3), 26–30.

Truman J E (1969), 'Methods available for avoiding SCC of austenitic stainless steels in potentially dangerous environments', in *Stainless Steels*, ISI publication 117, Iron and Steel Institute.

Truman J E (1971), 'The effects of composition and structure on the resistance to stress corrosion cracking of stainless steels', in *British Nuclear Energy Society Symposium on Effects of Environment on Materials Properties in Nuclear Systems*, paper no. 10, Institute of Civil Engineers.

Truman J E (1977), 'The influence of chloride content, pH and temperature on the occurrence of stress corrosion cracking with austenitic stainless steels', *Corrosion Science*, 17, 737–746.

Truman J E (1981), 'Stress corrosion cracking of martensitic and ferritic stainless steels', *International Metals Reviews*, 26(6), 301–349.

Truman J E and Perry R (1960), *British Corrosion Journal*, 1, 60.

Truman J E and Perry R (1965), 'The resistance of stress corrosion cracking of some Cr-Ni-Fe austenitic steels and alloys', *British Corrosion Journal*, 1(2), 60–66.

Tsai W-T and Chen M S (2000), 'Stress corrosion cracking behavior of 2205 duplex stainless steel in concentrated NaCl solution', *Corrosion Science*, 42, 545–559.

Turnbull A and Psaila-Dombrowski M (1992), 'A review of electrochemistry of relevance to environment-assisted cracking in light water reactors', *Corrosion Science*, 33, 1925–1966.

USNRC Regulatory Guide 1.36 (1973), Nonmetallic thermal insulation for austenitic stainless steel.

Warren D (1960), 'Chloride bearing cooling water and the stress corrosion cracking of austenitic stainless steel', in *Proceedings of 15th Annual Purdue Industrial Waste Conference*, Purdue University, 1.

Wasnik D, Kain V, Samajdar I, Verlinden B and De P K (2002), 'Resistance to sensitization and intergranular corrosion through extreme randomization of grain boundaries', *Acta Metallurgica*, 50, 4587–4601.

Wasnik D, Kain V, Samajdar I, Verlinden B and De P K (2003), 'Controlling grain boundary energy to make austenitic stainless steels resistant to intergranular stress corrosion cracking', *Journal of Materials Engineering and Performance*, 12, 402–407.

Watanabe Y and Kondo T (2000), 'Current and potential fluctuations in IGSCC of stainless steel', *Corrosion*, 56(12), 1250–1255.

Watanabe Y, Tonozuka T, Shoji T, Kondo T and Masuyama F (1999), 'Sensitization properties and IGSCC susceptibility of cerium modified stainless steel', *Corrosion 99*, Paper No. 453, NACE.

Watanabe Y, Kain V, Tonozuka T, Shoji T, Kondo T and Masuyama F (2000), 'Effect of Ce addition on the sensitization properties of stainless steels', *Scripta Materialia*, 42, 307–312.

Wheeler G C and Howells E (1960), 'A look at caustic stress corrosion', *Power*, 104(9), 86–87.

Wilms M E, Gadgil V J, Krougmen J M and Ijsseling F A (1994), 'The effect of sigma phase precipitation at 800°C on the corrosion resistance in sea water of a high alloyed duplex stainless steel', *Corrosion Science*, 36, 871–881.

Wilson I L, Pement F W and Aspden R G (1977), 'Stress corrosion on some stainless steels in elevated temperature aqueous environments', *CORROSION/77*, Paper no. 136, NACE.

Yeh T-K, Liang C-H, Yu M-S and Macdonald D D (1997), The effect of catalytic coatings on IGSCC mitigation for boiling water reactors operated under hydrogen water chemistry', *Proceedings of the 8th International Symposium on environmental Degradation of Materials in Nuclear Power Systems – water Reactors*, Amelia Island, GA, NACE International, 1, 551–558.

Yeh T-K, Lee M-Y and Tsai C-H (2002), 'Intergranular stress corrosion cracking of type 304 stainless steels treated with inhibitive chemical in simulated boiling water reactor environments', *Journal of Nuclear Science and Technology*, 39, 531–539.

Yeh T-K, Tsai C-H, Chen C-H, Cheng C-M, Sheng R.-F and Chu F (2004), 'Crack growth of ZrO_2 treated type 304 stainless steels in high temperature pure water', *International Water Chemistry Conference*, San Francisco, pp. 627–673.

Zhou Z F, Chalkova E, Lvov S N, Chou P and Pathania R (2007), 'Development of a hydrothermal deposition process for applying zirconia coatings on BWR materials for IGSCC mitigation', *Corrosion Science*, 49, 830–843.

Zucchi F, Trabanelli G and Demertzis G (1988), 'The intergranular stress corrosion cracking of a sensitized AISI 304 in NaF and NaCl solutions', *Corrosion Science*, 28, 69–79.

6

Factors affecting stress corrosion cracking (SCC) and fundamental mechanistic understanding of stainless steels

T. SHOJI, Z. LU and Q. PENG, Tohoku University, Japan

Abstract: The effects of metallurgical/material, environmental and mechanical factors and their synergistic contributions to stress corrosion cracking are summarized and analyzed. The contents are focused on stress corrosion cracking of materials in nuclear power plants and related key engineering parameters.

Key words: stress corrosion cracking, oxidation localization, oxidation acceleration, oxidation kinetics, crack growth model.

6.1   Introduction

Stress corrosion cracking (SCC) is defined as the growth of cracks due to the simultaneous action of a stress (nominally static and tensile) and a reactive environment [1]. SCC is the result of the combined and synergistic interactions of mechanical stress and corrosion/oxidation reactions [2]. The type of materials, types of loading, and types of environments affect stress corrosion cracking including initiation and propagation. Some contributing factors to SCC are summarized, emphasizing two main features of SCC: localization and acceleration of oxidation reactions [3].

6.2   Metallurgical/material factors

SCC is strongly affected by material chemistry and microstructure as a result of alloy compositions, forming methods, heat-treatment processes, and fabrication processes. Bulk alloy composition can affect passive film stability and phase distribution, for example, chromium in stainless steel and Ni-base alloys. Minor elements can cause local changes in passive film-forming elements, for example, carbon in stainless steels can cause sensitization. Impurity elements can segregate to grain boundaries and cause local differences in oxidation rate. Inclusions may affect local crack tip chemistry changes as the crack intersects them. Heterogeneity in material chemistry and microstructure are also found to affect SCC. Some examples

245

of stress corrosion cracking in (simulated) nuclear power plant environments are used to show the effects of metallurgical factors.

6.2.1   Grain boundary precipitation and chromium-depletion in stainless steels and stress corrosion cracking in high temperature water

Since the late 1960s, there have been many cases of stress corrosion cracking (SCC) in boiling water reactor (BWR) pressure boundary components, such as primary loop recirculation (PLR) pipes made with 304 stainless steel (SS) [4, 5], where the cracking was caused by typical weld sensitization in heat-affected zones due to carbide precipitation and chromium depletion along grain boundaries. Chromium carbide precipitation in stainless steel occurs in the temperature range 500–850°C, with the rate of precipitation controlled by chromium diffusion [6]. The degree of sensitization (DOS) can be measured by corrosion tests such as the Strauss test [7] or electrochemical potentiokinetic reactivation (EPR) tests [8, 9] which can be expressed by time/temperature/DOS diagrams or time/temperature/sensitization (TTS) diagrams. Stress corrosion cracking of sensitized stainless steel in simulated boiling water reactor environments is strongly affected by water chemistry such as dissolved oxygen or electrochemical potential, and concentration of impurities [4, 5].

6.2.2   The effect of strain-hardening on SCC of alloys in high temperature water

Low carbon stainless steels were used in boiling water reactors to replace the previously used 304 SS for mitigating SCC. However, stress corrosion cracking was also found in PLR pipes and core shrouds made of low carbon stainless steels in BWR plants [10–14]. Extensive analyses on these cracked components show that residual stress and strain due to weld shrinkage are crucial for the high crack growth rates observed in BWR plants. Figure 6.1 shows the schematic of recent SCC in BWR PLR piping, based on the review of original SCC data of BWR plant components [10, 11]. Several types of crack growth behavior were categorized [10, 11, 15]: growing in the heat affected zone (HAZ), stopping at fusion boundary, penetrating into weld metal, and changing the growing direction in the HAZ along the fusion line. It was believed that cracks initiated as transgranular cracks in the surface cold worked layer and then propagated in the hardened HAZ along grain boundaries (intergranular).

Figure 6.2(a) shows the Vickers hardness in the HAZ decreases with increasing distance from the fusion line. Figure 6.2(b) show the measured crack growth kinetics in 316L HAZ in simulated BWR environments. Recent

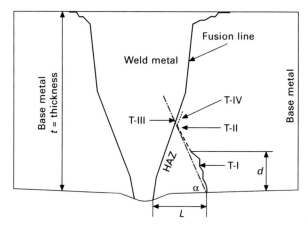

6.1 Schematic representation of several types of cracking behavior in PLR piping in BWR plants [10, 11, 15].

results showed that cold working or warm working increases the SCC growth rates of low carbon stainless steels in simulated BWR environments (Fig. 6.3) [10, 16–26].

Stress corrosion cracking growth rates of one-directionally (1D) warm-rolled 304L stainless steel specimens in T-L and L-T orientations were measured in oxygenated and deoxygenated pure water at 288°C [27]. More extensive intergranular SCC is observed in the specimen T-L orientation than in the specimen L-T orientation (Figs 6.4 and 6.5). In deoxygenated and oxygenated pure water environments at 288°C, SCC growth rates in the specimen T-L orientation are higher than those in the specimen L-T orientation. The corrosion fatigue crack growth rates during the in situ fatigue procedures are similar in both T-L and L-T specimens at each load ratio *R* (Fig. 6.6).

Figure 6.7 shows that the SCC growth rate of cold worked stainless steels increases with increasing yield strength in simulated PWR environments [10, 28, 29]. Results also show that CGR of cold-rolled Alloy 600 in a simulated pressurized water reactor primary water environment increases with increasing yield strength [30–34]. SCC growth rates in Alloy 600 HAZ is significantly higher than that in the base metal in simulated PWR environments [34, 35]. Strain-hardening in Alloy 600 HAZ is characterized by Vickers hardness measurements and misorientation measurement by electron back-scattering diffraction (EBSD) (Fig. 6.8).

6.2.3   Effects of metallurgical heterogeneity on SCC

Figure 6.9 shows a typical morphology of intergranular stress corrosion crack tip of cold worked 316NG stainless steel in simulated BWR water [36],

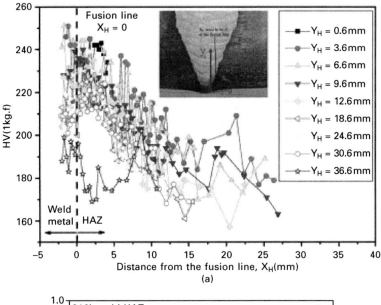

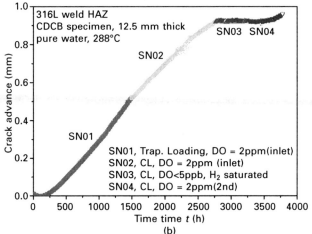

6.2 (a) The distribution of Vickers hardness (HV) in the HAZ of the 316L weldment; (b) the SCC crack advance vs time curve for the 316L weld HAZ specimen under various test conditions in pure water at 288°C [15].

with crack growth occurring preferentially along high angle boundaries. A heterogeneous microstructure such as the boundary between a large grain and small grains is observed, and is the favored path for intergranular stress corrosion cracking. Locally high misorientation at such boundaries has been revealed by kernel average misorientation (KAM) results by EBSD measurements.

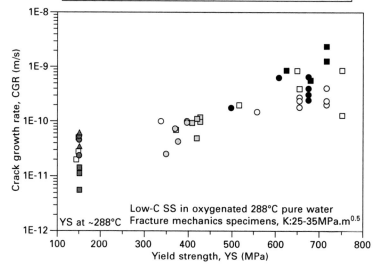

6.3 The effect of yield strength on the SCC growth rate on low carbon stainless steels in oxygenated pure water at 288°C [23].

SCC in the transition zone of Alloy 182-to-low alloy steel dissimilar metal weld (DMW) is another example of the effect of metallurgical heterogeneity. The transition zone has a composition gradient from the weld metal to the base metal across the dilution zone (DZ) in weld metal, the fusion boundary (FB) and the HAZ in the base metal. This change in composition causes a change in the microstructure, mechanical property and corrosion resistance. There is also a change in crystal microstructure from the bcc ferritic base metal to the fcc austenitic weld metal in the DMW. This in turn affects the mechanical properties of the DMW across the FB and may result in the formation of type II boundaries that run parallel to the fusion boundary in the weld metal within a distance of 100 μm [37, 38]. In addition, the formation of weld residual stress in the transition zone due to the microstructure heterogeneity and the resultant thermal mismatch result in the gradient of conditions across the FB [39, 40].

Since SCC in Alloy 182 weld metal in high temperature water was confirmed in recent years, concerns have been raised about SCC in the

(a) Optical picture of the whole

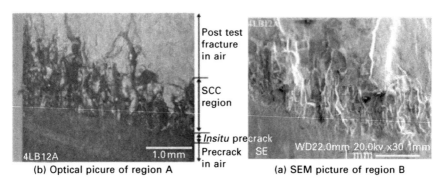

(b) Optical picure of region A

(a) SEM picture of region B

6.4 Optical and SEM morphologies of the fracture surface of WR 304L SS specimen T-L orientation (4LB12A) after SCC test: (a) optical picture of the whole, (b) optical picture of region A, and (c) SEM picture of region B [27].

transition zone in the Alloy 182-LAS DMW. To date, experimental works on the SCC behavior in the transition zone in high temperature water have been done by employing crack growth rate (CGR) tests and creviced bent beam (CBB) tests. One focus of the experiments is the synergistic effects of boundaries and water chemistry on the SCC behavior. The microstructure of the fusion and type II boundaries in the transition zone was investigated by electron backscattered diffraction (EBSD) and transmission electron microscopy (TEM). The EBSD analysis showed that both boundaries were high-angle boundaries (Fig. 6.10) [41]. TEM observation of the boundaries also found a high coverage of precipitates comprised of carbides of Ti and Nb on the type II boundary, but only a few precipitates of $Cr_{23}C_6$ on the FB [41]. The lower precipitate density on the FB is most likely due to the high solubility of carbon in the side of bcc-LAS.

For a SCC propagating perpendicular to the FB in the DZ in Alloy 182, the FB was found to be a barrier of the SCC propagation in high temperature oxygenated water with high purity or with doping of 30 ppb sulphate [42]. The fracture surface and the relative positions of SCC tip and FB shown in

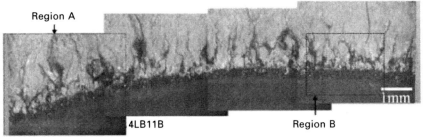

(a) optical picture of the whole

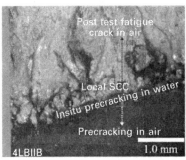

(b) optical picture of region A

(a) SEM picture of region B

6.5 Optical and SEM morphologies of the fracture surface of WR 304L SS specimen L-T orientation (4LB11B) after SCC test: (a) optical picture of the whole, (b) optical picture of region A, and (c) SEM picture of region B [27].

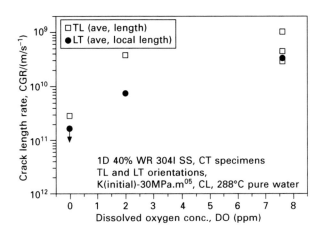

6.6 Stress corrosion cracking growth rates for 1DWR 304L SS T-L and L-T specimens in 288°C pure water with various DO concentrations.

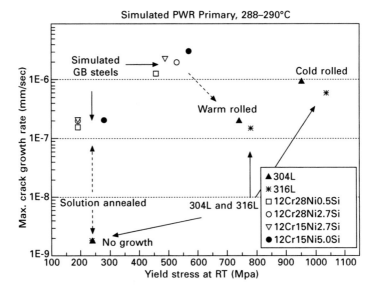

6.7 Experimental crack growth rates of strain-hardened stainless steels in simulated PWR pure water [10].

Fig. 6.11 (a) and (b) show that no further propagation of SCC in the LAS occurred during a SCC test of the 182-LAS DMW in pure water at 288°C and dissolved oxygen (DO) of 0.2 ppm, indicating the cessation of SCC propagation at the FB. Chloride has a much stronger effect on the SCC at the FB than sulphate. In 50 ppb chloride doped water at 288°C and DO of 2 ppm, the SCC in Alloy 182 weld metal can very easily cross the FB and further propagate into the HAZ and base metal of the adjacent LAS [43], as shown in Fig. 6.12.

Efforts were also made to investigate the effect of a higher sulphate concentration in water using CBB tests. In 100 ppb sulphate doped water at 288°C and 0.25 ppm DO, the tests revealed that a SCC propagated perpendicular to the FB in the DZ of Alloy 182 was arrested by pitting after it reached the FB (Fig. 6.13) [44]. The crack growth, however, was then reactivated from the pitting by localized oxidation along a grain boundary intersected with the pitting in the base metal, suggesting the synergistic effects of the grain boundary and the high sulphate concentration in water (see also in Fig. 6.4).

The CBB tests revealed that, compared to the FB, the type II boundary had a higher susceptibility to SCC propagation in high temperature oxygenated water [8, 9]. As shown in Fig. 6.14, a SCC propagated perpendicular to the FB deviated to the type II boundary after it intersected the type II boundary. The crack then propagated to the FB through the type I boundary, the normal boundaries perpendicular to the fusion boundary caused by epitaxial growth.

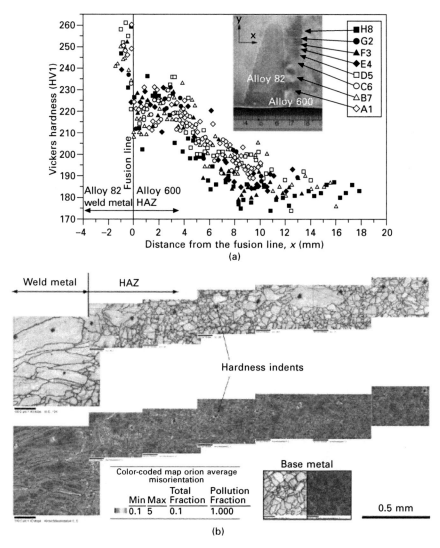

6.8 (a) The distribution of Vickers hardness and (b) the color-coded mapping of the grain average misorientation in the Alloy 600MA HAZ [34].

This suggests that for SCC growth perpendicular to the FB in the DZ, the role of the type II boundary is to link the crack growth to the FB. The high SCC susceptibility of type II boundary implies that it can be a potential propagation path in the DMW, and cause concern for the failure of the DMW along the type II boundary [41, 44].

The high SCC susceptibility of type II boundary in high temperature oxygenated water is most likely due to the high-angle misorientation at

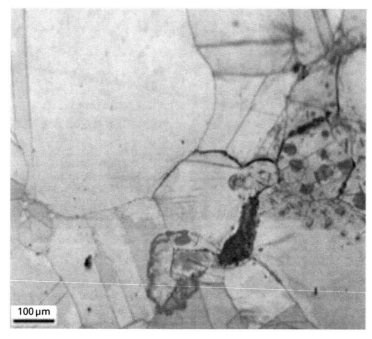

6.9 Observation of cracking path on the side surface of 3-directionally cold worked 316NG SS specimen 6LM212A (T-L orientation) [36].

the boundary and the high density of carbide precipitation on the boundary [41]. While a high-angle misorientation also exists at the fusion boundary, the enhanced corrosion of LAS by the galvanic effect may blunt the tip of a crack along the FB and thus lower the crack growth rate [41].

6.3   Environmental factors

The effects of environmental factors on SCC are often represented by a table of alloy-environment combinations in which SCC has been observed. The number of combinations has ncreased with the results of new SCC tests, refined crack growth monitoring, improved electrochemical control, and increased research activity [6]. The effects of environmental parameters such as temperature, dissolved oxygen and electrochemical potential on SCC in high temperature water are cited as examples. There are other environmental parameters such as solution pH, concentrations of aggressive ions and flow rate that affect SCC.

Early tests on the effects of temperature on SCC of sensitized 304 stainless steel (SS) were mainly performed by slow strain rate tests (SSRT) in water with relatively high conductivity due to intentionally added ions or unsatisfactory control of water chemistry [45–47]. Several kinds of temperature dependences

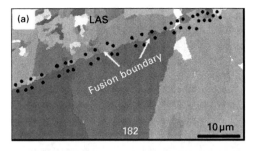

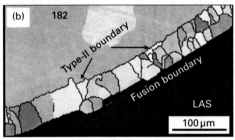

6.10 The misorientation maps of the FB region. (a) The misorientation along the FB. Black dots denote the points analyzed for the misorientation relationship. (b) The misorientation at the type II and type I boundaries. The black lines denote high-angle boundaries [41].

were obtained by different authors. Data from Ford and Povich [45] and Agrawal *et al.* [46] showed a monotonic increase of CGR with temperature from 200°C to 288°C for sensitized 304 SS in 8 ppm water, but a decrease of CGR after increasing temperature from 250°C to 288°C in 0.2 ppm water, by SSRT at strain rates of 2.1 × 10^{-7}/s and 4.0 × 10^{-7}/s. Ruther *et al.* [47] reported that maximum IGSCC measured by SSRT for lightly sensitized 304SS (EPR = 2 C/cm^2) occurred at temperatures between ~200°C and 250°C in 0.2 ppm DO pure water and the addition of sulfate increased the average CGRs at all temperatures and broadened the temperature ranges over which susceptibility occurred. Ruther *et al.*'s data generally showed a weak dependence of CGR on temperature between 200°C and 290°C, with an apparent activation energy of about 16 kJ/mol based on the later analyses by Andresen [48]. Weeks *et al.* [49] found that CGRs of furnace-sensitized 304 SS in pure water containing 22 ppm DO were quite dependent on strain rates of SSRT and showed a maximum at about 200°C. Hale [50] studied the effect of temperature on CGR of sensitized SS in 0.2 ppm DO pure water by testing 1TCT specimens under constant load at *K* of about 22 MPa.m$^{0.5}$. CGR showed a drastic increase when changing temperature from 100°C to 150°C. It seemed that the CGR was weakly dependent on temperature at 151–292°C, with data scattering larger than the change caused

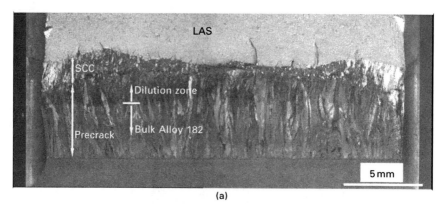

(a)

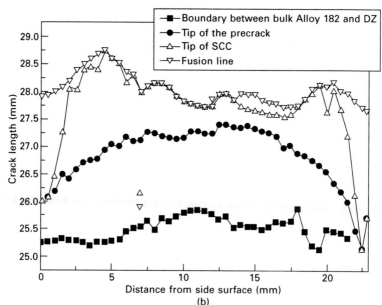

(b)

6.11 Observation of the fracture surface showing the cessation of crack propagation at the fusion boundary of an Alloy 182-LAS DMW following a SCC test in pure water at 288°C and DO = 0.2 ppm [42]: (a) outline of the fracture surface; (b) the relative positions of the fusion boundary, SCC tip and pre-crack tip on the fracture surface.

by the temperature change. CGR results by Magdowski and Speidel [51] on sensitized 304SS in pure water with a wide range of DO showed statistically a thermally activated temperature dependence with an apparent energy of about 46 kJ/mol between ~150°C and 288°C, while there seemed to be a change in crack growth mechanism below ~150°C. Such dependence was weakened due to quite a large range of DO and related parameters such as corrosion potential and pH, and the crack growth rate evaluation procedure,

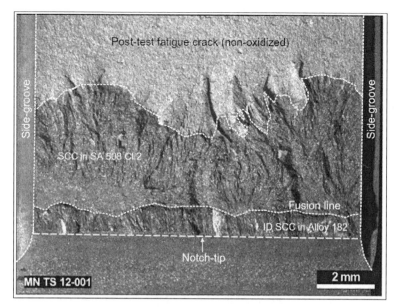

6.12 Fracture surface observation showing SCC in Alloy 182 and in the LAS HAZ and base metal following a test in high-temperature water with 2 ppm DO and 50 ppb of chloride [43].

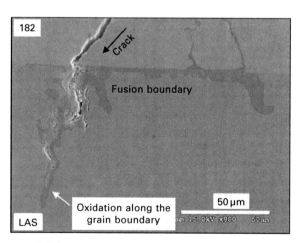

6.13 SEM observation of the SCC behavior in the FB region in a CBB specimen tested in 100 ppb sulphate doped water at 288°C and 0.25 ppm DO [44].

etc., as pointed out by Andresen [48]. Andresen [48] reported a CGR peak at about 200°C for sensitized 304SS in pure water containing 0.27 µs/cm H_2SO_4 and 0.2 ppm DO, by using 25 mm CT specimens under trapezoidal loading with 1000 s holding at $K_{max} = 33$ MPa.m$^{0.5}$ and $R = 0.5/0.01$ Hz unloading/

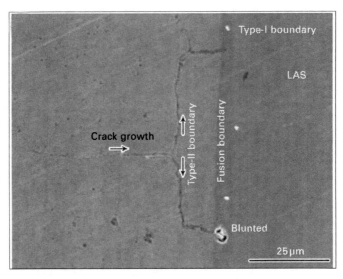

6.14 SEM observation showing the deviation of stress corrosion crack growth to the type II and type I boundaries before the crack reached the FB in a CBB specimen tested in 100 ppb sulphate doped water at 288°C and 0.25 ppm DO [43].

reloading. Jenssen and Jansson [52] reported a monotonic increase in CGR versus increasing temperature from 150°C to 288°C for sensitized 304SS in 0.5 ppm DO high purity water using CT specimens under quasi-constant K of 30 MPa.m$^{0.5}$. However, a CGR peak occurred near 150°C in sulfate solutions. Most of these reported data were obtained for sensitized stainless steels with relatively high carbon contents in oxygenated pure water or with a small amount of chemical additions. Andresen *et al.* [19] reported that the CGR of a cold worked 316L SS in hydrogenated pure water monotonically increases with temperature in the range 290°C to 340°C by using CT specimens.

The SCC growth for cold worked 316L stainless steel in oxygenated high purity water is controlled by thermally activated processes. The thermal activation energy tends to be higher in the higher temperature range [22, 53], implying that processes with a higher thermal activation energy at higher temperature are rate controlling. Based on the CGRs of CW316L SS in a similar environment and a similar temperature range, the thermal activation energies obtained under a higher K are lower than those obtained under a lower K (Fig. 6.15). These results suggest that the SCC growth rate of CW316L SS in oxygenated pure water could be controlled by multiple subprocesses. SCC growth of Ni-base alloys in simulated PWR primary water environments is thermally activated [31]. The effect of temperature has been included in the CGR disposition lines [54].

Further increase in temperature to >374°C and pressure to >22.05 MPa

CGR in high temperature pure water:
(1) FRI CW 316L SS, 760MPa (with 3D rolling)
■ 2ppm DO, K = 30MPa.m$^{0.5}$, Trap. Loading S2, S3, S4)
o 2.5ppm DO, K = 30MPa.m$^{0.5}$, Trap. Loading S11, S12, S13)
▲ 2ppm DO, K = 15MPa.m$^{0.5}$, CLSA08, SA09, SA10)
(2) Jenssen *et al.* 10th Degradation 2001
◇ Sera 304SS, DO = 500ppb
(3) Andresen *et al.* Corrosion 2001
✳ CW 316L SS, DH = 1.50ppm

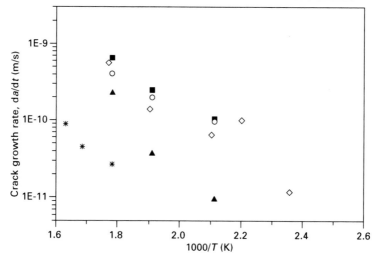

6.15 Summary of the effect of temperature on steady state crack growth rates for CW 316L SS in oxygenated pure water [22, 53], sensitized 304SS in oxygenated pure water in Ref. [52], and cold worked 316L SS in hydrogenated pure water in Ref. [19].

will cause a transition of water to supercritical. Since SCC of austenitic alloys in high temperature water has been acknowledged as an oxidation-driven process [29, 55], and that the oxidation in high temperature water is a thermally activated process, temperature is expected to have a strong effect on SCC. Supercritical water has significantly different properties than subcritical water. It acts like a dense gas and its density varies with temperature and pressure [56]. Temperature is expected to affect the SCC in supercritical water as well.

Stress corrosion crack growth in cold worked type 316L stainless steel in pure water across the subcritical–supercritical transition was investigated [57]. The results shown in Fig. 6.16 show that stress corrosion crack growth rates in the steel in pure water increase with temperature in the subcritical regime between 288°C and 360°C. The crack growth rate increase is consistent with an activation energy for crack growth of 26 kJ/mol. In the supercritical regime, the crack growth rate decreases with increasing temperature from 400°C to

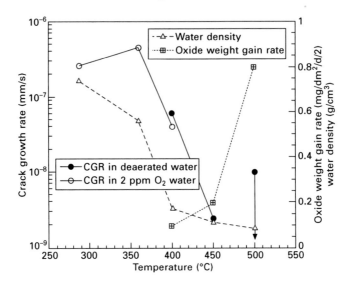

6.16 Crack growth rate vs temperature across the subcritical-supercritical line for unsensitized type 316L stainless steel in pure water. The rate of oxide weight gain over a period of approximately 500 h [58] and water density [54] are also shown for comparison.

450°C and then to 500°C. The very low crack growth rates in supercritical water at 450°C and 500°C suggest that the SCC is not a serious degradation mode in supercritical water for austenitic alloys. Figure 6.16 also shows an increased oxidation rate of stainless steel with the temperature in supercritical water [58]. The crack blunting by rapid oxidation in supercritical water was the cause of the low stress corrosion crack growth rate [57].

Electrochemical potential is affected by dissolved oxygen, and dissolved hydrogen concentration and has been found to have strong effects on SCC. It is well known that stress corrosion cracking of sensitized 304 SS is strongly affected by electrochemical potential [5]. SCC growth mitigated by decreasing electrode potential is less significant for cold or warm-worked low C stainless steels than for solution-annealed low C stainless steels and sensitized 304 SS in pure water at 288°C (Fig. 6.17) [36], consistent with the results of Toloczko *et al.* [59]. The crack growth rates (V_{CGR}) vs electrode potential (EP) are fitted using Eq. 6.1.

$$V_{CGR} = A_0 + B_0 \times 10^{(C_0 \times EP)} \qquad 6.1$$

where V_{CGR} is in m/s, E in V, A_0 in m/s, B_0 in m/s, and C_0 in 1/V.

The constants A_0, B_0 and C_0 in fitting Eq. 6.1 for crack growth rates of various materials in pure water at 288°C are summarized in Table 6.1.

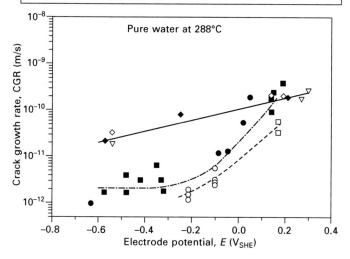

■ Sens, 304SS, Andresen *et al.*, JMS 2008, K = 30 MPa m$^{0.5}$
● Sens, 304SS, Kibochi *et al.*, Corrosion 100T, K-31 MPa m$^{0.5}$
□ 316NG SA, How *et al.*, PVP 2004, K = 31.2.32.2 MPa m$^{0.5}$
○ 316L SA, How *et al.*, PVP 2004, K = 27.8-33.3 MPa m$^{0.5}$
▽ 20% CW316L, Andresen *et al.*, JNM 2004, K = 27.5 MPa m$^{0.5}$
◇ 20% CW Alloy 600, Andresen *et al.*, JNM 2008, K = 30 MPa m$^{0.5}$
◆ 10% SDCR 316NG (T-L) FRI data, this work, K = 20 MPa m$^{0.5}$
—— Fit for FRI 3DCR 316NG
—·— Fit for Sens, 304SS
- - - Fit for low-C SS SA

6.17 Experimental SCC growth rate vs electrode potential for stainless steels and Alloy 600 in high temperature pure water at 288°C. The fitting curves based on the parameters in Table 6.1 are also plotted [36].

Table 6.1 Constants in fitting Eqs. (1)-(2) for CGR of sensitized 304SS, solution-annealed low-C SS, and 3DCR 316NG in pure water at 288°C [36]

Material Constants	A_0	B_0	C_0	Valid electrode potential range (V_{SHE})
Sensitized 304SS	2.11×10^{-12}	1.79×10^{-11}	6.15	-0.6-0.16
Low C SS SA	8.93×10^{-13}	7.74×10^{-12}	4.56	-0.25-0.18
3DCR 316NG	0	1.05×10^{-10}	1.21	-0.6-0.3

6.4   Mechanical factors

The mechanical factors are closely related to the specific alloy/environment combinations. Mechanical factors such as stress and strain, residual stress and strain, and strain rate have been found to be crucial parameters for SCC. The threshold conditions for SCC depend on the material, the environment and the load or a combination of these. The threshold stress intensity factor has often been used as an important threshold parameter for SCC, and has been described in several test standards such as ASTM E1681-95 [60]. The classical schematic description of the K threshold is modified to show detailed information that is used in further analysis of the threshold conditions for SCC of austenitic alloys in light water reactor coolant environments. (Fig. 6.18) [61].

The quantification of the effect of K is important especially in understanding the rate-controlling processes and in the crack growth rate disposition equations. Generally, crack growth rate in terms of K is expressed by a power law equation in an empirical way:

$$\text{CGR}(K) = C_0 * K^{n_k} \tag{6.2}$$

where C_0 is the prefix coefficient that is possibly affected by temperature, ECP, water conductivity and other variables [62] and n_k is the exponent in the CGR equation.

Some experimental crack growth rates vs stress intensity factor for non-sensitized austenitic stainless steels in simulated BWR oxygenated water environments are plotted in Fig. 6.19 [16, 20, 22, 23, 63, 64]. Andresen reported the value of n_k of about 2.3 or 2.5 for 20% cold worked 316L stainless steels in oxygenated pure water at 288°C [54]. The theoretical crack growth rate model based on the crack tip asymptotic fields and oxidation kinetics

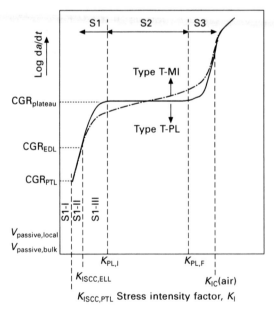

6.18 Schematic showing the K threshold for stress corrosion cracking [61].

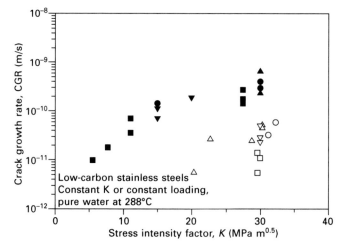

□ How *et al.*, SA, DO = 15.20ppm
○ How *et al.*, 316 MG SA, DO = 16-27ppm
△ How *et al.*, 316L SA, DO = 18.33ppm
▽ Tsubota *et al.*, 316L EA, DO = 8ppm DO
■ Andmon *et al.* CW316L, YS(200°C) = 550MPa, DO = 2ppm
● FRI data, 3D CW 316L, YS(RT) = 360MPa, DO = 2ppm
▲ FRI data, 3D CW 316L, YS(RT) = 360MPa, DO 7.5ppm
▼ Thia work, 3D CW, YS(300°C) = 354MPa, ELKO12A (T-L

6.19 Crack growth rate vs stress intensity factor for non-sensitized stainless steels in oxygenated pure water at 288°C [16, 20, 33, 63, 64].

has been used to predict the effect of K and the results are consistent with the experimental observation [25]. Besides K, other mechanical factors such as the change in K with time or crack advance, and applied loading mode are also found to affect stress corrosion behavior in high temperature water [65, 66].

6.5   Elemental mechanism and synergistic effects for complex stress corrosion cracking (SCC) systems

Stress corrosion cracking is a result of the combined and synergistic effects of materials, environments and mechanics. The individual processes in SCC are not isolated but inter-correlated. Various SCC mechanisms and controlling parameters have been proposed, such as slip-oxidation, oxidation penetration, stress-straining enhanced solid-state oxidation, hydrogen-related mechamism, and dealloy mechanism [1, 3, 5, 10, 31, 67]. Experimental approaches, analytical approaches, and numerical simulation approaches have been used to quantify the SCC sub-processes and controlling parameters. The general schematic of the oxidation/mechanics interactions for stress corrosion cracking in high temperature water is shown in Fig. 6.20 [25]. Depending on combinations of material/environment/loading conditions for SCC systems, the transient oxidation can take different kinetic laws, and the enhancement of crack tip oxidation can be realized via either physical degradation mode, physical-chemical degradation mode or both. Concerning the elemental mechanism and synergistic effects for complex SCC systems, systematic work has been carried out and more work, especially multi-scale modeling based on fundamental mechanisms, is required.

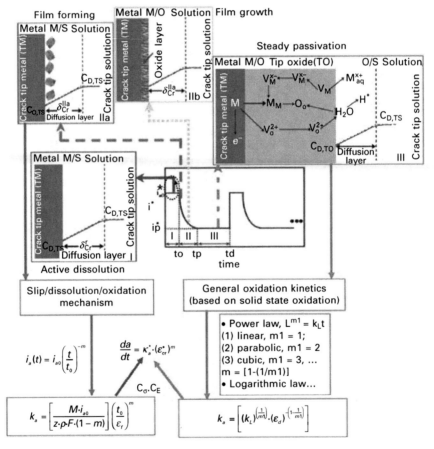

6.20 A schematic of the SCC sub-processes of austenitic alloys in high temperature water. [25].

6.6   Typical components and materials used in pressurized water reactors (PWR) and boiling water reactors (BWR)

Figure 6.21 shows typical materials used in PWR plants including primary and secondary circuits where various kinds of materials are used. The most popular materials are austenitic stainless steels such as austenitic 304 SS and 316 SS and their weld metals such as 308 SS and 309 SS, and Ni-base alloys such as Alloys 600 and 690, and their weld metals such as 182, 82, 152 and 52. In the core region, UO_2 as fuel, Zr-alloy as a fuel clad and B_4C as control rod are used. Among the most critical components are vessels, such as the pressure vessel, SG vessel and pressurizer vessel, where low alloy steels such as A533B and A508 are used. In some components, high strength alloys such as A286, X750 and 17-4PH SS are used. In condenser tubes, Ti or stainless steels tubes are used. Carbon steels are commonly used in secondary side steam piping as well as feedwater piping.

As can be seen clearly in Fig. 6.21, numerous kinds of materials are in use under different temperatures, chemical, mechanical and radiation conditions in primary, secondary and cooling systems. Most of the materials have greater or lesser susceptibility to SCC under some particular conditions. In particular, some of these materials such as Inconel 600 used in primary side high temperature water showed high SCC susceptibility under some metallurgical conditions, which is well known as PWSCC. The components using Alloy 600 are summarized below.

In Fig. 6.22, typical materials and components in BWR are summarized. Similar to the case of PWR, typical austenitic stainless steels 304, 304L, 316 and 316L, low alloy steels and their weld metals such 308 and 309 are used. Materials used for core region, steam turbine and condenser are very similar even though there are some differences in detail.

In some components, Alloy 600 and its weld metal 182 or 82 are used at CRD stub tubes. One of the most important points in the use of 304 stainless steel is to avoid thermal/weld sensitization during the fabrication and welding processes. Also, recent SCC events in BWR on 316L stainless steels were caused by hardened surface by machining, grinding or plastic forming where a high residual tensile stress can be associated with hardening. In some cases, a nano-grained region with a thickness of about a few microns is observed on the machined surface where SCC crack initiates and propagates along the nano-grains as IGSCC. After IGSCC propagates through the nano-grained layer, cracking mode changes to TGSCC because there is almost no grain boundary in front of the nano-grained IGSCC.

Figure 6.23 summarizes the component locations where Ni-alloys are used. As can be seen clearly, there are many components and parts where PWSCC should be taken into account as one of the possible degradation modes.

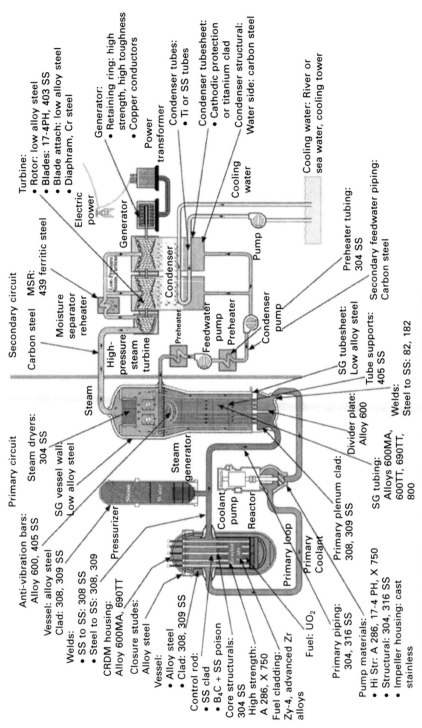

6.21 Typical materials in water cooled nuclear plants, e.g. PWRs [68].

Turbine:
• Rotor: low alloy steel
• Blades: 17-4PH, 403 SS
• Blade attach: low alloy steel
• Diaphram, Cr steel

Generator:
• Retaining ring: high strength, high toughness
• Copper conductors

Condenser tubes:
• Ti or SS tubes

Condenser tubesheet:
• Cathodic protection or titanium clad

Condenser structural: Water side: carbon steel

Cooling water: River or sea water, cooling tower

MSR: 439 ferritic steel

Preheater tubing: 304 SS

Secondary feedwater piping: Carbon steel

SG tubesheet: Low alloy steel

Tube supports: 405 SS

Welds: Steel to SS: 82, 182

Divider plate: Alloy 600

Primary plenum clad: 308, 309 SS

SG tubing: Alloys 600MA, 600TT, 690TT, 800

Primary piping: 304, 316 SS

Pump materials:
• Hi Str: A 286, 17-4 PH, X 750
• Structural: 304, 316 SS
• Impeller housing: cast stainless

Fuel: UO_2

Fuel cladding: Zy-4, advanced Zr alloys

Core structurals: 304 SS

Control rod:
• SS clad
• B_4C + SS poison

Vessel:
• Alloy steel
• Clad: 308, 309 SS

High strength: A 286, X 750

Closure studs: Alloy steel

CRDM housing: Alloy 600MA, 690TT

Welds:
• SS to SS: 308 SS
• Steel to SS: 308, 309

Vessel: alloy steel
Clad: 308, 309 SS

Anti-vibration bars: Alloy 600, 405 SS

Steam dryers: 304 SS

SG vessel wall: Low alloy steel

Carbon steel

Secondary circuit

Primary circuit

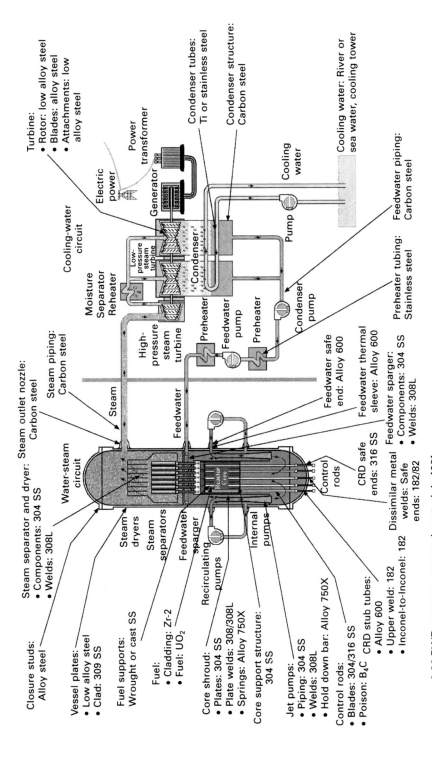

Closure studs:
Alloy steel

Steam separator and dryer: Steam outlet nozzle:
• Components: 304 SS   Carbon steel
• Welds: 308L

Vessel plates:
• Low alloy steel
• Clad: 309 SS

Fuel supports:
Wrought or cast SS

Fuel:
• Cladding: Zr-2
• Fuel: UO₂

Core shroud:
• Plates: 304 SS
• Plate welds: 308/308L
• Springs: Alloy 750X

Core support structure:
304 SS

Jet pumps:
• Piping: 304 SS
• Welds: 308L
• Hold down bar: Alloy 750X

Control rods:
• Blades: 304/316 SS
• Poison: B₄C   CRD stub tubes:
                 • Alloy 600
                 • Upper weld: 182
                 • Inconel-to-Inconel: 182

Turbine:
• Rotor: low alloy steel
• Blades: alloy steel
• Attachments: low
  alloy steel

Condenser tubes:
Ti or stainless steel

Condenser structure:
Carbon steel

Cooling water: River or
sea water, cooling tower

Feedwater piping:
Carbon steel

Preheater tubing:
Stainless steel

Feedwater safe
end: Alloy 600

Feedwater thermal
sleeve: Alloy 600

Feedwater sparger:
• Components: 304 SS
• Welds: 308L

CRD safe
ends: 316 SS

Dissimilar metal
welds: Safe
ends: 182/82

Control rods

Steam piping:
Carbon steel

6.22 BWR components and materials [68].

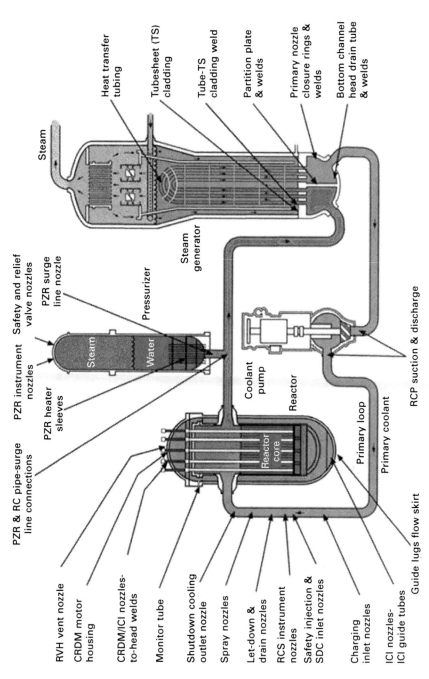

Heat transfer tubing

Tubesheet (TS) cladding

Tube-TS cladding weld

Partition plate & welds

Primary nozzle closure rings & welds

Bottom channel head drain tube & welds

Steam

Steam generator

Safety and relief valve nozzles

PZR surge line nozzle

Pressurizer

PZR instrument nozzles

PZR heater sleeves

Steam

Water

PZR & RC pipe-surge line connections

Coolant pump

Reactor

Reactor core

Primary loop

Primary coolant

RCP suction & discharge

RVH vent nozzle

CRDM motor housing

CRDM/ICI nozzles-to-head welds

Monitor tube

Shutdown cooling outlet nozzle

Spray nozzles

Let-down & drain nozzles

RCS instrument nozzles

Safety injection & SDC inlet nozzles

Charging inlet nozzles

ICI nozzles-ICI guide tubes

Guide lugs flow skirt

6.23 Locations of Ni-based alloys in the primary circuit of PWRs [68].

6.7   References

1. R.C. Newman, *Stress Corrosion Cracking Mechanisms*, Marcel Dekker, New York, 2002.
2. R.H. Jones, 'Stress corrosion cracking', in: *ASM Handbook, Vol. 13A*, ASM, Ohio, 2003, p. 356.
3. T. Shoji, Localized and accelerated oxidation and stress corrosion cracking – role of stress, strain, hydrogen and microstructures, Quantitative Micro-Nano (QMN) approach to predicting stress corrosion cracking in water-cooled nuclear plants, 2010, CDROM.
4. R.L. Jones, G.M. Gordon and G.H. Neils, Environmental degradation of materials in boiling water reactors, *Proc. 4th Int. Symp. on Environmental Degradation of Materials in Nuclear Power Systems – Water Reactors*, Newport Beach, CA, 1989, pp. 1–10.
5. P.L. Andresen and F.P. Ford, Life prediction by mechanistic modeling and system monitoring of environmental cracking of iron and nickel-alloys in aqueous systems, *Mat Sci Eng A* 103 (1988) 167–184.
6. R.H. Jones, Stress corrosion cracking, in: *ASM Handbook, Vol. 13A*, ASM, Ohio, 2003, pp. 346–366.
7. ASTM, A262 Practice E. Standard practices for detecting susceptibility to intergranular attack in austenitic stainless steels. *Annual Book of ASTM Standards*. American Society for Testing and Materials, 1998.
8. V. Čihal and J. Kubelka, Double loop recommendation of the State Research Institute for Materials Protection, Prague 1974, Tohoku Univ., Sendai, 1974.
9. V. Čihal, T. Shoji, V. Kain, K. Watanabe and R. Štefec, EPR – a comprehensive review, Tohoku Univ., Sendai, 2004.
10. T. Shoji, Progress in the mechanistic understanding of BWR SCC and its implication to the prediction of SCC growth behavior in plants, *Proc. 11th Int. Conf. on Environmental Degradation of Materials in Nuclear Power Systems – Water Reactors*, 10–14 August 2003, Skamania, Stevenson, WA. ANS, 2003, pp. 588–598.
11. S. Suzuki, K. Kumagayi, C. Shitara, J. Mizutani, A. Sakashita, H. Tohkuma and H. Yamashita, Damage evaluation of PLR pipe in nuclear power plant, *Maintenology* 3 (2004), 65–70.
12. T.M. Angeliu, P.L. Andresen, E. Hall, J.A. Sutliff, S. Sitzman and R.M. Horn, Intergranular stress corrosion cracking of unsensitized stainless steel in BWR environments, *Proc. 9th Int. Conf. on Environmental Degradation of Materials Nuclear Power Systems – Water Reactors*, TMS, 1999, pp. 311–317.
13. NISA, Documents presented at 5th structural integrity evaluation committee, 18 Feb. 2003.
14. U. Ehrnstén, P.A. Aaltonen, P. Nenonen, H.E. Hänninen, C. Jansson and T.M. Angeliu, Intergranular cracking of AISI316NG stainless steel in BWR environment, *Proc. 10th Int. Conf. on Environmental Degradation of Materials Nuclear Power Systems – Water Reactors*, NACE, 2001.
15. Z.P. Lu, T. Shoji, Y. Takeda, Y. Ito, A. Kai and N. Tsuchiya, Effects of loading mode and water chemistry on stress corrosion crack growth behavior of 316l HAZ and weld metal materials in high temperature pure water, *Corros Sci* 50 (2008) 625–638.
16. P.L. Andresen and M.M. Morra, IGSCC of non-sensitized stainless steels in high temperature water, *J Nucl Mater* 383 (2008) 97–111.
17. T. Shoji, G.F. Li, J.H. Kwon, S. Matsushima and Z.P. Lu, Quantification of yield strength effects on IGSCC of austenitic stainless steels in high temperature waters,

Proc. 11th Int. Conf. on Environmental Degradation of Materials in Nuclear Power Systems – Water Reactors, ANS, 2003, pp. 834–843.

18. M.M.L. Castaño, R.M.S. Garcia, V.G.D. Diego and B.D. Gòmez, Crack growth rate of hardened austenitic stainless steels in BWR and PWR environments, *Proc. 11th Int. Conf. on Environmental Degradation of Materials in Nuclear Power Systems – Water Reactors*, ANS, 2003, pp. 845–854.

19. P.L. Andresen, T.M. Angeliu and L.M. Young, Effect of martensite and hydrogen on SCC of stainless steels and Alloy 600, *Corrosion/2001*, Paper No. 228, 2001.

20. Z. Lu, T. Shoji, Y. Takeda, A. Kai and Y. Ito, Effects of loading mode and temperature on stress corrosion crack growth rates of a cold-worked type 316L stainless steel in oxygenated pure water, *Corrosion* 63 (2007) 1021–1032.

21. Z.P. Lu, T. Shoji, Y. Takeda, Y. Ito and S. Yamazaki, The dependency of the crack growth rate on the loading pattern and temperature in stress corrosion cracking of strain-hardened 316L stainless steels in a simulated BWR environment, *Corros Sci* 50 (2008) 698–712.

22. Z.P. Lu, T. Shoji, Y. Takeda, Y. Ito, A. Kai and S. Yamazaki, Transient and steady state crack growth kinetics for stress corrosion cracking of a cold worked 316L stainless steel in oxygenated pure water at different temperatures, *Corros Sci* 50 (2008) 561–575.

23. T. Shoji, Z.P. Lu, Y. Takeda, H. Murakami and C.Y. Fu, Deterministic prediction of stress corrosion crack growth rates in high temperature water by combination of interface oxidation kinetics and crack tip asymptotic field, *2008 ASME Pressure Vessels and Piping Division Conference* (ASME PVP 2008), ASME, 2008, PVP2008-61417.

24. T. Shoji, Z.P. Lu, N.K. Das, H. Murakami, Y. Takeda and T. Ismail, Modeling stress corrosion crack growth rates based upon the effect of stress/strain on crack tip interface degradation and oxidation reaction kinetics, ASME PVP 2009, PVP2009-77615.

25. T. Shoji, Z.P. Lu and H. Murakami, Formulating stress corrosion cracking growth rates by combination of crack tip mechanics and crack tip oxidation kinetics, *Corros Sci* 52 (2010) 769–779.

26. Z.P. Lu, K. Sakaguchi, K. Negishi, Y. Takeda, Y. Ito and T. Shoji, Quantifying the effects of strain-hardening and water chemistry on crack growth rates of 316L SS welds in high temperature water, *Proc. 14th Int. Symp. Environ. Degradation of Materials in Nuclear Power Systems – Water Reactors*, ANS, 2009, pp. 636–645.

27. Z.P. Lu, T. Shoji, T. Dan, Y.B. Qiu and T. Yonezawa, The effect of roll-processing orientation on stress corrosion cracking of warm-rolled 304L stainless steel in oxygenated and deoxygenated high temperature pure water, *Corros Sci* 52 (2010) 2547–2555.

28. Z.P. Lu, T. Shoji and S. Yamazaki, Effects of loading mode and water chemistry on stress corrosion cracking of 316L stainless steel in simulated PWR environments, *Proc. 14th Int. Conf. on Environmental Degradation of Mater. Nuclear Power Systems – Water Reactors*, ANS, 2009.

29. T. Shoji, K. Sakaguchi, Z.P. Lu, Y. Hasegawa, T. Kobayashi, K. Fujimoto and Y. Nomura, Effects of cold work and stress on oxidation and SCC behavior of stainless steels in PWR primary water environments, *Proc. International Symposium of Fontevraud 7*, SFEN, France, 2010, paper # A103 T103.

30. W.C. Moshier and C.M. Brown, Effect of cold work and processing orientation on stress corrosion cracking behavior of alloy 600, *Corrosion* 56 (2000) 307–320.

31. P.M. Scott and P. Combrade, On the mechanism of stress corrosion crack initiation and growth in alloy 600 exposed to PWR primary water, *Proc. of the 11th Int. Symp. Environmental Degradation Materials Nuclear Power Systems – Water Reactors*, ANS, 2003, pp. 29–38.

32. R.B. Rebak and Z. Szklarska-Smialowska, Effect of specimen thickness on cracking susceptibility of Alloy-600 in high-temperature water, *Corrosion* 51 (1995) 376–379.

33. R. Magdowski, F. Vaillant, C. Amzallag and M.O. Speidel, Stress corrosion crack growth rates of alloy 600 in simulated PWR coolant, *Proc. 8th Int. Conf. on Environmental Degradation of Materials Nuclear Power Systems – Water Reactors*, ANS, 1997, pp. 333–338.

34. S. Yamazaki, Z. P. Lu, Y. Ito, Y. Takeda and T. Shoji, The effect of prior deformation on stress corrosion cracking growth rates of alloy 600 materials in a simulated pressurized water reactor primary water, *Corros Sci* 50 (2008) 835–846.

35. G.A. Young, N. Lewis and D.S. Morton, The stress corrosion crack growth rate of alloy 600 heat affected zones exposed to high purity water, *Proc. Conf. on Vessel Head Penetration Inspection, Cracking and Repairs*, USNRC, 2003, pp. 309–335.

36. Z.P. Lu, T.S. Shoji, F.J. Meng, Y.B. Qiu, T.C. Dan and H. Xue, Effects of water chemistry and loading conditions on stress corrosion cracking of cold-rolled 316NG stainless steel in high temperature water, *Corros Sci* 51 (2011) 247–262.

37. T.W. Nelson, J.C. Lippold and M.J. Mills, *Sci Technol Weld Joi* 3 (1998) 249–255.

38. T.W. Nelson, J.C. Lippold and M.J. Mills, *Weld J* 78 (1999) 329S–337S.

39. H.T. Lee, S.L. Jeng, C.H. Yen and T.Y. Kuo, *J Nucl Mater* 335 (2004) 59–69.

40. J.W. Kim, K. Lee, J.S. Kim and T.S. Byun, *J Nucl Mater* 384 (2009) 212–221.

41. J. Hou, Q.J. Peng, Y. Takeda, J. Kuniya and T. Shoji, *Corros Sci* 52 (2010), 3949–3954.

42. Q.J. Peng, T. Shoji, S. Ritter and H.P. Seifert, *Proc. 12th Int. Conf. on Environmental Degradation of Materials in Nuclear Power System – Water Reactors*, TMS, 2005, pp. 589–599.

43. G.B. Naumov, *Handbook of Thermodynamic Data*, US Geological Survey, Menlo Park, CA, 1974.

44. ASTM G108, Standard test for electrochemical reactivation (EPR) for detecting sensitization of type AISI type 304 and 304L stainless steel, 2002.

45. F.P. Ford and M.J. Povich, *Corrosion* 35 (1979) 569–574.

46. A.K. Agrawal, G.A. Begley and R.W. Staehle, Stress corrosion of sensitized and quench annealed type 304 stainless steels in high purity water, *Corrosion/78*, NACE, 1978, Paper No. 187.

47. W.E. Ruther, W.K. Soppet and T.F. Kassner, *Corrosion* 44 (1988) 791–799.

48. P.L. Andresen, *Corrosion* 49 (1993) 714–725.

49. J.R. Weeks, B. Vyas and H.S. Isaacs, Environmental-factors influencing stress-corrosion cracking in boiling water-reactors, *Corros Sci* 25 (1985) 757–768.

50. D.A. Hale, *J Eng Mater-T ASME* 108 (1986) 44–49.

51. R. Magdowski and M.O. Speidel, Effect of temperature on stress corrosion crack growth in austenitic stainless steels exposed to water, Corrosion/90, NACE, 1990, Paper No. 291.

52. A. Jenssen, C. Jansson, Effects of temperature on crack growth rate in sensitized type 304 stainless steel in pure and sulfate bearing BWR environments, *Proc. 10th Int. Conf. on Environmental Degradation of Materials in Nuclear Power Systems – Water Reactors*, NACE, 2001.

53. Z.P. Lu, T. Shoji, Y. Takeda, Y. Ito, S. Yamazaki and N. Tsuchiya, *Proc. 13th Int. Conf. on Environmental Degradation of Materials in Nuclear Power Systems – Water Reactors*, NACE, 2007.

54. MRP115, Materials reliability program crack growth rates for evaluating primary water stress corrosion cracking (PWSCC) of alloy 82, 182, and 132 welds (MRP-115), EPRI, Palo Alto, CA, 2004. 1006696, vol. Nonproprietary version (MRP-115NP, 1006696-NP). 2002.

55. P.M. Scott, An overview of internal oxidation as a possible explanation of intergraular stress corrosion cracking of Alloy 600 in PWRs, *Proc. 9th Int. Conf. on Environmental Degradation of Materials in Nuclear Power Systems – Water Reactors*, TMS, 1999, pp. 3–12.

56. G.S. Was and S. Teysseyre, *Proc. 12th Int. Conf. on Environmental Degradation of Materials in Nuclear Power Systems – Water Reactors*, NACE, 2005.

57. Q.J. Peng, S. Teysseyre, P.L. Andresen and G.S. Was, Stress corrosion crack growth in type 316 stainless steel in supercritical water, *Corrosion* 63 (2007) 1033–1041.

58. G.S. Was, S. Teysseyre and Z. Jiao, Corrosion of austenitic alloys in supercritical water, *Corrosion* 62 (2006) 989–1005.

59. M. Toloczko, S.M. Bruemmer and P.L. Andresen, SCC crack growth of cold-worked 316LSS in BWR oxidizing and hydrogen water chemistry conditions. *Proc. 13th Int. Conf. on Environmental Degradation of Materials in Nuclear Power Systems – Water Reactors*, CNS, 2007.

60. ASTM, Standard test method for determining a threshold stress intensity factor for environment-assisted cracking of metallic materials under constant load, E1681-95, 1995, *Annual Book of ASTM Standards*, Vol. 03.01, ASTM.

61. T. Shoji, Z.P. Lu and Y. Takeda, Understanding the threshold conditions for stress corrosion cracking in light water reactor environments based on the deformation/oxidation mechanism, ASME PVP 2007/Creep8, ASME, 2007, PVP2007-26183.

62. R. Carter and R. Pathania, Technical basis for BWRVIP stainless steel crack growth correlations in BWRs, ASME PVP 2007, ASME, 2007, PVP2007-26618.

63. M. Itow, M. Kikuchi, N. Tanaka, J. Kuniya, M. Yamamoto, S. Suzuki, S. Namatame and T. Futami, SCC growth rates and reference curves for low carbon stainless steels in BWR environments, ASME PVP 2004, ASME, 2004, pp. 167–173.

64. M. Tsubota, Y. Katayama, Y. Saito and T. Kaneko, Stress corrosion cracking growth behavior of cold worked austenitic stainless steel in high temperature water, *Proc. 12th Int. Conf. on Environmental Degradation of Materials in Nuclear Power Systems – Water Reactors*, TMS, 2005, pp. 167–181.

65. T. Shoji, S. Suzuki and R.G. Ballinger, Theoretical prediction of SCC growth behavior-threshold and plateau growth rate, *Proc. 7th Int. Conf. on Environmental Degradation of Materials in Nuclear Power Systems – Water Reactors*, NACE, 1995, pp. 881–891.

66. P.L. Andresen and M.M. Morra, Effect of rising and failing K profiles on SCC growth rates in high-temperature water, *J Pressure Vessel Tech ASME* 129 (2007) 488–506.

67. N.K. Das, K. Suzuki, Y. Takeda, K. Ogawa and T. Shoji, Quantum chemical molecular dynamics study of stress corrosion cracking behavior for fcc Fe and Fe–Cr surfaces, *Corros Sci* 50 (2008) 1701–1706.

68. R.W. Staehle, Presented at Quantitative Micro-Nano(QMN) Approach to Predicting SCC of Fe-Cr-Ni Alloys, 12–18 June 2010, Sun Valley, Idaho, USA.

7

Stress corrosion cracking (SCC) of nickel-based alloys

R. B. REBAK, GE Global Research, USA

Abstract: Several families of commercial corrosion resistant high nickel alloys are used to handle aggressive industrial environments such as high temperature acids. Nickel alloys are resistant to cracking in hot chloride aqueous solutions. Environments that may cause stress corrosion cracking in nickel alloys include hot caustic solutions, high temperature water and hot and wet hydrofluoric acid. This chapter addresses the cracking behavior of nickel alloys in industries such as chemical process, nuclear power generation and in oil and gas exploration and production.

Key words: stress corrosion cracking, nickel alloys, alkalis, wet hydrofluoric acid, nuclear power, oil and gas, chemical process industry.

7.1   Introduction

Environmentally assisted cracking (EAC) is a general term that includes processes such as stress corrosion cracking (SCC), hydrogen embrittlement (HE), sulfide stress cracking (SSC), liquid metal embrittlement (LME), and corrosion fatigue (CF). In this chapter the terms EAC and SCC are generally interchangeable. EAC or SCC refers to a phenomenon by which a normally ductile metal loses its ductility (e.g. elongation to rupture) when it is subjected to mechanical stresses in the presence of a specific corroding environment. For EAC to occur, three affecting factors must be present simultaneously:

1. mechanical tensile stresses,
2. a susceptible metal microstructure and
3. a specific aggressive environment.

If any of these three factors is removed, EAC will not occur. That is, to mitigate the occurrence of EAC, engineers may, for example, eliminate residual stresses in a component, heat treat the material to produce a less susceptible microstructure, or limit its application to certain non-aggressive chemicals (environment). The term environment not only includes the chemical composition of the solution in contact with the component, but also other variables such as temperature and applied potential (or redox potential in the system).

Nickel alloys are in general more resistant than stainless steels to several

273

forms of EAC. For example, austenitic stainless steels (such as Type 304 or S30400) suffer chronic SCC in the presence of hot aqueous solutions containing chloride ions. Since chloride ions are ubiquitous in most industrial applications, the use of stainless steels components containing sometimes only minimal residual stresses is seriously limited because of the chloride cracking. On the other hand, nickel alloys (such as C-276) are practically immune to SCC in the presence of hot chloride solutions and therefore an excellent alternative to replace the troubled austenitic stainless steels. Nickel alloys may be prone to EAC in other environments such as hot caustic and hot wet hydrofluoric acid (Crum *et al.* 2000). The SCC of nickel alloys has been addressed before in other reviews (Sridhar and Cragnolino 1992, Rebak 2000a).

7.2   The family of nickel alloys

7.2.1   Properties

Most corrosion-resistant nickel-based alloys are solid solutions based in the element nickel (Ni). Some nickel alloys could be precipitation hardened (e.g. X-750, 625, 718). Even though Ni-based alloys in general contain a large proportion (sometimes up to 50%) of other alloying elements, nickel alloys still maintain the face centered cubic lattice structure (fcc, austenite, or gamma) from the nickel base element. As a consequence of the fcc structure, nickel-based alloys have excellent ductility, malleability and formability. Nickel alloys are also readily weldable. There are two large groups of the commercial Ni-based alloys. One group is designed to withstand high temperature and dry or gaseous corrosion while the other is dedicated mainly to lower temperature (aqueous and non-aqueous liquids) applications. Nickel-based alloys used for low temperature aqueous or condensed systems are generally known as corrosion-resistant alloys (CRA) and nickel alloys used for high temperature applications are known as heat-resistant alloys (HRA) or high temperature alloys (HTA). The practical industrial boundary between high and low temperature nickel alloys is in the order of 500°C (or approximately 1000°F). Most of the nickel alloys have a clear use either as CRA or HRA; however, a few alloys can be used for both applications (e.g. alloy 625 or N06625 and 718 or N07718).

7.2.2   Heat-resistant alloys

Unlike CRA, which are mostly selected for their capacity to resist corrosion in a given environment, most HRA need to play a dual role, namely, besides their capacity to withstand the corrosive aggressiveness of the environment, HRA also need to keep significant strength at high temperatures. In many instances, for example near and above 1000°C, alloy selection is dominated by how strong the alloy is in this temperature range rather than for its environmental degradation resistance. There are many different industrial high temperature aggressive environments. In general, for practical use, these high temperatures environments have been divided according to the most common causes of failure of a component in service. The most common failures are associated with the attack by a specific element such as oxygen (which causes oxidation), carbon (carburization and metal dusting), sulfur (sulfidation), halogen (e.g. chlorination) and nitrogen (nitridation). Other modes of failure such as molten metal attack and hot corrosion are less specific to a certain element in the atmosphere. Detailed descriptions of the mode of attack in these different environments are given elsewhere (Lai 1999, Gleeson 2000). The most common high temperature degradation mode is oxidation and the protection against oxidation in general is given by the formation of a chromium oxide scale on the surface of the component. In many instances, the presence of a small amount of aluminum or silicon in the alloy may improve the resistance against oxidation of a chromia-forming alloy. The degree of attack by other elements such as chlorine and sulfur will depend strongly on the partial pressure of oxygen in the environment (Dillon 1994). Due to their application, failures of HTA in service are not usually associated with EAC or SCC since they may not be in contact with a condensed phase promoting cracking. A few HTA such as Alloy 556 (R30556) are used in the handling of liquid metals (such as in liquid Zn galvanizing) and therefore may suffer embrittlement by liquid metal.

7.2.3   Chemical composition and corrosion behavior of CRA

Nickel alloys are highly resistant to corrosion, and in most environments nickel alloys outperform the most advanced stainless steels. One of the reasons is that nickel can be alloyed more heavily than iron. That is, larger amounts of specific elements can be dissolved purposely into nickel to tailor the alloy for a particular environment. In general, industrial corrosive environments can be divided into two broad categories, reducing and oxidizing. These terms refer to the range of electrode potential that the alloys experience, which is controlled by the cathodic reactions in the system. Thus, a reducing condition is generally controlled by the discharge of hydrogen from acids such as hydrochloric or sulfuric. An oxidizing environment has an electrode potential that is higher than the potential for hydrogen discharge. This potential may be established by cathodic reactions such as reduction of dissolved oxygen (O_2) from the atmosphere, chlorine gas (Cl_2), hydrogen peroxide (H_2O_2), chromate or chromic acid (CrO_4^{2-}), nitric acid (NO_3^-) and metallic ions in solution such as ferric (Fe^{3+}) and cupric (Cu^{2+}). Nickel alloys, similarly to other alloys, may suffer two main types of corrosion, uniform corrosion and localized corrosion. Uniform corrosion may happen under reducing conditions in the active region of potentials and also under oxidizing conditions in the form of a slow passive corrosion. Localized corrosion such as pitting and crevice corrosion generally occurs under oxidizing conditions. Stress corrosion cracking (SCC) or environmentally assisted cracking could occur at any electrochemical potential range.

From the chemical composition point of view, corrosion resistant Ni-based alloys can be grouped as:

- commercially pure nickel,
- Ni-Cu alloys,
- Ni-Mo alloys,
- Ni-Cr-Mo alloys and
- Ni-Cr-Fe-(Mo) alloys.

Table 7.1 gives the approximate chemical composition and the typical mechanical properties of the most familiar commercial wrought nickel-based alloys. A brief description of the corrosion behavior and application of each group of alloys is given below. More extended analyses are given elsewhere (Rebak 2000a, Agarwal 2000).

The main widely accepted application of commercially pure nickel is the handling of highly concentrated caustic solutions (alkalis). Nickel has lower corrosion rates in hot caustic solutions than alloyed nickel since alloying elements such as Cr and Mo dissolve preferentially from the alloy (dealloying) in hot caustic solutions. Nickel can also tolerate well cold reducing acids because of the slow discharge of hydrogen on its surface. Hot reducing acids and oxidizing acids corrode pure nickel rapidly. The main application of Ni-Cu alloys (or Monel alloys) is in the handling of pure hydrofluoric acid. However, if oxidants such as oxygen are present in hydrofluoric acid, Ni-Cu alloys may suffer intergranular attack (Rebak *et al.* 2001). Ni-Cu alloys are slightly more resistant to general corrosion than Ni-200 in hot reducing and oxidizing acids such as sulfuric acid and nitric acid. Ni-Mo alloys, commonly known as Hastelloy B type alloys, were specifically developed to withstand reducing HCl at all concentrations and temperatures. Besides more expensive materials such as tantalum, Ni-Mo alloys are the best alloys for hot hydrochloric acid (Agarwal 2000, Rebak 2000a, Rebak and Crook 2000a, 2000b, Agarwal and Kloewer 2001). Ni-Mo alloys are also used in the handling of other corrosive reducing environments such as dilute sulfuric, acetic, formic and hydrofluoric acids. B-2 has the lowest corrosion rate in boiling 10% sulfuric acid. However, Ni-Mo alloys perform poorly in oxidizing acids or, for example, in hydrochloric acid contaminated with ferric ions (Rebak and Crook 2000b).

There are many commercially available Ni-Cr-Mo alloys today. All

Table 7.1 Approximate chemical composition and applications of commercial Ni alloys. Typical mechanical properties (MPa) at ambient temperature for annealed plates

Alloy	UNS	Approximate composition	YS (0.2%)	UTS	ETF (%)	RH	Applications
Commercial Nickel							
Ni-200	N02200	99Ni-0.2Mn-0.2Fe	190	450	50	60 B	Strong caustic
Ni-301[A]	N03301	93Ni-4.5Al-0.6Ti	860	1170	25	35 C	Fasteners, springs
Ni-Cu Alloys							
Monel 400	N04400	67Ni-31.5 Cu-1.2Fe	270	540	43	68 B	Hydrofluoric acid
Monel K-500[A]	N05500	63Ni-30Cu-3Al-0.5Ti	700	1020	28	30 C	Fasteners, springs
Ni-Mo Alloys							
B-2	N10665	72Ni-28Mo	407	902	61	94 B	Hot hydrochloric
Hastelloy B-3	N10675	68.5Ni-28.5Mo-1.5Cr-1.5Fe-	400	885	58	NA	Reducing acids
Nimofer 6629 (B-4)	N10629	65Ni-28Mo-4Fe-1Cr-0.3Al	340	755	40	NA	Hydrochloric, sulfuric
Ni-Cr-Mo Alloys							
C-276	N10276	59Ni-16Cr-16Mo-4W-5Fe	347	741	67	89 B	Versatile CPI and pollution control
Inconel 625[A]	N06625	62Ni-21Cr-9Mo-3.7Nb	535	930	45	95 B	Aerospace, pollution control
Hastelloy C-22	N06022	59Ni-22Cr-13Mo-3W-3Fe	365	772	62	89 B	FGD, CPI, nuclear waste
C-22HS[A]	N07022	59Ni-21Cr-17Mo	1390	1590	20	30 C	Oil and Gas
Hastelloy C-2000	N06200	59Ni-23Cr-16Mo-1.6Cu	345	758	68	NA	CPI, oxidizing and reducing. Sulfuric
Nicrofer 5923hMo (59)	N06059	59Ni-23Cr-16Mo-1Fe	340	690	40	NA	Oxidizing and reducing acids, CPI
Inconel 686	N06686	46Ni-21Cr-16Mo-4W-5Fe	364	722	71	NA	Oxidizing and reducing acids. CPI
Ni-Cr-Fe-(Mo) Alloys							
Inconel 600	N06600	76Ni-15.5Cr-8Fe	275	640	45	75 B	Nuclear power
Inconel 690	N06690	58Ni-29Cr-9Fe	352	703	46	NA	Nuclear power
Inconel X-750[A]	N07750	73Ni-15Cr-7Fe-2.5Ti-1Nb-0.7Al	868	1270	25	36 C	Nuclear power

Table 7.1 Continued.

Alloy	UNS	Approximate composition	YS (0.2%)	UTS	ETF (%)	RH	Applications
Incoloy 800	N08800	33Ni-21Cr-40Fe-0.8(Al+Ti)	250	590	NA	83 B	Steam generator tubing
Incoloy 825	N08825	43Ni-21Cr-30Fe-3Mo-2.2Cu-1Ti	338	662	45	85 B	Oil and gas. Sulfuric, phosphoric
Incoloy 945[A]	N09945	50Ni-21Cr-18Fe-3Mo-3Nb-1.5Ti-0.3Al-2Cu	920	1194	28	40 C	Oil and Gas
Hastelloy G-30	N06030	44Ni-30Cr-15Fe-5Mo-2Cu-2.5W-4Co	317	689	64	NA	Nitric, phosphoric
Nicrofer 3033 (33)	R20033[D]	31Ni-33Cr-32Fe-1.6Mo-0.6Cu-0.4N	380	720	40	NA	Phosphoric acid

YS = yield stress (MPa), UTS = ultimate tensile strength (MPa), ETF = elongation to failure, RH = Rockwell hardness, CPI = chemical process industry, A = thermally aged, NA = not available, D = UNS starts with an R because it classified as a Cr-based alloy. Hastelloy is a trademark of Haynes International Inc., Inconel, Incoloy, and Monel are trademarks of Special Metals Corporation and Nicrofer and Nimofer are trademarks of Krupp-VDM.

these alloys were derived from the original C alloy (N10002), which was introduced to the market in cast form in 1932. The more advanced Ni-Cr-Mo alloys are Inconel 686, Nicrofer 5923 and Hastelloy C-2000 (Agarwal and Kloewer 2001). However, the more common Ni-Cr-Mo alloy in industrial applications is Hastelloy C-276, which was introduced in the market in the mid-1960s. Ni-Cr-Mo alloys are the most versatile nickel alloys since they contain molybdenum for protection against corrosion under reducing conditions and chromium, which protects against corrosion under oxidizing conditions (Rebak and Crook 2000b, Agarwal and Kloewer 2001). C-276 has low corrosion rates both in reducing conditions (boiling 10% sulfuric acid) and oxidizing conditions (boiling 10% nitric acid). One of the major applications of Ni-Cr-Mo alloys is in the presence of hot chloride containing solutions. Under these conditions, most of the stainless steels would suffer crevice corrosion, pitting corrosion and stress corrosion cracking. However, Ni-Cr-Mo alloys are highly resistant if not immune to chloride-induced attack in most industrial applications (Rebak 2000a).

The last group of nickel-based CRA is the group of Ni-Cr-Fe alloys. These alloys also may contain smaller amounts or molybdenum and/or copper such as in alloy N08825. Ni-Cr-Fe alloys in general are less resistant to corrosion than Ni-Cr-Mo alloys; however, they could be less expensive and therefore find a wide range of industrial applications where the use of stainless steels may be limited. The corrosion rate of alloy 600 in sulfuric acid is higher than the corrosion rate of alloy 825 since the latter contains small amounts of molybdenum and copper (Table 7.1), which are beneficial alloying elements for resistance to sulfuric acid. Also, alloy 825 has lower corrosion rate in nitric acid since it contains a larger amount of chromium. One of the most common applications of Ni-Cr-Fe-(Mo) alloys such as Alloy 33 and Hastelloy G-30 is in the industrial production of phosphoric acid and in highly oxidizing media such as nitric acid.

7.3   Environmental cracking behavior of nickel alloys

7.3.1   Environments that may cause cracking in nickel alloys

The main limitation in the application of nickel alloys is not EAC, because nickel alloys are less prone to suffer EAC than the more widely used austenitic stainless steels. In general, mill annealed nickel CRA have tensile strengths lower than 1000 MPa (Table 7.1) and large elongation to failure (>50%), that is, for example, they are not especially susceptible to failure mechanisms associated to hydrogen uptake. If the nickel alloy is being processed, for example by cold working or high temperature aging where they become less ductile, their susceptibility to hydrogen embrittlement may increase. Two of the specific environments that cause EAC in nickel CRA are hot caustic and hot wet hydrofluoric acid (Crum *et al.* 2000). These same two environments also produce EAC in austenitic stainless steels. Table 7.2 lists the different families of nickel alloys and the environments that were shown to cause their embrittlement. In the next few sections, the susceptibility

Table 7.2 Environments that may cause EAC in nickel alloys

Nickel alloys	Example UNS	Environments which may produce EAC
Commercial nickel	N02200	Molten metals, hydrogen embrittlement
Ni-Cu alloys	N04400	Hydrofluoric acid (especially in the vapor phase containing oxygen), mercury salts, ammonia, oil and gas (K-500)
Ni-Mo alloys	N10675	Cathodic and anodic acidic solutions (especially near welds), production of acetic acid, wet HF solutions, hydrogen embrittlement
Ni-Cr-Mo alloys	N10276	Hot caustic, super critical water, hot wet HF solutions
Ni-Cr-Fe alloys	N06600, N08825	Hot water, hot caustic, hot wet HF, high chloride high temperature

to cracking of Ni alloys is discussed by type of alloy and also by type of industrial application.

7.3.2   Commercial nickel

In general, there is an understanding that pure metals are less likely to suffer EAC than alloyed metals, mainly because the former have lower mechanical strength and great ductility. For example, it is claimed that commercially pure nickel is not highly susceptible to stress corrosion cracking, except in the heavily cold worked conditions in the presence of high temperature (>250°C) concentrated caustic solutions and liquid metal (Nelson 1987). C-ring specimens of 20%, 33% and 50% cold worked Ni-200 were free from cracking after testing for one week in 25% NaCl and in equivalent chloride concentration of $CaCl_2$ and $MgCl_2$ at 121°C, 149°C, 177°C, 204°C and 232°C (Kolts 1982). Laboratory testing has shown that commercially pure Ni may be susceptible to hydrogen embrittlement. Scully *et al.* (1991) tested Ni-200 specimens prepared from cold drawn rods in neutral 1.5M LiAlCl$_4$/ SOCl$_2$ solutions under cathodic applied potential using the slow strain rate technique. They reported intergranular and transgranular cracking in the tested materials and postulated that the presence of zero valence Li was responsible for the cracking (Scully *et al.* 1991). Wasynczuk *et al.* (1995) performed slow strain rate tests in the laboratory using both powder metallurgy (PM) and cast plus wrought (C+W) nickel strips. The composition of the Ni strips was similar to Ni-200 (Table 7.1) and the tests were conducted in a battery grade electrolyte of 1.5M LiAlCl$_4$/SOCl$_2$ under cathodic polarization. They reported intergranular SCC in the PM alloy and no cracking in the C+W alloy. Wasynczuk *et al.* (1995) also performed cathodic charging of hydrogen in 1 N H_2SO_4 solution for 24 hours of PM and C+W Ni strips and then bending in air. They reported that hydrogen charging produced intergranular cracking in both types of materials. Slow strain rate testing was also used to determine the effect of hydrogen on the cracking susceptibility of Ni-200 and Ni-270 specimens (Lee and Latanision 1987). For example, when Ni-200 specimens were deformed at 6.7×10^{-5} s^{-1} in air, the reduction in area was approximately 70%, but when the straining was carried out in 0.1 M N H_2SO_4 solution at −1000 mV SCE, the reduction in area was approximately 35%. In air the fracture mode of Ni-200 was dimpled but in the H_2SO_4 solution the fracture mode was a mixture of quasi cleavage and intergranular cracking showing the effects of hydrogen-induced cracking (Lee and Latanision 1987). Hydrogen-induced cracking of Ni may be exacerbated by the presence of sulfur at the grain boundaries (Lee and Latanision 1987). Ni-200 and 304 SS 1.6 mm diameter wires were stressed under constant load in a battery electrolyte melt containing $NaAlCl_4$ + 2% S at 300°C during typical battery cycling voltages of charge and discharge between 2 and 3 V. While the 304

SS wires failed within a few hours, the Ni-200 wires were free from cracking after more than 1400 hours of testing.

7.3.3   Ni-Cu alloys

Ni-Cu alloys were developed over 100 years ago as one of the first truly industrial Ni alloys. As in the case of Ni-200, alloy 400 does not seem to be highly susceptible to stress corrosion cracking (SCC) probably because of its low mechanical strength and high ductility (Table 7.1). Alloy 400 was found to be susceptible to SCC in acidic solutions containing mercury salts, in liquid mercury, in hydrofluoric acid and in fluosilicic acid (The International Nickel Company, 1968). The age hardenable alloy K-500 is used in seawater applications mainly as shafts and bolting due to its high strength; however K-500 may be susceptible to HE (Pound 1998). Alloy K-500 has also been used extensively for sour service in oil field applications (Krishnan *et al.* 2009). Several oil wells field cracking in K-500 were reported, mainly following an intergranular cracking mode. The failures may be attributed to a hydrogen embrittlement mechanism probably due to cathodic protection or because the alloy was coupled to less corrosion-resistant carbon steel (Krishnan *et al.* 2009).

7.3.4   Ni-Mo alloys

The main use of Ni-Mo alloys is in high temperature reducing acids. These alloys contain almost 30% Mo which makes them highly resistant to general corrosion in reducing acids. Because these alloys also contain approximately 70% Ni they are also resistant to SCC in hot concentrated chloride solutions. C-shaped specimens of 20%, 33% and 50% cold worked alloys B and B-2 were free from cracking after testing for one week in 25% NaCl and in equivalent chloride concentration of $CaCl_2$ and $MgCl_2$ at 121°C, 149°C, 177°C, 204°C and 232°C (Kolts 1982). That is, cracking did not occur even though for 50% cold worked alloy B-2 the hardness could be as high as HRC 49 (Kolts 1982). However, when alloy B-2 and, to a lesser extent alloy B-3, are exposed to temperatures in the range 550–850°C, they lose ductility due to a solid phase transformation which forms ordered intermetallic phases such as Ni_4Mo. The precipitation of these ordered phases changes the deformation mechanisms of the alloys making them susceptible to EAC such as hydrogen embrittlement (Agarwal *et al.* 1994, James *et al.* 1996). The kinetics of precipitation of the detrimental phases such as Ni_4Mo in alloy B-2 are highly sensitive to small chemistry variations and previous thermal processing and some compositions may form the detrimental ordered phases in the order of minutes, by exposure to the sensitive temperature range of 650–750°C, for example, in the heat affected zone (HAZ) during welding

(James *et al.* 1996). The alloy B-3 was developed as an improvement in the formation of the intermetallic and short ordering phases during welding and therefore increased resistance to SCC (James *et al.* 1996). It has been reported that B-2 alloy failed by intergranular stress corrosion cracking of the HAZ when exposed to organic solvents containing traces of sulfuric acid at 120°C (Takizawa and Sekine 1985). It has also been reported that B-2 alloy was prone to transgranular stress corrosion cracking in the presence of hydroiodic acid (HI) above 177°C (Kolts 1987). Stress corrosion cracking studies of B, B-2 and B-3 alloys in acidic solutions were carried out under laboratory and plant conditions (Nakahara and Shoji 1996). The effects of the electrochemical potential, cold work produced by drilling, and two different aging processes (that may simulate welding and the subsequent cooling cycle) were investigated. At anodic potentials (200 mV above the free corrosion potential), Nakahara and Shoji (1996) found transgranular fissuring in all three alloys both for mill annealed and aged materials. At cathodic potentials (100 mV and 400 mV below the free corrosion potential), they found intergranular cracking only for the aged (sensitized) alloys. Since the amount of intergranular brittle cracking increased at the lower applied cathodic potential, this environmentally induced cracking was attributed to hydrogen embrittlement (Nakahara and Shoji 1996).

7.3.5   Ni-Cr-Mo alloys

One of the major limitations of stainless steels is that these alloys are susceptible to chloride-induced localized attack such as crevice corrosion, pitting corrosion and stress corrosion cracking. Ni-Cr-Mo alloys such as C-276 are the most resistant Ni-based alloys to the classic chloride-induced localized corrosion that troubles and limits the use of the austenitic stainless steels. In some cases SCC was reported in high strength Ni-Cr-Mo materials; however, cracking only occurred in very aggressive conditions, such at temperatures higher than 200°C, pH lower than 4 and in the presence of hydrogen sulfide (Kolts 1987). U-bend specimens of C-2000, C-22 and C-276 alloys were free from cracking in boiling (154°C) 45% $MgCl_2$ solution after 1008 h of testing (Rebak 2000b). C-276 and C-4 alloy were free from cracking in a 25% NaCl solution at 232°C; however, these alloys were susceptible to cracking in a $MgCl_2$ solution of the same chloride content at the same temperature (Agarwal *et al.* 1994). C-22 alloy was immune to SCC in 20.4% $MgCl_2$ solution up to 232°C, even in the 50% cold worked condition and in the 50% cold worked plus aged at 500°C for 100 h condition (Rebak 2000b). Several alloys including stainless steels and Ni alloys were tested in the laboratory and in the Mont Blanc tunnel for bolt applications (Haselmair *et al.* 1994). The annual relative humidity and temperature averages in the tunnel are 72% and 18.5°C. The tunnel air also had gases including sulfur dioxide, nitrous oxides and hydrochloric acid. All the stainless steel type 303 bolts developed typical chloride-induced transgranular cracks after 36 months' exposure, but the Ni alloys C-4 and 625 were free from cracking (Haselmair *et al.* 1994).

When Ni-Cr-Mo alloys are aged at temperatures higher than 600°C for a long time (e.g. 1000 h at 650°C), long range ordering reactions and precipitation of tetrahedrally close packed (TCP) phases (μ, P, σ) may take place. The presence of the TCP phases produced by thermal aging may greatly reduce the ductility of Ni-Cr-Mo alloys. For example, for annealed C-276 alloy, the yield stress (YS) at room temperature is 360 MPa, the ultimate tensile stress (UTS) is 807 MPa, the elongation to rupture is 63%; however, for a C-276 alloy that was aged for 16 000 h at 760°C, the YS increases to 476 MPa, the UTS increases to 894 MPa and the elongation to rupture decreases to 10%. Figure 7.1 shows the effect of aging at 649°C for almost 2 years on the ductility of four Ni-Cr-Mo alloys. Alloys C-276 and 625 lose half of their ductility after aging for 4000 h. The alloy most resistant to aging is C-4, which still retains more than 40% ductility even after aging at 649°C for 16 000 h. The higher resistance of alloy C-4 to thermal aging effects could be related to its lower content of alloying elements.

It has been reported that thermally aged C-276 alloy was susceptible to hydrogen-induced cracking in environments containing hydrogen sulfide (H_2S) (Kane *et al.* 1977, Sridhar *et al.* 1980a, 1980b, Coyle *et al.* 1981).

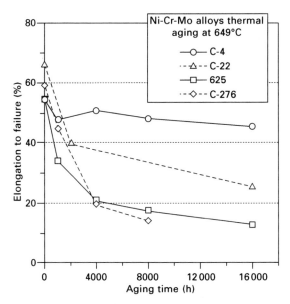

7.1 Effect of thermal aging at 649°C on the elongation to failure of four Ni-Cr-Mo alloys as a function of aging time.

Ni-Cr-Mo alloys were also found to suffer environmentally induced cracking in conditions associated to super critical water oxidation (SCWO). It has been reported that both C-276 (N10276) and Alloy 625 (N06625) suffered intergranular cracking when exposed to various aqueous solutions in the vicinity of the critical point of water (374°C) (Mitton *et al.* 1998, Kritzer *et al.* 1998, Alley and Bradley 2003). More details on SCWO appear in the next section.

7.3.6   Ni-Cr-Fe-(Mo) alloys

This is one of the largest groups of nickel-based alloys since it covers Inconel 600 (N06600), Incoloy 825 (N08825) and 800 (N08800) and Hastelloy G-30 (N06030) type alloys. Since Alloy 600 has been used to fabricate the tubes of steam generators in nuclear power plants, it has been by far the most studied nickel alloy regarding its stress corrosion cracking behavior, especially in hot water and caustic solutions. Alloy 600 has been found to suffer stress corrosion cracking in high temperature pure water (>300°C) both in service and in the laboratory. Due to its importance for the nuclear industry, the stress cracking of Alloys 600 and 690 in pure water and in caustic solutions has been extensively researched in the last three decades (Szklarska-Smialowska and Rebak 1996, Staehle and Gorman 2003, 2004) and thousands of technical papers have been published on this subject. The susceptibility to cracking of Alloys 600 and 690 depend strongly on environmental factors such as temperature, level of tensile stresses, deformation rate, presence of hydrogen gas, solution pH and electrochemical potential, and metallurgical factors, such as presence of minor alloying elements (impurities), the amount of cold work and heat treatment (intragranular or intergranular carbides). Cracking in Alloy 600 is predominantly intergranular (IGSCC) in nuclear service. In some cases (e.g. lead contamination from the secondary side), cracking may be transgranular (TGSCC). Further details are discussed in Section 7.4.4.

Alloy 690, which has double the amount of chromium in Alloy 600, has been found to be more resistant than Alloy 600 to high temperature cracking in pure water and in caustic solutions. Alloy 800 is also used in nuclear power generation. In steam generator applications, Alloy 800 is generally more resistant to cracking than Alloy 600, probably because of the intermediate Ni composition in Alloy 800. Precipitation hardened high strength Alloy X-750 is used in nuclear reactor internals, mainly as fasteners. Similar to Alloy 600, X-750 is susceptible to SCC in high temperature water typical of nuclear reactors. It has been reported that Alloy X-750 could also be susceptible to hydrogen embrittlement at temperatures below 150°C (Mills *et al.* 1999). Similarly, low temperature hydrogen induced cracking was also reported for Alloy 718 (Fournier *et al.* 1999).

Alloy 825 is more resistant to stress corrosion cracking in chloride solutions than 316 stainless steels (S31600) due to the higher content of nickel in Alloy 825. Slow strain rate tests and U-bend tests have shown that Alloy 825 was susceptible to transgranular stress corrosion cracking in 45% $MgCl_2$ solutions at temperatures above 146°C. Alloy 825 is used extensively in the oil and gas production in sour wells; however the performance of other nickel alloys such as C-276 and G-50 (N06950) is still superior of that of 825 (Hibner and Tassen 2000). These nickel alloys are used mainly in the cold worked condition for increased strength. Environmental factors that may affect the stress cracking performance of Alloy 825 (and other alloys) in oil and gas wells include temperature, amount of chloride and the presence of hydrogen sulfide gas (Hibner and Tassen 2000). Further discussion on the behavior of Ni alloys in oil and gas production can be found in Section 7.4.3. Data on the stress corrosion cracking behavior of G-30 alloy is scarce. It has been reported that G-30 components used in the industrial production of hydrofluoric acid suffered cracking (Rebak 2000b). U-bend specimens of G-30 alloy did not crack after exposure for 500 h in 45% $MgCl_2$ solution at 154°C. It has been found that G-30 as well as other nickel alloys would suffer cracking in the aggressive conditions encountered in super critical water oxidation (SCWO) treatments. Further discussion on the cracking susceptibility of Ni alloys appears in Section 7.4.2.

7.4   Resistance to stress corrosion cracking (SCC) by application

7.4.1   Chemical process industry (CPI)

Nickel-based corrosion-resistant alloys (CRA) are used widely in the chemical process industry in targeted applications. For example, Ni-Mo alloys (e.g., B-2) are used to handle hot reducing acids because they offer very low corrosion rates in these environments. Similarly, commercial nickel (e.g., Ni-200) is used to handle hot caustic solutions. Other alloys such as the C-family (e.g., C-276), which contain Ni, Cr and Mo, are versatile and can be used in almost every environment; however, their performance in hot reducing acids would be inferior to that of Ni-Mo alloys and in hot caustic it would have a higher corrosion rate than Ni-200. Unlike the austenitic stainless steels, Ni alloys are resistant to SCC in hot chloride environments. However, Ni alloys may suffer SCC in environments such as hot caustic and wet hydrofluoric acid environments (Table 7.2).

Caustic environments

Caustic environments are highly concentrated solutions (over 50%) of sodium hydroxide (NaOH) or caustic soda, potassium hydroxide (KOH) or caustic potash and calcium hydroxide ($Ca(OH)_2$) or caustic lime that may be found in the CPI as well as other industries such as oil refineries and pulp and paper (Crum and Shoemaker 2006, Rebak 2006). It is likely that the cracking susceptibility of Ni alloys is associated to a dealloying phenomenon (Rebak 2006). Figure 7.2 shows the cross section of a 0.6 mm thick sheet of C-276 that was in service for 10 months in a heat exchanger between water (lower part) and 50% NaOH plus traces of perchlorate at temperatures near 100°C. Cracking developed from the dealloyed layer exposed to the caustic solution. The best performing material in caustic environments is commercially pure Ni (Table 7.1). Large amounts of Mo in Ni alloys are detrimental and Cr seems to be a beneficial element only in high concentrations (Rebak 2006). When dealloying occurs, not only do Cr and Mo dissolve away leaving behind a porous pure Ni layer (such as in Fig. 7.2) but also the alloy is plated on the surface with pure Ni (Rebak 2006). Under slow strain rate conditions, C-276 alloy was susceptible to transgranular cracking in 50% NaOH at 147°C (Asphahani 1979). On the other hand, mill annealed and aged for 24 h at 677°C C-shape specimens of C-22 alloy did not exhibit cracking after immersion in 50% NaOH solution at 147°C for 720 h (Haynes International 2000).

Alloy 600, like other nickel based alloys, suffers stress corrosion cracking in hot caustic solutions (150–200°C). Laboratory SCC testing was performed using cylindrical slow strain rate specimens and spring loaded bend beam

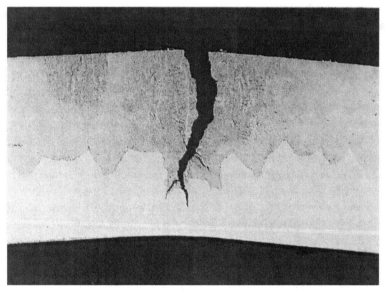

7.2 Crack formed on the dealloyed layer of a C-276 separator 0.6 mm thick sheet in a heat exchanger exposed for 10 months to 50% NaOH contaminated with traces of $NaClO_3$ at 100°C.

specimens of Alloys 600 and 800 in deaerated 10% NaOH solution at 550°F (288°C) (Cels 1976). SCC was reported in both alloys; however, Alloy 600 was more resistant to cracking than Alloy 800, probably because of the higher Ni content in the former (Cels 1976) (Table 7.1). Alloy 600 C-rings were tested for 1000 h in 10% NaOH solution and in 10% NaOH solution plus 7.5% silicon oxide (SiO_2) and 10% NaOH solution plus 15% sodium silicate (Na_2SiO_3) at 315°C (Navas *et al.* 1999). Intergranular cracks were reported in the specimens exposed to the three tested solutions; however, the cracks were longer in the pure NaOH solution suggesting that Si compounds had an inhibitive effect on the caustic cracking of Alloy 600. After the tests the specimens were covered with a layer of pure Ni (Navas *et al.* 1999). The fact that specimens tested in caustic solutions showed a re-deposited external layer of pure nickel has also been reported by other authors (Crum and Shoemaker 2006, Rebak 2006). U-bend specimens of Ni alloys such as 400, 600, 625, 690, 686, 800 and 825 were tested under various conditions in concentrated 10–90% NaOH solutions at temperatures ranging from 103°C to 210°C (Crum and Shoemaker 2006). In general the resistance to SCC increased with the amount of Ni in the alloy, even though there was a large variability in the results depending on the hydroxide concentration and temperature. Ni alloys performed substantially better than 300 series and high Cr-6 Mo austenitic stainless steels tested in parallel with the Ni alloys. Dealloying was also reported for some alloys in some of the tests (Crum and Shoemaker 2006). It was shown that Alloy 800 (N08800) was susceptible to caustic cracking (Mignone *et al.* 1990) and even more susceptible than Alloy 690, probably because of the higher Cr content of the latter (Yang *et al.* 2001). It is also apparent that Alloy 600 may be more resistant to caustic cracking than Alloy 690 because the former contains approximately 14% more Ni, which is the element that gives resistance in caustic environments (Staehle and Gorman 2004).

Wet hydrofluoric acid

Even though hydrofluoric (HF) acid may be classified as a weak acid, it is extremely corrosive and very few metallic materials are suitable for handling it. Pawel (1994) tested U-bends of Alloys 400 and C-276 for resistance to stress corrosion cracking as a function of acid concentration and temperature. Alloy 400 suffered cracking in HF environments in the vapor phase but not in the liquid phase. Alloy C-276 did not suffer cracking under any of the tested conditions (Pawel 1994). Internal crack-like penetrations from flat unstressed specimens exposed to HF were also reported for Alloys 400, 600 and 825 (Pawel 1994). Several Ni alloys were tested as a function of the HF concentration and temperature (Rebak 2000b, Rebak *et al.* 2001). Using U-bend specimens, it has been reported that the crack propagation rate in Alloy 400 exposed to the vapor phase of 20% HF for 240 h decreased as the temperature increased from 66°C to 93°C, probably because less oxygen was available in the vapor phase as the temperature increased (Rebak 2000b). In the same study, U-bends of Alloy 400 were found free from cracking while immersed in the liquid portion of 20% HF (Rebak 2000b, Rebak *et al.* 2001). It has also been reported that highly stressed alloy 400 suffers SCC in ammonia vapors at 300°C (Theus *et al.* 1982). Heat treatments that eliminate residual stresses and cold worked microstructures greatly reduce the susceptibility of alloy 400 to all types of environmentally induced cracking.

U-bend specimens of mill annealed B-3 (N10675) alloy were found to suffer stress corrosion cracking in the presence of vapor and liquid phase of a 20% HF solution at 66°C, 79°C and 93°C (Rebak 2000b). The cracking susceptibility of N10675 increased with the temperature and the liquid portion of 20% HF solution showed more aggressiveness than the vapor phase (Rebak 2000b).

Laboratory testing using U-bend specimens (ASTM G30) had shown that Ni-Cr-Mo alloys such as C-276, C-22 and C-2000 alloy were susceptible to SCC in wet HF in both the liquid and vapor phases (Rebak 2000b). Figure 7.3 shows the degradation appearance in Alloy C-22 after exposure to 20% HF at 93°C for 240 h. The attack in the liquid phase (Fig. 7.3(a)) was deeper than in the vapor phase (Fig. 7.3(b)). Some of the attack seems to be driven by stresses but it is known that HF also may promote internal attack on plain unstressed coupons. The most resistant of the Ni-Cr-Mo alloys to cracking in wet HF was C-2000 (N06200) probably because of the beneficial effect of 1.6% Cu content. Just in opposite behavior to Ni-Cu Alloy 400, Ni-Cr-Mo alloys were less susceptible to cracking in the vapor phase than in the liquid phase, suggesting that the presence of Cr is beneficial for HF vapor phase applications (Rebak 2000b). Alloy 600 was also prone to cracking and internal penetration in HF environments (Pawel 1994, Rebak *et al.* 2001).

7.4.2   Supercritical water oxidation

Supercritical water oxidation (SCWO) is a method that is being perfected for the treatment of dilute waste containing organic materials, including civilian and military hazardous waste (Latanision 1995, Bermejo and Cocero 2006, Mitton 2009). Organic compounds are highly miscible and inorganic salts have low solubility in water above its critical point (374°C and 221 atm for pure water). Toxic organic wastes turn into water, carbon dioxide, nitrogen and simple salts under SCWO conditions. The SCWO environments may range from acidic to alkaline and are extremely corrosive to most alloys. Ni-based alloys suffer cracking under certain conditions, mainly in subcritical conditions where the liquid phase is still stable but the temperature and pressure are high (just below critical). The aggressiveness is even higher if

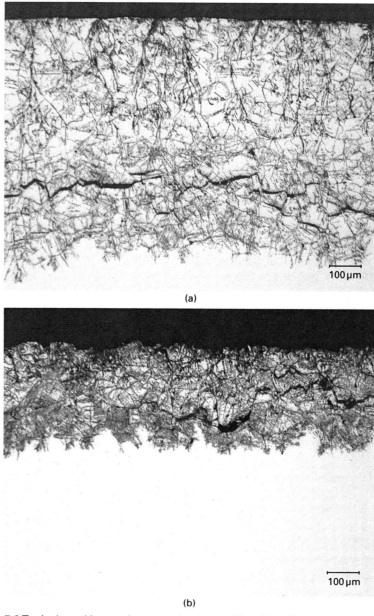

7.3 Typical cracking and penetration in an Alloy C-22 (N06022)
U-bend specimen exposed to a 20% HF solution at 93°C for 240 h: (a)
liquid phase, (b) vapor phase (magnification × 100).

the organic compounds contain chlorides (e.g., dichloromethane CH_2Cl_2 and
chloroprene C_4H_5Cl) because they decompose forming hydrochloric acid
(Schroer *et al.* 2005, Saito *et al.* 2006). A preheater tube of Alloy C-276

suffered intergranular cracking while feeding a waste of methylene chloride into a SCWO system; however, the cracking occurred under subcritical conditions (Latanision 1995). Cracking was also reported in an Alloy 625 tube exposed discontinuously for 300 h at 425°C to an aqueous stream containing HCl at pH = 2 (Mitton 2009). Cracking has also been reported for Alloy C-22 tubing exposed for 53 h at 350°C to a HCl feed of pH 0.48; however when the pH was 4.4 under the same testing conditions, cracking did not occur after testing for 70 h (Mitton 2009).

7.4.3   Oil and gas

A wide range of materials are used in the oil and gas industry as components for oilfield environments, depending on the aggressiveness of the environment. Until a couple a decades ago most of the material used in the production of gas and crude oil was carbon and low alloy steel since the operations were relatively easy and the commodity (petroleum products) was cheap (Kane 2006). However, crude production is becoming more difficult and stronger and more noble types of materials are needed. The newer materials include the corrosion-resistant alloys. Generally the corrosion-resistant alloys are separated into three categories:

- martensitic stainless steels,
- duplex stainless steel, and
- austenitic materials which include stainless steels and nickel-based alloys (Rhodes 2001).

Environmental factors affecting the corrosion behavior of the alloys include temperature, chloride concentration, partial pressure of hydrogen sulfide (H_2S) and carbon dioxide (CO_2), elemental sulfur, and the pH of the solution. The higher the temperature, chloride concentration and partial pressure of H_2S and the lower the pH the more aggressive the environment. Figure 7.4 gives a general guideline on the effect of both chloride concentration and partial pressure of H_2S on the materials that may be considered. The shape of Fig. 7.4 may change for other temperature and pH values. In general, two main factors are considered for the alloy selection process: mechanical properties and corrosion resistance (Craig 1995, Aberle and Agarwal 2008). Stronger alloys are preferred to minimize the section of the materials but at the same time the susceptibility to SCC generally increases with the strength of the alloy. For oil country tubular goods or production tubing, minimum requirements of yield stress are in the order of 550 MPa (80 ksi) for shallow wells and higher than 1100 MPa (160 ksi) for deep and high pressure wells (Rhodes 2001).

The corrosion resistance of stainless Ni alloys generally increases with increasing chromium and molybdenum contents. One general index to

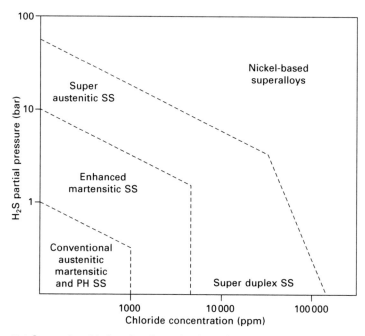

7.4 General guideline for the selection of materials for down hole oil and gas applications based on the chloride concentration and partial pressure of hydrogen sulfide. The materials selection will also depend on the temperature, pH of the environment and presence of carbon dioxide.

determine resistance to localized corrosion (and SCC) is the pitting resistance equivalent number (PREN = %Cr + 3.3%[Mo+W/2]). The higher the PREN the more resistant is the alloy to localized attack (Aberle and Agarwal 2008). The main factors affecting the cracking resistance of materials in an oil and gas well are the chemical composition, heat treatment, microstructure, strength of the material, and the environmental conditions (partial pressure of H_2S, pH, chloride concentration, oxygen, carbon dioxide, elemental sulfur, temperature, etc.). Nickel-based alloys are commonly used in sour wells where the hydrogen sulfide (H_2S) pressures and chloride concentrations are particularly high combined with temperatures at times exceeding 200°C. NACE International (The Corrosion Society) has issued two documents outlining testing, recommendations and applicability limits for materials for oil and gas applications. These documents are:

- TM0177-2005 'Standard Test Method Laboratory Testing of Metals for Resistance to Sulfide Stress Cracking and Stress Corrosion Cracking in H_2S Environments,' and
- MR0175/ISO 15156 'Materials Recommendation for Use in H_2S containing Environments in Oil and Gas Production.'

The TM0177 document specifies consistency in testing in order to compare data from different laboratories. The MR0175/ISO15156 recommends environmental and material limits for nickel alloys in two groups: (a) annealed and cold worked solid solution Ni-based alloys and (b) precipitation hardened Ni-based alloys. For example, MR0175/ISO15156 considers that a cold worked solid solution Ni-based alloy such as C-276 (N10276) is suitable for use up to 218°C (425°F) in an atmosphere containing up to 700 kPa (100 psi) partial pressure of hydrogen sulfide and any chloride concentration and the corresponding *in situ* pH provided the maximum hardness is lower than 40 HRC and the yield strength is lower than 1034 MPa (150 ksi). Craig (1995) notes that C-276 can be used at any chloride concentration (e.g., 100 000 ppm) and at any partial pressure of both H_2S and CO_2 (e.g., 10 000 psi) up to 260°C. However, the presence of elemental sulfur may cause catastrophic cracking in this alloy (Craig 1995). Other Ni-based alloys that are used in the oil and gas industry to resist SCC for sour wells applications include 825 (N08825), and higher end alloys such as Alloys 22 (N06022) and 59 (N06059) and the age hardenable alloys 625 (N06625), 718 (N07718), 925 (N09925), C-22HS (N07022) and 945 (N09945) (Table 7.1). Alloys such as C-276, 825, 22 and 59 can only be strengthened by cold working, so they may not be suitable for thick sections and intricate shapes and generally cold working does not produce a uniform microstructure. However, Alloys 625, 718, C-22HS and 945 can be aged hardened without large impact on the corrosion resistance (Rebak 2010). Strengthening of alloys such as 945 occurs by the precipitation in the austenitic matrix of small particles containing Nb, Ti and/or Al that interfere with the movement of dislocations. The alloy C-22HS (Table 7.1) is age hardenable by a long-range ordering reaction reaching high mechanical properties while still maintaining an elongation of approximately 20%.

A hybrid alloy between Alloys 925 and 718 was developed and tested for corrosion resistance in oil and gas applications (Mannan and Patel 2008). It is claimed that this age hardenable alloy 945 (Table 7.1) has excellent corrosion resistance with high strength properties. Pre-stressed to 100% of the yield stress C-ring specimens of high strength Alloy 945 were exposed NACE TM0177 method C solution (20% NaCl + 508 psi CO_2 + 508 psia H_2S) at 175°C and none of the specimens suffered cracking after 90 days of testing. Similarly, Alloy 945 tensile specimens stressed to 90% of the actual yield stress and galvanically coupled to steel that were exposed to TM0177 method A acidified solution at 24°C were free from cracking after 30 days of testing (Mannan and Patel 2008).

High strength age hardened Alloy C-22HS (Table 7.1) was tested using slow strain rate tests by triplicate at 4×10^{-6} s^{-1} for three different metallurgical conditions in deaerated 25% NaCl + 508 psig CO_2 + 508 psig H_2S at 205°C with and without elemental sulfur added. Figure 7.5 shows a plot generated

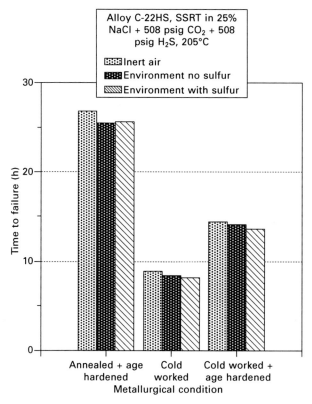

7.5 Effect of metallurgical condition of C-22HS on the time to failure in 25% NaCl + 508 psig CO_2 + 508 psig H_2S with and without elemental sulfur at 205°C

from table data published by Pike *et al.* (2010). In general, the longest time to failure was for the annealed + age hardened material and the shortest time to failure was for the cold worked material. It is interesting that some of the ductility is restored when the cold worked material is aged hardened (Fig. 7.5). For each one of the three metallurgical conditions (annealed + age hardened, cold worked and cold worked + age hardened) there was only a very small reduction in the time to rupture when the alloy was tested in the aggressive brine solution as compared with inert air. Also there was a very small reduction in the time to failure when the environment contained elemental sulfur. In general, the environmental effect was minimal and none of the specimens showed any secondary cracking (Pike *et al.* 2010). These data show that Alloy C-22HS has an excellent resistance to environmentally assisted cracking in simulated oil well conditions. The age hardenable Cu-Ni K-500 is also used in the oil and gas industry but it is limited to milder environments than the higher end alloys such as C-276 and 718 (Aberle and Agarwal 2008). Environmentally assisted cracking of materials in the production of petroleum products may be classified into three categories:

1. the well-known higher temperature stress corrosion cracking promoted by chlorides and that limits the application of austenitic stainless steels,
2. sulfide stress cracking mainly affecting martensitic materials and
3. hydrogen embrittlement generally associated with cathodic charging by coupling Ni alloys to carbon steel components (Rhodes 2001).

It is likely that most if not all the environmental fractures of Ni alloys are associated with the ingress of hydrogen into the alloys during service (Rebak 2010).

7.4.4   Nuclear power generation

Nickel alloys are used extensively as structural materials in nuclear power generation, mainly as steam generator tubing in pressurized water reactors (PWR) and as pressure vessel internals both in PWR and boiling water reactors (BWR) (Rebak 2009b, Allen and Busby 2009). The most common nickel alloys employed in nuclear power generation are Alloy 600 (N06600), Alloy 690 (N06690), and the weld metals alloys 82, 182, 152 and 52 (Table 7.1). Alloy 800 (N08800), which is used in some countries mainly for heat exchanger tubing, is listed as a Ni alloy even though it contains only about 33% Ni (Table 7.1). Alloys such as Inconel X-750 and 718 are also used but to a lower extent. The main mode of degradation or failure of the Ni alloys in nuclear power plants is stress corrosion cracking. Extensive research has been conducted in the last 40 years to determine the variables that control the occurrence of SCC in nuclear plants. The results from this research has been notably captured in a series of biannual conferences that are called *Environmental Degradation of Materials in Nuclear Power Systems – Water Reactors* that started in August 1983 in Myrtle Beach, SC. The latest *Environmental Degradation* meeting was the 14th in Virginia Beach, VA, in August 2009. One of the main failures of nickel alloys in power plants has been that of the PWR steam generator tubing made using Alloy 600, both from the primary and secondary sides of the tubing (Staehle and Gorman 2003). The cracking of steam generator tubing has been minimized lately by changes in the water chemistry, design, fabrication and tubing alloy material (e.g., using Alloy 690 instead of Alloy 600) (Staehle and Gorman 2003). SCC failures of other components were also reported, including the cracking of Alloy 600 reactor vessel head (RVH) penetrations (Pathania *et al.* 2002), Alloy X-750 bolts and springs (Jones 1996, Scott and Combrade 2003), and Alloys 82 and 182 welds (Paraventi and Moshier 2005).

Several factors control the susceptibility of Alloy 600 to SCC in the nuclear power industry including internal factors such as alloy composition and the microstructure of the alloy and external factors such as the redox potential in the system, the level of tensile stresses, temperature, the pH of the electrolyte, the presence of detrimental dissolved species such as lead, and the partial pressure and hydrogen. Staehle and Gorman (2004) identified seven principal variables to affect the depth of SCC penetration:

$$\text{Depth} - \text{SCC} = A \cdot [H^+]^n \cdot [x]^p \cdot \sigma^m \cdot e^{\frac{E-E_0}{b}} \cdot e^{\frac{Q}{R \cdot T}} \cdot t^q \qquad 7.1$$

where A represents the effect of the alloy composition and structure, the proton represents the effect of pH, x represents the effect of environmental species such as lead, sulfate, etc., σ represents the effect of stress, E in the first exponential represents the effect of the electrochemical potential and E_0 the effect of the alloy composition and environment, the last exponential represents the Arrhenius expression for the effect of the temperature (T) and t represents the effect of time (Staehle and Gorman 2004). These seven main variables (plus many others) are all interdependent and if one of them changes, the effect of all the other variables on SCC also changes.

Internal factors, chemical composition and microstructure

One of the most important factors affecting the susceptibility of Alloy 600 to SCC in high temperature water is its thermomechanical history and the amount, form and distribution of carbon in the matrix (Sridhar and Cragnolino 1992). Alloy 600 mill annealed (600MA) was found to be the most susceptible condition to SCC in PWR steam generator tubing service conditions. Alloy 600MA that is generally annealed at temperatures below 950°C has most of the carbide present in intragranular (or transgranular) sites. However, Alloy 600 thermally treated (600TT), which has at least 0.02% carbon and was exposed to annealing temperatures higher than 1000°C followed by a thermal exposure at 700°C to precipitate most of the carbides in intergranular form, was found to be highly resistant to SCC in high temperature water (Sridhar and Cragnolino 1992, Szklarska-Smialowska and Rebak 1996, Staehle and Gorman 2003). In the last two decades, instead of using the more SCC resistant Alloy 600TT, some plants have decided to combat Alloy 600 SCC by using Alloy 690TT (Staehle and Gorman 2003). It has been defended that by virtue of its higher Cr content, Alloy 690 is more resistant to SCC than Alloy 600. Furthermore, some plants designed in Germany prefer to use Alloy 800 nuclear grade (800NG) instead of 600TT or 690TT (Staehle and Gorman 2003). It has been suggested that alloys that contain intermediate amount of Ni (such Alloy 800 with 33% Ni, see Table 7.1) are more resistant to cracking in typical nuclear power high temperature pure water environments than high Ni containing alloys (Staehle 1996, Staehle and Gorman 2004). Figure 7.6 shows schematically how alloys with intermediate concentration

Ni such as Alloy 800 are both resistant to cracking in chloride containing solutions and in high temperature water. The schematic representation of Fig. 7.6 was first proposed by Henri Coriou in 1967 (Staehle 1996).

The presence of cold work greatly increases the susceptibility of Alloy 600 to SCC both for crack initiation and propagation (Cassagne and Gelpi 1993, Yamazaki *et al.* 2008). Paraventi and Moshier (2005) also reported an increase in SCC crack growth rate in Alloy 82 weld metal when the material was cold worked by 12%. Cold worked materials have higher yield stress and lower elongation to failure, and therefore appear less resistant to SCC.

The chemical composition of the alloy is also important since it is generally found that Alloy 690 (29% Cr) is more resistant to SCC in high temperature water than Alloy 600 (16% Cr) (Table 7.1). Andresen *et al.* (2008) reported that Alloy 690 and its weld metals (Alloy 52/152) have a SCC crack propagation rate that is approximately 100 to 400 times lower than for Alloys 600 and 182 tested under similar conditions in simulated primary water at 340–360°C. On the other hand, it was suggested that the

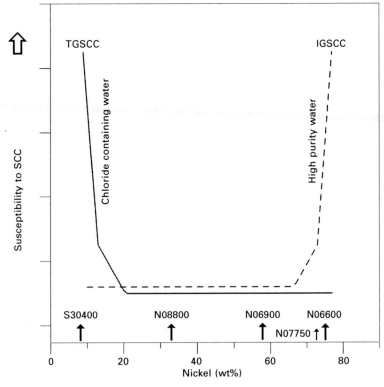

7.6 Schematic representation to show the effect of the content of nickel on the resistance of metallic alloys to stress corrosion cracking in the presence of chloride and pure water at high temperature.

higher Cr content in alloy 690 would make it less resistant to a hydrogen embrittlement mechanism than Alloy 600 (Symons 1997).

External factors, electrolyte composition, potential and temperature

Good water chemistry practices ameliorate the incidence of cracking of Alloy 600 tubing but do not totally eliminate it (Staehle and Gorman 2003). Some of the most important variables that control SCC from the electrolyte point of view are the pH of the solution, the presence of impurities, such as lead, and the presence of dissolved hydrogen and oxygen (which also control the electrochemical potential in the system (Staehle and Gorman 2004)). Cracking of Alloy 600 tubing from the secondary side tends to be more detrimental from occluded zones or crevices around the tubes where the electrolyte may evolve to become alkaline (higher pH values). The detrimental effect of lead (Pb) on the SCC of Ni alloys from the secondary side has been researched extensively for the last four decades (Castaño-Marin *et al.* 1993, Hwang *et al.* 1999, Staehle 2005). The presence of lead may change the mode of cracking of Alloy 600 and 690 in high temperature aqueous solutions from intergranular to transgranular (Hwang *et al.* 1999). The aggressive effect of Pb on the SCC of Ni alloys may also depend on other variables in the system such as metallurgical state of the alloys and pH of the electrolyte (Staehle 2005).

Temperature is a significant factor that controls both SCC initiation and propagation in Ni alloys such as Alloy 600 (Cassagne and Gelpi 1993, Economy *et al.* 1987, Staehle and Gorman 2004). The higher the temperature the less resistant is Alloy 600 to SCC. The Arrhenius activation energy seems to fluctuate approximately between 80 and 200 kJ/mol, probably depending on the values of the other affecting variables such as the amount of cold work, the presence of hydrogen, and the level of applied stresses (Cassagne and Gelpi 1993, Staehle 2005). The partial pressure of hydrogen is also an important variable controlling the susceptibility to SCC of Ni alloys. The hydrogen effect is through its effect on the electrochemical potential in the system. It was reported that the maximum susceptibility of Ni alloys (e.g. X-750 and 600) to SCC occurs for an intermediate partial pressure of hydrogen that gives rise to a narrow range of potential ($\sim$100 mV) corresponding to the transition of Ni metal to Ni oxide (NiO) (Morton *et al.* 2002). For a higher hydrogen partial pressure when the NiO is not allowed to form, the susceptibility to SCC decreases and for a lower partial pressure of hydrogen when the oxide that forms is protective, the susceptibility to SCC is also reduced. The partial pressure of hydrogen needed to produce the peak of maximum susceptibility is also a function of the other variables such as alloy composition, pH of the solution and temperature (Andresen *et al.* 2008). The fact that hydrogen gas depresses the electrochemical potential in the system has long been recognized as a measure to mitigate the occurrence of SCC. The susceptibility to environmental cracking of the structural materials increases with the potential in the system. That is, hydrogen gas is added to the water in the commercial power plants to maintain a low potential and therefore minimize the occurrence of cracking. It has been also found that if nanoparticles of platinum are applied to the wet parts of the alloy components, lower amounts of hydrogen gas are needed in the high temperature water to maintain a similar low protective potential (Hettiarachchi 2005).

SCC mechanisms for Ni alloys in nuclear reactor systems

During the last four decades many mechanisms were proposed to describe the initiation and propagation of SCC in Ni alloy power plant structural components (Rebak and Szklarska-Smialowska 1996, Scott and Combrade 2003, Young *et al.* 2005). However none of these mechanisms by themselves seems to be able to fully explain the intricacy of the variables that control the occurrence of SCC in the actual plants in the temperature range 250–360°C. The proposed mechanisms can be classified into three large groups:

1. oxidation/dissolution (with subgroups that may include slip step film rupture oxidation, internal oxidation and enhanced surface mobility),
2. hydrogen-assisted cracking (with subgroup mechanisms that may include hydrogen embrittlement and hydrogen enhanced plasticity) (Delafosse and Magnin 2001, Chêne and Brass 2004), and
3. mechanically oriented models including correlation between SCC and crack tip strain rate, vacancy condensation and creep-assisted grain boundary rupture.

All these mechanisms have some practical basis, for example, to explain the occurrence of cracking in some ranges of potential, temperature and/or applied stress. However, none of these mechanisms may have the predictive capabilities to explain the behavior of nickel alloy in power plants for time spans in the order of 60 years and longer.

7.4.5   Nuclear waste disposal

Because of its excellent resistance to stress corrosion cracking and other types of localized corrosion, C-22 (N06022) was selected by the US Department of Energy to fabricate the outer shell of the high level nuclear waste containers to be disposed permanently at the Yucca Mountain site (Gordon 2002, Cragnolino *et al.* 2002). C-22 has been extensively tested for its susceptibility to SCC in a variety of environments, mainly at GE Global Research, Southwest Research Institute and Lawrence Livermore National Laboratory (LLNL). This alloy was found extremely resistant to EAC in many different solutions at the corrosion potential, at all the tested temperatures from ambient to 110°C (Dunn *et al.* 2002, Estill *et al.* 2002, Andresen *et al.* 2003). Tests were carried out using cyclic loading, constant load, constant deformation and slow strain rate tests in solutions from 14 molal $MgCl_2$, to simulated concentrated groundwaters from pH 3 to 13. U-bend specimens of C-22 (N06022) and other nickel alloys such as C-4 (N06455), G-3 (N06985), 825 (N08825) and 625 (N06625) were used to characterize their stress corrosion cracking susceptibility in a variety of environments (Fix *et al.* 2004, Rebak 2009a). Gas tungsten arc welded (GTAW) and non-welded U-bend specimens were exposed for more than 5 years at the corrosion potential to the vapor and liquid phases of three different solutions (pH 2.8–10) simulating up to 1000 times the concentration of groundwater both at 60°C and 90°C. None of these alloys suffered any indication of environmentally induced cracking (Fix *et al.* 2004, Rebak 2009a).

Alloy C-22 was found susceptible to EAC when SSRT was performed on mill annealed specimens in hot simulated concentrated water (SCW) at anodic applied potentials (Estill *et al.* 2002, King *et al.* 2002). SCW is a multi-ionic alkaline solution approximately 1000 times more concentrated than a Yucca Mountain groundwater. It was initially assumed that the small amount of fluoride ions present in this solution (1400 ppm) contributed to the cracking of C-22 (Estill *et al.* 2002). However, it is now understood that the bicarbonate present in the SCW solution is the main reason for the SCC of Alloy 22 (Chiang *et al.* 2006). The susceptibility to cracking of C-22 was strongly dependent on the applied potential and the temperature of the solution. The highest susceptibility to EAC was found at around 90°C at +400 mV in the saturated silver chloride (SSC) electrode scale (Fig. 7.7). At the corrosion potential, C-22 was free from EAC even at 90°C (Fig. 7.8). Similarly, at anodic applied potentials, C-22 was free from EAC at ambient temperatures and as the temperature increased the time to failure in the tests decreased. The occurrence of EAC was related to the presence of an anodic peak in the polarization curve of the alloy in SCW environments. For example, at ambient temperatures, the peak is not present and EAC does not take place (King *et al.* 2004). It was demonstrated that the most aggressive species for EAC in SCW was bicarbonate, but that the presence of chloride in the bicarbonate solution enhances the aggressiveness of the environment (Chiang *et al.* 2006).

It has also been reported that Alloy C-22 (N06022) may suffer some embrittlement when it is slow strained under cathodic applied potentials (or currents) (King *et al.* 2004, Kesavan 1991, Scammon 1994). The maximum susceptibility to cracking under cathodic conditions seemed to occur at ambient temperatures suggesting a hydrogen-related failure mechanism.

7.7 Low magnification image of a slow strain rate test specimen after fracture showing secondary transgranular stress corrosion cracking. Specimen strained at $1.67 \times 10^{-6} s^{-1}$ in simulated concentrated water at 86°C at an applied potential of +400 mV in the saturated silver/silver chloride scale.

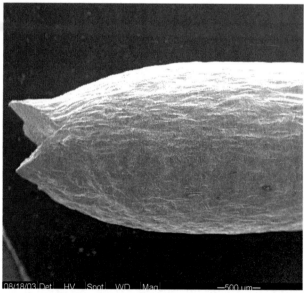

7.8 Low magnification image of a slow strain rate test specimen after fracture showing ductile failure (no secondary cracking). Specimen strained at $1.67 \times 10^{-6} s^{-1}$ in simulated concentrated water at 90°C at the corrosion potential (~ –140 mV in the saturated silver/silver chloride scale).

7.5   Conclusions

Since nickel can dissolve a large amount and variety of alloying elements, Ni alloys can be especially tailored for various applications. Nickel alloys are used in the chemical process industry, in nuclear power plants and in the harsh environments of supercritical water oxidation and oil and gas exploration. Some of the factors that greatly enhance the susceptibility of Ni alloys to stress corrosion cracking are temperature and a reduction in ductility by thermal aging or cold work. In general, Ni alloys are more resistant than stainless steels to stress corrosion cracking, mainly because Ni alloys are practically immune to SCC in hot chloride solutions. However, Ni alloys may be susceptible to cracking in high temperature water, the type of environments found in nuclear power reactors. Similarly to the stainless steels, Ni alloys are susceptible to cracking in hot caustic solutions and in hot and wet hydrofluoric acid solutions. Nickel alloys may also be susceptible to environmental cracking if they are cathodically polarized, especially in highly acidic environments.

7.6   References

Aberle D and Agarwal DC (2008), 'High Performance Corrosion Resistant Stainless Steels and Nickel Alloys for Oil & Gas Applications,' paper 08085, *Corrosion/2008* (Hosuton, TX: NACE International).

Agarwal D C (2000), 'Nickel and nickel alloys,' p. 831, in *Uhlig's Corrosion Handbook*, edited by Revie R W, John Wiley & Sons.

Agarwal D C and Kloewer J (2001), 'Nickel base alloys: corrosion challenges in the new millennium,' Paper 01325, *Corrosion/01* (Houston, TX: NACE International).

Agarwal D C, Heubner U, Kohler M and Herda W (1994), 'UNS N10629: a new Ni-28% Mo alloy,' *Materials Performance*, Vol. 33, No. 10, 64–68.

Allen T R and Busby J T (2009), 'Radiation damage concerns for extended light water reactor service,' *Journal of Metals*, Vol. 61, No. 7, 29–34.

Alley D W and Bradley S A (2003), 'Failure of a UNS N10276 reactor in supercritical water service,' Paper 03351, *Corrosion/2003* (Houston, TX: NACE International).

Andresen P L, Emigh P W, Young L M and Gordon G M (2003), 'Stress corrosion cracking growth rate behavior of alloy 22 (UNS N06022) in concentrated groundwater,' Paper 03683, *Corrosion/2003* (Houston, TX: NACE International).

Andresen P L, Hickling J, Ahluwalia A and Wilson J (2008), 'Effects of hydrogen on SCC growth rate of Ni alloys,' paper 08602, *Corrosion/2008* (Houston, TX: NACE International).

Asphahani A I (1979), 'Slow strain-rate technique and its applications to the environmental stress cracking of nickel-base and cobalt-base alloys,' in *Stress Corrosion Cracking – The Slow Strain-Rate Technique*, ASTM STP 665, p. 279 (Philadelphia: ASTM).

Bermejo M D and Cocero M J (2006), 'Supercritical water oxidation: a technical review,' *AIChE Journal*, Vol. 52, No. 11, pp. 3933–3951.

Cassagne T and Gelpi A (1993), 'Crack growth rate measurements on alloy 600 steam generator tubing in primary and hydrogenated AVT water,' in *Proceedings of the 6th International Symposium on Environmental Degradation of Materials in Nuclear*

Power Systems – Water Reactors, eds. Gold R E and Simonen E P, pp. 679–686 (Warrendale, PA: TMS).

Castaño-Marin M L, Gomez-Briceño D and Hernandez-Arroyo F (1993), 'Influence of lead contamination on the stress corrosion cracking of nickel alloys,' in *Proceedings of the 6th International Symposium on Environmental Degradation of Materials in Nuclear Power Systems – Water Reactors*, pp. 189–196 (Warrendale, PA: TMS).

Cels J R (1976), 'Stress corrosion cracking of stainless steels and nickel alloys at controlled potentials in 10% caustic soda solutions at 550°F,' *Journal of the Electrochemical Society*, Vol. 123, No. 8, 1152–1156.

Chêne J and Brass A M (2004), 'Role of temperature and strain rate on the hydrogen-induced intergranular rupture in alloy 600,' *Metallurgical and Materials Transactions A*, Vol. 35A, No. 2, 457–464.

Chiang K T, Dunn D S and Cragnolino G A (2006), 'The combined effect of bicarbonate and chloride ions on the stress corrosion cracking susceptibility of alloy 22,' Paper 06506, *Corrosion/2006*, (Houston, TX: NACE International).

Coyle R J, Kargol J A and Fiore N F (1981), 'The effect of aging on hydrogen embrittlement of a nickel alloy,' *Metallurgical and Materials Transactions A*, Vol. 12A, No. 4, 653–658.

Cragnolino G A, Dunn D S and Pan Y M (2002), 'Localized corrosion susceptibility of alloy 22 as a waste package container material,' in *Scientific Basis for Nuclear Waste Management XXV*, Vol. 713, pp. 53–60 (Warrendale, PA: Materials Research Society).

Craig B D (1995), 'Selection Guidelines for Corrosion Resistant Alloys in the Oil and Gas Industry,' Nickel Institute Technical Series #10073 (Toronto, Canada).

Crum J R and Shoemaker L E (2006), 'Corrosion resistance of nickel alloys in caustic solutions,' Paper 06219, *Corrosion/2006* (Houston, TX: NACE International).

Crum J R, Hibner E, Farr N C and Munasinghe D R (2000), 'Nickel-based alloys' Chapter 7, in *Casti Handbook of Stainless Steels and Nickel Alloys* (Edmonton, Alberta: CASTI Publishing).

Delafosse D and Magnin T (2001), 'Hydrogen induced plasticity in stress corrosion cracking of engineering systems,' *Engineering Fracture Mechanics*, Vol. 68, 693–729.

Dillon C P (1994), *Corrosion Control in the Chemical Process Industry* (Houston, TX: NACE International).

Dunn D S, Pan Y M and Cragnolino G A (2002), 'Stress corrosion cracking of nickel-chromium-molybdenum alloys in chloride solutions,' Paper 02425, *Corrosion/2002* (Houston, TX: NACE International).

Economy G, Jacko R J and Pement F W (1987), 'IGSCC behavior of alloy 600 steam generator tubing in water of steam tests above 360°C,' *Corrosion*, Vol. 43, No. 12, 727–734.

Estill J C, King K J, Fix D V, Spurlock D G, Hust G A, Gordon S R, McCright R D, Gordon G M and Rebak R B (2002), 'Susceptibility of alloy 22 to environmentally assisted cracking in Yucca Mountain relevant environments,' Paper 02535, *Corrosion/2002* (Houston, TX: NACE International).

Fix D V, Estill J C, Hust G A, Wong L L and Rebak R B (2004), 'Environmentally assisted cracking behavior of nickel alloys in simulated acidic and alkaline ground waters using U-bend specimens,' Paper 04549, *Corrosion/2004* (Houston, TX: NACE International).

Fournier L, Delafosse D and Magnin T (1999), 'Cathodic hydrogen embrittlement in alloy 718,' *Materials Science and Engineering*, Vol. A269, 111–119.

Gleeson B (2000), 'High-temperature corrosion of metallic alloys and coatings', in *Corrosion and Environmental Degradation, Volume II*, p. 173 (Weinheim, Germany: Wiley-VCH).

Gordon G M (2002), 'Corrosion considerations related to permanent disposal of high-level radioactive waste,' *Corrosion*, Vol. 58, No. 10, 811–825.

Haselmair H, Morach R and Boehni H (1994), 'Field and laboratory testing of high-alloy steels and nickel alloys used in fastenings in road tunnels,' *Corrosion*, Vol. 50, No. 2 160–168.

Haynes International (2000), Database (Kokomo, Indiana).

Hettiarachchi S (2005), 'Advances in electrochemical corrosion potential monitoring in boiling water reactors,' in *Proceedings of the 12th International Conference on Environmental Degradation of Materials in Nuclear Power Systems – Water Reactors*, 14–18 August 2005, pp. 3–15 (Warrendale, PA: TMS).

Hibner E L and Tassen C S (2000), 'Corrosion resistant OCTGs for a range of sour gas service conditions,' Paper 00149, *Corrosion/2000* (Houston, TX: NACE International).

Hwang S S, Kim H P, Lee D H, Kim U C and Kim J S (1999), 'The mode of stress corrosion cracking in Ni-base alloys in high temperature water containing lead,' *Journal of Nuclear Materials*, Vol. 275, 28–36.

James M M, Klarstrom D L and Saldanha B J (1996), 'Stress corrosion cracking of nickel-molybdenum alloys,' Paper 432, *Corrosion/96* (Houston, TX: NACE International).

Jones R L (1996), 'Critical corrosion issues and mitigation strategies impacting the operability of LWR's,' paper 103, *Corrosion/96* (Houston, TX: NACE International).

Kane R D (2006), 'Corrosion in petroleum production operations,' in *ASM Handbook Volume 13C*, edited by S D Cramer, B S Covino Jr. pp. 922–966 (Materials Park, OH: ASM International).

Kane R D, Watkins M, Jacobs D F and Hancock G L (1977), 'Factors influencing the embrittlement of cold worked high alloy materials in H_2S environments,' *Corrosion-NACE*, Vol. 33, No. 9, 309–320.

Kesavan S (1991), 'The kinetics of hydrogen evolution and absorption on high nickel alloys at elevated temperatures,' PhD Dissertation, The Ohio State University.

King K J, Estill J C and Rebak R B (2002), 'Characterization of the resistance of alloy 22 to stress corrosion cracking,' American Society of Mechanical Engineers, Pressure Vessels and Piping Division, PVP-449, pp. 103–109 (New York: ASME).

King K J, Wong L L, Estill J C and Rebak R B (2004), 'Slow strain rate testing of alloy 22 in simulated concentrated ground waters,' Paper 04548, *Corrosion/2004* (Houston, TX: NACE International).

Kolts J (1982), 'Temperature limits for stress corrosion cracking of selected stainless steels and nickel-base alloys in chloride-containing environments,' Paper 241, *Corrosion'82* (Houston, TX: NACE International).

Kolts J (1987), 'Environmental embrittlement of nickel-base alloys,' in *Metals Handbook Ninth Edition, Volume 13*, pp. 647–652 (Materials Park, OH: ASM International).

Krishnan K, Rooker J and Chitwood G B (2009), 'Case-history of environmental cracking failures with alloy K-500 for downhole completion tools,' Paper 09080, *Corrosion/2009* (Houston, TX: NACE International).

Kritzer P, Boukis N and Dinjus E (1998), 'Corrosion phenomena on alloy 625 in aqueous solutions containing sulfuric acid and oxygen under subcritical and supercritical conditions,' Paper 415, *Corrosion/98* (Houston, TX: NACE International).

Lai G Y (1999), *High-temperature Corrosion of Engineering Alloys* (Materials Park, OH: ASM International).

Latanision R M (1995), 'Corrosion science, corrosion engineering, and advanced technologies,' *Corrosion*, Vol. 51, No. 4, 270–283.

Lee T S F and Latanision R M (1987), 'Effects of grain boundary segregation and precipitation on the hydrogen susceptibility of nickel,' *Metallurgical and Materials Transactions A*, Vol. 18A, No. 9, 1653–1662.

Mannan S and Patel S (2008), 'A new high strength corrosion resistant alloy for oil and gas applications', Paper 08084, *Corrosion/2008* (Houston, TX: NACE International).

Mignone A, Maday M F, Borello A and Vittori M (1990), 'Effect of chemical composition on the SCC behavior of alloy 800,' *Corrosion*, Vol. 46, 57–65.

Mills W J, Lebo M R and Kearns J J (1999), 'Hydrogen embrittlement, grain boundary segregation, and stress corrosion cracking of alloy X-750 in low and high-temperature water,' *Metallurgical and Materials Transactions A*, Vol. 30A, No. 6, 1579–1596.

Mitton D B (2009), 'Supercritical water – material challenges,' Paper 09247, *Corrosion/2009* (Houston, TX: NACE International).

Mitton D B, Zhang S H, Quintana M S, Cline J A, Caputy N, Marrone P A and Latanision R M (1998), 'Corrosion mitigation in SCWO systems for hazardous waste disposal,' Paper 414, *Corrosion/98* (Houston, TX: NACE International).

Morton D S, Attanasio S A and Young G A (2002), 'Primary water SCC understanding and characterization through fundamental testing in the vicinity of the nickel/nickel oxide phase transition,' in *Proceedings of the 10th International Conference on Environmental Degradation of Materials in Nuclear Power Systems – Water Reactors*, 5–9 August 2001, Lake Tahoe, NV, p. 0122 (Houston TX: NACE International).

Nakahara M and Shoji T (1996), 'Stress corrosion cracking susceptibility of nickel-molybdenum alloys by slow strain rate and immersion testing,' *Corrosion*, Vol. 52, No. 8, 634–642.

Navas M, Gómez-Briceño D, García-Mazario M and McIlree A R (1999), 'Effect of silicon compounds on stress corrosion cracking of alloy 600 in caustic solutions,' *Corrosion*, Vol. 55, No. 7, 674–685.

Nelson J K (1987), 'Corrosion by alkalis and hypochlorite,' in *ASM International Metals Handbook*, 9th Edition, Volume 13 – Corrosion, p. 1174 (Metals Park, OH: ASM International).

Paraventi D J and Moshier W C (2005), 'The effect of cold work and dissolved hydrogen in the stress corrosion cracking of Alloy 82 and Alloy 182 weld metal,' in *Proceedings of the 12th International Conference on Environmental Degradation of Materials in Nuclear Power Systems – Water Reactors*, 14–18 August 2005, pp. 543–555 (Warrendale, PA: TMS).

Pathania R S, Tang H T and McIlree A R (2002), 'Overview of environmentally assisted cracking in LWR components,' paper 02507, *Corrosion/2002* (Houston, TX, NACE International).

Pawel S J (1994), 'Corrosion of high-alloy materials in aqueous hydrofluoric acid environments,' *Corrosion*, Vol. 50, 963–971.

Pike L M, Manning P E and Hibner E L (2010), 'A new high-strength, corrosion resistant alloy for oil and gas applications,' paper 10319, *Corrosion/2010* (Houston TX: NACE International).

Pound B G (1998), 'Effect of heat treatment on hydrogen trapping in alloy K-500,' *Corrosion*, Vol. 54, No. 12, 988–995.

Prakash J, Redey L and Vissers D R (1999), 'Corrosion studies of nickel-200 in high-temperature ZEBRA batteries at 300°C,' *Corrosion Science*, Vol. 41, 2075–2082.

Rebak R B (2000a), 'Corrosion of non-ferrous alloys. I. nickel-, cobalt-, copper, zirconium-

and titanium-based alloys' in *Corrosion and Environmental Degradation, Volume II*, p. 69 (Weinheim, Germany: Wiley-VCH).

Rebak R B (2000b), 'Environmentally assisted cracking in the chemical process industry: stress corrosion cracking of iron, nickel, and cobalt based alloys in chloride and wet HF services' in *Environmentally Assisted Cracking: Predictive Methods for Risk Assessment and Evaluation of Materials, Equipment and Structures*, ASTM STP 1401, p. 289 (West Conshohocken, PA: ASTM).

Rebak R B (2006), 'Industrial experience on the caustic cracking of stainless steels and nickel alloys – a review,' paper 06501, *Corrosion/2006* (Houston, TX: NACE International).

Rebak R B (2009a), 'Corrosion testing of nickel and titanium alloys for nuclear waste disposition,' *Corrosion*, Vol. 65, No. 4, 252–271.

Rebak R B (2009b), 'Environmental degradation of materials in light water reactors – stress corrosion cracking of nickel based alloys,' in *Proceedings of the 9th International Congress of Materials, Argentine Association of Materials (SAM)*, 19–23 October 2009, Buenos Aires.

Rebak R B (2010), 'Environmentally assisted cracking behavior of nickel alloys in oil and gas applications – a review,' *Eurocorr2010 – The European Federation of Corrosion*, Moscow, Russia, 13–17 September 2010.

Rebak R B and Crook P (2000a), 'Nickel alloys for corrosive environments,' *Advanced Materials and Processes*, Vol. 157, No. 2, 37–42.

Rebak R B and Crook P (2000b), 'Influence of alloying elements, temperature and electrolyte composition on the corrosion behavior or nickel based alloys,' Paper 00499, *Corrosion/2000* (Houston, TX: NACE International).

Rebak R B and Szklarska-Smialowska Z (1996), 'The mechanism of stress corrosion cracking of alloy 600 in high temperature water,' *Corrosion Science*, Vo. 38, No. 6, pp. 971–988.

Rebak R B, Dillman J R, Crook P and Shawber C V V (2001), 'Corrosion behavior of nickel alloys in wet hydrofluoric acid,' *Materials and Corrosion*, Vol. 52, 289–297.

Rhodes P R (2001), 'Environment-assisted cracking of corrosion-resistant alloys in oil and gas production environments: a review,' *Corrosion*, Vol. 57, 923–966.

Saito N, Tsuchiya Y, Akai Y, Omura H, Takada T and Hara N (2006), 'Corrosion performance of metals for supercritical water, oxidation-utilized organic waste processing reactors,' *Corrosion*, Vol. 62, No. 5, 383–394.

Scammon K M (1994), 'Hydrogen embrittlement of nickel based superalloy C-22,' MS Thesis, University of Central Florida.

Schroer C, Konys J, Novotny J and Hausselt J (2005), 'Material performance in chlorinated supercritical water systems,' *Corrosion/2005* Research Topical Symposium 'Corrosion Resistant Materials for Extreme Conditions,' pp. 117–142 (Houston, TX: NACE International)

Scott P M and Combrade P (2003), 'On the mechanism of stress corrosion crack initiation and growth in alloy 600 exposed to PWR primary water,' in *Proceedings of the 11th International Conference on Environmental Degradation of Materials in Nuclear Power Systems – Water Reactors*, 10–14 August 2003, Stevenson, WA, pp. 29–38 (Warrendale, PA: TMS).

Scully J R, Cieslak W R and Bovard F S (1991), 'Environmentally assisted cracking of cathodically polarized nickel in LiAlCl$_4$/SOCl$_2$ electrolyte,' *Journal of the Electrochemical Society*, Vol. 138, No. 8, 2229–2237.

Sridhar N and Cragnolino G A (1992), 'Stress-corrosion cracking of nickel-base alloys,' in *Stress Corrosion Cracking*, edited by R H Jones, p. 131 (Materials Park, OH: ASM International).

Sridhar N, Kargol J A and Fiore N F (1980a), 'Effect of low temperature aging on the hydrogen-induced crack growth in a Ni-base superalloy,' *Scripta Metallurgica*, Vol. 14, No. 2, 225–228.

Sridhar N, Kargol J A and Fiore N F (1980b), 'Hydrogen-induced crack growth in a Ni-base superalloy,' *Scripta Metallurgica*, Vol. 14, No. 11, 1257–1260.

Staehle R W (1996), 'Occurrence of modes and submodes of SCC,' in *Control of Corrosion on the Secondary Side of Steam Generators*, edited by R W Staehle, J A Gorman and A R McIlree, pp. 135–208 (Houston, TX: NACE).

Staehle R W (2005), 'Clues and issues in the SCC of high nickel alloys associated with dissolved lead,' in *Proceedings of the 12th International Conference on Environmental Degradation of Materials in Nuclear Power System – Water Reactors*, edited by T R Allen, P J King and L Nelson, pp. 1163–1210 (Warrendale, PA: The Minerals, Metals & Materials Society).

Staehle R W and Gorman J A (2003), 'Quantitative assessment of submodes of stress corrosion cracking on the secondary side of steam generator tubing in pressurized water reactors: part 1,' *Corrosion*, Vol. 59, 931–994.

Staehle R W and Gorman J A (2004), 'Quantitative assessment of submodes of stress corrosion cracking on the secondary side of steam generator tubing in pressurized water reactors: part 2,' *Corrosion*, Vol. 60, 5–63.

Symons D M (1997), 'Hydrogen embrittlement of Ni-Cr-Fe alloys,' *Metallurgical and Materials Transactions*,' Vol. 28A, No. 3, 655–663.

Szklarska-Smialowska Z and Rebak R B (1996), 'Stress corrosion cracking of alloy 600 in high-temperature aqueous solutions: influencing factors, mechanisms, and models,' in *Control of Corrosion on the Secondary Side of Steam Generators*, edited by R W Staehle, J A Gorman and A R McIlree, pp. 223–257 (Houston, TX: NACE).

Takizawa Y and Sekine I (1985), 'Stress corrosion cracking phenomena on Ni-Mo alloys in high temperature non aqueous solutions,' Paper 355, *Corrosion/85* (Houston, TX: NACE International).

The International Nickel Company (1968), *Corrosion Engineering Bulletin CEB-5* (New York: Inco).

Theus G J, Emanuelson R H and Russell J (1982), 'Stress corrosion cracking tests of Monel 400 steam generator tubing,' Paper 209, *Corrosion/82* (Houston, TX: NACE International).

Wasynczuk J A, Quinzio M V and Bittner H F (1995), 'Environmentally assisted cracking of nickel strip in $LiAlCl_4/SOCl_2$,' *Journal of the Electrochemical Society*, Vol. 142, No. 9, 2977–2985.

Yamazaki S, Lu Z, Ito Y, Takeda Y and Shoji T (2008), 'The effect of prior deformation on stress corrosion cracking growth rates of alloy 600 materials in a simulated pressurized water reactor primary water,' *Corrosion Science*, Vol. 50, 835–846.

Yang W, Lu Z, Juang D, Kong D, Zhao G and Congleton J (2001), 'Caustic stress corrosion cracking of nickel-rich, chromium-bearing alloys,' *Corrosion Science*, Vol. 43, No. 5, 963–977.

Young G A, Wilkening W W, Morton D S, Richey E and Lewis, N (2005), 'The mechanism and modeling of intergranular stresss corrosion cracking of nickel-chromium-iron alloys exposed to high purity water,' in *Proceedings of the 12th International Conference on Environmental Degradation of Materials in Nuclear Power Systems – Water Reactors*, edited by T R Allen, P J King and L Nelson, pp. 913–923 (Warrendale, PA: The Minerals, Metals & Materials Society).

8

Stress corrosion cracking (SCC) of aluminium alloys

M. BOBBY KANNAN, James Cook University, Australia,
P. BALA SRINIVASAN, Helmholtz-Zentrum Geesthacht,
Germany and V. S. RAJA, Indian Institute of Technology
Mumbai, India

Abstract: The stress corrosion cracking (SCC) behaviour of aluminium alloys has been studied for the past five decades and is still a research area of high interest due to the demand for higher strength aluminium alloys for fuel saving. This chapter brings out the general understanding of the SCC mechanism(s) and the critical metallurgical issues affecting the SCC behaviour of aluminium alloys. The developments made so far with regard to alloying and heat treatment of aluminium alloys for high SCC resistance are discussed. An overview of the available literature on the SCC of aluminium alloy weldments and aluminium alloy metal matrix composites is also presented.

Key words: aluminium alloys, stress corrosion cracking mechanisms, precipitates, weldment, composites.

8.1   Introduction

Aluminium alloys are extensively used in aircraft structures due to their high strength-to-weight ratio, excellent formability and good machinability. Although aluminium and its alloys possess good resistance to general corrosion due to the rapidly formed Al_2O_3 film, they are prone to pitting corrosion in aggressive environments (especially chloride-containing environment). More importantly, high strength Al-alloys are susceptible to another form of localized corrosion known as stress corrosion cracking (SCC). Out of the eight series of Al-alloys, SCC is most common in 2xxx, 7xxx and 5xxx (with high Mg) series Al-alloys. Speidel's detailed analysis of a large number of aircraft-component failures from 1960 to 1970 revealed that the components made of 7079-T6, 7075-T6 and 2024-T3 Al-alloys failed by SCC, and these alloys contributed to more than 90% of the service failures of all high-strength Al-alloys [1].

Numerous workers have studied the SCC behaviour of high strength Al-alloys [1–9]. From the metallurgy point of view, the SCC behaviour of Al-alloys is generally correlated to the chemistry and morphology of

307

grain boundary precipitates (GBPs) [10–12], the extent and distribution of dislocations [13–15] and segregation of alloying elements along the grain boundaries [16–18]. There are several review articles on SCC of Al-alloys [19–23].

SCC failures in peak-aged Al-alloys prompted research into heat-treatment procedures to successfully alter the microstructure of the alloy for attaining high SCC resistance. Overaging heat treatment was developed, which enhanced the SCC resistance of high strength Al-alloys. However, an increase in SCC resistance by such overaging treatments is usually associated with concomitant reduction in strength of these alloys. Later, Cina [24] claimed that high SCC resistance can be achieved without loss of strength through a novel heat treatment known as retrogression and re-aging (RRA). However, the disadvantage of such a treatment is its short retrogression time, which limits it application to only thin sheets. Hence, the original treatment has not become popular in industry. Recently, the use of multistep aging, together with controlled heating rate, to reach the final overaging temperature, has become popular for obtaining a combination of improved strength and SCC resistance [25–27].

This chapter broadly highlights the SCC mechanism(s) operating in high strength Al-alloys, the factors affecting the SCC behaviour of high strength Al-alloys, the SCC issues on Al-alloy weldments and Al-based metal matrix composites.

8.2   Stress corrosion cracking (SCC) mechanisms

Lynch has summarized the SCC mechanisms of metallic materials in Chapter 1 of this book. The main SCC mechanisms are: slip dissolution, film induced cleavage, hydride formation, hydrogen enhanced localized plasticity (HELP), adsorption induced dislocation emission (AIDE), hydrogen enhanced de-cohesion (HEDE) and corrosion enhanced localized plasticity (CELP) [3, 9, 22, 28]. However, the widely reported SCC mechanisms in high strength Al-alloys are anodic dissolution assisted cracking [1, 8, 20, 29] and hydrogen induced cracking [14, 15, 30, 31]. Burleigh [20] stated that 2xxx series alloys are more likely to undergo anodic dissolution assisted SCC and 7xxx series alloys undergo hydrogen induced cracking.

8.2.1   Anodic dissolution assisted cracking

Anodic dissolution assisted cracking is characteristically an intergranular mode of failure [1, 8, 20, 29]. According to electrochemical theory anodic dissolution assisted cracking requires a condition that makes the grain boundaries or adjacent regions anodic to the rest of the microstructure, so that the dissolution proceeds selectively along the boundaries. Choi *et al.*

[32] reported that SCC in 7039 Al-alloy was initiated by the dissolution of the GBPs and propagated by mechanical processes such as creep. A detailed study on the SCC behaviour of Al-Zn-Mg-Cu-Zr alloy carried out by Kannan and Raja [8, 33] showed that the coarse intermetallic particles can act as potential sites for the initiation of SCC (Fig. 8.1a). They have shown that the passive film breakdown in the vicinity of copper rich intermetallic caused the crack initiation in the alloy (Fig. 8.1b). Fracture surface analysis showed that the continuous nature of anodic grain boundary precipitates caused easy dissolution and crack propagation. They observed such a microstructure in the peak-aged Al-Zn-Mg-Cu-Zr alloy, and hence suggested that the alloy underwent intergranular stress corrosion cracking (IGSCC). Similarly, based

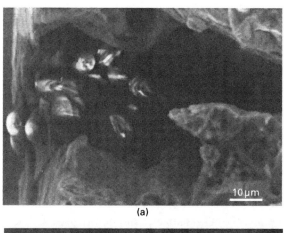

(a)

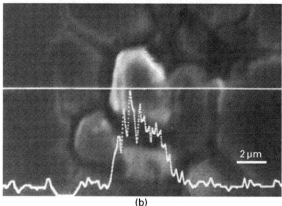

(b)

8.1 (a) SEM fractograph of 7010 Al-alloy shows cluster of coarse intermetallics associated with crack initiation; (b) SEM micrograph of 7010 Al-alloy exposed to chloride-containing solution shows passive film break down in the vicinity of coarse particle (WDX Cu scan shows the particle is rich in Cu) [8].

on the TEM micrographs of fracture zone of welded Al-Zn-Mg, Uesaki *et al.* [34] suggested that the main cause of fracture was due to these precipitates continuously lying along grain boundaries and their preferential dissolution in the test environment.

Sprowls [35] reported the existence of a difference between the open circuit potential for grain boundaries and grain interior in an Al-Cu system, suggesting that a local galvanic couple constituted by the Cu-lean boundaries and the relatively cathodic grain interiors would promote an intergranular corrosion. Later, Galvele and De Micheli [36] demonstrated a correlation between intergranular corrosion of Al-Cu and the difference in pitting potentials between Cu-lean boundaries and the grain interiors. Based on these concepts, Sugimoto *et al.* [37] explained the IGSCC for Al-4Cu and Al-Cu-Mg alloys. Their work showed that IGSCC occurred when the stressed samples were polarized anodic to the pitting potential of the grain boundaries. However, the samples polarized cathodic to the pitting potential of the grain boundaries did not fail.

Similar to 7xxx and 2xxx series alloys, Al-Li-Cu alloys tend to undergo intergranular SCC in chloride environments. The grain boundaries of Al-Li-Cu alloys consisting of strengthening phases $\theta'(Al_2Cu)$, $S''(Al_2CuMg)$ and $T1(Al_2LiCu)$ and solute depleted regions surrounding boundaries have been cited as the critical microstructural features responsible for SCC in Al-Li-Cu alloys. Wall and Stoner [38] correlated the SCC trend in Al-Li-Cu alloys to the electrochemical behaviour of the alloy using potentiodynamic polarization and scratching electrode techniques. Scratching electrode constant extension tests of stressed tensile samples have shown a change from no failure to a rapid failure due to a 10 mV change in applied potential. The potentials where these transitions occurred were linked to the electrochemistry of a copper depleted region along the grain boundaries.

8.2.2   Hydrogen induced cracking

Although in the past it has been suggested that anodic dissolution assisted SCC is the main mechanism for the failure of high strength Al alloys, later it was accepted that hydrogen induced cracking (HIC) also plays a role [14, 15, 30, 31].

It is generally know that pitting and pre-existing defects initiate SCC; however, it has also been reported that for the initiation of SCC a critical concentration of hydrogen should build up at the potential crack sites [31]. The lattice defects (vacancies, dislocations, grain boundaries) and precipitates provide a variety of trapping sites for hydrogen diffusion. Hydrogen traps have mechanistically been classified as reversible and irreversible traps, depending on the steepness of the energy barrier needed to be overcome by hydrogen to escape from the trap. Interestingly, a uniform distribution of irreversible traps is believed to provide a beneficial effect in alloy behaviour under embrittling conditions, by arresting hydrogen diffusion and thereby delaying its build-up at the crack sites [31]. Talianker and Cina [15] argued that diffusion of hydrogen to and along the grain boundary would be expected to be accelerated by the presence of dislocations adjacent to the grain boundaries. Gruhl [39] found that inhomogeneous slips induce hydrogen embrittlement and homogeneous slips inhibit hydrogen embrittlement by effectively reducing hydrogen transport to the grain boundaries. A recent study by Kannan *et al.* [40] on determining the true SCC susceptibility index by using glycerine as a non-corrosive environment showed that Al-Zn-Mg-Cu-Zr alloy underwent hydrogen embrittlement even in laboratory air (having a relative humidity of ~50%). Earlier, Speidel [1] has shown that the increase in relative humidity enhanced the SCC crack velocity significantly in 7075 Al-alloy (Fig. 8.2a). Bandyopadhyay *et al.* [41] also demonstrated that hydrogen charging deteriorates the mechanical properties of 2091 Al-alloy (Fig. 8.2b). Interestingly, they observed an increase in percentage reduction-in-area with increase in charging time. They suggested that in the case of hydrogen charged specimen, deformation occurred at a more localized region resulting in necking, which effectively increased the percentage reduction-in-area.

8.2.3   Synergetic effect

A few authors have claimed a synergistic effect of SCC and HIC in the failure of 7xxx and 5xxx series alloys. Magnin and Dubessy [42] suggested

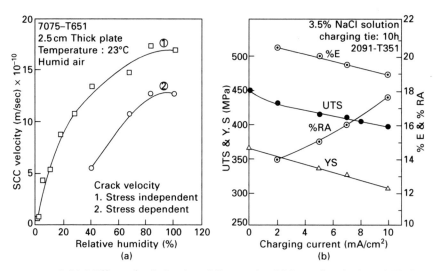

8.2 (a) Effect of relative humidity on the SCC crack velocity of 7075 Al-alloy [1]; (b) Effect of hydrogen charging on the mechanical properties of 2091 Al-Li alloy [41].

that anodic dissolution of grain boundary can lead to the discharge of hydrogen adjacent to the grain boundaries which in turn induce hydrogen embrittlement in Al-Zn-Mg alloys. Najjar *et al.* [5] also reported that both anodic dissolution and HIC operated simultaneously in 7050 alloy tested in 3.5 wt% NaCl solution. A schematic representation of a critical defect by localized dissolution around a metallurgical heterogeneity represented by Najjar *et al.* [5] is reproduced in Fig. 8.3. Interestingly, Gruhl [39] reported that larger precipitates in the grain boundary can act as trapping sites for atomic hydrogen to nucleate hydrogen bubbles. Hence, they improve the HIC resistance by lowering the hydrogen concentration in the crack tip below a critical hydrogen level required for HIC. Christodoulou and Flower [31] found that a critical size of approximately 20 nm for GBPs was required for nucleation of hydrogen bubble.

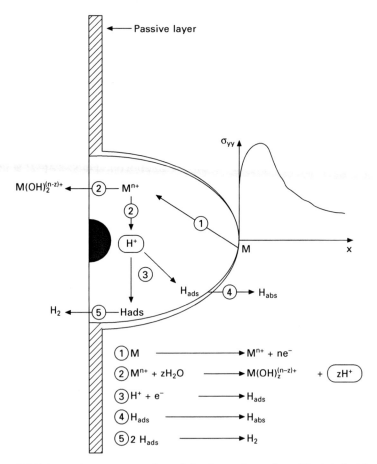

8.3 Schematic representation of the formation of a critical defect by localized anodic dissolution around a metallurgical heterogeneity [5].

8.2.4 Typical fracture modes

With regard to the SCC fracture modes, it is widely reported that high strength Al-alloys undergo intergranular fracture [43–49]. However, a few research teams have also observed transgranular cleavage-like characteristics [13, 49] and a mix-mode of fracture [4, 48]. Intergranular fracture occurs when the grain boundaries are weakened by selective dissolution of the anodic grain boundary precipitates and/or the hydrogen diffusion towards these grain boundaries. Forsyth [46] reported that intergranular attack was confined to the grain and sub-grain boundaries in Al-Zn-Mg-Cu-Zr under-aged alloy. The studies by Tsai and Chuang [45] on the SCC behaviour of 7475 peak-aged alloy in short-transverse and longitudinal directions, under cathodic charging conditions, showed a fracture surface with an equi-axed intergranular attack in the 7475 short-transverse specimen (Fig. 8.4a). On the other hand, the longitudinal specimen of this alloy exhibited a pancake shaped intergranular attack (Fig. 8.4b). Albrecht *et al.* [14] reported that in the pre-charged condition the fractured surface of 7075 alloy revealed brittle intergranular fracture only in the edge of the sample, whereas in the straining electrode test a dramatic increase in the extent of intergranular fracture was reported. Albrecht *et al.* [14] reported a predominantly transgranular cracking of 7075 alloy in the under-aged, peak-aged and over-aged conditions assessed under both cathodically charged and uncharged conditions. Tsai and Chuang [45] reported transgranular fracture of uncharged 7475 (short transverse) peak-aged alloy. Mix-mode (intergranular and transgranular) fracture has been reported by Gest and Troiano [48] in 7075 alloy in the peak-aged condition. Hardwick *et al.* [4] also reported a mix-mode of failure of under-aged 7050 alloy containing low Cu (0.01 wt%). They have suggested that the recrystallized grains undergo intergranular attack and the non-recrystallized grains undergo transgranular attack. Kannan and Raja [8, 49] have also observed intergranular attack of the recrystallized grains in the Al-Zn-Mg-Cu-Zr alloy, assessed by slow strain rate tensile and U-bend tests (Fig. 8.5). It is apparent that anodic-dissolution leads to intergranular cracking, whereas hydrogen diffusion promotes both intergranular and transgranular cracking.

8.3 Factors affecting stress corrosion cracking (SCC)

8.3.1 Composition

Aluminium alloys contain appreciable amounts of alloying elements such as Zn, Mg, Li or Cu. Precipitation and segregation of alloying elements along the grain boundaries play a critical role in the SCC behaviour. The alloying elements alter the electrochemical properties of a region or a phase in the alloys.

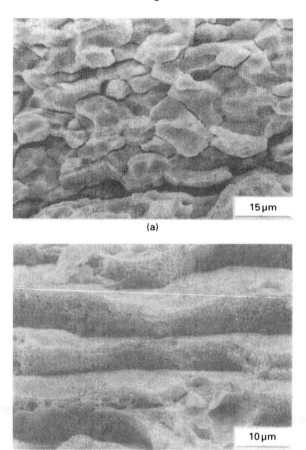

8.4 SEM fractographs of 7475 peak-aged alloy cathodically charged and tested in different orientation: (a) short-transverse direction shows intergranular cracking of equi-axed grains, (b) longitudinal direction shows intergranular cracking of pancake shaped grains [45].

Mg and Zn combine to form $MgZn_2$ precipitate known as η phase. Although fine $MgZn_2$ precipitates in the grains improve the strength of the alloy, they are anodic to the matrix and hence in certain heat-treated conditions where the precipitates are continuous along the grain boundary, the dissolution of these precipitates causes easy crack propagation under tensile stress, giving rise to SCC. In addition to precipitation, Mg also segregates along the grain boundaries, which again is detrimental in the context of hydrogen embrittlement [16, 18, 50]. The affinity of hydrogen atoms towards magnesium is larger than that towards aluminium and hence hydrogen preferentially interacts with magnesium to form magnesium hydride. Thus it is suggested that the segregation of magnesium along the grain boundaries enhances the amount

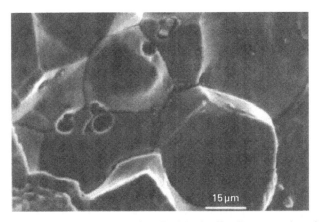

8.5 SEM fractographs of Al-Zn-Mg-Cu-Zr alloy tested under U-bend testing method, showing typical intergranular cracking of recrystallized grains [8].

of absorbed hydrogen, leading to hydrogen embrittlement. However, zinc and copper have a beneficial effect on the SCC resistance of Al-alloys [51, 52]. Ohasaki *et al.* [51] found that the addition of zinc (0.4–0.7%) enhances the SCC resistance of 2091 alloy (Fig. 8.6a). Basically, Cu alloying was carried out to improve the strength of Al-alloys. However, it was found that addition of Cu, higher than 1 wt%, to Al-Zn-Mg alloy imparts a high SCC resistance, primarily due to Cu substitution in the anodic η particles. This reduces the dissolution of Cu containing η particles, which in turn decreases the crack growth velocity. Sarkar *et al.* [52] reported that Cu addition of more than 1.6% also increased the number of partially coherent and incoherent precipitates and thus promoted homogeneous deformation. The homogeneous deformation in Cu-rich alloy is reported to be due to the looping mechanism of the strengthening precipitates in contrast to the shearing mechanism resulting in inhomogeneous deformation in the Cu-lean alloys. The pile-up of the effective number of dislocations decreased in Cu-rich alloy and hence the critical local stress necessary for a certain crack growth velocity increases with increase in Cu content (Fig. 8.6b).

Willey [53] reported that the addition of Sc significantly increases the strength of Al-alloys through precipitation hardening. In addition to the improvement in strength, Sc also inhibits recrystallization in Al-alloys [54]. It was further noticed that Sc in combination with Zr enhances the recrystallization inhibition effect significantly [55]. With regard to the SCC resistance, Braun [56] and Wang *et al.* [57] reported no beneficial effect of Sc addition to Al-Mg alloys. In contrast, Kannan and Raja [58] and Yi-Lei *et al.* [59] reported the beneficial effect of Sc addition towards the SCC resistance in Al-Zn-Mg-Cu alloy. Kannan and Raja [8, 58] showed that the 7010 alloy containing 0.25 wt% Sc exhibited a high SCC resistance even

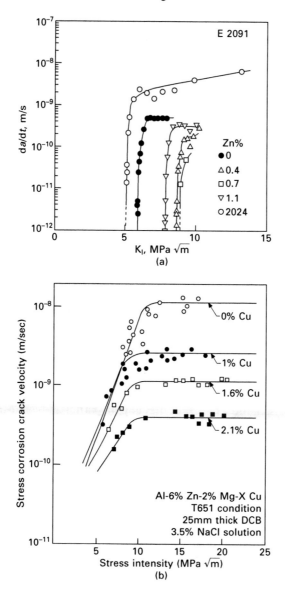

8.6 (a) Effect of zinc on SCC crack growth rate vs stress intensity of 2091 (Al-Li) alloy [51]; (b) effect of copper content (0.0–2.1 wt%) on the SCC behaviour of a peak-aged Al-6%Zn-2%Mg alloy [52].

in the peak-aged condition. They attributed the improvement in the SCC resistance to the sharp reduction in the recrystallized grains and change in the morphology of the grain boundary precipitates (i.e. continuous to broken network).

8.3.2   Heat treatment

Although heat treatment procedures are generally used to accelerate the precipitation kinetics for enhancing the strength of Al-alloys, they can also play a major role in altering the SCC behaviour by controlling the transformation and the micro-constituents. The role of various heat treatments towards the precipitation mechanism/morphology and their effects on the SCC behaviour are discussed in this section.

The high strength in aluminium alloys is achieved by the finely dispersed precipitates that form during aging heat treatments. The supersaturation of vacancies allows diffusion, and thus formation of the Guiner–Preston Zone (GPZ). It is generally agreed that GPZs appear initially at temperatures just below 100°C. In the precipitation process, the saturated solid solution first develops solid clusters which are involved in the formation of transition (non-equilibrium) precipitates. The final structure consists of equilibrium precipitates, whose contribution to precipitation strengthening becomes minimal. During artificial aging, the transitional (metastable) phase formed remains coherent with the solid solution matrix and thus contributes to the precipitation strengthening. With further heating, the precipitate grows, but even more importantly they become equilibrium phases, which are generally incoherent. Formation of equilibrium phases soften the materials and if continued further, produce the softest phase on annealed condition. The sequence of steps in the decomposition of supersaturated quaternary Al-Zn-Mg-Cu alloy solid solution may be summarized as:

$$\alpha \rightarrow \alpha + GPZ \rightarrow \alpha + \eta' \rightarrow \alpha + \eta \qquad 8.1$$

where α is the solid solution, η' is a semi-coherent intermediate $MgZn_2$ phase [60] and η is incoherent equilibrium $MgZn_2$ phase [61]. In 7xxx series alloys, the composition of Cu and Mg is chosen typically in such a way that during solution heat treatment at 460°C, the alloy lies in the single phase (i.e. α-Al) of the Al-Cu-Mg-Zn phase diagram. This would not allow the constituent S (Al_2CuMg) phase particles of the Al-Cu-Mg system to survive the typical homogenization and solution treatment temperatures, thus improving the fracture toughness of the alloy [62].

A typical aging effect on the SCC behaviour of high strength Al-alloys is shown in Fig. 8.7 [63]. In general, under-aged heat treatment leads to a very susceptible microstructure to hydrogen embrittlement. The coherent GPZs formed during under-aging are cut by passing dislocations during deformation, which leads to a local softening of the slip planes and concentration of slip bands [14]. It is believed that the extent of hydrogen dislocation transport and local hydrogen accumulation at the grain boundaries control the degree of embrittlement [4]. In the case of 7xxx alloys, peak-aging is reported to result in the precipitation of η phase along the grain boundaries in a continuous

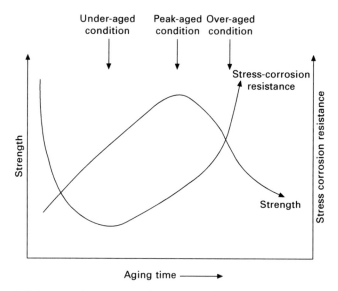

8.7 A general representation on the effect of aging time on the strength and SCC resistance of Al-alloys [63].

manner [8, 29]. These precipitates are reported to be anodic to the matrix and hence the selective dissolution in these regions favours intergranular SCC. Albrecht *et al.* [14] reported that in the peak-aged temper the matrix contains a mixture of GPZ and some coherent η′ precipitates. Under this condition, the slip distribution is suggested to be inhomogeneous at low plastic strain and hence leads to SCC.

Over-aged alloys are known to exhibit a better SCC resistance but with 10–15% compromise in their strength level. In a work on an Al-Li binary alloy, Christodoulou *et al.* [31] showed that the over-aged alloy was not susceptible to SCC, whereas the alloy in the peak-aged condition was highly susceptible. Albrecht *et al.* [14] reported that during over-aging semi-coherent precipitates are formed in the matrix, which results in homogeneous slip distribution and thus improving the SCC resistance of 7075 alloy. Peel and Poole [64] reported that the coarse grain boundary precipitates formed due to over-aging were responsible for the higher SCC resistance. However, recent studies by Kannan *et al.* [8, 29] reported that not only the change in the arrangement/morphology of the grain boundary precipitates (between the peak-aged and over-aged as shown in Fig. 8.8) was the reason for the improved SCC resistance, but the chemistry of the grain boundary precipitates also play an important role. The TEM-EDX studies showed enrichment of Cu in the grain boundary precipitates of over-aged alloy, which eventually decreases the dissolution due to its noble potential as compared to $MgZn_2$ precipitates and aluminium matrix [8].

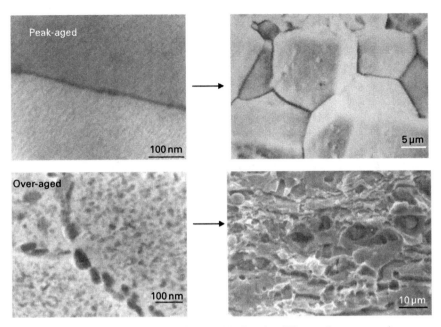

8.8 TEM micrographs of 7010 Al-alloy in different heat-treated conditions and the corresponding fracture surfaces show brittle failure for the alloy containing continuous network of GBPs and ductile failure for the alloy containing coarse broken network of GBPs [8].

Cina [24] claims that a higher SCC resistance without any loss in the UTS can be achieved through retrogression and re-aging (RRA) treatment. A typical RRA treatment involves subjecting the peak-aged material to a short retrogression treatment (from 10 s to 3 h) at elevated temperatures (180–260°C), quenching and subsequently repeating the peak-aging treatment i.e. 120°C for 24 h. The increase in the resistance to SCC is believed to be associated with the coarsening of η precipitates at the grain boundaries and the high strength resulting from the fine η′ precipitates in the matrix. However, Danh *et al.* [65] and Park [66] reported that the SCC resistance observed in RRA-treated alloys is due to the annihilation of dislocations during the retrogression step (i.e. heating at higher temperature). Komisarov *et al.* [67] also showed that the RRA treatment could improve the SCC resistance of 8090 alloy, which was attributed to the dissipation of dislocations with increase in retrogression time and temperature (Fig. 8.9).

In recent years, the use of multi-step aging together with controlled heating rate to reach the final over-aging temperature has become popular in order to obtain a combination of improved strength and fracture toughness [25–27]. The multi-step aging treatment is advantageous over RRA treatment, since the former can be applied to even thick plates, while the latter is restricted

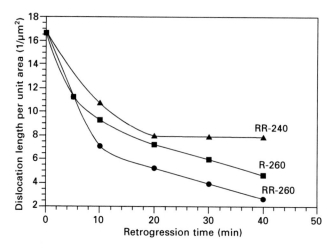

8.9 Effect of time and temperature of retrogression on density of dislocations of 8090 Al-alloy [67].

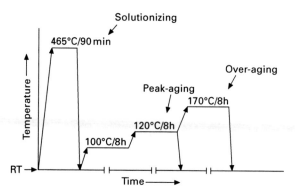

8.10 Schematic representation of multi-step aging treatment for 7010 Al-alloy [8].

to thin sheets because of the short retrogression treatment times at elevated temperature. Kannan *et al.* [8, 29] reported that through a multi-step over-aging treatment (Fig. 8.10), a higher ductility and superior SCC resistance can be accomplished in an Al-Zn-Mg-Cu-Zr alloy (Fig. 8.11).

8.3.3   Precipitate free zones

It is well known that precipitate free zones (PFZ) form during aging of high strength Al-alloys. Various factors such as aging temperature grain boundary mis-orientations and quench rate from solution treatment have an influence on the PFZ width. Although there exists a significant microstructural difference between the PFZ and the adjacent matrix, it is reported that the PFZ width of

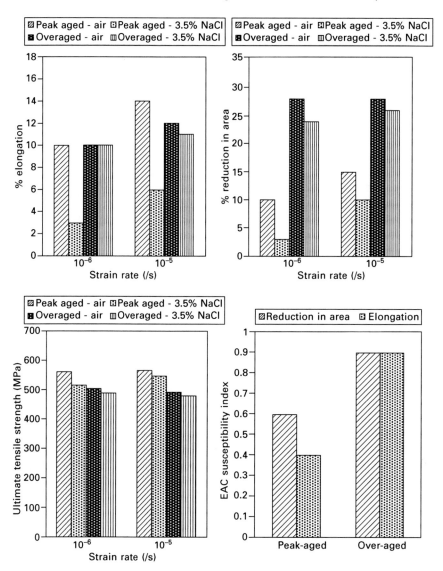

8.11 Slow strain rate testing result of peak-aged and over-aged 7010 Al-alloy showing high SCC resistance for over-aged alloy (i.e. the loss in ductility and mechanical strength for the over-aged alloy is low) [8].

ternary Al-Zn-Mg alloys has no influence on the SCC behaviour in aqueous sodium chloride solution [68]. However, the relative strengths of the PFZ and grain interior determine the extent of strain localization at grain boundaries and affect the fracture toughness [69].

8.3.4   Re-crystallization

Re-crystallization is a process which involves the nucleation and growth of relatively defect-free grains within deformed grains. The growth of re-crystallized grains occurs through the movement of high-angle grain boundaries. Re-crystallization generally occurs in hot-rolled Al-alloys. It is believed that fine particles (dispersoids) in Al-alloys inhibit re-cystallization

Kannan and Raja [8, 49] studied the influence of re-crystallized grain on the SCC behaviour of 7010 Al-alloy. They found that re-crystallized grains are more prone to SCC in chloride-containing solutions. They also reported that by inhibiting the re-crystallization, through scandium addition, the SCC resistance of 7010 alloy can be enhanced significantly, even in the peak-aged condition. Figure 8.12 shows the different modes of fracture observed in the partially re-crystallized and un-re-crystallized grain-structured alloys.

Ou *et al.* [70] claimed that through step-homogenization and step-quenching and aging treatment, both optimum strength and SCC resistance can be achieved for 7050 alloy. The sequence of the heat treatment is shown in Fig. 8.13. They reported that the fine and dense dispersoid distribution, generated by the step-homogenization treatment, effectively inhibited re-crystallization (Fig. 8.14) and also lowered the quench sensitivity of 7050 alloy. They

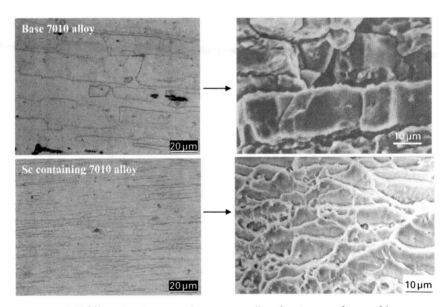

8.12 Microstructures and corresponding fracture surfaces of base and Sc containing 7010 Al-alloy in peak-aged condition showing intergranular fracture for base alloy (containing large pancake shaped grains) and mix-mode failure for Sc containing alloy (containing fine grains) [8].

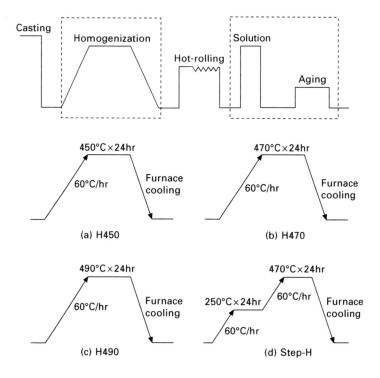

8.13 Schematic representation of different homogenization (H) treatments [70].

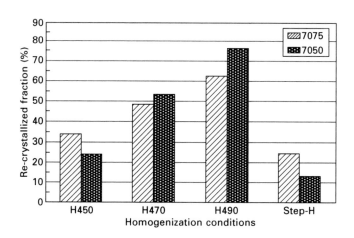

8.14 Re-crystallized fraction of 7075 and 7050 alloys as a function of homogenization conditions show Step-H alloy contains low fraction of recrystallized grains [70].

reported that the combination of step-homogenization and step-quenching provided not only an optimum strength in 7050 alloy but also a favourable morphology of grain boundary precipitates for high SCC resistance. They also found that such a treatment was not applicable for 7075 alloy because of its inherent high quench sensitivity.

8.3.5   Grain size and orientation

Tsai and Chuang [45] have studied the effect of grain size on the SCC behaviour of 7475 Al-alloy. They found that the grain refinement resulted in a more homogeneous slip mode and a smaller size of grain boundary precipitates (GBPs), which influenced the SCC resistance. It was suggested that a more homogeneous slip mode is beneficial for improving the SCC resistance. However, if the size of the GBPs was smaller than the critical precipitate size for nucleating hydrogen bubbles, then there is no beneficial effect due to either grain refinement or more homogeneous slip mode.

The grain orientation also influences the SCC behaviour of high strength Al-alloys. Intergranular SCC failures are very common in high strength Al-alloys containing pancake shape grains, typically occurring in rolled plates and extrusions (Fig. 8.15) [1]. The failures occur predominantly when the applied stresses are in the short-transverse direction. However, when the applied stresses are in the longitudinal or long-transverse directions, the intergranular crack path is not easy, hence leading to a high resistance to SCC. Although transgranular cracking can occur when there is no easy intergranular pathway, it is not that widely reported.

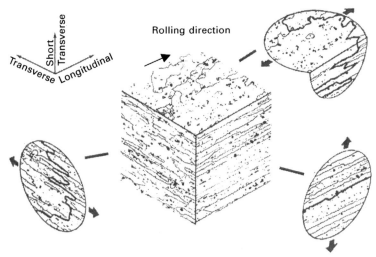

8.15 Effect of grain structure on the intergranular crack for sample stressed in different orientations [1].

8.3.6   Exfoliation corrosion

Exfoliation corrosion is a type of intergranular attack that appears in rolled and extruded products of Al-alloys, causing great losses of the surface material. Exfoliation corrosion occurs predominantly in Al-alloys that have marked directional structures. Exfoliation is experienced in automotive, aircraft and offshore structures [71]. The principal danger of exfoliation corrosion lies in the potential loss of an effective cross section. Figure 8.16 shows a schematic sequence illustrating the evolution of exfoliation corrosion [72]. McNaughtan *et al.* [73] studied the effect of forces due to exfoliation corrosion products on the SCC behaviour in high strength Al-alloys. They found that exfoliation and SCC are both strongly dependent on the rate of intergranular corrosion, which was controlled by the composition and distribution of grain boundary precipitates. They observed an inverse, linear relationship between corrosion product force generation during exfoliation and the K_{ISCC} values for SCC (Fig. 8.17) and this provided further evidence for the link between these two corrosion mechanisms.

8.4   Stress corrosion cracking (SCC) of weldments

Many engineering applications demand employment of joining processes for making components of desired size, shape and geometry. Even though

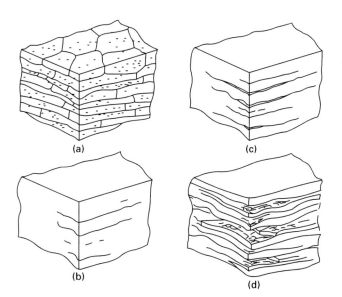

8.16 Schematic sequence illustrating the evolution of exfoliation in 2024 Al-alloy: (a) elongated grain structure and precipitation; (b) corrosion onset within grain boundaries; (c) delamination between grain layers; (d) complete loss of alloy thickness integrity [72].

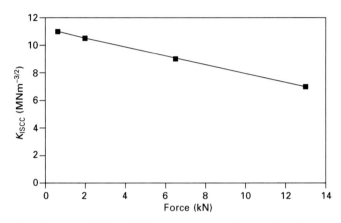

8.17 Inverse relationship between corrosion product force and K_{Iscc} in a high strength Al-Zn-Mg-Cu alloy [73].

aluminium alloys melt at lower temperatures than steels, they require higher heat inputs for welding due to their higher thermal conductivity. Gas tungsten arc (GTA), gas metal arc (GMA) and plasma arc (PA) welding processes are widely used for the joining of aluminium alloys. 5xxx and 6xxx series aluminium alloys are generally regarded as easily weldable by fusion welding processes when compared to the high strength aluminium alloys, i.e., 2xxx and 7xxx. With the advent of the solid state joining process, namely friction-stir welding (FSW) in 1991, the joining of high strength aluminium alloys has become relatively easier and this process has gained a wider acceptance, especially in the aircraft sector for the joining of heat-treatable aluminium alloys. The process is being employed in shipbuilding for the manufacture of panels of extruded sections with controlled distortion. Power beam processes such as electron beam (EB) and laser beam (LB) welding for the joining of age-hardenable aluminium alloys are also being contemplated in recent years.

In Al-alloy welds obtained by fusion welding processes (GTA/GMA/EB or LB), the weld metal consists of a cast microstructure. The composition of the weld metal is decided based on whether the joint is obtained autogenously or with use of an appropriate filler material. In addition to controlling the chemistry and microstructural features in the weld metal through dilution, the weld heat input also dictates the microstructural/phase transformations in the regions adjoining the weld metal. In the case of FSW, the weld nugget is obtained by an extensive plastic deformation and flow of material assisted by the frictional heat. Peak temperatures close to 480°C are common in FSW of aluminium alloys, and in some cases incipient melting in the weld nugget has also been reported. The regions adjoining the weld nugget also get affected by the thermomechanical effects. Thus, the changes that occur during welding of aluminium alloys are expected to alter the corrosion behaviour of the weldment significantly. Whilst the general and pitting corrosion behaviour of the different regions (weld metal, HAZ, parent alloy) of Al-alloy weldments have been studied extensively, the literature on the SCC behaviour of Al-alloy weldments is scarce.

Stress corrosion cracking of components made of aluminium alloy welds may be influenced by the prior/post-weld heat treatment, phase transformations/ micro-constituents in the weld metal and/or heat affected zone and the residual stresses that develop during welding. A couple of stress corrosion failures in GTA welded aerospace components made of an Al-Zn-Mg alloy were reported by Jha *et al.* [74, 75]. In the first case, cracking was observed along the fusion line of a circumferential weld, joining a nozzle press fitted to a cylindrical pressure vessel [74]. In the second, cracks were found in an adaptor made of a forged AFNOR 7020-T6 alloy used in the liquid propulsion system [75]. In both cases cracks were noticed in the components exposed to saline environment with a high humidity. The required tensile stresses were believed to have developed during different stages of manufacturing: (a) quenching from solutionizing temperature, (b) shrink fitting of the adaptor into the water tank and (c) GTA welding under conditions of high restraint. These residual stresses were augmented by the metallurgical factor, i.e. the formation of more electrochemically active $MgZn_2$ precipitates along the grain boundaries and that the synergistic effect of dissolution and stress had resulted in the dissolution-assisted intergranular fracture in the weld/ HAZ regions of the welded component.

Friction stir weldments of 6061-T651 Al-alloy have been reported to exhibit better SCC resistance than the parent metal [76]. A close examination of the data shows that this apparent resistance might be due to strain localization at the soft weld interface and the consequent higher strain rate suffered by this region than the remaining area. Hence a more detailed investigation is needed to substantiate this point. Similarly, 2219 Al-alloy in T87 temper and its friction stir weldment in the as-welded condition were reported to be not susceptible to SCC in 3.5% NaCl solution based on the SSRT tests performed at nominal strain rates of $10^{-6}\,s^{-1}$ and $10^{-7}\,s^{-1}$ [77]. Coarsening of intragranular strengthening precipitates in the thermomechanically affected and heat affected zones in the friction stir weldments of 2219 Al-alloy in T87 temper has been reported to reduce the hardness/mechanical properties of these regions. However, the SCC behaviour of a friction stir weldment of this alloy was found to be unaffected by the microstructural variations as was noticed from the results of the constant extension rate tensile tests performed at an extension rate of $2 \times 10^{-5}\,mm/s$ in 3.5% NaCl solution [78]. Similarly, 2219 Al-alloy in T87 temper and its friction stir weldment in the as-welded condition were reported to be not susceptible to SCC in 3.5% NaCl solution based on SSRT tests performed at nominal strain rates of

10^{-6} s^{-1} and 10^{-7} s^{-1} [79]. SCC failures observed in the heat affected zone regions of 7xxx series alloy weldments have been attributed to the degree of sensitization [80, 81]. Even though the term 'sensitization' in aluminium alloys is used to address the grain boundary precipitation in 5xxx alloys, in this case the authors seem to use this term to address a similar precipitation phenomenon in the 7xxx alloy.

A majority of the published literature suggests that the friction stir weldments of low and moderate strength Al-alloys, e.g. 5083, 2219 and 2195, possess a better SCC resistance when compared to the weldments of high strength alloys, e.g. 7075 and 7050 Al-alloys, on account of the favourable temper conditions, microstructural features, and corrosion behaviour of different regions of the weldment. However, the effect of strain rate has not been addressed well enough to comprehensively understand the SCC behaviour of these friction stir weldments.

It seems that the resistance of the weldments to SCC is influenced by the welding process/technique. Plasma arc welding of 2195 and 2219 Al-alloys were found to have a higher susceptibility to SCC than its FS weldment counterpart [82]. The stress corrosion crack initiation and propagation reported to be along the weld metal/partially melted zone interface in the PA weldment was attributed to the higher localized corrosion susceptibility. In the case of the FSW specimens, the fracture was found to originate at the root region of the weld, and propagate along the weld nugget/thermomechanically affected zone (TMAZ) interface. It was believed that the microstructural changes (especially the nature and distribution of precipitates at the PMZ/weld metal interface, also due to the employment of 4043 Al-alloy filler), which happened during welding, could be the reason for the stress corrosion crack initiation/propagation in these regions. The SCC resistance of friction stir weldments of 5454 Al-alloy in O and H43 conditions assessed by U-bend and slow strain rate tests was found to be better than that of its parent material and the GTA welds [83]. The SCC behaviour of a friction stir and GMA weldments of 5083 Al-alloy was assessed in ASTM EXCO solution and in 3.5% NaCl + 0.3 g/l H_2O_2 by SSRT tests [84]. The GMA weldment was reported to exhibit a much higher SCC susceptibility in both the environments, and the fracture was in the middle of the weld in the transverse tensile specimens SSRT tested. In the case of the FSW specimens, the fracture was outside the weld nugget region. The electrochemical tests suggested a better general and pitting corrosion resistance for the weld nugget region of the FSW specimens, which corroborated the SSRT SCC behaviour.

Power beam weldments of high and moderate strength aluminium alloys, in general, are reported to be susceptible to SCC in the as-welded condition. In some cases, an appropriate post-weld heat treatment has been found to be beneficial in retarding SCC and where possible, employment of suitable welding consumable may help in controlling the chemistry and microstructure of weldments. An electron beam weldment of alloy 7050-T7451 Al-alloy was found to be susceptible to SCC in 3.5% NaCl solution compared to its parent alloy, as assessed by constant load SCC tests at different stress levels. The higher susceptibility was attributed to the considerable level of residual stresses in the EB weldments [85]. In contrast, autogenous electron beam weldments of 5754 Al-alloy was found to possess very similar mechanical properties and SCC resistance as that of the parent alloy, which was attributed to the low strength levels of the weldment and also to the absence of any undesirable phase transformation that could cause higher corrosion susceptibility of the weld region [86].

Braun investigated the stress corrosion behaviour of AA6013 alloy and its Nd-YAG laser beam weldments, produced with silicon and magnesium containing filler powders, in 0.6 M NaCl + 0.06 M NaHCO$_3$ [87]. In the constant load tests performed at 170 and 270 MPa for 30 days, the parent alloy in both T4 and T6 tempers was found to be highly resistant to SCC. On the other hand, the as-welded laser joints in the T4 condition exhibited a high susceptibility to SCC, failing at a stress level of 100 MPa. The fracture was observed in the HAZ, at a location 1 mm away from the fusion boundary, regardless of the filler powder used. Based on the TEM observations, the formation of Si- and Mg-rich precipitates, possibly Mg$_2$Si at the grain boundaries in the HAZ, was reported to be responsible for the intergranular SCC (Fig. 8.18). However, the alloy welded in T6 temper and the T4 weldments after a T6 post-weld heat treatment were resistant to SCC at loads up to 170 MPa and 200 MPa, respectively.

8.18 Fractograph of 6013 Al-alloy joint made with aluminium alloy filler powder AlSi10Mg in the as-welded T4 condition, showing intergranular stress corrosion cracking [87].

Hatamleh *et al.* [88] produced friction stir welds from 7075 Al-alloy in T651 temper and evaluated the SCC behaviour of this weldment with and without laser shock peening and shot-peening treatments. The temper condition of the welded plates was modified from T651 to T7351 by artificial aging. The peening treatments were aimed at altering the state of residual stress at the surface in order to improve the SCC behaviour. The SSRT data summarized by the authors showed that there was no SCC of the weldments in 3.5% NaCl solution, even in the non-peened condition, which could be attributed to change in temper condition to a more stable/SCC resistant T7351. Yet another work from the same group [89] on the SCC of 2195 Al-alloy FSW joints showed very similar results as observed for the weldments of 7075 Al-alloy with none of the weldments showing SCC susceptibility. The beneficial effects of peening on the SCC resistance of the aluminium alloy welds could not be understood from these publications as even the un-peened alloy/weldment showed no signs of SCC. It appears that the strain rate (1×10^{-6} s^{-1}) employed in the above investigations was high, and that the test duration was too short to provide sufficient material-environment interaction to provide an accurate assessment. Unfortunately, there are no reports on the SCC behaviour of shot/laser peened weldments by employing lower strain rates, and hence the real influence of peening treatments on the SCC of aluminium alloy weldments still remains to be understood.

The limited available information on the SCC behaviour of aluminium alloy welds indicate that the alloys welded in the stabilized temper or that were given a post-welding heat treatment may resist the environmentally induced damage. Even though there are many publications on the surface treatments/coatings for improving the general/pitting corrosion resistance of aluminium alloys and a few on the laser treatment for combating SCC of aluminium alloys, the strategies for prevention of SCC of these weldments still remain an open topic for research.

There is growing interest in the joining of dissimilar aluminium alloys for industrial applications, and numerous works have been made on the microstructure-mechanical property correlation of dissimilar friction stir welded joints. The published information on the environmental cracking susceptibility of dissimilar weld joints is very limited. Srinivasan *et al.* [90] have investigated the SCC behaviour of a dissimilar friction stir weldment that comprised 6056 and 7075 Al-alloys. In the SSRT tests performed in air and in 3.5% NaCl solution at 10^{-6} s^{-1}, the fracture was observed in the TMZ/HAZ region of the 6056 Al-alloy, which was the weak region in the joint. Even though numerous pits were observed in the 7075 Al-alloy and also in the root region of the weld nugget, there was no SCC (Fig. 8.19). However, when the test was performed at a lower strain rate (10^{-7} s^{-1}), SCC was noticed in the 7075 parent alloy (Fig. 8.20), revealing the significant influence of strain rate in evaluating SCC of this dissimilar aluminium alloy weldment.

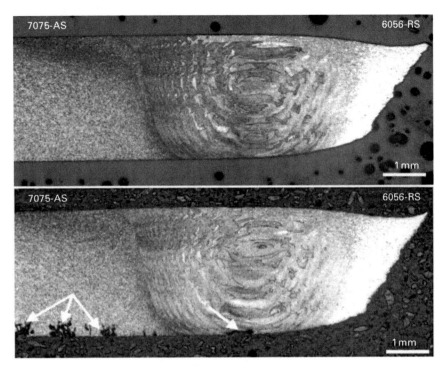

8.19 Optical macrograph showing the SSRT tested FSW specimens (a) in air and (b) in 3.5% NaCl solution [90]. (Note: failure location in both cases: TMAZ of 6056 Al-alloy; the arrows indicate pits in the weldment.)

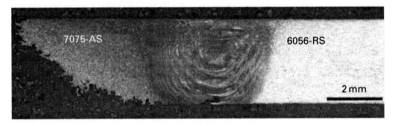

8.20 Optical macrograph showing the fracture location in a dissimilar friction stir weldment (7075-6056 Al-alloys) SSRT tested in 3.5% NaCl solution [90].

8.5   Stress corrosion cracking (SCC) of aluminium composites

Aluminium metal matrix composite systems, reinforced with secondary phase in the form of particulates, whiskers or fibres, are known for their better mechanical properties. The effect of incorporation of the secondary phases on the mechanical and tribological behaviour of these composite alloys has been extensively studied and also reviewed. The influence of reinforcements on the electrochemical degradation of the composite alloys in corrosive environments, especially those containing chlorides, has also been documented. In general, the corrosion behaviour of the composites is not adversely influenced by the chemically inert secondary phases. However, the reinforcements may have an indirect effect by way of influencing the aging characteristics of the composite alloy or by promoting the formation of metal/reinforcement interface or by influencing segregation of alloying elements during the processing. Unlike the case of monolithic aluminium alloys, there is no documentation on the corrosion/SCC failures of aluminium alloy composites in practical applications. However, there seems to be some research efforts to understand the SCC behaviour of aluminium composites. AA6061 alloy composites comprising SiC_w, Albite and Nextel 440 fibre as reinforcements have been examined for SCC behaviour by different techniques in different environments [91–94]. The composites with SiC_w and Albite were reported to be resistant to SCC, whilst the one with the Nextel fibre was found susceptible to SCC. Preferential attack of the interfacial precipitates along the length of the fibres was reported to be responsible for the higher SCC susceptibility of the Nextel composite. Similar degradation has been found to be responsible for SCC susceptibility of AA2024-boron fibre composites Sedriks *et al.* [95]. Furthermore, the fibre orientation to the stress axis seems to influence SCC. Whilst there was no SCC in the $Al_{18}B_4O_{33}/Al$ composite specimens loaded in the direction parallel to the extrusion direction, the specimens tested with the applied stress vertical to the extrusion direction showed the highest dissolution under stress and SCC susceptibility [96], suggesting orientation effects. In the double cantilever beam SCC experiments the authors observed the crack propagation along the interface between the whisker and matrix in the composite.

In a discontinuously reinforced AA2014 composite alloy containing 15 vol% Al_2O_3 particulates, the SCC susceptibility of the composite was attributed to the formation of higher crack density regions, promoted by secondary particle fracture and fracture at the matrix/particle interface (Fig. 8.21) [97]. SCC of 50 vol% Altex containing Al-Zn-Mg and Al-Zn-Mg-Ag alloy composites was reported to be influenced by the pits that originated owing to the setting up of a galvanic cell in the presence of Ti-rich and Fe-rich particles in the alloy at the fibre/matrix interfaces, and that it led to de-bonding at the fibre–matrix interfaces (Fig. 8.22). Even though the role of hydrogen in the SCC process was speculated by the authors, there was no experimental evidence to substantiate the claims [98]. AA2024 alloy composite containing 15 vol% SiC particulates of 10 µm size showed a high resistance to SCC in 3.5% NaCl solution when tested by SSRT and DCB tests at open circuit potential [99]. Fine microstructure in the matrix

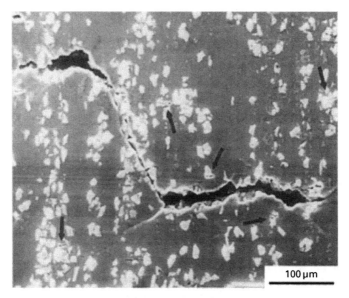

8.21 Surface of 2014 Al-composite alloy showing cracks at the particles and particle/matrix interface [97].

8.22 De-bonding at the fibre–matrix interfaces and corrosion product at these regions of Al-Zn-Mg composite [98].

and the hindering effect of secondary phase particles located at the grain boundaries to the intergranular cracking were reported to be the factors for the observed behaviour. Reduction in macro stress in the composites by annealing at higher temperatures was reported to increase the SCC resistance markedly [100].

The available literature on the SCC aluminium alloy composites, in general, suggests that the SCC susceptibility of aluminium alloy matrix composites is governed by the processing route, orientation of the secondary phase, the interface/bonding between the reinforcement and the matrix and residual stress of the composite alloy. However, it is necessary to gather information on the SCC failures of composites in actual service conditions, and use the field information to frame research plans to investigate different alloy systems with different reinforcements in appropriate test environments. Such documentation could be useful in selecting composite alloys for improved performance in specific applications.

8.6   Conclusions

A significant amount of work has been carried out to understand the SCC behaviour of Al-alloys. Anodic dissolution assisted cracking and hydrogen induced cracking are the two widely cited SCC mechanisms in Al-alloys. Various heat treatment procedures like over-aging, retrogression and re-aging and multi-step aging have been developed to improve the SCC resistance of Al-alloys. Grain size and orientation, precipitation along grain boundary and dislocations influence the SCC behaviour of Al-alloys. Notably, inhibiting re-crystallization by scandium alloying has shown to enhance the SCC resistance of Al-alloys. The limited literature on the SCC of aluminium alloy weldments suggests the environmental cracking susceptibility of these weldments is governed by the transformations that happen during welding in the different regions of weldment, and the residual stresses also contribute to the failure. However, there is large scope for a better understanding of the SCC behaviour of weldments and composite alloys by appropriate selection of SCC test method/procedures/conditions.

8.7   References

[1]   Speidel O.M. (1975), 'Stress corrosion cracking of aluminum alloys', *Metall. Trans. A*, 6A, p. 631.

[2]   Holroyd N.J.H. and Hardie D. (1981), 'Strain-rate effects in the environmentally assisted fracture of a commercial high-strength aluminium alloy (7049)', *Corros. Sci.*, 21, p. 129.

[3]   Lynch S.P. (1982), 'Mechanisms of environmentally assisted cracking in Al-Zn–Mg single crystals', *Corros. Sci.*, 22, p. 925.

[4]   Hardwick D.A., Thompson A.W. and Bernstein I.M. (1988), 'The effect of copper content and heat treatment on the hydrogen embrittlement of 7050-type alloys', *Corros. Sci.*, 28, p. 1127.

[5]   Najjar D., Magnin T. and Warner T.J. (1997), 'Influence of critical surface defects and localized competition between anodic dissolution and hydrogen effects during

stress corrosion cracking of a 7050 aluminium alloy', *Mater. Sci. Eng. A*, 238A, p. 293.

[6] Forsyth P.J.E. (1999), 'Initiation of intergranular cracking and other corrosion assisted fracture paths in two precipitation hardening aluminium alloys', *Mater. Sci. Technol.*, 15, p. 301.

[7] Cooper K.R., Young L.M., Gangloff R.P. and Kelly R.G. (2000), 'The electrode potential dependence of environment-assisted cracking of AA 7050', *Mater. Sci. Forum*, 331–337, p. 1625.

[8] Bobby Kannan M. (2005), Role of multistep aging and scandium addition on the environmentally assisted cracking and exfoliation behaviour of 7010 Al alloy, PhD thesis, Indian Institute of Technology Bombay, Mumbai, India.

[9] Knight S. (2008), Stress Corrosion Cracking of Al-Zn-Mg-Cu Alloy, Effects of Heat-Treatment, Environment, and Alloy Composition, PhD thesis, Monash University, Melbourne Australia.

[10] Buchheit R.G., Wall F.D., Stoner G.E. and Moran J.P. (1995), 'Anodic dissolution-based mechanism for the rapid cracking, pre-exposure phenomenon demonstrated by aluminium-lithium-copper alloys', *Corrosion*, 51, p. 417.

[11] Puiggali M., Zielinski A., Olive J.M., Renauld E., Desjardins D. and Cid M. (1998), 'Effect of microstructure on stress corrosion cracking of an Al-Zn-Mg-Cu alloy', *Corros. Sci.*, 40, p. 805.

[12] Liu X., Frankel G.S., Zoofan B. and Rokhlin S.I. (2004), 'Effect of applied tensile stress on intergranular corrosion of AA2024-T3', *Corros. Sci.*, 46, p. 405.

[13] Jacobs A.J. (1965), 'The role of dislocation in stress corrosion cracking of 7075 aluminum alloy', *Trans. ASM*, 58, p. 579.

[14] Albrecht J., Thompson A.W. and Berntein I.M. (1979), 'The role of microstructure in hydrogen-assisted fracture of 7075 aluminum', *Metall. Trans. A*, 10A, p. 1759.

[15] Talianker M. and Cina B. (1989), 'Retrogression and reaging and the role of dislocations in the stress corrosion of 7000-type aluminum alloys', *Metall. Trans. A*, 20A, p. 2087.

[16] Vishwanadham R.K., Sun T.S. and Green J.A.S. (1980), 'Grain boundary segregation in Al-Zn-Mg alloys – implications to stress corrosion cracking', *Metall. Trans.*, 11, p. 85.

[17] Scamans G.M., Holroyd N.J.H. and Tuck C.D.S. (1987), 'The role of magnesium segregation in the intergranular stress corrosion cracking of aluminium alloys', *Corros. Sci.*, 27, p. 329.

[18] Song R.G., Tseng M.K., Zhang B.J, Liu J., Jin Z.H. and Shin K.S. (1996), 'Grain boundary segregation and hydrogen-induced fracture in 7050 aluminium alloy', *Acta Mater.*, 44, p. 3241.

[19] Speidel M.O. and Hyatt M.V. (1972), 'Stress-corrosion cracking of high-strength aluminum alloys' in *Advances in Corrosion Science and Technology, Vol. 2* (Eds M.G. Fontana and R.W. Staehle), New York: Plenum Press, pp. 115–335.

[20] Burleigh T.D. (1991), 'The postulated mechanisms for stress corrosion cracking of aluminum alloys – a review of the literature 1980–1989', *Corrosion*, 47, p. 89.

[21] Speidel M.O. (1984), 'Hydrogen embrittlement and stress corrosion cracking of aluminum alloys' in *Hydrogen Embrittlement and Stress Corrosion Cracking: a Troiano Festschrift* (Eds R. Gibala and R.F. Hehemann), Materials Park, OH: American Society for Metals.

[22] Lynch S.P. (2003), 'Mechanisms of hydrogen assisted cracking – a review' in *Hydrogen Effects on Material Behavior and Corrosion Deformation Interactions* (Ed. R.H. Jones), Warrendale, PA: The Minerals, Metals and Materials Society (TMS), pp. 449–466.

[23] Braun R. (2007), 'Environmentally assisted cracking oa aluminium alloys', *Mat. Wiss. Werkstofftech.*, 38, p. 674.

[24] Cina B. (1974), Reducing the Susceptibility of Alloys, Particularly Aluminium Alloys, to Stress Corrosion Cracking, US Patent 3856584.

[25] Hatch J.E. (1984), *Aluminum Properties and Physical Metallurgy*, Materials Park, OH: ASM, p. 270.

[26] Polmear I.J. (1989), *Light Alloys* (Metallurgy and Material Science Series), Burlington, MA: Butterworth-Heinemann, p. 101.

[27] Mukhopadhyay A.K. (1997), 'Development of reproducible and increased strength properties in thick extrusions of low-alloy Al-Zn-Mg-Cu based AA 7075', *Metall. Mater. Trans. A*, 28A, p. 2429.

[28] Lynch S.P. (1989), 'Metallographic contributions to understanding mechanisms of environmentally assisted cracking', *Metallography*, 23, p. 147.

[29] Bobby-Kannan M., Raja V.S., Raman R. and Mukhopadhyay A.K. (2003), 'Influence of multistep aging on the stress corrosion cracking behavior of Al alloy (7010)', *Corrosion*, 59, p. 881.

[30] Bobby Kannan M. and Raja V.S. (2006), 'Evaluation of hydrogen embrittlement susceptibility of over aged 7010 Al alloy', *J. Mater. Sci.*, 41, p. 5495.

[31] Christodoulou L. and Flower H.M. (1980), 'Characterization of trapped hydrogen in exfoliation corroded aluminium alloy 2024', *Acta Metall.*, 28, p. 481.

[32] Choi Y., Kim H.C. and Pyun S.I. (1984), 'Stress corrosion cracking of Al-Zn-Mg alloy AA 7039 by slow strain rate method', *J. Mater. Sci.*, 19, p. 1517.

[33] Bobby Kannan M. and Raja V.S. (2007), 'Role of coarse intermetallic particles on the stress corrosion cracking behavior of peak aged and over aged Al-Zn-Mg-Cu-Zr alloy during slow strain rate testing', *J. Mater. Sci.*, 42, p. 5458.

[34] Uesaki K., Kawakami T. and Takechi H. (1985), *Kenkyu Kiyo-Anan Kogyo Koto Senmon Gakko*, 21, p. 11.

[35] Sprowls D.O. (1996), 'Evaluation of stress corrosion cracking', in *Stress Corrosion Cracking – Material Performance and Evaluation* (Ed. R.H. Jones), Materials Park, OH: ASM, p. 394.

[36] Galvele J.R. and De Micheli S.M. (1970), 'Mechanism of intergranular corrosion of Al-Cu alloys', *Corros. Sci.*, 10, p. 795.

[37] Sugimoto K., Hoshino K., Kageyama M., Kageyama S. and Sawada Y. (1975), *Corros. Sci.*, 15, p. 709.

[38] Wall F.D. and Stoner G.E. (1997), 'The evaluation of the critical electrochemical potentials influencing environmentally assisted cracking of Al-Li-Cu alloys in selected environments', *Corros. Sci.*, 39, p. 835.

[39] Gruhl W. (1984), *Z. Metallkd.*, 75, p. 819.

[40] Bobby Kannan M., Raja V.S. and Mukhopadhyay A.K. (2004), 'Determination of true stress corrosion cracking index of a high strength Al alloy using glycerin as the non-corrosive atmosphere', *Scripta Materialia*, 51, p. 1075.

[41] Bandyopadhyay A., Ambat R. and Dwarakadasa E.S. (1992), 'Effect of hydrogen charging on the mechanical properties of medium strength aluminium alloys 2091 and 2014', *Bull. Mater. Sci.*, 15, p. 311.

[42] Magnin T. and Dubessy C. (1985), *Mem. Etud. Sci. Rev. Metall.*, 82, p. 559.

[43] Doig P., Flewitt P.E.J. and Edington J.W. (1977), 'The stress corrosion susceptibility of 7075 Al-Zn-Mg-Cu alloys tempered from T6 to an overaged T7X', *Corrosion*, 33, p. 217.

[44] Lee S., Pyun S. and Chu Y. (1991), 'A critical evaluation of the stress-corrosion cracking mechanism in high-strength aluminum alloys', *Metall. Trans. A*, 22A, p. 2407.

[45] Tsai T.C. and Chuang T.H. (1997), 'Role of grain size on the stress corrosion cracking of 7475 aluminum alloys', *Mater. Sci. Eng. A*, 225A, p. 135.

[46] Forsyth P.J.E. (1999), 'Initiation of intergranular cracking and other corrosion assisted fracture paths in two precipitation hardening aluminium alloys', *Mater. Sci. Technol.*, 15, p. 301.

[47] Yang J. and Ou B. (2001), 'Influence of microstructure on the mechanical properties and stress corrosion susceptibility of 7050 Al- alloy', *Scand. J. Metallurgy*, 30, p. 158.

[48] Gest R.J. and Troiano A.R. (1974), 'Stress corrosion and hydrogen embrittlement in an aluminum alloy', *Corrosion*, 30, p. 274.

[49] Bobby Kannan M. and Raja V.S. (2010), 'Enhancing stress corrosion cracking resistance in Al-Zn-Mg-Cu-Zr alloy through inhibiting recrystallization', *Eng. Fracture Mech.*, 77, p. 249.

[50] Tseng M.K. and Marcus H.L. (1981), 'An investigation of the effect of gaseous corrosion fatigue on the J_{IC} of aluminum alloys', *Scripta Metall.*, 15, p. 427.

[51] Ohasaki S., Dobayashi K., Iino M. and Sakamoto T. (1996), 'Fracture toughness and stress corrosion cracking of aluminium-lithium alloys 2090 and 2091', *Corros. Sci.*, 38, p. 793.

[52] Sarkar B., Marek M. and Starke Jr. E.A. (1981), 'The effect of copper content and heat treatment on the stress corrosion characteristics of Al-6Zn-2Mg-X Cu alloys', *Metall. Trans. A*, 12A, p. 1939.

[53] Willey L.A.: US Patent 3,619,181, 1971.

[54] Elagin V.I., Zakharov V.V., Petrova A.A. and Vyshegorodtseva E.V. (1983), 'Effect of scandium on the structure and properties of aluminium-zinc-magnesium alloys', *Russian Metallurgy*, 4, p. 180.

[55] Riddle Y.W. and Sanders T.H. (2000), 'Aluminium alloys their physical and mechanical properties', *7th International Conference ICAA7, Charlottesville, VA*, 9–14 April.

[56] Braun R. (1995), 'Evaluation of the stress corrosion cracking behaviour of damage-tolerant Al-Li sheet using the slow strain rate testing technique', *Mater. Sci. Eng.*, 190, p. 143.

[57] Wang Z., Wang P., Kumar K.S. and Briant C.L. (2001), 'Chemistry and electrochemistry of corrosion and stress corrosion Cracking', *A Symposium Honoring the Contributions of R. W. Staehle, Proceedings of Symposium*, (Ed. R.H. Jones) New Orleans, LA, p. 573.

[58] Bobby Kannan M. and Raja V.S. (2007), 'Influence of heat treatment and scandium addition on the electrochemical corrosion behavior of Al-Zn-Mg-Cu-Zr (7010) alloy', *Metall. Mater. Trans. A*, 38A, p. 2843.

[59] Yi-Lei W., Froes F.H., Chenggong L. and Alex A. (1999), 'Microalloying of Sc, Ni, and Ce in an advanced Al-Zn-Mg-Cu alloy', *Metall. Mater. Trans. A*, 30A, p. 1017.

[60] Gjonnes J. and Simensen C.J. (1970), 'An electron microscope investigation of the microstructure in an aluminium-zinc-magnesium alloy', *Acta Metall.*, 18, p. 881.

[61] Mondolfo L.F. (1971), 'Structure of the aluminium:magnesium:zinc alloys', *Metals Mater*, 5, p. 95.

[62] Mukhopadhyay A.K., Reddy G.M., Prasad K.S., Kamat S.V., Dutta A. and Mondal C. (2001), 'Influence of scandium addition on the microstructure and properties of AA7010 alloy sheets', *J.T. Staley Honorary Symposium on Al Alloys* (Ed. M. Tiryakioglu), ASM Materials Solution Conference and Exposition (Indianapolis, USA), p. 63.

[63] Shreir L.L. (ed.) (1976), *Corrosion*, 2nd edn. Newnes-Butterworth, Section 1:49.

[64] Peel C.J. and Poole P. (1985), 'The stress corrosion resistance of high strength Al-Zn-Mg alloys', *Proceedings of the International Conference on Fatigue, Corrosion Cracking, Fracture Mechanics and Failure Analysis* (Ed. V.S. Goel), p. 147.

[65] Danh N.C., Rajan K. and Wallace W. (1983), 'A TEM study of microstructural changes during retrogression and reaging in 7075 aluminum', *Metall. Trans.*, 14A, p. 1843.

[66] Park J.K. (1988), 'Influence of retrogression and reaging treatments on the strength and stress corrosion resistance of aluminium alloy 7075-T6', *Mat. Sci. Eng.*, A103, p. 223.

[67] Komisarov V., Talianker M. and Cina B. (1996), 'The effect of retrogression and reaging on the resistance to stress corrosion of an 8090 type aluminium alloy', *Mater. Sci. and Eng.*, A221, p. 113.

[68] Holroyd N.J.H. (1990), 'Environment-induced cracking of high-strength aluminum alloys' in *Proceedings of Environment-induced cracking of metals* (Eds R.P. Gangloff and M.B. Ives), Houston; TX: NACE, p. 311.

[69] Vasudevan A.K. and Doherty R.D. (1987), 'Grain boundary ductile fracture in precipitation hardened aluminum alloys', *Acta Metallurgica*, 35, p. 1193.

[70] Ou B., Yang J. and Wie M. (2007), 'Effect of homogenization and aging treatment on mechanical properties and stress-corrosion cracking of 7050 alloys', *Metall. Mater. Trans.*, 38A, p. 1760.

[71] Onoro J. and Ranninger C. (1995), 'Exfoliation corrosion behaviour of welded high strength aluminium alloys', *Br. Corros. J.*, 30, p. 203.

[72] Posada M., Murr L.K., Niou C.S., Roberson D., Little D., Arrowood R. and George D. (1997), 'Exfoliation and related microstructures in 2024 aluminum body skins on aging aircraft', *Materials Characterization*, 38, p. 259.

[73] McNaughtan D., Worsfold W. and Robinson M.J. (2004), 'Corrosion product force measurements in the study of exfoliation and stress corrosion cracking in high strength aluminium alloys', *Corros. Sci.*, 45, p. 2377.

[74] Jha A.K., Murty S.V.S.N., Diwakar V. and Sree Kumar K. (2003), 'Metallurgical analysis of cracking in weldment of propellant tank', *Engineering Failure Analysis*, 10, p. 265.

[75] Jha A.K., Naga Shirisha G., Sreekumar K., Mittal M.C. and Ninan K.N. (2008), 'Stress corrosion cracking in aluminum alloy AFNOR 7020-T6 water tank adaptor for liquid propulsion system', *Engineering Failure Analysis*, 15, p. 787.

[76] Lim S., Kim S., Lee C.G. and Kim S. (2005), 'Stress corrosion cracking behaviour of friction stir welded Al 6061-T651', *Metall. Mater. Trans.*, 36A, p. 1977.

[77] Bala Srinivasan P., Arora K.S., Dietzel W., Pandey S. and Schaper M.K. (2009), 'Characterisation of microstructure, mechanical properties and corrosion behaviour of an AA2219 friction stir weldment', *J. Alloys Compounds*, 492, p. 631.

[78] Kramer L.S., Blair T.P., Blough S.D., Fisher J.J., Jr. and Pickens J.R. (2002), 'Stress-corrosion cracking susceptibility of various product forms of aluminium alloy 2519', *J. Mater. Eng. Performance*, 11, p. 645.

[79] Paglia C.S. and Buchheit R.G. (2006), 'Microstructure, microchemistry and environmental cracking susceptibility of friction stir welded 2219-T87', *Mater. Sci. Eng.*, A429, p. 107.

[80] Paglia C.S., Carroll M.C., Pitts B.C., Reynolds T. and Buchheit R.G. (2002), 'Strength, corrosion and environmentally assisted cracking of a 7075-T6 friction stir weld', *Materials Science Forum*, 396–402, p. 1677.

[81] Lumsden J., Pollock G. and Mahoney M. (2003), 'The effect of thermal treatments on the corrosion behaviour of friction stir welded 7050 and 7075 aluminium alloys', *Materials Science Forum*, 426–432, p. 2867.

[82] Hu W. and Meletis E.I. (2000), 'Corrosion and environment assisted cracking behaviour of friction stir welded Al2195 and Al2219 alloys', *Materials Science Forum*, 331–337, p. 1683.

[83] Frankel G.S. and Xia Z. (1999), 'Localized corrosion and stress corrosion cracking resistance of friction stir welded Al alloy 5454', *Corrosion*, 55, p. 139.

[84] Zucchi F., Trabanelli G. and Grassi V. (2001), 'Pitting and stress corrosion cracking resistance of friction stir welded AA5083', *Mater. Corros.*, 52, p. 853.

[85] Ciompi E. and Lanciotti A. (1999), 'Susceptibility of 7050-T7451 electron beam welded specimens to stress corrosion', *Engineering Fracture Mechanics*, 62, p. 463.

[86] Czechowski M. (2003), 'Mechanical properties and stress corrosion of electron beam melting through joint of AlMg alloy', *Adv. Mater. Sci.*, 4, p. 9.

[87] Braun R. (2006), 'Nd:YAG laser butt welding of AA6013 using silicon and magnesium containing filler powders', *Mater. Sci. Eng.*, A426, p. 250.

[88] Hatamleh O., Singh P.M. and Garmestani H. (2008), 'Corrosion susceptibility of peened friction stir welded 7075 aluminium alloy joints', *Corros. Sci.*, 51, p. 135.

[89] Hatamleh O., Singh P.M. and Garmestani H. (2009), 'Stress corrosion cracking behaviour of peened friction stir welded 2195 aluminium alloy joints', *J. Mater. Eng. Performance*, 18, p. 406.

[90] Bala Srinivasan P., Dietzel W., Zettler R., dos Santos J.F. and Sivan V. (2005), 'Stress corrosion cracking susceptibility of friction stir welded AA7075 – AA6056 dissimilar joint', *Mater. Sci. Eng.*, A392, p. 292.

[91] McIntyre J.F., Le A.H., Golledge S.L. and Conrad R.K. (1987), 'Corrosion behaviour of SiC reinforced aluminium alloys', Report No. NSWC-TR-87-326, United States Naval Surface Warfare Centre.

[92] Cao L., Li G.F. and Yao C.K. (1989), In: *Proceedings of the First Japan International SAMPE Symposium*, 28 November–1 December, pp. 1066–107.

[93] Sharma C. (2001), 'A study on stress corrosion behaviour of Al6061/Albite composite in higher temperature acidic medium using autoclave', *Corros. Sci.*, 43, p. 1877.

[94] Berkeley D.W., Sallam H.E.M. and Hashemi H.N. (1998), 'The effect of pH on the mechanism of corrosion and stress corrosion and degradation of mechanical properties of AA6061 and Nextel 440 fibre reinforced AA6061 composite', *Corros. Sci.*, 40, p. 141.

[95] Sedriks A.J., Green J.A.S. and Novak D.L. (1971), *Metall. Trans.*, A2, p. 871.

[96] Hu S.J., Chen C.S., Xu L.X., Yao C.K. and Zhao L.C. (2002), 'Effect of whisker

orientation on the stress corrosion cracking behaviour of aluminium borate whisker reinforced pure Al composite', *Mater. Lett.*, 56, p. 642.

[97]   Singh P.M. and Lewandowski J.J. (1996), 'Stress corrosion cracking of discontinuously reinforced aluminium (DRA) alloy during slow strain rate testing', *J. Mater. Sci. Lett.*, 15, p. 490.

[98]   Winkler S.L. and Flower H.M. (2004), 'Stress corrosion cracking of cast 7XXX aluminium fibre reinforced composites', *Corros. Sci.*, 46, p. 903.

[99]   Yao H.Y. (1999), 'Effect of particulate reinforcing on stress corrosion cracking performance of a SiC_p-2024 aluminium matrix composite', *J. Compos. Mater.*, 33, p. 962.

[100]   Hu J., Lou R.S., Yao C.K. and Zhao L.C. (2001), 'Effect of annealing treatment on the stress corrosion cracking behaviour of SiC whisker reinforced aluminium composite', *Mater. Chem. Phys.*, 70, p. 160.

9

Stress corrosion cracking (SCC) of magnesium alloys

A. ATRENS, The University of Queensland, Australia,
W. DIETZEL and P. BALA SRINIVASAN,
Helmholtz-Zentrum Geesthacht, Germany, N. WINZER,
Fraunhofer Institute for Mechanics of Materials IWM, Germany
and M. BOBBY KANNAN, James Cook University, Australia

Abstract: Stress corrosion cracking (SCC) of Mg alloys is intergranular (IGSCC) or transgranular (TGSCC). A continuous or nearly continuous second phase, typically along grain boundaries, causes IGSCC by micro-galvanic corrosion of the adjacent Mg matrix. IGSCC is expected in all such alloys, typical of most creep resistant alloys, because each known second phase has a more positive corrosion potential than the matrix α-Mg; the degree of severity depends on the electrochemical properties of the second phase; these electrochemical properties need to be studied. Nearly continuous second phases can be avoided by Mg alloy design. TGSCC is most likely caused by an interaction of hydrogen (H) with the microstructure. A study of H-trap interactions is needed to understand this damage mechanism, and to design alloys resistant to TGSCC. Understanding is urgently needed if wrought alloys are to be used safely in service, because prior research indicates that many Mg alloys have a threshold stress for SCC of about half the yield stress in common environments including high-purity water.

Key words: magnesium, corrosion, stress corrosion cracking.

9.1 Introduction

9.1.1 Aims

This chapter builds on our reviews on magnesium (Mg) corrosion [1–3] and Mg stress corrosion cracking (SCC) [4, 5]. Mg SCC [4–7] can be characterised by the threshold stress, σ_{SCC}, the threshold stress intensity factor, K_{1SCC}, and the stress corrosion crack velocity, V_{SCC}. Mg TGSCC is most likely caused by an embrittlement mechanism (Fig. 9.1):

(1) an embrittled region forms ahead of the crack tip,
(2) there is a crack propagation,
(3) the crack stops as it enters the parent material, and
(4) the process recurs when an embrittled region has re-formed.

341

9.1 Model for TGSCC by an embrittlement mechanism. Reprinted from [8].

9.1.2   Mg alloys

Polmear [9] provides a good introduction to Mg physical metallurgy. Mg has low density, a hexagonal close packed crystal structure and a melting point of 650°C. Mg alloys have been use since the beginning of the twentieth century. Cd has complete solubility. All other alloying elements form intermetallic phases and result in either eutectic or peritectic phase diagrams. Major alloying elements include Al, Zn, Ce, Y, Ag, Th, Zr and rare earth (RE) elements. Al improves castability and strength. Increasing Al content typically increases the volume fraction of the β-phase ($Mg_{17}Al_{12}$). Zn is commonly used in combination with either Al or Zr. Zr is a grain refiner in Mg alloys which do not contain Al and Mn. Zr is contained in almost all Mg-Zn alloys. Creep properties are typically improved by RE elements by the formation of intermetallic phases. Mg alloys are often designated following the ASTM standard B275-94, e.g. AZ91 is the designation of the most common casting alloy, containing 9%Al and 1%Zn. The first two letters indicate the major alloying elements; the nominal composition, in wt%, is given by the two numbers. The designation and the corresponding alloying element are as follows: A, Aluminium; E, Rare earths; H, Thorium; K, Zirconium; M, Manganese; Q, Silver; R, Chromium; S, Silicon; W, Yttrium and Z, Zinc. WE43 refers to an alloy that contains 4% Y and 3% rare earth elements. Table 9.1 presents some common M alloys and their compositions [10].

Table 9.1 Compositions of common Mg alloys [10]

	Al	Zn	Mn	Zr	RE	Ag	Si	Y	Sr	Cu
High pressure die casting alloys										
AE42	4.0		0.10		2.5					
AJ52x	5.0								1.95	
AM50A	4.9		0.26							
AS41A	4.2		0.20				1.0			
AZ91D	8.7	0.7	0.13							
Sand and gravity die casting alloys										
AM100A	10.0		0.1							
AZ81A	7.6	0.7	0.13							
EZ33A		2.7		0.6	3.3					
WE54A				0.7	3.0			5.2		
ZE41A		4.2		0.7	1.2					
ZK51A		4.6		0.7						
Wrought alloys										
AZ31B	3.0	1.0	0.2							
AZ61A	6.5	1.0	0.15							
ZC71		6.5	0.5							1.25
ZK21A		2.3		0.45						
ZK61A		6.0		0.7						
ZM21		2.0	0.5							

9.1.3   Mg corrosion

There are two classes of Mg alloys in relation to corrosion [2, 4]: (i) Zr-containing alloys and (ii) Zr-free alloys. Zr reacts with impurity elements removing them from the melt. Zr-containing alloys are typically high purity with low corrosion rates and often contain alloying elements like Ag, RE, Th, etc., that allow attainment of good mechanical properties. These alloys tend to be expensive and their applications are restricted to applications where their properties are more important than cost, such as in the aerospace industry. The most common of the Zr-free alloys are the alloys containing Al, such as AZxx, AMxx and ASxx. Their corrosion behaviour is highly dependent on their impurity content. High purity alloys have corrosion rates comparable with those of the Zr-containing alloys.

For Mg in an aqueous solution like 3% NaCl, corrosion occurs at breaks in a partially protective surface film [1, 2] by the electrochemical anodic partial reactions (Eqs 9.1–9.3) and the chemical reaction (Eq. 9.4). The product formation reaction, Eq. 9.5, tends to repair the surface film. Hydrogen evolution occurs by: (a) the cathodic partial reaction, Eq. 9.3 that balances the anodic partial reactions, Eqs 9.1 and 9.2; and (b) by the reaction of Mg^+ with water, Eq. 9.4. H^+ are consumed; the pH increase favours the Mg hydroxide film by the precipitation reaction, Eq. 9.5. The overall reaction, Eq. 9.6, produces one hydrogen molecule for each Mg atom dissolved.

$$Mg = Mg^+ + e \qquad \text{(anodic reaction)} \qquad 9.1$$

$$kMg^+ = kMg^{2+} + ke \qquad \text{(anodic reaction)} \qquad 9.2$$

$$(1 + k)H^+ + (1 + k)e = (1 + k)(½)H_2 \qquad \text{(cathodic reaction)} \qquad 9.3$$

$$(1 - k)Mg^+ + (1 - k)H^+ \qquad \text{(chemical reaction)} \qquad 9.4$$
$$= (1 - k)Mg^{2+} + (½)(1 - k)H_2$$

$$Mg^{2+} + 2OH^- = Mg(OH)_2 \qquad \text{(product formation)} \qquad 9.5$$

$$Mg + 2H_2O = Mg(OH)_2 + H_2 \qquad \begin{array}{l}\text{(overall reaction in} \\ \text{steady state)}\end{array} \qquad 9.6$$

Song *et al.* [1, 2, 11, 12, 20] provided a comprehensive analysis of the negative difference effect based on this reaction sequence. Anodic polarisation increases the H evolved by Eq. 9.4. The negative open circuit potential causes a high hydrogen fugacity, f_H. If Eq. 9.3 is in equilibrium at a potential E, in a solution of a particular pH, then the Nernst equation [13, 14] provides an estimate of the H fugacity, f_H:

$$f_H = k_1 \exp (k_2[E_H - E]) \qquad 9.7$$

where k_1 and k_2 are constants. E_H is the reference equilibrium potential of the H evolution reaction, and is equal to the standard potential for a solution with pH = 0 and f_H = 1 atmosphere. The constant k_1 is determined by f_H = 1 atmosphere when $E = E_H$. Eq. 9.7 estimates the equilibrium hydrogen fugacity at the freely corroding Mg surface, if the corrosion potential is substituted into Eq. 9.7. The equilibrium H fugacity, f_H, is large because $[E_H - E]$ is large, over 1000 mV. The H fugacity at the surface of Mg, corroding in an aqueous solution, is many orders of magnitude larger than that for freely corroding steel. Corrosion is important [1–4, 15–19], and there is much research on Mg corrosion [20–48]. Our research [1–4, 11, 12, 15, 17–26] has shown that Mg corrosion, in a typical solution like 3% NaCl, is controlled by: (i) the composition of the α-Mg matrix, (ii) the composition and amount of the second phases, and (iii) the distribution of the second phases. No alloying element produces a solid solution Mg alloy with a corrosion rate lower than that of pure Mg in solutions such as 3% NaCl [1–3], in contrast to Cr in Fe-Cr alloys in which a Cr content greater than 11.5% changes the surface film and drastically increases the corrosion resistance [49, 50]. Thus, high purity (HP) Mg should be used as a standard to compare the corrosion of Mg alloys. All second phases tend to cause micro-galvanic acceleration of the corrosion of the α-Mg matrix [1–3, 17, 18, 25], so a multi-phase alloy has typically a corrosion rate greater than that of pure Mg. Second phases associated with the impurity elements Fe, Ni, Co and Cu are particularly effective in accelerating micro-galvanic corrosion; corrosion rates are high for Mg alloys with concentrations of impurity elements above their tolerance limits, as much as 100× greater [1–3, 18, 21]. Study of alloying influences requires HP alloys; HP [1, 21, 23] means that the alloy contains less than the concentration dependent tolerance level of the impurity elements [21]. The second phase can act as a barrier if it is continuous and has a corrosion rate lower than pure Mg [18, 25]; otherwise the corrosion rate is accelerated, even for second phase particles as small as 40 nm.

9.1.4   Background reading

Our reviews [4, 5] and our research [34, 51–55] elucidated the critical aspects of Mg SCC. This chapter provides an update, based on our understanding of Mg corrosion [1–3, 140], SCC [7, 56–66], HE [67, 68], CF [69, 70], diffusion [68, 71] and passivity [72, 73].

9.2   Alloy influences

9.2.1   Pure Mg

Pure Mg is susceptible to SCC [74–79]. Winzer *et al.* [79] found TGSCC for pure Mg in 5 g/L NaCl. Meletis and Hochman [76] reported crystallographic TGSCC for 99.9% pure Mg in a chloride-chromate solution. Fracture was cleavage-like, consisting of flat, parallel facets on $\{2\bar{2}03\}$ planes separated by steps also on $\{2\bar{2}03\}$ planes. The cleavage was attributed to a reduction in the surface energy due to atomic H or to Mg hydride. Stampella *et al.* [78] reported SCC for commercial purity (99.5%) Mg and HP (99.95%) Mg in deaerated, pH = 10, 10^{-3} M Na_2SO_4. They proposed SCC occurred by cleavage facilitated by atomic H in solid solution; the crack was exclusively TGSCC for fine grain (0.025 mm) commercially pure Mg, whereas larger grain (0.075 mm) HP Mg produced mixed TGSCC and IGSCC.

The research of Lynch and Trevena [75] was in stark contrast as they observed concave features on opposing fracture surfaces, so that the opposing fracture surfaces were not matching. They studied SCC of cast 99.99% pure Mg in aqueous 3.3%NaCl + 2%K_2CrO_4, by cantilever bending of specimens at various deflection rates up to high rates. They proposed SCC occurred by localised micro-void coalescence resulting from dislocations at crack tips, due to weakening of inter-atomic bonds by adsorbed H, in a mechanism similar to that for liquid metal embrittlement. They suggested tubular voids were nucleated at intersections of $\{0001\}$ and $\{10\bar{1}X\}$ slip bands, resulting in a fluted fracture surface parallel to $\{10\bar{1}X\}$ planes (Fig. 9.2). There was fractographic evidence of SCC for crack velocities between 10^{-8} and 5×10^{-2} m/s. They proposed SCC was induced by H adsorption, since insufficient time was available for H diffusion or localised dissolution at the fastest crack velocities.

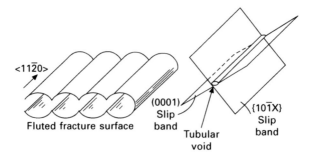

9.2 Formation of fluted fracture topography by localised micro-void coalescence. Reprinted from [75].

Table 9.2 SCC threshold stress for common Mg alloys [4]

Alloy, environment	σ_{SCC}
HP Mg, 0.5%KHF$_2$	60%YS
Mg2Mn, 0.5%KHF$_2$	50%YS
MgMnCe, air, 0.001N NaCl, 0.01Na$_2$SO$_4$	85%YS
ZK60A-T5, rural atmosphere	50%YS
QE22, rural atmosphere	70–80%YS
HK31, rural atmosphere	70–80%YS
HM21, rural atmosphere	70–80%YS
HP AM60, distilled water	40–50%YS
HP AS41, distilled water	40–50%YS
AZ31, rural atmosphere	40%YS
AZ61, costal atmosphere	50%YS
AZ63-T6, rural atmosphere	60%YS
HP AZ91, distilled water	40–50%YS

9.2.2   Alloy composition

Table 9.2 presents values of the SCC threshold stress for common Mg alloys [4].

Aluminium

Al in Mg alloys increases strength and fluidity in casting. Mg-Al alloys are susceptible to SCC [78, 80–86] in air, distilled water and chloride-containing solutions. Busk [80] reported that susceptibility increased as the Al content increased from 1% to 8%. Miller [84] showed SCC in distilled water for AZ91, AM60 and AS41. The SCC threshold stress was ~50% of yield strength for all three alloys. There was the same SCC behaviour in high purity AZ91 and AZ91-0.01%Fe (AZ91 containing a high iron content of 0.01%). Arnon and Aghion [85] found SCC in AZ31 in 0.9% NaCl. Chen *et al.* [86] observed TGSCC for AZ91 in 0.5 M MgCl$_2$, with the threshold stress ~1/3 yield stress. Kannan and Raman [87] found small decreases of strength and ductility for AZ91 in a simulated body fluid using CERT; these decreases usually indicate SCC although the decreases were claimed to be insignificant.

Zinc

Miller [74] reported that Zn increased SCC susceptibility. Mg-Zn alloys containing rare earths, such as the ZExx alloys, are considered to have moderate SCC susceptibility relative to Mg-Al-Zn alloys. The AZxx alloys contain both Al and Zn and are considered particularly susceptible in air and chloride-containing solutions. They are also the most common, which is one reason why many studies have focused on Mg-Al alloys [88–93, 127], Mg-Al-Zn alloys [81, 83, 94–98] and pure Mg [75–78]. Ben-Hamu [99] showed that Mg-Zn-Mn alloys were susceptible in 3.5 wt% NaCl saturated with $Mg(OH)_2$.

Manganese

Mg-Mn alloys were considered immune to SCC in the atmosphere, chloride solutions and chloride-chromate solutions [74], but were susceptible (i) in the atmosphere and in distilled water [100], (ii) in solutions containing chloride and sulphate ions [100], and (iii) Mn-2%Mn-0.5%Ce showed SCC in distilled water and 0.5% KHF [93]. Timonova [101] stated that addition of Mn or Zn to Mg-8%Al decreased susceptibility; however, addition of both elements increased susceptibility.

Rare earths

Mg alloys incorporate rare-earth (RE) elements [102] to improve (i) creep resistance, which is primarily achieved by RE-containing phases along grain boundaries [103, 104], (ii) castability, (iii) age hardening [105], and corrosion resistance [106–108]. Rokhlin [102] reported that Nd addition to Mg-Zn-Zr increased SCC resistance. Kannan *et al.* [53] found SCC for three RE alloys (ZE41, QE22 and EV31A) in 0.5 wt% NaCl and in distilled water. TGSCC in ZE41 and QE22 in distilled water was consistent with HE, whereas IGSCC in ZE41, QE22 and EV31A in 0.5 wt% NaCl was attributed to micro-galvanic corrosion associated with the second phase along grain boundaries (Fig. 9.3).

Iron

Fe is present as an impurity in Mg alloys and significantly increases the corrosion rate [1, 2, 21, 77] above the tolerance limit. Perryman [93]

9.3 SCC of RE containing Mg alloys in 3.5% NaCl. (a) ZE41:
predominantly IGSCC with isolated TGSCC, arrows; (b) EV31: IGSCC
with some TGSCC, arrow; and (c) QE22: IGSCC and TGSCC, arrow
[53]. The arrows point to transgranular features. Reprinted from [53].

reported that the SCC susceptibility in distilled water of Mg-5Al+0.13%Fe was higher than for Mg-5Al+0.0019% Fe. Similarly, Pelensky and Gallaccio [82] reported that SCC susceptibility for Mg-5Al alloys increased with Fe concentration and Pardue *et al.* [98] reported that the fraction of TGSCC in AZ61 increased with Fe concentration. In contrast, Fairman and Bray [83] found that Fe has minimal effect on SCC. Similarly, Miller [84] found equivalent SCC susceptibility in distilled water for low purity (0.010%Fe) and HP AZ91 and Timonova [101] reported that Fe had no effect on the SCC of Mg-Al-Zn-Mn alloys.

Other elements

Reports are varied on other elements. Li, Ag, Nd, Pb, Cu, Ni, Sn and Th have little influence on SCC susceptibility [74, 77]. Busk [80] stated that cast Mg-Zr alloys have negligible SCC whereas the *ASM Handbook* [77] states that alloys containing Zr and REs have intermediate SCC susceptibility in atmospheric environments. Cd, Ce and Sn may increase susceptibility in certain alloys [77, 101]. Conversely, Rokhlin [109] reported that addition of Cd and Nd to Mg-Zn-Zr increased SCC resistance.

9.2.3   Mg-Al alloys

Winzer *et al.* [34] found SCC for AZ91, AZ31 and AM30 in distilled water using constant extension rate tests (CERT) and linearly increasing stress tests (LIST) [54, 63] (Fig. 9.4). Thus water itself is the key environment factor and thus a TGSCC tendency is expected in all aqueous solutions unless there is clear contrary evidence. AZ91 is a two-phase alloy containing an α-Mg matrix and significant amounts of the β-phase. AZ31 is largely a single-phase alloy with an α-matrix of composition similar to that of the α-phase of AZ91. AM30 is also a single α-phase alloy, with Al content similar to AZ31, but AM30 contains Mn instead of Zn.

The threshold stress, σ_{SCC}, was measured using a DC potential drop (DCPD) technique (Fig. 9.5). The threshold stress was 55–75 MPa for AZ91, 105–170 MPa for AZ31 and 130–140 MPa for AM30. SCC susceptibility increased with decreasing strain rate. The low σ_{SCC} of AZ91 is attributed to: (i) the tendency for the β-particles to fracture; and (ii) their behaviour as reversible H traps, which enhance H transport within the matrix. This implies that the increasing susceptibility of Mg alloys to SCC with increasing Al concentration is related to β-particles. Generalisation implies low SCC resistance for two-phase Mg alloys if there is an H influence on the fracture of the second phase.

The stress corrosion crack velocities for AM30 ($V_c = 3.6 \times 10^{-10} - 9.3 \times 10^{-10}$ m/s) were slower than for AZ91 ($V_c = 1.6 \times 10^{-9} - 1.2 \times 10^{-8}$ m/s)

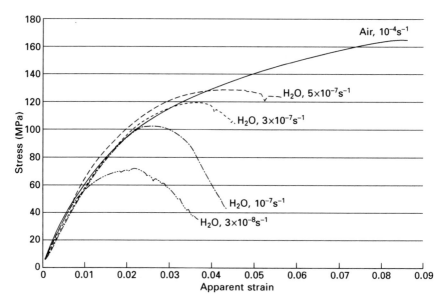

9.4 Typical stress vs apparent strain curves for AZ91 in distilled water and air under CERT conditions showing a decrease in apparent ductility with decreasing applied strain rate indicative of stress corrosion cracking. Reprinted from [34].

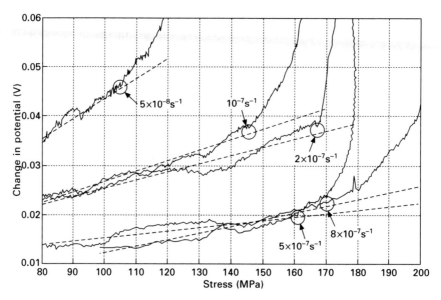

9.5 DCPD vs stress curves for AZ31 in distilled water and air under CERT conditions showing threshold stress values. The technique works equally well for LIST and similar threshold values are measured. Reprinted from [34].

and AZ31 ($V_c = 1.2 \times 10^{-9} - 6.7 \times 10^{-9}$ m/s). This is consistent with lower H diffusivity in the α-phase in the absence of Zn.

9.2.4   Cast alloys

Research at Dow [110] showed that the SCC susceptibility for sand-cast AZ63 in a rural atmosphere was lower than for extruded or rolled AZ61. This may be due to the residual stresses resulting from extrusion or rolling. Stephens *et al.* [111] reported that cast AZ91E-T6 was highly susceptible to SCC in 3.5% NaCl. SCC resistance is reduced by residual tensile stress [77]. Timonova *et al.* [112] found reduced SCC resistance for Mg-Y-Zn alloys in NaCl following hot deformation, and that resistance was partly recovered by annealing, which was attributed to the removal of residual tensile stresses.

9.2.5   Crack path

It can be rationalised from the literature that (i) there are different mechanisms for IGSCC and TGSCC, and (ii) the operative mechanism is determined by the microstructure and the environment. Pardue *et al.* [98] proposed that TGSCC is discontinuous and involves alternating fracture and dissolution, whereas IGSCC is continuous and completely electrochemical. Stampella *et al.* [78] observed TGSCC for fine grained (25 μm) 99.5% commercial-purity Mg and TGSCC-IGSCC for large grained (75 μm) 99.95% HP Mg. IGSCC was attributed to the large grains being subject to higher stresses across grain boundaries, due to more concentrated dislocation pile-ups. In contrast, Meletis and Hochman [76] reported TGSCC for 99.9% purity Mg, heat-treated and furnace cooled to produce large grains. Stampella *et al.* [78] used a dilute 10^{-3} Na_2SO_4 solution whilst Meletis and Hochman [76] used a chloride-chromate solution, suggesting that crack morphology is influenced by the environment. This is supported by Perryman [93] who showed that Mg-5Al evinced TGSCC in saturated $MgCO_3$ solution, 0.5% KF solution, 0.5% KHF solution and 0.5% HF solution, whilst IGSCC in 0.05% potassium chromate solution; and Mears *et al.* [113] found IGSCC for Mg-6.5Al-1Zn in the pH 5.0 chloride-bichromate solution (3.5%NaCl + 2.0%$K_2Cr_2O_7$) and TGSCC in the pH 8.1 chloride-chromate solution (3.5%NaCl + 4.0%K_2CrO_4.).

Priest *et al.* [81] and Fairman and Bray [97] found IGSCC for Mg-Al-Zn and Mg-Al alloys, heat treated to produce fine grains (~5 μm) and heavy $Mg_{17}Al_{12}$ precipitation at grain boundaries. Fairman and Bray [97] stated that Mg-Al alloys are more inclined to IGSCC if the Al content is greater than 6% when IGSCC occurred by preferential corrosion of the Mg matrix adjacent to the continuous $Mg_{17}Al_{12}$ precipitate at grain boundaries (Fig. 9.6). Miller [74] and Pardue *et al.* [98] attribute all IGSCC to this micro-galvanic

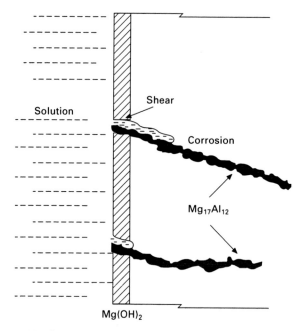

9.6 Preferential corrosion of metal matrix adjacent to $Mg_{17}Al_{12}$
[97]. A continuous or nearly continuous second phase, typically
along grain boundaries, causes IGSCC by micro-galvanic corrosion
of the adjacent Mg matrix. IGSCC is expected in all such alloys,
typical of most creep resistant alloys. All Mg alloys with nearly
continuous second phases are expected to be prone to IGSCC
because each known second phase has a more positive corrosion
potential than the matrix α-Mg; the degree of severity depends
on the electrochemical properties of the second phase; these
electrochemical properties need to be studied. Nearly continuous
second phases can be avoided by appropriate Mg alloy design.
Reprinted from [97].

corrosion. Priest *et al.* [81] found that maximum grain size for IGSCC was
~28 μm regardless of the presence of $Mg_{17}Al_{12}$. They proposed that, where
both IGSCC and TGSCC are possible, due to the presence of $Mg_{17}Al_{12}$ and a
source of H, increasing the grain size causes a transition to TGSCC because
the crack may propagate along a more direct path, or HE might be easier
with large grains.

Fracture is ductile in air for pure Mg and Mg alloys. In contrast, Meletis
and Hochman [76] reported cleavage for SCC of pure Mg in NaCl + K_2CrO_4
solution. Cleavage consisted of long, parallel facets separated by perpendicular
steps with jogs at edges. The facets and steps corresponded to $\{2\bar{2}03\}$
planes, with the ledges of each step oriented in the $<10\bar{1}1>$ direction. Similar
fractography was reported by Pugh and co-workers [89, 91, 92] for Mg-7.5Al
in NaCl + K_2CrO_4 solution, but the cleavage facets corresponded to $\{31\bar{4}0\}$

planes. Pugh and co-workers also observed (i) opposite fracture surfaces were interlocking, and (ii) fine parallel markings ~1 μm apart, interpreted as crack arrest markings, within some cleavage-like facets. Fairman and West [95] reported similar markings. Quasi-cleavage fracture for SCC of Mg alloys has also been reported by Wearmouth *et al.* [88] and Nozaki *et al.* [94]. In marked contrast is the fracture observed by Lynch and Trevena [75] for pure Mg in a NaCl + K_2CrO_4 solution (Fig. 9.2), which was ductile on the micro-scale.

Winzer *et al.* [55] found different SCC fracture morphologies for Mg alloys for different combinations of alloy and environment. The diversity of SCC fracture morphologies is attributed to various SCC mechanisms. For example, Winzer *et al.* [55] found that the fracture surfaces for AZ31 consisted of relatively smooth regions containing small, elongated dimples, whereas those for AM30 consisted of cleavage-like markings.

Fractography has contributed to the discussion regarding hydrides. Early works [8, 76, 89, 91, 92] proposed that Mg SCC involved delayed hydrogen cracking (DHC). The plausibility of this mechanism was reinforced by the cleavage-like fracture surfaces, as such features were also related to hydrides in other metals.

9.2.6   Microstructure

The cooling rate from elevated temperatures influences SCC by the precipitation of β-particles ($Mg_{17}Al_{12}$) at grain boundaries. $Mg_{17}Al_{12}$ occurs in Mg alloys with more than 2% Al [114] and $Mg_{17}Al_{12}$ is precipitated by slow cooling from the solution treatment temperature [74] and slow cooling of castings [18]. Thus, Priest *et al.* [81] found TGSCCC in chloride-chromate solutions for Mg-6Al-1Zn heat-treated for 24 h at 345°C and water quenched, whereas there was IGSCC for furnace-cooled samples, which had $Mg_{17}Al_{12}$ precipitates. Similarly, Pardue *et al.* [98] found IGSCC in a chloride-chromate solution for wrought AZ61 heat treated at 345, 425 or 480°C for 24 h and furnace cooled to have a high concentration of $Mg_{17}Al_{12}$ at grain boundaries, whereas there was a significant TGSCC for samples water quenched from the heat treatment temperature. Similarly Kiszka [115] found IGSCC in humid air for Mg-14Li with more than 1% Al, rapidly cooled after hot-working at 370°C. The IGSCC susceptibility was associated with the Al content and the formation of a second phase during rapid cooling.

Stress relief by T4 and T5 heat treatment was noted by Busk [80] to restore the threshold stress reduced by residual tensile stresses. Similarly, Timonova [101] stated that relieving residual stresses in MA5 by annealing significantly increased SCC resistance, with resistance increasing with annealing temperature.

Precipitation heat treatment was found by Speidel *et al.* [116] to have no influence on the threshold stress intensity factor and stage 2 crack velocity for all high strength Mg alloys, including ZKxx. This would imply that SCC of high strength Mg alloys does not depend on aging and also does not depend on slip morphology as is the case for high strength steels and high strength aluminium alloys.

9.2.7   Welding

Engineering structures are often fabricated using welding. Welding changes the microstructure, and the weld metal can have a composition significantly different from the parent metal. There is extensive research on Mg welding [117–119]. Laser beam (LB) welding has the following advantages: (i) high welding speed; (ii) narrow joints with reduced heat affected zone (HAZ) and distortion; and (iii) a high joint efficiency (joint strength/base metal strength) approaching 100% [117], whereas conventional welding, such as tungsten inert gas (TIG) welding, has a joint efficiency of ~70–90% [118]. Friction stir welding (FSW) [120] achieves a metallic bond at temperatures below melting and thereby avoids melting-related issues [121].

A few researchers [122–124] have studied SCC of welded Mg. Winzer *et al.* [124] reported high SCC susceptibility of TIG welds of AZ31 sheet. SCC initiated at the interface between the weld and the HAZ. Kannan *et al.* [122] also reported high SCC susceptibility of LB-welded AZ31. SCC occurred in the fusion boundary (Fig. 9.7(a)). SCC was attributed to galvanic corrosion between the weld and the base metal. The fracture was mixed IGSCC and TGSCC (Fig. 9.7(b)). Kannan *et al.* [123] also reported high SCC susceptibility of FSW AZ31. Srinivasan *et al.* [125, 126] found that plasma electrolytic oxidation (PEO) coating did not improve SCC resistance for LB-welded AZ31 and FSW AZ61.

9.3   Influence of loading

9.3.1   Fracture mechanics

Usually, stress corrosion cracks are considered macroscopically brittle, i.e. they occur at stresses below general yield and propagate in an essentially elastic body, even though local plasticity may be necessary for cracking. Linear elastic fracture mechanics is used to study SCC. The crack tip stress intensity factor in the opening mode, K_I, represents the stress field in the vicinity of the crack tip, and the mechanical driving force for crack growth. The stress intensity factor at the crack tip is given by:

$$K_I = Y \sigma \sqrt{\pi a} \qquad\qquad 9.8$$

where Y is a geometric factor, σ is the applied stress and a is the crack

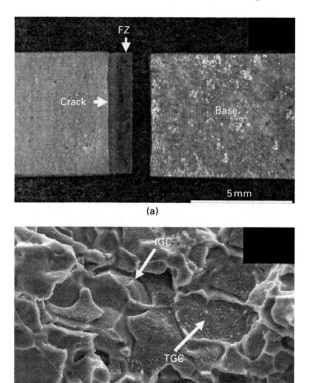

9.7 (a) overview of LB-AZ31 sample after SCC test showing SCC at the interface between weld and base material, and (b) the fracture surface showed IGSCC and TGSCC.

length. In air, fracture is expected when K_I exceeds a critical value, the fracture toughness, K_{Ic}. Figure 9.8 provides typical stress corrosion crack velocity data [116]. The threshold for SCC, K_{ISCC}, is the stress intensity factor above which occurs the first measurable crack extension; the growth rates are typically above 10^{-10} m/s. The typical curve has three stages, although region 3 is not always observed; for example, region 3 is absent from Fig. 9.8 [116]. During stage 1, the crack velocity increases rapidly as the stress intensity factor increases above K_{ISCC}. During stage 2, the crack velocity is essentially constant. In stage 3, as the stress intensity factor approaches K_{Ic}, the crack propagation mechanism becomes dominated by ductile tearing and rapid crack growth ensues until failure occurs [7, 116]. Ebtehai *et al.* [127] have also shown that environmental conditions can influence the steady-state crack velocity and the threshold values.

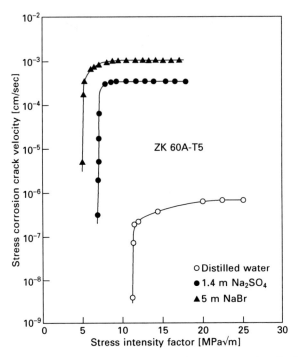

9.8 Effect of the stress intensity factor on crack velocity for ZK60 alloy in various environments. Reprinted from [116].

K_{ISCC} is useful as it predicts the combinations of stress and flaw size that leads to SCC. K_{ISCC} may be used as a design criterion for ensuring no SCC in service, provided that the stress, minimum detectable flaw size, and environmental conditions are well defined, and that the service loads are essentially sustained, i.e. that cyclic loading is not significant. Figure 9.9 illustrates how K_{ISCC} can be used to assess structural integrity. In an inert environment, the mechanical limits are the yield stress, σ_y, and the fracture toughness, K_{1C}. In a SCC environment, these limits may be reduced to the threshold stress intensity factor, K_{1SCC}, and the threshold stress, σ_{SCC}. These two parameters define a region without SCC. For small cracks, the K concept is non-conservative. A combination of the SCC threshold stress, σ_{SCC} and K_{ISCC} may serve as a design criterion defining a minimum crack size for the use of the K concept.

9.3.2   Stressing

The threshold stress σ_{SCC} is typically approximately half the yield strength [4, 77, 88, 127] (Table 9.1). K_{1SCC} may be determined from data as in Fig. 9.8. σ_{SCC} may be determined by a potential drop technique associated with LIST

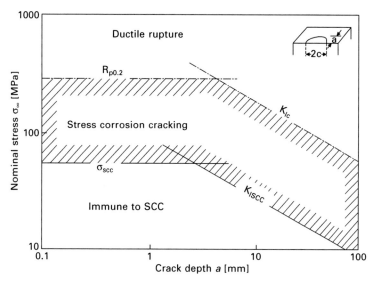

9.9 Mechanical limits of a typical system involving an inert and a SCC environment. Reprinted from [4].

or CERT as in Fig. 9.5. A correlation between σ_{SCC} and K_{1SCC} was found by Wearmouth *et al.* [88] for Mg-7Al in 3.5% NaCl + 2% K_2CrO_4; and by Ebtehaj *et al.* [127] for Mg-9Al in aqueous solutions with 35 g/L NaCl and various K_2CrO_4 concentrations. Ebtehaj *et al.* [127] investigated the influence of strain rate on SCC susceptibility for cast Mg-9Al. They proposed that the SCC mechanism involved diffusion of cathodically generated H, and that the SCC susceptibility is defined by opposing effects relating to H ingress as suggested by Wearmouth *et al.* [88]. At slow strain rates, film integrity was maintained preventing H ingress and ductile fracture ensued. As the strain rate was increased, the effectiveness of repassivation was reduced, allowing H to ingress more freely. At high strain rates, ductile tearing occurred before embrittlement, because insufficient time was available for H ingress. Maximum susceptibility occurred at an intermediate strain rate (Fig. 9.10). Furthermore, the strain rate corresponding to maximum susceptibility was dependent on the balance between active and passive corrosion (as defined by the ratio of chloride and chromate ion concentrations) with this effect becoming less significant at high strain rates. As the chromate concentration increased, the tendency for repassivation increased, so higher strain rates were required to overcome repassivation and the overall resistance to cracking increased. This explains the locations of the minima with respect to strain rate for the curves representing 5, 20 and 35 g/L K_2CrO_4 concentration.

Figure 9.4 in contrast showed that SCC susceptibility for AZ91 in distilled water [34] continued to increase with decreasing strain rate. At these strain rates, the influence of repassivation at the crack tip is negligible. In contrast,

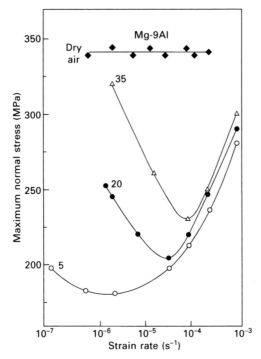

9.10 Effect of strain rate on SCC susceptibility for 5, 20 and 35 g/L K_2CrO_4 and 5 g/L NaCl. Reprinted from [127].

the data of Fig. 9.10 used strongly passivating solutions. This indicates that the occurrence of maximum SCC susceptibility at intermediate strain rates for Mg alloys in strongly passivating solutions (Fig. 9.10), is characteristic of these environments rather than a characteristic of Mg alloys. Nozaki *et al.* [94] also found that SCC susceptibility increased continuously as the strain rate was decreased from $8.3 \times 10^{-4} s^{-1}$ to $8.3 \times 10^{-7} s^{-1}$ for AZ31B in distilled water (Fig. 9.11). SCC susceptibility was characterised by a SCC susceptibility index, I_{SCC}, given by

$$I_{SCC} = \{(E_{oil} - E_{SCC})/E_{oil}\} \times 100\% \qquad\qquad 9.9$$

where E_{SCC} and E_{oil} are the area under the stress–strain curve in the SCC solution and in oil, respectively, the latter being inert. I_{SCC} is high for a system with high SCC susceptibility whereas I_{SCC} tends to zero for a system with low susceptibility. Figure 9.10 shows that susceptibility in distilled water increased from 0 to 85% as the strain rate was decreased from $8.3 \times 10^{-4} s^{-1}$ to $8.3 \times 10^{-7} s^{-1}$. For NaCl concentrations of 4 and 8%, susceptibility was around 90%, independent of strain rate.

Makar *et al.* [114] found also that the SCC velocity increased continuously with increasing loading rate for rapidly solidified (RS) Mg-1Al and RS

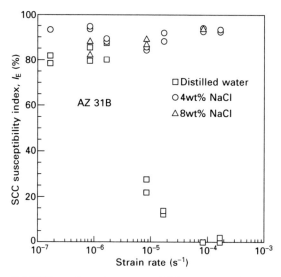

9.11 Effect of strain rate on SCC susceptibility index for AZ31B in distilled water and NaCl solutions. Reprinted from [94].

Mg9Al in 0.21 M K_2CrO_4 + 0.6 M NaCl. Markar *et al.* indicated that ductile tearing became increasingly dominant, as the loading rate increased from 4.8×10^{-3} mm·s^{-1} to 8.9×10^{-3} mm·s^{-1}, above which the fracture surfaces were mostly ductile. This could explain the high rates and ductile tearing mechanism observed by Lynch and Tevena [75].

9.3.3   Incremental loading

Winzer *et al.* [54] compared the linearly increasing stress test (LIST) and the constant extension rate test (CERT) in the evaluation of TGSCC of AZ91 in distilled water and 5 g/L NaCl. The LIST apparatus [63] is illustrated in Fig. 9.12. The specimen is attached to one end of the lever arm. To the opposite end of the arm a known mass is attached such that the tensile load applied to the specimen increases linearly as the distance between the fulcrum and the mass is increased by means of a screw thread and synchronous motor. LIST is load controlled, whereas CERT is extension controlled. They are essentially identical until SCC initiation. Thereafter, LIST ends as soon as a critical crack size is reached, whereas CERT can take much longer as typically CERT only ends when the final ligament suffers ductile rupture.

The LIST and CERT techniques are both useful in identifying SCC occurrence. Coupled with a technique for identifying crack initiation, both LIST and CERT can measure the threshold stress (Fig. 9.5), and crack velocity from the crack size divided by the time for cracking. The increased crack

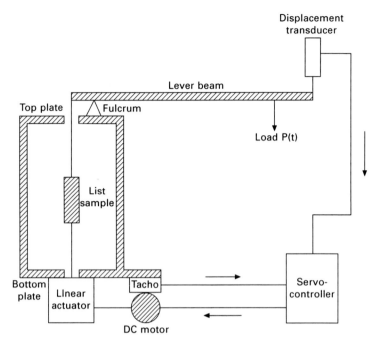

9.12 Schematic of LIST apparatus. Reprinted from [63].

propagation time under CERT conditions may be important in determining the mechanisms for SCC in Mg alloys. However, the stress corrosion crack growth occurs under conditions of decreasing load, and the fractography towards the end of a CERT may not be typical of that in the earlier stages of SCC.

9.4   Environmental influences

9.4.1   Gaseous hydrogen

Mg alloys have shown SCC in gaseous H [34, 92, 127]. Figure 9.13 shows the influence of H_2 pre-charging. The specimen fractured just above the yield stress, σ_Y, and without the apparent plastic strain or reduction in load that typically characterises SCC of AZ91 in aqueous solutions under CERT conditions (see Fig. 9.4). The control sample, which was exposed to gaseous Ar at 3 MPa for 15 h at 300°C, had a UTS and ductility comparable to that of the sample tested in air. Thus HE was responsible for the large reductions in UTS and elongation-to-failure for the specimen pre-charged in gaseous H_2.

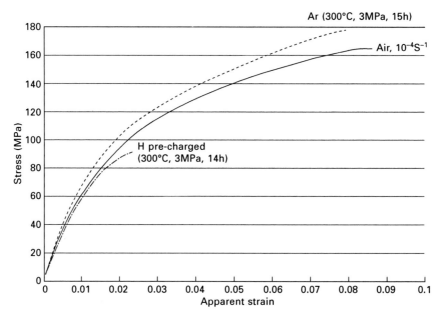

9.13 Stress vs apparent strain curves for AZ91 pre-charged in gaseous H_2 and Ar at 3 MPa and 300°C and fractured rapidly in air, compared with a non-charged specimen fractured in air. Reprinted from [34].

9.4.2   Air

Dry air is generally considered inert for Mg [75, 78, 83], whereas SCC is reported in outdoor atmospheres. Pelensky and Gallaccio [82] reported SCC for all AZxx alloys in outdoor exposure, including marine and rural atmospheres, with susceptibility increasing with increasing periods of rain, humidity or high temperature. Furthermore, AZ61 was susceptible to SCC at 98 to 100% relative humidity. Similarly, Loose and Barbian [128] reported SCC of AZ61 in annealed and as-rolled conditions in rural and coastal atmospheres. Kiszka [115] reported SCC for rapidly cooled Mg-14Li alloys containing 1% or 1.5% Al in humid air. In contrast, Pelensky and Gallacio [82] reported that M1 (Mg-1.2Mn) was immune to SCC in marine air.

9.4.3   Solution composition

A vast literature deals with SCC in solutions containing chloride and chromate ions. Chromate ions inhibit corrosion by protective film growth. Chloride ions cause film breakdown and H evolution, which facilitates SCC [88, 114, 127]. Ebtehaj *et al.* [127] reported that SCC was a maximum when the ratio of chloride ions to chromate ions was approximately 1:2. In these solutions, SCC required a balance between active and passive behaviour: chromate-only solutions result in complete passivity preventing localised corrosion which was essential for H ingress; whereas chloride-only solutions result in excessive general corrosion which outran crack growth. Maximum susceptibility occurred at intermediate chloride/chromate ratios. This was contradicted by Pelensky and Gallaccio [82], who found SCC susceptibility of AZ61 increased as the concentration of K_2CrO_4 increased from 3 to 200 g/L in 35g/L NaCl and, similarly, susceptibility increased as the concentration of NaCl increased from 40 to 200 g/L in 5 g/L K_2CrO_4 solution.

However, the relationship between strain rate and environment varies with system. Winzer *et al.* [34] found that SCC susceptibility of AZ91, AZ31 and AM30 in distilled water increased continuously with decreasing strain rate (Fig. 9.4). Nozaki *et al.* [94] similarly reported an increase in susceptibility index from 0 to 85% for distilled water for strain rates between $8.3 \times 10^{-5} s^{-1}$ and $8.3 \times 10^{-7} s^{-1}$ (Fig. 9.11), but no significant variation in the SCC susceptibility index in NaCl solutions with concentrations between 2 and 8%.

Fairman and Bray [83] found that additions of $NaNO_3$ or Na_2CO_3 to 4%NaCl + 4%Na_2CrO_4 increased SCC resistance for pure Mg and various Mg-Al, Mg-Al-Fe and Mg-Al-Zn alloys, heat treated to ensure TGSCC. They proposed that $NaNO_3$ or Na_2CO_3 improved repair of surface film defects. Similarly, inhibition of SCC for Mg-6Al in chloride-chromate solutions by NO_3^- ions was found by Frankenthal [129].

Timonova [101] reported SCC for MA3 in sodium carbonate solution, with susceptibility constant for concentrations between 0.005 and 0.15 M, and increasing with concentrations between 0.15 and 1.0 M.

Pelensky and Gallaccio [82] carried out an extensive assessment of SCC of pure Mg and Mg-Mn, Mg-Al, Mg-Al-Zn, Mg-Zn-Zr, Mg-Al-Mn and Mg-Li alloys in air, water and numerous aqueous solutions. Most alloy-environment combinations resulted in SCC. The following trends were observed:

- Mg-5Al failed in distilled water.
- ZK60A-T5 (Mg-5.5Zn-0.45Zr) failed rapidly in distilled water and sea-water.
- ZK60A-T5 stressed to 90% of its yield strength failed rapidly in KCl, CsCl, NaBr, NaCl and NaI solutions.
- AZ61 failed in Na_2CO_3 solution with susceptibility constant for concentrations between 0.265 and 15.9 g/L, and increasing with concentration between 15.9 and 53 g/L.
- Susceptibility of AZ61 and Al-1.5Mn in NaCl increased with solution concentration.
- AZ61 was susceptible in NaCl + K_2CrO_4, K_2CrO_4 + Na_2SO_4, Na_2SO_4, $NaNO_3$, Na_2CO_3, NaCl, $NaCH_3COO$ (in order of decreasing susceptibility).

- High purity Mg and Mg-2Mn failed in KHF_2.
- AZ61 did not fail in $K_2CrO_4 + NaCH_3COO$, $K_2CrO_4 + Na_2CO_3$ or $K_2CrO_4 + NaNO_3$ solutions.

9.5   Mechanisms

Mg SCC has been generally attributed to (i) continuous crack propagation by dissolution at the crack tip, or (ii) discontinuous crack propagation by a series of mechanical fractures at the crack tip (Fig. 9.1) [4, 5, 74, 78, 130]. Lynch and Travena [75] suggested that the mechanism is similar to that occurring during liquid metal embrittlement (LME). Dissolution mechanisms include preferential attack, film rupture and tunnelling, whilst mechanical mechanisms include cleavage, HE and LME. IGSCC occurs typically by micro-galvanic acceleration of the Mg matrix adjacent to cathodic grain boundary precipitates (Figs 9.3 and 9.6). For TGSCC, dissolution cannot explain the crack propagation rates and fracture surfaces, particularly the interlocking fracture surfaces of Pugh and co-workers [8, 89, 91, 92]. There is considerable evidence for H enhanced decohesion (HEDE) and H enhanced localised plasticity (HELP). H adsorption induced plasticity may also play a role, particularly at the high crack velocities reported by Lynch and Trevena [75].

9.5.1   Micro-galvanic corrosion

Dissolution of the Mg matrix adjacent to $Mg_{17}Al_{12}$ grain boundary precipitates causes IGSCC in Mg-Al alloys (Fig. 9.3). The precipitates appear during slow cooling of alloys with Al concentrations greater than 2% [97]. Cracking is driven by the potential difference between $Mg_{17}Al_{12}$ and the matrix, which is ~ 300 mV [2, 97]. Stress pulls apart opposite crack surfaces and allows solution access to the crack tip. Pardue *et al.* [98] showed continuous cracking, assuming a continuous grain boundary precipitate, as indicated by a lack of signals in acoustic emission studies. Fairman and Bray [97] reported some discontinuities in IGSCC, attributing growth discontinuities to crack intersection with unfavourably oriented grains. A similar IGSCC mechanism is expected for other Mg alloys, because all second phases tend to accelerate the corrosion of the adjacent Mg matrix [2].

9.5.2   Dissolution

Logan [131] proposed an alternative dissolution mechanism. He postulated that surface film rupture produces an electrochemical cell between the anodic film-free area and the cathodic filmed area. Stress concentration at the crack tip was postulated to prevent the film from reforming such that crack propagation is continuous. Logan [131] proposed that TGSCC was electrochemical because of the observation that cathodic polarisation prevented SCC. However, Logan calculated, from the Faraday law, that the observed crack propagation rates (10×10^{-6} m/s) required an effective current density of 14 A/cm^2. Similarly Pugh *et al.* [132] observed crack velocities between 6×10^{-3} and 40×10^{-6} m/s for Mg-7.6Al in a chloride-chromate solution, which corresponded to current densities between 8 and 58 A/cm^2. Such current densities were considered by Pugh *et al.* to be prohibitively high.

9.5.3   Cleavage

Mg TGSCC has been associated with discontinuous cleavage. Pardue *et al.* [98] and Chakrapani and Pugh [89] showed that stages of crack propagation coincided with discrete spikes in acoustic emission. Unstressed specimens in solution emitted a steady acoustic signal, which was associated with H evolution from pitted areas, whilst specimens undergoing plastic deformation in air emitted no signal. Stressed specimens in solution emitted discrete acoustic signals superimposed on the continuous H signal, indicating discontinuous crack advance. Pardue *et al.* [98] also observed that peaks of electrochemical anodic current corresponded to crack advance and the exposure of fresh metal.

Chakrapani and Pugh [89] determined, from measurements of the distance between acoustic emission spikes, that the average crack velocities in Mg-7.6Al were between 5 and 30×10^{-6} m/s. Similarly, using a travelling microscope Pugh *et al.* [132] determined that, for the same alloy, average crack velocities were between 6 and 40×10^{-6} m/s. Such velocities could not be reconciled with dissolution models according to the Faraday law, adding support to a mechanical fracture model.

Various workers [8, 76, 89, 91] have reported that TGSCC in Mg results in fracture surfaces consisting of flat, parallel facets separated by perpendicular steps, consistent with cleavage. The steps and facets are generally parallel to the direction of crack propagation and change direction at grain boundaries [76]. Opposite fracture surfaces are matching and interlocking, which is also consistent with the occurrence of cleavage and is difficult to explain by a dissolution model [8]. The fracture surfaces show numerous jogs resulting from overlap of parallel cleavage cracks, further evidencing discontinuous propagation. That there can be a ductile component of the fracture mechanism was shown by Lynch and Trevena [75] who observed concave features on opposing fracture surfaces for higher crack propagation velocities, such that the opposing fracture surfaces were not interlocking.

9.5.4   Role of H

Chakrapani and Pugh proposed that TGSCC for hot-rolled Mg-7.5Al in NaCl-K_2CrO_4 solution occurred by discontinuous cleavage evidenced by the stepped fracture surface topography; they also observed that H evolution invariably occurred at corrosion pits [89, 91]. Subsequent work [92] proposed that the cleavage-like fracture surfaces could be due to HE. Specimens stressed in gaseous H, or exposed to aqueous solution prior to stress application, exhibited a loss of ductility and cleavage-like fracture surfaces. The role of H was further evidenced by (i) inert gas fusion methods, which showed that the H concentration of the specimens progressively increased with time, and (ii) the fact that vacuum annealing partially reversed the effects of H exposure.

Chakrapani and Pugh [92] inferred that the mechanism for TGSCC growth was repeated cycles of H diffusion to the crack tip region followed by brittle fracture. This was supported by discontinuous acoustic emission signals. It was also suggested that the role of H in the brittle fracture could be to form brittle hydrides or to produce decohesion. They inferred that, since SCC fractures tend to occur on $\{31\bar{4}0\}$ planes, these planes may correspond to the habit or cleavage planes of a hydride. Fracture surfaces for SCC and pure HE systems were different: the latter tended to be flatter and without the pleated/stepped structure. It was speculated that this could be related to H fugacity and H entry kinetics, and the fact that, for SCC conditions, dissolution occurs at the crack tip.

Meletis and Hochman [76] reported that TGSCC initiation and propagation for pure Mg in a chloride-chromate solution was accompanied by H evolution. Using electron channeling pattern analysis and SEM photogrammetry they showed that cracks propagated primarily by cleavage on $\{2\bar{2}03\}$ planes as evidenced by parallel facets. These facets were separated by steps also on $\{2\bar{2}03\}$ planes. The authors proposed that cleavage occurs by reduction in surface energy of $\{2\bar{2}03\}$ planes by HE resulting from preferential H accumulation or hydride formation on these planes. H may be absorbed from the solution at the crack tip or transported to the region by dislocation motion.

Bursle and Pugh [8] also concluded that TGSCC in Mg-Al alloys occurs by discontinuous cleavage induced by HE. They rejected the adsorption model (reduction of metal inter-atomic bond strength at the crack tip by the interacting of absorbed ions) on the basis that adsorption would only affect a few atomic layers ahead of the crack tip, and therefore would not cause discontinuous crack propagation involving the distances between crack arrest markings observed by SEM. Adsorption induced propagation would be macroscopically continuous at a rate determined by the transport of ions to the crack tip. Dealloying models were also rejected on the basis that solutions that cause dealloying are not usually associated with TGSCC and there has been no correlation reported between dealloying and cleavage.

9.5.5   H adsorption

Lynch and Travena [75] carried out metallographic and fractographic observations of crack growth in pure Mg in dry air, aqueous and liquid metal environments. Crack growth in dry air at ambient temperatures was macroscopically brittle, but fracture surfaces were microscopically fluted or dimpled. Fluted fractures and dimpled intercrystalline facets were also produced by SCC and LME, with flutes and dimples smaller and shallower than those produced in air. Cleavage fractures were also observed after SCC and LME. The close similarities between SCC and LME suggested that absorbed H (rather than solute H, hydrides or localised dissolution) was responsible for SCC.

9.5.6   H diffusion

The speed of H diffusion in Mg is critical for several of the mechanisms proposed for Mg TGSCC [8, 75, 114]. Literature data [133–135] are presented in Fig. 9.14 [51]. Renner and Grabke [133] hydrided samples of a Mg-Ce alloy; the samples were of significant size (~ cms). The Ce in the alloy reacted with H to form hydrides in a surface layer of thickness ξ, which was related to the diffusion coefficient of H in the alloy. This data [133], when extrapolated to ambient temperature, indicated a low diffusion coefficient, ~10^{-13} m²/s, so that doubts were expressed [75] about the viability of any mechanism for Mg SCC, in which H transport is required some distance ahead of the crack tip.

Nishimura *et al.* [134] carried out permeation experiments through 99.9% pure Mg sheets of thickness 0.6 mm to 2.9 mm. Pure H_2 gas was the charging medium. The H permeation flux was measured with a calibrated vacuum gauge on the exit side of the membrane that was maintained at a pressure of ~10^{-7} Pa. Schimmel [135] estimated the H diffusion coefficient from molecular dynamics simulations. There is a reasonable agreement among these three estimates of the H diffusion coefficient (Fig. 9.14), so that it is possible to put a line through all these data, and extrapolate to 23°C. This yields an estimate of the diffusion coefficient at ambient temperature to be ~10^{-9} m²/s to ~10^{-5} cm²/s. This value of the H diffusion coefficient is sufficient to allow significant H transport ahead of a stress corrosion crack in Mg at ambient temperature.

Dietzel *et al.* [136] estimated the H diffusion coefficient in AZ91 to be $D_{H\_in\_AZ91} = 2 \times 10^{-13}$ m²/s at RT based on modelling the CERT results of Fig. 9.4 for AZ91 in distilled water [34]. In the model, the stress–strain behaviour of the cylindrical specimen was represented by a bundle of parallel fibres oriented parallel to the direction of the applied force. The fibre-bundle model assumed that each individual fibre followed the same

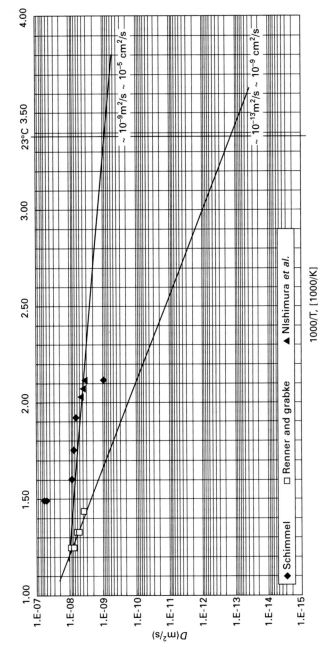

9.14 Literature data for the diffusion coefficient of H in Mg, $D_{H\_in\_Mg}$. Reprinted from [51].

stress–strain curve as the bulk material. For an environment leading to HE it was assumed that H diffused into the material and reduced the strain-to-failure of individual fibres and that fracture of a particular fibre occurred once the combination of applied strain and local H concentration reached a critical value. Crack initiation and growth were treated as a sequence of failure events at individual fibres. It was assumed that H was generated by the corrosion reaction inside pits at the specimen surface. These pits were created by mechanically straining the specimen and thus rupturing the hydroxide layer on the surface, where part of the hydrogen thus generated diffused into the material. The model produced fracture surfaces similar to those produced by CERT, and stress–strain curves were generated which reflected the influence of the applied strain rate on the H induced fracture of the Mg alloy. The model had two unknown parameters: $D_{H\_in\_AZ91}$ and x_H, (which relates to the fracture strain in the presence of H). Best fit was obtained for $D_{H\_in\_AZ91} = 2 \times 10^{-13}$ m^2/s.

An alternative approach to the available H diffusion data is presented in Fig. 9.15 [137]. Separate best-fit lines are drawn through the diffusion data for Mg-2%Ce [133] and the data for pure Mg [134, 135]. These lines have slopes similar to each other and to data for other close packed lattices (e.g., hcp Zr and Ti, and fcc Ni, Cu, Au). Included on this figure as 'AZ91 (sim)' is the modelling estimate from the work of Dietzel *et al.* [136] of $D_{H\_in\_AZ91} = 2 \times 10^{-13}$ m^2/s. This value is in good agreement with lower shelf values predicted for H diffusion in Mg at room temperature (Fig. 9.25), by extrapolating the elevated temperatures measurements of Renner and Grabke [133].

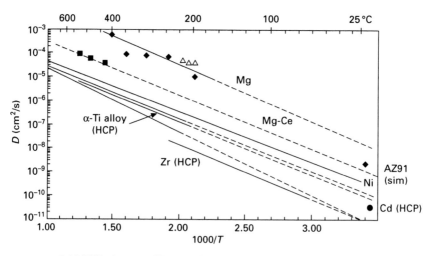

9.15 Diffusion coefficient of H in Mg and other hexagonal metals. Reprinted with permission from [137].

9.5.7   Multiple mechanisms

Recent work has indicated multiple TGSCC mechanisms. Winzer *et al.* [34, 54, 55, 124] found, for AZ31, AM30 and AZ91 in water, tested with CERT, TGSCC initiated by highly localised dissolution, with a transition to HE at some critical crack length. The HE mechanism is dependent on microstructure. TGSCC initiation in LIST is directly by HE.

9.5.8   β particles

In two phase α + β Mg-Al alloys like AZ91, β particles are critical for TGSCC. Winzer *et al.* [34, 55] found that TGSCC propagation for AZ91 involved crack nucleation within β particles ahead of the crack (Fig. 9.16). Crack nucleation at β particles could be due to: (i) the inherently low fracture toughness of the β particles; (ii) a reduction in the fracture toughness of β particles by internal H (by HEDE or transformation to MgH_2); or (iii) the synergistic effects of dislocation pile-ups at the α–β interface, as observed by Wang *et al.* [138], and H-enhanced dislocation mobility. Using the same AZ91 as Winzer *et al.* [34, 55], Chen *et al.* [40, 45] showed that cathodic polarisation in Na_2SO_4, in the absence of an external stress, resulted in fracture of the β at the surface, indicating that β particle fracture is indeed due to H and not due to their inherently low toughness.

9.5.9   H trapping

Mg TGSCC is most likely a form of HE. This implies (i) that the durability of Mg alloys is dependent on H diffusivity, and (ii) that microstructure influences need to be understood in terms of H transport and trapping. Winzer *et al.*

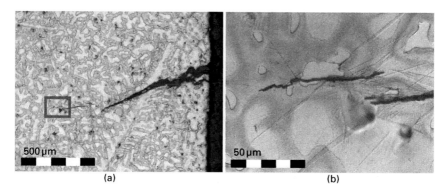

9.16 Micrographs of secondary crack in AZ91 tested in distilled water at $3 \times 10^{-7} s^{-1}$ showing crack nucleation within a β particle. Reprinted from [35].

[55] showed that small alloy changes result in significant variations in the mechanism. The fracture surface for AZ31 was characterised by relatively smooth regions containing small, elongated dimples, whereas for AM30 was cleavage-like (Fig. 9.17). Moreover, the crack velocities for AM30 were lower than those for AZ31. These differences may be related to the influences of Zn and second phase particles on H diffusivity such that: (i) different HE mechanisms occur in AZ31 and AM30; or (ii) the same HE mechanism occurs, with the difference in H diffusivity changing the crack velocity and fracture surface morphology. There is no understanding of H trapping by microstructure features in Mg alloys. This represents a substantial gap in understanding, particularly since there is evidence to suggest that key microstructure features (i.e. β particles) can act as H trapping and crack nucleation sites.

9.6   Recommendations to avoid stress corrosion cracking (SCC)

This topic is discussed in our prior review [4] and summarised here. The general principle to avoid SCC is to avoid the use of a susceptible alloy under load in an environment that produces SCC. For Mg alloys, research has shown that common alloys are susceptible in distilled water, with the threshold stress about half the yield stress. Research is underway to provide the scientific understanding to allow the production of Mg alloys more resistant to SCC. In the meantime, the authors wish to offer some general guidelines.

- It is a general trend in SCC prevention to start with a design that avoids concentration of stresses at the start, or during use, or through increasing susceptibility to different forms of degradation in service, especially

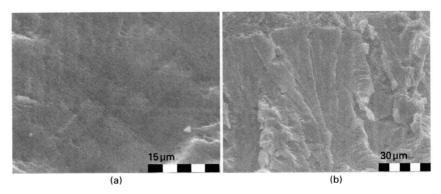

(a)   (b)

9.17 Morphology of SCC fracture surface for AZ31 (a) and AM30 (b). Reprinted from [55].

localised corrosion, galvanic corrosion, (corrosion) fatigue and erosion-corrosion.

- SCC is not expected to be an issue in dry atmospheres provided the relative humidity is less than 90% and provided that there are no crevices. For crevices, capillary condensation can cause the formation of a liquid in the crevices at lower values of relative humidity. It would be prudent to fill crevices with a corrosion-inhibited putty. Furthermore, it should be noted that crevices are part of the microstructure of diecast alloys.

- SCC occurs for many alloy + environment combinations and there are only some references to alloy + environment combinations where there is information regarding SCC resistance [4]. Moreover, such data needs to be approached with the appropriate degree of critical evaluation. The one situation that does seem to merit more attention is the use of Mg-Mn alloys, including the use of Mg-2Mn as cladding for higher strength Mg alloys.

- The total stress in service (stress from the service loading + the fabrication stress + the residual stress) should be below the threshold stress, which, in the absence of other data could be estimated to be ~ 0.4 times the tensile yield strength.

- Bolted or riveted joints can also produce high local stresses that can cause SCC, so that attention should be given to joint design and construction. Attention should also be paid to tensile residual stresses from welding.

- The prime function of anodising, other coatings, surface treatments and inhibitors, is to prevent the Mg alloy from contacting the environment. There would be no SCC if there were no contact with the environment.

- There should be appropriate inspection and maintenance programs to avoid SCC and precursors, for example a concentration of stresses by localised corrosion. For corrosion protection by protective coatings, routine maintenance is essential since scratches could create favourable sites for initiation of SCC.

9.7   Conclusions

IGSCC of Mg alloys is well understood. A continuous second phase along the grain boundaries, a microstructure typical of many creep-resistant Mg alloys, causes IGSCC by micro-galvanic corrosion. In contrast, TGSCC occurs through the α-phase Mg matrix. TGSCC occurs for common Mg alloys in environments such as distilled water and dilute chloride solutions.

Mechanistic studies have established the main components for TGSCC. Pugh and co-workers [8, 89, 91, 92, 132] provided convincing evidence for a brittle cleavage type mechanism involving H. Particularly noteworthy are the stepped and facetted interlocking fracture surfaces. In contrast, the fractography of Lynch and Trevena [75] indicates some plasticity, particularly at higher loading rates and faster crack velocities. The slow strain rate testing of both Ebtehaj *et al.* [127] and Stampella *et al.* [78] also indicate a mechanism involving strain induced film rupture leading to corrosion and H production, with crack advance due to H. Makar *et al.* [114] confirmed the fractography, the importance of strain rate and H, but they propose a brittle hydride model.

A key aspect of the environment is that TGSCC is associated with conditions leading to the local breakdown of a partially protective surface film. The environment can cause film breakdown, e.g. localised corrosion by chloride ions. However, loading can also cause film breakdown as indicated by the facts that SCC occurs for (1) pure Mg in a dilute sulphate solution and (2) Mg alloys (AZ91, AM60, AS41, ZK60A-T5) in distilled water. The role of the film is consistent with a barrier preventing H ingress, the H being produced as part of the Mg corrosion reaction.

All two-phase alloys, like AZ91, are expected to have poor TGSCC properties if there is H facilitated fracture of large second-phase particles. In contrast, there is no information regarding H-influenced fracture of the small precipitates of precipitation hardening, although there are promising indications, extrapolating from their positive influence for some heat treatment conditions in Al alloys. SCC resistance is much higher for Al alloys in the over-aged condition and for particular double aging conditions [139].

AM30 has a TGSCC propagation velocity significantly lower than that of AZ31 and AZ91 [34, 35, 54, 55], consistent with a lower effective H diffusivity, probably due to the absence of Zn in AM30. This again demonstrates the necessity to understand the influence of alloying on SCC propagation velocity and H diffusivity.

The next major advance in understanding Mg TGSCC requires an understanding of the microstructure influences on SCC. Included needs to be an understanding of the effective H diffusion rate to the fracture process zone. This requires an understanding of H-trap interactions in Mg alloys.

Mg alloys are starting to be used in structural applications and new classes of wrought Mg alloys are emerging. These include high-performance high-strength heat-treatable wrought alloys. Promising alloy systems include Mg-Zn-Φ, Mg-Sn-Λ and Mg-RE-Σ. Their SCC behaviour has not been characterised. A significant SCC risk would negate much of the potential for Mg alloys in critical service components, such as stents (catastrophic fracture due to SCC in a heart artery would probably be fatal) or stressed Mg-alloy auto components exposed to road spray.

The prior work on pure Mg, Mg-Al alloys and Zr-containing alloys showed that SCC is a significant issue, and that SCC can occur for a load ~50% of the yield stress for many combinations of alloy + environment. Since the prior research indicates that all existing Mg alloys are likely to be susceptible, guidelines are needed to ensure safe application of Mg alloys in service. An urgent task is to delineate safe operational limits for common alloys in likely service environments.

Table 9.1 indicates that it would be conservative to assume that the threshold stress was ~50%YS, unless there is other convincing data for the particular alloy + environment combination. It would be prudent to apply this recommendation for the common Mg alloys.

9.8   Acknowledgements

This research was supported by an Australian Research Council (ARC) Linkage grant in collaboration with General Motors Corporation USA and by the Australian Research Council, Centre of Excellence, Design of Light Alloys in Australia. Atrens wishes to thank GKSS-Forschungszentrum Geesthacht GmbH for their support that allowed him to work at GKSS as a visiting scientist.

9.9   References

1   G Song, A Atrens, Corrosion mechanisms of magnesium alloys, *Advanced Engin Materials* 1 (1999) 11.

2   GL Song, A Atrens, Understanding magnesium corrosion mechanism: a framework for improved alloy performance, *Advanced Engineering Materials* 5 (2003) 837.

3   G Song, A Atrens, Recent insights into the mechanism of magnesium corrosion and research suggestions, *Advanced Engineering Materials* 9 (2007) 177–183.

4   N Winzer, A Atrens, G Song, E Ghali, W Dietzel, KU Kainer, N Hort, C Blawert, A critical review of the stress corrosion cracking (SCC) of magnesium alloys, *Advanced Engineering Materials* 7 (2005) 659–693.

5   A Atrens, N Winzer, W Dietzel, Stress corrosion cracking of magnesium alloys, *Advanced Engineering Materials* (2009), doi: 10:1002/adem.200900287.

6   A Atrens, ZF Wang, Stress corrosion cracking, *Materials Forum*, 19 (1995) 9.

7   W Dietzel, *Encyclopedia of Materials: Science and Technology*, KHJ Buschow, RW Cahn, MC Flemings, B Ilschner, EJ Kramer, S Mahajan (eds), Elsevier Science Ltd., Amsterdam (2001), 8883.

8   AJ Bursle, EN Pugh, in *Mechanisms of Environment Sensitive Cracking of Materials*, PR Swann, FP Ford, ARC Westwood (eds), Materials Society (London) (1977), 471.

9   IJ Polmear, *Light Alloys*, Arnold (1981).

10 KU Kainer, PB Srinivasan, C Blawert, W Dietzel, Corrosion of magnesium alloys: *in Shreir's Corrosion Handbook*, Elsevier (2009) 2011–2041.

11 G Song, A Atrens, D StJohn, X Wu, J Nairn, The anodic dissolution of magnesium in chloride and sulphate solutions, *Corrosion Science* 39 (1997) 1981.

12 G Song, A Atrens, X Wu, B Zhang, Corrosion behaviour of AZ21, AZ501 and AZ91 in sodium chloride, *Corrosion Science* 40 (1998) 1769.

13 M Pourbaix, *Atlas of Electrochemical Equilibria in Aqueous Solutions*, NACE, Houston, TX, CEBELCOR, Brussels (1974).

14   DA Jones, *Principles and Prevention of Corrosion*, Prentice-Hall (1996).

15   MC Zhao, P Schmutz, S Brunner, M Liu, G Song, A Atrens, An exploratory study of the corrosion of Mg alloys during interrupted salt spray testing, *Corrosion Science* 51 (2009) 1277–1292.

16   B Zberg, PJ Uggowitzer, JF Loffler, *Nature Materials* 8 (2009) 887–891.

17   MC Zhao, M Liu, G Song, A Atrens, Influence of pH and chloride ion concentration on the corrosion of Mg alloy ZE41, *Corrosion Science* 50 (2008) 3168.

18   MC Zhao, M Liu, G Song, A Atrens, Influence of the β-phase morphology on the corrosion of the Mg Alloy AZ91, *Corrosion Science* 50 (2008) 1939–1953.

19   M Liu, S Zanna, H Ardelean, I Frateur, P Schmutz, G Song, A Atrens, P Marcus, A first quantitative XPS study of the surface films formed, by exposure to water, on Mg and on the Mg-Al intermetallics: Al_3Mg_2 and $Mg_{17}Al_{12}$, *Corrosion Science* 51 (2009) 1115–1127.

20   A Atrens, W Dietzel, The negative difference effect and unipositive Mg^+, *Advanced Engineering Materials* 9 (2007) 292–297.

21   M Liu, PJ Uggowitzer, AV Nagasekhar, P Schmutz, M Easton, G Song, A Atrens, Calculated phase diagrams and the corrosion of die-cast Mg-Al alloys, *Corrosion Science* 51 (2009) 602.

22   M Liu, D Qiu, MC Zhao, G Song, A Atrens, The effect of crystallographic orientation on the active corrosion of pure magnesium, *Scripta Materialia* 58 (2008) 421–424.

23   JX Jia, GL Song, A Atrens, Influence of geometry on galvanic corrosion of AZ91D coupled to steel, *Corrosion Science* 48 (2006) 2133–2153.

24   JX Jia, A Atrens, G Song, T Muster, Simulation of galvanic corrosion of magnesium coupled to a steel fastener in NaCl solution, *Materials and Corrosion* 56 (2005) 468–474.

25   GL Song, A Atrens, M Dargusch, Influence of microstructure on the corrosion of diecast AZ91D, *Corrosion Science* 41 (1999) 249–273.

26   A Seyeux, M Liu, P Schmutz, G Song, A Atrens, P Marcus, ToF-SIMS depth profile of the surface film on pure magnesium formed by immersion in pure water and the identification of magnesium hydride, *Corrosion Science* 51 (2009) 1883–1886.

27   G Song, A Atrens, DH St John in *Magnesium Technology*, J Hryn (ed.), TMS, New Orleans (2001) 255.

28   Z Shi, G Song, A Atrens, Influence of the β phase on the corrosion performance of anodised coatings on magnesium–aluminium alloys, *Corrosion Science* 47 (2005) 2760.

29   MB Haroush, CB Hamu, D Eliezer, L Wagner, The relation between microstructure and corrosion behavior of AZ80 Mg alloy following different extrusion temperatures, *Corrosion Science* 50 (2008) 1766–1778.

30   G Williams, HN McMurray, Localized corrosion of magnesium in chloride-containing electrolyte studied by a scanning vibrating electrode technique, *Journal of the Electrochemical Society* 155 (2008) C340–C349.

31   W Zhou, NN Aug, Y Sun, Effect of antimony, bismuth and calcium addition on corrosion and electrochemical behaviour of AZ91 magnesium alloy, *Corrosion Science* 51 (2009) 403–408.

32   WC Neil, M Forsyth, PC Howlett, CR Hutchinson, BRW Hinton, Corrosion of magnesium alloy ZE41 – the role of microstructural features, *Corrosion Science* 51 (2009) 387–394.

33   MP Staiger, AM Pietak, J Huadmai, G Dias, Magnesium and its alloys as orthopedic biomaterials: a review, *Biomaterials* 27 (2006) 1728–1734.

34   N Winzer, A Atrens, W Dietzel, VS Raja, G Song, KU Kainer, Characterisation of stress corrosion cracking (SCC) of Mg-Al Alloys, *Materials Science and Engineering A* 488 (2008) 339–351.

35   N Winzer, A Atrens, W Dietzel, G Song, KU Kainer, Comparison of the linearly increasing stress test and the constant extension rate test in the evaluation of transgranular stress corrosion cracking of magnesium, *Materials Science and Engineering A* 472 (2008) 97–106.

36   RG Song, C Blawert, W Dietzel, A Atrens, A study of the stress corrosion cracking and hydrogen embrittlement of AZ31 magnesium alloy, *Materials Science and Engineering* 399 (2005) 308–317.

37   SB Abhijeet, R Balasubramaniam, M Gupta, Corrosion behaviour of Mg–Cu and Mg–Mo composites in 3.5% NaCl, *Corrosion Science* 50 (2008) 2423–2428.

38   J Zhang, D Zhang, Z Tian, J Wang, K Liu, H Lu, D Tang, J Meng, Microstructures, tensile properties and corrosion behavior of die-cast Mg–4Al-based alloys containing La and/or Ce, *Materials Science and Engineering A* 489 (2008) 113–119.

39   E Zhang, W He, H Du, K Yang, Microstructure, mechanical properties and corrosion properties of Mg–Zn–Y alloys with low Zn content, *Materials Science and Engineering* A488 (2008) 102–111.

40   J Chen, J Wang, E Han, W Ke, Effect of hydrogen on stress corrosion cracking of magnesium alloy in 0.1 M Na_2SO_4 solution, *Materials Science and Engineering A* 488 (2008) 428–434.

41   M Jönsson, D Persson, C Leygraf, Atmospheric corrosion of field-exposed magnesium alloy AZ91D, *Corrosion Science* 50 (2008) 1406–1413.

42   MB Kannan, RK Singh Raman, *In vitro* degradation and mechanical integrity of calcium-containing magnesium alloys in modified-simulated body fluid, *Biomaterials* 29 (2008) 2306–2314.

43   J Chen, J Dong, J Wang, E Han, W Ke, Effect of magnesium hydride on the corrosion behavior of an AZ91 magnesium alloy in sodium chloride solution, *Corrosion Science* 50 (2008) 3610–3614.

44   X Zhou, Y Huang, Z Wei, Q Chen, F Gan, Improvement of corrosion resistance of AZ91D magnesium alloy by holmium addition, *Corrosion Science* 48 (2006) 4223–4233.

45   J Chen, J Wang, E Han, J Dong, W Ke, States and transport of hydrogen in the corrosion process of an AZ91 magnesium alloy in aqueous solution, *Corrosion Science* 50 (2008) 1292–1305.

46   LJ Liu, M Schlesinger, Corrosion of magnesium and its alloys, *Corrosion Science* 51 (2009) 1733–1737.

47   N Birbilis, MA Easton, AD Sudholz, SM Zhu, MA Gibson, On the corrosion of binary rare earth alloys, *Corrosion Science* 51 (2009) 683–689.

48   W Liu, F Cao, L Chang, Z Zhang, J Zhang, Effect of rare earth element Ce and La on corrosion behavior of AM60 magnesium alloy, *Corrosion Science* 51 (2009) 1334–1343.

49   P Bruesch, K Muller, A Atrens, H Neff, Corrosion of stainless steels in chloride solution: an XPS investigation of passive films, *Applied Physics A* 38 (1985) 1–18.

50   P Bruesch, A Atrens, K Muller, H Neff, Corrosion of rust-free steels – an XPS study of passive films, *Helvetica Physica Acta* 57 (1984) 487.

51   A Atrens, N Winzer, GL Song, W Dietzel, C Blawert, Stress corrosion cracking and hydrogen diffusion in magnesium, *Advanced Engineering Materials* 8 (2006) 749–751.

52 N Winzer, A Atrens, W Dietzel, G Song, KU Kainer, Evaluation of the delayed hydride cracking mechanism for transgranular stress corrosion cracking of magnesium alloys, *Materials Science and Engineering A* 466 (2007) 18–31.

53 MB Kannan, W Dietzel, C Blawert, A Atrens, P Lyon, Stress corrosion cracking of rare-earth-containing magnesium alloys ZE41, QE22, and Elektron 21 (EV31A) compared with AZ80, *Materials Science and Engineering A* 480 (2008) 529–539.

54 N Winzer, A Atrens, W Dietzel, G Song, KU Kainer, Comparison of the linearly increasing stress test and the constant extension rate test in the evaluation of transgranular stress corrosion cracking of magnesium, *Materials Science and Engineering A* 472 (2008) 97–106.

55 N Winzer, A Atrens, W Dietzel, G Song, KU Kainer, The fractography of stress corrosion cracking (SCC) of Mg-Al alloys, *Metallurgical and Materials Transactions A* 39 (2008) 1157.

56 E Gamboa, A Atrens, Material influence on the stress corrosion cracking of rock bolts, *Engineering Failure Analysis* 12 (2005) 201.

57 E Gamboa, A Atrens, Environmental influence on the stress corrosion cracking of rock bolts, *Engineering Failure Analysis* 10 (2003) 521.

58 JQ Wang, A Atrens, DR Cousens, PM Kelly, C Nockolds, S Bulcock, Measurement of grain boundary composition for X52 pipeline steel, *Acta Materialia* 46 (1998) 5677.

59 A Oehlert and A Atrens, SCC propagation in Aermet 100, *J Materials Science* 33 (1998) 775–781.

60 A Oehlert, A Atrens, Initiation and propagation of stress corrosion cracking in AISI 4340 and 3.5NiCrMoV rotor steel in constant load tests, *Corrosion Science* 38 (1996) 1159.

61 ZF Wang, A Atrens, Initiation of stress corrosion cracking for pipeline steels in a carbonate-bicarbonate solution, *Metallurgical and Materials Transactions A* 27 (1996) 2686.

62 A Oehlert, A Atrens, Room temperature creep of high strength steels, *Acta Metallurgica et Materialia* 42 (1994) 1493.

63 A Atrens, CC Brosnan, S Ramamurthy, A Oehlert and IO Smith, Linearly increasing stress test (LIST) for SCC research, *Measurement Science and Technology*, 4 (1993) 1281.

64 RM Rieck, A Atrens, IO Smith, The role of crack tip strain rate in the stress corrosion cracking of high strength steels in water, *Met Trans*, 20A (1989) 889.

65 RG Song, W Dietzel, BJ Zhang, WJ Liu, MK Tseng, A Atrens, Stress corrosion cracking and hydrogen embrittlement of an Al–Zn–Mg–Cu alloy, *Acta Materialia* 52 (2004) 4727–4743.

66 S Ramamurthy, A Atrens, The influence of applied stress rate on the stress corrosion cracking of 4340 and 3.5NiCrMoV steels in distilled water at 30°C, *Corrosion Science* 52 (2010) 1042–1051.

67 CD Cann, A Atrens, A metallographic study of the terminal solubility of hydrogen in zirconium at low hydrogen concentrations, *Journal of Nuclear Materials* 88 (1980) 42.

68 A Atrens, D Mezzanotte, NF Fiore, MA Genshaw, Electrochemical studies of hydrogen diffusion and permeability in Ni, *Corrosion Science* 20 (1980) 673.

69 A Atrens, W Hoffelner, TW Duering, J Allison, Subsurface crack initiation in high cycle fatigue in Ti-6Al-4V and a typical martensitic stainless steel, *Scripta Metallurgica* 17 (1983) 601.

70 R Zeng, E Han, W Ke, W Dietzel, KU Kainer, A Atrens, Influence of microstructure on tensile properties and fatigue crack growth in extruded magnesium alloy AM60, *International Journal of Fatigue* 32 (2010) 411–419.

71 IG Ritchie, A Atrens, The diffusion of oxygen in alpha-zirconium, *Journal of Nuclear Materials* 67 (1977) 254.

72 S Jin, A Atrens, ESCA – Studies of the structure and composition of the passive film formed on stainless steels by various immersion times in 0.1 M NaCl solution, *Applied Physics A* 42 (1987) 149.

73 AS Lim, A Atrens, ESCA studies of Fe-Ti, *Applied Physics A* 54 (1992) 500.

74 WK Miller, in *Stress Corrosion Cracking: Materials Performance and Evaluation*, RH Jones (ed.), ASM International, USA (1992) 251.

75 SP Lynch, P Trevena, Stress corrosion cracking and liquid metal embrittlement in pure magnesium, *Corrosion* 44 (1988) 113–123.

76 EI Meletis, RF Hochman, Crystallography of stress corrosion cracking in pure magnesium, *Corrosion* 40 (1984) 39–48.

77 *ASM Specialty Handbook: Magnesium and Magnesium Alloys*, ASM International, USA (1999) 211.

78 RS Stampella, RPM Procter, V Ashworth, Environmentally induced cracking of magnesium, *Corrosion Science* 24 (1984) 325–341.

79 N Winzer, G Song, A Atrens, W Dietzel, C Blawert, Stress corrosion cracking of magnesium, *Corros Prevention* Australasian Corrosion Association, (2005) 37.

80 RS Busk, *Magnesium Products Design*, Marcel Dekker, USA (1986) 256.

81 DK Priest, FH Beck, MG Fontana, Stress corrosion mechanism in a magnesium-base alloy, *Transactions of the American Society for Metals* 47 (1955) 473–492.

82 MA Pelensky, A Gallaccio, *Stress Corrosion Testing, STP425*, ASTM (1967), 107.

83 L Fairman, HJ Bray, Transgranular SCC in magnesium alloys, *Corrosion Science* 11 (1971) 533–541.

84 WK Miller, *Mat Res Soc Symp Proc* 125 (1988) 253.

85 A Arnon, E Aghion, Stress corrosion cracking of nano/sub-micron E906 magnesium alloy, *Advanced Engineering Materials* 10 (2008) 742–745.

86 J Chen, M Ai, J Wang, E Han, W Ke, Stress corrosion cracking behaviours of AZ91 magnesium alloy in deicer solution using constant load, *Materials Science and Engineering A* 515 (2009) 79–84.

87 MB Kannan, RKS Raman, *In vitro* degradation and mechanical integrity of calcium-containing magnesium alloys in modified-simulated body fluid, *Biomaterials* 29 (2008) 2306–2314.

88 WR Wearmouth, GP Dean, RN Parkins, Role of stress in stress corrosion cracking of a Mg-Al alloy, *Corrosion* 29 (1973) 251–260.

89 DG Chakrapani, EN Pugh, The transgranular SCC of a Mg-Al alloy: crystallographic, fractographic and acoustic-emission studies, *Metallurgical Transactions* 6A (1975) 1155–1163.

90 G Oryall, D Tromans, Transgranular stress corrosion cracking of solution treated and quenched Mg-86 Al alloy, *Corrosion* 27 (1971) 334–341.

91 DG Chakrapani, EN Pugh, On the fractography of transgranular stress corrosion failures in a Mg-Al alloy, *Corrosion*, 31 (1975) 247–251.

92 DG Chakrapani, EN Pugh, Hydrogen embrittlement in a Mg-Al alloy, *Metallurgical Transaction* 7A (1976) 173–178.

93 ECW Perryman, Stress corrosion of magnesium alloys, *Journal of the Institute of Metals* 79 (1951) 621–642.

94   T Nozaki, S Hanaki, M Yamashita, H Uchida, *Proceedings of the 13th Asian-Pacific Corrosion Control Conference*, Osaka (2003), K-15.

95   L Fairman, JM West, Stress corrosion cracking of a magnesium alloy, *Corrosion Science* 5 (1965) 711–716.

96   A Moccari, CR Shastry, An investigation of stress corrosion cracking in Mg AZ61 alloy in 3.5% NaCl + 2% $K_2 CrO_4$ aqueous solution at room temperature, *J. Mater. Technol. (Zeitschrift fur Werkstofftechnik)* 10 (1979) 119–123.

97   L Fairman, HJ Bray, *British Corrosion Journal* 6 (1971) 170–174.

98   WM Pardue, FH Beck, MG Fontana, Propagation of stress-corrosion cracking in a magnesium-base alloy as determined by several techniques *Transactions of the American Society for Metals* 54 (1961) 539–548.

99   G Ben-Hamu, D Eliezer, W Dietzel, KS Shin, Stress corrosion cracking of new Mg–Zn–Mn wrought alloys containing Si, *Corrosion Science* 50 (2008) 1505–1517.

100 ND Tomashov, VN Modestova, *Intercrystalline Corrosion and Corrosion of Metals Under Stress*, IA Levin (ed.), L. Hill, London (1962), 251–262.

101 MA Timonova, *Intercrystalline Corrosion and Corrosion of Metals Under Stress*, IA Levin (ed.), L. Hill, London (1962), 263.

102 LL Rokhlin, *Magnesium Alloys Containing Rare Earth Metals*, Taylor and Francis, London (2003).

103 JF Nie, X Gao, SM Zhu, Enhanced age hardening response and creep resistance of Mg–Gd alloys containing Zn, *Scr. Mater.* 53 (2005) 1049–1053.

104 C Sanchez, C Nussbaum, P Azavant, H Octor, Elevated temperature behaviour of rapidly solidified magnesium alloys containing rare earths, *Materials Science and Engineering A* 221 (1996) 48–57.

105 P Lyon, T Wilks, I Syed, The influence of alloying elements and heat treatment upon the properties of elektron 21 (EV31A) alloy, in *Magnesium Technology*, NR Neelameggham, HI Kaplan, BR Powell (eds), Warrendale, PA (2005), 303.

106 J Chang, X Guo, P Fu, L Peng, W Ding, Effect of heat treatment on corrosion and electrochemical behaviour of Mg–3Nd–0.2Zn–0.4Zr (wt.%) alloy, *Electrochim. Acta* 52 (2007) 3160–3167.

107 JH Nordlien, K Nisancioglu, S Ono, N Masuko, Morphology and structure of water-formed oxides on ternary MgAl alloys, *Journal of the Electrochemical Society* 144 (1997) 461–466.

108 S Krishnamurthy, M Khobaib, E Robertson, FH Froes, Corrosion behaviour of rapidly solidified Mg-Nd and Mg-Y alloys, *Materials Science and Engineering* 99 (1988) 507–511.

109 LL Rokhlin, *Magnesium Alloys Containing Rare Earth Metals*, Taylor and Francis (2003), 221.

110 *Exterior Stress Corrosion Resistance of Commercial Magnesium Alloys, Report Mt 19622*, Dow Chemical, USA (1966).

111 RI Stephens, CD Schrader, DL Goodenberger, KB Lease, V V Ogarevic, SN Perov, *Society of Automotive Engineers, No 930752*, USA (1993).

112 MA Timonova, LI D'yalchenko, YM Dolzhanskii, MB Al'tman, NV Sakharova, AA Blyablin, *Protection of Metals* 19 (1983) 99–102.

113 RB Mears, RH Brown, EH Dix, in *Symposium on Stress-Corrosion Cracking of Metals*, American Society for Testing Materials (1945) 323.

114 GL Makar, J Kruger, K Sieradzki, Stress corrosion cracking of rapidly solidified magnesium alloys, *Corrosion Science* 34 (1993) 1311–1342.

115 JC Kiszka, Stress corrosion tests of some wrought magnesium-lithium base alloys, *Materials Protection* 4 (1965) 28–29.

116 MO Speidel, MJ Blackburn, TR Beck, JA Feeney, *Corrosion Fatigue: Chemistry, Mechanics and Microstructure*, NACE-2 (1972), 324.

117 A Munitz, C Cotler, H Shaham, G Kohn, Electron beam welding of magnesium AZ91D plates, *Weld J* 79 (2000) 202s–208s.

118 J Matsumoto, M Kobayashi, M Hotta, Arc welding of magnesium casting alloy AM60, *Weld Int* 4 (1990) 23–28.

119 X Cao, M Jahazi, JP Immarigeon, W Wallace, A review of laser welding techniques for magnesium alloys, *Journal of Materials Processing Technology* 171 (2006) 188–204.

120 WM Thomas, US Patent No 5,460,317.

121 R Zettler, AC Blanco, JF dos Santos, S Marya, The effect of process parameters and tool geometry on thermal field development and weld formation in friction stir welding of the alloy AZ31 and AZ61, in *Magnesium Technology*, NR Neelameggham, HI Daplan, BR Powell (eds), TMS, The Minerals, Metals & Materials Society (2005), pp 409–423.

122 MB Kannan, W Dietzel, C Blawert, S Riekehr, M Kocak, Stress corrosion cracking behavior of Nd:YAG laser butt welded AZ31 Mg sheet, *Materials Science and Engineering A* 444 (2007) 220–226.

123 MB Kannan, W Dietzel, R Zeng, R Zettler, JF dos Santos, A study on the SCC susceptibility of friction stir welded AZ31 Mg sheet, *Materials Science and Engineering A* 460–461 (2007) 243–250.

124 N Winzer, P Xu, S Bender, T Gross, WES Unger, CE Cross, Stress corrosion cracking of gas-tungsten arc welds in continuous-cast AZ31 Mg alloy sheet, *Corrosion Science* 51 (2009) 1950–1963.

125 PB Srinivasan, R Zettler, C Blawert, W Dietzel, A study on the effect of plasma electrolytic oxidation on the stress corrosion cracking behaviour of a wrought AZ61 magnesium alloy and its friction stir weldment, *Materials Characterization* 60 (2009) 389–396.

126 PB Srinivasan, S Riekehr, C Blawert, W Dietzel, M Kocak, Slow strain rate stress corrosion cracking behaviour of as-welded and plasma electrolytic oxidation treated AZ31HP magnesium alloy autogenous laser beam weldment, *Materials Science and Engineering A* 517 (2009) 197–203.

127 K Ebtehaj, D Hardie, RN Parkins, The influence of chloride-chromate solution composition on the stress corrosion cracking of a Mg-Al alloy, *Corrosion Science* 28 (1993) 811–821.

128 WS Loose, HA Barbian, in *Symposium on Stress Corrosion Cracking of Metals*, American Society for Testing Materials, USA (1945) 273.

129 RP Frankenthal, The inhibition of pitting and stress corrosion cracking of Mg-Al Alloys by NO_3^-, *Corrosion Science* 7 (1967) 61–62.

130 D Hardie, The environment-induced cracking of hexagonal metals: magnesium, titanium and zirconium, in *Environment induced cracking of metals, NACE 10*, RP Gangloff, MB Ives (eds), NACE (1990) 347–361.

131 HL Logan, Film rupture mechanism of stress corrosion, *Journal of Research of the National Bureau of Standards* 48 (1952), 99–105.

132 EH Pugh, JAS Green, PW Slattery, *Fracture 1969: The Proceedings of the Second International Conference on Fracture*, PL Pratt (ed.), Chapman and Hall, London (1969), 387.

133 J Renner, HJ Grabke, Bestimmung von Diffusionskoeffizienten bei der Hydrierung von Legierungen, *Z Metallkunde* 69 (1978) 639.

134 C Nishimura, M Komaki, M Amano, Hydrogen permeation through magnesium, *Journal of Alloys and Compounds* 293–295 (1999) 329.

135 HG Schimmel, Towards a hydrogen-driven society?, PhD thesis, Technical University of Delft (2004).

136 W Dietzel, M Pfuff, N Winzer, Testing and mesoscale modelling of hydrogen assisted cracking of magnesium, *Engng Fract Mech* 77 (2009) 257.

137 SP Lynch, unpublished work, Defence Science and Technology Organisation (2009), presented at the International Symposium on Stress Corrosion Cracking in Structural Materials at Ambient Temperatures, Padova, Italy, 30 August–4 September, 2009.

138 RM Wang, A Eliezer, EM Gutman, An investigation on the microstructure of an AM50 magnesium alloy, *Materials Science and Engineering* 355A (2003) 201–207.

139 RG Song, W Dietzel, BJ Zhang, WJ Liu, M Tseng, A Atrens, *Acta Mater.* 52 (2004) 4727.

140 Z Shi, A Atrens, An innovative specimen configuration for the study of Mg corrosion, *Corrosion Science* 53 (2011) 226–246.

10

Stress corrosion cracking (SCC) and hydrogen-assisted cracking in titanium alloys

I. CHATTORAJ, Council of Scientific and Industrial Research (CSIR), India

Abstract: By virtue of their excellent physical, mechanical and chemical properties, titanium alloys have tremendous application potential. Apart from the high cost of production and fabrication, one of the hindrances to its application has been corrosion, especially under-performance in severe environments due to stress corrosion. This chapter provides a summary of the present knowledge on stress corrosion cracking of titanium and its alloys with due emphasis on hydrogen enhanced cracking.

Key words: titanium alloys, stress corrosion cracking, hydrogen embrittlement, hydride, crack tip.

10.1 Introduction

10.1.1 Popularity and utility

Titanium alloys were first considered on account of their very favourable specific strengths and therefore were found to be suitable for the aerospace industries. Ti-6Al-4V, which continues to be the workhorse of the titanium industry, has been used in airframes, landing gears, empennage, and wings and even in gas turbine engines. Only at temperatures below 300°C do carbon fibre reinforced plastics have a higher specific strength than titanium alloys, with the latter surpassing all other commercial materials at all other application temperatures in terms of specific strength. However, the other important advantages of titanium alloys have also been realized and are being exploited in a number of non-aerospace applications. These properties are excellent corrosion resistance, good creep resistance, good high temperature strength, low thermal coefficient of expansion, biocompatibility and relative abundance of the metal in the earth's crust. Counterbalancing these obvious advantages of titanium alloys is the high component cost arising out of the high cost of metal extraction and high cost of shaping and forming. Even in the aerospace industries, which has been largely performance driven, the cost of production is expected to have a prominent role in materials selection in the future.

381

10.1.2 Different titanium grades and their uses

Pure commercial (CP) grade titanium, as well as a number of alloys of titanium exhibit hexagonal close packed (hcp) crystal structure with suitable adjustments of lattice structure and bond strength depending on alloying, collectively referred to as α *alloys*. At higher temperatures (above 882°C (the β transus temperature) for the pure alloy) the body-centred cubic (bcc) structure is stable and is referred to as the β *phase*. All alloys of titanium exist in either of these two phases or as a combination of these two phases. Additionally, precipitate phases, which may be intrinsic like ω phase, intermetallics, carbides and nitrides, or extrinsic like hydrides formed due to hydrogen entry, may be present in varying amounts, usually small. The major phases present in the alloy at the application temperature has a profound influence on its physical and chemical properties and therefore decide its application suitability. The alloying elements of titanium are classified as neutral, as α-stabilizer or as β-stabilizer depending on their effect on the β transus temperature. Aluminium is the most important α-stabilizer, with oxygen, nitrogen and carbon being others. Vanadium is the most important commercially used β-stabilizer, with molybdenum being another important element of this category. Tin and zirconium are considered neutral elements with regard to the β-transus but not with regard to mechanical properties, which they influence.

Thus titanium alloys are classified as α, α–β or β alloys, depending on the majority phase in the alloy. It should be noted that often the α alloys may have varying amounts of β phase and likewise β alloys can have small amounts of α phase in them. Thus a sub-classification of the alloys is often made which recognizes near-α alloys (those having small amounts of β) and metastable β alloys (those with small amounts of α phase at room temperature (RT)). The properties of these alloy systems are obviously influenced by the nature, fraction and distribution of these phases, which in turn relates to the crystal structure of and alloy partitioning in each of these phases. In relation to the α phase, the β phase exhibits much higher strengths through aging, better deformability, much higher diffusivity for interstitial elements, and isotropic properties. However, β alloys have lower creep resistance, higher density owing to the heavier alloying elements (as opposed to aluminium in α alloys), and are generally costlier.

α alloys and near-α alloys

α alloys find several elevated temperature applications in the aerospace industries owing to their good creep resistance. The primary alloy for this application is Ti-6Al-2Sn-4Zr-2Mo (Boyer, 1996) which is used for gas turbine components like blades, rotors and discs up to temperatures of 540°C. IMI829 and IMI834 also find high temperature applications. Ti-3Al-2.5V is used in hydraulic tubings while Ti-5Al-2.5V is used for cryogenic applications.

α–β alloys

The classic Ti-6Al-4V has a number of aerospace and non-aerospace applications. This stems from its combination of the desirable properties of both the phases. Steam turbine blades, pressure vessels, electromagnetic cooking wares, and golf club heads have been manufactured using this alloy (Yamada, 1996). The same alloy also found a number of medical applications especially for surgical implants until it was observed that vanadium is not bio-compatible. Vanadium replaced grades like Ti-6Al-7Nb and Ti-5Al-2.5Fe have been developed to address this issue (Balazic et al., 2007).

β alloys

β titanium alloys represent a versatile class of materials. While they offer highest specific strengths, toughness and fracture resistance, they are limited by a small processing window and higher costs. Ti-13V-11Cr-3Al was the first commercially significant β alloy. Five popular alloys of this group today are the Timetal 21S, Beta C, Ti-10-2-3, BT 22 and Ti 17. The first four are used in structural applications and the last in gas turbines. Beta C is used for springs including return springs for brakes. Timetal 21S is a corrosion resistant alloy which is resistant to hot hydraulic fluids.

10.1.3 Proliferation and recent advancements in titanium alloys

The popularity of titanium and its alloys in the last 20 years can be gauged by looking at the patents that have been filed in titanium-related areas and topics as shown by the histogram in Fig. 10.1. It should be cautioned that many of the patents relate to titanium used in steels and as an alloying element in other alloy systems. Figure 10.2 provides a worldwide distribution of the patents on titanium-related topics. Clearly there has been significant Japanese activity on titanium in the last 20 years.

Future titanium alloys would aim to attain lower costs of production while maintaining the definite property advantage they have. Timet has developed low cost Timetal 62s (Ti-6Al-1.7Fe-0.1Si) and Timetal LCB (Ti-6.8 Mo-4.5Fe-1.5Al), the former as a replacement for Ti-6Al-4V and the latter with automotive springs in mind. A surgical implant alloy TMZF (Ti-12Mo-6Zr-2Fe) has low modulus, good strength and corrosion resistance. Alloys for the energy industry, specifically for tubulars for oil exploration, have been developed, like the Ti-6Al-4V-Ru and Ti-3Al-2.5V-Ru (Schutz and Watkins,

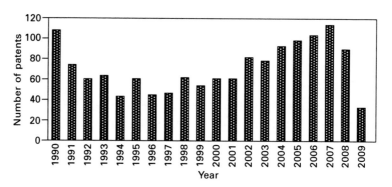

10.1 Year-wise distribution of patents in Ti-related areas (data obtained using Delphion).

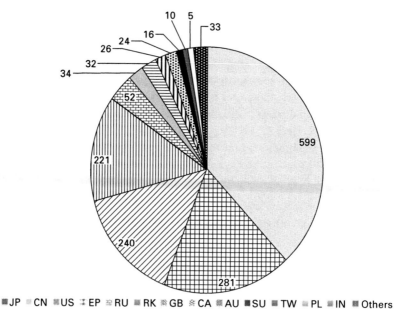

■JP ▨CN ▨US ░EP ▨RU ▦RK ▨GB ▨CA ▨AU ■SU ▨TW ▨PL ▨IN ▨Others

10.2 Country-wise distribution of patents in Ti-related areas over the last 20 years (data obtained using Delphion).

1998). A Ti-5Al-1Sn-1Zr-1V-0.8Mo (Ti-5111) near-α alloy was developed by the US Navy for fastener applications (Been and Faller, 1999).

The developments in Japan have been towards non-aerospace applications, more specifically for chemical processing and power generation components. NKK has developed a superplastic alloy (Ti-4.5Al-3V-2Fe-2Mo). High corrosion resistant titanium alloys have been developed based on small additions of (i) palladium, (ii) ruthenium and nickel, and (iii) palladium, nickel, chromium and ruthenium (Yamada, 1996). Cold deformable β alloys that have been developed in Japan include Ti-22V-4Al, Ti-20V-4Al-1Sn and Ti-16V-4Sn-4Nb. The TIX-80 (Ti-0.5Fe-0.1Nl) is a low cost high strength alloy, developed for the consumer goods market. Dispersion strengthening has been utilized in Ti-6Al-4V-10Cr-1.3C alloy, which contains fine crystallized TiC particles in a β titanium matrix.

10.2   Corrosion resistance of titanium alloys

Titanium alloys are very corrosion resistant in ambient environments owing to their ability to maintain a passive film in non-conducive and aggressive environments like brackish waters. Protective oxide films on titanium are formed in the presence of water, even though this may only be present in trace amounts. Thus, if titanium is exposed to highly oxidizing environments in the complete absence of water, rapid oxidation can occur and a violent, often pyrophoric, reaction results. The nature, thickness and composition of the passive film formed depend on the environmental conditions. In most aqueous environments the passive film is TiO_2, Ti_2O_3 or TiO. The amount of Ti^{4+} in the passive film is reported to increase with potential while that of Ti^{3+} and Ti^{2+} decreases. The passive film of the order of 10 nm thickness is very resistant to most aggressive media and even HF. Titanium in its bare state is highly reactive and spontaneously forms the passive films in oxidizing environments in the presence of traces of moisture. This passive film thus has self-healing capabilities. The Pourbaix diagram for the Ti–H_2O system (Fig. 10.3) shows that the passive film is stable for the entire pH range for oxidizing and mildly reducing potentials. Additionally at reducing potentials TiH_2 film is formed which is also protective to general corrosion, but is not desired on account of its deleterious effects in terms of ductility loss to the material and as it is concurrent with hydrogen introduction into the alloy that causes hydrogen embrittlement (HE). The corrosion resistance of titanium can be further enhanced by molybdenum additions (Chu, 1972); however, the practical difficulty of avoiding molybdenum segregations in large castings, on account of the widely differing melting points of the two elements, limits its use.

The high temperature resistance of titanium alloys also stems from oxide films. At high temperatures a TiO_2 or rutile film is formed while at relatively lower temperatures an amorphous Ti_2O_3 (anatase) film is formed, or a mixture of rutile and anatase is formed.

It is also very resistant to localized corrosion up to 70°C, regardless of salinity or acidity of the environment. Titanium lies only below platinum, gold and silver in the galvanic series. As a result titanium is practically immune from galvanic corrosion except under reducing potentials and environments which exclude passive film formation. Titanium alloys have lower susceptibility to crevice corrosion in brine on account of the hydrolysable Ti^{3+} that forms

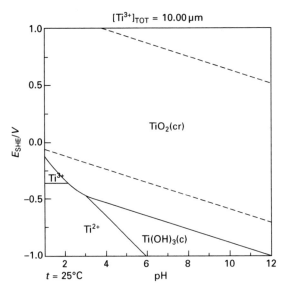

10.3 Ti–water Pourbaix diagram.

as a product of the anodic dissolution, which therefore sustains the crevice reaction. Titanium alloys are very resistant to erosion corrosion, much more than stainless steels and are one of the few materials which are highly resistant to cavitation attack. It is evident that titanium alloys possibly have the best ambient temperature resistance to various corrosion phenomena amongst the commercial alloys.

10.3 Stress corrosion cracking (SCC) of titanium alloys

10.3.1 Passive film breakdown

For passivating alloys the first step in stress corrosion cracking is the breakdown of the protective passive film. This can be studied through anodic current transients and atom force microscopy (Kolman *et al.*, 1998), as well as several other surface characterization techniques. A current transient that lasts longer indicates an inability to repassivate quickly. In stress corrosion cracking the most important events occur at or near the crack tip; thus, the crack tip passivity or lack thereof has special significance in SCC. This is to a large extent determined by the crack geometries for two major reasons. Firstly, crack geometries determine crack tip stress intensities; secondly, the crack geometry always has an effect on electrochemical kinetics thereby imposing different crack chemistry (in terms of pH and presence of ionic species) compared to the bulk. It was observed that the anodic transients for compact tension (CT) specimens were much longer than those for circumferentially notched samples.

The passive film formed on titanium and its alloys are chemically protective, but there is evidence from the studies of different groups that the passive film itself can induce tensile stresses (Lu *et al.*, 2000; Guo *et al.*, 2003). Laser moiré interferometric study (Lu *et al.*, 2000) on Grade 2 alloy showed that a passive film-induced stress was present. These film-induced stresses can have significant magnitude (up to 60 MPa at anodic overpotentials). These studies were conducted in methanol and addition of water to the solution generally led to a decrease in the film-induced stress.

10.3.2   Crack initiation and propagation

The anodic dissolution process and consequent crack advance is well understood. A new twist to the anodic dissolution process has been provided by certain plastic events reported (Zhang and Vereecken, 1990; Gao *et al.*, 2000). In Ti-6Al-4V alloys, SCC studies in methanol revealed secondary cracks along slip traces indicating a preferential dissolution along slip planes. It was proposed that, in the first place, the dislocation movement caused passive film breakdown and secondly, caused an increase in the reactivity of the slip line atoms. This would promote preferential dissolution along slip lines. In a study carried out in methanol on Ti-24Al-11Nb, it was found that the process of crack/notch tip dissolution led to dislocation emission and mobility (Gao *et al.*, 2000). The immediate neighbourhood of the crack tip was often found to be devoid of dislocations and a dislocation free zone (DFZ) resulted. The continuous emission and motion of the dislocations away from the crack tip resulted in local thinning and subsequent rupture of the thin ligament and the crack advanced. Here too, the passive film-induced stress was suspected to enhance the dislocation activity at lower applied stresses.

As with other aspects of SCC and HE there are divergent views on crack initiation and propagation. In a CP titanium alloy, hydrogen embrittlement possibly through hydride formation was thought to be the main mechanism of cracking (Simbi and Scully, 1997).

10.3.3   Microstructural effects

In the single phase as well as dual phase titanium alloys, the final microstructure after aging has a profound influence on SCC susceptibility. Single phase Beta C as well as dual phase Beta C (isothermally aged for short times) with low fractions of α were immune to SCC in NaCl (Somerday and Gangloff, 1998). However Beta C alloys subjected to higher aging times showed an initial transgranular fracture which transformed to intergranular fracture as the crack length increased or concomitantly the stress intensity increased.

The susceptibility of CP titanium to SCC has been found to be a function of the interstitials present (Simbi and Scully, 1995). High iron containing and low oxygen containing titanium were found not to be susceptible to SCC in seawater. For low iron containing titanium, the susceptibility to SCC increased with oxygen content.

In addition to the presence of interstitials, the susceptibility of titanium alloys to SCC is dependent on texture and grain shape and size (Gregory and Brokmeier, 1995). In Ti-6Al-4V the susceptibility to SCC can be correlated to the orientation of basal planes, which are sensitive to cleavage.

10.4   Hydrogen degradation of titanium alloys

10.4.1   Hydrogen titanium interaction

Titanium and hydrogen form a simple eutectoid with the α + hydride phases forming directly from the β phase. The strong β stabilizing effect of hydrogen results in a decrease of the α-to-β transformation temperature from 882°C to a eutectoid temperature of 300°C (Tal-Gutelmacher and Eliezer, 2005). The terminal hydrogen solubility in the β phase is about 50 at% at above 600°C while in the α phase, the maximum terminal hydrogen solubility is only about 7 at% at 300°C and decreases rapidly with decreasing temperature. In titanium alloys, like in other group IV transition metals, hydrogen tends to occupy tetrahedral interstitial sites. Due to the relatively open BCC structure of the β phase, hydrogen has a much higher solubility as well as much higher diffusivity in β alloys compared to α alloys. This is also facilitated by the presence of twelve tetrahedral and six octahedral sites per Ti atom in the β alloys compared to only four tetrahedral and two octahedral interstitial sites per Ti atom in the hcp α alloys.

Three different kinds of hydrides have been reported for titanium alloys (Tal-Gutelmacher and Eliezer, 2004, 2005). The δ hydrides (TiH$_x$) have a face-centred cubic (fcc) lattice with the hydrogen atoms occupying the tetrahedral interstitial sites (CaF$_2$ structure). This hydride is non-stoichiometric with x, varying from 1.5 to 1.99. At high hydrogen concentrations ($x \geq 1.99$), the δ hydride transforms into ε hydride, with a face-centered tetragonal (fct) structure ($c/a \leq 1$ at temperatures below 37°C). At low hydrogen concentrations (less than 3 at%), the metastable γ hydride forms, with an fct structure (c/a higher than 1). In γ hydrides, the hydrogen atoms occupy one-half of the tetrahedral interstitial sites.

In pure titanium and its alloys, hydrides form as lenticular plates, with different habit planes depending on the hydrogen charging conditions with the most common habit plane being $\{10\bar{1}0\}$, which is explicable from the fact that hydride formation and consequent lattice volume changes have to be accommodated both elastically as well as plastically through slip, and

{10$\bar{1}$0} are the preferred prism planes for slip. The mechanism of hydride formation has been explained as a martensitic phase transformation whereby the hcp lattice converts to an fcc lattice (Xiao, 1992).

10.4.2   Hydrogen absorption and trapping

The entry, permeation and trapping of hydrogen in different titanium alloys have been studied (Pound, 1991, 1994, 1997). The permeation techniques have allowed determination of the hydrogen diffusivities, while thermal desorption curves (Eliezer *et al.*, 2006a) have revealed information on the extent and nature of hydrogen trapping, while certain sophisticated techniques like secondary ion mass spectroscopy, SIMS (Hamada *et al.*, 2009) have provided insight on hydrogen partitioning to different phases and traps. The hydrogen absorption during electrochemical charging of titanium was reported to follow a linear law at low current densities and a parabolic law at higher current densities (Mizuno, 1977). Moreover the behaviour of hydrogen absorption changed from the parabolic rate to a linear rate with increasing temperature. The linear rate law corresponds to a surface reaction being rate limiting, the surface reaction possibly being the adsorption or absorption reaction of hydrogen on the metal surface. The parabolic rate law was thought to be due to diffusion through the hydride being rate controlling.

In grade 2 titanium, the trapping of H was felt to be at interstitial nitrogen with some trapping at grain boundaries (Pound, 1991). In Ti-6Al-4V, irrespective of hydrogen fugacity, one irreversible trapping site was identified (Eliezer *et al.*, 2006a), probably the hydride formed in this low hydrogen solubility alloy. The microstructure of the two phase alloys (α–β), including the volume fraction and the distribution and asperity of the phases, plays an important role in determining the hydrogen absorption and transport. The hydrogen concentration absorbed in the fully lamellar alloy is reported to be always higher than in the duplex microstructure, irrespective of the charging conditions.

In β alloys, hydrogen trapping has been studied in some detail (Pound, 1994, 1997; Eliezer *et al.*, 2006b). It was reported that on aging the hydrogen trapping was enhanced in Beta 21S and Beta C while the Ti-15V-3Cr-3Sn alloy remained unaffected by aging and had a relatively low hydrogen trapping constant. This was found to correlate directly with the HE susceptibility of these alloys. Moreover, it has been reported (Eliezer *et al.*, 2006b) for Beta 21S that the method of hydrogen introduction affects the hydrogen trapping and distribution in the alloy. The direct evidence for hydrogen partitioning can be obtained through SIMS (Hamada *et al.*, 2009) as shown in Fig. 10.4. For a CP titanium, SIMS revealed hydrogen presence at intragranular hydrides as well as at the grain boundaries.

There are a few other interesting and important consequences of hydrogen

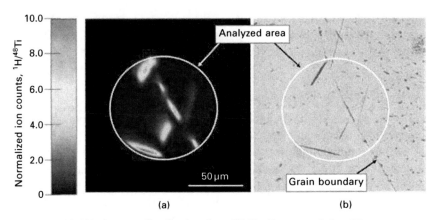

10.4 Hydrogen distribution in a CP Ti alloy containing 34 ppm hydrogen. (a) SIMS image showing H presence near grain boundary and at hydrides; (b) corresponding optical image of the analyzed region. From Hamada *et al.* (2009).

entry into titanium alloys. The β transus is decreased due to hydrogen entry due to its strong β stabilizing effect. This is used sometimes for thermomechanical treatments of titanium alloys, where temporary alloying with hydrogen enhances the deformable β phase, thereby reducing the flow stresses. Additionally there is evidence of hydrogen promoting dynamic recrystallization (Zhao *et al.*, 2010).

10.4.3   Hydrogen degradation of α and near α alloys

In commercial purity alloys (Grade 1 and 2), low hydrogen fugacities caused no embrittlement. With increasing hydrogen fugacities, at a critical value, fast brittle fracture occurs (Clarke *et al.*, 1997). This critical hydrogen content depends inversely on the material strength. It also depends on material texture and presence of residual β phase. The dependence on texture stems from the fact that a high availability of basal planes would promote fast fracture. In commercial purity titanium, increasing impurity caused an increasing susceptibility to HE. For instance, Grade 3 Ti which has higher oxygen content and iron content, was significantly more susceptible than Grade 2, although the difference in the impurity levels is in fractions of weight percentages (Wang *et al.*, 1998). This difference in susceptibility can be attributed to the lower solubility of hydrogen in Grade 3 which resulted in internal hydrides in this alloy but none in Grade 2 after electrochemical charging at cathodic overpotentials. The comparison was extended to Grade 4 alloys (Briant *et al.*, 2002) which has higher oxygen than either Grade 2 or Grade 3, but lower iron than Grade 3. Out of these three alloys it was found that only Grade 3 was susceptible during low fugacity electrochemical hydrogen charging. Thus there seemed to be a detrimental influence of iron on HE susceptibility. This impurity effect was attributed to the ease of hydride nucleation due to the presence of impurities and not due to reduced interstitial availability, which would effectively reduce hydrogen solubility. Gaseous charging of the Grade 2 alloy, however, resulted in significant hydride formation, mostly at the grain boundaries. This naturally led to hydrogen embrittlement when tested at room temperature. In this particular study the hydrogen charging was carried out at 475°C and three different amounts of hydrogen was charged into the samples. It was observed that hydrides became coarser with increasing hydrogen content in the lattice. In addition to that, at very high hydrogen contents, the alloy on observation at room temperature showed a Widmanstatten type structure, very unlike that of the samples with lower hydrogen content. Reference to the Ti-H phase diagram is necessary. The hydrogen would be in solid solution at 475°C. The lowest hydrogen content (600 ppm) sample would remain single phase, but those with higher hydrogen content would have varying amounts of β at 475°C. The hydrides would precipitate out on cooling to room temperature. The hydrides for the lower hydrogen content samples nucleated at the grain boundaries. It is suggested that for the exception, that is the sample having the highest hydrogen content, a martensitic transformation occurred, resulting in very different microstructure.

The early reports (Boyer and Spurr, 1978; Williams, 1976; Meyn, 1974) of hydrogen initiated crack propagation in α and near-α alloys were attributed to the diffusion of hydrogen to crack tips under the influence of the hydrostatic stress field at the crack tip and subsequent hydride formation at the crack tip (Westlake, 1969). As a refinement to this hypothesis, it was proposed (Hack and Leverant, 1980) that large hydrostatic stresses can result from long, blocked slip bands. Hydrogen permeation to these slip bands results in cleavage and the consequent cleavage facets are along the slip planes. *In-situ* TEM studies of the hydrogen embrittlement of Ti-4% Al alloys provided important information on the hydrogen influenced fracture (Shih *et al.*, 1988). It was observed that depending on the stress intensity, either the hydrogen enhanced localized plasticity (HELP) mechanism (at high stress intensities) or the brittle hydride cleavage mechanism (at low stress intensities) would operate. This is consistent with the fact that hydrogen accumulation is required ahead of the crack tip for hydride formation, and this would be possible only when the crack velocities are low (at low stress intensities). The same study observed that the hydrides formed were non-reversible, that is, removal of neither hydrogen nor stresses caused a reversion. This lack of reversion was attributed to the fact that the volume accommodation needed during hydride formation is brought about by plastic processes. It was also conclusively shown that dislocation mobility was enhanced on hydrogenation of α alloys. This corroborates the operation of a HELP mechanism at high stress intensities. At high stress intensities, the high crack velocities preclude significant hydrogen concentration at crack tips for the 'hydride formation and cleavage' mechanism to operate. Instead, crack advance and failure occurs by strain localization. In methanol-iodine mixtures, the cracking of CP titanium was attributed to hydrogen embrittlement (Hollis and Scully, 1993); hydride formation was not reported and it is likely that HE was due to the HELP mechanism, although not claimed by the authors.

The presence of small amounts of β, especially in the near-α alloys, can significantly alter the HE response and mechanism. This was shown for Grade-2 and Grade-12 alloys (Clarke *et al.*, 1994), the former being essentially an α alloy and the latter having small fractions of β. Under similar hydrogen charging condition, Grade-2 demonstrated ductile failure up to 500 wppm hydrogen levels, while in Grade-12 fast unstable fracture was observed at much lower hydrogen levels, but such fracture was always preceded by slow crack growth. Above 700 wppm hydrogen, Grade-2 failed in a brittle fashion, indicative of hydride formation. Various authors have attributed the slow crack growth rate observed prior to fast failure in titanium alloys to creep. It is not clear if this is a misplaced attribution brought on by the fact that hydrogen enhanced strain localization (following the HELP mechanism) may be misconstrued as creep.

In a Ti-Al-Zr alloy, HE was studied through impact testing (Liu *et al.*, 2005). This essentially single phase alloy was cathodically charged and quite high hydrogen introduction (up to 800 ppm) was possible. The hydride that resulted was δ-hydride and the Charpy impact energy decreased dramatically above a critical hydrogen concentration of 300 ppm. This hydride is incoherent with the matrix and has a ~ 23% larger specific volume and as a result stress field developed near the hydrides owing to the volume mismatch. Hydrides near the crack tips resulted in a superposition of the crack tip stress field and the hydride stress field. As a result, either the hydride grew towards the crack tip or the crack advanced into the hydride. Cleavage fracture resulted at the hydrides due to suppression of plastic deformation processes.

10.4.4   Hydrogen degradation of α–β alloys

The most important titanium alloy in terms of usage is definitely the Ti-6Al-4V. This alloy will be taken as the representative alloy for this binary alloy class and discussed in some details in this section. The reader may recollect that this class has an α matrix with dispersed β or a dispersed eutectoid structure. The crack initiation, crack propagation and extent of HE depend on a number of factors including phase fraction, stress intensity, grain shape, size and distribution, minor alloying and of course the amount of hydrogen. The crack growth during slow strain rate tests carried out on a hydrogen pre-charged Ti-6Al-4V alloy could be divided into a slow crack growth stage and a fast unstable crack growth stage. The slow crack growth was observed to be transgranular at lower hydrogen concentrations and intergranular for higher hydrogen concentrations (500 ppm). Interestingly, it was found that for lower hydrogen levels, there was an enhancement in threshold stress intensity, although the slow crack growth rate was increased by hydrogen. Most of the hydrogen under these conditions was present in the interstitial sites and do not form hydrides. For higher hydrogen contents, the solubility limits in the α phase would be exceeded. Hydrogen in solid solution would still exist in the β phase, but the α–β interface would become especially susceptible owing to either hydride precipitation or due to hydrogen supersaturation. Consequently, crack propagation at higher hydrogen concentration would essentially follow the α–β interface resulting in intergranular cracking.

The much higher diffusivity and solubility of hydrogen in the β phase implies that the continuity of this phase fraction would be important in hydrogen transport and consequent embrittlement. The duplex alloy with continuous β phase, would have hydrogen available at the crack tip. Hydrogen diffusivity is also increased with temperature. It was reported (Gu and Hardie, 1997) for this alloy that with increasing hydrogen content the slow crack initiation changed from within the α phase to the α–β interface with resulting change in the fracture mode from transgranular to intergranular. At low levels of hydrogen (below 90 ppm) it was surprisingly observed that the threshold for slow crack growth was enhanced by hydrogen presence. This was thought of as due to an inhibition by dissolved hydrogen of cleavage cracking. The fast crack growth predominated at higher hydrogen content. This was thought to be due to an enrichment of hydrogen at the crack tip as well at α–β interfaces which facilitated crack growth in an intergranular fashion.

In-situ hydrogenation during crack growth was used to study the effect of hydrogen availability on cracking kinetics in a Ti-6Al-4V alloy having equiaxed α grains with intergranular β phase (Yeh and Huang, 1997). The extent of the stable crack growth regime (Stage II) was found to be enhanced by temperature of testing. This can be explained by the decreased yield strength of the alloy with increasing temperature, enabling easy crack blunting, which more than compensated for the enhanced H diffusion. When the study was extended to include the stress intensity effects (Yeh and Huang, 1998), it was observed (Fig. 10.5) that the stage II crack growth rates were higher for lower initial applied stress intensities (40–50 MPa m$^{1/2}$) although the threshold stress intensity for crack initiation was also higher as compared to that for the higher initial applied stress intensities ($\geq$ 50 MPa m$^{1/2}$). The higher crack growth velocities were associated with brittle fracture. This apparent anomaly in crack growth rates vis-à-vis the applied stress intensity was rationalized as follows. An increasing intensity would increase the plastic zone size ahead of the crack tip. The consequent

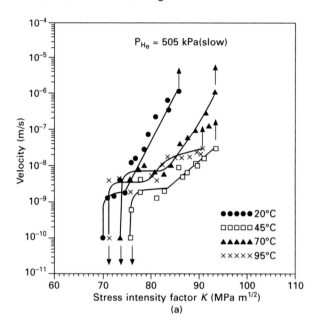

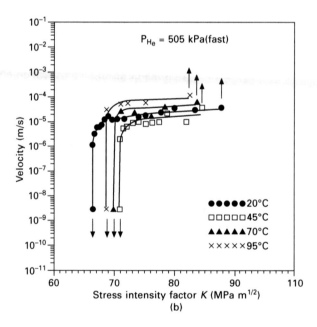

10.5 Crack growth rate in Ti-6Al-4V at 505 kPa H_2 from 20 to 95°C: (a) specimens tested under higher initial stress intensities, $K_{ini} > 55$ MPa $m^{1/2}$ and (b) specimens tested under lower initial stress intensities, $40 < K_{ini} < 50$ MPa $m^{1/2}$. From Yeh and Huang (1998).

reduction in hydrostatic tensile stress would mean that the hydrogen entry would be reduced leading to less hydride formation at the crack tip. In such circumstances the crack rate would be controlled by hydrogen induced strain localization and not by hydride cracking. This would mean that brittle cracking would occur under conditions of reduced plastic zone size (lower initial applied stress intensities) or with higher hydrogen availability. The hypothesis was proved by charging with high hydrogen fugacities at high applied stress intensities, when a fracture mode transition from relatively ductile to brittle failure occurred. Other reports (Hardie and Ouyang, 1999) throw up similar anomalies with regard to hydrogen content influence on brittle cracking for Ti-6Al-4V (Fig. 10.6), while establishing the importance of crack tip stress intensity in HE. Smooth specimens with negligible initial stress intensities when tested in embrittling environment at low strain rates, demonstrated a reduction in ductility, but introduction of stress intensity by using CT specimens caused similar reduction of ductilities at one-tenth the amount of hydrogen. It was observed that brittle cracking was the final failure mode; this may be preceded by slow crack growth for low hydrogen concentrations. It was also reported that the threshold for slow crack growth (K_s) increased with hydrogen content while that for fast crack growth (K_H) increased with hydrogen content. The increase in K_s is thought to be due either to hydrogen reducing creep rate or a crack tip blunting brought on by hydrogen induced plasticity at the crack tip.

Such anomalies in hydrogen-induced behaviour have also been reported with regard to fracture toughness for another α–β alloy, Ti-8Al-1Mo-1V

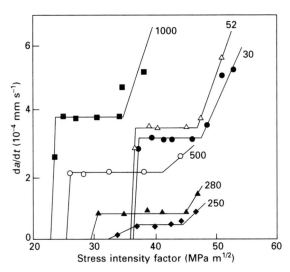

10.6 Crack growth rate (d*a*/d*t*) as a function of applied stress intensity for different hydrogen content (in ppm). From Hardie and Ouyang (1999).

(Orman and Picton, 1974). In this alloy, for hydrogen greater than 65 ppm, an increase in K_{IC} and K_{ISCC} were observed with increasing hydrogen content, although the corresponding values for the unhydrogenated alloy could not be achieved by hydrogenation. On removal of hydrogen below 65 ppm there was an increase in K_{IC} and K_{ISCC} with lowering of hydrogen, thus a property minima was observed around 65 ppm hydrogen. No satisfactory explanation for this effect was provided by the researchers; however, these findings may be explained along similar lines as indicated earlier (Yeh and Huang, 1997, 1998; Hardie and Ouyang, 1999; Gu and Hardie 1997).

The effect of the β phase distribution in Ti-6Al-4V was examined (Tal-Gutelmacher et al., 2004) by producing two different microstructures, one duplex and the other fully lamellar (beta annealed) by varying the heat treatment, as shown in Fig. 10.7. The hydrogen absorption in the beta annealed structure was observed to be significantly higher. Hydrogen introduction caused lattice strains in both the phases of the beta annealed alloy as was evident from X-ray diffraction studies, but negligible lattice strain in the duplex alloy, which corroborates the high hydrogen absorption in the former. TEM observations revealed a high density of dislocations inside the β lamellae as well as in the α regions in the beta annealed alloy. No hydride precipitation was observed in any of the alloys. A similar study on the effect of microstructure was carried out for Ti-8Al-1Mo-2V (Rusli, 2008). Two microstructures, one having equiaxed α grains with β grains distributed discontinuously and another with coarse Widmanstatten structure having colonies of α separated by continuous β network, were produced by suitable heat treatments. The equiaxed alloy did not demonstrate slow crack growth during constant hold loads close to the respective tensile strengths, nor was there a significant decrease in tensile strength on straining to failure in hydrogen. The failure was essentially by overload with characteristic ductile microvoids with some evidence of cracking along α–β interfaces. The Widmanstatten alloy, on the other hand, demonstrated slow crack growth and much reduced tensile strength. The crack proceeded along the α–β interfaces (Fig. 10.8) and only small regions of overload failures were observed. It was hypothesized that hydride formed at the interfaces were the sites of crack propagation.

10.4.5   Hydrogen degradation of β and near-β alloys

The high solubility of hydrogen in the β phase would retard hydride formation to very high hydrogen concentrations. This benefit in terms of HE resistance is offset somewhat by generally higher strength of these alloys. Additionally these alloys have to be aged for improvement of mechanical properties, and the consequent microstructural changes can also influence HE. It has been observed that even the soluble hydrogen exerts a definite influence on

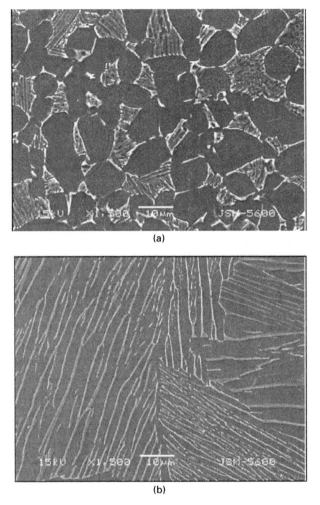

(a)

(b)

10.7 Ti-6Al-4V alloy showing (a) duplex microstructure; (b) lamellar microstructure with continuous β. From Tal-Gutelmacher *et al.* (2004).

mechanical properties. However, in terms of environment severity, these alloy grades are the most resistant to HE of the three classes of titanium alloys. The important phenomena related to HE of β alloys are as follows (Tal-Gutelmacher and Eliezer, 2004). The degradation of mechanical properties is usually through the formation of the brittle δ-hydride. Hydrogenation of these alloys can cause a transformation of the α phase to β at elevated temperatures. As the α phase is the main source of strengthening in these alloys, this transformation causes a decrease in the alloy strength. It was also observed that at hydrogen contents much lower than that required for hydride precipitation, significant mechanical property degradation was possible.

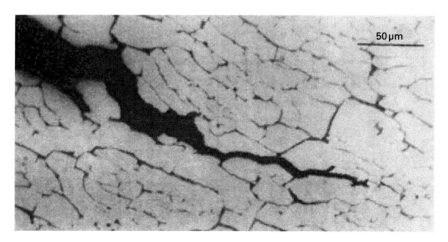

10.8 Hydrogen-induced crack propagation along α–β interface in a Ti-8Al-1Mo-2V alloy. From Rusli (2008).

Formation of the fcc δ-hydride was responsible for embrittlement of the Ti-30Mo alloy (Shih and Birnbaum, 1986) On the other hand, significant embrittlement of the same alloy has been reported for hydrogen contents (up to 1800 ppm) that do not produce hydride precipitation in the β phase (Gerberich *et al.*, 1981). Comparative studies on hydrogen embrittlement resistance of Beta 21S alloy (Ti-15Mo-3Nb-3Al) and 15-3 (Ti-15V-3Cr-3Al-3Sn) were carried out (Young and Scully, 1994). These alloys were heat treated to their respective solution heat treated (SHT), peak aged (PA) and duplex-aged (DA) structures. The embrittlement tests were carried out with different notch-geometry samples thereby enabling studies of the constraint (notch tip stress concentration) effects on HE. The hydrogen absorption by 15-3 was many times greater than that for 21S. Neither of the alloys showed any hydride formation at the hydrogen charging conditions used. In spite of the greater hydrogen absorption by 15-3, it demonstrated much higher HE resistance in the SHT condition. The HE resistance of both alloys decreased with aging (PA and DA conditions) but even in the aged conditions the 15-3 demonstrated better resistance. The fracture in 15-3 for all heat treatment conditions and all constraints tested was essentially ductile microvoid, although at higher hydrogen concentration in the DA alloy transgranular ductile tearing was observed. Distinct changes in the fracture mode were observed in the aged 21S alloy with increasing hydrogen content as well as with increasing constraint. At high hydrogen content, distinct cleavage facets were observed. XRD studies indicated that there was significant lattice dilation of the β phase while the α phase remained unaffected, suggesting that hydrogen partitioning occurred predominantly into the β phase. In aged 21S alloys, grain boundary α precipitation was observed. This led to the postulation that in these alloys with fine grain boundary α, shearing of the α precipitates would lead to planar slip with subsequent concentration of hydrogen on the slip bands thereby sustaining the process of fracture.

Another study (Somerday *et al.*, 2000) reported that cracking in Beta C and Beta 21S alloys could be observed at relatively high loading rates. A classical HE hypothesis is that the extent of embrittlement would be inversely proportional to the bulk strain rate or the crack tip strain rate. The situation in titanium alloys is slightly different especially in aqueous environments. This is due to the ease of oxide barrier formation, which is stable even at cathodic overpotentials. Thus a sufficient strain rate at the crack tip is required to break up surface films and allow enough time for hydrogen entry before surface repassivation occurs. An oscillation in the crack growth rate was observed for both alloys which could be reconciled to its dependence on the crack tip stress intensity and the crack tip strain rates, which fluctuated with crack advance.

Comparison of a near-β (Ti-10V-2Fe-3Al) and stable β (Alloy-C (Ti-35V-15Cr)) alloys with regard to hydrogen influence on mechanical properties was carried out (Christ *et al.*, 2003). The stable β alloy showed a significant decrease in ductility on hydrogenation along with an increase in the tensile strength, a similarly pronounced decrease in fatigue cracking threshold, and an increase in the ductile-brittle transition temperature [DBTT]. By contrast, the near-β alloy showed only slight decrease in ductility along with a softening, and surprisingly the fatigue crack threshold was found to increase on hydrogenation (Fig. 10.9). This is reasoned as due to the extrinsic hydrogen effect of suppression of the α precipitation leading to a lowered yield strength and therefore lower local stress amplitudes at the crack tip.

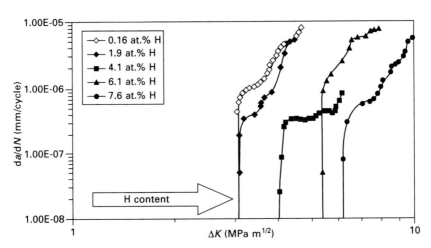

10.9 Effect of hydrogen on the fatigue crack propagation rate of near-β Ti-10V-2Fe-3Al. From Christ *et al.* (2003).

The hydrogen effect on DBTT in Beta 21S alloys was explained as due to hydrogen enhanced decohesion (Teter *et al.*, 2001). This explanation was a result of elimination of the other possible mechanisms of HE as stated below. This study on Beta 21S used *in-situ* TEM studies to show that hydrogen enhanced dislocation motion occurred in these alloys; however, at and above critical hydrogen concentrations there was limited evidence of plastic deformation to substantiate a HELP mechanism. Another important observation was that hydrogen caused a suppression of ω phase in these alloys. The fracture surfaces (Fig. 10.10) varied from microvoid coalescence at low hydrogen values, to quasi-cleavage with some localized plasticity above a critical hydrogen concentration. Neither was hydride formation observed, thereby dismissing a hydride-based mechanism of ductility loss.

In metastable β alloys, like Beta C, some α coexist with β. At a critical hydrogen concentration the α phase transforms to body centred tetragonal (bct) hydrides, and to δ-hydrides on further increase in hydrogen concentration in the matrix (Alvarez *et al.*, 2004). This transformation was stress-induced as it was observed only in front of crack tips. For low hydrogen concentrations, the hydrogen was in solid solution in α and β phases; this would facilitate HELP. Hydrogen can accumulate at grain boundaries leading to some intergranular features, but the fracture would be predominantly transgranular with slip band

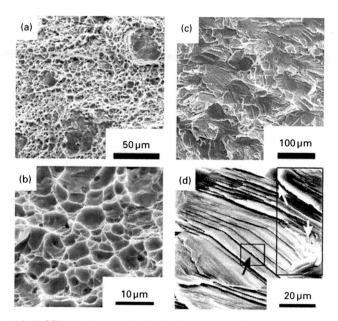

10.10 SEM fractographs of the fracture surfaces of specimens tested at 298 K. (a, b) H/M = 0.014; (c, d) H/M = 0.21. The white arrows indicate regions of plasticity on the fracture surface and the black arrows in (d) indicate a small microvoid. From Teter *et al.* (2001).

markings. For higher hydrogen contents, stress-induced hydrides would form near crack tips. The hydride formation is accompanied by volume increase and consequent compressive stresses, which should temporarily arrest the cracks. But the crack tip stresses may be sufficient to cleave the cracks and advance, or the cracks are cleaved when the load is increased. Crack arrest markings were observed in these studies. As the crack propagated, its velocity increased and a situation may have occurred where the crack velocity was too rapid to allow hydride formation. Under such situations the HELP mechanism would be triggered which would result in flat slip planes with ductile tear ridges separating them, as was reported.

10.4.6 Alloy class specific hydrogen degradation: summing up

The low solubility of hydrogen in the HCP α phase means that for α and near-α alloys, hydrogen embrittlement is invariably linked to hydride formation. During ambient electrochemical charging, no phase changes are observed and surface hydrides and internal hydrides are formed. But elevated temperature gaseous charging, followed by subsequent cooling to room temperature, causes excursions into the β phase field and martensitic transformations on cooling are reported. Such phase changes have a further adverse effect on HE resistance.

In the two-phase alloys, irrespective of whether it belongs to the near-α, α–β, near β or metastable-β, the volume fractions of the phases, their shape and the continuity of the β phase play important roles in hydrogen degradation. The much higher solubility of hydrogen in the β phase would indicate an inherent resistance to HE as hydride precipitation is postponed to high hydrogen contents. But the much higher diffusivity of hydrogen through β implies that a continuous β phase provides easy conduit for hydrogen to reach the α phase and the α–β interfaces and then precipitate there. Thus the α–β interfaces are often observed to be the crack pathways in these alloys. The situation is further complicated by β stabilization by hydrogen.

In pure β alloys, much higher hydrogen contents are required for hydride precipitation. This is to some extent counterbalanced by the high hydrogen diffusivity. The high amounts of soluble hydrogen can have a profound effect on the fracture processes, often invoking the HELP mechanism or, less often, the hydrogen enhanced decohesion. The former leads to slip oriented fractures and the latter to cleavage or quasi cleavage.

10.4.7 HE of titanium alloys in biomedical applications

There is increasing demand for titanium alloys in biomedical applications like implants, prosthetic devices, fasteners, orthodontic appliances, and others, on account of their excellent corrosion resistance, non-toxicity and bio-compatibility, low thermal coefficient and high specific strength. All three classes of titanium alloys discussed earlier are used in biomedical applications.

Under no-load situations titanium alloys are very resistant to chemical attack in physiological media due to the regenerative power of the oxide film on titanium. Under *in-vivo* conditions, these alloys see fairly severe environments, and alloys used in load-bearing implants undergo high cyclic loads. Such loads activate the oxide layer. Most tests on biomedical alloys are carried out in simulated physiological solutions like Ringer's solution, Hank's solution or brine. In reality (*in-vivo*), implants experience protein containing serums as their environment which has a definite effect on corrosion resistance (Fleck and Eifler, 2010).

A substantial literature is available on the studies carried out in fluoride solutions (Yokoyama *et al.*, 2004, 2006; Ogawa *et al.*, 2005; Kaneko *et al.*, 2004). This is essentially to simulate oral cavity situations, where fluoride is expected in toothpastes and dental rinses. The fluoride ranges from 100–10 000 ppm and the pH ranges from 3.5 to neutral. The acidic conditions are generated by the metabolic activities of oral bacteria. Therefore conducive conditions for both SCC as well as HE exist. Tests on CP titanium showed that the time to fracture under constant load situations were lower for acidified phosphate fluoride environments as compared to neutral fluoride environments (Yokoyama *et al.*, 2004). The above findings are indicative of HE as the degrading mechanism. Thermal desorption curves showed (Yokoyama *et al.*, 2004, 2006) a high amount of hydrogen accumulation in the alloy exposed to acidified solution. Similar desorption studies carried out in acidic fluoride solutions on α and β alloys (Ogawa *et al.*, 2005) showed two desorption peaks for the α alloy; and the hydrogen resided mostly in a surface hydride layer. In the β alloy the hydrogen was found to be uniformly distributed, which is hardly surprising considering the higher diffusivity and solubility of hydrogen in β. It was observed (Yokoyama *et al.*, 2006) that while α alloys absorb hydrogen even in neutral solutions, α–β and β alloys do not do so even after long exposures. The comparative performance of Ni-Ti, β-Ti, SS and Co-Cr-Ni alloys for orthodontic applications has been reported (Kaneko *et al.*, 2004). The titanium-containing alloys were found to be degraded in comparison to the other two on account of significant hydrogen absorption by both. All these reports would indicate that for orthodontic applications titanium alloys are possibly not very suitable.

In prosthetic devices, fretting and crevice corrosion of a Ti-6Al-4V alloy was observed (Rodrigues *et al.*, 2009) along with γ-hydride formation. The gait simulation loading seemed to decrease lifetime as compared to normal cyclic loading (Leinenbach *et al.*, 2005).

10.4.8   Hydrogen-enhanced fatigue in titanium alloys

Corrosion fatigue and hydrogen-enhanced fatigue are truly not within the scope of this chapter, but as a matter of continuity, it is felt necessary to discuss some of the relevant reports. Cyclic loading under conditions of simultaneous hydrogen entry, or on pre-charged samples are discussed briefly.

In Ti-6Al-4V, it was observed (Ding *et al.*, 2009) that the fatigue crack growth rates in gaseous hydrogen was less than that in air at lower load ratios, but this approached that for air at higher load ratios. For Ti-15V-3Cr-3Al-3Sn alloys, the fatigue crack growth rates [FCGRs] under hydrogen charging were also found to be slightly lower than those in air at medium ΔK and approached those for air at low ΔK (Nakasa and Satoh, 1996). A reasonable explanation for this phenomenon was not forthcoming. For a two-phase alloy [Ti-24Al-11Nb] the FCGRs for hydrogen and argon atmospheres were compared (Christ *et al.*, 1996). No differences in the FCGRs were observed during elevated temperature testing, although significant embrittlement due to hydrogen was observed at RT through tensile tests. This was possibly due to the presence of hydrogen in solution in the matrix at elevated temperatures. Two possibilities for this observed decrease in FCGR under hydrogen charging for different titanium alloys are phase transformation induced crack closure and introduction of a compressive stress field at the crack tip due to hydride formation. Neither of these theories has been conclusively proven.

In Ti-13V-11Cr-3Al-3Sn alloys, the FCGRs are higher in hydrogen atmospheres compared to those in air at low ΔK but approach those of air at high ΔK (Nakasa and Satoh, 1996). This was attributed to hydrogen enhanced decohesion or cleavage from fractographic observations, which did not reveal any hydride phase formation. Once again, one is at a loss to explain why the FGCR approaches that in air at high ΔK, since higher loads should facilitate further decohesion and not retard it.

There seems to be no significant effect of hydrogen on the fatigue properties especially on FCGR and even on the threshold values for titanium alloys, even when the alloy may show significant hydrogen degradation under constant loading or monotonically increasing loading. This is also observed for the metastable β alloy Ti-3Al-8V-4Mo-4Zr (Christ *et al.*, 1996), which showed loss in ductility on hydrogen charging, but no effect of hydrogen on the fatigue life. This apparent inertness to hydrogen effects in cyclic loading is baffling at first glance. But it has to be realized that in cyclic loading, the stress field ahead of the crack tip (most samples are pre-cracked) fluctuates in magnitude and even in sign (+/−). Even when the load ratio is positive so that the remote load cycling is between positive stresses, the plastic zone ahead of the crack tip can see substantial compressive stresses in the unload cycle. Thus the stress-induced hydride formation as well as hydrogen diffusion ahead of the crack tip may be affected in a complex, non-conducive fashion.

Whether this is the reason for a relative insensitivity to hydrogen in cyclic loading of titanium alloys has not been explored.

10.5   Conclusions

Titanium alloys are poised for rapid commercialization. But their cost of production and some chinks in their corrosion resistance, especially in the presence of hydrogen, are obstacles that have to be overcome. *Per se*, titanium alloys do not have poor resistance to hydrogen, weight by weight they can sustain many times more hydrogen presence than most commercial alloys. But the high cost of the alloys limit their use to very severe environments where the hydrogen contents are substantial. It is under these conditions that titanium alloys are found to be prone to degradation.

The inherent corrosion resistance of titanium alloys is due to a very protective passive oxide layer. The SCC in titanium and its alloys therefore occurs as a result of compromise of this passive film. One environment to which it is especially susceptible is acidified fluorides. This is partly due to the formation of non-protective fluoride corrosion products and partly due to the hydrogen entry during the corrosion process. In most reported cases of SCC of titanium alloys, hydrogen invariably has a role. Even in anodic conditions, the cathodic reaction is generally hydrogen liberation and hydrogen finds its way into the titanium alloy. Titanium alloys have very high solubility limits for hydrogen, especially if β phase is present and low amounts of hydrogen do not cause much problem.

The hydrogen degradation of titanium alloys is definitely class specific, with the pure β alloys (bcc) showing the most resistance owing to their high solubility limits for hydrogen and the α alloys (hcp) least so. However, β alloys have to be age hardened in actual practice which often results in varying fractions of α phase which provides strengthening. This complicates the hydrogen response of aged β alloys, and α–β alloys. The hydrogen can move easily and quicken through the β phase without precipitating as hydrides, but once they move into the α phase, precipitation of hydrides are inevitable which often occurs at α–β interfaces. It is no surprise that cracks often follow these interfaces. It is also obvious that the continuity of the β phase would have a direct correlation with the hydrogen response of these alloys.

The hydride precipitation is accompanied with an increase in volume; therefore tensile stresses especially at a crack tip tend to promote its formation due to the significant stress intensification at the crack tip. In electrochemical charging, local acidification at the crack tip enhances hydrogen production even in neutral conditions. The high crack tip stresses also facilitate passive film breakdown. Thus a sample with a constraint (notch, pre-crack, etc.) is always found to embrittle at significantly lower hydrogen levels compared to a smooth sample. It is not surprising to find ambiguity of hydrogen effects on fatigue of titanium alloys as cyclic loading does not provide a sustained crack tip tensile stress for activating the hydride precipitation.

Although in titanium alloys hydride formation is the main reason for hydrogen degradation of the mechanical properties of the alloy, it is by no means the only reason. The HELP mechanism as well as decohesion mechanism have been invoked to explain loss in ductility in which no hydride formation was observed, the former for situations resulting in transgranular fracture with slip markings, and the latter for situations demonstrating cleavage or quasi-cleavage. Such situations are naturally more prevalent in β alloys where the high hydrogen solubility postpones hydride precipitation to very high hydrogen contents. It is worthwhile noting that hydrides formed at the crack tips would redissolve after crack advance so that hydrides may not be observed on *ex-situ* observation of the fracture surface. However, this would be possible only if the hydride-induced volume expansion is accommodated elastically and no plastic deformation is involved, which is rarely the case. Hydrogen-enhanced dislocation mobility has been observed in titanium alloys, but in some cases this did not result in significant plastic signatures to justify a HELP-induced embrittlement. On the other hand the creation of a hydride is followed by a compressive stress at the crack tip which would reduce plastic processes and enhance hydride cleaving. The mechanism of decohesion is possible and is reported in titanium alloys as very high hydrogen concentrations are possible in these alloys unlike other commercial alloys, sufficient for decohesion or for sufficient reduction in interatomic bond strengths. But even in titanium alloys, decohesion is often invoked for the last stages of fracture, when crack growth velocities are too high to sustain either a hydride cleavage mechanism or a HELP mechanism.

It is evident from these discussions that the SCC of titanium alloys, especially when hydrogen plays a part, is not a closed chapter. The issues which are unresolved are many, including crack tip strain rate effects, apparent anomaly on fatigue crack growth threshold, the applicable hydrogen degradation mechanism, the role of certain interstitials, and others. For increased application of titanium alloys, addressing the SCC and HE issues in general and these issues in particular are required.

10.6   Acknowledgements

The author would like to acknowledge the help of Dr. S.K. Pal, Dr. N.G. Goswami and Mr. S.N. Hembram, of National Metallurgical Laboratory, Jamshedpur, in preparing this article.

10.7   References

Alvarez A-M, Robertson IM and Birnbaum HK (2004), 'Hydrogen embrittlement of a metastable β-titanium alloy', *Acta Mater.*, 52, 4161–4175.

Balazic M, Kopac J, Jackson MJ and Ahmed W (2007), 'Titanium and titanium alloy applications in medicine', *Int. J. Nano & Biomater.*, 1, 3–34.

Been J and Faller K (1999), 'Using Ti-5111 for marine fastener applications', *JOM*, 51, 21–24.

Boyer RR (1996), 'An overview on the use of titanium in the aerospace industry', *Mater. Sci. Engr.*, A213, 103–114.

Boyer RR and Spurr WF (1978), 'Characteristics of sustained-load cracking and hydrogen effects in Ti-6Al-4V', *Metall. Trans.*, 9A, 23–29.

Briant CL, Wang ZF and Chollocoop N (2002), 'Hydrogen embrittlement of commercial purity titanium', *Corros. Sci.*, 44, 1875–1888.

Christ H-J, Alvarez A-M, Birnbaum HK and Robertson IM (1996), 'The influence of hydrogen on the fatigue behaviour of the beta-titanium alloy Ti-3Al-8V-6Cr-4Mo-4Zr', *Fatigue Fract. Engng. Mater. Struct.*, 19, 1421–1434.

Christ H-J, Senemmar A, Decker M and Prüßner K (2003), 'Effect of hydrogen on mechanical properties of β-titanium alloys', *Sādhanā*, 28, 453–465.

Chu HP (1972), 'Fracture characteristic of titanium alloys in air and seawater environment', *Engr. Frac. Mech.*, 4, 107–117.

Clarke CF, Hardie D and Ikeda BM (1994), 'The effect of hydrogen content on the fracture of pre-cracked titanium specimens', *Corros. Sci.*, 36, 487–509.

Clarke CF, Hardie D and Ikeda BM (1997), 'Hydrogen-induced cracking of commercial purity titanium', *Corros. Sci.*, 39, 1545–1559.

Ding YS, Tsay LW and Chen C (2009), 'The effects of hydrogen on fatigue crack growth behaviour of Ti-6Al-4V and Ti-4.5Al-3V-2Mo-2Fe alloys', *Corros. Sci.*, 51, 1413–1419.

Eliezer D, Tal-Gutelmacher E, Cross CE and Boellinghaus T (2006a), 'Hydrogen absorption and desorption in a duplex-annealed Ti-6Al-4V alloy during exposure to different hydrogen-containing environments', *Mater. Sci. Engr.*, A433, 298–304.

Eliezer D, Tal-Gutelmacher E, Cross CE and Boellinghaus T (2006b), 'Hydrogen trapping in β-21S titanium alloy', *Mater. Sci. Engr.*, A421, 200–207.

Fleck C and Eifler D (2010), 'Corrosion, fatigue and corrosion fatigue behaviour of metal implant materials, especially titanium alloys', *Inter. J. Fatigue*, 32, 929–935.

Gao KW, Chu WY, Gu B, Zhang TC and Qiao LJ (2000), '*In-situ* transmission electron microscopic observation of corrosion-enhanced dislocation emission and crack initiation of stress corrosion', *Corrosion*, 56, 515–522.

Gerberich WW, Moody NR, Jensen CL, Hayman C and Jatavallabhula K (1981), *Hydrogen Effects in Metals*, Warrendale PA, TMS of AIIME, 731–745.

Gregory JK and Brokmeier H-G (1995), 'The relationship between crystallographic texture and salt water cracking susceptibility in Ti-6Al-4V', *Mater. Sci. Engr.*, A203, 365–372.

Gu J and Hardie D (1997), 'Effect of hydrogen on the tensile ductility of Ti6Al4V', *J. Mater. Sci.*, 32, 609–617.

Guo XZ, Gao KW, Chu WY and Qiao LJ (2003), 'Correlation between passive film-induced stress and stress corrosion cracking of α-Ti in a methanol solution at various potentials', *Mater. Sci. Engr.*, 346, 1–7.

Hack JE and Leverant GR (1980), 'A model for hydrogen-assisted crack initiation on planar shear bands in near-alpha titanium alloys', *Scripta Mater.*, 14, 437–441.

Hamada S, Ohnishi K, Nishikawa H, Oda Y and Noguchi HJ (2009), 'SIMS analysis of low content hydrogen in commercially pure titanium', *J. Mater. Sci.*, 44, 5692–5696.

Hardie D and Ouyang S (1999), 'Effect of hydrogen and strain rate upon the ductility of mill-annealed Ti6Al4V', *Corros. Sci.*, 41, 155–177.

Hollis AC and Scully JC (1993), 'The stress corrosion cracking and hydrogen embrittlement of titanium in methanol-iodine solutions', *Corros. Sci.*, 34, 837–850.

Kaneko K, Yokoyama K, Moriyama K, Asaoka K and Sakai J (2004), *Angle Orthodontist*, 74, 487–495.

Kolman DG, Gaudett MA and Scully JR (1998), 'Modeling of anodic current transients resulting from oxide rupture of plastically strained β+α titanium', *J. Electrochem. Soc.*, 145, 1829–1840.

Leinenbach C, Schwilling B and Eifler D (2005), 'Cyclic deformation behaviour and fatigue induced surface damage of titanium alloys in simulated physiologica media', *Mater. Sci. Engr.*, C25, 321–329.

Liu YZ, Zu XT, Li C, Qiu SY, Li WJ and Huang XQ (2005), 'Hydrogen embrittlement of a Ti-Al-Zr alloy evaluated by impact test method', *Scripta Mater.*, 52, 821–825.

Lu H, Gao KW, Qiao LJ, Wang YB and Chu WY (2000), 'Stress corrosion cracking caused by passive film-induced tensile stress', *Corrosion*, 56, 1112–1118.

Meyn DA (1974), 'Effect of hydrogen on fracture and inert-environment sustained load cracking resistance of α–β Ti alloys', *Metall. Trans.*, 5, 2405–2414.

Mizuno T (1977), *Boshoku-Gijutsu*, 26, 185–192.

Nakasa K and Satoh H (1996), 'The effect of hydrogen-charging on the fatigue crack propagation behavior of β-titanium alloys', *Corros. Sci.*, 38, 457–468.

Ogawa T, Yokoyama K, Asaoka K and Sakai J (2005), 'Distribution and thermal desorption behavior of hydrogen in titanium alloys immersed in acidic fluoride solutions', *J. Alloys Comp.*, 396, 269–274.

Orman S and Picton G (1974), 'The role of hydrogen in the stress corrosion cracking of titanium alloys', *Corros. Sci.*, 14, 451–459.

Pound BG (1991), 'Hydrogen ingress in titanium', *Corrosion*, 47, 99–104.

Pound BG (1994), 'The effect of aging on hydrogen trapping in β-titanium alloys', *Acta Metall. Mater.*, 42, 1551–1559.

Pound BG (1997), 'Hydrogen trapping in aged β-titanium alloys', *Acta Mater.*, 45, 2059–2068.

Rodrigues DC, Urban RM, Jacobs JJ and Gilbert JL (2009), '*In vivo* severe corrosion and hydrogen embrittlement of retrieved modular body titanium hip-implants', *J. Biomed. Mater. Res.*, 88B, 206–210.

Rusli RH (2008), 'Role of hydrogen environment induced hydrogen embrittlement of Ti-8Al-1Mo-2V alloy', *Mater. Sci. Engr.*, A494, 143–146.

Schutz RW and Watkins HB (1998), 'Recent development in titanium alloy applications in the energy industry', *Mater. Sci. Engr.*, A243, 305–315.

Shih DS and Birnbaum HK (1986), 'Evidence of FCC titanium hydride formation in β titanium alloy: an X-ray diffraction study', *Scripta Metall.*, 20, 1261–1264.

Shih DS, Robertson IM and Birnbaum HK (1988), 'Hydrogen embrittlement of α titanium: *in situ* TEM studies', *Acta Metall.*, 36, 111–124.

Simbi DJ and Scully JC (1995), 'The effect of iron and three interstitial elements on the stress corrosion cracking of titanium in seawater', *Corros. Sci.*, 37, 1325–1330.

Simbi DJ and Scully JC (1997), 'Aspects of fracture morphology exhibited in the stress corrosion behavior of Ti-155 in methanol-hydrochloric acid solution', *Corrosion*, 53, 298–305.

Somerday BP and Gangloff RP (1998), 'Effect of strength on environment-assisted cracking of Ti-8V-6Cr-4Mo-4Zr-3Al in aqueous NaCl Part 1: Age hardening vs. work hardening', *Mater. Sci. Engr.*, A254, 166–178.

Somerday BP, Young LM and Gangloff RP (2000), 'Crack tip mechanics effects on environment-assisted cracking of beta-titanium alloys in aqueous NaCl', *Fatigue Fract. Engng. Mater. Struct.*, 23, 39–58.

Tal-Gutelmacher E and Eliezer D (2004), 'Hydrogen based degradation of titanium based alloys', *Mater. Trans.*, 45, 1594–1600.

Tal-Gutelmacher E and Eliezer D (2005), 'The hydrogen embrittlement of titanium-based alloys' *JOM*, 57, 46–49.

Tal-Gutelmacher E, Eliezer D and Eylon D (2004), 'The effects of low fugacity hydrogen in duplex- and beta-annealed Ti-6Al-4V alloy', *Mater. Sci. Engr.*, A381, 230–236.

Teter DF, Robertson IM and Birnbaum HK (2001), 'The effects of hydrogen on the deformation and fracture of β-titanium', *Acta Mater.*, 49 4313–4323.

Wang ZF, Briant CL and Kumar KS (1998), 'Hydrogen embrittlement of grade 2 and grade 3 titanium in 6% sodium chloride solution', *Corrosion*, 54, 553–560.

Westlake DG (1969), *Trans. ASM*, 62, 1000–1006.

Williams DN (1976), 'Effects of hydrogen in titanium alloys on subcritical crack growth under sustained load', *Mater. Sci. Engr.*, 24, 53–63.

Xiao HZ (1992), 'On the mechanism of hydride formation in a-Ti alloys', *Scripta Metall. Mater.*, 27, 571–576.

Yamada M (1996), 'An overview on the development of titanium alloys for non-aerospace application in Japan', *Mater. Sci. Engr.*, A213, 8–15.

Yeh M-S and Huang J-H (1997), 'Internal hydrogen induced subcritical crack growth in Ti-6Al-4V', *Scripta Mater.*, 36, 1415–1421.

Yeh M-S and Huang J-H (1998), 'Hydrogen-induced subcritical crack growth in Ti-6Al-4V alloy', *Mater. Sci. Engr.*, A242, 96–107.

Yokoyama K, Kaneko K, Miyamoto Y, Asaoka K, Sakai J and Nagumo M (2004), 'Fracture associated with hydrogen absorption of sustained tensile-loaded titanium in acid and neutral fluoride solutions', *J. Biomed. Mater. Res.*, 68A, 150–158.

Yokoyama K, Ogawa T, Asaoka K and Sakai J (2006), Hydrogen absorption of titanium and nickel-titanium alloys during long-term immersion in neutral fluoride solution', *J. Biomed. Mater. Res.*, 78B, 204–210.

Young Jr. GA and Scully JR (1994), 'Hydrogen embrittlement of solution heat-treated and aged β-titanium alloys Ti-15%V-3%Al-3%Sn and Ti-15%Mo-3%Nb-3%Al', *Corrosion*, 50, 919–933.

Zhang XG and Vereecken J (1990), 'Stress corrosion cracking mechanism of Ti-6Al-4V in acidic methanol', *Corrosion*, 46, 136–141.

Zhao JW, Ding H, Hou HL and Li ZQ (2010), 'Influence of hydrogen content on hot deformation behavior and microstructural evolution of Ti600 alloy', *J. Alloys Comp.*, 491, 673–678.

11

Stress corrosion cracking (SCC) of copper and copper-based alloys

M. BOBBY KANNAN, James Cook University, Australia
and P. K. SHUKLA, Southwest Research Institute, USA

Abstract: This chapter summarizes stress corrosion cracking of copper and copper-based alloys in different chemical, thermal, and electrochemical environments. The chapter begins with description of different copper-based alloys and their common application. A description of different operating SCC mechanisms for the copper-based alloys is provided. Information regarding the chemical, thermal and electrochemical conditions that could induce SCC of the copper and copper-based alloys and additional details on the SCC mechanism of particular copper-based alloys in specific chemical conditions are also detailed. The role of secondary phase particles on SCC of copper-based alloys is discussed. SCC mitigation strategies for the copper-based alloys are discussed. A summary of the chapter is provided in the end. The aim of this chapter is to develop an understanding of SCC susceptibility of different copper-based alloys in different chemical environments.

Key words: brass, tin bronze, cupronickel, aluminum bronze, secondary-phase particles.

11.1   Introduction

Copper and its alloys have found use in various industrial applications. Copper is alloyed with zinc, tin, nickel, aluminum, and silicon to produce various types of brass, bronze, cupronickel, aluminum bronze, and silicon bronze, respectively. A list of commercially used copper alloys, their composition, and applications is provided in Table 11.1. Although copper and its alloys are resistant to atmospheric corrosion, they are prone to localized degradation such as stress corrosion cracking in industrial environments. Several copper-based alloys used in manufacturing heat exchange tubes, gears for pumps, valves, control rods, impellers, turbines and compressor blades have undergone SCC. Agarwal and Bapat (2009) reported SCC in cupronickel alloy structures in marine environment containing a small amount of ammonia and sulphides ions. Rao and Nair (1998) reported that admiralty brass condenser tubes, used in a nuclear power plant cooled by fresh water, underwent SCC due to the presence of nitrate-reducing bacteria in the cooling water resulting in the formation of ammonia, causing failure.

Table 11.1 List of commercially used copper alloys

Name	Composition	Applications
Brass	Brass is predominantly a mixture of copper and zinc. Most commercially used cartridge's brass composition is 70 wt% Cu and 30 wt% Zn. Brass made with 60 wt% Cu and 40 wt% Zn is known as Muntz metal. Sometimes, small amounts of lead and iron are also added.	Good for cold-working, radiators, hardware, electricals, external casing of bullets, and condensers, evaporators and heat exchanger tubes.
Tin bronze	Commercially used tin bronze contains approximately 88 wt% Cu and 9–10 wt% Sn. Tin bronzes are also alloyed with phosphorous in the range of 0.15–0.5 wt%, lead in the range of 0.05–0.25 wt%, and zinc in the range of 0.05–0.3 wt%.	Commonly used in marine pumps, valves, bearings, brushings, piston rings, steam fittings, gears, and high-pressure pumps.
Cupronickel	Two forms of cupronickel are most commonly used: (i) alloy with 70 wt% Cu and 30 wt% Ni and (ii) alloy with 88.3 wt% Cu, 10 wt% Ni, 1.3% wt Fe and 0.4 wt% Mn.	Applications include piping, heat exchangers and condensers in seawater systems as well as marine hardware, and sometimes for the propellers, crankshafts and hulls of premium tugboats, and for condenser tubes in steam-power plants.
Aluminum bronze	Aluminum bronzes predominantly contain aluminum and copper. For different alloys, the amount of aluminum varies between 5 and 11 wt%. Iron, nickel, manganese, and zinc are also added to different aluminum bronzes. The amount of iron varies between 0.5 and 6 wt%, nickel varies between 0.8 and 6 wt%, manganese varies between 0.5 and 2 wt%, and zinc is usually 0.5 wt%.	Applications include plain bearings and landing gear components on aircraft, engine components, underwater fastenings in naval architecture, and seawater ship propellers, inland water supply, and tools used for oil and petrochemical industries.
Silicon bronze	Silicon bronze usually contains about 96% copper. The amount of silicon varies between 1 and 3 wt% of silicon; about 1 wt% of iron, and less that 1 wt% of nickel, manganese or tin can also be present.	Used in the chemical processing industry.

11.2   Stress corrosion cracking (SCC) mechanisms

Most of the SCC failure mechanisms for copper-based alloys have been predominantly investigated for ammonia-brass systems (Pugh *et al.*, 1969;

Pugh, 1971). Two failure mechanisms that have found wide acceptance on SCC of copper-based alloys are: (a) the passive-film rupture and transient dissolution theory also referred to as film-rupture theory, and (b) the de-alloying or selective dissolution theory.

In the passive-film rupture theory, a passive film formed on a metal surface ruptures due to tensile force. The rupture of the passive film results in the formation of cracks which become conduits for a chemical environment in contact with the base metal. The metal surface exposed to the chemical environment undergoes rapid localized dissolution as repassivation is inhibited by the plastic deformation at the crack tip. Intergranular SCC is believed to occur due to passive-film rupture and transient dissolution mechanism in copper-zinc alloys. Chen *et al.* (2005) reported that aluminum bronze (Al-Cu-Zn) underwent SCC in fluoride-containing solutions through the film-rupture mechanism.

According to de-alloying, i.e., selective dissolution of the alloying component, e.g., Zn in Cu-Zn alloys, is the primary factor in the SCC process. For example, in the case of Cu-Zn alloys, preferential dissolution of Zn occurs in the region of the crack tip, or at the grain boundaries. Domiaty and Alhajji (1997) found that cupronickel alloy Cu-Ni (90–10) underwent SCC in seawater polluted with sulphide ions due to selective dissolution of copper. Chen *et al.* (2005) also reported that de-alloying resulted in SCC of aluminum brass (Al-Cu-Zn) in fluoride-containing aqueous solutions. The Pourbaix diagrams for pure metals can be useful in predicting the susceptibility to de-alloying, and in evaluating the likelihood of selective dissolution and hence the SCC behavior.

11.3   Stress corrosion cracking (SCC) of copper and copper-based alloys

11.3.1   Copper

Hard copper tubes are generally employed in a cooling system. Kuznicka and Junik's (2007) failure analysis of a copper tube used in a cooling system suggested that the failure was due to SCC. They found that copper metal processed into tubes in hard temper was sensitive to SCC in a wet aerated environment containing small quantities of ammonium, nitrites and nitrates. SCC initiation was reported to proceed by creation of shallow pits where the protective passive layer of CuO had been destroyed (see Fig. 11.1). Kuznicka and Junik (2007) suggested that the SCC occurred due to the passive-film rupture theory mechanism, i.e., by local anodic dissolution of copper and by repeatable formation and cracking of oxide layer, and as a result the fractured sample exhibited an intergranular mode of failure. Similarly, Mori *et al.* (2005) reported intergranular SCC of semi-hard deoxidized high phosphorous

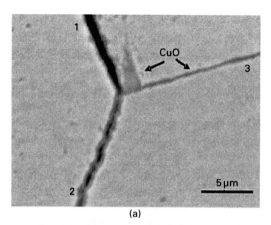

(a)

(b)

11.1 (a) Micrograph of copper showing CuO film formed along a grain boundary: (1) ruptured, (2) microcracks and (3) un-attacked, (b) intergranular cracking of copper due to repeated formation and cracking of CuO film (Kuznicka and Junik, 2007).

copper (DHP-Cu) in nitrite solutions. They suggested that for initiation of intergranular SCC in this corrosion system a tenorite layer is a prerequisite, because the cracks propagate by a combined process of chemical dissolution of copper and cracking of tenorite layer.

11.3.2   Brass

Brass is an alloy of copper and zinc. The proportions of copper and zinc are varied to create a range of brasses with varying properties. It is well known that Cu-Zn alloys are susceptible to SCC in ammonia-releasing environment. In fact, the SCC problem of brasses was first discovered in brass cartridge cases used for rifle ammunition during the 1920s in the Indian Army. The high residual stress from cold formed brass cases together with traces of ammonia in the atmosphere had caused SCC.

Hoar and Booker (1965) examined the SCC of α-brass in solutions containing cupric, sulphate and ammonia, over a wide range of pH (4.0–9.0). They observed very rapid cracking of the α-brass within only a very narrow pH range 7.0–7.3 of the solution containing 0.05 M Cu^{2+} and 1.0 M NH_4^+ ions. Parkins and Holroyd (1982) reported that brass (70 Cu–30 Zn) is susceptible to SCC in sodium acetate, formate and hydroxide at only a certain range of potential and pH. They observed intergranular cracking in the alloy when the employed potential and pH fell within the region (Pourbaix diagram) where Cu_2O is stable, whereas when the potential was above the Cu_2O region, the alloy underwent transgranular cracking. Recently, Nishimura and Yoshida (2008) work on SCC behavior of Cu-Zn (70–30) alloy in Mattsson solution [$(NH_4)_2SO_4$, NH_4OH, $CuSO_4$]. They suggested that the alloy undergoes transgranular SCC at pH 10.0 and intergranular SCC at pH 7.0. Nishimura and Yoshida (2008) reported that corrosion current density is higher at pH 7 than at pH 10. They also argued that the higher current density at pH 7 result in transgranular SCC of the alloy, whereas intergranular SCC of the alloy at pH 10 was due to the lower corrosion current density.

Uhlig *et al.*'s (1975) investigation into the SCC behavior of brass (63 Cu–37 Zn) alloy in $(NH_4)_2SO_4$ revealed that Cu^{2+} ions play a critical role. By increasing the copper concentration, the critical potential of the alloy shifts towards active direction, consequently making the alloy more susceptible to SCC (see Fig. 11.2). They also found that heavy metals such as cobalt, cadmium and nickel influence the alloy SCC susceptibility (see Fig. 11.2). A linear relationship between the stress corrosion crack propagation rate of α-brass and Cu^{2+} concentration in the cupric-sodium nitrite solution was reported by Giordano *et al.* (1997) as seen in Fig. 11.3. Increase in Cu^{2+} concentration increased the stress corrosion crack propagation rate in all four brasses. Another conclusion one could draw from Fig. 11.3 is that the increase in zinc content in the alloy also increases the stress corrosion crack propagation rate. Similarly, Farina *et al.* (2005) reported that zinc content strongly influences the SCC behavior of Cu-Zn alloys in 1 M $Cu(NO_3)_2$, i.e., the increase in zinc content increased the crack propagation rate of the alloy (see Fig. 11.4).

Shih and Tzou (1993) studied the SCC of Cu-Zn (70–30) brass in fluoride solution (pH 6.2) at 20°C. The authors reported that the alloy is prone to intergranular SCC in the fluoride solution, with an approximate threshold concentration of about 1 ppm or above at open circuit potential (see Fig. 11.5). Yu and Parkins (1987) reported the SCC susceptibility of brass (70 Cu-30 Zn) in various concentrations of sodium nitrite solutions. Alvarez *et al.* (2005) also observed SCC in α-brass and also pure copper in nitrite solutions. They reported that passivity rupture was a necessary condition for

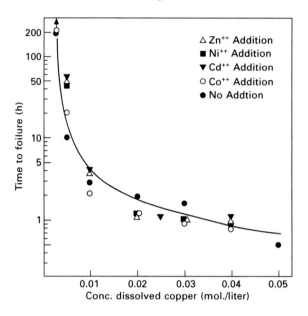

11.2 Effect of dissolved copper in the presence of metal sulfates on the time to failure of brass (63 Cu–37 Zn) in 1 M $(NH_4)_2SO_4$, pH 6.5 at room temperature (Uhlig *et al.*, 1975).

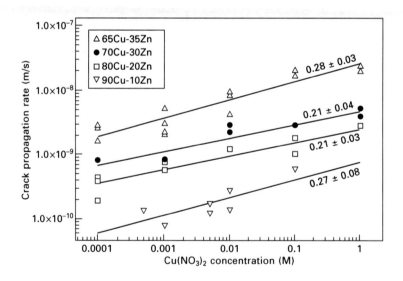

11.3 Stress corrosion crack propagation rates for four brass alloys, in $Cu(NO_3)_2$ solution and in mixtures of $Cu(NO_3)_2$ and $NaNO_3$ with total NO_3^- equal to 2 M (Giordano *et al.*, 1997).

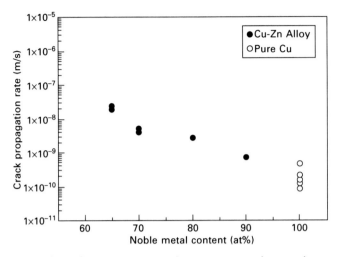

11.4 Effect of zinc content on the stress corrosion crack propagation rate of Cu-Zn alloy in 1 M Cu(NO₃)₂ (Farina *et al.*, 2005).

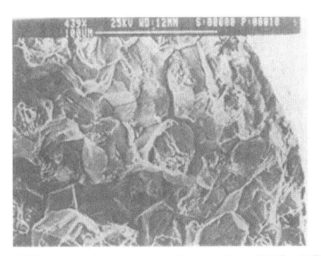

11.5 Intergranular stress corrosion cracking of 70 Cu–30 Zn brass in 1000 ppm fluoride solution at pH 6.2 (Shih and Tzou, 1993).

transgranular SCC of α-brass and copper. The SCC behavior of the alloy also depended on the applied potential. SCC occurred when the potential was equal to or higher than a certain critical value (E_c) at which the passivity breakdown takes place. The E_c for α-brasses and copper increased towards the noble direction with the increase in pH of the solution (see Fig. 11.6a). However, the E_c decreased as the zinc content increased (see Fig. 11.6b), which explains why addition of zinc increased the SCC susceptibility of brass. Alvarez *et al.* (2005) suggested that below the critical potential, E_c, a thin

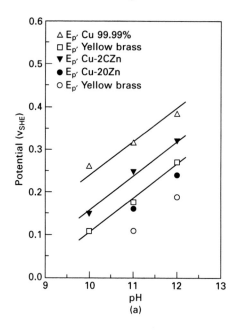

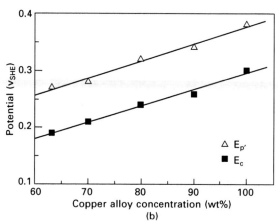

11.6 Pseudo-pitting potential (E_p) and critical potential (E_c) for α-brasses and copper in a 1 M NaNO$_2$ solution (a) as a function of the solution pH, and (b) as a function of the copper alloy concentration at pH 12 (Alvarez *et al.*, 2005).

adherent film (Cu$_2$O) protects the metal, whereas above E_c non-protective dark Cu$_2$O precipitates form. Alvarez *et al.* (2005) have also observed extensive secondary cracking at high potentials (see Fig. 11.7).

11.7 Lateral view of fracture Cu–20Zn alloy in 1 M NaNO$_2$ solution at pH 12, and at 0.32V_{SHE} reveals extensive secondary cracking (Alvarez *et al.*, 2005).

11.3.3 Aluminum-copper alloy

Aluminum-copper alloys are of two types: (i) alloys containing copper and aluminum (known as aluminum bronze) and (ii) alloys containing copper, aluminum and zinc (known as aluminum brass).

Ateya *et al.* (1994) reported that aluminum-bronze (93 Cu–7 Al) is susceptible to SCC in chloride-containing solution only when the samples are held at a high anodic potential. However, they observed that the alloy was not susceptible to SCC at corrosion potential. Accordingly, the failure mode of the alloy changed from ductile to brittle as the applied anodic potential increased. They attributed the high resistance of aluminum-bronze at near open corrosion potential to the formation of aluminum oxide film on the alloy. Although at high anodic potential aluminum oxide was detected, cuprous chloride was also found. Hence, they suggested de-alloying (copper dissolution) was the SCC operating mechanism.

Aluminum-bronze is resistant to SCC in a wide concentration of caustic solutions (0.1–4 M) (Ashour *et al.*, 2001). The high SCC resistance of the alloy was attributed to the thin layer of Al$_2$O$_3$. Interestingly, Ashour *et al.*, (2001) found that de-aluminification of the alloy had not affected the alloy SCC resistance. However, dealuminizing in aluminum-silicon bronze alloy in 3 wt% HF acid caused intergranular SCC (Hoffman *et al.*, 2003).

The studies of Torchio (1981) and Mazza and Torchio (1983) on the SCC of aluminum brass in acidic chloride-citrate solutions suggest that de-alloying leads to SCC of the alloy. Chen *et al.* (2005) reported that the aluminum brass and aluminum bronze suffer SCC in fluoride environment.

Interestingly, aluminum brass was susceptible to SCC even at the corrosion potential, whereas aluminum bronze was not susceptible to SCC at corrosion potential but only at anodic potentials. This behavior convinced the authors to suggest that the de-alloying mechanism is responsible for such cracking in aluminum brass and the film-rupture mechanism in aluminum bronze.

11.3.4   Cupronickel alloy

Cupronickel alloys are generally considered superior to other copper-base alloys in terms of their SCC resistance (Brown, 1975). However, the literature indicates that cupronickel alloys do undergo SCC in ammonia/amine, sulfide and nitrate environments (Pugh *et al.*, 1968; Graf, 1969; Islam *et al.*, 1991). Recently, Agarwal and Bapat (2009) reported that increase in Ni content (90/10 to 70/30 Cu-Ni) enhanced the SCC resistance in ammonia and sulfide environments.

Domiaty and Alhajji (1997) reported that 90/10 Cu/Ni alloy is prone to SCC in sulfide-containing seawater. The SCC severity increases as the concentration of sulfide increases, but within the range of 100–1000 ppm, because at concentrations below 100 ppm the corrosion of the material was less, and above 1000 ppm the corrosion of the material was high. The authors proposed two different SCC mechanisms: (i) sulfide stress cracking associated with the anodic dissolution in low sulfide concentration range, and (ii) hydrogen embrittlement in high sulfide concentration range. They further suggested that the synergistic interaction between sulfide and stress enhanced the alloy's susceptibility to hydrogen-induced SCC in the high sulfide concentration range.

Popplewell (1973) reported that Fe addition to Cu-Ni (90–10) alloy enhanced the SCC resistance alloy in moist ammonia environment. However, it was found that only solutionized Fe showed beneficial effect, whereas Fe precipitation accelerates the crack growth. The author suggested that under solutionized condition doping (by Fe) of the Cu_2O corrosion product film occurs and hence there is a reduction of the Cu_2O film growth rate ahead of the advancing crack, enhancing the SCC resistance. In the case of precipitated Fe, the free Fe is not available for doping the corrosion product film. The precipitates also further accelerates the crack growth rate due to galvanic effect.

Cu-Ni tubes are used at refiners where they are exposed to monoethanolamine (MEA), which is used in aqueous solutions for scrubbing weakly acidic gases. Shalaby *et al.* (2007) investigated the SCC behavior of Cu-Ni (70–30) alloy in 18 wt% MEA solution. They found that the alloy undergoes transgranular SCC under anodic polarization in aerated MEA solution at 90°C and pH 9. SCC occurred at both active–passive transition and transpassive regions.

They suggested that de-alloying of copper at the crack tip was responsible for the SCC susceptibility of the alloy.

Agarwal (2002, 2003) and Agarwal and Bapat (2009) investigated the SCC susceptibility of Cu-Ni alloy in ammonia environment. It was reported that the alloy underwent SCC in pure ammonia solution. Although the SCC was initiated by an intergranular mechanism, the crack propagation was transgranular. Increasing the concentration of ammonia increased the SCC susceptibility of the alloy. Interestingly, it was found that combined addition of NaCl and $MgCl_2$ to ammonia solution enhanced the SCC resistance of the alloy. Another study by Agarwal and coworkers (2005) on the individual effect of NaCl and $MgCl_2$ in ammonia solution towards the SCC behavior of Cu-Ni alloy revealed that NaCl has no noticeable effect on the SCC behavior of the alloy; however, $MgCl_2$ has a significant effect. The authors reported that 7.5 wt% $MgCl_2$ is the critical concentration beyond which all of the SCC caused by 10 wt% ammonia solution is essentially non-existent.

11.4   Role of secondary phase particles

Secondary phase particles in copper-based alloys could potentially initiate SCC when the particles undergo selective dissolution or induce galvanic corrosion. To the authors' surprise, the literature on the role of secondary phase particles on the SCC behavior of copper-based alloys is limited. The available literature suggests that secondary phase particles along the grain boundaries influence the SCC behavior, whereas the precipitates within the grains have no effect. Fonlupt et al. (2005) reported that the SCC susceptibility of nickel-aluminum bronze in synthetic seawater increases with the increase in the amount of secondary phase particles ($CuAl_9Ni_3Fe_2$) in the grain boundaries. Dissolution of secondary phase particles initiated the crack shown in Fig. 11.8. However, secondary phase particles, such as Cu-sulfide and P-oxide within the grains of semi-hard deoxidized high phosphorous copper (DHP-Cu) did not influence the SCC behavior of the alloy in nitrite solution (Mori et al., 2005). In fact, the alloy predominantly underwent intergranular cracking (see Fig. 11.9).

11.5   Stress corrosion cracking (SCC) mitigation strategies

11.5.1   Heat treatment

SCC in copper-based alloys can be mitigated by annealing the alloy to relieve residual tensile stresses. The annealing temperatures for different copper-based alloys are listed in Table 11.2 (Brown, 1975). The annealing time for these alloys is typically 30–60 minutes. In fact, a post-annealing test can be

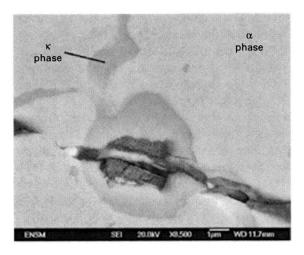

11.8 A fine crack associated with the dissolution of secondary phase particles (Fonlupt *et al.*, 2005).

conducted to evaluate the effectiveness of the annealing process in relieving residual tensile stresses. The post annealing test has been standardized as ASTM B154-05 (ASTM International, 2005).

11.5.2   Inhibitors

Inhibitors, cathodic and anodic, have been widely used to mitigate SCC of copper and copper-based alloys. Sircar *et al.* (1983) reported that addition of ammonium chloride, cobalt sulfate, cesium sulfate and ethylene diamine to a Mattsson solution [$(NH_4)_2SO_4$, NH_4OH, $CuSO_4$] produces an inhibitive effect on the SCC of α-brass. Based on the cracking time and cracking mode, it was suggested that the inhibitors are adsorbed on the metal surface at preferential sites and influence the cathodic reactions (cathodic inhibitors).

Luo and Yu (1996) investigated the effectiveness of SCC inhibition of brass (70 Cu–30 Zn) with additions of arsenic anions (AsO_2^- and $HAsO_4^{2-}$) into 1 M $NaNO_2$ solution. They found that arsenic anions were effective in suppressing the anodic dissolution behavior (anodic inhibitors) and inhibiting the SCC. The effectiveness increased with increase in the arsenic concentration (see Fig. 11.10). They observed that arsenic has the effect of inhibiting both crack initiation and crack propagation. They suggested the chemisorption mechanism for the inhibitive effect of arsenic. They also reported that arsenic anions suppressed the anodic dissolution behavior and also inhibit the SCC of pure copper in 1 M $NaNO_2$.

Shih and Tzou (1993) found that small addition (300 pm) of 1,2,3-benzotriazole (BTA) inhibits the alloy SCC susceptibility of brass (70 Cu–30

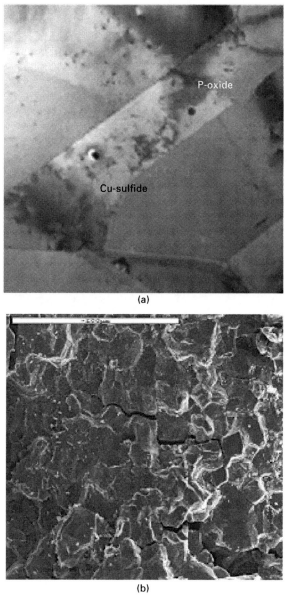

11.9 DHP-Cu: (a) TEM micrograph showing secondary phase particles within grains, and (b) intergranular SCC (Mori *et al.*, 2005).

Zn) in fluoride solution (pH 6.2, 20°C). Infrared spectra suggested that the Cu-BTA complex formed on the surface of the alloy inhibited the SCC. Anodic polarization of the alloy accelerated SCC while cathodic potentials prevented SCC. They noticed that the alloy was polarized cathodically in

Table 11.2 Stress relief annealing temperatures for different copper-based alloys (Brown, 1975)

Alloy name	Temperature (°F)
Brass	260–500
Muntz metal	190–375
Cupronickel (30 wt% Ni and 70 wt% Cu)	426–800
Aluminum bronze	600–1112
Silicon bronze	371–700

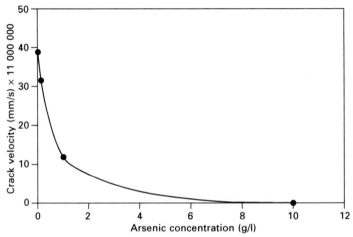

11.10 The effect of arsenic concentration, added as arsenite in 1 M NaNO$_2$ on the average crack velocity of brass A at –50 mV$_{SCE}$ (Luo and Yu, 1996).

the presence of BTA, which suggested that BTA is a cathodic inhibitor for this alloy (see Fig. 11.11).

Recently, Allam and Ashour (2009) found that the SCC susceptibility of Cu-Zn (67–33) alloy in 3.5% NaCl and 0.1 M NaNO$_2$ can be reduced by adding a small concentration (0.1 M) of di-sodium biphosphate (DSHP). The fracture mode changed from transgranular cracking to ductile failure due to the addition of DSHP. The reduction in SCC was observed under both open circuit potential and controlled potential of 300 mV$_{SHE}$, which is also reflected in the electrochemical polarization curves (see Fig. 11.12). The authors reported that formation of zinc phosfate on the alloy surface inhibited the dezincification process and hence reduced the SCC susceptibility of Cu-Zn alloy.

11.6   Conclusions

Generally, copper and copper-based alloys are susceptible to SCC in ammonia, nitrite, nitrate, sulfide, fluorides, and other environments. Nickel addition

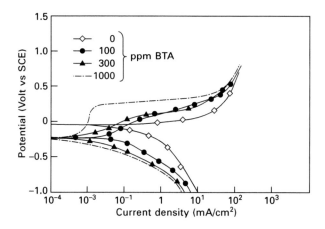

11.11 Polarization curves of 70/30 brass in 1000 ppm fluoride solution containing various BTA concentrations (Shih and Tzou, 1993).

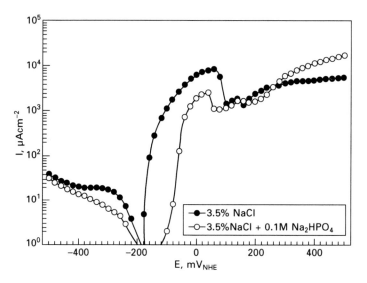

11.12 Effect of 0.1 M DSHP in 3.5% NaCl on the potentiodynamic polarization behavior of Cu-Zn alloy (Allam and Ashour, 2009).

to copper enhances the SCC resistance, while zinc addition decreases it. The pH of the solution and the electrochemical potentials (both anodic and cathodic) of the materials have shown to influence the SCC behavior significantly. Passive-film rupture and de-alloying are the two most widely reported SCC mechanisms for copper and copper-based alloys. Depending on the alloy composition and environment, de-alloying can be due to selective dissolution of zinc, copper or aluminium. Heat treatment and inhibitors have been successfully used to mitigate the SCC susceptibility of copper and copper-based alloys.

11.7   References

Agarwal, D.C. (2002), 'Stress corrosion in copper-nickel alloys: influence of ammonia', *British Corrosion Journal*, 37, 267–275.

Agarwal, D.C. (2003), 'Effect of ammonia concentration on environment-assisted failure in a low-nickel copper alloy', *Practical Failure Analysis*, 3, 58–68.

Agarwal, D.C. and A.M. Bapat (2009), 'Effect of ammonia and sulphide environment on 90/10 and 70/30 cupronickel alloy', *Journal of Failure Analysis and Prevention*, 9, 444–460.

Agarwal, D.C., R. Vishwakarma, M.B. Deshmukh, S. Kurian, S. Sarin, and K. Wadhwa (2005), 'Mitigation of ammonia-induced SCC in a cupronickel alloy by additions of $MgCl_2$ part 2: role of $MgCl_2$ concentration in a cupronickel alloy exposed to 10% ammonia', *Journal of Failure Analysis and Prevention*, 5, 70–78.

Allam, N.K. and E.A. Ashour (2009), 'Electrochemical and stress corrosion cracking behavior of 67Cu–33Zn alloy in aqueous electrolytes containing chloride and nitrite ions: effect of di-sodium hydrogen phosphate (DSHP)', *Materials Science and Engineering*, B 156, 84–89.

Alvarez, M.G., P. Lapitz, S.A. Fernández, and J.R. Galvele (2005), 'Passivity breakdown and stress corrosion cracking of α-brass in sodium nitrite solutions', *Corrosion Science*, 47, 1643–1652.

Ashour, E.A., E.A. Abd El Meguid, and B.G. Ateya (2001), 'Technical note: stress corrosion behavior of alpha aluminum bronze in concentrated alkali solutions', *Corrosion*, 57, 749–752.

Ateya, B.G., E.A. Ashour, and S.M. Sayed (1994), 'Stress corrosion behavior of α-aluminum bronze in saline water', *Corrosion*, 50, 20–25.

ASTM International (2005), *Standard test method for mercurous nitrate for copper alloys: B154–05*, West Conshohocken, PA: ASTM International.

Brown, B.F. (1975), *Stress corrosion cracking control plans: copper alloys*, National Technical Information Service, Office of Naval Research, Washington D.C., U.S. Department of Commerce.

Chen, Y.Y., R.J. Tzou, Y.S. Chang, L.H. Wang, J.C. Oung, and H.C. Shih (2005), 'Two distinct fracture modes of copper alloys in fluoride environments', *Corrosion Science*, 47, 79–93.

Domiaty, A. El. and J.N. Alhajji (1997), 'The susceptibility of 90Cu–10Ni alloy to stress corrosion cracking in seawater polluted by sulfide ions', *Journal of Materials Engineering and Performance*, 6, 534–544.

Farina, S.B., G.S. Duffó, and J.R. Galvele (2005), 'Stress corrosion cracking of copper and silver, specific effect of the metal cations', *Corrosion Science*, 47, 239–245.

Fonlupt, S., B. Bayle, D. Delafosse, and J-L. Heuze (2005), 'Role of second phases in the stress corrosion cracking of a nickel–aluminium bronze in saline water', *Corrosion Science*, 47, 2792–2806.

Giordano, C.M., G.S. Duffo, and J.R. Galvene (1997), 'The effect of Cu^{2+} concentration on the stress corrosion cracking susceptibility of α-brass in cupric nitrate solutions', *Corrosion Science*, 39, 1915–1923.

Graf, L. (1969), in *Fundamental Aspects of Stress Cracking*, Ed. R.W. Staehle, Houston, TX, NACE International.

Hoar, T.P. and C.J.L. Booker (1965), 'The electrochemistry of the stress corrosion cracking of alpha brass', *Corrosion Science*, 5, 821–840.

Hoffman, J.J., J.W. Slusser, and J.L.O'Leary (2003), 'Stress corrosion cracking susceptibility of aluminum–silicon–bronze', *Proceedings of the CORROSION 2003 Conference*, Paper No. 03510.

Islam, M., W.T. Riad, S. Al-Kharraz, and S. Abo-Namous (1991), 'Stress corrosion cracking behavior of 90/10 Cu–Ni alloy in sodium sulfide solutions', *Corrosion*, 47, 260–268.

Kuznicka, B. and K. Junik (2007), 'Intergranular stress corrosion cracking of copper – a case study', *Corrosion Science*, 49, 3905–3916.

Luo, X. and J. Yu (1996), 'The inhibition of the stress corrosion cracking of 70/30 brass in nitrite solution with additions of alloying arsenic and arsenic anions', *Corrosion Science*, 38, 767–780.

Mazza, F. and S. Torchio (1983), 'Factors influencing the susceptibility to intergranular attack, stress corrosion cracking and de-alloying attack of aluminum brass', *Corrosion Science*, 23, 1053–1061.

Mori, G., D. Scherer, and S. Schwentenwe (2005), 'Intergranular stress corrosion cracking of copper in nitrite solutions', *Corrosion Science*, 47, 2099–2124.

Nishimura, R. and T. Yoshida (2008), 'Stress corrosion cracking of Cu–30% Zn alloy in Mattsson's solutions at pH 7.0 and 10.0 using constant load method – a proposal of SCC mechanism', *Corrosion Science*, 50, 1205–1213.

Parkins, R.N. and N.J.H. Holroyd (1982), 'Stress corrosion cracking of 70/30 brass in acetate, formate, tartrate, and hydroxide solutions', *Corrosion*, 39, 245–255.

Popplewell, J.M. (1973), 'The effect of iron on the stress corrosion resistance of 90/10 cupro–nickel ammonical environment', *Corrosion Science*, 13, 593–603.

Pugh, E.N. (1971), The mechanism of stress corrosion cracking of alpha-brass in aqueous ammonia in *Theory of Stress Corrosion Cracking in Alloys*, Ed. by J.C. Scully, NATO Science Affair Division.

Pugh, E.N., J.V. Craig, and W.G. Montague (1968), 'Factors influencing the path of stress-corrosion cracking in alpha-phase copper alloys exposed to aqueous ammonia environments', *ASM Transactions Quarterly*, 61, 468–473.

Pugh, E.N., J.V. Craig, and A.J. Serdikas (1969), 'The stress corrosion cracking of copper, silver, and gold alloys', in *Proc. Conf. Fundamental Aspects of Stress Corrosion Cracking*, Ed. R.W. Staehle, A.J. Forty, and D. Van Rooyen, Houston, TX, NACE International.

Rao, T.S. and K.V.K. Nair (1998), 'Microbilogically influenced stress corrosion cracking failure of admiralty brass condenser tubes in nuclear power plant cooled by freshwater', *Corrosion Science*, 40, 1821–1836.

Shalaby, H.M., A. Husain, A.A. Hasan and A.Y. Abdullah (2007), 'Electrochemical behaviour and stress corrosion cracking of 70:30 copper–nickel alloy in monoethanolamine solutions', *Corrosion Engineering, Science and Technology*, 42, 64–72.

Shih, H.C. and R.J. Tzou (1993), 'Studies of the inhibiting effect of the 1,2,3-benzotriazole on stress-corrosion cracking of 70/30 brass in fluoride environments', *Corrosion Science*, 35, 479–488.

Sircar, S.C., U.K. Chatterjee, and G.M. Sherbini (1983), 'The inhibitive effect of extraneous ion addition on stress corrosion cracking of alpha-brass', *Corrosion Science*, 23, 777–787.

Torchio, S. (1981), 'The influence of arsenic and phosphorus on the stress corrosion cracking of aluminum brass in chloride–citrate solution', *Corrosion Science*, 21, 425–437.

Uhlig, H., K. Gupta, and W. Liang (1975), 'Critical potentials for stress corrosion cracking of 63–37 brass in ammoniacal and tartrate solutions', *Journal of the Electrochemical Society*, 122, 343–350.

Yu, J. and R.N. Parkins (1987), 'Stress corrosion crack propagation in α-brass and copper exposed to sodium nitrite solutions', *Corrosion Science*, 27, 159–182.

12

Stress corrosion cracking (SCC) of austenitic stainless and ferritic steel weldments

H. SHAIKH, T. ANITA, A. POONGUZHALI,
R. K. DAYAL and B. RAJ, Indira Gandhi
Centre for Atomic Research, India

Abstract: Welded steels are widely used in structural engineering. However, welding introduces potentially unwelcome microstructural changes, stresses and contaminants which can weaken the metal and make it prone to stress corrosion cracking. This chapter reviews the mechanisms by which these changes occur during welding and the ways they can be minimised.

Key words: stress corrosion cracking (SCC), ferritic steels, austenitic stainless steels, welds, weldments.

12.1   Introduction

Austenitic stainless steels and Cr-Mo ferritic steels are chosen as the major structural materials in a wide range of industries in view of their adequate high temperature properties, which include high creep strength, resistance to low cycle fatigue and creep–fatigue interactions, and good resistance to environmental sensitive cracking. Their choice is based on their better weldability and availability of code data on mechanical properties [1]. However, their use involves fabrication of components by welding. A welded joint is required to perform either equal to or better than the base metal it joins. However, in practice, this objective is never achieved since the welding process itself introduces features, which degrade the mechanical and corrosion properties of the welded joints as compared to the wrought base metal. Despite shielding by a gas or by slag, the weld metal can get contaminated by, for example, slag inclusions or tungsten inclusions. The fast cooling rates associated with the weld metal causes the formation of a dendritic structure besides straining of the weld metal. Also, several metallurgical transformations can take place in the weld metal during this cooling. Apart from weld contamination and metallurgical changes, improper welding procedures can leave behind a host of defects, such as porosities, undercuts or microfissures, in the weld metal. All these detrimental features do not augur well from the point of view of mechanical and corrosion properties. Because of their microstructural and compositional heterogeneities, weld

427

metals are inherently more prone to corrosion than the unaffected base metal, though the basic corrosion mechanisms are the same.

The various regions of a weldment are schematically illustrated in Fig. 12.1. The composite zone, also known as mixed zone or fusion zone, is the conventionally fused region in the weldment where the filler material has been diluted with material from the base metal. The composition of this zone depends on the compositions of filler and base material. The extent of dilution is a function of heat input, joint geometry, position and the welding process. In an autogeneous weld metal, where no filler metal is employed, the solidified material composition is very close to that of the base metal. However, in case of welding using a 'matching' consumable, the composition of the weld metal tends to be different from the base metal. It is frequently necessary to weld a joint with a consumable with composition entirely different from the base metal it joins, thereby resulting in problems of galvanic corrosion. Next to the mixed zone is the narrow region of unmixed zone where the base metal is melted but not fused with the filler metal and its composition is close to the base metal. In fact, the unmixed zone can be considered to be the equivalent of an autogeneous weld. The occurrence of an unmixed zone depends on a number of factors such as welding process, welding parameters, composition of filler metal and its physical properties. The extent of unmixed zone in a gas tungsten arc weld (GTAW) is less than in shielded metal arc welding (SMAW), gas metal arc welding (GMAW) and submerged arc welding (SAW) thus indicating that metal transfer from the filler rod and the weld pool motion play a very significant role in the formation of unmixed zones. The width of the unmixed zone varies with distance from the surface to the root of the weld, as shown in Fig. 12.2 for the C-22 and Inconel 625 filler metals used for welding AL6XN alloy [2]. The difference in the width and extent of unmixed zone with location is

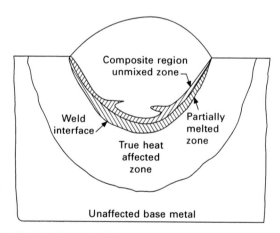

12.1 A schematic illustrating the various regions of a weldment.

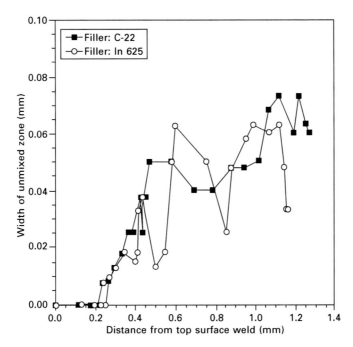

12.2 Change in the width of the unmixed zone as function of distance from root to the surface of the weld.

because the surface of the weld pool is more turbulent as compared to that at the root thus causing differences in mixing of filler and base materials. The heat affected zone (HAZ) in a weldment is situated in the base metal adjacent to the fusion line. It experiences different thermal cycles depending on the distance from the fusion line. Each thermal cycle is characterised by a heating and cooling cycle and a peak temperature. The subsequent sections deal with the stress corrosion cracking behaviour of the various regions of austenitic stainless steel and ferritic steel weld joints.

12.2   Effect of welding defects on weld metal corrosion

Welding defects such as slag and other entrapments, cracks, microfissures, porosities, inclusions, lack of penetration and fusion, etc., oxide and other scales, etc., deteriorate the corrosion properties of weld metal. These defects normally cause early pitting and crevice attacks in the weld metal [3, 4]. Sulphide inclusions are the most susceptible sites for pitting and crevice attack; however, other non-metallic inclusions are also capable of causing pit nucleation. Also, these defects act as regions of stress concentration which aid in faster initiation of SCC. Non-metallic inclusions, such as sulphides, are undesirable from a SCC point of view. Influence of non-metallic inclusions in reducing the SCC life is directly related to the easy initiation of pitting corrosion at inclusion sites [5]. In acid solutions, and hence in occluded cells, sulphide inclusions dissolve to form H_2S, which has an accelerating effect on corrosion of steel [6]. H_2S is reported to accelerate both the anodic and cathodic processes [5]. In many service applications, SCC initiated from pits which act as stress raisers [7, 8]. In these cases, the induction times for SCC were longer than those for pitting [7]. Apparently, the non-metallic inclusions do not participate directly in crack nucleation, but their presence is undesirable, as they give rise to pitting. Clarke and Gordon [9] reported that cracking nucleated from the crevice corrosion attack around an included particle. Non-metallic inclusions also play a role in hydrogen embrittlement (HE) of stainless steel welds. Surface inclusions facilitate entry of hydrogen into the weld and thus induce cracking [5]. Bulk inclusions may act as trap sites for hydrogen and thus assist in nucleation and development of internal crevices and cracks [5]. The shape of the sulphide inclusion influences the crack initiation time. Elongated sulphide particles may increase hydrogen entry six-fold into the metal as compared to spherical inclusions [5].The non-metallic inclusion also cause stress concentration in the material. The extent of stress concentration would depend on the shape of the inclusion [5]. Sharp edged inclusions act as more effective notches than spherical or elliptically shaped inclusions. Apart from the shape, the ratio of thermal expansion coefficients and modulus of elasticity of the inclusion to that of the matrix also contribute to a lesser extent in determining the magnitude of stress concentration [5]. Apart from sulphide inclusions, weld metals can contain other inclusions resulting from the oxidation and deoxidation reactions in the molten weld metal. They may also contain slag inclusions, which could result from slag entrapment in the solidified weld bead. The extent of slag entrapment is dictated by the flux coating on the welding consumable. Rutile coatings give ease of slag detachment and good bead shape. Out-of-position welding can also cause slag entrapment.

In designing welding consumables, it is common practice to adjust weld metal composition by adding alloying elements in the flux. During welding, these elements may not get mixed well in the molten pool thus leaving regions rich in these elements which may act as nucleation sites for localised corrosion attack. For example, in offshore applications, the electrode coating contains ferro-managanese particles, which if not homogeneously mixed in the weld metal leaves areas harder than the adjoining matrix [10]. These hard areas provide sites for initiation of SCC in sour service. Certain measures during welding, such as increase in welding current, reducing travel speed or the width of weaving, may help in mitigating the problem.

Porosities can lead to faster initiation of pitting attack or SCC since they act as sites for stress concentration [11]. Appropriate choice of heat input could reduce the amount of porosities in the weld metal. Cracks and microfissures also contribute to reduction in localised corrosion resistance of weld metal. The weldment may undergo cracking during welding, or immediately after welding, or during service or during post-weld heat treatment (PWHT) in the solidified weld metal, HAZ of base metal or HAZ of weld metal. These cracks may act as sites for crack initiation or as crevices. In stainless steels and Ni-base alloys, microfissuring and hot cracking is a major problem. C-Mn steels and low alloy steels normally undergo cold cracking and reheat cracking. Undercuts, lack of fusion and lack of penetration are preferential sites for corrosion attacks. Judicious selection of the weld joint design, welding parameters, consumables and welding process may avoid corrosion problems associated with such faulty joints. Certain joint designs, like lap and stake joints may result in crevice corrosion attack and should be avoided and instead full penetration butt joints are preferred [12]. In some applications, it is common practice to use welding inserts, which may not get properly fused in the weld metal and leave regions, which may become prone to crevice attack. In addition to crevice corrosion problems, the crevices thus formed will act as stress raisers and cause SCC. In welding of steels, if the electrodes are overbaked, the amount of hydrogen, which increases bead penetration, is reduced thus leading to lack of penetration defects in the weld metal.

12.3   Stress corrosion cracking (SCC) of austenitic stainless steel weld metal

A general definition of the phenomenon is that SCC is the fracture of a material by the simultaneous action of a tensile stress and a corrosive environment. It is a synergistic process in the sense that the time-to-fracture, the decrease in load-bearing capacity, and other effects manifested in the phenomenon, are different from similar effects by stress or corrosion acting alone. The phenomenon is of great industrial significance since stresses are invariably induced during the fabrication of plant components and structures and it is not always possible to remove the stresses or to make reliable measurements of them in the engineering structures. A number of factors influence the corrosion properties of austenitic stainless steel weld metal. These include the solidification process, micro-segregation, alloying element partitioning, formation of multi-phase microstructure, heat input, high temperature aging, welding defects, etc. These aspects are discussed in detail below.

12.3.1   Microstructure evolution in an austenitic stainless steel weld metal

The welding of austenitic stainless steel has two major problems, namely hot cracking of weld metal and sensitisation of the HAZ. Hot cracking is due to the inability of the solidifying weld metal to bear the strains, arising during solidification in a certain temperature range. This is normally encountered in a fully austenitic weld deposit. Hot cracking is a common phenomenon in alloys which freeze over a large temperature range. In austenitic stainless steel welds, this interdendritic cracking is due to the presence of elements such as sulphur, phosphorous, silicon, oxygen, tantalum, titanium and niobium which form low melting phases along the interdendritic boundaries. The problem of hot cracking could be overcome by (i) reduction of the S+P+B content of the weld to less than 0.01% (Fig. 12.3) [13] and (ii) addition of Mn to form high melting MnS instead of low melting FeS. However, the most common method adopted is by intentionally rendering the austenitic stainless steel weld deposit inhomogeneous by retaining some amount of high temperature δ-ferrite to room temperature [14]. δ-ferrite has a higher solubility for impurity elements like sulphur and phosphorous in its lattice than in austenite. Thus, its presence reduces the amount of these impurities in the last-to-solidify melt [15].

δ-ferrite can be retained by appropriate choice of the filler metal composition which could be done in consultation with the 70% iron isopleth (Fig. 12.4) [13, 16]. This diagram determines the position of the alloy composition with respect to the liquidus minimum. The figure shows that four distinct solidification modes are normally possible: austenitic (A), austenitic-ferritic

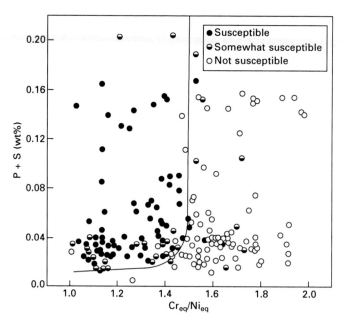

12.3 Solidification cracking behaviour in austenitic stainless steel welds as a function of Schaeffler Cr_{eq}/Ni_{eq} ratio and PCS levels [13].

or primary austenitic (AF), ferritic-austenitic or primary ferritic (FA) and ferritic (F). The approximate composition ranges in which these modes occur are indicated in Fig. 12.4. Alloys solidifying in the A mode remain unchanged to low temperatures, while those solidifying as AF would form some eutectic ferrite. Compositions that solidify in the FA and F modes pass through the austenite-δ-ferrite two-phase region and may re-enter the single-phase austenite field. This is due to the asymmetry of the two-phase field towards the primary ferritic side of the diagram, as seen in Fig. 12.4. Thus, alloys such as types 304 and 316 stainless steels that are fully austenitic at room temperature enter this two-phase region after AF/FA solidification and may undergo solid-state transformation to a fully austenitic structure. For higher ratios of chromium over nickel, the equilibrium structure at room temperature may retain considerable amounts of δ-ferrite as in duplex stainless steels. In all the solidification modes, the δ-ferrite is enriched in ferritisers like Cr, Mo and Si, while the austenite is enriched in austenitisers, such as Ni, Mn, C, N, etc. (Fig. 12.5) [17]. Koseki *et al.* [18] reported that in fully austenitic stainless steels welds, the interdendritic regions are slightly enriched in both Cr and Ni; while in AF mode weld metal significant enrichment of Cr and depletion of Ni occur in the interdendritic regions. Ferrite nucleates in the Cr-rich and Ni-depleted regions as a non-equilibrium phase. When

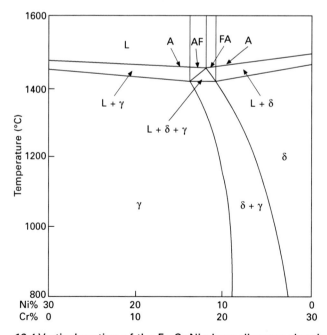

12.4 Vertical section of the Fe-Cr-Ni phase diagram showing the variation of solidification mode with composition, for a constant iron content of 70% [13].

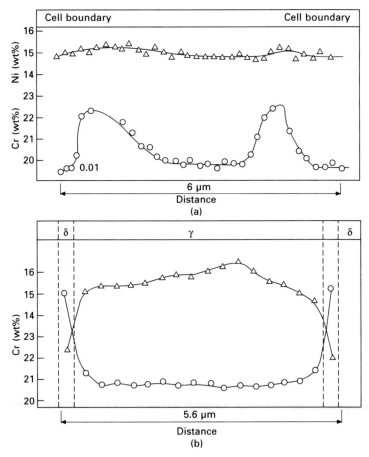

12.5 Effect of heat input on partitioning of Cr, Mo and Ni in type 316L weld metal during (a) primary austenitic and (b) ferritic solidification modes [17].

FA and F mode solidification takes place, the dendrite core is significantly enriched in Cr and depleted in Ni. The segregation of Cr to ferrite and Ni to austenite during solidification plays a major role in stabilising the ferrite during subsequent solid-state transformation. The extent of partitioning of alloying elements depends on heat input [19]. Apart from partitioning of alloying elements, impurities, such as S and P, segregate to the δ-ferrite/austenite (δ/γ) interface. The extent of elemental partitioning and segregation depends on the solidification mode and heat input [16, 19].

A minimum δ-ferrite content is necessary to ensure hot cracking resistance in these welds [20, 21], while an upper limit on the δ-ferrite content determines the propensity to embrittlement due to secondary phases, e.g. σ phase, etc., formed during elevated temperature service [22]. At cryogenic temperatures, the toughness of the austenitic SS welds is strongly influenced by the δ-ferrite content [23]. The desired amount of ferrite can be achieved by adjusting the ratio of ferrite promoters (such as chromium) to austenite promoters (such as nickel) in the electrode [24].

The use of welded components of austenitic stainless steels at high temperature would cause time-dependent degradation owing to the transformation of the δ-ferrite in the weld metal to a number of intermetallic phases. Small quantities of these intermetallic phases can cause large variations in the mechanical and corrosion properties of the weld metal. Hence, a knowledge of the type, amount and physical parameters, such as size, shape and morphology, of these phases is essential in designing the optimum operational parameters to achieve the desired lifetime of welded austenitic stainless steel components.

Investigations on the transformation kinetics of δ-ferrite have indicated that δ-ferrite transformed to austenite, carbides, σ, χ, Lave's, R phases, etc., on exposure to high temperatures [25–28]. No two investigators have reported matching results for the transformation of δ-ferrite because of the presence of a large number of variables in the weld metal. The differences in the transformation kinetics of δ-ferrite arise due to variations in chemical composition of the weld metal, temperature of exposure, size and shape of δ-ferrite, etc. For example, Thomas and Yapp [29] reported that at 973 K, δ-ferrite disappeared in 100 hours in a manual metal arc (MMA) weld of type 316 stainless steel, while Gill *et al.* [30] reported total transformation of δ-ferrite below 20 hours in a gas tungsten arc (GTA) weld of type 316L stainless steel. σ phase is known to nucleate and grow easily in austenitic stainless steel welds containing low concentration of carbon [30–32]. This was because the critical concentration of molybdenum and chromium necessary to form σ phase, after carbide precipitation is over, would be more in a low carbon stainless steel weld than in a high carbon stainless steel weld [31]. Leitnaker [32] reported that in an austenitic stainless steel, alloys containing δ-ferrite such as are used as weld deposits are protected against the transformation of δ-ferrite to σ phase during aging by the presence of carbon plus nitrogen in a weight percent 0.015–0.030 times the volume percent δ-ferrite present in the alloy. The formation of χ phase upon aging is controlled by controlling the Mo content. Zhao *et al.* [33] reported that the amount and size of δ-ferrite is reduced with the increase of nitrogen in the shielding gas composition, thus indicating a reduction in σ phase transformation. Apart from the carbon and nitrogen contents, the amount of molybdenum and chromium present in the weld metal would also govern the transformation kinetics of δ-ferrite. Figure 12.6 shows the T-T-T diagram for two weld metals, one with 17 Cr and 2 Mo and the other with 19 Cr and 3 Mo. It is seen that the δ-ferrite in the weld metal with higher Cr and Mo transforms faster [34]. The effect of other alloying elements, such as Mn,

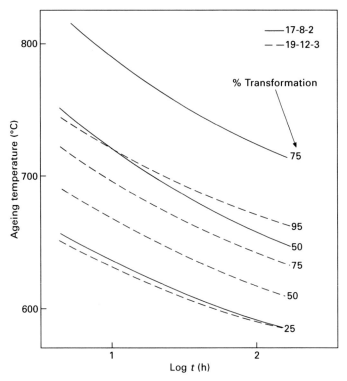

12.6 Transformation kinetics of δ-ferrite in weld metals with different chemical compositions [34].

Ti, Si, Nb, Ni, etc., has been expressed by Hull [35] in terms of equivalent chromium content (ECC) as:

$$\begin{aligned}
ECC = {} & [\%Cr] + 0.31\,[\%Mn] + 1.76\,[\%Mo] + 0.97\,[\%W] + 2.02 \\
& [\%V] + 1.58\,[\%Si] + 2.44\,[\%Ti] + 1.70\,[\%Nb] + 1.22 \\
& [\%Ta] - 0.266\,[\%Ni] - 0.177\,[\%Co]
\end{aligned}$$

According to Hull, the embrittlement due to σ and χ phases takes place if ECC is less than 17.8%. However, Hull did not take into account the influence of carbon on the transformation of δ-ferrite. Gill *et al.* [36] accounted for the influence of carbon by normalising ECC against the carbon content of the weld metal.

Figure 12.7 shows that the rate of transformation of δ-ferrite increases with increasing temperature [30]. An interdependence between the growth kinetics of σ phase and carbide phase was reported by Gill *et al.* [30]. Microstructurally, low temperature aging (< 1023 K) reduces δ-ferrite to austenite and carbides [26, 34, 37, 38]. However, at higher temperatures, δ-ferrite dendrites transformed directly to σ phase [27, 39]. The compositions of δ-ferrite, σ and χ phases were close to each other [27], thus indicating

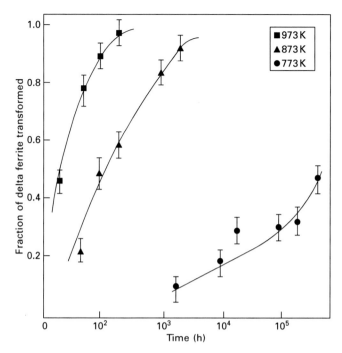

12.7 Effect of temperature on transformation kinetics of δ-ferrite [30].

that a crystallographic transformation with a small compositional adjustment was necessary for the reaction $\delta \rightarrow \sigma + \chi$ to proceed.

Aging of type 308 stainless steel weld at 400°C or 475°C for times up to 5000 h showed a number of changes in the as-welded microstructure. Unlike the specimens aged at temperatures greater than 550°C, aging at temperatures less than 550°C showed no evidence of ferrite to sigma phase transformation. However. $M_{23}C_6$ carbide precipitation at the austenite/ferrite interface was found to occur. During the initial stages of aging, a fine scale spinodal decomposition of ferrite into iron rich a and chromium rich a' phases was observed within the ferrite [40–42]. This decomposition of ferrite in the weld is similar to the observations made in some ferritic steels [43–48]. This particular transformation has been found to embrittle the ferritic steel. Within the ferrite, in addition to the spinodal decomposition, abundant precipitation of G-phase was observed. G-phase is a nickel-rich silicide that has been identified in both austenitic stainless steel welds and castings [41, 49–52]. Microstructural changes observed in specimens aged at 400°C and 475°C were similar except the kinetics of the transformation were faster at 475°C. Abe *et al.* [53] reported spinoidal decomposition of ferrite into iron-rich a and chromium-rich a' phases during isothermal aging of a type 316L stainless steel weld metal at 608 K. Abe and Watanabe reported that the age-hardening rate of δ-ferrite was faster for type 316L stainless steel weld metal with primary austenite solidification mode (AF mode) than that with primary ferrite solidification mode (FA mode) in the initial stage of the aging up to 2000 h at 608 K [54].

12.3.2   SCC of base and weld metals of austenitic stainless steels

The beneficial effect of δ-ferrite in controlling hot cracking is offset by the fact that in some corrosive media, it can cause severely localised corrosion attacks. However, δ-ferrite is not harmful under all conditions of environmental corrosivity. The attack on δ-ferrite is controlled by material composition and does not solely depend on its mere presence in the austenite matrix. Also, preferential weld metal corrosion involving δ-ferrite takes two principal forms: attack on the ferrite or attack along the DF/A interface. Both these types of attack can occur in Mo-containing and Mo-free austenitic SS weld metals. But attack along the interface is more common in the former while attack on the δ-ferrite is often encountered in the latter.

The difference in SCC behaviour of base and weld metals of austenitic stainless steels has been a subject of much disagreement in the literature. The differences in SCC behaviour between the base and weld metals depend on the chemical composition, environment and testing techniques. Weld metal of austenitic stainless steel possessed equal or better SCC resistance than the base metal of similar composition when tested by slow strain rate technique (SSRT) in boiling 45% $MgCl_2$ solution [55–57]. The better SCC resistance of the weld metal was attributed to the cathodic protection offered to the austenite by the corroding δ/γ interfaces [58]. Tests on type 304 stainless steel in 1 N HCl at room temperature showed a much higher SCC resistance for the base metal [59]. Raja and Rao [60] reported that the base metal of type 316 stainless steel had better SCC resistance than its autogeneous weld in a 5 N H_2SO_4 + 0.5 N NaCl solution at room temperature. Shaikh et al. [61] reported lower K_{ISCC} and higher plateau crack growth rates for type 316N stainless steel weld metal compared with type 316LN base metal (Fig. 12.8). They attributed the poorer SCC resistance of the weld metal to its higher yield strength and higher microscopic defect density. Shaikh et al. [62] reported that a sensitised HAZ was the weakest link in a weld joint from a SCC point of view. For a non-sensitised weldment, they reported failure in the weld metal. Baeslack III et al. [59] reported that the differences in the SCC susceptibilities of base and weld metals of type 304 stainless steel decreased with increasing strain rate of testing using the SSRT technique (Fig. 12.9). The decreasing difference on increasing the strain rate was because the higher strain rate of testing would be beyond the most susceptible range of strain rates for SCC.

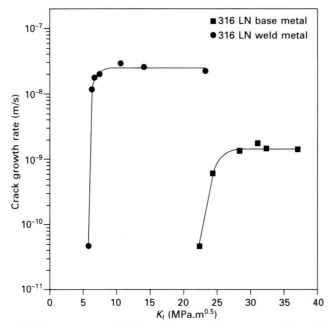

12.8 Comparison of SCC resistance of base and weld metals [61].

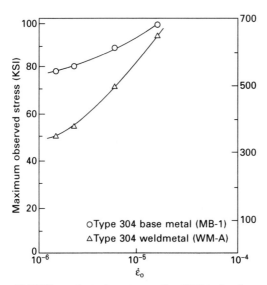

12.9 Effect of strain rate on the SCC behaviour of base and weld metals [59].

The presence of δ-ferrite can appreciably alter both the SCC resistance and the crack morphology of the weld metal. The SCC resistance of the weld metal depends on the δ-ferrite content, its distribution and the solidification mode. Duplex weld metal of type 304 stainless steel, which solidifies in the primary ferritic solidification mode, has SCC resistance similar to that of the base metal in boiling 45% $MgCl_2$ solution and failure occurred by interphase–interface cracking [63]. Duplex weld metal of type 304 stainless steel, which solidifies in the primary austenitic solidification mode, has SCC resistance similar to that of the base metal in boiling 45% $MgCl_2$ solution and failure occurred by a mixed transgranular-intergranular mode [63]. Fully austenitic weld metal has the most degraded SCC property and fails by intergranular SCC due to extensive segregation of S and P at the grain boundaries [64]. In NaCl and HCl solutions, both at ambient and high temperatures, the weld metal failed by stress-assisted dissolution (SAD) of δ-ferrite (Fig. 12.10) and SCC of austenite (Fig. 12.11) [59, 62]. In boiling 45% $MgCl_2$ solution, the failure occurred due to cracking of the DF/A interface (Fig. 12.12) and SCC of austenite [56, 57, 59]. The amount, morphology and continuity of the δ-ferrite network influences the SCC resistance of the weld metal. A continuous network of δ-ferrite was most harmful for SCC of weld metal as it provided a continuous path for crack propagation (Fig. 12.13) [59, 65]. Shaikh et al. [62] reported that on polarising the weldments anodic to the critical cracking potential (CCP), a decrease in SCC resistance was observed, while slight cathodic polarisation prevented failure (Fig. 12.14). This suggested that dissolution-mechanism was operative during SCC of type 316 stainless steel weldments. Krishnan and Rao [66] reported improvement in SCC resistance of the weld metal on cathodic polarization.

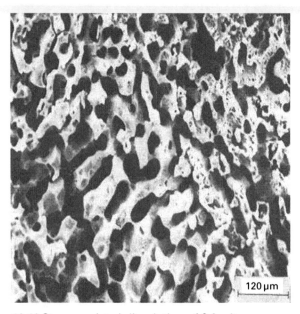

120 µm

12.10 Stress-assisted dissolution of δ-ferrite.

12.11 Transgranular SCC of austenite.

12.12 δ/γ interphase–interface cracking.

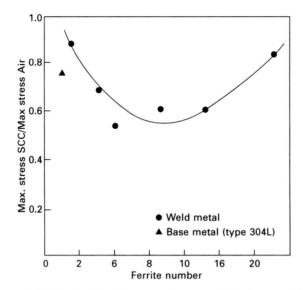

12.13 Effect of ferrite content on the SCC of austenitic SS weld [59].

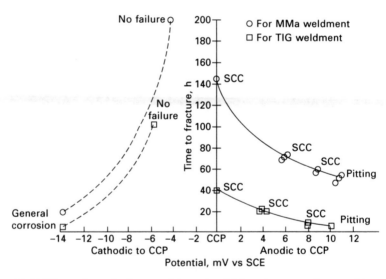

12.14 Effect of applied potential on the SCC resistance of austenitic SS [62].

Baeslack III *et al.* [63] reported that increasing the nitrogen content in 18Cr-8Ni stainless steel weld metal promoted primary austenitic solidification mode and reduced its SCC resistance. However, Nage and Raja [67] reported that in high temperature water at 288°C, type 904L stainless steel welds, obtained without nitrogen in the shielding gas, failed at the weld fusion zone.

Small additions of nitrogen (0.5 vol%) in the argon shielding gas imparted significant SCC resistance to the weld fusion zone such that the failure occurred in the base metal. Li and Congleton [68] reported that the transition zone of low alloy steel to stainless steel transition joint had a higher susceptibility to SCC than both the stainless steel and the low alloy steel parts. The SCC in this zone was mainly intergranular in the austenitic layer but transgranular cracking occurred both at the interface and in the low alloy steel part of the joint. Li *et al.* [69] reported that PWHT at 560°C, 620°C and 680°C for 20 h and also at 620°C for 5 and 50 h of the A508-309L/308L transition weld did not significantly affect the susceptibility of the as-welded joint to SCC in simulated PWR primary water at 292°C.

Sensitisation of duplex weld metals of austenitic stainless steels is not commonly heard of since chromium depletion due to $M_{23}C_6$ precipitation at δ/γ interfaces does not readily occur. This is because any chromium depletion occurring due to $M_{23}C_6$ precipitation gets healed due to rapid diffusion of chromium from the chromium-rich δ-ferrite phase. However, this immunity from sensitisation and IGC can be severely compromised especially on heat treatment in the temperature range of transformation. The severity of this problem depends on the carbon content as well as the extent to which the duplex structure is lost through the dissolution/transformation of δ-ferrite. Sensitisation data on weld metal of austenitic stainless steel is rare. Hamada and Yamauchi [70] reported that type 308 stainless steel weld overlays on low alloy steel became sensitised on post-weld heat treatment (PWHT) at 873 K. They evaluated the effect of δ-ferrite distribution on the sensitisation behaviour by using a microstructural parameter. Parvathavarthini *et al.* [71], in a very detailed study, reported that type 316N stainless steel weld metal was sensitised in the temperature range 625–725°C, when tested by ASTM A262 Practice E test (Fig. 12.15). They determined a critical cooling rate (CCR) of 1°C/h above which there was no risk of sensitisation. Figure 12.26 also shows that type 316L stainless steel weld metal was sensitised in the temperature range 625–675°C and the CCR was 160°C/h. They reported that the failure in Practice E test was only due to chromium depletion in type 316N stainless steel weld metal, while in type 316L stainless steel weld metal the loss of ductility was also due to σ phase precipitation.

Effect of heat input

Heat input affects the microstructure and hence the microstructure-sensitive properties of the weld metal. Increasing the heat input decreases the cooling rate of the weld metal, which increases the partitioning of the alloying elements and segregation of S and P, coarsens the δ-ferrite dendrites and increases the mean spacing between its secondary arms (Fig. 12.16) [19, 58, 72]. This is because decrease in the cooling rate of the weld metal increases

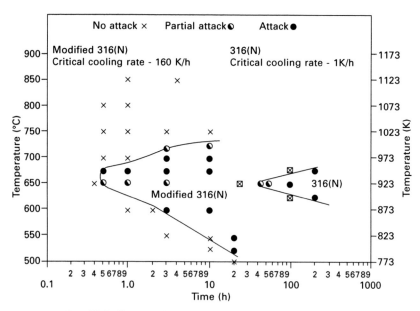

12.15 TTS diagram for types 316 N and type 316 L stainless steel weld metals [71].

solid-state diffusion and aids the redistribution of the solute between the austenite and ferrite phases. The size and shape of δ-ferrite in the weld metal strongly depends on heat input. The shape of the δ-ferrite also governs the transformation kinetics of δ-ferrite. Kokawa *et al.* reported early σ phase precipitation in vermicular ferrite than in lacy ferrite at 973 K [28].

Very little work has been done on the influence of heat input on the SCC susceptibility of welds of austenitic stainless steel. Sensitivity to intergranular corrosion (IGC) is increased with increasing heat input and it would seem the risk of intergranular SCC (IGSCC) is also promoted at higher heat inputs [58]. To a lesser extent, a similar adverse effect of high heat input has been found also for transgranular SCC (TGSCC) [69]. However, Franco *et al.* [73] reported an increase in SCC resistance of weld metal of AISI type 304 stainless steel with increasing heat input in boiling 45% MgCl$_2$ solution. Krishnan and Rao [65] reported that welds deposited by the low heat input (GTAW) process showed better SCC resistance than their SAW counterparts because they had a finer ferrite network. Lu *et al.* [74] reported that the GTAW process made the type 304 stainless steel weld metal and HAZ more sensitive to SCC than the base metal, but the laser beam welding (LBW) process improved the SCC resistance of the weld metal. In their studies on the SCC behaviour of multi-pass welds of type 316N stainless steel welds made by the manual metal arc welding (MMAW) process, Anita *et al.* [72] reported an improvement in the SCC resistance with increasing heat input

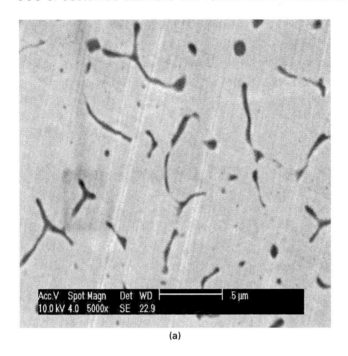

(a)

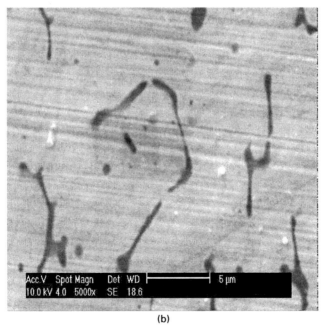

(b)

12.16 Micrographs showing coarsening of δ-ferrite and increasing interdenritic spacing with increasing heat input: (a) 3.07 kJ/cm, (b) 4.8 kJ/cm (c) 7.4 kJ/cm [72].

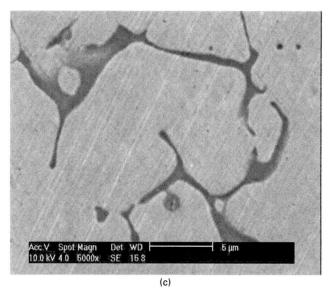

(c)

12.16 Continued.

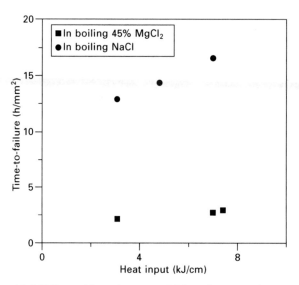

12.17 Effect of heat input on SCC resistance of type 316N stainless steel weld metal [72].

(Fig. 12.17). They attributed it to the reheating of the previous passes which led to homogenisation of the microchemistry and decrease in the matrix hardness due to thermal recovery.

Effect of secondary phases

The transformation products of δ-ferrite, such as σ, χ, μ, carbide phases, etc., which are formed during high temperature exposure or during post-weld heat treatment (PWHT), have their own effects on the SCC behaviour of austenitic stainless steel weld metals. Krishnan and Rao [75] reported a decrease in SCC resistance of weld deposits of types 308Cb and 309Cb stainless steels on PWHT at 600 and 800°C due to the formation of σ phase. However, they reported an improvement in the SCC resistance on PWHT of 1000°C due to globularisation of δ-ferrite. SCC studies in boiling 45% $MgCl_2$ solution on aged weld metal of type 316L stainless steel showed that, on aging at 873 K, the SCC resistance of the weld metal was governed by the occurrence of two complementary processes of matrix hardening and softening [76]. Maximum SCC resistance was observed for weld metal aged for 200 h (Fig. 12.18) because effects of matrix softening, caused by processes such as dissolution of ferrite network, overwhelmingly dominated the effects of matrix hardening, caused by factors like σ phase precipitation. Deterioration of SCC resistance was observed beyond 200 h aging because the effects of matrix hardening dominated [76].

Kim and Ballinger [77] measured the SCC growth rates for type 316L stainless steel weld metal in the as-welded condition and after aging at 673 K for 1000 and 5000 h in high purity water containing 300 ppb of oxygen at

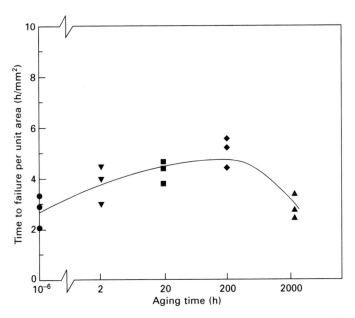

12.18 Effect of high temperature aging on SCC resistance of 316L stainless steel weld in boiling 45% $MgCl_2$ solution [76].

561 K. Crack growth rates for material in the as-welded and aged for 5000 h at 673 K were found to be generally within the scatter band for wrought material, although the aged material data fell at the high end of the scatter band.

Effect of residual stresses

The residual stresses in a component or structure are stresses caused by incompatible internal permanent strains. Allowing for residual stresses in the assessment of service performance varies according to the failure mechanism. During welding, residual stresses are retained in the component due to faster cooling rates associated with the weld metal. These stresses increase the dissolution rate of the material, thus deteriorating its corrosion properties. The major impact of residual stresses is felt on the SCC properties of the material. Many SCC failures have been reported on welded components during storage [78]. During service, residual stresses add on to the service load and accelerate SCC failures. Hence, the control and management of residual stresses assumes significance.

The extent of residual stresses induced into the weld is dependent on a number of parameters such as thickness of plate, extent of restraint, welding procedure, welding sequence, etc. [79–82]. Figure 12.19 shows the residual pattern developed in a type 304 stainless steel plate of dimensions 100 mm × 100 mm × 6 mm having a triangular groove of included angle 75° up to a depth of 4 mm. The general features of austenitic welds appear to be longitudinal tensile stresses in and near the weld, but the normal and transverse stresses can be positive or negative [79].

Post-weld heat treatment (PWHT) is most commonly employed to rid welded components of their residual stresses. In austenitic stainless steel components, PWHT can be of substantial benefit in avoiding IGSCC. In this regard, a stabilising anneal at 1143–1223 K is frequently applied to welded components in the petrochemical industry to guard against polythionic acid attack. For chloride-induced SCC, cracking occurs where environmental conditions are adverse and threshold stress for this form of failure is low. Service experience shows that when loaded components suffer chloride SCC, cracking preferentially occurs in weld areas, strongly implying weld residual stresses to be a significant factor. Hence, PWHT is advisable, at least for critical applications. The high coefficient of thermal expansion and the extremely low limits of elasticity for austenitic stainless steel would mean that residual stresses would readily arise during cooling from peak PWHT temperature, unless the operation is carefully controlled. Heating to over 1203 K is necessary to achieve maximum stress relief, and these high temperatures can pose critical problems such as control of distortion, scaling etc. Slow heating and cooling is preferred, subject to avoidance of sensitisation during

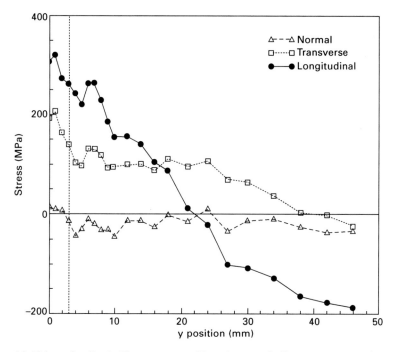

12.19 Longitudinal, *X*, transverse, *Y*, and normal, *Z*, components of residual stress measured along a line from the centre of the weld transversely to the edge of the plate. Typical uncertainties in stress are +40 MPa [79].

the cooling part of the cycle. Heat treatment at intermediate temperature, say 823–1023 K, represents an easier operation but only about 60% residual stress will be relieved. However, the problems of sensitisation and formation of intermetallic brittle phases such as σ phase rule out this heat treatment cycle. Low temperature stress relief at 673 K has been advocated to avoid SCC even though less than 40% stress relief is achieved [83]. Due to the above considerations, it may not be possible to guarantee that PWHT can avoid chloride SCC under the said service condition [84, 85]. Elimination of chloride ions, minimising operating temperature and appropriate design to eliminate crevices could prevent SCC of austenitic stainless steel weld joints in preference to PWHT. This is not to say that PWHT should be overlooked. PWHT is definitely beneficial in countering chloride SCC and should be resorted to when other properties, like creep and fatigue, render them essential.

Residual stresses in as-welded structures may be minimised by appropriate selection of materials, welding process and parameters, structural geometry and fabrication sequence. Residual stresses may be reduced by various special welding techniques including low stress non-distortion welding

(LSND), last pass heat sink welding (LPHSW) or inter-run peening. They may be relaxed by thermal processes including PWHT and creep in service, or by mechanical processes including proof testing and vibratory stress relief. Several processes, including laser shock processing (LSP), low plasticity burnishing (LPB), hole expansion techniques, and shot peening exist for manipulating the magnitude and type of residual stresses. Different stress relief treatments are appropriate in different applications. The effectiveness of the treatment may be reduced or the residual stresses may be increased if the treatment is not applied properly. Specialised processes are available for inducing beneficial compressive residual stresses, including peening, shot blasting, induction heating stress improvement (IHSI), low plasticity burnishing (LPB) and mechanical stress improvement procedures (MSIP).

12.3.3   Stress corrosion cracking (SCC) of heat affected zone

The more often encountered phenomenon that degrades the corrosion resistance of an austenitic stainless steel weld joint is the sensitisation of the HAZ. Sensitisation of austenitic stainless steel occurs in the temperature range 723–1123 K. During this high temperature exposure, depletion of Cr to less than 12% occurs in the region around the grain boundary, due to the precipitation of a continuous network of $M_{23}C_6$ carbides. Sensitisation makes the steel susceptible to intergranular attack. The probability of the HAZ being sensitised during welding would depend on the time it spends in the sensitisation temperature range. Figure 12.20 illustrates the maximum time spent by four different regions of the HAZ in the sensitisation temperature range. The region that experiences a peak temperature of 1123 K is sensitised the most. Below 723 K, carbide precipitation is too sluggish to cause any concern. The residence time spent by the HAZ in this temperature range depends on the heat input, which in turn governs the heating and cooling rates. Figure 12.21 shows that for a butt weld configuration for thicker sections, increasing heat input increases the time spent by the HAZ in the sensitisation temperature range. Also, the use of higher interpass temperatures increases the probability of sensitisation of HAZ [86].

Figure 12.22 illustrates the microstructure of a sensitised austenitic stainless steel after etching in oxalic acid. Optical microscopy reveals any of the three structures: step, ditch and dual. Step structure corresponds to clean grain boundaries, ditch structure means continuous carbide depletion and dual structure means discontinuous carbide precipitation. Normally step and dual structures are screened to be non-sensitised as per ASTM standards. This could be a mistake since dual structure could result in a fully ditched structure during low temperature service due to low temperature sensitisation

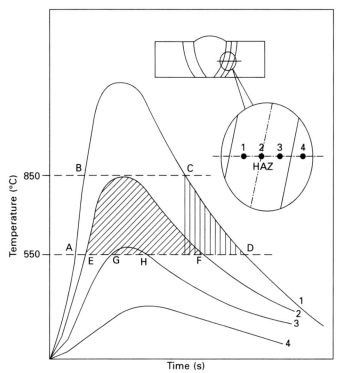

12.20 A schematic diagram indicating the effect of thermal cycling in inducing sensitisation in HAZ [86].

(LTS), as in boiling water reactors (BWRs) where LTS is known to cause IGSCC of nuclear reactor components.

Figure 12.23 illustrates the relationship between $M_{23}C_6$ precipitation and IGC. It is seen that the curve indicating IGC is offset to the right of the carbide precipitation curve. This indicates a time lag between onset of carbide precipitation and susceptibility to IGC. This is because, although carbide precipitation would occur, the regions adjacent to the carbides at grain boundaries have to be sufficiently depleted of Cr (< 12%) for IGC to occur [87]. The kinetics of the chromium carbide precipitation, and, hence, the resultant sensitisation could be predicted from a time-temperature-sensitisation diagram (TTS) (Fig. 12.24) [88]. These curves represent sensitisation during isothermal heat treatments. The TTS diagram indicates the possibility of sensitisation occurring in austenitic stainless steels. If a cooling curve, which is superimposed on the TTS diagram, cuts through it, it would indicate that the steel would become sensitised. As the TTS diagram moves to the right, it would indicate improvement in the resistance to sensitisation. It would also mean slower cooling rates could be tolerated during welding without the risk of the HAZ being sensitised. Figure 12.24 shows that on lowering

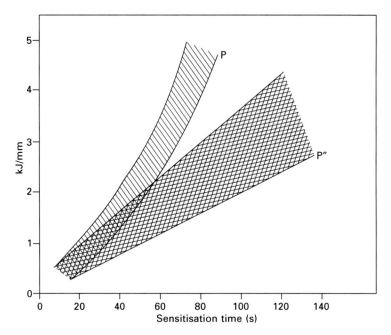

12.21 Relationship between heat input and maximum time spent by HAZ for welds made with 373 K and 300 K interpass temperatures [86].

the carbon content in type 316 stainless steel, the TTS diagram is moved to the right. Addition of nitrogen to type 316L stainless steel moves the TTS diagram further to the right. However, sensitisation during welding can also be predicted by a continuous cooling sensitisation (CCS) diagram (Fig. 12.25) [89].

Sensitisation of austenitic stainless steel requires the precipitation of Cr-rich carbides along grain boundaries, which makes carbon and Cr the predominant compositional variables for sensitisation. By reducing the carbon content in stainless steel, the TTS curve is displaced towards longer times because carbon concentration in austenite becomes insufficient to form Cr-carbide readily (Fig. 12.26) [90]. The limit of C content for which steel is not sensitive to intergranular corrosion (IGC) is closely connected with the presence of other alloying elements such as Cr, Mo, Ni, N, Mn, B, Si, as well as Ti and Nb in stabilised steel. The detrimental effect of C on sensitisation can be reduced by the addition of stabilising elements like Ti and Nb. These elements form TiC and NbC, which results in reduction of carbon available for Cr-carbide precipitation.

Cr has a pronounced effect on the passivation characteristics of stainless steel. With higher Cr contents, time to reach resistance limit of Cr depletion at the grain boundaries is shifted to longer times (Fig. 12.27). Higher Cr

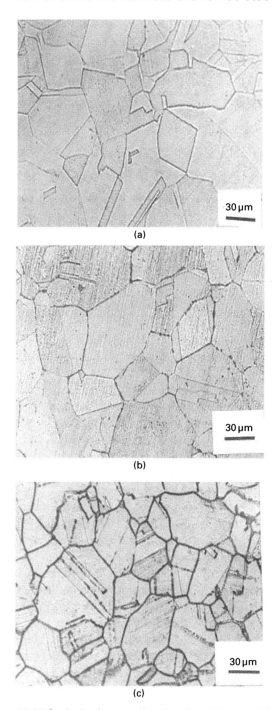

12.22 Optical micrographs showing (a) step, (b) dual and (c) ditch structures.

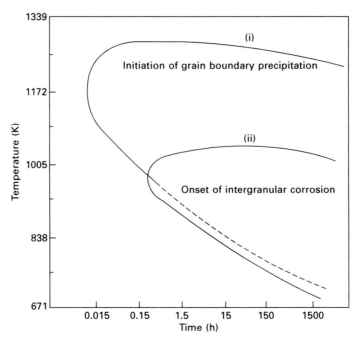

12.23 Relationship between $M_{23}C_6$ precipitation and IGC [87].

contents facilitate the diffusion of Cr into the depleted grain boundary area [90]. Ni is required in austenitic stainless steel to stabilise the austenite and must be increased with increasing Cr concentration. Increasing the bulk Ni content decreases the solubility and increases the diffusivity of C. This effect is much more pronounced when Ni content is greater than 20%. It is generally recommended that in 25/20 Cr-Ni steel, C content should be less than 0.02% to guarantee resistance to IGC. Mo reduces the solubility of C in austenite. Carbide precipitation is accelerated at higher temperatures whereas at lower temperatures it is slowed down (Fig. 12.28) [90]. When Mo is present, it is also incorporated in $M_{23}C_6$. Therefore, in addition to Cr depletion, Mo depletion is also revealed. In Mo-containing Cr-Ni austenitic stainless steel, $(Fe,Cr)_{23}C_6$ is precipitated first at 1023 K to 1123 K. With prolonged aging, Mo is also incorporated as $(Fe,Cr)_{21} Mo_2C_6$ which is finally converted to χ phase. With increasing Mo contents, $M_{23}C_6$ precipitation and IGC become increasingly influenced by the precipitation of intermetallic phases. The influence of Mn is of special importance because in fully austenitic welds, this element is added. Mn reduces the carbon activity and increases its solubility. Carbide precipitation is slowed down and hence it appears to inhibit carbide precipitation [91]. Boron retards the precipitation of Cr-carbide but depending upon the heat treatment it promotes IGC.

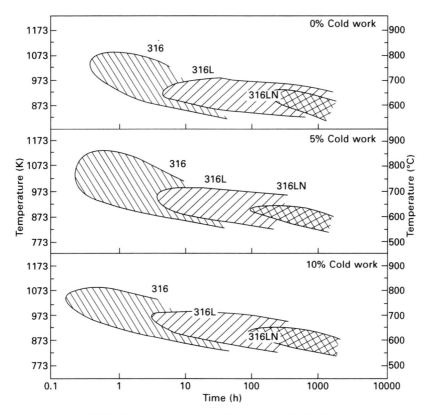

12.24 TTS diagrams for type 316, 361 L and 316 LN stainless steels [88].

Si promotes IGC of high purity and commercial stainless steel. Steels containing Mo were found to be much more sensitive to Si additions. The increased susceptibility to IGC in highly oxidising solution is due to the segregation of Si to grain boundaries [90]. Figure 12.29 represents the influence of Si on the kinetics of sensitisation. One of the alloying additions studied extensively in recent years is nitrogen. Its effect is quite complex and is dependent on the presence of other alloying additions. N content up to 0.16 wt% is reported to improve sensitisation resistance by retarding the precipitation and growth of $Cr_{23}C_6$ [92]. Beyond 0.16 wt%, the solubility limit of nitrogen in austenite is exceeded and Cr2N precipitates, thus reducing the sensitisation resistance of the steel. Dayal *et al.* [93] have established that as the nitrogen content increases, time required for sensitisation at the nose temperature increases from 0.5 h (316 stainless steel) to as much as 80 h (316 LN stainless steel) indicating the beneficial effect of N. Based on numerous data in the literature, the effect of chemical composition on sensitisation behaviour of austenitic stainless steels has been described by

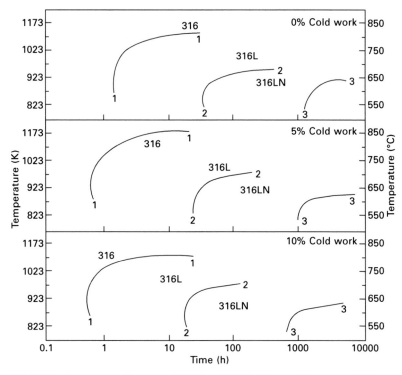

12.25 Continuous cooling sensitisation (CCS) diagram for types 316, 316L and 316LN stainless steels [89].

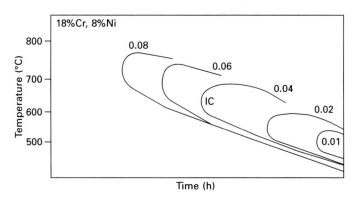

12.26 Influence of carbon on sensitisation kinetics [90].

an effective chromium content, Cr^{eff}, by giving proper weightings to various elements as [88, 93]:

$$Cr^{eff} = Cr + 1.45Mo - 0.19Ni - 100C + 0.13Mn - 0.22Si - 0.51Al - 0.2Co + 0.01Cu + 0.61Ti + 0.34V - 0.22W + 9.2N$$

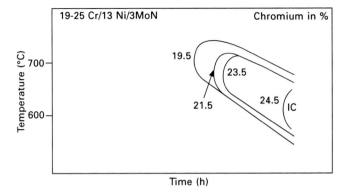

12.27 Influence of chromium on sensitisation kinetics [90].

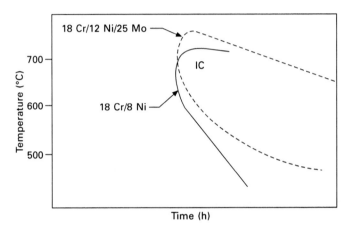

12.28 Influence of molybdenum on sensitisation kinetics [90].

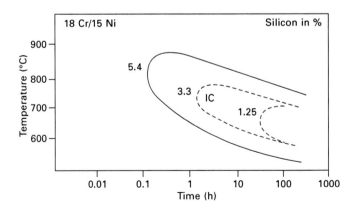

12.29 Influence of silicon on sensitisation kinetics [90].

Table 12.1 relates Cr^{eff} to the critical cooling rates. It is seen that as Cr^{eff} increases, the critical cooling rate, above which sensitisation will ot occur, decreases.

As the grain size decreases, the resistance to sensitisation increases [94]. This is because the grain boundary area available for Cr-carbide precipitation increases which would cause the effective depletion of Cr per unit grain boundary area to reduce. Also, Cr has to diffuse smaller distances to heal the Cr-depleted regions to desensitise the material. Cold work is reported to increase the sensitisation kinetics of austenitic stainless steel at moderate cold works and decrease the kinetics at higher cold works. Parvathavarthini and Dayal [89] reported that TTS diagrams of cold worked materials are shifted to the left (shorter times) with increasing cold work (Fig. 12.24), and below that of the as-received material. The nose temperature corresponding to maximum rate of sensitisation is also shifted to lower temperatures with increasing cold work. Desensitisation is faster at higher levels of cold work especially at high aging temperatures.

A sensitised HAZ could fail by TGSCC or IGSCC. The most commonly encountered environments that cause failure in austenitic stainless steel are those containing chlorides. It is very difficult to specify chloride levels below which SCC will not occur since other environmental factors such as oxygen content and pH of environment play a role in the failure. SCC in weld sensitised HAZ has been observed in coastal atmospheres [84, 95]. Many failures due to sensitisation have been reported in high temperature water in nuclear reactors [96]. IGSCC has been observed mainly in small diameter pipes where the stresses are expected to be very high. The cracks were mostly circumferential due to axial stresses. An important aspect to be considered to avoid IGSCC in nuclear reactors is low temperature sensitisation (LTS), wherein sensitisation occurs at temperatures lower than 723 K, for example at boiling water reactor (BWR) operating temperatures of 561 K [97]. In his exhaustive review, Hanninen reported that Japanese workers have carried out detailed investigations which involved heat treatment of weld HAZ for 10 000 h at 623 K and testing for IGC and IGSCC [98]. They found that

Table 12.1 Dependence of critical cooling rates on Cr^{eff} [93]

% CW	Cr^{eff} 316 SS	Cr^{eff} 316L SS	Cr^{eff} 316LN SS
	12.57	14.36	16.03
0	365	17	0.43
5	710	22	0.54
10	765	27	0.73
15	515	27	0.76
20	815	26	0.93
25	790	18	0.97

normal grade of austenitic stainless steel were more prone to LTS than nuclear grade austenitic stainless steel. Kain *et al.* [99] investigated the effects of low-temperature sensitisation of non-sensitised austenitic stainless steel on SCC in nuclear power reactors. Types 304, 304L and 304LN developed martensite after 15% cold working. Heat treatment of these cold worked steels at 773 K led to sensitisation of grain boundaries. Types 316L and 316LN did not develop martensite upon cold rolling due to its chemical composition suppressing the martensite transformation temperature, hence these were not sensitised at 773 K. Type 304 stainless steel in both the as-received and the solution annealed conditions and type 304LN stainless steel with 0.15% nitrogen developed LTS when heated at 500°C for 11 days, but the degree of sensitisation was not sufficient to make it susceptible to IGSCC in BWR simulated environment [100].

The combined role of oxygen and chloride ions in causing IGSCC of BWR piping is shown in Fig. 12.30. It is seen that under the operating condition of BWRs, IGSCC occurs in sensitized steels. The problem of IGSCC in BWRs can be overcome by resorting to modifications in welding processes, or by altering the material or by modifying the environment. Modifications in welding process include:

- *Last pass heat sink welding*: A TIG welding arc is used as the heat source to heat the outer surface while simultaneously melting the filler metal.

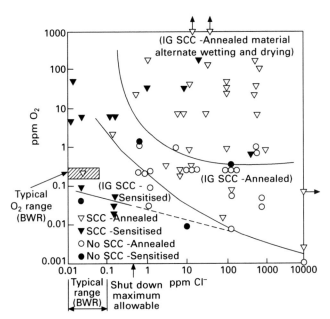

12.30 Effect of chloride and oxygen ions on the SCC of austenitic stainless steel.

During the process, the inner surface is flushed with water, thus cooling it. A temperature difference is thus established between the outer and inner surfaces. The resulting thermal stresses produce localised plasticity inducing compressive stresses on the inner surface of the pipe [101].

- *Induction heat stress improvement*: In this process, the weld area in the pipe is inductively heated from outside. The pipe is simultaneously cooled from inside with water. Just as in the above process, compressive stresses are induced on the inner surface of the pipe [101].

- *Mechanical stress improvement*: In this process, the pipe is radially compressed a slight amount on one side of the weld by means of hydraulic jaws to produce a permanent deformation. The deformation involved is less than 2%. The resulting curvature reduces the tensile stresses, produced by welding, on the root side of the weld area and produces compressive stresses in both the axial and radial directions [101].

- *Solution annealing treatment*: Helps in dissolving the grain boundary carbide network and evens out the material composition.

- *Corrosion resistant cladding*: The weld joint is deposited with weld overlays which have a duplex microstructure. In the BWR environment, the weld overlay may crack but complete resistance to IGSCC is ensured for the pipe.

- *Alternate pipe material*: Choose better materials for the application, such as types 304 LN SS, which is resistant to sensitisation and hence to IGC.

Cold working enhances sensitisation [90, 91, 99, 102–107], which, in turn, accelerates high temperature SCC [98]. Briant and Ritter [108] showed that deformation which induced martensite caused rapid sensitisation at temperatures below 600°C, led to extensive transgranular corrosion, and produced rapid healing. They also observed that deformation alone noticeably increased the kinetics of sensitisation only at temperatures where undeformed samples were readily sensitised. García *et al.* [109] reported that at lower levels of cold work, type 304 stainless steel showed higher tendency to IGSCC in boiling 42% $MgCl_2$ solution, while for higher deformations there was a greater susceptibility to TGSCC. However, Muraleedharan *et al.* [110] reported a transgranular to intergranular transition in cracking mode in the solution annealed steel while aging of the steel was found to inhibit the above transition, when the SCC test was conducted in $MgCl_2$ solution boiling at 428 K. Abu Elazm *et al.* [111] reported that type 304 stainless steel with a high degree of sensitisation showed only intergranular cracking in boiling 45% $MgCl_2$ solution at regions of lower stress in the constant-strain samples and transgranular cracking at regions of high stress. Nakano *et al.* [112] reported that cold worked type 304 stainless steel showed maximum susceptibility to irradiation-assisted SCC as compared to sensitised material in oxygenated high purity water at 561 K.

Sensitisation results in weaker passive film in the regions along the grain boundary compared with solution annealed material. Hence, it possesses lower threshold stress parameters (K_{ISCC} and J_{ISCC}) and higher plateau crack growth rate (PCGR) compared with solution annealed material (Fig. 12.31) [113, 114]. The susceptibility to IGSCC does not increase with increasing degree of sensitisation, as seen in Fig. 12.32 [115]. Instead, for a given electrochemical potentiokinetic reactivation (EPR) charge value, IGSCC susceptibility increased with decreasing sensitising temperature. This suggested that the IGSCC susceptibility of the austenitic stainless steel was dependent on the temperature of aging and is governed by the ratio of width to depth of the depleted zone. Smaller width to depth ratio promotes IGSCC.

Interfaces control many properties in engineering materials, several of which are critical to the integrity of the engineering structure. In single-phase austenitic alloys, grain boundaries are often the weak link, displaying susceptibility to creep, corrosion and stress corrosion cracking. As such, grain boundary structure control affords the opportunity to improve the overall performance of alloys in a variety of applications. The role of coincident site lattice boundary (CSLB) enhancement and grain boundary connectivity on the response of an alloy to stress and the environment has been examined by some authors [116, 117]. Wasnik et al. [116] showed improvement in resistance to IGSCC with extreme randomisation of grain boundaries. Yang

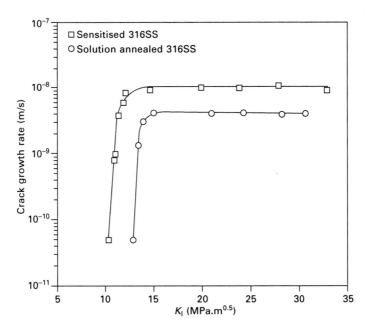

12.31 Effect of sensitisation on SCC resistance of type 316 stainless steel [115].

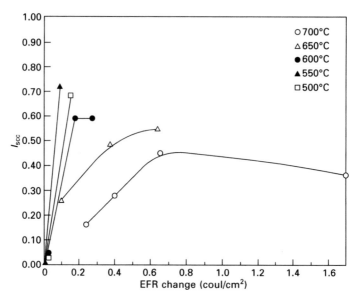

12.32 Correlation between IGSCC and DOS [116].

et al. [117] showed that a combination of laser surface melting and annealing on type 304 stainless steel resulted in a high frequency of twin boundaries and consequent discontinuity of random boundary network, which led to an improvement in resistance to intergranular corrosion. The maximum CSLB density that could be reached was 88.6% under optimal processing conditions of 1220 K and 28 h. Shaikh *et al.* [118] reported transgranular initiation in thermally aged type 316LN stainless steel in boiling acidified concentrated NaCl solution. They attributed it to dominating presence of Σ3 grain boundaries in the steel. Abdeljawad [119] reported that fcc polycrystalline materials with the Σ3 grain boundaries are characterised by their strength and resistivity to crack nucleation as evidenced by the higher local stresses and accumulation of dislocation densities at the grain boundary regions.

12.4   Welding issues in ferritic steels

Chromium-molybdenum steels are widely used in the power and process industries, combining high strengths at both ambient and high temperatures with adequate toughness. The availability of these properties in low alloy steels makes them economically attractive. These steels contain chromium contents up to 12% with usually about 1% Mo. The requirements of these steels with high Cr content have been necessitated for high-temperature operation in corrosive environments such as those of the nuclear and petrochemical industries [120, 121]. The properties are controlled by microstructure, which, in turn, depends on the heat treatment. The heat treatments given are also simple and therefore economically viable. The steels most commonly used vary in chromium content from 0.5 to 12% and all of them contain some molybdenum, usually up to a maximum of 1%. In recent years, 9Cr-1Mo steels has become a favoured tubing material for fast reactor steam generators because of its low thermal expansion, high resistance to chloride-induced SCC, acceptable mechanical properties at service temperatures, and easy control of microstructure [122]. The material undergoes degradation in properties due to long-term exposures to elevated temperatures and to aggressive environments. It is also susceptible to intergranular SCC in environments where hydrogen is produced [123, 124]. These steels are readily weldable only if recommended welding practices such as pre-heating and post-weld heating are followed, failing which a host of cracking problems [125], such as delayed cracking, hot cracking and reheat cracking, and sensitisation [126], could result.

The failure of a ferritic steel component is initiated during welding or during service in a localised microscopic region of the weldment [127]. This is due to the significant differences in the behaviour of the various regions of the weldment because of the heterogeneity in their microstructure [128, 129]. The variations in the microstructure and microchemistry of the different regions in ferritic steel weldment arise due to the differences in the thermal cycles experienced by these regions [125]. The major causes of failure of ferritic steel weldments have been identified to be δ-ferrite formation, segregation of impurity elements, overcoarsening of austenite near the weld metals, and formation of intercritical regions [127, 130, 131]. In addition, sensitisation, and as a result intergranular corrosion, has been reported under certain conditions such as improper PWHT.

12.4.1   Microstructure evolution in the weld joint

The pseudo-equilibrium diagram for the Fe-Cr-0.1%C system is given in Fig. 12.33 [132]. The phase fields crossed during heating or cooling by a particular alloy are dependent on the chemical composition, which in turn can be expressed as the net chromium equivalent [133]. Patriarca *et al.* [133] proposed a net chromium equivalent (NCE), which is expressed as

$$NCE = \%Cr + 6 * \% \ Si + 4 * \% \ Mo + 1.5 * \% \ W + 11 * \% \ V$$
$$+ \ 5 * \% \ Nb + 12 * \% \ Al + 8 * \% \ Ti - 40 * \% \ C$$
$$- \ 2 * \% \ Mn - 4 * \% \ Ni - 2 * \% \ Co - 30 * \% \ N - \% \ Cu.$$

Figure 12.34 shows the microstructure of the normalised and tempered base metal of 9Cr-1Mo steel. The structure of this base metal is reported to contain martensite laths with $M_{23}C_6$ carbides at lath boundaries, and uniform dispersion of fine, acicular M_2X, where X could be carbide or nitride, within

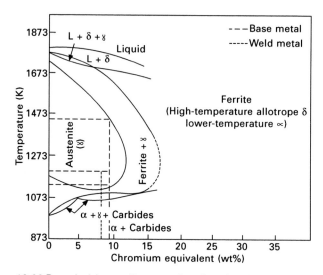

12.33 Pseudo-binary diagram showing the dependence of various phases on net chromium equivalent [132].

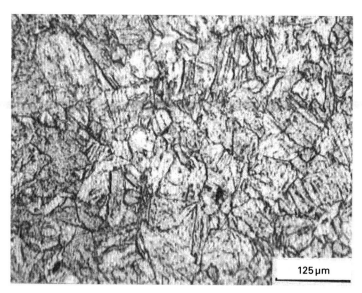

12.34 Microstructure of base metal of 9Cr-1Mo steel.

the ferrite [134]. Figure 12.35 shows that the microstructure of the weld metal of 9Cr-1Mo ferritic steel contained only the martensite phase [135]. In multipass welding, the martensite in the previous beads could get tempered, leading to the precipitation of δ-ferrite and carbide phases. Vijayalakshmi *et al.* [135] had reported the presence of prior austenite grains elongated in the

125 µm

12.35 Microstructure of weld metal of 9Cr-1Mo steel.

direction of cooling. The solidification of the liquid metals could proceed by two different routes depending on the chemical composition and the cooling rate: either 100% austenite, which subsequently could undergo a solid-state transformation to martensite, or a mixture of austenite with small amounts of δ-ferrite [135]. Solidification of the liquid metal into the duplex δ + γ phase field has been reported in a series of steels with Cr content in the range of 9–12%, with 9% being the borderline [136]. Apart from the cooling rate, which is governed by the thickness of the weld section, and the degree of repartitioning of Cr into the liquid, the solidification of 9Cr-1Mo steel into the duplex phase field depends critically on the amount of ferrite stabilisers, particularly Si [135]. Comparison of Figs 12.34 and 12.35 indicates that the weld metal has a larger grain size than the base metal. This could be attributed to the higher temperatures that the multipass weld metals experience during welding and also the longer time spent by the weld metal in the austenitising temperature range during cooling of the weld metal.

The four main HAZ microstructures observed in the 9Cr-1Mo weld joint depend mainly on the peak temperatures attained in the different regions during welding [137, 138]. The regions closest to the fusion boundary experience peak temperatures above AC4 (boundary between γ and γ + δ fields). At this temperature, Laha *et al.* [139] reported the formation of δ-ferrite along the austenite grain boundaries. The resultant structure is coarse-grained martensite with δ-ferrite. They reported an austenitic grain size of about 80 µm. After this region, where the peak temperatures experienced by HAZ lie between AC3 and AC4, the carbides, which impede grain growth, dissolve. Saroja *et al.* [140] reported that total solutionising of the microstructure occured above 1323 K. Laha *et al.* [139] reported an austenite grain size of about 60 μm in this region. On cooling, this region would transform to coarse-grained martensite (Fig. 12.36). The grain size decreases with increasing distance from the fusion line as the peak temperature experienced decreases [138, 141]. This would result in the formation of fine-grained martensite (Fig. 12.37). When the HAZ experiences a peak temperature between AC1 and AC3, only partial transformation to austenite takes place. Consequently, the microstructure resulting after cooling would be a mixture of austenite transformed products surrounding the volumes of non-transformed ferrite, which would have been tempered during the thermal cycle. This region is known as the intercritical region (Fig. 12.38).

12.4.2   Hydrogen embrittlement of steel weldments

Environment-assisted failures, caused mainly by hydrogen, have included failures of fossil fuel boiler tubes [142, 143], retaining ring of generator rotors [143], water-side components of condensers [144], among ethers. Exposure of carbon steel components to caustics at temperatures above 50–80°C, in the presence of residual stress, leads to caustic SCC and such failures have been experienced in petrochemical plants and petroleum refineries [145].

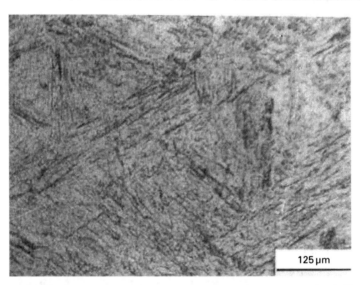

125 μm

12.36 Microstructure showing coarse grained martensite in HAZ of 9Cr-1Mo steel.

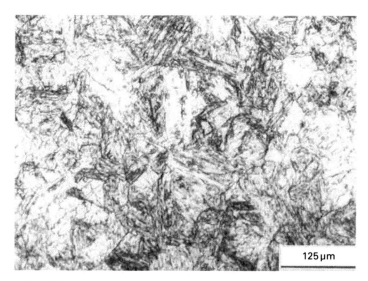

12.37 Microstructure showing fine-grained martensite in HAZ of 9Cr-1Mo steel.

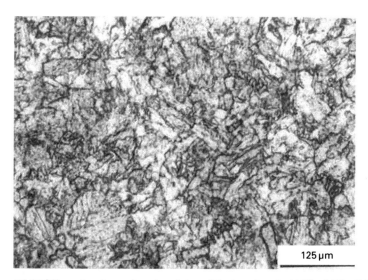

12.38 Microstructure showing intercritical region in HAZ of 9Cr-1Mo steel.

Traces of caustics can become concentrated in boiler feed water and cause caustic SCC [145]. Blowouts in buried pipelines of API 5L_46 steel due to high pH intergranular SCC in the presence of a considerable concentration of carbonates and bicarbonates in the soil has been reported [146]. Seifert and Ritter [147] reported that different reactor pressure vessel (RPV) steels, their weld metals and HAZ showed a very low susceptibility to SCC crack growth under simulated boiling water reactor (BWR), normal water (NWC) and hydrogen water chemistry (HWC) conditions in the temperature range of 150–288°C on periodical partial unloading, constant and ripple loading with pre-cracked fracture mechanics specimens.

Cold cracking or hydrogen-assisted cracking is probably the most widely encountered cracking problem in steel welds. It occurs mainly in the coarse grained heat affected zone (CGHAZ), though in high alloy steels, where welding is carried out using consumables of matching composition, weld metal is also susceptible to this form of cracking. Further, in steels of very low carbon content like HSLA steels, weld metal is more susceptible to cracking than the HAZ. As the name suggests, it is caused by hydrogen that is introduced into the molten metal during the welding process and retained in the weld at ambient temperature even after many hours or even days of welding. For the cracking to occur the following three conditions have to be met: (a) sufficient amount of hydrogen, (b) a susceptible microstructure, and (c) presence of sufficient restraint.

The main source of hydrogen in the weld metal is the moisture in the electrode coating (in the case of the SMAW process) or in the powder flux (in the SAW process). Thus, this type of cracking is less severe in other welding processes like GTAW and GMAW where no flux is used. However, moisture can either be present as an impurity in the shielding gas or gain entry into the arc from the atmosphere due to poor shielding. Other sources of hydrogen are the organic material used in cellulose-coated electrodes, grease or other organic products that may be present in the dirty joint surfaces or even corrosion products.

Hydrogen enters the weld metal through the welding arc where gaseous hydrogen, moisture or other compounds containing hydrogen dissociate to form atomic hydrogen that dissolves in the weld metal. As the weld solidifies, the hydrogen in the weld metal becomes supersaturated (the solubility of hydrogen in iron is very low and it decreases with temperature). However, due to high cooling rate, not all supersaturated hydrogen can diffuse out of the weld metal during cooling. Some of these hydrogen atoms get trapped at various defects like grain boundaries, inclusions, etc., where they combine to form hydrogen molecules. Hydrogen thus trapped is called residual hydrogen and this does not contribute to cracking. The atomic hydrogen that diffuses through the matrix is called diffusible hydrogen. Hydrogen in this form assists plastic deformation in the regions of stress concentration like notches and root of the welds, thus leading the cracking.

Susceptibility to cracking is strongly influenced by the microstructure. In general, martensitic microstructure is the most susceptible and ferritic the least. The structure of the HAZ depends on both alloy composition and heat input (cooling time $t_{8/5}$). Composition is represented by a parameter called carbon equivalent (CE) which normalises the effect of different alloying elements on the structure into a single parameter. A large number of equations are available to determine CE of steels, which are valid for different ranges of alloy composition. CE_{IIw}, formulated by the International Institute of Welding, and P_{CM}, originated in Japan, which are used in the discussion that follows are defined below.

$$CE_{IIw} = C + Mn/6 + Cu + Ni/15 + Cr+Mo+V/5$$

$$P_{CM} = C + Si/30 + Mn+Cu+Cr/20 + Ni/60 + Mo/15 + V/10 + 5B$$

Welding heat input is best represented by cooling rate of the weld, as the efficiency of different welding processes differs significantly. Time required for the weld to cool from 800°C to 500°C ($t_{8/5}$) is the most widely used parameter to represent the cooling rate as the austenite to ferrite transformation take place in this temperature range. Another parameter employed is t_{100}, the time required for weld to cool down to 100°C. Hardness of the HAZ is also used as a parameter that indicates the susceptibility of the welds to cracking. As a rule of thumb, it is assumed that steel is not susceptible to cracking if its CE_{IIw} is <0.4 and if its HAZ hardness is <350 VHN. Restraint imposed by different parts of a component during fabrication can significantly influence the stress state in the actual weld joints. Restraint of the joint increases with increase in thickness of the weld and hence thickness is often used as a parameter in empirical equations that are available to determine the critical pre-heat temperature to avoid cracking.

It was initially believed that cold cracking could be prevented in the steel if the maximum hardness of the HAZ was kept below a critical value. This critical hardness was considered as indepenedent of CE or the alloy composition. In many C-Mn steels, maximum hardness decreases with cooling rate, and, hence, the main emphasis was to reduce the cooling rate to obtain hardness below the critical value. This was achieved by pre-heating the job before welding, so that after welding it cools slowly, thus resulting in a microstructure of lower hardness. This approach was often referred to as the hardness control approach in the literature and was developed by The Welding Institute, UK. A nomogram based on this approach is shown in Fig. 12.39. It considers four factors, namely, combined thickness, diffusible hydrogen content of the electrode (HD), carbon equivalent (CE_{IIw}) and weld heat input. The weld metal hydrogen content is divided into four groups: A with HD levels > 15 ml/100 g (high), B with 15 > HD > 10 (medium), C with 10 > HD > 5 (low), and D HD < 5 (very low). CE values valid for these four levels of HD are shown separately in the nomogram. However, later it was shown by Graville that this approach is valid only for steels of limited alloy content. He divided the steels into three broad categories based on carbon content and CE and proposed a diagram as shown in Fig. 12.40. According to this diagram, steels falling in Zone I (low carbon and

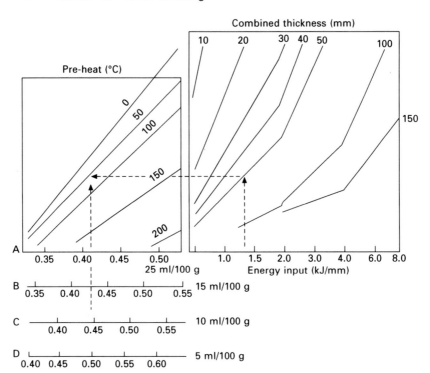

12.39 Nomogram for choosing safe pre-heat temperature to avoid HAC.

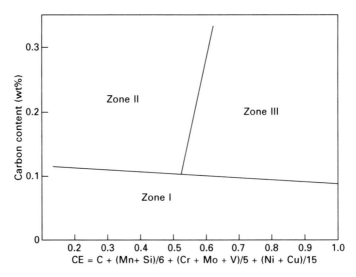

12.40 Graville diagram for classifying steels according to their cracking susceptibility.

low alloy content) are least susceptible to cracking, which occurs only under conditions of high hydrogen content and restraint. For alloys falling in Zone II, hardness varies with cooling rate, and hardness control approach can be applied to determine the pre-heat temperature to prevent cracking. For high alloy and high carbon steels falling in Zone III, their hardenability will be too high to alter by varying the cooling rate in normal welding conditions.

For steels falling in Zones I Zone III, pre-heat temperature is determined by hydrogen control approach. In this approach, it is assumed that cracking can be avoided by reducing the diffusible hydrogen content in the weld. In addition to reducing HAZ hardness, pre-heating helps in lowering the diffusible hydrogen content of the steel by giving more time for removal of hydrogen at high temperatures where diffusion is much faster than at ambient temperature. The time taken by the weld to cool down to 100°C is also considered in determining the pre-heat temperature. Based on this approach, Ito and Bessyo defined a cracking parameter P_w, which considers the effect of composition, hydrogen content and restraint separately:

$$P_W = P_{CM} + H_{JIS}/60 + R_F/40 \times 10^4$$

P_{CM} has been defined earlier. H_{JIS} is the diffusible hydrogen content as measured by glycerin method and R_F is the restraint factor ($N\ mm^{-1}\ mm^{-1}$), which is defined as force/unit length of weld required to contract or expand a gap by unit length in the direction perpendicular to the weld. R_F is given by

$$R_F = \eta r_f h$$

where r_f is a restraint coefficient ($<400\ N\ mm^{-2}\ mm^{-1}$ for normal welds), h is the plate thickness in mm and η a correction factor for the plate thickness. The correction factor η varies with the plate thickness and for thicknesses in the range of 5–100 mm, it is estimated using the following equation

$$\eta = \exp[(5.6-h)/80.3]$$

The critical pre-heat temperature ($T_{critical}$) above which no cracking takes place can be estimated from the parameter P_w using the following relation:

$$T_{critical}\ (°C) = 1440\ P_w - 392$$

Minimum cooling time for the weld to cool down to 100°C to avoid cracking (t_{100}) can also be calculated from P_w using the following relation:

$$t_{100}(s) = 14 \times 10^{-4}(P_w - 0.28)^2$$

There are many more equations available in the literature to determine either the pre-heat temperature or time for cooling down to 100°C. Most are determined using the data derived from different cracking tests which determine the pre-heat temperature in severe restraint conditions and hence are conservative in nature.

Though hydrogen-assisted cracking is associated mainly with CGHAZ, it can also take place in the weld metal if favourable conditions exist. One such condition is the very high hydrogen levels in the welds. This is the case when electrodes with cellulose coatings are used for welding. It has been found that cracking can take place in weld even if its hardness is as low as 200 VHN. Another condition that favours weld metal hydrogen cracking is the use of welding electrodes of composition matching that of the base metal, especially in the case of alloy steels. Unlike in C-Mn steel welds, both microstructure and hardness of the weld metal and HAZ would be similar, making both HAZ and weld metal susceptible to cracking.

Cracking in weld metal can also occur if the C content in the base metal is very low and the situation encountered in the welding of HSLA steels where C contents is kept low and strengthening is achieved by precipitation and grain refinement. CE and hardness of the weld metal is often higher that that of the base metal, and thus weld metal becomes more prone to cracking.

Classification of hydrogen damage

There are four major forms of hydrogen damage [148]. They are (i) creation of internal flaws, (ii) hydrogen attack, (iii) hydride formation, and (iv) hydrogen embrittlement. These are briefly discussed below.
Creation of internal flaws

Blistering

Hydrogen-induced blistering (Fig. 12.41) is observed in low strength alloys, such as ferritic steels and highly tempered martensitic alloys during electropolishing, pickling, corrosion, etc. Hydrogen is absorbed into the metal and accumulates just below the surface at voids, laminations, inclusions/matrix interfaces. This can build up enough pressure to lift up and bulge out the exterior layer of the material so that it resembles a blister. This type of failure is often encountered in petroleum drilling, refining and storage equipment, in vessels handling hydrocarbons and H_2S, and in alkylation units where HF is often used as catalyst. Steel plates used as cathodes in electrolysis also suffer this damage.

Shatter cracks, flakes and fish eye

Heavy steel forgings often contain a number of hairline cracks, called flakes, in the centre part. These are formed during cooling after forging and not during cooling after solidification. Moisture and additives are the main source of hydrogen. Flakes are formed due to recombination of hydrogen at inclusions and micropores.

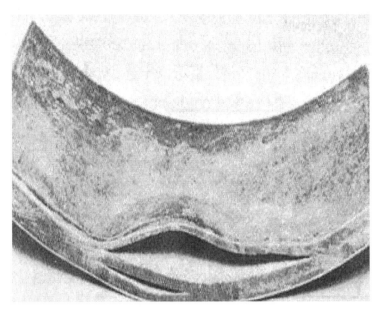

12.41 Hydrogen blistering in steel.

Hydrogen attack

When steel structures are exposed to high pressure hydrogen at high temperatures, hydrogen dissociates on the steel surface to form atomic hydrogen, which readily diffuses into the steel. The hydrogen then reacts with the carbon or metal carbides to form CH_4. The large size of methane prevents its diffusion. This results in high internal pressure which can cause blistering or intergranular cracking [149, 150]. At still higher temperatures, carbon can diffuse to the surface, and form methane with hydrogen on the surface, leading to decarburisation of the material. Decarburisation results in reduction in load bearing ability of the material. This is a serious problem for the design and operation of refinery equipment in hydrogen service. It is of serious concern in hydrotreating, reforming and hydrocracking units at temperatures above 533 K and hydrogen partial pressures above 689 kPa [151]. Figure 12.42 shows the Nelson diagram which indicates the temperature-hydrogen partial pressure combinations in which various steels can be used. These curves are based on long-term refinery experience rather than laboratory studies [152].

Hydride formation

A number of transition, rare earth, alkaline earth metals and alloys form brittle hydrides and are prone to embrittlement. The precipitations of these

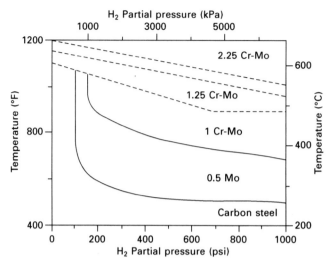

12.42 Nelson diagram for hydrogen attack.

hydrides can cause sufficient stress to produce failure even without application of external stress. Susceptibility is dependent on a number of variables such as hydrogen content, hydride orientation, distribution, morphology, strain rate and temperature. These metal hydrides are not only brittle and less dense than the metal matrix but precipitate at grain boundaries, twin boundaries, preferred crystallographic planes, etc., and dictate the initiation and growth stages of fracture. A classic example is hydride cracking of zircaloys in nuclear reactors. Ingress of hydrogen into the zircaloy lattice causes hydriding and embrittlement of the clad. The three sources of hydrogen are: (i) aqueous corrosion, (ii) hydrogen overpressure in the coolant, and (iii) radiolytic decomposition of water. When the hydrogen solubility limit in zircaloy is exceeded, excess hydrogen precipitates as hydride. The hydrides preferably precipitate perpendicular to the basal pole of the hcp-structured zircaloy. Hydrides impair the mechanical properties of the clad, eg. hydrides oriented in the radial direction reduce the ductility of the clad while circumferntial hydrides have only marginal effect. The problem of hydriding is overcome by texture control in the clad so that the basal plane is oriented in the radial direction by a fabrication process known as 'pilgering'.

Hydrogen embrittlement

This is the classic embrittlement process in which hydrogen atoms reduce the bond strength between atoms and cause loss of load bearing ability of the materials and also reduced ductility. The premature failure by hydrogen embrittlement is a function of applied stress and time.

Hydrogen embrittlement in industries

Steels containing Cr and Mo, which are extensively used in steam generators, are susceptible to hydrogen embrittlement (HE) at the weld metal/HAZ interface [153]. The effect of tempering temperature on the susceptibility of tested steel to stress corrosion cracking is connected with the effect on solubility of hydrogen in steel [154]. The microstructure and strength level of steel affect the susceptibility to hydrogen embrittlement. Most studies have shown that quenched and tempered martensite or bainite is most resistant to hydrogen, followed by spheroidised structures which demonstrate intermediate resistance, and untempered martensite which is the least resistant structure to hydrogen [155]. Tempering at high temperatures increases the amount of carbide precipitates that trap hydrogen, decreasing its diffusivity and reducing the rate of hydrogen cracking [156]. Performing a proper PWHT on the welded steel reduces the risk of HE since the weld joint is softened and toughness of the weld joint improved. Less aggressive media such as boiler water (deionised, oxygen-free) can also result in HSC in unalloyed steels at high temperature (100–200°C). The hydrogen is generated by the formation of magnetite (Fe_3O_4) on the steel. Several failures were observed in the HAZ of repair welded boilers after a few years of service resulting in cracking. The reason for the failures was the fact that PWHT after repair welding was overlooked which resulted in increased hardness of HAZ [157].

In the oil and gas industry, sour water containing H_2S has caused spontaneous cracking of steel components containing hard welds. Failures are especially likely with the use of submerged arc welding for pressure vessel construction and the weld metal had significantly higher hardness which led to transverse cracking in weld deposits. Whenever corrosion of steel takes place (aqueous H_2S corrosion, sour water or pickling solution) the cathodic reaction is the generation of atomic hydrogen from cathodic areas. When hydrogen recombination poisons such as suphide, arsenic or cyanide are present, atomic hydrogen concentration builds up at the surface and is absorbed within the material. If sufficient concentration is built up, it can lead to blistering at the surface in low strength steel. If residual stresses, arising due to fabrication processes like welding, are present, then hydrogen stress cracking (HSC) takes place. For example in sour water service, HSC has occurred in carbon steel vessels containing hard weld [158, 159]. Cracking is typically transgranular and contains sulphide corrosion products as shown in Fig. 12.43. Martensite formation leads to increase in hardness. In sour gas application, the hardened HAZ may undergo sulphide SCC. It is specified that a maximum hardness of HRC 22 is adequate to avoid this problem [160].

It has been reported that 9–12%Cr creep strength enhanced ferritic steels can be susceptible to SCC in the fully hardened condition due to sensitisation and there are a few case histories in the literature which attributes some failures of 9Cr-1Mo steel to sensitisation [161–165]. Intergranular oxide

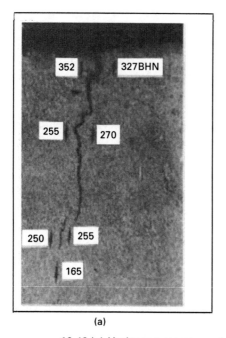 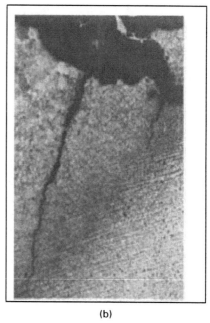

(a)   (b)

12.43 (a) Hydrogen stress cracking of a hard weld of a carbon steel vessel in sour water service; (b) hydrogen stress cracking of a hard HAZ next to weld in A516-70.

penetration in caustic environment and poor SCC resistance of as-welded structures are attributed to the sensitisation of the weldments. Henry has reported failure of dissimilar metal joint connecting T91 nipples to T22 safe end in a replacement SH outlet header and final SH assemblies header nipples in a coal-fired subcritical utility boiler [161]. Figure 12.44 shows the T91 to T22 safe end cracking on the T91 side of the weld. In some welds cracking had developed 360° around the weld to a depth greater than 50% of the total wall thickness. This failure has been attributed to sensitisation during cooling of the weld and the subsequent susceptibility to IGSCC through exposure to moisture/dampness. Even traces of aggressive contaminants like sulphur species can lead to transgranular hydrogen stress cracking in the as-welded (hardened) condition. Failure of cadmium plated carbon steel socket head cap screws, generally used in valve assemblies, are reported to fail during service due to absorption of hydrogen during the plating process and subsequent insufficient baking of the screws [166].

Control measures for SCC

Protection against SCC has to start from the design stage itself, starting with material selection, avoidance of crevices and any artifact that can cause

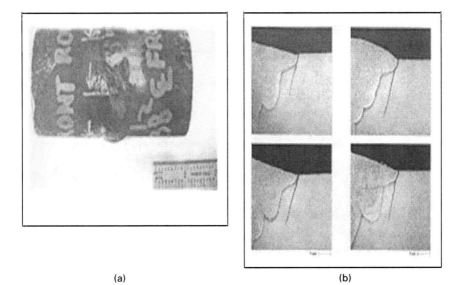

(a)   (b)

12.44 Cracking of dissimilar metal joint connecting T91 nipples to T22 safe end of a coal-fired subcritical utility boiler. Cracking was present on the T91 side of the weld. Cracking had developed 360° around the weld to a depth greater than 50% of the wall thickness.

stagnation of water, and use of fabrication procedures that produce no or low residual stresses. Control of environment to remove/minimise the damage-causing anionic species helps control/eliminate SCC. Surface modification to induce surface compressive stresses would help delay/eliminate SCC. In the case of stainless steel, low heat input welding procedures to avoid sensitisation, and use of low chloride containing, dye penetrant, hydrotest fluid and insulation would help control SCC. Pickling and passivation after fabrication to remove iron contamination and develop a passive film is recommended. Fabricated components should be stored in low humidity atmosphere. For ferritic steels, hardness and strength levels should be controlled by resorting to post-weld heat treatments.

Control measures for hydrogen embrittlement

1. Avoid hydride formers for use in hydrogen atmosphere.
2. Avoid galvanic couples, galvanic coatings, stray currents.
3. Avoid pickling and cathodic cleaning of susceptible materials:
    ➤   use mechanical means
    ➤   use vapour degreasing or solvent cleaning
    ➤   if pickling is necessary, use inhibited acid.
4. Avoid electroplating of HE susceptible alloys.

5. Modify the environment:
    ➤ add inibitors to reduce general corrosion and H evolution
    ➤ valeronitrile, benzonitrile, naphtonitrile, dibenzyl sulfoxide
    ➤ avoid hydrogen recombination poison.
6. Metallurgical modifications:
    ➤ use alloys with high SFE and low hydrogen solubility
    ➤ use clean (inclusion-free) steel
    ➤ reduce S content and other trace impurities to improve fracture behaviour.
7. Design modifications:
    ➤ avoid stress raisers in the design
    ➤ decrease residual stress by proper heat treatment
    ➤ lower the operating stress below threshold value for HE.

12.5   Conclusions

This Chapter deals with the SCC behaviour of austenitic stainless steel and ferritic steel welds. Defects and residual stresses are harmful to SCC resistance. The SCC behaviour of austenitic stainless steel welds depends on the content of δ-ferrite and other secondary phases formed during high temperature service, and on heat input. A continuous network of δ-ferrite is harmful to SCC resistance of the austenitic stainless steel weld metal. Similarly, σ phase deteriorates SCC resistance. In multipass welds, increasing heat input is beneficial for SCC resistance of austenitic stainless steel weld metal, as long as secondary phases other than δ-ferrite do not precipitate. Sensitisation is harmful to the SCC resistance of austenitic stainless steel weld joints. A sensitised HAZ is the most susceptible region of an austenitic stainless steel weld joint from a SCC point of view.

Hydrogen-assisted cracking during welding has been discussed with respect to the chemical composition and carbon equivalent. Environment-induced failure of ferritic steel weld joint occurs due to hydrogen produced in the corrosion reaction. The HAZ of a ferritic steel weld joint is most susceptible to hydrogen embrittlement. Appropriate pre-heat and post-weld heat treatments considerably improve the resistance to hydrogen embrittlement.

12.6   References

1. S. L. Mannan, S. C. Chetal, Baldev Raj and S. B. Bhoje, *Proceedings of the Seminar on Materials R & D for PFBR*, Eds.: S. L. Mannan and M. D. Mathew, IGCAR, Kalpakkam, (2003) 9
2. C. D. Lundin, *WRC Progress Report* (1994) 64
3. R. H. Jones and R. E. Ritter, *Materials Handbook*, Vol. 13, *Corrosion*, ASM International, OH (1987) 145
4. T. Rogne, J. M. Drugli and S. Valen, *Corrosion*, 48 (1992) 864

5. S. Szklarska-Smialowska and E. Lunarska, *Werkstoffe und Korrosion*, 32 (1981) 478
6. H. Holtan and H. Sigurdsson, *Werkstoffe und Korrosion*, 32 (1981) 478
7. S. Szklarska-Smialowska and J. Gust, *Corrosion Science*, 19 (1979) 753
8. R. L. Shamakian, A. R. Troiano and R. F. Hehemann, *Corrosion*, 36 (1980) 279
9. W. L. Clarke and G. M. Gordon, *Corrosion*, 29 (1973) 1
10. D. Kotecki and D. G. Howden, *WRC Bulletin No. 184* (1973)
11. H. T. Shirley, *Journal of Iron and Steel Institute*, 174 (1954) 242
12. D. N. Noble, *Welding and Metal Fabrication*, 59 (1991) 295
13. V. Kujanpaa, N. Suutala, T. Takalo and T. Moisio, *Welding Research International*, 9 (1979) 55
14. F. C. Hull, *Welding Journal*, 46 (1967) 399-s
15. D. Peckner and I. M. Bernstein, *Handbook of Stainless Steel*, McGraw-Hill, New York (1977)
16. J. C. Lippold and W. F. Savage, *Welding Journal*, 59 (1980) 362-s
17. J. A. Brooks and A. W. Thompson, *International Metals Review*, 36 (1991) 16
18. T. Koseki, T. Matsumiya, W. Yamada and T. Ogawa, *Metallurgical and Materials Transactions A*, 25 (1994) 1309
19. T. P. S. Gill, V. Seetharaman and J. B. Gnanamoorthy, *Corrosion*, 44 (1988) 511
20. C. D. Lundin and C. P. D. Chou, *WRC Bulletin No. 289* (1983) 1
21. C. D. Lundin, W. T. Delong and D. F. Spond, *Welding Journal*, 54 (1975) 241-s
22. J. M. Vitek and S. A. David, *Welding Journal*, 65 (1986) 106s
23. E. R. Szumachowski and H. F. Reid, *Welding Journal*, 57 (1978) 325s
24. J. Lefebvre, *Welding in the World*, 13 (1993) 21
25. V. S. Raghunathan, V. Seetharaman, S. Venkadesan and P. Rodriguez, *Metallurgical Transaction*, 10A (1979) 1683
26. H. Shaikh, M. G. Pujar, N. Sivaibarasi, P. V. Sivaprasad and H. S. Khatak, *Materials Science and Technology*, 10 (1994) 1096
27. G. F. Slattery, S. R. Keown and M. E. Lambert, *Metals Technology*, 10 (1983) 373
28. H. Kokawa, T. Kuwana and A. Yamamoto, *Welding Journal*, 68 (1989) 92-s
29. R. G. Thomas and D. Yapp, *Welding Journal*, 57 (1978) 361-s
30. T. P. S. Gill, M. Vijayalakshmi, P. Rodriguez and K. A. Padmanabhan, *Metallurgical Transactions*, 20A (1989) 1115
31. B. Weiss and R. Stickler, *Metallurgical Transactions*, 3 (1972) 851
32. J. M. Leitnaker, United States Patent US4265983, 1981
33. L. Zhao, Z Tian, Y. Peng, Y. Qi and Y. Wang, *Journal of Iron and Steel Research International*, 14 (2007) 259
34. R. G. Thomas and S. R. Keown, *Proceedings of the International Conference on Mechanical Behaviour and Nuclear Applications of Stainless Steels at Elevated Temperature, Varese, Italy, 1981*, The Metals Society, London (1981) 30
35. F. C. Hull, *Welding Journal*, 52 (1973) 104-s
36. T. P. S. Gill, M. Vijayalakshmi, J. B. Gnanamoorthy and K. A. Padmanabhan, *Welding Journal*, 65 (1986) 122-s
37. J. K. Lai and J. R. Haigh, *Welding Journal*, 58 (1979) 1-s
38. R. A. Farrar, *Journal of Materials Science*, 22 (1987) 363
39. R. J. Gray R. T. King and V. K. Sikka, *Journal of Metals*, 30 (1978) 18

40. S. A. David, J. M. Vitek, J. R. Keiser and W. C. Oliver, *Welding Journal*, 66 (1987) 235

41. J. M. Vitek, S. A. David, D. J. Alexander, J. R. Keiser and R. K. Nanstad, *Acta Metallurgica et Materialia*, 39 (1991) 503

42. M. Shafy, *Egyptian Journal of Solids*, 29 (2006) 151

43. R. M. Fisher, E. J. Dulis and K. G. Carroll, *Transactions AIME*, 197 (1953) 690

44. R. O. Williams and H. W. Paxton. *Journal of Iron and Steel Institute*, 185 (1957) 358

45. D. Chandra and L. H. Schwartz, *Metallurgical Transactions A*, 2A (1971) 511

46. J. Nishizawa, M. Hasebe and M. Ko, *Acta Metallurgica*, 27 (1979) 817

47. T. J. Nichol, A. Dotra and G. Aggun, *Metallurgical Transactions A*, 11 A (1980) 573

48. S. S. Brenner, M. K. Miller and W. A. Soffa, *Scripta Metallurgica*, 16 (1982) 831

49. J. M. Vitek, *Metallurgical Transactions A*, 18A (1987) 154

50. H. M. Chung and O. K. Chopra, in *Environmental Degradation of Materials in Nuclear Power Systems – Water Reactors*, ed. G. J. Theus and J. R. Weeks, TMS-AIME, Warrendale, PA (1988) 359

51. M. K. Miller and J. Bentleyin, *Environmental Degradation of Materials in Nuclear Power Systems – Water Reactors*, ed. G. J. Theus and J. R. Weeks, TMS-AIME, Warrendale, PA (1988) 341

52. M. Vrinat, R. Cozar and Y. Meyzaud, *Scripta Metallurgica*, 20 (1986) 1101

53. H. Abe, K. Shimizu and Y. Watanabe, *Journal of Power and Energy Systems*, 2 (2008) 2

54. H. Abe and Y. Watanabe, *Metallurgical and Materials Transactions A*, 39 (2008) 1392

55. D. H. Sherman, D. J. Duquette and W. F. Savage, *Corrosion*, 31 (1975) 376

56. F. Stalder and D. J. Duquette, *Corrosion*, 33 (1977) 67

57. W. A. Baeslack III, W. F. Savage and D. J. Duquette, *Metallurgical Transactions A*, 10A (1979) 1429

58. T. G. Gooch, *Welding in the World*, 22 (1984) 64

59. W. A. Baeslack III, D. J. Duquette and W. F. Savage, *Corrosion*, 35 (1979) 45

60. K. S. Raja and K. P. Rao, *Corrosion*, 48 (1992) 634

61. H. Shaikh, G. George, F. Schneider, K. Mummert and H. S. Khatak, *Werkstoffe und Korrosion*, 51 (2000) 719

62. H. Shaikh, H. S. Khatak and J. B. Gnanamoorthy, *Werkstoffe und Korrosion*, 38 (1987) 183

63. W. A. Baeslack III, W. F. Savage and D. J. Duquette, *Welding Journal*, 58 (1979) 83-s

64. A. Garner, *Materials Performance*, 21 (1982) 9

65. K. N. Krishnan and K. Prasad Rao, *Materials Science and Engineering A*, 142 (1991) 79

66. K. N. Krishnan and K. Prasad Rao, *Werkstoffe und Korrosion*, 41 (1990) 178

67. D. D. Nage and V. S. Raja, *Corrosion Science*, 48 (2006) 2317

68. G. F. Li and J. Congleton, *Corrosion Science*, 42 (2000) 1005

69. G. F. Li, E. A. Charles and J. Congleton, *Corrosion Science*, 43 (2001) 1963

70. I. Hamada and K. Yamauchi, *Metallurgical and Materials Transactions A*, 33A (2002) 1743

71. N. Parvathavarthini, R. K. Dayal, H. S. Khatak, V. Shankar and V. Shanmugam, *Journal of Nuclear Materials*, 355 (2006) 68

72. T. Anita, H. Shaikh, H. S. Khatak and G. Amarendra, *Corrosion*, 60 (2004) 873
73. C. V. Franco, R. P. Barbosa, A. E. Martinelli and A. J. A. Buschinelli, *Werkstoffe und Korrosion*, 49 (1998) 496
74. B. T. Lu, Z. K. Chen, J. L. Luo, B. M. Patchett and Z. H. Xu, *Electrochimica Acta*, 50 (2005) 1391
75. K. N. Krishnan and K. P. Rao, *Proceedings of the International Conference on Stainless Steel*, Indian Institute of Metals, New Delhi (1992) 419
76. H. Shaikh, H. S. Khatak, S. K. Seshadri, J. B. Gnanamoorthy and P. Rodriguez, *Metallurgical and Materials Transactions A*, 26A (1995) 1859
77. J. H. Kim and R. G. Ballinger, *Corrosion*, 64 (2008) 645
78. H. S. Khatak, V. Seetharaman and J. B. Gnanamoorthy, *Practical Metallography*, 20 (1983) 570
79. P. J. Webster, G. Mills, X. D. Wang, W. P. Kang and T. M. Holden, *The Journal of Strain Analysis for Engineering Design*, 30 (1995) 35
80. A. M. Paradowska, J. W. H. Price, T. R. Finlayson, U. Lienert, P. Walls and R. Ibrahim, *Journal of Physics: Condensed Matter*, 21 (2009) 124213
81. K.-S. Lee, W. Kim, J.-G. Lee, C.-Y. Park, J.-S. Yang, T.-R. Kim and J.-H. Park, *Journal of Mechanical Science and Technology*, 23 (2009) 2948
82. M. Mochizuki, M. Hayashi and T. Hattori, *Journal of Pressure Vessel Technology*, 122 (2000) 27
83. C. L. Cole and J. D. Jones, *Proceedings of Conference on Stainless Steels*, ISI Publication 117, (1969) 71
84. C. Edeleanu, *Corrosion Technology*, 4 (1957) 49
85. T. G. Gooch, *Proceedings of the Conference on The Influence of Welding and Welds on Corrosion Behaviour of Constructions*, IIW Annual Assembly, Tel Aviv (1975) 1.52
86. T. P. S. Gill, *Proceedings of the Corrosion Management Course*, The Indian Institute of Metals, Kalpakkam, October 1995, Paper No. L-6
87. C. J. Novak, *Handbook of Stainless Steels*, McGraw-Hill, New York (1977)
88. N. Parvathavarthini and R. K. Dayal, *Journal of Nuclear Materials*, 305 (2002) 209
89. N. Parvathavarthini, R. K. Dayal, J. B. Gnanamoorthy and S. K. Seshadri, *Journal of Nuclear Materials*, 168 (1989) 83
90. E. Folkhard, *Welding Metallurgy of Stainless Steels*, Springer-Verlag, Vienna (1988)
91. T. A. Mozhi, W. A. T. Clarke, K. Nishimoto, W. B. John and D. D. McDonald, *Corrosion*, 41 (1985) 555
92. N. Parvathavarthini, R. K. Dayal and J. B. Gnanamoorthy, *Corrosion*, 208 (1994) 251
93. R. K. Dayal, N. Parvathavarthini and Baldev Raj, *International Materials Reviews*, 50 (2005) 129
94. R. Singh, S. G. Chowdhury, B. Ravi Kumar, S. K. Das, P. K. De and I. Chattoraj, *Scripta Materialia*, 57 (2007) 185
95. T. V. Vinoy, H. Shaikh, H. S. Khatak, J. B. Gnanamoorthy and Baldev Raj, *Practical Metallography*, 34 (1997) 527
96. C. F. Cheng, *Journal of Nuclear Materials*, 57 (1975) 11
97. M. J. Povich, *Corrosion*, 34 (1978) 269
98. H. E. Hanninen, *International Metals Review*, 24 (1979) 85
99. V. Kain, K. Chandra, K. N. Adhe and P. K. De, *Journal of Nuclear Materials*, 334 (2004) 115

100. V. Kain, R. Samantaray, S. Acharya, P.K. De and V.S. Raja, *Environment-induced Cracking of Materials* (2008) 163

101. J. Schmidt, D. Pellkofer and E. Weiss, *Nuclear Engineering and Design*, 174 (1997) 301

102. R. S. Dutta, P. K. De and H. S. Gadiyar, *Corrosion Science*, 34 (1993) 51

103. S. Pednekar and S. Smialowska, *Corrosion*, 36 (1980) 565

104. C. L. Briant and A. M. Ritter, *Scripta Metallurgica*, 13 (1979) 177

105. S. M. Bruemmer, L. A. Charlot and D. G. Atteridge, *Corrosion*, 44 (1988) 427

106. B. K. Shah, A. K. Sinha, P. K. Rastogi and P. G. Kulkarni, *Materials Science and Technology*, 6 (1990) 157

107. S. K. Mannan, R. K. Dayal, M. Vijayalakshmi and N. Parvathavarthini, *Journal of Nuclear Materials*, 126 (1984) 1

108. C. L. Briant and A. M. Ritter, *Metallurgical Transactions A*, 11 (1980) 2009

109. C. García, F. Martín, P. De Tiedra, J. A. Heredero and M. L. Aparicio, *Corrosion Science*, 43 (2001) 1519

110. P. Muraleedharan, J. B Gnanamoorthy and P. Rodriguez, *Corrosion Science*, 38 (1996) 1187

111. A. Abou-Elazm, R. Abdel-Karim, I. Elmahallawi and R. Rashad, *Corrosion Science*, 51 (2009) 203

112. J. Nakano, Y. Miwa, T. Tsukada, S. Endo and K. Hide, *Journal of Nuclear Materials*, 367–370 (2007) 940

113. H. S. Khatak, J. B. Gnanamoorthy and P. Rodriguez, *Metallurgical and Materials Transactions*, 27A (1996) 1313

114. T. V. Vinoy, H. Shaikh, H. S. Khatak, N. Sivaibharasi and J. B. Gnanamoorthy, *Journal of Nuclear Materials*, 238 (1996) 278

115. P. Muraleedharan, J. B. Gnanamoorthy and P. Rodriguez, *Corrosion*, 52 (1996) 790

116. D. N. Wasnik, I. Samajdar, V. Kain, P. K. De and B. Verlinden, *Journal of Materials Engineering and Performance*, 12 (2003) 402

117. S. Yang, Z. J. Wang, H. Kokawa and Y. S. Sato, *Journal of Material Science*, 42 (2007) 847

118. H. Shaikh, T. Anita, R. K. Dayal and H. S. Khatak, *Corrosion Science*, 52 (2010) 1146

119. F. F. Abdeljawad, MS Thesis, North Carolina State University, 2005

120. J. A. G. Holmes, *Nuclear Engineering*, 23 (1981) 23

121. F. Pollard, *Proceedings of the International Conference on Production, Fabrication, Properties and Applications of Ferritic Steels for High Temperature Applications*, Ed. A. K. Khare, ASM International, Materials Park, OH (1981) 153

122. J. Orr and S. J. Sanderson, *Proceedings of the Tropical Conference on Ferritic Alloys for Use in Nuclear Energy Technologies*, TMS-AIME, Warrendale, PA (1984) 261

123. C. A. Hippsley and N. P. Howarth, *Materials Science and Technology*, 4 (1988) 791

124. S. K. Banerjee, C. J. McMohan, Jr. and H. C. Feng, *Metallurgical Transactions A*, 9A (1978) 237

125. R. Menon and K. K. Khan, Report No. ORNL/sub/81-07685/02&77, Oak Ridge National Laboratory, Oak Ridge, TN

126. B. Paulson, *Corrosion Science*, 18 (1978) 371

127. M. Dewitt and C. Coussement, *Materials at High Temperature*, 9 (1991) 178

128. K. Laha, K. B. S. Rao and S. L. Mannan, *Materials Science and Engineering*, 129A (1990) 183

129. J. G. Zhang, F. W. Noble and B. L. Eyre, *Materials Science and Technology*, 7 (1991) 315

130. H. K. D. H. Bhadeshia, S. A. David and J. M. Vitek, *Materials Science and Technology*, 7 (1991) 50

131. G. S. Kim, J. E. Indacochea and T. D. Spry, *Materials Science and Technology*, 7 (1991) 42

132. S. J. Sanderson, *Proceedings of the International Conference on Production, Fabrication, Properties and Applications of Ferritic Steels for High Temperature Applications*, Ed. A. K. Khare, ASM International, Materials Park, OH (1981) 85

133. P. Patriarca, S. D. Harkness, J. M. Duke and L. R. Cooper, *Nuclear Technology*, 28 (1976) 516

134. S. Saroja, M. Vijayalakshmi and V. S. Raghunathan, *Materials Transactions of the Japan Institute of Metals*, 34 (1993) 901

135. M. Vijayalakshmi, S. Saroja, V. Thomas Paul, R. Mythili and V. S. Raghunathan, *Metallurgical Transactions A*, 30A (1999) 161

136. J. N. Soo, CEGB Report no. RD/L/R, CEGB 1918

137. R. S. Fidler and D. J. Gooch, *Proceedings of the International Conference on Ferritic Steels for Steam Generators*, Eds. S. F. Pugh and E. A. Little, British Nuclear Energy Society, London (1978) 128

138. F. B. Pickering and A. D. Vassiliou, *Metals Technology*, 7 (1980) 409

139. K. Laha, K. S. Chandravathi, K. B. S. Rao and S. L. Mannan, *International Journal of Pressure Vessel and Piping*, 62 (1995) 303

140. S. Saroja, M. Vijayalakshmi and V. S. Raghunathan, *Journal of Materials Science*, 27 (1992) 2389

141. J. N. Soo, *Proceedings of the Specialist Meeting on the Mechanical Properties of Structural Materials Including Environmental Effects*, IAEA and IWGFR, Chester, England (1983) 579

142. W. L. Weiss, *Handbook of Case Histories in Failure Analysis*, Vol. 2, Ed. K. A. Easaklul, ASM International, Materials Park, OH, (1993) 271

143. R. K. Dayal and N. Parvathavarthini, *Sadahana*, 28 (2003) 431

144. *Metal Handbook, Vol. 13, Corrosion*, ASM International, Materials Park, OH (1987)

145. R. K. Dayal, H. Shaikh and N. Parvathavarthini, *Weld Cracking in Ferrous Alloys*, Ed. Raman Singh, Woodhead Publishing Limited, Cambridge (2009) 477

146. C. Manfredi and J. L. Otegui, *Engineering Failure Analysis*, 9 (2002) 495

147. H. P. Seifert and S. Ritter, *Journal of Nuclear Materials*, 372 (2008) 114

148. T. K. G. Namboodhiri, *Proceedings of the National Workshop on SCC*, RRC, Kalpakkam, (1980) 121

149. G. Sorell and M. J. Humphries, *Materials Performance*, 17 (1978) 33

150. A. R. Ciuffreda and W. R. Rowland, *Proceedings of API*, 37 (III) (1957) 116

151. R. D. Merrick and A. R. Ciuffreda, *Proceedings of API*, 61 (III) (1982) 101

152. R. Chiba, K. Ohnishi, K. Ishii and K. Maeda, *Corroison*, 41 (1985) 415

153. N. Parvathavarthini, R. K. Dayal and T. P. S. Gill, *Indian Welding Journal*, 28 (1995)

154. M. Schutze, *Corrosion and Environmental Degradation*, Eds R. W. Cahn, P. Haasen, E. J. Kramer, Wiley-VCH, (2000)

155. B. Pawlowski, A. Mazur and S. Gorczvca, *Corrosion Science*, 32 (1991) 685

156. A. W. Thompson, Effect of metallurgical variables on environmental fracture of engineering materials. In: *Environment-Sensitive Fracture of Engineering Materials*, Ed. Z. A. Foroulis, The Metallurgical Society of AIME, Warrendale (1977) 379

157. B. London, D. V. Nelson and J. C. Shyne, *Metallurgical Transaction A*, 19A (1988) 2497

158. J. Gutzeit, R. D. Merrick, L. R. Scharfstein, *Metals Handbook, Vol. 13, Corrosion*, P.1262

159. E. L. Hildebrand, *Proc. API*, 50 (III), (1970) 593

160. J. A. Brooks and A. W. Thompson, *International Metals Review*, 36 (1991) 16

161. J. F. Henry, ALSTOM, IIW, AWS, Technical lectures on Cr-Mo steels, January/February 2006

162. B. Poulson, *Corrosion Science*, 18 (1978) 371

163. B. Poulson, *Proceedings of the International Conference on Ferritic Steels for Steam Generators*, Eds. S. F. Pugh and E. A. Little, British Nuclear Energy Society, London (1978), Paper No: 68, 413–418

164. G. J. Bignold, *Proceedings of the International Conference on Ferritic Steels for Steam Generators*, Eds. S. F. Pugh and E. A. Little, British Nuclear Energy Society, London (1978), Paper No: 58, 346

165. D. Robinson, Clarke Chapman report no. ML/73/68/Dec. 1973

166. G. M. Tanner, *Handbook of Case Histories in Failure Analysis*, Vol. 1, Ed. K. A. Easaklul, ASM International, Materials Park, OH (1993) 332

13

Stress corrosion cracking (SCC) in polymer composites

J. K. LIM, Chonbuk National University, South Korea

Abstract: Polymer composite materials used as structural components can be subjected to a wide variety of different loading conditions or environmental conditions. This chapter describes typical examples of stress corrosion cracking (SCC) related to loading and environmental degradation. It analyzes subcritical crack growth and fracture behavior through microstructure characterization and strength evaluation, fiber orientation on deformation and fracture, fracture toughness, fatigue behavior and the use of scanning electron microscopy (SEM) in the analysis of fiber reinforced polymer composites.

Key words: stress corrosion cracking (SCC), fiber reinforced polymer (FRP) composites.

13.1   Introduction

Engineering plastics, such as advanced polymer composites, have been developed intensively over the past decade because of the great potential benefits of these materials in such areas as aerospace and automotive structural applications and in electronic packaging. Polymer composite materials used as structural components can be subjected to a wide variety of different loading or environmental conditions such as humidity levels, temperature cycles, sunlight, wind, dust or other contamination and oxygen concentration as shown in Fig. 13.1. If critical values of these loading and environmental conditions are exceeded, certain failure events, e.g., the initiation of fatigue cracking or the formation of wear particles, can take place in the material, finally resulting in modulus degradation or structural strength reduction. Any given combination of environmental factors can culminate in specific chemical and physical conditions, thereby determining the material–environment interactions, and the ultimate performance of the material.

The common mechanisms of moisture penetration into composite materials are capillary flow along the fiber–matrix interface, followed by diffusion from the interface into the bulk resin and transport by microcracks. Each of these mechanisms becomes active only after the occurrence of specific damage to the composite. When joints are exposed to a wet and hot environment, the absorption of water plays an important role in the degradation of the joints.

485

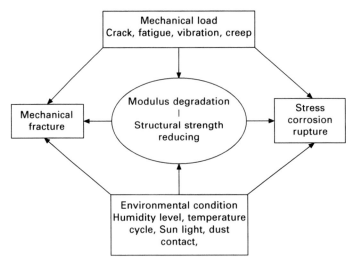

13.1 Schematic diagram showing relationship between loading and environmental conditions and fracture of polymer composites.

Huneke and Porey [1] studied the electrical performance of glass fiber/epoxy laminates. Laminates with silane treatment were susceptible to water absorption as determined by the electrochemical potential for a conductive growth along the glass–resin interface, but the laminate without any silane treatment was most susceptible. Schrader and Block [2] studied the lifetime of pyrex/epoxy/pyrex block adhesive joints under a static load in hot water. Ishida and Koenig [3] showed that extensive polysiloxane networks must be hydrolyzed by moisture into low molecular weight oligomers before any desorption is detected.

This chapter describes typical examples of stress corrosion cracking (SCC) and environmental degradation. It will analyze subcritical crack growth and fracture behavior through microsturcture characterization and strength evaluation, fiber orientation on deformation and fracture, fracture toughness, fatigue behavior and the use of scanning electron microscopy (SEM) in the analysis of fiber reinforced polymer composites. The chapter consists of five sections discussing:

- SCC of short fiber reinforced polymer injection moldings
- SCC evaluation of glass fiber reinforced polymer (GFRP) composites in synthetic sea water
- fatigue crack propagation mechanism of GFRP in synthetic sea water
- aging and crack propagation of natural fiber reinforced polymer composites related to water absorption behavior and the effect of absorption cycles on tensile properties and fracture toughness
- aging of biodegradable composites based on natural fiber and PLA in a hygrothermal environment.

13.2   Stress corrosion cracking (SCC) of short fiber reinforced polymer injection moldings

Thermoplastic injection moldings are used in many applications with the requirement for ever higher molding quality. There are still many problems to be solved in molding technology. In particular, any multiple gating or splitting of the melt flow gives rise to points within the structure where the flow fronts will recombine and become welded together. In the molding of very complex pieces such as automotive grill and electronic components, a multiplicity of weld lines is generated [4–6]. The weld line looks somewhat like a crack and, indeed, does lead to reduced strength [7, 8]. Formation of a weld line in injection molded products is practically unavoidable. They should therefore be designed not to occur in high stress areas. Short fiber reinforced thermoplastics are also known to be anisotropic at weld lines because of the variation of fiber orientation induced by mold filling [9, 10].

The mechanical behavior of composite materials is also affected by environmental and service conditions. The effect of temperature on the mechanical properties of polymer composites is derived from mechanical stresses which can be induced in composites containing constituents with different degrees of thermal expansion [11]. Such internal stresses may be present in the structure at room temperature and can change their magnitude according to environmental temperature changes. Moisture penetration into composite materials is conducted by one major mechanism, namely, diffusion. This mechanism involves direct diffusion of water molecules into the matrix and the interface of the matrix and fibers, in some cases into the fibers. The combination of temperature and moisture can have a significant effect on the mechanical properties of short fiber reinforced thermoplastics, particularly in the region of the weld lines in a component. This section reviews the effects of water absorption, strain rate and temperature on the mechanical properties and environmentally assisted cracking of short fiber reinforced polymer composite materials using the slow strain rate test (SSRT) method, SEM and light transmission microscopy (LM).

13.2.1   Microstructure of parent and weld materials

The polymer composite materials discussed in this section are polycarbonate (PC) with an amorphous structure typical of a general engineering plastic, including 30% short glass fibers and polyphenylene sulfide (PPS) with a crystalline structure. The diameter of the glass fiber is 13 micrometer and its length is 200–400 micrometer. The mold conditions were: an injection temperature of 320°C and a mold temperature of 115°C for the PC; and an injection temperature of 320°C and a mold temperature of 135°C for the PPS.

Studies were carried out in the environmental conditions of air–air, air–water and water–water. Air–air means that the specimens were both held and subsequently tested in air, air–water means that the specimen was held in air and then tested in water after absorption of over 100 hours. Water–water means that the specimens were both held and tested in water. A slow strain rate test (SSRT) was carried out from 10^{-4} sec^{-1} to 10^{-7} sec^{-1} and the temperature from 20°C to 80°C. Figure 13.2 provides a schematic drawing of a slow strain rate tester.

The fiber orientation pattern is highly influenced by mold geometry, processing conditions and rheological properties of the material. The rheological properties are affected by the viscosity and by other factors such as fiber aspect ratio, injection temperature and pressure. Figure 13.3 shows photomicrographs taken of the fiber orientation of the PC parent. The detailed characterization of the fiber morphology of injection flow was obtained by means of LM from sheets at orientation parallel and perpendicular to the injection flow direction.

Reinforcements such as glass fibers need to show the right orientation. When examining for flow and orientation, it is necessary to microtome the specimen parallel to the direction of flow. Figure 13.3 (a) and (b) show the

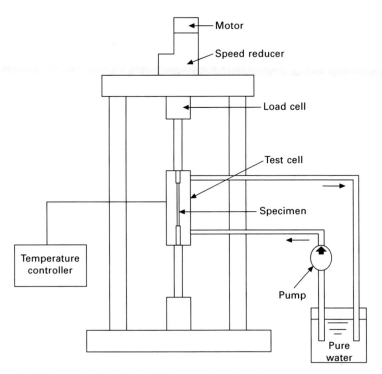

13.2 Schematic drawing of slow strain rate tester.

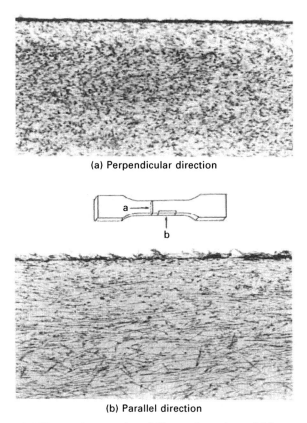

(a) Perpendicular direction

(b) Parallel direction

13.3 Photomicrographs of fiber orientation of PC parent.

PC parent with perpendicular and parallel orientation of glass fiber to the injection flow direction, respectively. Fiber orientation is perpendicular to the mold filled direction (MFD) and is longitudinal to the injection direction. It is also confirmed in the parallel direction. Figure 13.4 shows photomicrographs taken of fiber orientation of the PC weld in the x-z plane. The fiber orientation in the weld line is perpendicular to the MFD, that is, nearly in the direction of the x-y plane.

The flow front is spherical during injection, due to melt freezing effects in the direct vicinity of the mold surface. When two flow fronts profile during cavity filling, fibers in the flow front change to a perpendicular direction whilst the following fibers are injected at an angle of nearly 45 degrees. The top part of the mold surface resembles a flower shape. We therefore called the appearance of the fiber pattern at weld points the flower-like pattern [12]. But in the y-z plane the fiber orientation makes a fountain or volcano-like pattern [13] at the weld line. The mold thickness is 3 mm in the y-z plane and contrasted with the 13 mm dimension in the x-z plane. So the flow front is changed in the perpendicular direction only.

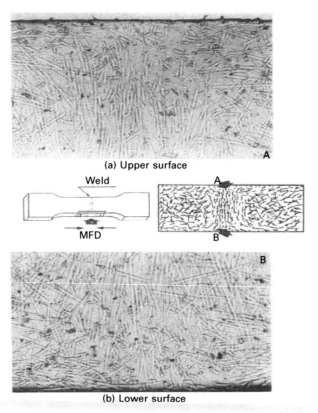

(a) Upper surface

Weld

MFD

A

B

(b) Lower surface

13.4 Photomicrographs of fiber orientation of PC weld in x-z plane.

13.2.2   Effect of temperature on strength

The air temperature on the earth's surface can vary over a wide range, bounded by lowest and highest records of −88°C and 58°C. The effect of temperature on strength was tested within the range of −60°C to +100°C. Lower temperatures were controlled by liquid nitrogen and high temperatures by a heating coil. Figure 13.5 shows the effect of temperature on the tensile strength of PC and PPS. Tensile strength at room temperature is lower than at −30°C. It is reasonable to assume that interface force of matrix and fiber at −30°C is stronger than that at room temperature, which is the effect of thermal shrinkage force at −30°C. The effect of temperature on tensile strength in the parent is greater than that of the weld because the strength of the parent depends on the matrix whilst the weld is dependent upon fiber orientation. The parent seems to be affected by the matrix rather than the glass fiber, but the weld is affected by the glass fiber rather than the matrix.

Figure 13.6 shows the fracture surfaces of PC parent fractured at −60°C and +80°C. The matrix produces voids and shows microcracks at a temperature

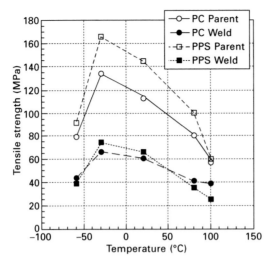

13.5 Relation between tensile strength and temperature.

(a) Test temperature, −60°C

(b) Test temperature, +80°C

13.6 Comparison of fracture surface depending on test temperature of PC parent.

of −60°C without debonding, but there is debonding between matrix and fiber at a temperature of +80°C. The fracture mechanism of this composite can be guessed as follows. The matrix breaks first followed by fiber breaking at low temperatures but the fibers break first followed by the matrix at high temperatures.

Figure 13.7 shows the fracture surface of a PPS parent fractured at −60°C. There is no debonding of the fiber and the matrix because of the effect of shrinkage at such low temperatures. Figure 13.7(b) shows the magnitude of embrittlement in the matrix. Figure 13.8 shows the fracture surface of a PPS parent fractured at +100°C. There is no apparent debonding of the fiber and the matrix, but delamination can be seen because of the softening of the matrix at high temperature. Figure 13.9 shows that the fracture surface at −60°C has a brittle and cleavage-like appearance but Fig. 13.10 shows less damage at 100°C. These results suggest that the matrix is brittle and is broken first at low temperature but that the glass fibers are broken first at high temperature. Figure 13.11 shows the fracture surface of a PPS weld at −60°C. The matrix is brittle and the fibers have shrunk. Both affect the

(a) × 300

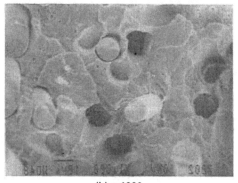

(b) × 1000

13.7 SEM fractography of PPS parent at temperature of 60°C.

(a) × 300

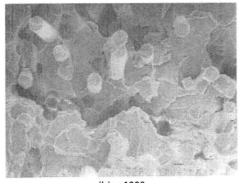

(b) × 1000

13.8 SEM fractography of PPS parent at temperature of 100°C.

strength of the PPS weld. Figure 13.12 shows the fracture surface of a PPS weld at +100°C. The matrix of PPS is soft, meaning that it fractures after the glass fibers break.

13.2.3   Effect of environment on the mechanical properties of polymers

The effect of water on the performance of composite materials is generally discussed in terms of short- or long-term effects, of reversible or irreversible effects, and of chemical or physical effects. We have examined the effect of exposure time in water on the tensile strength and the fracture mechanism of short glass fiber filled polymer composites. The various moisture levels were achieved by immersion in water at room temperature.

Figure 13.13 shows the relation between water content (%) and exposure time (hours) for PC and PPS. The water content increases linearly with time to 100 hours, and the material is saturated after 200 hours. The water content is 0.14% and 0.22% respectively at room temperature, and increases with

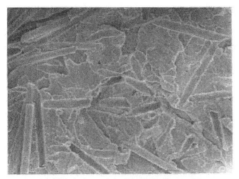

(a) Test temperature (–60°C, × 300)

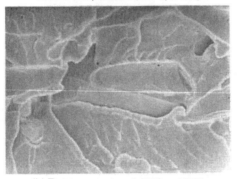

(b) Test temperature (–60°C, × 1000)

13.9 SEM fractography of PC weld at temperature of –60°C.

water temperature. Figure 13.14 shows the relation between water content and water exposure time of PC and the PPS parent at room temperature and +50°C. The water content also increases linearly with time to 100 hours and the material is saturated after 200 hours. The water content of PC increases with water temperature by up to 57%, but the water content of PPS does not change with the increase in water temperature. This result shows that the cohesive force between the matrix and fiber of PPS is stronger than that of PC.

Turning to material strength, the tensile strength of the PC parent is 114 MPa, but that of the weld line is 65 MPa, almost half of the parent. The mechanical properties of injection molded parts are affected by the microstructural anisotropies that are generated by the spatial variation of glass fiber orientation. Strength seems to be affected by variations in the matrix and the relation of glass fiber and matrix. Figure 13.15 shows the relation between tensile strength and water exposure time. Water exposure time reduces the tensile strength of both parent and weld material by nearly 10%. Figure 13.16 shows that the effect of water content on strength of PPS is less than that of PC.

(a) Test temperature (100°C, × 300)

(b) Test temperature (100°C, × 1000)

13.10 SEM fractography of PC weld at temperature of 100°C.

Figure 13.17 shows the relation between tensile strength and strain rate. The tensile strength is increased with the strain rate in three stages, that is, a viscoelasticity effect in the first stage, an insensitive second stage and a creep effect in the last stage. Figure 13.18 shows the relation between tensile strength and strain rate in air–air and water–water conditions. The strength in air–air is greater than in water–water conditions. In addition, the effect of PC parent on strain rate is greater than that of weld material.

Figure 13.19 shows the relationship between load and elongation of parent and weld material in different conditions at room temperature. All are nearly linear to maximum load. The elastic modulus of an immersed specimen in water is greater than that of a dried specimen in air. Immersion seems to affect the debonding of matrix and fiber due to swelling. Figure 13.20 shows the curves of load and elongation at a temperature of 80°C. These curves illustrate the time to fracture at maximum load. In particular, it takes longer to fracture at maximum load at 80°C in air–air than in water–water conditions at 80°C. Figure 13.21 shows the curves in air–air. These curves are linear not only for the parent but also the weld material

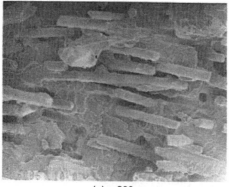

(a) × 300

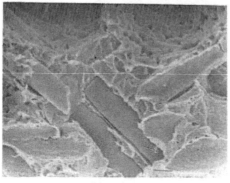

(b) × 1000

13.11 SEM fractography of PPS weld at temperature of –60°C.

in air–air conditions. All fractured at maximum load. As shown in Figure 13.22, however, the curves in water–water conditions are different according to water temperature. Tensile load is decreased with the increase in water temperature and elongation is long for both parent and weld. Figure 13.23 shows the relation between tensile strength and temperature. The strength of parent and weld is decreased linearly with increase in temperature. The parent is more sensitive than the weld material to temperature.

13.2.4   Fractography and environmentally assisted cracking

Figure 13.24 shows the fracture surface of the parent at a temperature of 20°C in (a) air–air, (b) air–water and (c) water–water. In Figure 13.24(a) the matrix fractured first followed by the glass fiber, but (b) shows the debonding of matrix and fiber, with matrix and fiber breaking at same time. In Fig 13.24(c) the fiber appears to break first followed by the matrix, but

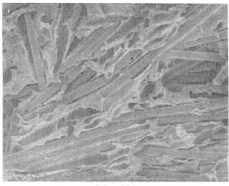

(a) × 300

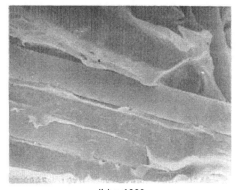

(b) × 1000

13.12 SEM fractography of PPS weld at temperature of 100°C.

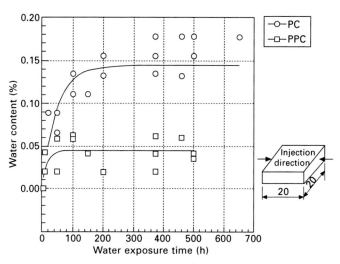

13.13 Relation between water content and water exposure time of PC and PPS parent.

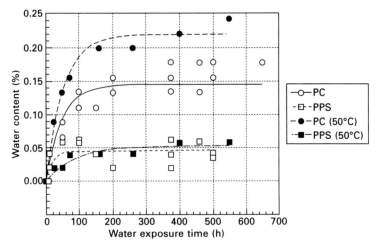

13.14 Relation between water content and water exposure time of PC and PPS parent at room temperature and +50°C.

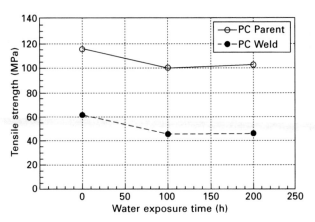

13.15 Relation between tensile strength and water exposure time of PC parent and weld.

it is not possible to see the internal cracking. As shown in (b) and (c), the cohesive force in the interface of matrix and fibers is weak due to the effect of water in the glass fiber composite materials. Figure 13.25 shows the fracture surface of parent at a temperature of 80°C in (a) air–air, (b) air–water and (c) water–water. Whilst (a) shows the debonding of matrix and fiber which does not appear in (b) and (c), (b) and (c) show internal cracking because of the expansion of the matrix at 80°C in air–water and water–water. The fractographic examples of glass fiber polymer composite in water at ambient and high temperatures demonstrate that an important feature is the hackle zone in the matrix and the mirror zone in the fiber. Figure 13.26 shows the

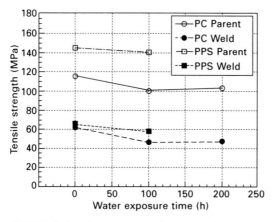

13.16 Relation between tensile strength and water exposure time of PC and PPS parent and weld.

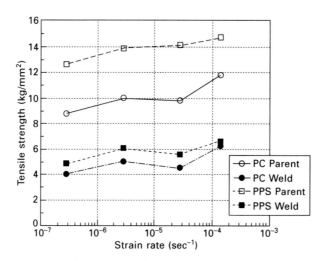

13.17 Tensile strength of PC and PPS at various strain rates in air–air.

fracture surface of weld materials at a temperature of 20°C in (a) air–air, (b) air–water and (c) water–water. Figure 13.26(a) shows a brittle fracture surface compared to (b) and (c) which show greater ductility.

Figure 13.27 shows weld materials at a temperature of 80°C in (a) air–air, (b) air–water and (c) water–water. The fracture surface of (a) is different from that of (b) and (c). The weld materials show the same hackle zone detected in parent with immersion at both ambient and high temperatures. Water content changes the matrix structure of PC and fracture mechanism of matrix and fiber.

Figure 13.28 shows the effect of temperature and water content on environmentally assisted cracking, showing surface cracking in (a) air–air,

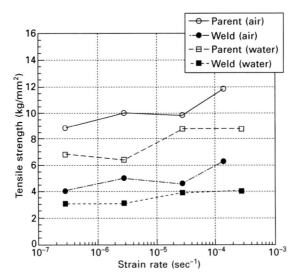

13.18 Tensile strength of PC at various strain rates in air–air and water–water.

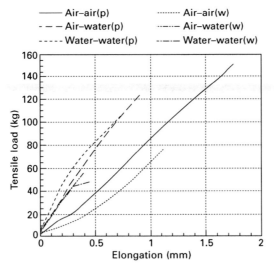

13.19 Relation between tensile strength and elongation at room temperature.

(b) air–water and (c) water–water. There is no surface cracking in air–air but small and large cracks respectively in air-water and water–water. The surface cracking of polymer composites induced in ambient and high temperature water conditions is similar to stress corrosion cracking (SCC) in metals. The observed results of surface cracking are summarized in Table 13.1.

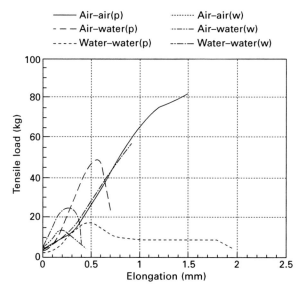

13.20 Relation between tensile strength and elongation at environmental temperature of 80°C.

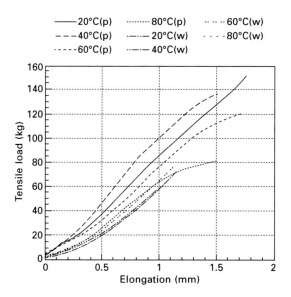

13.21 Relation between tensile strength and elongation in air–air condition.

This dotted line shows the point at which environmental and temperature conditions instigate SCC in PC polymers. Initially the fracture is created in the fibers due to swelling of the matrix and the gradual increase of stress on

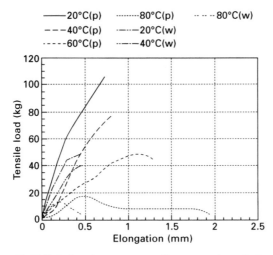

13.22 Relation between tensile strength and elongation in water–water condition.

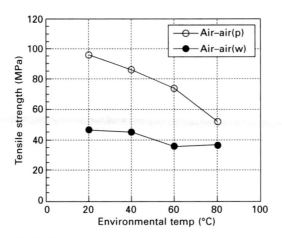

13.23 The effect of environmental temperature on tensile strength in air–air condition.

the fibers. Fracture then travels into the surrounding matrix and river lines are formed from adjacent fibers.

13.2.5   Summary

The tensile strength of polymer composites is dependent upon temperature. Tensile strength at room temperature is lower than that at –30°C. It is reasonable to assume that interface force of matrix and fiber at –30°C is stronger than that at room temperature due to the effect of thermal shrinkage force. The

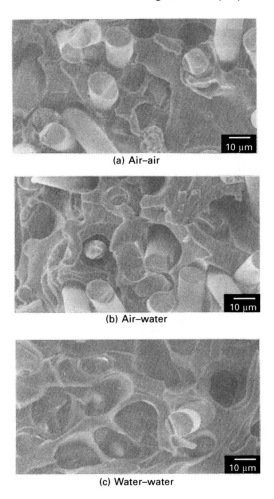

(a) Air–air

(b) Air–water

(c) Water–water

13.24 SEM fractography of PC parent at environmental temperature of 20°C.

fracture mechanism of polymer composites also varies with temperature. At low temperatures the matrix breaks first followed by the fibers, whereas the fibers break first followed by the matrix at high temperature. Exposure to water and elevated temperature reduces tensile strength, with saturation after 100 hours. Environmentally assisted cracking is focused around the planer zone in polymer matrix and the mist or hackle zone in glass fiber composites. It is affected by exposure to water and temperature. Surface cracking was observed at a temperature of 60°C in water–water and over 80°C in air–water.

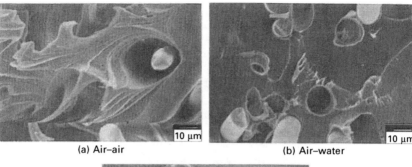

(a) Air–air   (b) Air–water

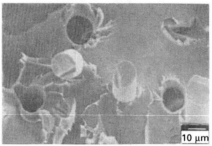

(c) Water–water

13.25 SEM fractography of PC parent at environmental temperature of 80°C.

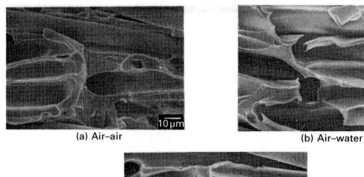

(a) Air–air   (b) Air–water

(c) Water–water

13.26 SEM fractography of PC weld at environmental temperature of 20°C.

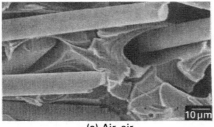

(a) Air–air

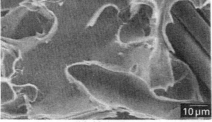

(b) Air–water

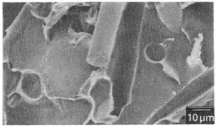

(c) Water–water

13.27 SEM fractography of PC weld at environmental temperature of 80°C.

13.3   Stress corrosion cracking (SCC) evaluation of glass fiber reinforced plastics (GFRPs) in synthetic sea water

Composite materials are widely used in many industrial applications which require light-weight, high strength materials, including marine engineering. The mechanical behavior of glass fiber reinforced plastics (GFRP) composites is affected both by service load and environmental parameters such as sea water. A likely cause of failure in GFRP composites will be by SCC in sea water conditions, but this has not been studied much. Studies on SCC of GFRP composites have been performed in strong acid solutions, because SCC occurs easily in these conditions [14] and failures have been reported in corrosive acid environments [15]. Recently, the probability of SCC of GFRP composites in sea water has been reported [16]. Accordingly, we

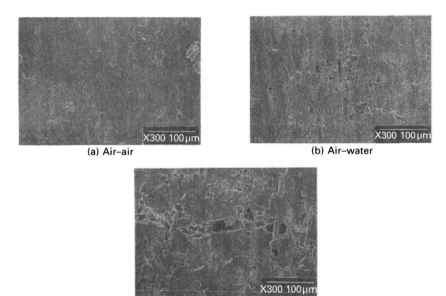

(a) Air–air

(b) Air–water

(c) Water–water

13.28 Environmental surface cracking of PC parent at environmental temperature of 80°C.

Table 13.1 Environmentally assisted cracking of PC parent

Conditions/Temp. (°C)	20	40	60	80
In air–air	X	X	X	X
In air–water	X	X	X	O
In water–water	X	X	O	O

X: no crack in gage length
O: surface crack in gage length

investigated the tensile properties and possibility of SCC of a chopped strand glass mat/polyester composite in synthetic sea water. Tensile properties are increased with pH concentration increase because of hardening of the resin. Specimens exposed to sea water and tested at particular strain rates showed evidence of SCC in the co-planar, mirror and hackle zone.

13.3.1　SCC evaluation of GFRP

Our studies were based on GFRP composites produced by the hand lay up (HLU) process. Vinylester type epoxy acrylate resin (that is, an unsaturated polyester) and E-glass fiber (CSM) were used as the matrix material and reinforcement, respectively. In the preparation of GFRP plate, one surface mat (SM) was placed on the ground first and four CSM were laid up and one SM was put on the upper surface. This plate was allowed to sit at room temperature for 24 hours followed by a post cure of 2 hours at 120°C. The final thickness of the plate was 3.2 mm. The resulting sheets were finally machined into specimens with the dimensions indicated by JIS K7054 [17].

Two types of specimens were used in this test: a dry specimen which was cured for 2 hours at 50°C before testing, and a wet specimen which was immersed in sea water for 4 months. The sea water absorption rate of the wet specimen was 0.45%. The composition of the synthetic sea water was as recommended by ASTM D 1141 [18] with the pH concentration varied between pH 6.0, 8.2 and 10.0. The sea water was circulated to prevent it from stagnation and the pH concentration measured every 12 hours. Slow strain rate tests were also carried out at air and sea water conditions. The strain rates varied from 1×10^{-4} sec^{-1} to 1×10^{-7} sec^{-1}. Parameters are shown in Table 13.2.

A typical example of the load-displacement curves obtained from SSRT is shown in Fig. 13.29 with a strain rate of 1×10^{-6} sec^{-1}. For the AA material, it was found that the load-displacement curve showed a linear response at first up to P_{max} followed by a sudden drop as the load arrived at P_{max}. We think that this behavior just before P_{max} is due to the nonlinear relationship between the fiber and resin because of the plasticity of the resin. However, for the AS82 and SS82 materials, this ductile behavior is less compared with the AA material because the resin is hardened by sea water as shown in Fig. 13.30.

On the other hand, if we compare mechanical properties with pH concentration, it can be seen that P_{max} and displacement are decreased with pH concentration increase as shown in Fig. 13.31, because the hardening level is increased with pH. There is a wide difference between tensile properties compared with the values of fracture strain energy (E_f) which are calculated by the area under the load-displacement curve as shown in Fig. 13.32. Figure 12.33 illustrates a typical stress corrosion fracture surface [15] of fiber for the SS82 material with strain rate of 1×10^{-7} sec^{-1}. The region on the left is a mirror zone where the stress corrosion crack for glass fiber is initiated whilst on the right there is a hackle zone. Another feature of stress corrosion cracking is the planar nature of the fracture surface. In the CSM type of fiber reinforced polyester composites, stress corrosion cracking produces

Table 13.2 Specimens tested under each condition

Environment	Synthetic sea water		
Specimen	pH 6.0	pH 8.2	pH 10.0
Dry specimen	AS60	AS82	AS10	AA
Wet specimen	SS60	SS82	SS10	

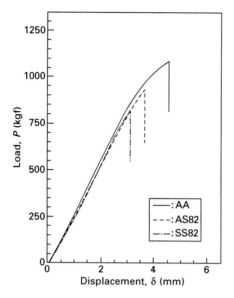

13.29 Load-displacement curves (pH8.2, $1 \times 10^{-6}\,\text{sec}^{-1}$).

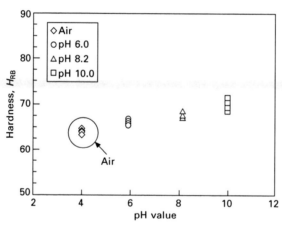

13.30 Relation between hardness of resin and pH value in synthetic sea water (100 kgf).

a characteristic planar fracture. Obviously, if bundles of fibers lie parallel to the plane of cracking, then fracture may deviate along the fiber–matrix interface. However, fibers inclined at small angles to the fracture plane fail in a co-planar position to the matrix fracture with minimal pull-out.

Figure 13.34 shows a SEM fractograph by SSRT for SS materials. In this fractograph, a fully co-planar fracture surface is observed at 1×10^{-7} sec^{-1} for the SS82 material and at $1 \times 10^{-6}\,\text{sec}^{-1}$, $1 \times 10^{-7}\,\text{sec}^{-1}$ for the SS10

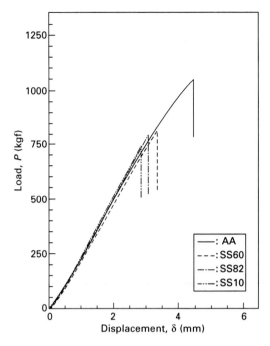

13.31 Load-displacement curves of wet specimens ($1 \times 10^{-7}\,sec^{-1}$).

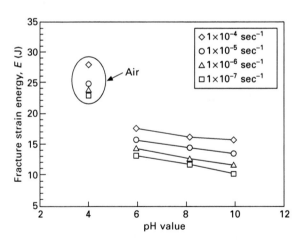

13.32 Relation between fracture strain energy and pH value of wet specimen tested in synthetic sea water.

material. Fiber pull-out is observed at $1 \times 10^{-6}\,sec^{-1}$ for the SS60 material. It is well known that SCC of glass fiber occurs in strong acid solutions because of an exchange process between Na^+, K^+ ions in the glass and H^+ ions in the environment [19]. In addition, it has been reported that the net-shaped

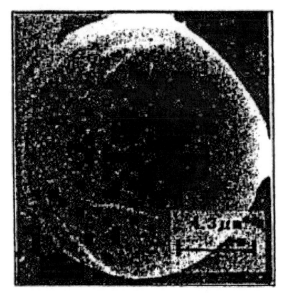

13.33 SEM photograph of fiber fracture surface (SS82, $1 \times 10^{-7} \mathrm{sec}^{-1}$, $\times 7000$).

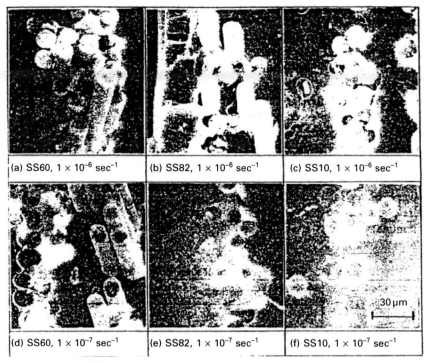

| (a) SS60, $1 \times 10^{-6} \mathrm{sec}^{-1}$ | (b) SS82, $1 \times 10^{-6} \mathrm{sec}^{-1}$ | (c) SS10, $1 \times 10^{-6} \mathrm{sec}^{-1}$ |
| (d) SS60, $1 \times 10^{-7} \mathrm{sec}^{-1}$ | (e) SS82, $1 \times 10^{-7} \mathrm{sec}^{-1}$ | (f) SS10, $1 \times 10^{-7} \mathrm{sec}^{-1}$ |

13.34 SEM photographs of wet specimen tested in synthetic sea water ($\times 1000$).

structures of silica are rapidly corroded when the concentration of OH⁻ is in
a range from pH 9.5 to pH 10.0 [20]. Therefore, we think that SCC appears
in weak alkaline sea water (pH 8.2, pH 10.0) because OH⁻ concentration
is high but that SCC is less likely in weak acid sea water (pH 6.0) because
the metallic ions do not dissolve easily in the environment. From these test
results, it can be seen that the probability of fracture by SCC is increased
with strain rate decrease and with increase in pH concentration of sea water.
The region of SCC failure obtained from the SSRT is shown in Table 13.3.
The dotted line shows the border of the SCC area.

13.3.2   Summary

In this section we investigated tensile properties, the effect of pH concentration
on the tensile fracture behavior of GFRP by SSRT and conditions affecting
SCC. The results from SSRT show a load-displacement curve of GFRP
demonstrating ductile behavior just before P_{max} in air, and linear behavior
in sea water because of hardening of the resin. Sea water decreases the
interfacial strength between fiber and resin. The tensile properties of GFRP
are considerable decreased in sea water as compared to air. Because the resin
is hardened and the interfacial strength between fiber and resin is decreased
with pH concentration increase, the tensile properties of GFRP in sea water
are decreased a little with no change in stiffness. Wet specimens tested at a
particular strain rate showed evidence of SCC in the co-planar, mirror and
hackle zones. For example, the strain rates are 1×10^{-5} sec^{-1}, 1×10^{-6} sec^{-1}
and 1×10^{-7} sec^{-1} for SS10, SS82 and SS60, respectively.

13.4   Fatigue crack propagation mechanism of glass fiber reinforced plastics (GFRP) in synthetic sea water

The fatigue failure mechanism in fiber reinforced composites is quite different
from those in monolithic, homogeneous materials such as metals. Fatigue
failure in metals, for example, occurs as a result of initiation and growth of a

Table 12.3 Conditions of SCC in glass fiber in synthetic sea water

Strain rate	Dry specimens				Wet specimens		
	AA	AS60	AS82	AS10	SS60	SS82	SS10
1×10^{-4} (sec^{-1})	X	X	X	X	X	X	X
1×10^{-5} (sec^{-1})	X	X	X	X	X	X	Δ
1×10^{-6} (sec^{-1})	X	X	X	X	X	Δ	O
1×10^{-7} (sec^{-1})	X	X	X	X	Δ	O	O

O: SCC Δ: quasi-SCC X: No SCC

principal crack. Fiber reinforced laminate composites, on the other hand, can sustain a variety of subcritical damage, such as matrix fracture, fiber–matrix debonding, fiber pull-out and fiber crack bridging [21]. The fatigue crack propagation mechanism depends on the type of composites, the function of the structure and relevant conditions such as stress and environment. Under many circumstances, composites are superior to metals in their fatigue resistance. However, our present knowledge of the fatigue and fatigue damage in composites is incomplete and more research is required. The failure of GFRP composites in marine structures such as floating structures, harbor facilities, yachts and minesweepers occurs due to the combined action of service load and sea water. However, the is little research on their fracture behavior in sea water [22].

The study discussed in this section examine laminates made from chopped strand mat type E-glass, which is used as a common resin, and those made from polyester resin which has been used for many years in applications requiring resistance to chemical attack. Bending fatigue tests ($R = -1$) were performed on dry and wet specimens in air and sea water respectively. The pH concentration of sea water was maintained at 6.0, 8.2 and 10.0 with wet specimens immersed in sea water for 4 months. Throughout the bending tests, fatigue cracks, both in dry and wet specimens tested in air or sea water, occurred at the beginning of the cycle.

13.4.1   Sea water absorption of GFRP

During preparation of the GFRP plates, the filaments were surface-treated with an epoxy-compatible silane finish by the manufacturers. One surface mat (SM) was placed on the ground, four CSM, in turn, laid up and finally one SM put on the upper surface. This plate was left at room temperature for 24 hours followed by a post cure for 2 hours at 120°C. The plates had a final thickness of 3.2 mm and a fiber content of 30–35 wt%. The specimens were machined in the form of bars 76.2 mm long by 25.4 mm wide by 3.2 mm thick. Two types of specimens were used: a dry specimen cured for 2 hours at 50 ± 3°C before testing, and a wet specimen immersed to be saturated in synthetic sea water for 4 months.

Synthetic sea water recommended by ASTM D 1141 [23] was used in the test with pH concentrations of 6.0, 8.2 and 10.0. The specimens were soaked in synthetic sea water at room temperature and weighed at various time intervals to determine the rate of sea water absorption. The sea water absorption rate does not change according to pH concentration of synthetic sea water and saturated at 0.45 wt% as shown in Fig. 13.35. Environmental testing was done using a Lucite-walled chamber installed around the testing apparatus for measurements of the immersed specimens. A more detailed description of the test procedure can be found in ASTM D 570 [24].

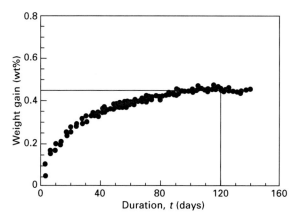

13.35 Diffusion behavior of synthetic sea water into GFRP at room temperature.

Double-edge notched specimens were used in bending fatigue testing. The fatigue testing was performed on a repeated torsion and bending fatigue testing machine programmed to cycle sinusoidally in $R = -1$ at 30 Hz. Crack length was measured using a traveling microscope positioned in front of the specimen. The fracture surface was also evaluated using SEM. Upon crack initiation at the machined notch, the testing machine was stopped at regular intervals to record the crack length, a, and the corresponding number of cycles, N. The values of a are then plotted versus N and best-fit curve drawn. From such a graph, the crack growth rate, da/dN, is obtained by measuring the slope of the curve at various values of a. In this study, we use the stress intensity factor, K, recommended by JSMS:

$$K = f(a/w) \cdot \rho_b \sqrt{\pi a}$$

$$F(a/w) = 1.98 + 0.36(a/w) - 2.12(a/w)^2 + 3.42\,(a/w)^3$$

where ρ_b is applied bending stress and $f(a/w)$ is a parameter depending on the specimen and crack geometries. For $R = -1$, the K_{min} is zero because the fatigue crack is closed. The stress intensity factor range (ΔK) is $\Delta K = K_{max}$.

13.4.2   Fatigue life

The S-N curves obtained under air and synthetic sea water with dry and wet specimens are shown in Fig. 13.36. The fatigue limit is not observed in a range up to 10^7 cycles. This is a common feature of GFRP [25]. The fatigue lives of AS and SS materials are considerably less than that of AA and the fatigue life of GFRP in synthetic sea water decreases with the pH concentration increase. This behavior is due to the hardening of resin in

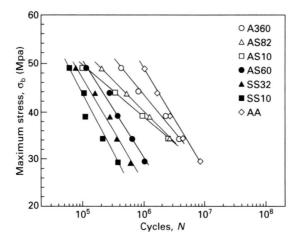

13.36 S-N curves of dry and wet specimens tested in air or synthetic sea water.

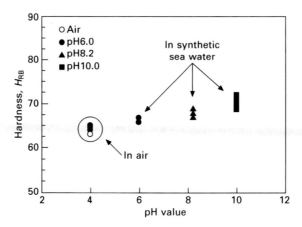

13.37 Relation between hardness of resin and pH value in synthetic sea water (980N).

synthetic sea water as pH concentration increases as shown in Fig. 13.37. In addition, it has been reported that glass fibers are rapidly corroded when the concentration of OH⁻ ranges from pH⁻ 9.5 to pH 10.0 [26, 27].

13.4.3 Fatigue crack propagation behavior

Three typical *a-N* curves of AA, AS82 and SS82 obtained under a repeated bending fatigue test ($R = -1$) are shown in Fig. 13.38. If we look at the whole fatigue crack propagation behavior, the curves can be divided into two regions: region 1 indicates crack deceleration and region 2 represents

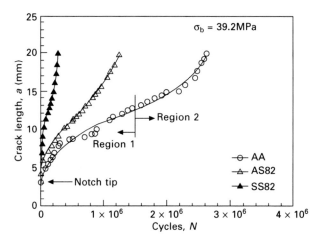

13.38 Crack length versus number of cycles.

crack reacceleration. It has been known that these regions occur under a compressive fatigue process [28]. Consequently, fatigue crack growth can be explained in the following way. The crack that has been initiated at the beginning of testing grows rapidly under the applied stress and then visibly slows down because of the axial cracks which reduce the magnitude of the stress field at the crack tip and cause deceleration of the advancing crack. Subsequent axial cracking reaches its limit and this reaccelerates fatigue crack growth because the ligament of the specimen is reduced, and then instable crack growth occurs. From Fig. 13.38, it can be seen that the fatigue crack growth mode of GFRP in synthetic sea water is the same as the mode in air and the process of AS82, SS82 is faster than that of AA.

Figure 13.39 shows the *a-N* curves according to the pH concentration of synthetic sea water for wet specimens. In this figure, it appears that crack propagation is faster as the pH concentration increases. This behavior is due to the hardening of the resin as shown in Fig. 13.37 and is caused by degradation of the interfacial strength between fiber and resin as pH concentration increases [22, 29]. Crack growth rate (*da/dN*) as a function of ΔK can be obtained at consecutive positions along *a-N* curves. A typical log-log plot of *da/dN* versus ΔK for AA materials is shown schematically in Fig. 13.40.

These curves have a V-shape that can be divided into two major regions and confirm the existence of a transition point where the fatigue crack growth changes from deceleration to acceleration. In this figure, we think that the change depends on the load because the shift to a large ΔK value occurs as the applied stress increases. On the other hand, the change is independent of environmental conditions because the transition in SS materials is located at the same ΔK values as AA materials as shown in Fig. 13.41. Figure 13.42

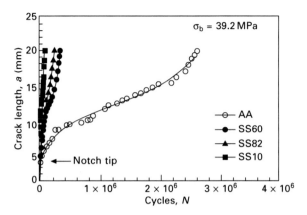

13.39 Crack length versus number of cycles.

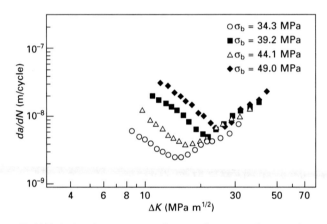

13.40 Relation between crack growth rate and stress intensity factor range of dry specimen tested in air.

shows the relation between da/dN and ΔK for SS materials as $\rho_b = 39.2$ MPa. It can be seen that the fatigue crack growth rate in alkaline sea water is very high because the rate of SS10 is higher than that of any other specimens and the transition point is independent of pH concentration in sea water.

13.4.4 Fatigue crack propagation mechanism

In general, the fatigue crack propagation behavior for conventional materials can be divided into three regions [30]. The behavior in region 1 exhibits a fatigue-threshold cyclic stress intensity factor range, ΔK_{th}, below which cracks do not propagate under cyclic stress. Region 2 represents the stable crack propagation behavior and in region 3 the fatigue crack growth rate is higher than that predicted for region 2. However, in this study we can

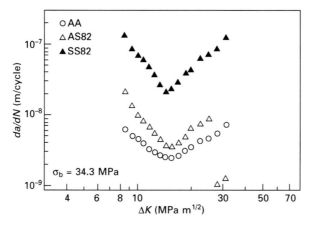

13.41 Relation between crack growth rate and stress intensity factor range (ρ_b = 34.3 MPa).

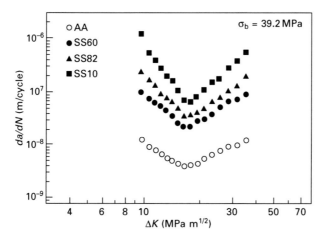

13.42 Relation between crack growth rate and stress intensity factor range (ρ_b = 39.2 MPa).

observe something remarkable about the crack propagation behavior, that is, the behavior can be divided into two regions and confirms the existence of a transition point. From these experimental observations we can describe the fatigue crack propagation process as shown in Fig. 13.43. In this figure, (a) is a schematic diagram in cross section of the specimen and describes its state before testing. The hatched portion on the left side of the cross section is a pre-notch 3 mm in depth and the right side is a central position of the specimen. At the beginning of test, the fatigue crack initiates and propagates from the notch tip and the crack proceeds with a visible increase in velocity because the axial cracks, attributed to elastic buckling of fibers, are very

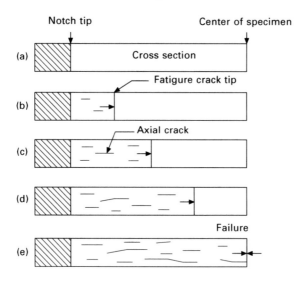

13.43 Fatigue crack growth and model for axial cracking.

small and few as shown in (b). Continued cycling produces many cracks in the axial direction perpendicular to the cross section as shown in (c) and these cracks constrain fatigue crack propagation. After a number of cycles, the axial cracking reaches its limit and this reaccelerates the fatigue crack growth because the ligament of the specimen is reduced as shown in (d), and then complete failure results from the unstable fatigue crack growth as shown in (e).

13.4.5   Summary

Through this test we have ascertained a particular mode of fatigue crack propagation for CSM type E-glass fiber reinforced polyester resin composite tested in air and synthetic sea water. The fatigue crack propagation behavior is composed of deceleration and acceleration processes. The transition point of fatigue crack propagation depends on the load, while the point is independent of environmental conditions. But as its pH concentration increases, the sea water promotes axial cracks such as debonding and delamination and further decelerates and accelerates cracking. Moreover, it is found that the fatigue life of GFRP in synthetic sea water is considerably less than its life in air and is decreased with increase in pH concentration. This behavior is due to the hardening of the resin and is caused by reduction of interfacial strength between fiber and resin as pH concentration increases.

13.5   Aging crack propagation mechanisms of natural fiber reinforced polymer composites

Growing environmental awareness has resulted in a renewed interest in the use of natural materials for different applications [31–33]. Tougher environmental regulation by governments has forced industries like the automotive, packaging and construction industries to search for new materials that can substitute the traditional composite materials consisting of a plastic matrix and inorganic filler as reinforcement. The most used composites currently are glass fiber filled thermoplastics, but different inorganic fillers like aramid or carbon fibers are also widely used. However, inorganic fibers present several disadvantages, for example their non-biodegradability, the abrasion that they produce in the processing equipment and the health problems that they cause to workers due to the skin irritation they cause during processing and handling.

Sisal fiber is one of the most widely used natural fibers in industry and is very easily cultivated [32]. Nearly 4.5 million tons of sisal fibers are produced every year throughout the world. Tanzania and Brazil are the two main producing countries [32]. At present, sisal fiber is mainly used as ropes for the marine and agricultural industries [34]. Other applications of sisal fibers include twines, cord, upholstery, padding and mat making, fishing nets, fancy articles such as purses, wall hangings, table mats, etc. A new potential application is for manufacture of composite corrugated roofing panels that are strong and cheap with good fire resistance [33].

Unfortunately, several disadvantages of natural fibers, such as thermal and mechanical degradation during processing, can make them undesirable for certain applications [35]. Lack of good interfacial adhesion and poor resistance to moisture absorption make the use of natural fiber reinforced composites less attractive. This problem can be overcome by treating these fibers with suitable chemicals [36]. However, several considerations have to be taken into account in the design of natural fibers composites. One of the most important issues is the degradation behavior of the composites exposed to environmental conditions such as humidity, sunlight or microorganisms. The poor resistance of the fibers to water absorption can have undesirable effects on the mechanical properties and the dimensional stability of natural fiber composites [37]. Therefore, it is important to study in detail their water absorption behavior in order to estimate the durability of natural fibers composites exposed to water.

Moisture penetration into composite materials is conducted by three different mechanisms [38]. The main process consists of diffusion of water molecules inside the microgaps between polymer chains. The other common mechanisms are capillary transport into the gaps and flaws at the interfaces between fibers and polymer, because of incomplete wettability, and impregnation and transport by microcracks in the matrix, formed during the compounding process [39]. In spite of the fact that all three mechanisms are active jointly in case of moisture exposure of the composite materials, the overall effect can be modeled conveniently considering only the diffusional mechanism.

This section discusses the water absorption behavior of sisal fiber reinforced composites. It reviews the durability of sisal fiber composites in distilled water and exposed to controlled cycles of wetting and drying. The durability of sisal composites was measured as the loss of strength over time.

13.5.1 Natural fiber reinforced polymer composites and water absorption

The natural sisal fibers used in this investigation were extracted from five-year-old sisal plant grown in southern China. The diameter of the fiber varied from 100 to 250 μm. The average diameter of the fiber of about 140 μm was calculated from the fraction of fibers of different diameter in a sample. The sisal fiber contains 57% cellulose, 16% hemicelluloses and 11% lignin by weight. The tensile strength, modulus and failure strain of the fiber were 400–700 MPa, 9–20 GPa and 5–14%, respectively. The materials used were reinforced by two plies of woven roving sisal fiber with a fiber volume fraction of about 60%.

To produce a higher fiber matrix fiber modification with silane was applied. For this procedure the fibers first had to be dewaxed in an alcoholic solution for 24 hours and then washed with distilled water. The subsequent silane treatment similarly was carried out in an alcoholic solution with a silane content of 2 wt% for 24 hours at 23°C. The coupling agent used was a γ-glycidoxy propyltrimethoxy-silane with the chemical constitution of $CH_2CHCH_2O(CH_2)_3Si(OCH_3)_3$. The silane application was finished with a four-hour drying process in a vacuum oven at 75°C.

For the preparation of the composites, two different kinds of resin were used: an epoxy and vinyl-ester resin. The tested composites were made of woven roving fibers which were embedded in a resin by using the resin transfer molding (RTM) technique. The sisal fibers were aligned in a Teflon mold and the liquid resin mixture poured in. Curing was performed at 120°C for 2 hours. Rectangular specimens 76.2 mm long, 25.4 mm wide and 3.2 mm thick were cut from the composites. The specimens were then immersed in distilled water at 20 ± 1°C. At regular intervals, the specimens were removed from distilled water, wiped with filter paper to remove surface water and weighed with a precision analytical balance with 0.1 mg resolution. This study reviews the results of tests on samples which were aged for 400 days under water or exposed to cycles of wetting for 9 days until saturation and dried for 1 day at 50°C to simulate the natural extremes of weather.

The durability of the composites is discussed using the results of tensile and fracture toughness tests carried out before and after 400 days aging or five cycles of wetting and drying, and observations of fracture surface from a scanning electron microscope. Two kinds of tests were performed on the dried samples in order to determine the influence of the water absorption cycles on the mechanical properties following ASTM D 638-90 and D 5045-96. The ¼ T compact tension (CT) specimens were used for the fracture toughness tests with 32.0 mm width. The standard dumbbell-shaped specimens were used for tensile tests with 25.0 mm gage length.

13.5.2   Water absorption behavior

The amount of water absorbed in the composites was calculated by the weight difference between the samples exposed to distilled water and the wetting and drying cycled samples. Figure 13.44 shows the percentage of water absorbed plotted against time for the aged and cycled samples. Figure 13.44(a) is for the epoxy and vinyl-ester composite samples aged for 9600 hours (400 days). Figure 13.44(b) shows five cycled specimens immersed for 216 hours (9 days) and then dried for 24 hours (1 day). In both cases the samples absorbed water very rapidly during the first stages (0–260 h for epoxy and 0–150 h for vinyl-ester) reaching a certain value, the saturation points, where no more water was absorbed and the content of water in the composites remained the same. The hydrophilic character of natural fibers is responsible for the water absorbed. The matrix has little effect on the amount of water absorbed by pure resin composites, such as an epoxy and vinyl-ester [38], as shown in Fig. 13.44.

The cycles of wetting and drying also have an influence on the water absorption curves. Figure 13.44(b) shows the absorption curves for the woven sisal fiber reinforced epoxy and vinyl-ester composites at different cyclic times. It can be concluded that the cycling process increases the water uptake of the composite materials. When the wetting and drying process is increased, the amount of water uptake is slightly increased in the same wetting period. For the epoxy composites, the absorption rate is increased by 1.0 wt% as the cycling process moved from the first to the fifth wetting cycle. For vinyl-ester, the rate is almost the same. At the fifth drying cycle the absorption rate is decreased by 1.28 wt% for the composites with epoxy and vinyl-ester, as shown in Fig. 13.44(b). This weight reduction may be due to the extracted cellulosic fibers which form the polymeric matrix [37].

13.5.3   Effect of absorption cycles on tensile properties

Tensile tests were performed on the epoxy and vinyl-ester samples before (0 cycle) and after water absorption during the cycled stages. Figure 13.45(a) and

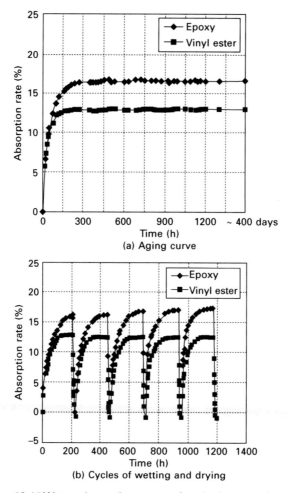

13.44 Water absorption curves for sisal composites at 20°C.

(b) present these results for the epoxy and vinyl-ester composites respectively. In general, the tensile strength of these materials decreases after moisture uptake, due to the effect of the water molecules which change the structure and properties of the fiber, matrix and the interface between them [39, 40]. Once the moisture penetrates inside the composite materials, the fibers tend to swell. The matrix structure can also be affected by the water uptake processes such as chain reorientation and shrinkage. The aging under conditions of cyclic wetting and drying may also lead to the degradation of natural fibers by a hydrolysis mechanism [41]. Water absorption and their resulting effects contribute to the loss of compatibility between fibers and matrix, which results in debonding and weakening of the interface adhesion.

Considering stress at maximum load, values are gradually reduced after cyclic water absorption for the epoxy as shown in Fig. 13.45(a). This effect is particularly evident for the composites cycled four and five times, where the maximum stress is drastically reduced. For the vinyl-ester, the values of maximum stress are decreased with cyclic water uptake as shown in Fig. 13.45(b), but the maximum stresses of cyclic composites are distributed within the limits of 27.0–29.8 MPa without wide dispersion. The changes in the strain at failure with cyclic aging are more complex. It has been reported that water molecules act as a plasticizer agent in the composite material, which should lead to an increase in the maximum strain for the composites after water absorption [40]. Such an increase is shown for composites with the epoxy resin immersed in water for four and five cycles as shown in Fig.

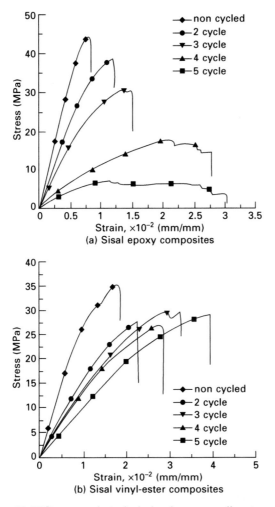

13.45 Stress and strain behaviors according to various cyclic times.

13.45(a). The maximum strain is 3.1% after the fifth cycle. For the vinyl-ester composites, the maximum strain is also gradually increased after cyclic water absorption with the increase in cyclic times as shown in Fig. 13.45(b). In this case, the plasticizer effect of water in the composite structures is superior to the degradation effect, so the maximum strain is increased in spite of indistinct stress behavior at cyclic composites.

13.5.4   Effect of absorption cycles on fracture toughness

The critical stress intensity factor in mode I fracture (K_{IC}) as a function of the wetting and drying cycle times was evaluated with the composites immersed for 400 days (9600 h) for the epoxy and vinyl-ester. The notches were machined perpendicularly to the direction of woven fibers. The K_{IC} values demonstrate a decrease in inclination with increasing cyclic times of wetting and drying for the epoxy and vinyl-ester.

The two curves of the average K_{IC} values show a parallel downward trend with the increase in the number of cycles as shown in Fig. 13.46. For composites that have a 45°-oriented notch to the direction of woven fibers, the K_{IC} values distribute within the similar limit band to perpendicular oriented composites for the epoxy and vinyl-ester. The K_{IC} values of composites aged 400 days distribute within similar region for four or five cycles. The effect of crack orientation on K_{IC} values is not clear. The K_{IC} values of the epoxy and vinyl-ester ranged from 0.64 to 2.66 MPa.m$^{1/2}$ and from 0.22 to 1.75 MPa.m$^{1/2}$.

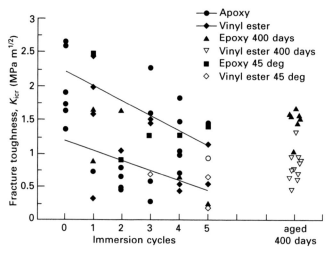

13.46 Distribution of critical stress intensity according to immersed cyclic times.

13.5.5   Fracture mechanism in absorption cycles

SEM photographs clearly show the degradation of the fibers through the water absorption cycles. Figure 13.47 shows the composite structure before and after water absorption cycles, where the loss of adhesion between fiber and matrix, characterized by the appearance of voids, can be noticed. The fibers appear seriously degraded with complete lack of resin layer and the microfibers can be clearly observed. The ruptured ends of the sisal fibers were not planar with some parts of the pulled-out sisal fibers. This is the feature of microcell multi-cracking followed by interfacial debonding. The fiber and resin interface seems to be moderate because interfacial debonding took place at the sisal fiber contour.

From Fig. 13.47, the fiber section can be seen to be neither circular nor regular. A bundle of microcells was pulled away, leaving a hole. A pulled-out microcell can be seen. The microcell was covered with a thick layer of cuticle. The cuticle was very flexible and very deformed, and its adhesion

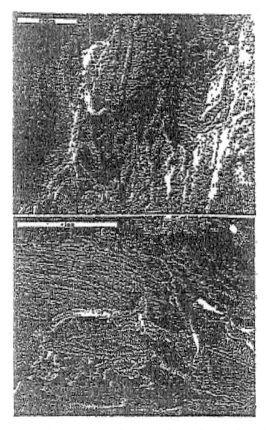

13.47 Photographs of composites after cycles of wetting and drying.

to the microcell was not strong. The microcell was ruptured by decoiling, splitting, torsion, etc. The decoiling is a specific characteristic of sisal fibers. In fact, inside the wall of microcells, there exists many short aligned microfibrillar particles which are oriented in a direction different than that of sisal fiber [42]. Therefore, the microcell wall is also a composite structure of lignocellulosic material reinforced by helical microfibrillar bands of cellulose. It was observed that some microcells, particularly near the fiber and resin interface, were compressed transversely and so split longitudinally along their length.

13.5.6   Summary

The study presented here has described the influence of wetting and drying cycles on the mechanical properties of woven sisal fiber reinforced epoxy and vinyl-ester composites. The water absorption characteristics of sisal composites were studied with cycled and aged conditions. Water uptake of the epoxy composites was found to increase with cycle times. Mechanical properties are dramatically affected by the water absorption cycles. Water-saturated samples present poor mechanical properties such as lower values of maximum strength and extreme elongation. The K_{IC} values demonstrate decrease in inclination with increasing cyclic times of wetting and drying for the epoxy and vinyl-ester. The K_{IC} values of composites aged 400 days distribute within a similar region to those exposed to the four or five cycles. The fibers appear seriously degraded with complete lack of resin layer and the microfibrillars can be clearly observed.

13.6   Aging of biodegradable composites based on natural fiber and polylactic acid (PLA)

Waste plastic is a major environmental problem. Environmental awareness, new rules and legislation are forcing industries to seek new materials which are more environmentally friendly. A remarkable result of the efforts is the development of polylactic acid (PLA), a biodegradable polymer which was developed and came into use largely during the 1990s [43]. Plant fibers from agricultural crops are renewable materials which have potential for creating green products and replacing synthetic materials such as glass fiber, carbon fiber and plastic fibers. A review article by Bledzki and Gassan [44] summarized a number of natural fiber applications in polymer/bio-fiber composites. The combination of natural fiber and fully biodegradable polymer could produce a fully biodegradable composite. This section focuses on the investigation of china jute fiber reinforced PLA composites. Composites with three different fiber volume fractions were fabricated by film stacking using the hot press method. Mechanical properties and the microstructure of the composites were observed. Fiber surface treatment by silane was applied in order to investigate the effect on the properties of the composites. The moisture absorption rates in a hygrothermal environment were measured.

The moisture absorption test is generally used for quality control purposes and to measure the degradation of the quality of composite materials [45]. Due to the intrinsic moisture absorption property of natural fiber and the degradability of PLA in the environment, the durability of such composites is a major concern. However, up to now most research has focused on improving material fabrication and mechanical properties. There is little research on the aging behavior of nature fiber reinforced polylactide in hygrothermal environments.

Aging is a multi-scale chemical and physical process. Natural aging tests are very representative because they make it possible to reproduce identical conditions of exposure [46]: however, they are time consuming and may last for years. That is why accelerated aging tests are used to investigate aging behaviors in a short period. Hygrothermal aging is such a method, providing aggressive conditions for an accelerated aging test. In this research, the hygrothermal aging behaviors of short jute fiber reinforced polylactide are investigated in a simulated working environment where the material might be used in practice, for example, as automobile interior panels. The automotive industry requires material to be lightweight but with high strength, and to be easily recycled after service. Jute fiber reinforced PLA potentially meets these requirements.

13.6.1   Biodegradable composite fabrication and hygrothermal environment

The jute fibers in this study were removed from the jute stem by wetting. The original fiber length of 2.5–4 m was cut into short fibers 5–10 mm long. The short fibers were divided into two groups. One was treated with a 1.5% aqueous silane agent solution for 24 hours at room temperature. The fiber was then dried for 24 hours in an oven at 102°C. Another group of fibers was only cleaned in water and dried in the same conditions as described above. The fiber density was measured as 0.92 g/cm^3 by the Archimedes method.

Biodegradable PLA film produced by Cargill Dow LLC was used in this study. Some important properties of the as-received PLA film were: thickness: 0.3 mm; tensile strength: 35 MPa; elastic modulus: 3.5 GPa; elongation: 6%. The film was cut into sheets as large as the moulds for tensile test, bending test and impact test, respectively. Composite plates with three different fiber volume fractions of 30, 40 and 50% were prepared. The thicknesses of the composite plates were designed to be 4 mm for the tensile and bending tests, and 10 mm for the impact test.

The general process of composite fabrication is indicated in Fig. 13.48.

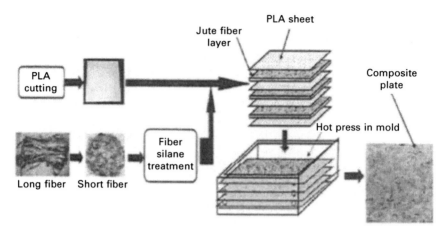

13.48 Fabrication procedure of composite plate.

PLA film and the dried short fiber were placed into the mold uniformly layer on layer by controlling the weight of each layer. Subsequently, the mold together with the materials were placed onto the heating panel of the hot press machine. In accordance with the melting point of the polymer, the temperature was set to 170°C and was controlled automatically. A 1.3 MPa pressure was applied and maintained for 10 minutes after the temperature rose to 170°C. Then the mold was removed from the heating panel. The composite plate was removed after it was cooled down to room temperature. The samples were cut from plates with different fiber volume fractions. They were placed into a sealed plastic box with water inside so as to create a saturated humid environment. The box together with the samples were placed into an oven. The temperature was set to 70°C. The weight of each sample was measured periodically to determine the moisture absorption rates.

Hygrothermal aging is an accelerated aging test in order to investigate aging behavior during a short period. The test simulates an extreme working environment for natural jute fiber/PLA composite such as an automobile interior. When an automobile is exposed to sunlight in summer, the temperature inside the car can rise to nearly 70°C along with high humidity. As a result, the conditions were set to 70°C in saturated water vapor. To investigate the effect of coating on the moisture uptake, two samples were carefully coated with 0.1 mm thick polypropylene plastic adhesive tape and exposed to vapor. Two uncoated samples were directly exposed to vapor.

13.6.2   Mechanical properties on moisture absorption rate

Tensile strength, flexural strength and impact strength are indicated in Figs 13.49, 13.50 and 13.51. The tensile strength of the as-received PLA film is 35 MPa. The tensile strengths of composites are from 39.3 MPa to 42.8 MPa,

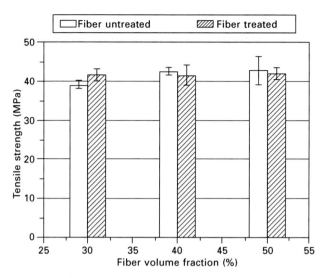

13.49 Tensile strength vs fiber volume fraction.

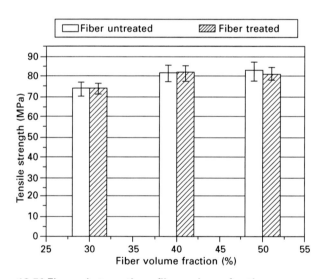

13.50 Flexural strength vs fiber volume fraction.

which vary with the fiber volume fractions. The strengths exhibit 12.3% to 21.7% increases compared with pure PLA. Theoretically, the strength should change with the fiber volume fraction according to the rule of mixture [47]. However, the test results have shown minor differences among the composites with different fiber volume fractions.

The flexural strengths, as indicated in Fig. 13.50, range from 74 MPa to 82.8 MPa. When the fiber volume fraction increases from 30% to 40%,

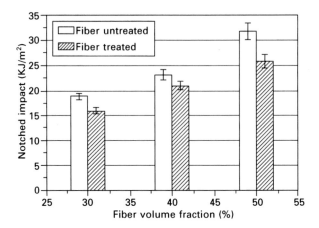

13.51 Notched impact strength vs fiber volume fraction.

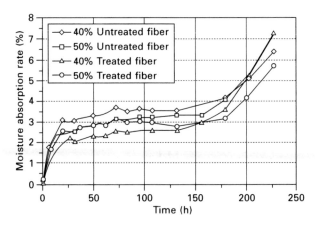

13.52 Moisture absorption rate at 70°C saturated humid environment.

the flexural strength increased from 74 MPa to around 82 MPa, indicating a 10.8% increase, whereas the strengths of composites with 50% fiber volume fraction showed nearly the same values as those with 40% fiber fraction.

The moisture absorption rate in a 70°C hygrothermal environment is plotted in Fig. 13.52. Three stages can be observed from the curves. The first stage lasted for around 25 hours and corresponds to the quick water absorption. The second stage, which appears as a plateau, corresponds to the saturated stage. The saturated water absorption rate, which is usually defined as maximum moisture content by Ficken's second law of diffusion, ranges from 3.0% to 3.6%. These values are much higher than that of neat PLA. Wang and Sun [48] reported that the water absorption of neat PLA leveled at approximately 1%. The higher water absorption rate is attributed to the hydrophilic property of natural fibers due to the hydroxyl group on the cellulose molecules. H-bond could be formed between the hydroxyl group and water molecules. As can be seen in Fig. 13.52, the third stage shows rapid increase of moisture absorption after 180 hours' exposure. At this stage, the appearance of samples changed remarkably. The resin matrix cracked into scraps exposing the natural fiber, as shown in Fig. 13.53(d). Since more fibers were directly exposed to moisture, they would absorb more water.

All of the composites with silane treated fibers have a lower moisture absorption rate than that of composite with untreated fibers. Herrera-Franco and Valadez-Gonzalez [49] studied the mechanism of natural fiber treatment with silane. The grafting of silane onto cellulose molecule causes an intermolecular condensation reaction between Si-OH groups. The number of hydroxyl groups is reduced to some extent. As a result, the water absorption rate of the composite with silane treated fibers may drop slightly.

In contrast to tensile and flexural properties, the impact strength increases significantly with the increase in fiber volume fraction. The impact strength of neat PLA as-received is 12.8 KJ/m^2. As indicated in Fig. 13.51, the maximum values occurred when the fiber volume fraction is 50%. The maximum impact strength of the composite is 31.8 KJ/m^2 and 25.9 KJ/m^2, respectively. They demonstrated increases of 148% and 102%. In this study, silane treatment did not significantly improve mechanical properties, although silane is commonly used as a coupling agent to improve the bonding conditions between matrix and fibers.

Figure 13.53(a) shows the fracture surface of an impact sample with 50%

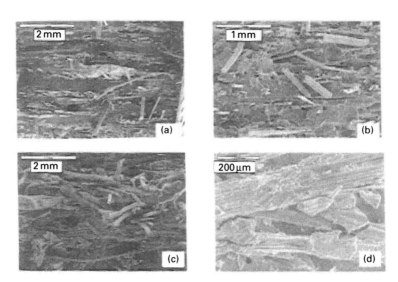

13.53 SEM observation of the microstructure: (a) sandwich structure; (b) uniformly distributed single fiber; (c) fiber bundles; (d) morphology after aging.

fiber. Delamination was observed when the bending test was conducted. Fig. 13.53(b) and (c) show tensile fracture surfaces of samples with 30% and 50% fiber content, respectively. For the 30% fiber content tensile specimen, single fibers are very well bonded to the resin matrix. For the 50% fiber content tensile specimen, fibers exist in bundles and are not well bonded to the matrix. This may explain why strength only increases slightly with the increase in fiber content. Fiber pull-out is the main failure mode in both of the composites. This is an indicator of fiber–matrix interfacial failure. Fiber pull-out has significant effect on impact strength because the pull-out process consumes more energy [47]. This may account for the great improvement in impact strength.

The tensile strength values of coated and uncoated specimens with respect to aging time are shown in Fig. 13.54. Specimens without aging have an average tensile strength of 42.6 MPa. For coated specimens, the tensile strength (41.32 MPa) does not show a significant difference after 24 h aging compared to that of specimens without aging, whereas the tensile strength of uncoated specimen is 36.39 MPa, which is 85.4% that of specimens without aging. Although there are no significant defects observed in this period, the whitened surface of the uncoated sample implies that some changes in the surface layer have occurred which may cause the strength decrease after aging as shown in Fig. 13.53(d). After 72 h aging, the tensile strength has badly deteriorated both for coated and uncoated specimens. The load-displacement curve for aged samples is indicated in Fig. 13.55. The results reveal that the load bearing capacity and elongation of the samples have greatly decreased with aging time.

Molecular weights of PLA before and after aging were tested by GPC.

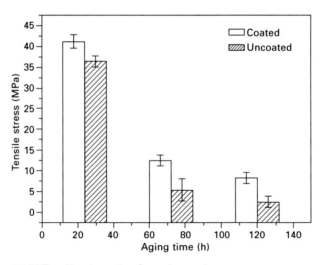

13.54 Tensile strength of aged samples.

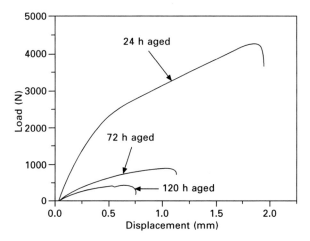

13.55 Load-displacement curve of aged samples.

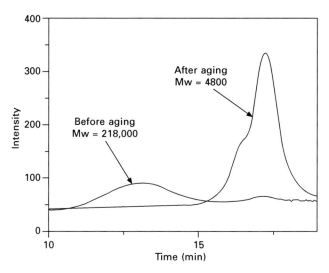

13.56 Gel permission chromatography of PLA before and after aging.

Chromatography showed that the average molecular weight (Mw) of reaction products after aging was significantly reduced, as indicated in Fig. 13.56. The Mw of PLA is 218 000 before aging whereas the Mw of the products after aging is 4800. Ahmad-Faris *et al.* investigated the hydrolysis rate of PLA in high-pressure steam [50]. They found that the molecular weight of hydrolytic products strongly depends on the time and temperature at which the hydrolytic reaction occurs. At higher temperature, the hydrolysis reaction of PLA has a higher rate. Hydrolytic products also have lower molecular weights. A reaction between water and PLA takes place and the water molecules are permanently absorbed by PLA and transform into hydrolytic products.

Aging is a chemical and physical interaction process. Hydrolysis of PLA is a chemical process. It results in molecular chain breakage and causes a decrease in strength. According to observations of the microstructure, defects occurring in sequence during the aging process include pores, microcracks and delamination. During the first stage of aging, moisture absorption was dominated mainly by capillary transport of natural fibers as well as flaws and gaps between the matrix and fibers. Moisture is transported along natural fibers by capillary action. Hydrolysis of PLA matrix took place through the interaction of moisture and temperature at the interface of fiber and matrix. The resulting breakage of the molecular chain of the PLA resulted in debonding of fiber and matrix which created more fresh surfaces able to absorb moisture. Debonding then caused microcracks between fiber layers and resin layers. This is the reason the microcracks were found to be parallel to the layers of fiber and matrix. Once the microcracks occurred, more surface area was exposed to moisture. Therefore it resulted in the abrupt increase in moisture absorption. At this stage, the moisture absorption process was dominated by interaction of hydrolysis and cracking. On the one hand, the water molecules were absorbed physically by the matrix and fibers; on the other hand, they were absorbed chemically by PLA molecules. The fact that the occurrence of microcracks and the inflexion on the curve of moisture absorption rate appeared simultaneously can be explained by this mechanism. By the end of the aging period, the matrix was severely degraded to scraps. Finally, relaxation of the composites occurred in the whole structure.

13.6.3   Summary

This section has reviewed the mechanical properties and moisture absorption rates of a fully biodegradable composite reinforced by short jute fiber. With the fiber content increases, tensile strengths vary from 39.3 MPa to 42.8 MPa and flexural strengths from 74 MPa to 82.8 MPa. They exhibit only minor differences with fiber content increase because higher fiber content resulted in poorer bonding. Silane treatment has no distinct effect on the mechanical properties. However, it could reduce the moisture absorption rate. The maximum moisture absorption rates were tested to be from 3% to 3.6%, depending on the fiber content and treatment. The material demonstrated poor durability in the testing conditions. These results suggest that some precautions should be taken to avoid direct exposure to a humid environment.

The main defects during the aging process include pores, microcracks, delamination and complete structure relaxation. PP film coating on the samples can slow the moisture absorption rate. The deterioration of the microstructure can also be effectively retarded by coating. Tensile test results show that significant strength decrease of the samples occurs in a hygrothermal environment when exposed for long periods.

13.7   References

[1]   Huneke J.J. and Porey J.K., *Proceedings of 37th Annual Technical Conf.*, RP/C Institute, SPI, session 2-b, 1982.

[2]   Schrader M.E. and Block A.J., *Polym. Sci., Part C*, 34, 1971, p. 286.

[3]   Ishida I. and Koenig J.L., *J. Polym. Sci., Poly. Phys. Ed.*, 18, 1989, p. 1931.

[4]   Grafton P., *Handbook of Plastics and Elastomers*, McGraw-Hill, New York, 1975, p. 124.

[5]   Tomari K., Tonogai S. and Harada T., *Polymer Eng. and Science*, 30(15), 1990, p. 931.

[6]   Tadmore Z. and Gogas C.G., *Principles of Polymer Processing*, John Wiley, New York, 1979, p. 603.

[7]   Malgurnera S.C. and Manisali A., *Polymer Eng. and Science*, 21 (10), 1981, pp. 586–592.

[8]   Hubbauer P., *Plast. Eng.*, 1973, p. 37.

[9]   Hobbs S.Y., *Polymer Eng. and Science*, 14(9), 1974, p. 621.

[10]   Matsuoka T., Takabatake J.I, Inoue Y. and Takahasi H., *Polymer Eng. and Science*, 30(16), 1990, p. 957.

[11]   Camahort J.L., Renhack E.H. and Coons W.C., 'Effects of thermal cycling environment on graphite/epoxy composites', in ASTM STP 602, 1976, p. 37.

[12]   Lim J.K. and Shoji T., *ASTM/JSME Joint Conference*, San Jose, CA, 1992, April 8–12.

[13]   Hagerman E., *Plast. Eng.*, 1973, p. 67.

[14]   Hogg P.J., Hull D. and Legg M.J., Failure of GFRP in corrosive environment, in *Composite Structures*, Applied Science, London, 1981, pp. 106–122.

[15]   Price J.N., Stress corrosion cracking in glass reinforced composites, in *Fractography and Failure Mechanisms of Polymers and Composites*, Elsevier Applied Science, London and New York, 1989, pp. 495–531.

[16]   Aveston J., Kelly A. and Sillwood J.M., Long term strength of glass reinforced plastics in wet environments, in *Advances in Composite Materials*, I.C. C. M. 3, Bunsell A.R. *et al.* Pergamon Press, Oxford, 1980, pp. 556–568.

[17]   JIS K704, Tensile testing method of GFRP.

[18]   ASTM D 1141, 1975, Standard specification for substitute ocean water.

[19]   Caddock B.D., Evans K.E. and Hull D., in *Proceedings of the Institution of Mechanical Engineers*, Fiber Reinforced Composites, Liverpool, 1986, p. 55.

[20]   Schmitz G.K. and Metcalfe A.G., Stress corrosion of E-glass Fibers, *Ind. Eng. Chem., Res. & Dev.*, 5(1), 1966, pp. 1–8.

[21]   Hahn H.T. and Lorenzo L., in *Advances in Fracture Research*, ICF6, New Delhi, India, Pergamon Press, Oxford, vol. 1, 1984, p. 549.

[22]   Kim Y.J. and Lim J.K., A study on properties of corrosion fracture surfaces of GFRP in synthetic sea water, *KSME International Journal*, 11(3), 1997, pp. 249–254.

[23]   ASTM D 1141, 1975, Standard Specification for Substitute Ocean Water.

[24]   ASTM D 570, 1981, Standard Test Method for Water Absorption of Plastics.

[25]   Dharan C.K.H., Fatigue failure mechanism in a unidirectionally reinforced composite material, *Fatigue of Composite Material*, ASTM STP 569, 1975, pp. 171–188.

[26] Schmitz G.K. and Metcalfe A.G., Stress corrosion of E-glass fibers, *Ind. Eng. Chem., Res & Dev.*, 5(1), 1966, pp. 1–8.

[27] Ishai O., Environmental effects on deformation, strength and degradation of unidirectinal glass-fiber reinforced plastics, *Polym. Eng. Sci.*, 15(7), 1975, pp. 486–490.

[28] Kunz S.C. and Beaumont P.W.R., Microcrack growth in graphite fiber-epoxy resin systems during compressive fatigue, *Fatigue of Composite Materials*, ASTM STP 569, 1975, pp. 71–91.

[29] Kim Y.J. and Lim J.K., Synthetic sea water and strain rate effects on tensile properties of E-Glass/polyester composites, *Korean Journal of Materials Research*, 2(2), 1992, pp. 133–142.

[30] Knott J.F. *Fundamentals of Fracture Mechanics*, Tower Press, 1979, pp. 246–251.

[31] Mohanty A.K., Misra M. and Drzal L.T., *J. Polym. Environ.* 10(1), 2002, pp. 19–26.

[32] Li Y., Mai Y.W. and Ye L., *Compos. Sci. Technol.*, 60, 2000, pp. 2037–2055.

[33] Savastano H.J., Agopyan V., Nolasco A.M. and Pimental L., *Const. Build. Mater.*, 13, 1999, pp. 433–438.

[34] Dweib M.A., Hu B., O'Donnell A., Shenton H.W. and Wool R.P., *Compos. Struct.*, 63, 2004, pp. 147–157.

[35] Dalaprasad G., Pradeep P., Mathew G., Pavithran C. and Thomas S., *Compos. Sci. Technol.*, 60, 2000, pp. 2967–2977.

[36] Bisanda E.T.N., *Appl. Compos. Mater.*, 7, 2000, pp. 331–339.

[37] Lu X., Zhang M.Q., Rong M.Z., Yue D.L. and Yang G.C., *Compos. Sci. Technol.*, 64, 2004, pp. 1301–1310.

[38] Espert A., Vilaplana F. and Karlsson S., *Compos. Part A*, 35, 2004, pp. 1267–1276.

[39] Lin Q., Zhou X. and Dai G., *J. Appl. Polym. Sci.*, 85(14), 2002, pp. 2824–2832.

[40] Joseph P.V., Rabello M.S., Mattoso L.H.C., Joseph K. and Thomas S., *Compos. Sci. Technol.*, 62, 2002, pp. 1357–1372.

[41] Joseph K., Thomas S. and Pavithran C., *Compos. Sci. Technol.*, 53, 1995, pp. 99–110.

[42] Bai S.L., Li R.K.Y., Wu L.C.M., Zeng H.M. and Mai Y.W., *J. Mater. Sci. Lett.*, 17, 1998, pp. 1805–1807.

[43] Kulinski Z. and Piordowska E., *Polymer*, 46, 2005, pp. 10290–10300.

[44] Bledzki A.K. and Gassan J., *Prog. Polym. Sci.*, 24, 1999, pp. 221–274.

[45] Lemana Z., Sapuana S.M., Saifola A.M. *et al.*, Moisture absorption behavior of sugar palm fiber reinforced epoxy composites. *Mater Design*, 29, 2008, pp. 1666–1670.

[46] Boubakri A., Elleuch K., Guermazi N. and Ayedi H.F., Investigation on hygrothermal aging of thermoplastic polyurethane material. *Mater Design*, 30, 2009, pp. 3958–3965.

[47] Harris B., *Engineering Composite Materials*, 2nd edn, Chemical Industry Press, Beijing, China.

[48] Wang H. and Sun X.Z., *Journal of Polymer Sci.*, 82, 2005, pp. 1761–1767.

[49] Herrera-Franco P.J. and Valadez-Gonzalez A., *Composites Part B*, 36, 2005, pp. 597–608.

[50] Ahmad-Faris M.A., Haruo N. and Yoshihito S., Evaluation of kinetics parameters for poly(lactic acid) hydrolysis under high-pressure steam. *Polym. Degrad. Stab.*, 93, 2008, pp. 1053–1058.

Part IV
Environmentally assisted cracking problems in various industries

14

Stress corrosion cracking (SCC) in boilers and cooling water systems

M. J. ESMACHER, GE Water & Process Technologies, USA

Abstract: This chapter reviews the common causes of stress corrosion cracking (SCC) damage in boiler system components and in water-cooled heat exchangers/condensers. Key influencing factors that cause SCC include alloy selection, fabrication history, water chemistry impact, design aspects, operational characteristics, and maintenance/repair issues. These key factors can influence the onset of cracking in boiler and heat exchanger systems and will be discussed. Monitoring strategies for detecting SCC, as well as a review of precautions that can be taken to mitigate the potential causes of SCC, will be evaluated.

Key words: stress corrosion cracking (SCC), boiler tube cracking, heat exchanger tube cracking.

14.1 Overview of stress corrosion cracking (SCC) in water systems

Stress corrosion cracking (SCC) is produced by the simultaneous interaction of three essential factors: tensile stress (residual or applied), a susceptible alloy, and an aqueous environment unique to that particular alloy. The absence of any one of these individual factors essentially eliminates the susceptibility of the component to SCC damage.

Because of the relative lack of general corrosion on most metal surfaces, stress corrosion crack growth can initiate and propagate to total fracture with little or no warning. This happens in materials that are considered fairly corrosion resistant in the boiler water or cooling water system in which they are exposed. Inspection of the fracture surfaces often reveals severe embrittlement, with hundreds or thousands of small cracks being present at the time a component fails.

From a historical perspective, over the past century three types of SCC damage in boilers and cooling water systems are considered 'classic' failure mechanisms. In boilers, the 'caustic embrittlement' noted in early, riveted boiler drum designs, has been attributed to have caused numerous catastrophic ruptures in steam locomotives and steam-powered ships. In ambient water environments, 'season cracking' was noted over 100 years ago in brass metal during the rainy season in the tropics. The cracking occurred in brass cartridge cases stored in the tropics in the presence of ammonia during heavy rainfall (driven by nearby decaying plant and animal matter). Similar ammonia SCC damage can be seen when a brass heat exchanger is put into storage without proper flushing to remove plant or other organic matter that can decompose and produce ammonia. Lastly, chloride SCC of 300 series austenitic stainless steel alloys became more prevalent following World War II when stainless steels found more widespread use. Alloys like 304SS and 316SS continue to be subject to cracking and fracture today due to a lack of understanding of application limitations in terms of specific chloride concentration, pH, dissolved oxygen or metal temperatures in the given environment.

These classic examples clearly illustrate that cracking damage related to SCC in boiler and cooling water systems has been around for a long time, but perhaps not fully understood. Over time, the unique alloy/environment combinations that exhibit SCC behavior in industrial water systems have become fairly well documented. Also, the type of residual/applied tensile stresses that are required to induce brittle cracking in a particular alloy/ environment combination is now better understood. Thus a design engineer can consult guidelines for an alloy or construction material that will help predict potential susceptibility to SCC damage. An engineer can modify the alloy choice based on anticipated upsets in boiler water or cooling water chemistry. However, despite these advances and precautions, catastrophic component fracture in boiler and cooling water systems can still develop through SCC for a number of reasons:

• unanticipated concentration of aggressive ions in the water that has 'embrittling characteristics' on a stressed metal surface,
• improper manufacturing or design execution in fabrication/welding, or
• inadequate material selection for a given operating environment.

14.2 Stress corrosion cracking (SCC) in boiler water systems

14.2.1 Carbon and low-alloy steel in boiler systems

During the heart of the Industrial Revolution, when steam power was driving unprecedented growth in manufacturing processes and transportation, boiler explosions were an unfortunate consequence of shoddy manufacturing processes or poor water chemistry control. Poor quality manufacturing methods often left boiler units with leaks that could concentrate caustic boiler water salts of specific concentration depending upon the level of free caustic alkalinity in the boiler feedwater. This in turn would lead to 'caustic embrittlement' of the steel. For example, drum leaks often originated at the spaces (or gaps) created by riveted construction, or from poor joint sealing at tube-to-drum connections.

To put the level of carnage in perspective, in the 1850s some 50 000 Americans died every year from boiler explosions, and boiler accidents were on the order of one every four days (Nichols, 1996). The establishment of the ASME boiler code in 1915 brought standardization to boiler manufacturing that greatly reduced boiler accidents through improved assembly practices and eventual elimination of riveted joints in favor of welding processes.

What was not fully understood at the time was that operating with any type of leak in a boiler could result in producing a key concentrating factor for caustic salt accumulation at the point where the escaping water flashes to steam. In essence, a leak path concentrates the free caustic (sodium hydroxide alkalinity) that is commonly present in boiler feedwater right at the point where residual fabrication stresses are high. The use of welded drum construction helped to eliminate inherent leaks characteristic of riveted drum design. However, cracks at tube-to-drum connection leaks were still commonplace.

During the 1930s and into the 1940s, advances in boiler water treatment to mitigate caustic SCC were discovered and implemented. Also the ability to monitor the effectiveness of water treatment, via a caustic embrittlement detector, allowed boiler operators to adjust water chemistry to virtually eliminate the problem of cracking in carbon steel at areas where concentrated caustic could form. For example, prior to 1939, an average of 30 steam locomotive boilers per year from fleet service at Chesapeake & Ohio Railway operations had to be repaired due to caustic cracking (Berk and Schroeder, 1943). However, the subsequent introduction of the addition of sodium nitrate inhibitor to boiler feedwater essentially eliminated all the caustic cracking problems in this fleet of steam locomotive boilers.

In addition to the use of nitrate additions to boiler water to reduce inherent embrittling characteristics of certain types of boiler feedwater, use of coordinated pH phosphate control schemes were being researched by Whirl and Purcell in the early 1940s, in order to counteract the presence of free caustic alkalinity via the phosphate treatment. Thus, from a historical perspective, the use of boiler water treatment to control the propensity for brittle cracking from caustic SCC is well founded, and instances of cracked drums or tubing have been greatly minimized.

Mechanism for caustic SCC in carbon and low-alloy steels

As indicated above, the mechanism for caustic SCC in carbon and low-alloy steels in boilers typically involves intergranular (intercrystalline) branched cracks, which initiate in areas of high residual tensile stress. These cracks form in the presence of concentrated hydroxide solutions (principally sodium hydroxide) that may form in areas of evaporative concentration. Sodium hydroxide concentrations above 20% are required to initiate cracking within a realistic time frame in stressed boiler components (Gabrielli, 1990). In reference to the NACE Caustic Soda Service Graph (see Fig. 14.1), the actual concentration of sodium hydroxide to produce cracking can be observed to be a function of temperature. In this graph, regions of susceptibility to caustic

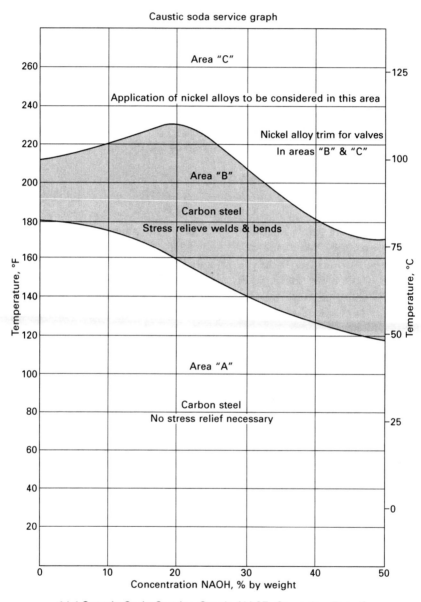

14.1 Caustic Soda Service Graph, NACE, Corrosion Data Survey, Metals Section, Sixth Edition, 1985. Reprinted with permission of NACE International®. All rights reserved.

SCC are indicated in which stress-relieving welds and bends should be considered, or upgrading to nickel-based alloys (at higher temperatures and concentration ranges) is required to avoid brittle cracking (Graver, 1985).

Because bulk boiler water chemistry does not typically have caustic concentrations above 20%, caustic SCC can only develop in specialized circumstances to create localized caustic concentration to achieve these levels. An example of a boiler tube cracked due to localized caustic concentration at a leak site is shown in Fig. 14.2. In this example, as water leaked out via a loose rolled-in tube joint at a boiler drum, it flashed to steam. At that point, dissolved caustic alkalinity in the boiler water can concentrate sufficient caustic via evaporative concentration to produce branched, predominately intergranular stress corrosion cracking in the steel microstructure (see Fig. 14.3). Under these circumstances, a failure analysis that includes cross-sectional metallography will typically identify crack initiation on the fireside surface, just past the roll leak, where the wet, caustic salts concentrate.

If left unattended, these types of rolled-in tube cracking issues can in turn create cracks in the ligament spacing between the boreholes in the drum. This can subsequently lead to drum fracture via link up of these small cracks and possible catastrophic boiler drum rupture. Thus, from a maintenance perspective, it is not good engineering practice to 'run with a leak', especially if the boiler water has embrittling characteristics.

14.2 Photograph of caustic embrittlement crack (caustic SCC) on the external surface of a boiler tube, where water leaked at the tube roll joint area at the boiler drum.

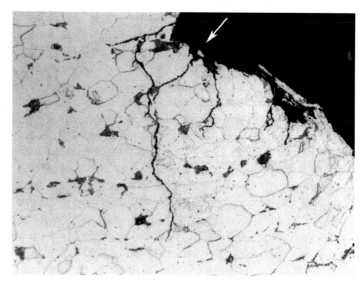

14.3 Photomicrograph of caustic embrittlement cracking (caustic SCC) in the microstructure of a steel boiler tube. 200×, nital etch.

Monitoring for caustic SCC in carbon and low-alloy steels

As discussed above, in the 1940s, a detector was developed to identify the susceptibility of stressed carbon steel to caustic SCC in boiler water environments. This 'embrittlement detector' was developed by the US Bureau of Mines and is covered by ASTM D807. Figure 14.4 shows a cut-away view of the steel block and how the steel test bar is configured inside of the device. This device allows for the simultaneous application of applied tensile stress and evaporative concentration of boiler salts via leaking of the boiler water at the test bar surface. Figure 14.5 shows a test bar that has failed a ductility test after exposure to boiler water that has embrittling characteristics.

The advantage to having a proven technique to monitor caustic SCC potential in a boiler water environment is that it can be of significant assistance in troubleshooting an operating system in which metal cracking has been identified. In addition, analytical testing to determine the precise boiler water chemistry via a water analysis can direct a suitable boiler water treatment approach that can reduce the possibility of caustic embrittlement, should any leak or other concentration mechanism develop in the boiler. For example, in lower-pressure boilers, a sufficient ratio of sodium nitrate to sodium hydroxide (0.2–0.5 ratio, depending upon boiler operating pressure) is required to impart caustic cracking inhibition (Betz Laboratories, 1991). Thus, if sodium nitrate is fed at the correct ratio, a steel test bar will be crack-free after a 30–90 day test with the embrittlement test device.

In a similar fashion, coordinated phosphate/pH control can be utilized to

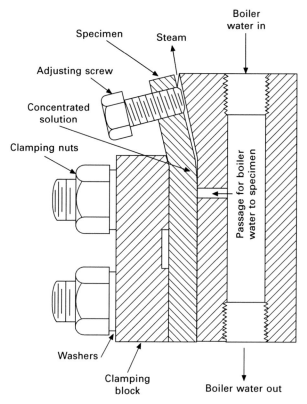

14.4 Caustic embrittlement detector schematic. Source: Betz Laboratories (1991).

minimize the presence of 'free' caustic alkalinity, controlled by appropriate phosphate chemical additions, blowdown rates, etc. Figure 14.6 shows a typical control range for a coordinated pH phosphate program, depending upon the operating pressure. This coordinated pH phosphate treatment approach is typically used in higher-pressure boiler systems that employ high purity feedwater.

In modern boilers, another common instance of caustic SCC in carbon and low-alloy steels involves inadvertent carryover of water droplets into steam lines or superheaters where only high purity steam is typically flowing (Kotwica, 1995). Areas of high residual stress such as welds (Figs 14.7 and 14.8) or tube bends (Figs 14.9 and 14.10) can rapidly crack when the moisture that does carry over into the steam phase flashes to steam and leaves behind moist, concentrated caustic salts at the metal surface (Esmacher, 2002). Avoidance of this type of caustic SCC damage involves vigilance in maintaining high steam purity at all times (which can be verified with a sodium analyzer in the steam). In addition to verifying that proper steam drum water levels are

14.5 Embrittlement test bar, carbon steel, after removal from caustic embrittlement test device, and failure upon bend test. Source: Betz Laboratories (1991).

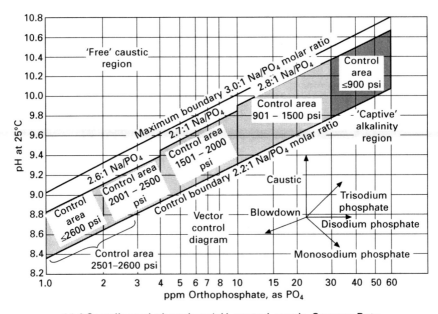

14.6 Coordinated phosphate/pH control graph. Source: Betz Laboratories (1991).

maintained below a certain level, making sure that cyclone separators, mist eliminators, and other mechanical devices in the upper part of a steam drum are in place and intact are of utmost importance.

In addition to the corrective action of minimizing carryover into superheaters that can produce caustic SCC, residual tensile stress at bends and welds can be significantly reduced at the time of fabrication via a stress relief anneal.

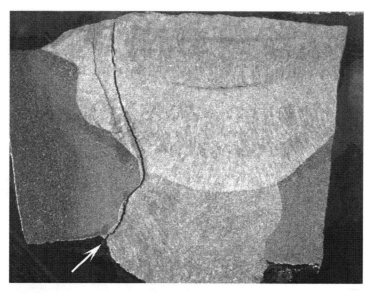

14.7 Steam line, carbon steel, caustic SCC at weld due to carryover inside line. Metallographic cross-section. Nital etch.

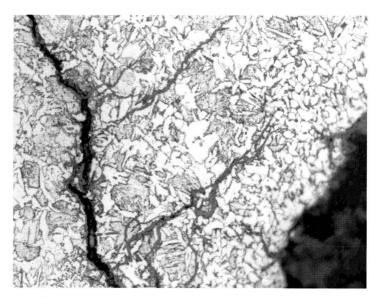

14.8 Steam line, carbon steel, caustic SCC at weld due to carryover inside line. Metallographic cross-section. Nital etch.

For example, in one case of caustic SCC of a carbon steel superheater bend due to carryover, the residual tensile hoop stress was measured to be 19 000 psi (131 MPa) in the bend that cracked. This residual tensile stress was high

14.9 Superheater tube at a bend, carbon steel, caustic SCC at bend due to carryover in steam.

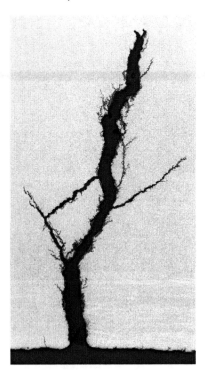

14.10 Superheater tube, carbon steel, caustic SCC in the microstructure at a bend due to carryover in steam, 50×, Nital etch.

because the stress relief step was not carried out as specified (Barer and Peters, 1970).

14.2.2   Stainless steel alloys in boiler and steam systems

Analogous to crack susceptibility of ferritic carbon steel alloys, austenitic stainless steel alloys in boiler environments are subject to rapid crack growth rates when exposed to dilute aqueous chlorides and/or concentrated caustic salts (sodium hydroxide, potassium hydroxide, etc.). These salts can form due to wetting/drying in the boiler environment. Softened water make-up used to provide boiler feedwater can be a source of chlorides. Another source of chlorides can be cooling water leaking into the steam-side of a condenser. If chlorides are present in conjunction with high dissolved oxygen content in the water, stressed austenitic stainless steels can experience severe cracking. An example of this can be seen in the upper areas of deaerator units fabricated from 304SS or 316SS, where brittle cracking can develop when chloride salts concentrate on the metal surface, and dissolved oxygen is present (see Figs 14.11 and 14.12).

In the absence of dissolved oxygen in boiler feedwater via deaeration and implementing appropriate oxygen scavengers (e.g., <10 ppb dissolved oxygen

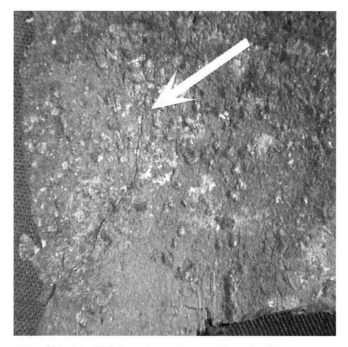

14.11 Chloride SCC in a deaerator vent line, 304SS.

14.12 Chloride SCC in a deaerator vent line, 304SS microstructure. 50×, oxalic acid etch.

levels), and with no appreciable chloride levels (e.g., <10 ppb chlorides) present via utilization of high purity reverse osmosis (RO) or demineralized boiler feedwater make-up, the risk of chloride SCC in austenitic stainless steel is greatly reduced in boiler systems. For example, reducing dissolved oxygen levels to less than 0.2 ppm has been shown effective in preventing chloride SCC up to chloride concentration levels of 1000 ppm at metal temperatures up to 300°C (McIntyre, 1987).

With regard to cracking in boiler systems where concentrated caustic salts can form, the instances of stainless steel caustic SCC damage can often be seen related to unintended carryover of moisture into the steam phase. The prevention of caustic SCC in stainless steel superheater tubes, expansion joints, and other components in steam service requires vigilance in maintaining a high purity steam environment. This is achieved by reducing or eliminating carryover and/or maintaining high purity water for desuperheating or attemporation (Esmacher, 1993). Failure to maintain a high purity steam environment can result in catastrophic caustic SCC of austenitic stainless steel within 24 hours of an upset event. Also, repeat failure of replacement components frequently occurs until the steam system is completely flushed of any residual caustic salts (Esmacher, 2001). Steam expansion joints fabricated from austenitic stainless steel are often subject to brittle failure via caustic SCC when steam purity guidelines are neglected (see Figs 14.13 and 14.14).

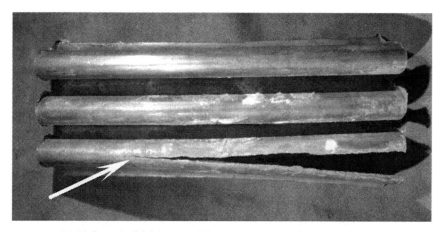

14.13 Caustic SCC in a stainless steel expansion joint due to carryover in steam.

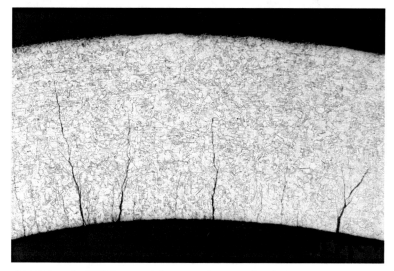

14.14 Caustic SCC in a stainless steel expansion joint due to carryover in steam. 200×, oxalic acid etch.

14.2.3   Other alloys in boiler and steam systems

In addition to steel alloys, other alloys that may undergo SCC failure in boiler water and steam environments include material used to construct various components such as valves, feedwater heaters and condensers. These include high nickel-content alloys, such as Monel and Incoloy or Inconel, as well as copper-based alloys (such as admiralty brass and copper-nickel alloys).

With regard to copper-based alloys, the main risk is exposure to ammonia and oxygen in areas where steam condensation is present. In addition, moist atmospheres that contain sulfur dioxide (SO_2) within specific concentration ranges have been noted to induce SCC in brass condenser tube applications (Howell, 1993).

There is also a risk of caustic SCC in high nickel content-based alloys in boiler and steam environments. However, the probability of SCC damage in these more corrosion resistant alloys is considerably lower when compared to carbon/low-alloy steel grades or 300 series austenitic stainless steel alloys. Nickel-based alloys, such as Inconel 600 will undergo caustic SCC at elevated temperatures, such as caustic-contaminated boiler water or steam environments that approach or exceed 600°F (315°C) (Loginow, 1979).

When SCC failures do develop in feedwater heaters and condensers, possible failure areas include U-bend locations where high residual forming stresses have not been relieved by an appropriate stress-relief anneal following fabrication (Syrett et al., 2006). Other high-risk areas for feedwater heater SCC include locations on the steam-side at or near the desuperheating zone, where the condensing steam can create variable wetting/drying regions dependent upon plant operating conditions (via steam load variations, peaking operations, etc.). To reduce SCC tendency in feedwater heaters and condenser tubing, specifying low residual stress in 300 series stainless steel tubing (below 5000 psi (34.5 MPa) maximum tensile hoop stress) should be considered as per EPRI Report GS-6913, Project 2504-5 guidelines (Janikowski, 2008),

14.3   Stress corrosion cracking (SCC) in cooling water systems

There are several important environments/alloy combinations that are frequent problems for SCC in cooling water systems, including: (1) ammonia/copper alloys and (2) chlorides/austenitic stainless steels. In addition, there are other cooling water chemistry conditions that can produce SCC damage, including nitrate SCC of carbon steel, etc. These various environment/alloy combinations will be considered below. In addition, corrective actions to minimize crack susceptibility in piping and heat exchanger operations, and monitoring options, are evaluated.

14.3.1 SCC of brass in cooling water

As was outlined previously, brass 'season cracking' is a relatively outdated term but one that relates to ammonia-induced SCC. While all copper-based alloys have certain susceptibility to ammonia SCC, there is an acute sensitivity for stressed copper-zinc alloys (i.e., 70/30 Cu-Zn brass) alloys to crack when exposed to a water environment contaminated with ammonia. Table 14.1 shows a relative ranking of the copper-based alloys versus susceptibility to ammonia SCC (McIntyre and Dillon, 1985).

Table 14.1 Relative ranking of copper and its alloys to ammoniacal SCC

Class	Class description	CDA number
Class 1	Very low susceptibility	
	Cupro-nickel 90-10	706
	Cupro-nickel 70-30	715
	ETP copper	110
Class 2	Low susceptibility	
	DLP copper	124
	DHP copper	122
Class 3	Intermediate susceptibility	
	Red brass	230
	Commercial bronze	220
	Aluminum bronze	614
	Silicon bronze	655
	Phosphor bronze	510
	Nickel silver	745
Class 4	High susceptibility	
	Leaded brass	360
	Naval brass	464
	Admiralty brass	443
	Yellow brass	270
	Manganese bronze	675
	Aluminum brass	687
	Muntz metal	280
	Cartridge brass	260

Source: McIntyre, D.R. and Dillon, C.P., 1985, *Guidelines for Preventing Stress Corrosion Cracking in the Chemical Process Industries, MTI Publication No. 15*, Materials Technology Institute of the Chemical Process Industries. Reprinted with permission of MTI®. Hyperlink http://www.mti-global.org, www.mti-global.org

There are important considerations to review with brass alloys when they are exposed to ammonia in cooling water streams. These include: crack morphology, water chemistry influence, and tensile stress conditions. These factors will be considered below.

Crack morphology in brass alloys

In many cases, stress corrosion cracks grow in an intergranular fashion, that is, they grow between the crystals or grains, hence the term 'intergranular' cracking. However, ammonia-induced cracks in brass do not always grow along grain boundaries. In many cases, the cracks are observed to propagate in a transgranular fashion. This was later discovered to be a function of pH in ammoniacal solutions (Fig. 14.15), with rapid intergranular cracking being favored at near-neutral (pH 6–8) conditions, and transgranular cracking more dominant in alkaline and acidic ranges (Warke, 2002). In some cases, intergranular SCC is seen to initiate on a brass heat exchanger tube surface, followed by a shift to transgranular crack propagation (see Figs 14.16 and 14.17).

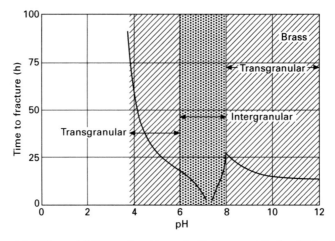

14.15 Brass SCC as a function of pH. Source: Warke, W.R., *Metals Handbook*, 10th edn, *Volume 11: Failure Analysis and Prevention* (1985). Reprinted with permission of ASM International®. All rights reserved.

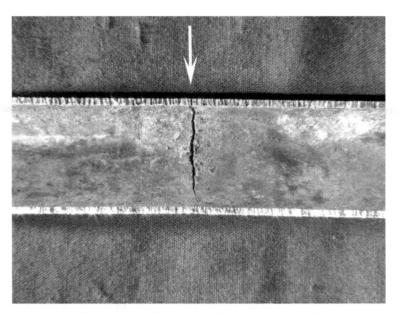

14.16 Brass SCC example. Cracks developed on the cooling waterside of steam condenser tube due to ammonia contamination.

Chemistry influence on brass-SCC

This type of crack morphology and propagation, as discussed above, underscores the dependence of SCC on the specific chemical environment

14.17 Brass SCC due to ammonia, 200×. Cracking initiating on the cooling water-side of a steam condenser tube.

that is present on the metal surface or in the developing crack front. Given the importance of surface chemistry reactions, the natural question arises, 'how much ammonia content is necessary to cause brittle SCC damage in a brass alloy?' Unfortunately, no precise guidelines exist as to what is a 'safe' ammonia level. Assuming that sufficient residual forming stresses are present, the ammonia content required could be quite low.

In one study involving brass fittings researched by Fontana (1986), as little as 1 ppm of ammonia was required to cause cracking damage. In another well-documented case, trace ammonia from bird droppings was sufficient to cause SCC of cold-drawn 70-30 brass tubing (Peters *et al.*, 1965). In addition to ammonia content in the environment, the presence of oxidizing substances such as dissolved oxygen, nitrates, or cupric/ferric ions are generally required to drive SCC damage in aqueous ammoniacal solutions (Warke, 2002).

On occasion, open recirculating cooling water systems are required to operate with significant ammonia contamination, especially during water re-use situations. The relative amount of ammonia contamination in a re-use cooling water system that will produce ammonia SCC of copper alloys can be quite low, depending upon the type of copper alloy in the system. For example, copper-nickel alloys show a slightly higher tolerance for ammonia contamination than copper-zinc alloys, like admiralty brass (see Table 14.2) (Thompson *et al.*, 2006).

Table 14.2 Resistance of copper alloys to SCC caused by ammonia

Copper alloy	Maximum safe level of ammonia in water (mg/L as NH3)
Admiralty (70/30 Cu/Zn)	< 0.2
90/10 Copper/nickel	< 10
70/30 Copper/nickel	< 20

Source: Thompson, K.W., Christofferson, W., Robinette, D., Curl, J. and Baker, L., *Characterizing and Managing Salinity Loadings in Reclaimed Water*, Copyright 2006. AWWA Research Foundation. All Rights Reserved. Hyperlink http://www. WaterResearchFoundation.org, www.WaterResearchFoundation.org

Stress influence on brass SCC

The sensitivity of crack growth to corrosion reactions that take place on susceptible (deformed) metal surfaces in the presence of specific corrosive species is an important factor to consider in predicting SCC susceptibility. Although some amount of residual or applied stress is definitely required for SCC to become operative, the threshold stress that triggers the onset of ammonia-induced SCC in brass can be extremely low. For example, cracking can be triggered at stress levels approaching only 1% of the tensile strength of the alloy (Warke, 2002). Naturally, the higher the stress level the shorter the time to failure. Another important factor is temperature, with increasing temperatures causing shorter time to failure of brass exposed to ammoniacal solutions (McIntyre and Dillon, 1985).

Prevention of brass SCC may involve substituting copper alloys that have better overall resistance to damage (such as alloys below 15% zinc content), modification of the environment to preclude SCC agents (such as ammonia), or consideration of reduction of residual stresses in manufactured components via stress-relief annealing. With regard to the latter, depending upon the alloy, short heating (30 minutes to 1 hour) between 300 and 500°F (150 and 260°C) can be useful to reduce residual stress levels below the point at which SCC will not develop in copper-based alloys. As a verification step of the effectiveness of the thermal stress relief process in copper-based alloys, a mercurous nitrate evaluation (ASTM B154-05, 'Standard Test Method for Mercurous Nitrate Test for Copper Alloys') or an ammonia test (ASTM B858-06, 'Standard Test Method for Ammonia Vapor Test for Determining Susceptibility to Stress Corrosion Cracking in Copper Alloys') can be used to screen the parts for SCC potential, and adjustments made to the thermal stress relief process (Ricksecker, 1961).

14.3.2   SCC of austenitic stainless steel alloys in cooling water

In cooling water systems, chloride SCC of 300 series (austenitic) stainless steels is a well-known mechanism that produces branched cracking and leaks in supply lines and heat exchangers (Hargrave, 2004). Commonly used grades such as types 304 and 316 SS are susceptible to failure by exposure to chloride ions in the presence of tensile stress (residual and applied) at metal temperatures in excess of 140°F (60°C). The relationship between SCC susceptibility of 300 series stainless steel alloys as a function of chloride content of an aqueous environment versus metal temperature is shown in Fig. 14.18 (McIntyre, 1987).

Not all austenitic stainless steel grades are equally susceptible to chloride SCC. For example, there is a direct relationship between nickel concentration in an austenitic stainless steel (Fig. 14.19), and susceptibility to chloride SCC, referred to as the Copson curve (Copson, 1959). Thus, the stainless steel nickel content with the greatest potential for chloride SCC is 8%, which is the minimum nickel value for type 304SS. Therefore, cracking susceptibility can be reduced by use of alloys with very low or very high nickel contents, such as found in ferritic/super-ferritic stainless, duplex stainless steel, and super-austenitic stainless steel alloys. Duplex stainless steel alloys have been shown to be cost effective and considerably more resistant to chloride SCC as compared to 300 series austenitic stainless steels. In general, duplex stainless steels show excellent resistance to chloride SCC in neutral pH cooling waters, but not in acidic media (Munster, 1991). Also, super-ferritic stainless steels such as 'SeaCure' that have low nickel contents are also very resistant to chloride SCC, and have found widespread use in steam condenser

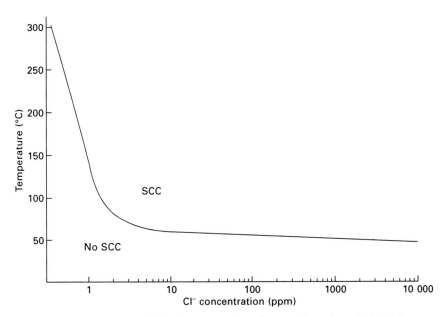

14.18 Chloride SCC of types 304/316 SS as a function of chloride concentration and temperature. Source: McIntyre (1987), reprinted with permission of MTI®.

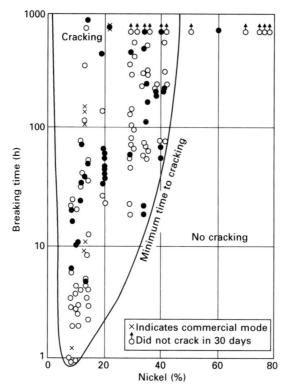

14.19 Copson curve showing the effect of stainless steel nickel content on chloride SCC susceptibility in a boiling magnesium chloride solution. Source: Copson (1959).

applications (Janikowski, 2007). Alternatively, stainless steel alloys with very high nickel levels, such as found in the higher-grade super-austenitic alloys, can markedly reduce chloride SCC susceptibility, but tend to be more costly.

Crack morphology and mechanism in Cl SCC

Both transgranular and intergranular modes of chloride SCC have been observed in austenitic stainless steels. However, branched transgranular SCC is the more common cracking mode. On occasion when intergranular crack propagation is observed, the main cause of intergranular chloride SCC may be related to carbide precipitation (sensitization) near a weld, or caused by sensitization related to excessive carbon content coupled with a non-optimal heat treatment. Intergranular chloride SCC in a sensitized stainless steel occurs readily, even at lower temperatures, as low as ambient (Nakahara, 1993). For example, a sensitized 300 series stainless steel bolt in a reverse osmosis (RO) water system pipe clamp fractured after less than one year in service, displaying both intergranular and transgranular chloride SCC, (see Figs 14.20 and 14.21). In this case, the bolt did not meet the 316 or 316L carbon requirement (0.08%–0.03% carbon maximum, respectively). Thus, the excessive carbon content in the stainless steel in this case (0.13% carbon), combined with an incorrect heat treatment within the sensitization range (500–800°C), produced bolts that cracked rapidly via chloride SCC under ambient, atmospheric exposure conditions.

Chloride SCC is considered a type of localized corrosion that progresses in two stages: initiation and propagation (Okada, 1998). Similar to pitting and crevice corrosion, SCC begins with establishment of a localized chemical environment suitable for cracking prior to the initiation process. Crack initiation in austenitic stainless steels is believed to be more dependent upon the initial establishment of a suitable local chemistry at the stressed metal surface than the actual stress concentration. The impact of local chemistry interaction can be seen in some cases where pitting precedes the generation of stress corrosion cracks. Cracks can also develop within crevice environments where concentration effects are enhanced.

Water chemistry and temperature influence in Cl SCC

In plant environments, low levels of chlorides in cooling water can often concentrate inside crevices or underneath deposits. Chlorides can also concentrate through evaporation. Thus, in cooling water environments it is

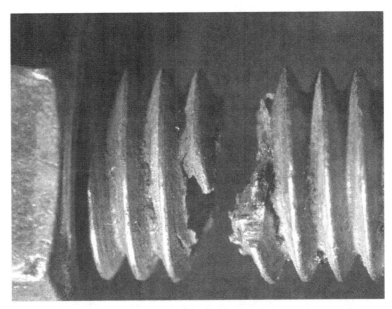

14.20 Bolt fracture due to chloride SCC, sensitized 300 series SS.

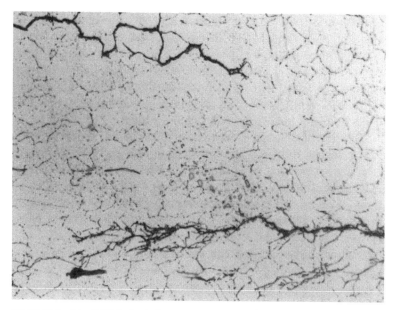

14.21 Chloride SCC in bolt fracture case, sensitized stainless steel microstructure. Note both intergranular and transgranular cracking observed. 1000×, oxalic acid etch.

often difficult to predict what is the 'safe' threshold chloride content at which chloride SCC will not occur in 300-series stainless steel alloys under heat exchanger operating conditions. In laboratory studies, chloride-induced SCC in austenitic stainless steels has been documented at concentrations below 10 ppm chloride (McIntyre and Dillon, 1985). Thus, from a practical viewpoint, there is no safe threshold limit at which chloride SCC can develop assuming sufficient residual/applied stress loading is present in an environment that can concentrate chlorides at the metal surface.

In the absence of concentrating mechanisms and with the practice of maintaining good water flow rates in 'normal' water environments, it has been asserted that austenitic stainless steels may withstand exposure up to a maximum of 1000 ppm chloride without SCC developing (Muraleedharan, 2002). However, actual operating success at this level of chloride may be impaired by slow flow rates or stagnant conditions, high operating temperatures, the presence of crevices (created by fouling or other design factors), high level of fabrication/applied stress loading, and the actual composition of the austenitic stainless used in the heat exchanger manufacture.

A key factor in any discussion of chloride content in cooling water versus the risk of chloride SCC is the metal temperature in the heat exchanger. In essence, there will always be some risk of SCC in 300 series austenitic stainless steels at elevated temperatures above 140°F (60°C), even when chloride levels are relatively low. Based on plant experience with austenitic SS vessels and piping, cracking almost always occurs when temperatures are above 130–140°F (55–60°C) (McIntyre and Dillon, 1985).

Measures to control susceptibility to transgranular stress corrosion cracking (TGSCC) of austenitic stainless steels include reducing the temperatures and lowering chloride concentration by using a higher purity make-up water source for the cooling water. Also, eliminating crevices that can concentrate chlorides can help reduce chloride SCC susceptibility. For example, during design/ fabrication, crevices at tube joints or connection points in a heat exchanger can be minimized. Also, control of water-side fouling on a heat transfer surface by maintaining good flow velocities, and adding deposit control agents to minimize scale build-up (or biocides to limit microbiological growth), can reduce cracks that tend to grow via crevice concentration of chlorides.

Fabrication/design considerations that impact Cl SCC

For 300 series SS alloys, the threshold stress for chloride SCC is relatively low so it can be assumed that sufficient stress is always present, whether from applied loads, thermal cycling, or residual stresses from welding and forming (Warke, 2002). Welded areas, which contain residual stresses, are particularly prone to chloride SCC unless stress-relief annealing (to relieve residual tensile strain) or shot-peening (to induce compressive surface stress) is employed. Unfortunately, post-fabrication heat treatment with 300 series austenitic stainless steel alloys must be approached with caution. The temperature range for annealing may cross over into the sensitization range (500–800°C), in which carbide precipitation can dramatically increase susceptibility to intergranular SCC.

Heat exchanger design considerations

Water flow designed to be on the inside of the tubing in a heat exchanger bundle (tube-side) is the best way to prevent chloride concentration. Also, when water is circulated on the shell side of a heat exchanger, there is a risk that under high heat transfer, concentrated chlorides can develop via trapped steam/water conditions. Shell-side cooling water flow can also be inherently low velocity (on the order of one foot per second (0.3 m/sec), or less), which makes deposition control extremely difficult. Consequently, chloride accumulation under deposits or in the vicinity of baffles or support plates can develop when water is circulated on the shell-side of heat exchangers. This is why tube-side water circulation design is much better, achieving on average four to six feet per second (1.2–1.8 m/sec) flow rates, with limited risk of fouling to produce corrosive under-deposit crevice conditions at the metal surface.

Vertical heat exchangers with water on the shell side are especially prone to SCC because of the potential for a waterline to exist a few inches below the upper tubesheet, or at some point in the upper region of the tube bundle. In these upper regions, chlorides can concentrate at a waterline through wetting/drying cycles via evaporative concentration. Excessive temperatures can exist in the vapor space if the hot process stream enters the upper tubesheet. A design for a heat exchanger should take into account the inlet and outlet temperatures of a process stream, as well as the location of the process inlet/outlets versus the cooling water pathways. For example, venting of the top tubesheet in a vertical condenser that has water circulating on the shell-side can be one way to insure that the top tubesheet is continually flushed with flowing water. This design practice minimizes concentration effects. The impact of a horizontal waterline that created a repeated wet/dry concentration mechanism is shown in Figs 14.22 and 14.23. In this case, the chloride SCC of the 304SS developed only within the band where excessive chloride concentration was taking place via the wetting/drying condition.

Maximum tube metal temperatures should be maintained below the point at which water-formed scale and deposits are driven to precipitate. While this depends upon the relative hardness of the cooling water, the temperature range at which the onset of fouling can become severe is estimated to be between 50°C and 60°C (Dillon, 1994). Once a heat transfer surface is fouled, chloride concentration under the deposit formation can increase to the point where Cl SCC can develop in 300 series SS alloys.

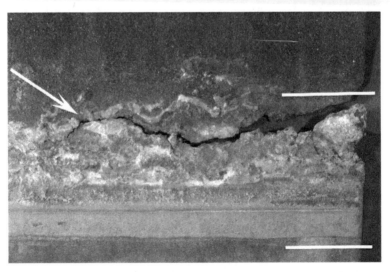

14.22 Cl SCC of 304SS at a horizontal waterline that fluctuated within a band (between the white lines) and caused excessive chloride concentration.

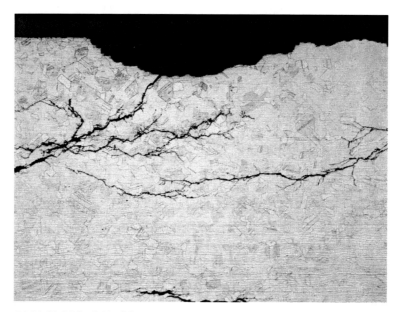

14.23 Cl SCC of 304SS at a horizontal waterline, metallographic cross-section. 100×, oxalic acid etch.

In addition to being aware of the cooling water chemistry as it pertains to scale-forming potential under high temperatures, maintaining good flow rates is essential to avoid fouling of heat transfer surfaces, and prevent under-deposit Cl SCC. Unfortunately, cooling water flow rates during a heat exchanger operation can be inadvertently impaired, leading to fouling and subsequent Cl SCC under deposits that can form. For example, reduced cooling water flow rate in one heat exchanger was observed to have caused excessive calcium carbonate fouling, which in turn caused chlorides to concentrate under-deposit to a sufficient level to cause SCC of the 304SS heat exchanger tubing (see Figs 14.24 and 14.25).

In summary, there are several ways that chloride SCC of austenitic stainless steels can be avoided. Chloride SCC susceptibility can be based on conditions dictated by the interplay of the water-side environment, design/operation, stress, and material selection. Table 14.3 shows possible corrective actions that can be considered in preventing chloride SCC of austenitic stainless steel heat exchangers (Nakahara, 1993).

14.3.3   SCC in other cooling water environments

Although the two most common SCC problems in cooling water systems have been reviewed (brass SCC in ammonia and chloride SCC in 300 series SS alloys), there are other, less common SCC mechanisms that can develop

14.24 Severe calcium carbonate scale build-up inside a 304SS heat exchanger tube, under low cooling water flow conditions.

14.25 Chloride SCC in the 304SS heat exchanger tube that was subject to under-deposit crevice concentration of chlorides on the cooling water-side of the heat exchanger. 100×, oxalic acid etch.

Table 14.3 Prevention of chloride SCC in austenitic stainless steel heat exchangers

Condition	Corrective action
Environment	Lower temperature below 60°C (140°F), the typical metal temperature threshold for CI-SCC
	Lower chloride concentration (example: substitute high purity water source for cooling water via reverse osmosis or demineralization)
	Raise pH to delay onset of cracking
	Reduce dissolved oxygen levels, or metal cations in solution (that can drive cathodic reduction reactions) Minimize scale, deposit, sediment or biofilm formation that can concentrate chlorides
Design/operation	Eliminate crevices at connection points that can concentrate chlorides
	Improve environment to avoid wet/dry concentration at heat transfer surfaces. Design for tube-side cooling water flow. Considering venting top tubesheets in vertical heat exchangers with shell-side water flow
	Maintain a minimum water flow rate to keep surfaces clean and do not throttle flow back on cooling water to meet process effluent temperature objectives
Stress	Minimize residual tensile stress in fabrication and welding. Consider heat treatment but avoid sensitization temperature range, or use a stabilized grade of austenitic stainless steel
	Avoid severe thermal fluctuations or pressure transients that can induce severe applied stresses
	Shot-peen areas to induce compressive stresses to mitigate high residual tensile stress
Material selection	Use austenitic SS alloys that have higher Ni content (above 45% Ni), or select ferritic or duplex SS alloys with lower Ni content
	Apply protective coating (or plating) if feasible to block chloride access to the austenitic stainless steel

Source: Derived from Nakahara (1993).

in cooling water systems. For example, intergranular nitrate SCC of carbon steel can develop if nitrate-contaminated cooling water is circulated in a cooling water system.

Historically, plant operations that produce nitrate-related chemicals (fertilizer plants, etc.) could be subject to such nitrate SCC of carbon steel. In addition, nitrate SCC of carbon steel can take place in closed loop cooling systems if sufficient concentration of nitrates becomes prevalent. One way that nitrate concentration can accumulate in a closed-loop cooling water system is unexpected conversion of sodium nitrite corrosion inhibitor to nitrates due to microbiological degradation via nitrifying bacteria in the system. Welded connections in carbon steel that have not been stress-relief annealed tend to be more susceptible to nitrate SCC damage, as a result of high residual tensile welding stresses.

Also, carbon content of the steel can play a role in nitrate SCC, with the threshold stress for cracking being low at 0.05% carbon, but increasing likelihood of cracking seen at a carbon content of 0.10% carbon (Warke, 2002). Although medium-carbon stainless steels (carbon content in excess of 0.18% C) are thought to be less susceptible to nitrate SCC, decarburized surfaces in these types of steels can be susceptible to crack initiation, which in turn leads to through-wall fracture of the component (McIntyre and Dillon, 1985).

14.4   Stress corrosion cracking (SCC) monitoring strategies

The use of constant-strain test strips, like an ASTM G 30 U-bend Test Coupon, can provide a convenient method to monitor the likelihood of stress corrosion cracking propensity in a cooling water system. In the cold-bent position, the stress-state simulated by the U-bend and bolting arrangement can adequately simulate the stresses present in field-fabricated piping, deformed areas (elbows, flanges, etc.), and other high residual stress regions of the plant. The U-bend coupon can be made of the suitable alloy that is desired to be studied, and after immersion, can be removed for visual and microscopic study via metallographic cross-sectioning. In addition, a weld bead can be placed in the center of the U-bend, to simulate the combined action of deformation stresses and residual stresses related to welding. Other pre-stressed metal test coupons can be used for SCC monitoring, including C-rings, cup specimens, etc. (McIntyre and Dillon, 1985).

14.4.1   Field inspection/maintenance

Monitoring field equipment such as heat exchangers and piping in cooling water systems for SCC damage via visual examination can be challenging because in most cases, little to no metal corrosion loss from the surface is realized. Conventional non-destructive testing (NDT) methodologies are geared to detecting pitting or metal thinning, and are thus not suitable to detect fine cracks that may form on the metal surface due to SCC. If crack patterns are not readily visible, dye penetrant (DP) testing is very useful to identify the tell-tale sign of SCC, branched interconnected cracking. These cracks can typically originate in areas of deformation (bends, rolled-in tube ends, flanges, etc.) or welded areas where residual stresses can be very high. For ferritic steel components, wet magnetic florescent particle testing can be also used to identify regions suspected to be cracked. Other NDT methodologies such as ultrasonic testing (UT), acoustic emission (AE), eddy current (ED) and radiography testing (RT) may only work if the cracking is severe, as these techniques may lack the sensitivity to isolate and identify fine crack growth below the metal surfaces that are being scanned.

Once a component has been identified to be cracked via SCC, repair and maintenance of the affected area can be problematic. For example, due to the branched interconnected nature of a stress corrosion cracking network, it is not uncommon for cracks to 'run' when heat is applied to weld-repair an area. Thus, it is often more suitable to use a grinding disk on thick-wall components to 'chase' cracks to reveal the depth, then repeat a dye-penetrant procedure before weld repairs are attempted.

It may be required to cut away an entire area from an assembly and have it replaced in-kind in order to repair a component that has been damaged via SCC. It is important to note that any weld-repair process should be followed by a suitable post-weld heat treament (PWHT) procedure. A PWHT procedure may be required in order to counteract the presence of high residual tensile stresses that are induced by welding, as the molten weld pool cools, shrinks, and contracts, leaving the adjoining surfaces in a high tensile strain condition. For example, a typical PWHT procedure to provide a thermal stress relief for carbon steel to minimize SCC would be to hold an annealing temperature of 1150–1250°F (620–677°C) for a minimum of 1 hour hold time (Thielsch, 1984). Regrettably, lack of attention to residual stresses created by weld repair procedures can often result in repeat cracking of the component by the same SCC mechanism.

With regard to preventative maintenance, precluding the causative agent from coming in contact (or concentrating) on the metal surface is essential to reduce the risks of SCC damage from occurring or repeating. For example, obtaining a water sample to analyze for the contamination of the water source by potential causative agents (free caustic alkalinity, ammonia, chlorides, nitrates, etc.) is considered a best practice in troubleshooting a water system that has experienced SCC damage. In addition, determining how design or process conditions allow a concentration mechanism to develop that elevates chloride salts or other causative agents to a threshold level sufficient to crack metal components can help direct corrective actions. Lastly, when unexpected contamination of a water source is confirmed to be unavoidable, alternative material selection options can become an opportunity to upgrade the piping or tubing in the water system to avoid crack susceptibility.

14.5   References

Barer, R.D. and Peters, B.F. (1970), *Why Metals Fail*, Gordon and Breach Science Publishers.

Berk, A.A. and Schroeder, W.C. (1943), 'A practical way to prevent embrittlement cracking', *Transactions of the ASME*, August, p. 702.

Betz Laboratories (1991), *Handbook of Industrial Water Treatment*, 9th edn. Available from: http://www.gewater.com/handbook/index.jsp

Copson, H.R. (1959), 'Effect of composition on stress corrosion cracking of some alloys containing nickel', in *Physical Metallurgy of Stress corrosion Fracture*, Rhodin, T.N. (ed.), New York, Interscience, p. 247.

Dillon, C.P. (1994), *Corrosion Control in the Chemical Process Industries*, 2nd *edn*, MTI Publication No. 45, NACE International.

Esmacher, M. (1993), 'Stress corrosion cracking of stainless steel superheater tubing', *Handbook of Case Histories in Failure Analysis: Volume 2*, ASM International.

Esmacher, M. (2001), 'Stress corrosion cracking of stainless steel components in steam service', Paper No. 01496, NACE 2001 Corrosion Conference.

Esmacher, M. (2002), 'The impact of water chemistry on boiler tube failures', Paper No. 02-55, International Water Conference.

Fontana, M.G. (1986), *Corrosion Engineering*, McGraw-Hill.

Gabrielli, F. (1990), 'Caustic corrosion in industrial boilers', *Corrosion/90*, Paper No. 191, NACE, Houston.

Graver, D.L. (ed.) (1985), *NACE Corrosion Data Survey, Metals Section*, 6th edn. NACE, p. 176.

Hargrave, R.E. (2004), 'Behavior of 300-series stainless steel heat exchangers in cooling water service', Paper No. 04080, NACE 2004 Corrosion Conference.

Howell, A. (1993), 'Sodium sulfite decomposition in boilers and subsequent condenser tube damage', Paper No. 46, NACE 1993 Corrosion Conference.

Janikowski, D. (2007), 'Selecting tubing materials for power generation heat exchangers', Power-Gen Int'l Conference, New Orleans, LA.

Janikowski, D. (2008), 'Stainless steel feedwater and condenser tubing – expectations, results, choices', Power-Gen Int'l Conference, Orlando, FL.

Kotwica, D. (1995), 'Deposit related failures of boiler superheater tubing and steam piping', Paper No. 615, NACE 1995 Corrosion Conference.

Loginow, A.W. (1979), 'Failure analysis of expansion bellows in high pressure steam', *Materials Performance*, 18, 10.

McIntyre, D.R. (1987), *Experience Survey: Stress Corrosion Cracking of Austenitic Stainless Steels in Water*, MTI Publication No. 27, Materials Technology Institute of the Chemical Process Industries.

McIntyre, D.R., and Dillon, C.P. (1985), *Guidelines for Preventing Stress Corrosion Cracking in the Chemical Process Industries*, MTI Publication No. 15, Materials Technology Institute of the Chemical Process Industries.

Munster, R.L. (1991), 'Use of duplex stainless steels to avoid stress corrosion cracking (SCC) due to cooling waters, and external corrosion', Paper No. 300, NACE 1991 Corrosion Conference.

Muraleedharan, P. (2002), 'Metallurgical influences on stress corrosion cracking', *Corrosion of Austenitic Stainless Steels Mechanism, Mitigation and Monitoring*, ASM Narosa Publishing House, New Delhi, India, p. 143.

Nakahara, M. (1993), 'Preventing stress corrosion cracking of austenitic stainless steels in chemical plants', NiDI Technical Series No.10 066. Available from: http://www.nickelinstitute.org/index.cfm/ci_id/3170/la_id/1/document/1/re_id/0

Nichols, D. (1996), 'ASME Boiler Code became constitution for steam age', *Power Engineering*, February.

Okada, H. (1998), 'Mechanistic understanding and prevention of localized corrosion', *Materials Performance*, December.

Peters, B.F., *et al.* (1965), 'Stress corrosion cracking in marine service', *Materials Protection*, 4 (5), 24–37.

Ricksecker, R. (1961), 'Corrosion of wrought copper and copper alloys', in *Metals Handbook*, 8th edn, *Volume 1: Property and Selection of Metals*, ASM International.

Syrett, B.C., Jonas O. and Mancini, J.M. (2006), 'Corrosion in the condensate-feedwater system', in *Corrosion: Environments and Industries*, Vol. 13C, *ASM Handbook*, ASM International, pp. 447–460.

Thielsch, H. (1984), *Defects and Failures in Pressure Vessels and Piping*, Reinhold Pub. Co.

Thompson, K., Christofferson, W., Robinette, D., Curl, J. and Baker, L. (2006), *Characterizing and Managing Salinity Loadings in Reclaimed Water*, AWWA Research Foundation.

Warke, W.R. (2002), *Metals Handbook*, 10th edn, *Volume 11: Failure Analysis and Prevention*, 'Stress corrosion cracking', ASM International.

15

Environmentally assisted cracking (EAC) in oil and gas production

M. IANNUZZI, Det Norske Veritas, Norway

Abstract: With conventional oil and gas reserves dwindling, the oil and gas industry is transitioning to more aggressive environments, many rich in hydrogen sulfide, carbon dioxide, chlorides, and elemental sulfur at elevated temperatures and total pressures. Under these conditions, the management of environmentally assisted cracking becomes one of the main challenges, especially during the materials selection process, which is guided by the use of industry standards such as NACE MR0175/ISO15156. Environmentally assisted cracking (EAC) is a common cause of failure during oil and gas production, especially for intermediate sour service conditions, or when the requirements of NACE MR0175/ISO15156 are not fully addressed during design. This chapter describes the EAC mechanisms encountered in oil and gas production and the performance of low grade and corrosion resistant alloys. Standards and recommended practices for materials selection and design are also discussed.

Key words: corrosion resistant alloys, environmentally assisted cracking, sulfide stress cracking, ISO 15156, NACE MR-0175.

15.1   Introduction

With conventional oil and gas reserves dwindling, the oil and gas industry is transitioning to more aggressive environments, many rich in hydrogen sulfide (H_2S), carbon dioxide (CO_2), chlorides, and elemental sulfur (S^0) at elevated temperatures and total pressures. Under these conditions, the management of environmentally assisted cracking (EAC) becomes one of the main challenges, especially during the materials selection process, and it is not an uncommon cause of failure during oil and gas production. For this reason, it is crucial to understand EAC mechanisms and kinetics and to quantify the risks of catastrophic failures during the life of the reservoir.

Typical EAC failure modes in oil and gas drilling and production include sulfide stress cracking (SSC), hydrogen stress cracking (HSC), hydrogen induced cracking (HIC) (previously known as stepwise cracking or SWC), stress oriented hydrogen induced cracking (SOHIC), and stress corrosion cracking (SCC). While the first three modes are common to most systems, from carbon and low alloy steels to the highly corrosion resistant Ni-based alloys, the last one has been observed only in carbon and low alloy steels.

570

There are a number of standards, recommended practices, and technical documents available for materials selection and design in sour service. In this regard, the international standard ISO 15156 (NACE MR0175) looks into SSC and HIC performance of carbon and low alloy steels (CS and LAS, respectively) and SSC, HEC, SCC of corrosion resistant alloys (CRAs) in H_2S containing environments and it is used worldwide to qualify materials and components. Likewise, the European Federation of Corrosion (EFC) publications 16 and 17 (EFC, 2002a, 2002b) complement the ISO standard and are implemented extensively by most oil and gas companies to determine the suitability of materials for sour service.

This chapter looks into the main EAC mechanisms encountered in oil and gas production and the performance of carbon and low alloy steels as well as CRAs. Standards and recommended practices for materials selection and design are also discussed.

15.2   Overview of oil and gas production

Since there are a number of definitions and expressions unique to the oil and gas industry that may be confusing or even misleading if not defined properly, they will be explained below. The objective is to have a uniform understanding of the terminology that will be used throughout this chapter, which might differ from similar concepts described in other sections of this book. Table 15.1 summarizes common oil and gas terminology.

15.2.1 Terminology

Sweet and sour gas

The definition of sweet and sour gas was first introduced in API Book 29 'Corrosion of Oil and Gas Well Equipment' during the 1950s (now referred to as API G12900). Today, ISO standard ISO15156 (NACE MR0175) (NACE, 2003) defines 'sweet gas' as any gas free of hydrogen sulfide. In contrast, 'sour service' is used to identify environments containing measurable amounts of H_2S. The EFC utilizes the term H_2S service rather than sour service to identify environments that contain measurable amounts of H_2S. In this regard, H_2S can be measured in the gas stream down to 0.5 ppm using techniques described in the ASTM D 4810 standard (EFC, 2002b).

Shallow, deep and ultra-deep water

The first modern offshore operation started production in 1977 at a recorded water depth of 124 m (406 ft). Nevertheless, offshore drilling dates back to 1869, when a patent was granted to T.F. Rowland for his offshore drilling rig

Table 15.1 Summary of typical oil and gas terminology

Term	Definition
Sweet service	Any gas free of H_2S.
Sour service	Environments containing measurable amounts of $H_2S(g)$. $H_2S(g)$ can be measured in the gas stream down to 0.5 ppm.
Shallow, deep, and ultra-deep water	Shallow water: water depths up to 300 m. Deep water: water depths from 300 to 1500 m. Ultra-deep water: water depths above 1500 m. Note: Water depths ranging from 2000 to 3000 m are not uncommon in Brazil and the Gulf of Mexico
Shallow, deep, and ultra-deep wells	Shallow wells: <3000 m deep Deep wells: from 3000 to 6100 m deep Ultra-deep wells: > 6100 m deep Record: 10 421 m as of December 2006
High pressure high temperature and extreme high pressure high temperature reservoirs (HPHT and xHPHT, respectively)	HPHT Bottom-hole temperature > 200°C Bottom-hole pressure > 103 MPa xHPHT Bottom-hole temperature > 230°C Bottom-hole pressure > 206 MPa
Extended reach drilling (ERD)	ERD is an extreme form of directional drilling that achieves horizontal well departures beyond conventional, or achieves challenging geometries in terms of horizontal versus vertical offsets. Stress state: High torque low tension Length: comparable with ultra deep drilling (i.e. > 6100 m) Depth: record set to 4600 m
Ultra deep drilling (UDD)	UDD refers to the drilling of ultra-deep wells with total vertical depths (TVDs) of more than 6100 m with little or no horizontal deviation. Stress state: Low torque high tension. Length = TVD: > 6100 m exploratory wells of up to 15 000 m have been drilled.

design. In addition, offshore drilling was performed in the Gulf of Mexico (GOM) from barges in the coastal waters during the 1940s and 1950s. Since then, the industry has moved into increasingly deeper waters. Offshore operations are commonly classified according to the water depth and can be divided into three main categories: (i) shallow water, (ii) deep water, and (iii) ultra-deep water. In this context, the terminology 'shallow water' refers to platforms operating in water depths lower than 300 m (990 ft) (Leffler *et al.*, 2003). The term 'deep water' refers to wells drilled in water depths ranging from 300 to 1500 m (990 to 4900 ft). Most of the current StatoilHydro fields off the coast of Norway operate in water depths in the range of 350–400 m

(1100–1300 ft) (Bell *et al.*, 2005). The term 'ultra-deep water' refers to wells drilled in water depths greater than 1500 m (4900 ft) (Leffler *et al.*, 2003). Typical regions where ultra-deep water operations are gaining momentum include the coast of Brazil, the GOM, West Africa, South China, and the North Sea (Bell *et al.*, 2005).

Shallow, deep and ultra-deep wells

Oil and gas wells are commonly classified based on their total vertical depth (TVD), which is broadly divided into three main categories: (i) shallow, (ii) deep, and (iii) ultra-deep (Fig. 15.1). Shallow wells represent the conventional oil and gas reserves that are now scarce. Shallow wells refer to wells less than 3000 m (10 000 ft) deep (Jellison *et al.*, 2008). Shallow wells can be either sweet or sour depending on the properties of the reservoir.

Deep wells enclose most of the newer fields in production today (Jellison *et al.*, 2008; Morsi, 1989). Deep wells refer to wells with depths ranging

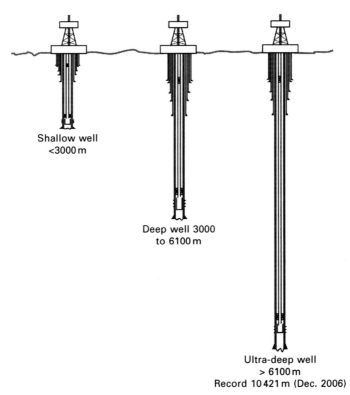

Shallow well
<3000 m

Deep well 3000
to 6100 m

Ultra-deep well
> 6100 m
Record 10 421 m (Dec. 2006)

15.1 Evolution of TVD from shallow wells to deep and ultra-deep reservoirs.

from 3000 to 6100 m (10 000–20 000 ft). As with shallow wells, deep wells can be either sweet or sour depending on the properties of the reservoir (Jellison *et al.*, 2008). High concentrations of CO_2 and H_2S are not uncommon (Watkins and Greer, 1976).

Ultra-deep wells are wells deeper than 6100 m (20 000 ft) (Jellison *et al.*, 2008). According to Jellison *et al.* the deepest well successfully drilled as of December 2006 had a TVD of 10 421 m (34 819 ft). Exploratory wells of more than 12 100 m (40 000 ft) have been reported. Moreover, reservoirs deeper than 15 000 m (about 50 000 ft) have been discovered in the Gulf of Mexico and Brazil (Jellison *et al.*, 2008). Ultra-deep wells are associated with high H_2S and CO_2 environments. Environments with more than 25% H_2S and 10% CO_2 have been encountered (Formigli, 2008; Hamby, 1975; Heacox and Stroup, 1976; Jellison *et al.*, 2008; Morsi, 1989; Watkins and Greer, 1976). Elemental sulfur is also likely to be present if the concentration of H_2S is higher than 5–10% in the gas phase (EFC, 2002b; Rhodes, 2001).

High pressure-high temperature and extreme high pressure-high temperature wells

Temperature and pressure increase dramatically with the depth of the well. Oil and gas fields with temperatures and pressures above 230°C (445°F) and 160 MPa (23 000 psi), respectively, have been reported (Jellison *et al.*, 2008; Shaughnessy *et al.*, 2003; SPE, 1975). Although there is not a universal definition, the term high pressure-high temperature (HPHT) is used by operators to indicate downhole temperatures in excess of 150°C (approximately 300°F) and surface shut-in pressures greater than 70 MPa (10 000 psi). Wells with downhole temperatures exceeding 220°C (425°F) and pressures above 103 MPa (15 000 psi) are referred to as ultra- or extreme-HPHT (xHPHT or X-HPHT) wells (Suryanarayana and Bjorkevoll, 2009).

In addition to temperature and pressure, a third parameter that is used to identify HPHT conditions is the so-called geothermal gradient (Suryanarayana and Bjorkevoll, 2009). In this regard, geothermal gradients above 0.046°C/m (0.014°F/ft) are a good indication of HPHT environments in which thermal effects merit serious consideration.

15.2.2   Drilling and completion

Exploration, drilling, completion, and production are the four main stages in conventional petroleum operations. Drilling is defined as the process in which a hole is made through the subsurface rocks to allow hydrocarbons to flow to the surface (Perrin, 1999). Drilling is considered the most expensive upstream operation and the one with the highest associated risk (Jellison *et al.*, 2008; Khan, 2007; Rocha *et al.*, 2003; Shaughnessy *et al.*,

2003). The purpose of drilling can vary depending on the well, which can be categorized as: (i) exploration wells, (ii) appraisal or confirmation wells, and (iii) development wells.

During the exploration stage seismic surveys provide information related to the likelihood of finding hydrocarbons based on the structural properties of the rock formation (Khan, 2007; Perrin, 1999). The presence of hydrocarbons is subsequently confirmed by exploratory drilling. The prime objective of this stage is to define the reservoir capacity, production rates and nature of the fluids (i.e. oil, gas, and/or water) in the reservoir and the main characteristics of the downhole environment (Khan, 2007; Perrin, 1999).

The objective of appraisal wells is to refine or complete the data obtained during exploration. Additionally, during this stage the off-wellbore properties are determined. The number of appraisal wells could be limited by safety concerns depending on the nature of the reservoir (Perrin, 1999).

The main objective of development wells is to bring the field into production. During production, three types of development wells can be drilled. The most important and numerous type of well is the so-called production well from where the hydrocarbons are extracted. Injection wells, in contrast, are less abundant but crucial to maintain reservoir pressure and to assure the flow of hydrocarbons to the surface during secondary recovery. Finally, observation wells are sometimes drilled to monitor the conditions of the reservoir during the production stage (Perrin, 1999).

Casing is necessary once the well grows deeper into the formation. Casing is usually required for depths greater than 1000 m. Casing is used to stabilize the well and to prevent the sides from collapsing into the well. In addition, casing protects fresh water containing formations from oil, gas and brine contamination and it prevents the production from being diluted due to waters originated in other formations in the well (Perrin, 1999). On top of the casing a blowout preventer is installed to shut down and regain control of the reservoir in the event of uncontrollable pressures (Perrin, 1999).

After a well has been drilled and the presence of hydrocarbons confirmed, the well needs to be completed. Well completion is defined as a single operation involving the installation of production casing and equipment in order to bring the well into production from one or more zones (Perrin, 1999). A typical well prognosis is shown in Fig. 15.2.

With the rare exception of *tubingless* completions, small diameter tubing is run into the well to just above the bottom to guide the oil/gas and water/brine to the surface (Greer and Holland, 1981, Hope, 2008, Watkins and Greer, 1976). Except for the case of open-hole completions, the tubing string isolates the casing from the hydrocarbons, preventing or reducing the risk of corrosion of the casing by the produced fluids. Likewise, the tubing string provides more effective pressure control and management. Because the casing is cemented in place, it is very difficult or impossible to repair. The tubing

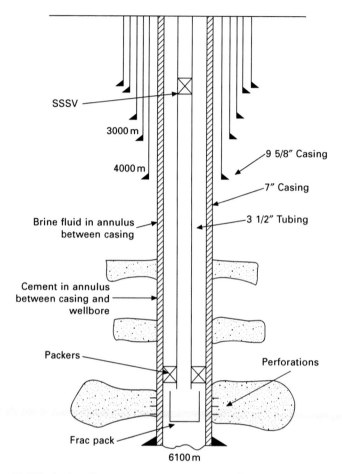

15.2 Typical well prognosis showing the different casing diameters and lengths as well as the main components of a well; adapted from Leffler *et al.* (2003).

string, in contrast, is suspended in the well and it can be retrieved from the well for inspection and repair (Khan, 2007).

A completion, annular, or packer fluid is commonly injected in the annular space between the casing and the production tubing to further protect against corrosion and prevent accidental inflow of brine and hydrocarbons through the annulus. Inhibited petrodiesel is a typical packer fluid for shallow wells and relatively low pressure systems. In contrast, high density brines are typically formulated for deep and ultra-deep wells (Khan, 2007).

An expandable device named production packer is used near the bottom of the tubing string to keep the tubing string central in the well and to prevent the produced fluids from flowing vertically (i.e. up) through the annular space between the tubing and the casing (Perrin, 1999).

Subsurface safety valves (SSSVs) are used in the tubing to stop the flow of hydrocarbons during an emergency (Leffler *et al.*, 2003). There are different types of valve designs, but the most common SSSV are held open by pressure and a drop in well pressure will automatically close the valve. In contrast, surface-controlled SSSVs are operated from the topside facilities via hydraulic lines that run down the wellbore (Perrin, 1999).

15.2.3   Production and surface equipment

Even though there are significant design differences between on-shore and submerged systems, the main functions of wellheads and surface equipment are similar (Leffler *et al.*, 2003). The wellhead is a permanent valve assembly at the top of the well. It is located at the ground level in on-shore wells and at the seabed in submerged systems. It usually consists of two main components: (i) the casing heads and (ii) the tubing head and hangers.

The lower casing heads seal the annular space between the casing strings and contain the casing hangers for the top of each casing string. Casing hangers are used to suspend the casing strings in place. There is a casing head and a casing hanger for each casing string (Perrin, 1999).

Wellhead or surface equipment is attached to the top of the tubing and casing. For both on- and offshore operations wellhead equipment includes: casing heads, tubing heads, wellhead valve assembly or 'Christmas trees' (industry jargon) for gas and naturally flowing oil wells or artificial lift pumps when the reservoir pressure is not enough to allow natural flow to the surface, stuffing boxes and pressure gauges (Leffler *et al.*, 2003; Perrin, 1999).

Christmas trees are essential surface components use to control the flow from the well when the hydrocarbons flow to the surface by their own pressure. Applications that use Christmas trees include all gas wells and, less commonly, oil wells in the early stages of field development (Khan, 2007). Wellhead valve assembly consist of a casing head housing, casing and tubing head spools, as well as production valves and chokes (Khan, 2007). Although Christmas tree design varies widely, they all have at least one master valve to regulate flow. A tree swab and a pressure gauge sit on top of the wellhead valve assembly. A flowline valve is usually located to the side of the tree. Submerged Christmas tress can be either wet or dry. Whereas wet trees are directly exposed to seawater, dry trees are encapsulated in an air-fitted chamber (Leffler *et al.*, 2003).

15.2.4   Common downhole environments

The environment found in deep and ultra-deep wells will affect corrosion resistance and materials performance. The main parameters affecting materials in service are:

- H_2S and CO_2 partial pressures (the gas phase),
- chloride concentration (the water phase),
- the presence of elemental sulfur,
- pH (determined by the amount of H_2S and CO_2 in the gas phase and the presence of buffers),
- temperature and pressure,
- the presence of non-production chemicals, and
- contaminants such as oxygen and liquid metals.

Since understanding the effects of such variables is critical for defining the performance in service, typical compositions of the gas and water phases in HPHT wells are discussed below. The implications for corrosion resistance will be discussed in a later section.

The gas phase

The gas phase in a HPHT reservoir is usually composed of a mixture of H_2S and CO_2. While a moderately sour gas could have between 20 and 100 ppm of H_2S, the CO_2 concentration of such system could be as high as 1–3% (NACE, 2000). The expression *extreme* H_2S environment indicates more than 1.5% H_2S in the gas phase (Morsi, 1989; NACE, 2000; Rhodes, 2001; Smith *et al.*, 1970). In addition to H_2S, extreme environments usually have concentrations of $CO_2(g)$ greater than 3% (NACE, 2000; Rhodes, 2001). Elemental sulfur can also be found if the concentration of H_2S is greater than 5–10% (EFC, 2002b; Rhodes, 2001; Smith *et al.*, 1970; Smith and Craig, 2005). Finally, traces of Hg and organic acids (mostly acetic acid, HAc) are not uncommon in the gas phase of deep and ultra-deep wells (EFC, 2002b).

It is important to note that while the concentration of H_2S and CO_2 in the gas phase can be higher than 1.5 and 3%, respectively; it is the amount of gas dissolved in the liquid that directly affects the corrosivity of the water phase. The concentration of dissolved H_2S and CO_2 in water is a fraction of their concentration in the gas phase. Solubility is a direct function of the temperature and the total pressure of the reservoir as well as the concentration of other species in the brine (Rhodes, 2001).

The water phase

The composition of the water present in a reservoir plays a critical role in determining corrosion behaviour. Regardless of the type of reservoir, the composition of the water changes significantly with the life of the well. While new wells usually have low concentration of dissolved solids, the concentration increases as the well ages and highly concentrated brines are often produced (EFC, 2002b). In addition, there are important differences in water composition between oil and gas reservoirs (EFC, 2002b).

Water in oil reservoirs originates from any geological formation adjacent to the well. The specific composition of the produced water is site-specific and since its nature varies between geological layers it may also change significantly during the life of a field (EFC, 2002b). Nevertheless, water extracted from oil fields tends to be highly concentrated with high chloride and bicarbonate levels, resulting in pH values ranging from 4.5 to more than 6 (Choi, 2009; Craig, 1993, 1997; EFC, 2002b).

In oil reservoirs, water can also be injected during secondary recovery to increase the pressure of the field and force the flow of hydrocarbons to the surface (Perrin, 1999). The composition of the injection water depends on the location of the field. Key requirements are availability and compatibility with the reservoir rock (EFC, 2002b; Perrin, 1999). Oxygen can be introduced to the well during injection of seawater for secondary recovery. In this regard, Mack and Carminati have reported oxygen concentrations in the 100 ppb range (Rebak, 2010). The addition of oxygen to the downhole environment can have deleterious consequences, in terms or uniform and localized corrosion, for tubular and other components if not designed to handle oxygen contamination.

Gas reservoirs are usually water-saturated when produced. The amount of water produced by a gas reservoir is controlled by the water-carrying capacity of the gas. In most cases, this is enough to produce the extensive wetting of the production tubing due to the formation of thin condensed water films (EFC, 2002b).

Water in gas reservoirs originates from condensation, a consequence of a decrease in temperature as the gas moves up to the surface, or due to water entrapped in geological layers adjacent to the well. Condensed water has a very low concentration of dissolved solids and, in the absence of dissolved bicarbonate and sulfide and the presence of H_2S and CO_2, it can achieve very acidic pH values (i.e. pH values below 3.5) (Asphahani *et al.*, 1989; Rhodes, 1996; Wilhelm and Kane, 1986). In addition to condensed water, *produced* water can be carried out to the surface. As the reservoir ages or as the gas is produced from deeper layers, the amount of produced water increases and eventually becomes predominant (Asphahani *et al.*, 1989; Rhodes, 1996; Wilhelm and Kane, 1986). The composition of produced water is usually similar to that found in oil fields (EFC, 2002b).

In practice, there are reservoir models that assess formation and production characteristics; however, there are no reliable means for predicting the expected composition of the produced or injected water for either an oil or a gas reservoir. Therefore, careful downhole and surface sampling during exploratory drilling operations is necessary.

NACE International has published a survey of end-user experience on CRAs used in sour oil and gas production in which materials and service conditions are listed alongside service experience (NACE, 2000). In this survey, a wide spectrum of chloride concentrations was reported, ranging from 2000 to 200 000 ppm. In one particular case, Exxon Mobil reported 400 000 ppm of total dissolved solids for their extremely corrosive deep Mobile Bay fields (NACE, 2000).

As explained in the EFC-17, once water reaches the surface, its final composition depends on various factors, including:

- equilibrium with the hydrocarbon phase at the surface temperatures and pressures
- mixing with water from other wells or fields
- precipitation of scaling ions
- additions and return of production chemicals
- oxygen intake from different sources.

There are key differences between the composition of the outlet water in *liquid* and *gas* streams for separators (EFC, 2002b). Water in the liquid stream will not change composition dramatically between the vessel inlet and outlet. In contrast, water in the gas stream is present as vapour which condenses as the streams cools down. In general, unless the water in the inlet vessel is buffered, the pH of condensed water in the gas stream could be very low (EFC, 2002b).

15.3   Environmentally assisted cracking (EAC) mechanisms common to oil and gas production

In this chapter the term environmentally assisted cracking is used to describe a diverse number of corrosion phenomena in which ductile materials (i.e., including metals, ceramic or polymers) show brittle failures during tensile loading at stress levels well below their ultimate stress when exposed to certain environments (Galvele, 1999).

Since there have been numerous ways to refer to similar hydrogen-induced corrosion mechanisms, in this chapter the terminology accepted by ISO 15156 and ASTM G193-10 will be adopted. According to ISO 1516, typical EAC failure modes in oil and gas drilling and production include sulfide stress cracking (SSC), hydrogen stress cracking (HSC), hydrogen induced cracking (HIC) (previously known as stepwise cracking or SWC), stress oriented hydrogen induced cracking (SOHIC), and stress corrosion cracking (SCC). Other types of EAC that occur under very special environmental conditions are caustic, carbon monoxide, and liquid metal embrittlement, which are not going to be discussed in this chapter (Craig, 1993).

In oil and gas drilling and production operations, the thermal and environmental gradients along the depth of a well introduce a gradient in EAC susceptibility and mechanisms. Figure 15.3 summarizes possible

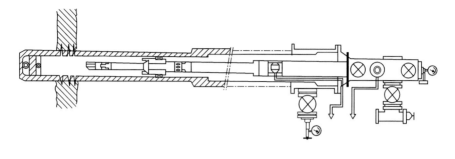

Temperature profile

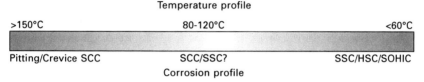

>150°C   80-120°C   <60°C

Pitting/Crevice SCC   SCC/SSC?   SSC/HSC/SOHIC

Corrosion profile

15.3 Temperature gradient and possible corrosion mechanisms along an ultra-deep well.

cracking modes along a sour HPHT reservoir. A brief description of the most common failure mechanisms typical of drilling and production environments is presented below.

15.3.1   Sulfide stress cracking (SSC)

SSC is a form of hydrogen embrittlement cracking involving embrittlement of the metal by atomic hydrogen produced by acid corrosion on the metal surface (Kermani, 2000; NACE, 2004b; Rhodes, 1996). High strength metallic materials and hard weld zones are, in general, more susceptible to this type of failure (NACE, 2004b).

In SSC, H_2S is the main source of hydrogen, which differentiates SSC from other hydrogen embrittlement mechanisms. The presence of H_2S and CO_2 in the gas phase also lowers the pH of the environment to values below the depassivation pH of the alloy, increasing the rate of proton discharge. Depassivation can also occur inside pits and crevices and, therefore, localized corrosion can be a precursor for SSC in some systems. Likewise, the sulfide present in the solution causes a greater amount of hydrogen to enter the metal, given that sulfur has been shown to act as a hydrogen recombination poison (Rhodes, 1996, 2001).

SSC is a cathodic process and it is accelerated by an applied cathodic polarization. As a result, SSC can be further exacerbated due to galvanic coupling to a less noble metal (EFC, 2002b; Rhodes, 1996, 2001). In SSC of carbon steel and martensitic CRAs brittle cracks form by solid-state reactions between the hydrogen atoms and the metal lattice (Rhodes, 2001). Nevertheless, the initiation of this process in CRAs can involve the initiation of local or general anodic corrosion by the presence of chlorides and low pH.

The SSC susceptibility of CRAs is a function of the stability of the passive film of the material and the pH and chloride content of the environment. The worst cases of SSC occur at lower temperatures close to the wellhead (e.g., between room temperature and 60°C). However, synergism between passivity, micro-creep, diffusion and hydrogen charging can lead to failures at higher temperatures (e.g., failures at 80–120°C have been reported for duplex stainless steels) (Fig. 15.3) (EFC, 2002b). Ferritic and martensitic microstructures are prone to SSC and industry and international standards, such as ISO 15156, DNV RP F101, and EFC-16 and 17, limit their maximum hardness depending upon the H_2S, pH and temperature of the environment (Craig, 1993,1997; Rhodes, 2001). Figure 15.4 illustrates the potential and pH regions where SSC can occur for carbon and low alloy stainless steels in typical production environments.

15.3.2   Hydrogen stress cracking (HSC)

Some authors make an explicit distinction between HSC and SSC although they are both particular modes of hydrogen embrittlement cracking. In this

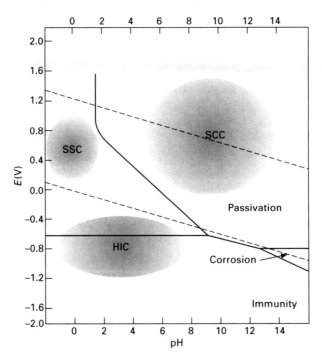

15.4 EAC modes as a function of pH and electrochemical potential for carbon steel.

regard, HSC is associated with external cathodic charging and, therefore, cathodically initiated in the absence of anodic dissolution induced by galvanic couple to more active materials (e.g., high strength UNS N10276 coupled to steel) (Kane *et al.*, 1977a, 1977b). SSC, in contrast, may involve cathodic initiation but it is always accompanied by corrosion (Rhodes, 2001).

15.3.3   Hydrogen blistering

Hydrogen blistering is a process strongly influenced by the cleanliness and impurity content of the alloy. Hydrogen blistering occurs where inclusions or voids are present in the metal. Atomic hydrogen can diffuse to the defects and recombine to give molecular hydrogen. Since molecular hydrogen cannot diffuse, the concentration and local pressure of hydrogen gas increases within the defects and it may be sufficiently high to cause local yielding, which ultimately leads to the formation of blisters (Kermani, 2000). Hydrogen blistering has been associated with lower strength steels (Kermani, 2000).

15.3.4   Hydrogen induced cracking (HIC)

HIC or SWC is a form of hydrogen embrittlement common to carbon steels that originates from the propagation and linking of small laminar cracks in a step-like manner resulting from the pressurization of trap sites by hydrogen. In contrast to SSC and HSC, no externally applied stress is needed for the formation of hydrogen-induced cracks. As the concentration of hydrogen increases, the strain of the internal hydrogen pressure inside laminar cracks increases causing the linkage of adjacent cracks in the through-thickness (i.e. transverse) direction between the individual planar cracks (Craig, 1993; Kermani, 2000). Figure 15.4 illustrates the potential and pH regions where HIC can occur for carbon and low alloy stainless steels in typical production environments.

15.3.5   Stress oriented hydrogen induced cracking (SOHIC)

SOHIC results from the combination of two different hydrogen-induced damage mechanisms: HIC and SSC (Hay, 2000; Kermani, 2000; Cayard and Kane, 1997). When the material is subject to stress, small laminar HIC cracks become lined up in the transverse direction and step cracks form between them. SOHIC is generally related, but not limited, to notches, defects and heat affected zones (HAZs) in welded components made of carbon steel (Hay, 2000, Kermani, 2000). It is important to highlight the occurrence of SOHIC given that alloys immune to either SSC or HIC can still suffer from this type of hydrogen embrittlement under certain types of environmental and stress/strain conditions (EFC, 2002b). This can further extend the region of susceptibility shown in Fig. 15.4 by combining the SSC and HIC zones.

15.3.6   Stress corrosion cracking (SCC)

Stress corrosion cracking became a significant challenge for the oil and gas industry in the 1950s during the development of the sour fields in western Canada (Craig, 1993). In contrast to SSC and other hydrogen embrittlement mechanisms, SCC is a form of localized corrosion involving dissolution due to mechanical or residual micro-creep depassivation, in other words, SCC is an anodic process that requires the presence of a passive film (Galvele, 1999; Kermani, 2000; NACE, 2004a, 2004b). SCC is controlled mainly by the stability of the passive film and it is, therefore, sensitive to the pH, temperature and halide anion content of the environment (Galvele, 1999). Figure 15.4 illustrates the potential and pH regions where SCC can occur in typical production environments, which is, for CRAs, usually much more extended than the region of SSC and HIC susceptibility.

Since SCC is an anodic process, the application of a cathodic potential can mitigate and prevent its occurrence (Rhodes, 2001). While SSC occurs at lower temperatures where hydrogen diffusion is minimized, SCC can occur at the highest service temperature (i.e. >80°C) (EFC, 2002b). Whereas ferritic and martensitic microstructures tend to be insensitive to SCC, austenitic microstructures are intrinsically susceptible to SCC possibly due to the high micro-creep associated with FCC lattices (EFC, 2002b). Martensitic CRAs can also suffer SCC when manufactured at high strength levels (i.e. above 690 MPa) and/or contain Ni (e.g., modified 13Cr and 15Cr alloys).

15.4   Materials for casing, tubing and other well components

In this chapter the term carbon steel (CS) is used to describe any alloy of carbon and iron containing up to 2% carbon, up to 1.65% manganese, and residual quantities of other elements, except those intentionally added in specific quantities for deoxidation (e.g., silicon and aluminium). Likewise, the term low alloy steel (LAS) refers to steels with a total alloying element content of less than approximately 5%, but above the amount specified for carbon steel. LAS are the most common alloys used in the oil and gas industry today.

There exists some contradiction in the definition of CRAs between ISO 15156 and API 6L standards. In this chapter, the term CRAs is used to describe alloys intended to be resistant to general and localized corrosion (i.e., pitting and crevice corrosion and EAC) in oilfield environments that are corrosive to carbon steels (ASM International, 2001a, 2001b; Craig,

1993; EFC, 2002b). Today CRAs are used for several downhole applications including seal assemblies in packers, SSSV components, valve components, seating nipples and tubulars when highly corrosive environments and high flow rates are encountered (Craig, 1993, 1997; Galis and Damman, 1991; Heisler and Mortiz, 1975; Hope, 2008; Mahmund and Chung, 1990; Schillmoller, 1988). CRAs are also popular in surface and subsea equipment such as valves, actuators and Christmas trees (Schillmoller, 1988, 1989), or where steels and inhibitors are not effective or not cost effective (e.g., high production rate, deep and offshore wells).

The objective of this section is to summarize the main families of CS, LAS and CRAs common in oil and gas drilling and production environments. The alloys to be considered include: CS and LAS, martensitic stainless steels (MSS), duplex and super duplex stainless steels (DSS and SDSS, respectively), nickel-based alloys (Ni alloys), titanium alloys (Ti alloys), and nickel-cobalt (Ni-Co) alloys. The following section is not intended to be a comprehensive discussion about carbon steels and CRAs but rather a summary of the most important properties with respect to mechanical and corrosion resistance for use in typical oil and gas production environments, with focus on sour HPHT conditions. For more information, other chapters in this book may be consulted. A list of common alloys and their nominal compositions is presented in Table 15.2.

15.4.1   Carbon and low alloy steel

The large variety of carbon and low alloy steels represent the most common choice for oil and gas production even today. Metallurgical and mechanical requirements for carbon steels for tubing and casing are covered by the American Petroleum Institute (API) Specification 5CT, which is equivalent to the ISO standard 11960 (API, 2001; ISO, 2004). The API 5CT/ISO 11960 regulations are divided into four main groups depending on the grade of the steel. Group 1 covers grades H, J and N; Group 2 includes all casing and tubing grades C, L and T; Group 3 is applicable to carbon steels grade P, and finally Group 4 includes carbon steels grade Q for casing applications (API, 2001; ISO, 2004).

Groups 1–3 are only limited by the impurity content, which gives manufacturers certain flexibility but can result in a large variation in mechanical and corrosion properties (Craig, 1993; Tuttle, 1997). Carbon-manganese steels are used for low strength requirements. High strength and improved corrosion performance can be achieved by micro alloying with Cr, Mo, Ti, B or V in various combinations (Craig, 1993; NACE, 2004a; Schillmoller, 1989, Tuttle, 1997).

The most common LAS grades for tubing and casing include ISO 11960 grades H40, J55, K55, M65, C-75, L-80, C-90, T-95 and C-95 (Wilhelm

Table 15.2 Nominal composition of common CRAs for oil and gas drilling and production

Type	Alloy name	UNS No.	Composition (wt%)											
			Fe	Ni	Cr	Mo	W	Cu	Co	Ti	C$_{max}$	V	Al	Other
Martensitic	410S	S41008	Bal.		13						0.08			
	M13Cr		Bal	4	13	1					0.02			
	S13Cr		Bal.	5	13	2					0.01			
	15Cr		Bal.	6	15	2					0.03			
Duplex SS	22Cr	S31803	Bal.	5.5	22	3					0.03			N:0.2
	25Cr		Bal.	6.5	25	3					0.03			N:0.2
Austenitic SS	316L	S31603	Bal.	12	17	2.5					0.03			
	27-7Mo	S31277	Bal.	27	22	7					0.02			0.35
Ni-base alloys	825	N08825	30	42	22	3		2			0.03			
	925	N09925	22	42	21.5	3		2		2.2	0.03			
	28	N08028	34	31	27	3		1						
	G-3	N06985	19.5	Bal.	22	7		2			0.015			
	C-276	N10276	5	Bal.	15.5	16	4				0.02			
Ti alloys	ASTM Grade 2	R50400	0.3 max							Bal.	0.1			O:0.25, N:0.03
	ASTM Grade 28	R56323	0.25 max							Bal.	0.08	2.0-3.0	2.5-3.5	Ru:0.08-0.14 O:0.15 max
Ni-Co alloys	MP35N	R300035	1.0	33-37	19-21	9-10.5			Bal.		0.025			Si:0.025

and Kane, 1986). Grades L-80, C-90 and T-95 are the only ones having a hardness limitation imposed by specification (Craig, 1993). Likewise, ISO 11960 grade Q-125, which belongs to API Spec 5CT/ISO 11960 Group 4, is the only alloy that requires impact testing and has a minimum limit of Charpy V-notch energy. For temperatures above 65°C only grades N-80 Type Q and C95 are recommended by ISO 15156-2 (NACE MR0175-2) for use in sour service; for temperatures greater than 80°C only grades N-80 and P-110 are suggested by the standard, and, finally, for temperatures above 105°C only the alloy Q-125 is acceptable (API, 2001; Craig, 1993; ISO, 2004).

There are numerous specialty steel grades not listed in API 5CT/ISO 11960. Typical examples of these grades include grades S-95, HC-95 and V-150. These steels are usually sought for their tensile pull capacity. However, their corrosion resistance, in particular their SSC resistance, is significantly lower than other steels conforming to API 5CT/ISO 11960 (Craig, 1993, Wilhelm and Kane, 1986). Ultra-high strength steels such as Grades 140 and 150 can fail by hydrogen cracking even in the absence of H_2S (Craig, 1993).

Microstructure is the most important parameter determining corrosion resistance at an equivalent strength and environmental conditions. LAS for tubular components have a martensitic structure that provides sufficient strength and allows for heat treatments (e.g. tempering) that can maintain strength levels at about 655–760 MPa (Craig, 1993). Cracking resistance is usually improved by good as-quenched microstructures and tempered martensite formed by relatively high temperature heat treatments (Craig, 1993; Wilhelm and Kane, 1986). In this regard, API requires 90% martensite in grades C-90 and T-95 (API, 2001; Craig, 1993; ISO, 2004; Wilhelm and Kane, 1986).

A general rule of thumb suggests that the higher the yield strength of the material the more susceptible to SSC it becomes, which is true for carbon steels. CA and LAS steels are not usually capable of sustaining environments with more than 0.34 kPa H_2S at 25°C with yield strength (YS) exceeding 760 MPa (Wilhelm and Kane, 1986). However, some exceptions have been reported recently indicating that uninhibited carbon steels could be used in certain conditions in sour environments in oil and gas production systems given the formation of a protective FeS protective layer (Johnson *et al.*, 2008).

According to ISO 15156-2 (NACE MR0175-2), high strength CS and LAS steels with less than 1% Ni (and not cold worked) are limited to HRC 22 in order to prevent SSC and other hydrogen embrittlement mechanisms (NACE, 2004a). Although the hardness limitation is necessary, it is, by no means, sufficient to ensure the suitability of the alloy for service. Typically, LAS with more than 1% Ni have been proven more susceptible to SCC even at hardness levels equal to or below HRC 22 (Craig, 1993; Tuttle, 1997; Wilhelm and Kane, 1986).

15.4.2   Martensitic stainless steels (MSS)

MSS are Fe-Cr-C-(Ni-Mo) alloys that have a body-centered tetragonal (bct) crystal structure (martensitic) in the hardened condition. They are ferromagnetic, hardenable by heat treatments, and generally resistant to corrosion only in relatively mild environments (ASM International, 2001b). Chromium content is in the range of 10.5 to 18% and carbon content can exceed 1.2%, although for materials commonly used for tubular goods and other oil and gas applications carbon content is normally 0.020% or lower. Small amounts of nickel can be added to improve corrosion resistance in some media and to improve toughness. The most common MSS alloys are shown in Table 15.2.

Martensitic stainless steels are specified when the application requires tensile strength, creep and fatigue strength properties, in combination with moderate corrosion resistance and heat resistance. In general, MSS are prone to SSC and SCC in several environments (Hay, 2000, 2001; Mack, 2008; Rhodes, 1996). Typical YS requirements for sour gas services are 550 MPa (80 ksi), with higher strength desirable but constrained as in the case of low alloy steels to minimize the risk of SSC (NACE, 2004b). One of the most common alloys used in oil and gas production is UNS4200 (or 13Cr-80) with YS in the range of 550 to 565 MPa (80–96 ksi) (NACE, 2004b).

Another common family of MSS alloys very popular in oil and gas production is the so-called modified 13Cr (M13Cr). The modified 13Cr family is a group of MSS alloys with very low carbon content and modified by the addition of Ni and Mo to improve corrosion resistance and mechanical properties. Modified 13Cr materials are not generic and their performance depends on their composition, which differs from supplier to supplier. The most common of this family is the M13Cr-95 alloy with a YS of 565 MPa (95 ksi) (NACE, 2000; Rhodes, 2001). Other alloys such as the super or hyper M13Cr-110 are also becoming more common in downhole applications (Kimura et al., 2005; Rhodes, 2001; Turnbull and Griffiths, 2003). Even though the ISO 15156-3 (MR0175-3) standard does not list all the special or custom MSSs, some of these alloys have been used successfully in slightly sour service (up to 10 kPa H_2S) with no evidence of cracking or corrosion (NACE, 2004b).

15.4.3   Duplex and super-duplex stainless steels

Duplex and super-duplex stainless steels (DSS and SDSS, respectively) are two-phase alloys based on the Fe-Cr-Ni system having a Cr content in the 22–25% range (SDSS having 25% or more Cr) with additions of Mo, N and W to tailor the desire microstructure and corrosion and mechanical properties (ASM International, 2001a). The microstructure of duplex and super-duplex steels is composed of austenitic (γ) and ferritic phases (α), both having more than 12 wt% Cr. In this regard, it is usually recommended that the optimal α/γ ratio should be close to 50% so that the amount of ferrite equals the amount of austenite (ASM International, 2001a; Tuck *et al.*, 1994). By combining the advantages of ferritic and austenitic stainless steels, DSSs have good SCC resistance (due to the ferrite) but some susceptibility to SSC and HSC due to the ferrite, adequate toughness, ductility and good general and localized corrosion resistance (due to the austenite and ferrite depending on the environment) (Tuck *et al.*, 1994).

DSS and SDSS are characterized by their high tensile yield strengths obtained by heat treatment and enhanced further by cold working during metallurgical processing, offering designers the possibility of using a thin-wall material (which can lead to major reductions in weight) with adequate pressure containing and load-bearing capacity. However, the high alloy content and the presence of a ferritic matrix render duplex stainless steels susceptible to embrittlement and loss of mechanical strength, particularly toughness, through prolonged exposure to elevated temperatures (Tuck *et al.*, 1994). For this reason, the upper temperature of application is generally less than 300°C. YSs are normally greater than 450 MPa (65 ksi) in the solution-annealed condition (SA) but YS can surpass 1110 MPa (160 ksi) when cold-worked (CW). Typical alloys used in the oil and gas industry include: S39274, S31803, S32550, S2205, S2507 and U-50M (i.e., a cast product with 25–45% ferrite) (Table 15.2) (NACE, 2000, 2004b,).

15.4.4   Precipitation-hardenable stainless steels

Most CRAs can be strengthened only by cold work, which limits the shape and dimensions of components. Precipitation hardenable alloys are used to obtain a uniform distribution of mechanical properties in complex and/ or thick samples. Precipitation hardenable alloys are strengthened by heat treatment. Three main types of precipitation-hardenable stainless steels are commonly utilized in oilfield applications: (i) martensitic, (ii) semi-austenitic, and (iii) austenitic. Typical alloys include J92180, S13800, S15700 and S45500. These alloys are designated by their microstructural response to cooling from an austenizing heat treatment (NACE, 2000). As in aluminum alloys, the precipitation-hardening response is created by the addition of elements with limited solubility in the matrix such as Cu, Al, Ti and Mo (NACE, 2000). Precipitation hardening increases strength in addition to any martensitic transformation.

A broad range of mechanical properties can be produced by different combinations of heat treatments. Room temperature YS can range from 280 to 1240 MPa (40–180 ksi), depending on composition and mechanical treatments. Strength levels are usually limited to 830 MPa (120 ksi) due to

SSC (low temperature) and SCC (high temperature due to Ni content and often low Mo) considerations which severely limits their use particularly for components involving high tensile stress. Nevertheless, some high strength components for specialized applications have been used in sulfide-containing environments (NACE, 2000).

15.4.5   Ni-based alloys

The most common Ni-based alloys for oil and gas applications are austenitic Fe-Cr-Ni-Mo alloys where Ni is the main alloying element. Today, the UNS designation system is commonly implemented to distinguish between Ni alloys and austenitic stainless steels. Alloy numbers starting with N belong to the Ni-based family whereas alloys starting with S indicate an austenitic stainless steel (NACE, 2004b; Rebak, 2010; Rhodes, 2001). Typical applications for Ni-based alloys include rods, production tubing, valves, landing nipples and tool joints.

Ni-based alloys with sufficient alloy addition of Mo and Cr are highly resistant to corrosion and environmentally assisted cracking in severe H_2S-CO_2 environments at high temperatures and chloride contents. As with steels, they can be broadly classified as solid solution and precipitation-hardenable alloys. Solid solution Ni-alloys such as N06022 (alloy C-22) and N10276 (alloy C-276) are part of the C family of Ni-based alloys (Ni-Cr-Mo), which are generally used in the solution annealed or the annealed and cold worked condition (NACE, 2004b; Rebak, 2010; Rhodes, 2001). Precipitation-hardenable Ni-alloys, such as N06625 or N09925, contain niobium and are generally used as solution-annealed, solution-annealed plus aging, hot-worked and aged, or cold-work and aged condition (NACE, 2004b; Rhodes, 2001). In all cases, CW is necessary to achieve the highest strengths. Table 15.2 summarizes typical Ni-based alloys common to oil and gas production. Newer alloys, such as Alloy 945 and C-22HS, have been introduced to the market recently and promise excellent resistance to localized corrosion and EAC (Rebak, 2010).

The room temperature YS of solid-solution Ni-based alloys can range from 210 to 1380 MPa (30–200 ksi), depending on composition and the degree of cold working. Mechanical properties of precipitation-hardenable Ni-based alloys can exceed 1400 MPa (200 ksi) (NACE, 2000, 2004b).

Most of the limitations of Ni-based alloys are related mainly to cost, extended lead times and manufacturing challenges. High-Ni alloys are more difficult to cast and prone to cast defects such as hot tears, cracking, porosity and gassing. Long aging of heavily cold worked alloys such as UNS N10276 and UNS N06625 at approximately 315°C can lead to lower hydrogen embrittlement (HE) resistance (Rhodes, 2001 and Kane *et al.*, 1977a). Likewise, heat treatments of precipitation-hardenable Ni-alloys from about 260 to 650°C can also lead to a significant decrease in HE resistance (Rhodes, 2001).

15.4.6   Titanium alloys

Titanium alloys for oil and gas production combine a very good resistance to localized corrosion with lightweight and excellent mechanical properties (Table 15.2), (Ueda *et al.*, 1993). The mechanical properties of Ti alloys vary considerably between alpha, alpha/beta, and beta alloys. Alpha alloys have the lowest YS, which varies between 170 and 480 MPa (25–75 ksi). Alpha/beta alloys have YSs ranging from 480 MPa (75 ksi) to more than 1200 MPa (175 ksi) depending upon heat treatment and the degree of cold work. Beta alloys have minimum YSs in the range of 790 MPa (115 ksi) but can attain more than 1400 MPa (200 ksi). Current recommended practices for oil country tubular goods (OCTG) limit the YS for sour services to 1400 MPa (NACE, 2000). However, no clear explanation justifying such limitation is given in the standard.

Ti alloys have a number of manufacturing and operational challenges including elevated cost, weldability, low fatigue strength and hydrogen embrittlement issues. The cost of Ti drilling pipe has been calculated to be 17 times higher than conventional high strength steels, which limits the implementation of Ti alloys in the drilling phase (Jellison *et al.*, 2008). Weldability is limited and welding requires the used of an inert environment or an inert cover gas. The presence of even small amounts of alpha phase in beta Ti alloys greatly reduces its resistance to HSC; which impedes the use of fully alpha Ti alloys from oil and gas service. Ti alloys are also very susceptible to hydrofluoric acid and fluorides (F^- attacks the Ti passive film), concentrated HCl, and anhydrous methanol (EFC, 2002b) therefore must not be exposed to commonly used downhole stimulation acids with high HCl or with HF. Hydrogen uptake can form brittle hydride phase, which kinetics can be increased by: coupling to less corrosion resistant alloys cathodic protection, tensile loading or residual stresses, pH < 2 or >12, and H_2S under cathodic polarization (Ueda *et al.*, 1993). Finally, similarly to some stainless steels and nickel alloys, titanium alloys can suffer SCC in the presence of trace amounts of Hg, which is induced by an amalgam formation due to the liquid metal (EFC, 2002b).

15.4.7   Ni-Co alloys

The Ni-Co family of alloys contains at least 35 wt% Co with Ni and Cr percentages that can vary from as little as 1.5 wt% to as much as 37 wt% (NACE, 2000). In addition to Co, Ni, and Cr, Mo can be added in concentrations ranging from 0.5 to 11 wt%.

The most popular alloy is UNS R30035 (MP35N), which is produced by double vacuum melting and contains 35 wt% Ni, 35 wt% Co, 20 wt% Cr, and 10 wt% Mo (Heacox and Stroup, 1976). UNS R30035 combines ultra-high strength with good ductility and excellent corrosion resistance. UNS R30035 has an austenitic face centered cubic (fcc) structure at high temperatures, which is retained even at room temperature given the high activation energy for the fcc to hexagonal close packed (hcp) transformation. Cold work can induce the formation of coherent hcp platelets in the metastable fcc matrix. This coherency strengthens the alloy by straining the lattice. The amount of strengthening depends strongly on the number of hcp platelets and, to a lesser extent, on the temperature of the fcc to hcp transformation. Further strengthening can be achieved by aging, which results in the precipitation of a coherent Co_3Mo compound at the interface between the fcc matrix and the hcp platelets (Heacox and Stroup, 1976).

Ni-Co alloys are generally used in the solution annealed (SA), SA and aged, hot-worked (HW) and aged, or CW and aged condition. The properties of UNS R30035 can be controlled over a wide range by adjusting the amount of cold work. In the SA condition, a minimum YS of 827 MPa (120 ksi) is usually reported. YS as high as 2260 MPa (328 ksi) can be achieved by a 70% cold work reduction. Pipe and tubing components are usually manufactured with a minimum YS of 1516 MPa (220 ksi) (Heacox and Stroup, 1976; NACE, 2000).

UNS R30035 offers excellent resistance to general and localized corrosion as well as outstanding performance in H_2S environments (Heacox and Stroup, 1976). As such, UNS R30035 is generally used for applications requiring ultra-high strength, good ductility and superior corrosion resistance. Currently, UNS R30035 is used as armour material for logging cables even at extreme concentrations of H_2S. Other applications include underwater structures, oceanographic cables, springs, fasteners, bearing components, as well as biomedical applications (Heacox and Stroup, 1976).

The use of UNS R30035 as OCTG has been proposed but it has not gained mainstream attention thus far given that it is not economic or available in large quantities required for OCTG applications. In this regard, the main limitation of Ni-Co alloys is cost. Likewise, UNS R30035 can become susceptible to hydrogen embrittlement by prolonged aging of high strength parts at temperatures between 260 and 650°C.

15.5   Corrosivity of sour high pressure/high temperature (HPHT) reservoirs

In downhole environments in the absence of oxygen contamination and elemental sulfur (S^0) the only cathodic reaction is the electrochemical reduction of protons to hydrogen, H^0, which may then recombine and form $H_2(g)$ or dissolve into the metal as atomic hydrogen (Rhodes, 2001). The adsorption of H_2S at the surface is widely recognized as a promoter for H absorption into the lattice in Fe-based alloys (EFC, 2002a, 2002b; Rhodes, 2001). The kinetics of the cathodic reaction depends on the electrochemical potential (E), temperature, pH, and the concentration of weak acids that could donate protons, such as H_2S and H_2CO_3 (formed due to the presence of CO_2), as well as acetic acid (HAc) (Rhodes, 2001).

The anodic reactions depend on the type of alloy but they can be summarized based on the dissolution of the main alloying elements: Ni, Fe and Cr. In this regard, Ni can react with H_2S to form NiS, Fe can form FeS precipitates, and Cr can react with H_2O to form Cr_2O_3 (Rhodes, 2001; Suryanarayana and Bjorkevoll, 2009).

In the absence of S^0 and O_2, solution corrosivity increases with incremental concentrations of H_2S, H^+ and Cl^- as well as temperature (Rhodes, 2001). Both Cl^- and H_2S promote general corrosion and the synergistic effect between H_2S and Cl^- (and S^0 if available) enhances localized corrosion. Loss of passivity can also occur under cathodic conditions if the pH of the environment is sufficiently acidic so that the solution pH is below the depassivation pH of the alloy (Rhodes, 2001).

At constant temperature, SCC is promoted by a cathodic shift in the pitting and crevice potentials (E_P and E_{CREV}, respectively) and SSC by rapid dissolution kinetics. In this regard, there is a critical pH value, pHd, where the open circuit potential (OCP) shifts from the active to passive state, which represents a boundary for SCC and SSC modes.

For downhole conditions, the initiation of SCC at anodic potentials is generally observed to be associated with pits and crevices. Thus, localized corrosion acts as a precursor for cracking in some CRAs in most cases. In this regard, Cragnolino *et al.* (2001) have shown that for stainless steel 316L (UNS S31603) in aggressive chloride environments at elevated temperature, SCC only took place below the re-passivation potential (E_{REP}), suggesting that localized corrosion is a prerequisite for SCC. The authors also suggested that E_{REP} could be used to monitor the risk of SCC (Cragnolino *et al.*, 2001). In this regard, a decrease in the difference between OCP and E_P or E_{CREV} is usually a clear indication of an increased susceptibility to SCC (Szklarska-Smialowska, 2005).

The role of temperature requires a separate mention. While an increase in temperature increases the rate of uniform dissolution and the risk of SCC, the susceptibility to SSC and other HEC phenomena decrease with an increase in solution temperature (however, in some environments there is evidence that SCC of MSS can be more severe at intermediate temperatures) (Cayard and Kane, 1998). For MSS and possibly for DSS, this seems to be related to a decreasing contribution of absorbed H at elevated temperatures (EFC, 2002b; Rhodes, 2001). In general, for highly alloyed steels having more than 22% Cr, increasing temperature increases the susceptibility to pitting and crevice corrosion resulting in a lower resistance to SCC.

15.5.1   The role of elemental sulfur (S^0)

Elemental sulfur is an oxidant that is frequently dissolved in high pressure gas with more than 5–10% H_2S (EFC, 2002b; Rhodes, 2001; Smith *et al.*, 1970; Smith and Craig, 2005). Nevertheless, S^0 has also been found in gas streams having less than 5% H_2S (NACE, 2000). The total concentration of dissolved sulfur varies greatly from reservoir to reservoir. In many H_2S-containing gas wells the concentration of S^0 usually ranges between 50 and 100 ppm, which is well below the solubility limit. The solubility of S^0 in the gas phase depends on several environmental factors, increasing with incremental concentrations of H_2S, total pressure and temperature, and decreasing in the presence of a liquid hydrocarbon phase (hydrocarbons are effective solvents for S^0) (EFC, 2002b; NACE, 2000, 2004b; Smith and Craig, 2005). Reservoirs saturated in S^0 have also been encountered (NACE, 2000).

As in the case of oxygen, the presence of S^0 has been shown to have a marked negative effect on the corrosion performance of low and high grade CRAs (NACE, 2000; Smith *et al.*, 1970; Smith and Craig, 2005). S^0 has been shown to reduce E_P, E_{CREV}, and E_{REP} of most CRAs, suggesting an increased susceptibility to localized corrosion. Given that for stainless steels and higher grade CRAs stress corrosion cracking appears to be directly related to or initiated at localized corrosion sites (i.e., pits and crevices) the presence of S^0 has a direct negative impact on SCC resistance as well (Rhodes, 2001). In this regard, a survey conducted by NACE shows that the presence of even small amounts of S^0 (actual concentrations were not always reported) systematically resulted in SCC of CRAs in service (NACE, 2000). The presence of S^0 also has a marked effect on general corrosion rates as discussed in EFC-17 (EFC, 2002b).

15.6   Environmentally assisted cracking (EAC) performance of typical alloys for tubing and casing

An overview of the SCC and SSC performance of MSS, DSS and Ni-based alloys in sour HPHT oilfield environments is presented in this section.

15.6.1   MSS

Under oil and gas field exposure conditions, most MSS are normally passive. Passivity is usually present in low chloride, moderate pH regions (i.e. pH values above 3.50), as well as highly concentrated brines if they have higher pH values as a consequence of HCO_3^- buffering effects (Turnbull and Griffiths, 2003). SCC is more likely to occur due to cracking at pits and crevices than because of general corrosion. Likewise, SCC is considered H-assisted due to the formation of H_2S inside pits (Rhodes, 2001; Turnbull and Griffiths, 2003; Turnbull and Saenz de Santa Maria, 1989; Tuttle, 1997).

Under laboratory conditions, however, MSS have been shown to be prone to SSC and SCC under a variety of conditions. Most testing in H_2S solutions has been conducted at room temperature in severe solutions (i.e., high H_2S 1 atm and low pH 2.70) where an increased corrosion and SSC is expected (Rhodes, 1996, 2001; Schillmoller, 1989; Turnbull and Griffiths, 2003; Turnbull and Saenz de Santa Maria, 1989; Tuttle, 1997). At room temperature, the pHd (i.e. the pH that determines the transition between active dissolution and passivity) is the critical parameter used to determine the boundary between SSC and SCC (Rhodes, 1996, 2001; Schillmoller, 1989; Turnbull and Griffiths, 2003; Turnbull and Saenz de Santa Maria, 1989; Tuttle, 1997).

To differentiate between SSC and SCC, susceptibility domains are commonly used. Susceptibility domains are defined by adding electrochemical data such as OCP, E_P and E_{REP} values to E vs pH diagrams (Rhodes, 2001) (Fig. 15.5). The domain of active corrosion shown on such plots can be used to separate SSC and SCC regions based on pH values. At pH values between 3 and 4, SSC is possible if the OCP drifts into the active region. However, SSC can

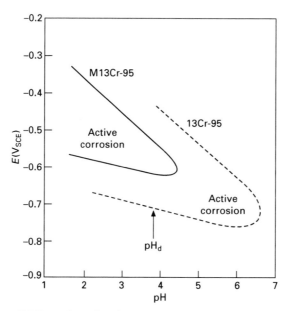

15.5 Domains of active dissolution and localized corrosion for 13Cr and modified 13Cr MSS indicating the occurrence of SSC or SCC, respectively; adapted from Rhodes (2001).

still occur even with a passive film present or when active dissolution occurs in isolated areas of the surface (Rhodes, 2001). In all cases presented by Hara and Asahi, SCC occurred at pH values greater than 4.9 under passive conditions (Hara and Asahi, 2000).

There is agreement that an increase in the partial pressure of H_2S (P_{H2S}), chloride content and temperature decreases the SCC resistance of MSS. In this regard, 13Cr MSS can be exposed to temperatures up to 140°C if the chloride content of the environment is kept below 25 000 ppm with pH_2S lower than 10 kPa (1.5 ksi) at pH values above 3.50. The maximum service temperature drops to less than 50°C if the chloride content is 200 000 ppm (Galis and Damman, 1991). Limited information, however, is available regarding the effects of other variables such as galvanic coupling, the effects of crevices, the presence of S^0 and oxygen contamination (Rhodes, 1996; Rhodes *et al.*, 2007).

For MSS there is also a compromise between YS and the maximum allowable P_{H2S}, which is a function of alloy composition. Keeping P_{H2S} constant, SCC resistance decreases with increasing yield strength (YS). Likewise, at a fixed stress level, the critical P_{H2S} value decreases with increasing YS (Rhodes *et al.*, 2007; Schillmoller, 1989).

15.6.2   DSS

While DSS are resistant to SCC in H_2S and O_2-free boiling NaCl solutions, SCC was found in de-aerated boiling $CaCl_2$ and MgCl solutions (Rhodes, 2001; Rhodes *et al.*, 2007). Interestingly, no relationship between H_2S SCC susceptibility and mechanical strength has been found. The morphology of the SCC attack is usually transgranular (TG) across both phases and intergranular (IG) along the austenite/ferrite interphase.

In sour environments, three modes of failure are encountered: (i) SSC, (ii) low temperature SCC, and (iii) high temperature SCC (Rhodes, 2001; Rhodes *et al.*, 2007). SSC is mostly related to uniform dissolution and it is therefore similar to SSC of MSS. The low temperature SCC mode has a maximum susceptibility at 80°C. The high temperature SCC mode manifests at temperatures above 120°C, leading to a maximum susceptibly to EAC in this intermediate temperature range. The low temperature mode is more sensitive to the H_2S concentration and it ends up defining the suitability of the alloy for service. Most tests are recommended for the low temperature mode (Rhodes, 2001; Rhodes *et al.*, 2007).

According to Rhodes, both SSC and low temperature SCC are considered H-enhanced as in the case of SSC/SCC described for MSS. The main difference in mechanisms is related to the crack morphology. While selective corrosion of austenite occurs during SSC in NACE solutions, selective dissolution of the ferrite occurs during SCC at both low and high temperatures (Rhodes, 2001; Rhodes *et al.*, 2007).

As in the previous case, domain diagrams can be used to determine the susceptibility of DSS to SSC and SCC. In this regard, E vs pH diagrams are similar to those of MSS but with a much-extended passive region (Fig. 15.6). Under normal exposure conditions, DSS and SDSS have been reported immune to SSC at room temperature up to pH 2.7. However, if a cathodic shift was introduced loss of passivity and SSC have been reported for 22Cr DSS. SCC on statically loaded specimens occurred only if localized corrosion was present. At pH 4, a potential range for maximum SCC susceptibility between –0.38 and –0.51 V vs SCE was identified under very aggressive testing conditions (i.e., 22Cr DSS samples were pre-attacked potentiostatically) (Rhodes, 2001; Rhodes *et al.*, 2007; Roberge, 2000; Tuck *et al.*, 1994).

Finally, DSS and SDSS can suffer from hydrogen embrittlement in the form of HSC caused by cathodic protection in subsea systems (e.g., manifolds, connections, valve bodies and bonnets, trees, piping, and other structural components). In this regard, HSC can be minimized in service by following the recommendation of the DNV-RP-F112 during the design stage of the project. However, in its current stage DNV-RP-F112 appears to be too conservative in terms of allowed austenitic spacing and strain criteria for welded components in the region where residual stresses and strain needs to be considered (L_{RES}) (DNV, 2008).

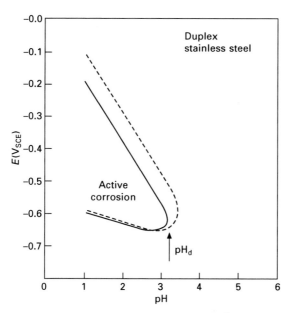

15.6 Domains of active dissolution and localized corrosion for two DSSs indicating the occurrence of SSC or SCC respectively; adapted from Rhodes (2001).

15.6.3 Austenitic stainless steels and Ni-based alloys

Most Ni-based alloys are used in wells containing more than 10% H_2S at temperatures that can exceed 200°C with expected service lives of more than 20 years. CW products, in the form of OCTG or production tubing, have been the primary application of Ni-based alloys for the oil and gas industry due to strength requirements in HPHT applications. The ISO 15156-3 (MR0175-3) standard lists more than 40 commercial compositions including austenitic and Ni-based alloys. Ni-based alloys with more than 40% Ni are highly resistant to the classical chloride-induced SCC mechanisms (NACE, 2000, 2004b) but can be susceptible (Kane *et al.*, 1979). Alloy C-276 (UNS N10276) is the most common Ni-based system and UNS S31600 the most popular austenitic stainless steel for HPHT applications (Tuttle, 1997; Ueda and Kudo, 1991; Wilhelm and Kane, 1986).

Limited data exist on SCC resistance on H_2S/CO_2 + chloride-containing environments, especially at lower temperatures (i.e. 80–180°C). The main limitation for corrosion testing is usually the extended induction time for both pitting and SCC initiation what makes tests in simulated conditions impractical. As a consequence, the majority of the work reported in the literature uses extremely aggressive environments, which in many cases could be far more corrosive than those expected in current service conditions

SCC mechanisms of Ni-based alloys and most austenitic stainless steels are similar to the classical SCC studies of stainless alloys in boiling $MgCl_2$. A synergism between H_2S and Cl^- extends the domain of SCC to lower temperatures, chloride concentrations and more cathodic potentials (Rhodes, 2001). For SA and CW Ni-based alloys, a transgranular path is usually found (but not exclusively), whereas an intergranular SCC mode has also been observed on precipitation-hardenable Ni-based systems with σ-phase precipitation along grain boundaries. The presence of σ-phase is related to overheating or prolonged well aging. In all cases SCC nucleates at pits or crevices and it is not observed when coupled to lower grade alloys and carbon steel (Rhodes, 2001).

An increase in temperature lowers the SCC resistance of the high-grade alloys (i.e. alloys having more than 30 wt% Ni). Nevertheless, a low temperature mode similar to the one described for DSS is suspected for lower Ni content alloys and UNS S31600 with a maximum susceptibility around 80°C. H_2S is expected to accelerate anodic dissolution rates in pits and cracks, promoting acidification and H adsorption (Aberle, 2008; Rebak, 2010; Rhodes, 2001; Rhodes *et al.*, 2007). Finally, solution corrosivity dramatically increases with additions of S^0, oxygen, and acetic acid, which reduces the resistance to pitting corrosion and, hence, increases SCC susceptibility (EFC, 2002b; NACE, 2000; Rhodes, 2001).

15.7   Qualification of materials for oil- and gas-field applications

Premature failures related to sour gas operations caused concerns for the oil and gas industry in the early 1950s. As a result, NACE International organized its first symposium on SSC in 1952 (Craig, 1993). It was not until 1966 that NACE published its first work specific to materials selection in sour systems (NACE Standard 1F166). In 1975 this work provided the basis for the Materials Recommendation Standard (MR) 01-75 (Craig, 1993). The standard was fully updated in 1976 as a consequence of a serious accident in Texas that lead legislators to require nearly all oil and gas equipment and operations to conform to NACE MR 0175. The ISO committee approved the NACE standard in 2003, obtaining the status of international standard and becoming ISO-15156 (MR0175) (NACE, 2003). Since then NACE International, currently via the Task Group (TG) 299 working in conjunction with the ISO Maintenance Panel, reviews and updates the standard systematically.

In addition to NACE, the European Federation of Corrosion publishes two key documents, the EFC Publications 16 and 17 that complement ISO 15156 and are also widely used by oil and gas companies worldwide to qualify materials for oil and gas production environments. The EFC-16 and EFC-17 are also updated periodically by the EFC via the Working Party 13 (EFC, 2002a, 2002b).

Today, the qualification of materials for oil and gas production is based on a combination of standard tests, fitness for purpose evaluations, and, perhaps more importantly, experience. For example, the boundaries or limits of application are typically presented in maps such as that shown in Fig. 15.7. Craig has also summarized a number of alloy selection diagrams that can be used to estimate the application limits of a number of CRAs in service conditions including 13Cr MSS, several duplex steels and Ni-based alloys (Craig, 1997).

In addition to these maps, the ISO15156-3 (NACE MR0175-3) standard lists a number of environments where different CRAs could be used in service without further testing. The standard also includes a procedure for laboratories and oil and gas companies to request the NACE/ISO committee to include new environment/material combinations in future revisions of the standard. However, it is important to highlight that qualification to ISO 15156 does not imply suitability for sour service or immunity to corrosion (NACE, 2003). It only indicates that within the guidelines of the standard no SSC should occur. Therefore, the ISO 15156 standard cannot be used to make any prediction about corrosion kinetics and lifetime estimations.

Although the ISO 15156 standard has been revised systematically over the last three decades, several limitations are cited in the literature, restricting its use in certain circumstances. The main areas of disagreement can be

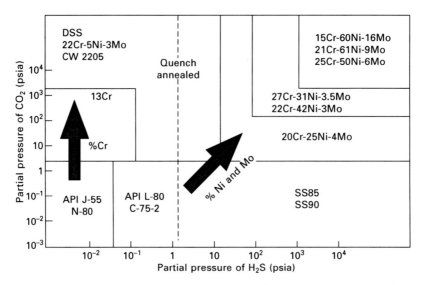

15.7 Typical materials selection chart for low alloy steels and CRAs; adapted from Sumitomo Metals.

summarized as follows (Kermani, 2000, Mack, 2008, Smith and Craig, 2005):

- The environments included in the standard are usually too conservative and most of the forecasted environments lie well outside this range.
- There is reported evidence that materials listed as acceptable in the ISO 15156 standard failed at less than 100% YS when examined in laboratory conditions (Mack, 2008).
- The standard is limited to low temperature (i.e. below 66°C) SSC.
- Problems associated with high temperature SCC are not addressed by the standard.

For these reasons, testing and fitness-for-purpose investigations are considered mandatory for most HPHT sour environments for which limited experience exists (Hay, 2000, 2001; Mack, 2008). The most common testing methodologies for qualifying CRAs for severe downhole environments are summarized below.

15.7.1   Statically and quasi-statically loaded tests

U-bend tests

U-bend testing as described by the ASTM G30 standard is usually an economic screening test that induces high degrees of plastic deformation. The main caveat of U-bend testing is stress relaxation during prolonged exposure at high temperatures (EFC, 2002a, 2002b; Hay, 2000; Hay, 2001; Mack, 2008). MSSs, DSSs and some precipitation-hardened SS are more susceptible to stress relaxation than other CRAs (Hay, 2000, 2001; Mack, 2008). The EFC-17 recommends that U-bend testing should not be considered as the only screening test for materials selection and evaluation (EFC, 2002b).

C-ring tests

C-rings are statically loaded specimens that are generally used for the evaluation of tubular products and sometimes forgings. They are loaded to cause the outer surface to be in tension, although loading of the internal surface is possible and it is becoming more frequent (NACE, 2005). The NACE TM-0177 Method C, ISO 7539-5 and ASTM G38 standards are the most frequently implemented testing procedures. It is important to remark that the equations for stressing the C-ring coupons given in NACE Test Method TM-0177 do not compensate for the strong dependence of the non-linear portion of the σ-ε plot above the material yield point and effects of temperature. Therefore, deflections applied at room temperature before testing may not reproduce the same strain in the specimen when heated to the testing temperature (NACE, 2005). To account for this effect, Mack recently presented a number of plots summarizing the temperature dependence of the Young's modulus of common CRAs as well as the changes in YS with temperature (Mack, 2008). Another approach is to measure the actual stress–strain behavior using strain gauges during heating. Finally, as with U-bend specimens, stress relaxation can occur during C-ring testing even at room temperature. DSSs have also been shown to be more susceptible to this effect.

Standard tensile testing

The procedures for standard tensile testing are described in NACE TM0177 Method A Section 8. The methodology is used to evaluate the EAC resistance of carbon steels and CRAs under uniaxial tensile loading. The method provides a well-defined stress state and susceptibility is determined by measuring the time to failure of the test coupon. Stress thresholds can, in principle, be determined when testing a large number of samples under various loads. The test environments proposed by NACE TM0177 are: (i) 5.0 wt% NaCl + 0.5 wt% glacial acetic acid (CH_3COOH) at pH about 2.70 (known as NACE Solution A), (ii) 5.0 wt% NaCl + 2.5% CH_3COOH + 2.5 wt% sodium acetate at a pH of around 3.50 to 3.60 (known as NACE Test Solution B), or (iii) 5 mM sodium acetate with a chloride, H_2S partial pressure, and pH specified by the user or purchaser. Loading can be applied by a constant load (dead-weight) device or a proof ring. The degree of loading depends on the type of environment. Typical stress levels are 80% YS for Test Solution A, 90% for Test Solution B, and user defined for Solution C. The standard duration of the test is 720 h or at tensile test specimen failure, whichever occurs first (NACE, 2005).

Four-point bend testing

Four-point loaded bent specimens are prepared following either the ASTM G39 or the ISO 7539-2 standards. The ASTM G-39 also proposes a double beam configuration, which can be implemented for HPHT testing as suggested by the EFC-17 (EFC, 2002b). Specimens are loaded such that the service-wetted surface is in tension. The bent beams methods appear to show more passes than other tests methods, possibly due to stress relaxation (Mack, 2008). Typical stress levels depend on the solution chosen for testing. In general, aggressive environments (e.g., NACE Test Solution A at pH 2.70) require lower loading, which is usually about 80% of YS. As pH increases above 3.50 loading can be above 90% YS.

Standard double cantilever beam (DCB) testing

DBC testing is a fracture mechanics approach that gained popularity for EAC testing of carbon steels. However, this approach has not been extensively implemented for CRAs to date (Mack, 2008). The NACE TM0177 Method D standard describes the implementation of DCB testing for corrosion studies (NACE, 2005). The standard duration of a DCB test for SSC is two weeks. K_{IFINAL} should not be considered as the threshold stress intensity for the initiation of crack propagation unless sufficient crack growth has taken place on the test specimen during exposure. Generally, the measured value for K_{IFINAL} will decrease with increasing YS. Likewise, the minimum value required for design considerations will increase with increasing YS as high yield usually implies a higher design stress (Mack, 2008). Including the effects of strain aging (well aging) in DBC tests is desirable.

15.7.2   Dynamic loading

Slow strain rate testing (SSRT) (constant extension rate testing – CERT)

The SSRT method is described in the ISO 7539-7 standard. This method of threshold stress determination still requires validation for use in sour service environments. Moreover, its usability to predict SSC is subject to interpretation (Mack, 2008). Nevertheless, SSRT has been successfully used in other environments, such as the nuclear and aeronautical industries, to determine SCC susceptibility and to rank families of alloys.

EAC susceptibility of CRAs depends on strain rate. As described in

NACE TM0198, the choice of strain rate ensures that the test is sufficiently discriminating, reasonably rapid, and gives acceptable repeatability and reproducibility. Experience suggests that a strain rate of 1×10^{-6} s^{-1} gives satisfactory results for many systems, including M13Cr steels. When more rapid assessment is required, a strain rate of 4×10^{-6} s^{-1} may be adopted. This strain gives satisfactory results for nickel and austenitic alloys. However, it may lead to reduced repeatability and reproducibility for other systems (NACE, 2004c).

The main criticism of the SSRT methodology is that specimens are strained to failure in all cases, making the testing conditions too aggressive. The surface finish is another critical variable that has been shown to greatly affect the results of SSRT (Mack, 2008).

15.7.3   Pass/fail criterion

Both the ISO 15156 and the EFC-17 determine that the pass/failure criterion for statically loaded specimens is the absence of crack initiation as observed upon magnification using optical microscopy. The stress requirement of $\sigma/\sigma_y = 0.9$ proposed by the EFC-17 is more aggressive than the one suggested in ISO 15156 (σ/specified minimum YS = 0.8) (EFC, 2002a, 2002b; NACE, 2005; Rhodes, 2001).

In contrast, there is no agreement on a pass/fail criterion for fracture mechanics tests for CRAs. An acceptance criterion of $K_{IFINAL} = 35$ MPa$\sqrt{m}$ has been proposed for super-duplex stainless steels when tested in NACE solution. However, this threshold is not necessarily suitable for other CRAs or other environments (Rhodes, 2001).

The threshold or acceptance criterion for SSRT is still in doubt. A measure of the reduction in nominal area (R_{An}) is proposed by EFC-17. In all cases, the loss of ductility as measured by R_{An} is greater than that measured by the loss in nominal elongation (ε_n). A common pass criterion valid only for Ni-based alloys is $R_{An} > 0.8$, coupled with the absence of visible cracking on the gauge section (Rhodes, 2001).

15.8   The future of materials selection for oil and gas production

Many researchers agree that current trends in materials engineering for extreme oil-field environments are on the optimization of the materials selection process (Choi, 2009; Craig, 1997; Kermani, 2000; Schillmoller, 1989; Wilhelm and Kane, 1986). MSS and super MSS are the focus of oil and gas industries today given their combination of mechanical properties, cost and availability. Nevertheless, DSS, SDSS and lower grade Ni-based alloys are still required for more aggressive conditions.

Research is ongoing and continues to gain a better understanding concerning the mechanisms of SCC and SSC in sour HPHT environments. In this regard, the main area of uncertainty is related to the initial stages of EAC and the transition from pits and crevices to cracks. Improved accelerated laboratory tests to better simulate service conditions are desirable, especially to shorten the extremely long exposure times required for higher grade CRAs. Finally, the current testing procedures to determine the influence of S^0 on EAC performance are unsatisfactory. Better and more consistent methodologies for adding S^0 into the test cells are needed.

15.9   References

Aberle D (2008), High performance corrosion resistant stainless steels and nickel alloys for oil & gas applications. *CORROSION/08*. New Orleans, LA: NACE International.

API (2001), API Specification 5CT/ISO 11960: 'Specification for Casing and Tubing', Washington, DC: American Petroleum Institute.

ASM International (2001a), Classification of Stainless Steels: Duplex Stainless Steels. *Metals Handbook*. Metals Park, OH: ASM International.

ASM International (2001b), Classification of Stainless Steels: Martensitic Stainless Steels. *Metals Handbook*. Metals Park, OH: ASM International.

Asphahani A, Prouheze J and Petersen G (1989), Corrosion-resistant alloys for hot, deep sour wells: properties, experience, and future trends In: Engineers S O P (ed.), *Offshore Technology Conference*. Houston, TX.

Bell J M, Chin Y D and Hanrahan S (2005), State-of-the-art of ultra deepwater production technologies. *Offshore Technology Conference*. Houston, TX: Society of Petroleum Engineers.

Cayard M S and Kane R D (1997), An exploratory examination of the effect of SOHIC on the fracture resistance of carbon steels, *CORROSION/97*, NACE International T8 Symposium, Houston, TX, March 1997.

Cayard M S and Kane R D (1998), Serviceability of 13Cr tubulars in oil and gas production environments, *CORROSION/98*, NACE International, San Diego, CA, March 1998.

Choi H J (2009), Materials selection for smart well completions in conjuction with expandable casing technology. *SPE Saudi Arabia Section Technical Symposium and Exhibition*. Al-Khobar, Saudi Arabia: Society of Petroleum Engineers.

Cragnolino G A, Dunn D S, Pan Y-M and Sridhar N (2001), The critical potential for the stress corrosion cracking of Fe-Ni-Cr alloys and its mechanical implications. In: Jones R H (ed.), *Chemistry and Electrochemistry of Corrosion and Stress Corrosion Cracking: A Symposium Honoring the Contributions of R. W. Staehle*. New Orleans, LA: The Minerals, Metals and Materials Society.

Craig B D (1993), *Sour-gas Design Considerations*, Richardson, TX, Society of Petroleum Engineers.

Craig B (1997), Selection guidelines for corrosion resistant alloys in the oil and gas industry. In: Patton C, Craig B, Drake D, Hilliard H and Lewis R (eds), *SPE Reprint Series No 46: Corrosion*. Richardson, TX: Society of Petroleum Engineers.

Det Norske Veritas (DNV) (2008), Recommended Practice DNV-RP-F112, HSC of Duplex and Super Duplex Stainless Steels, Høvik, Norway.

EFC (2002a), European Federation of Corrosion Publications No 16, 'Guidelines on Materials Requirements for Carbon and Low Alloy Steels for H₂S-containing Environments in Oil and Gas Production', London: Maney Publishing on behalf of The Institute of Materials.

EFC (2002b), European Federation of Corrosion Publications No. 17, 'Corrosion Resistant Alloys for Oil and Gas Production: Guidance on General Requirements and Test Methods for H₂S Service', London: Maney Publishing on behalf of The Institute of Materials.

Formigli J. (2008), Santos Basin pre-salt cluster: how to make production development technical and economically feasible. In: Rio Oil & Gas, 2008 Rio de Janeiro, Brazil. Instituto Brasilero de Petróleo, gás e biocombustíveis.

Galis M F and Damman J J (1991), 13% Cr Steels for slightly sour service. *CORROSION/91*. Houston, TX: NACE International.

Galvele J R (1999), W.R. Whitney Award Lecture: Past, present and future of stress corrosion cracking. *Corrosion*, 55, 723–731.

Greer J and Holland W (1981), High-strength heavy-wall casing for deep, sour gas wells. *J. Petrol. Tech.*, 33, 2389–2398.

Hamby T (1975), Deep, high-pressure sour gas wells – an industry challenge. *American Petroleoum Institute Annual Meeting Papers – Divison of Production*. Dallas, TX.

Hara T and Asahi H (2000), Conditions under which cracks occur in modified 13% chromium steel in wet hydrogen sulfide environments. *Corrosion*, 56, 533.

Hay M G (2000), Fit-for-purpose materials testing for sour gas service. *CORROSION/00*. Orlando, FL: NACE International.

Hay M G (2001), Fitness-for-purpose material testing for sour gas service – an overview. *Corrosion*, 57, 236–252.

Heacox R and Stroup J (1976), A material for tubing and casing applications in deep, sour gas wells. *J. Petrol. Tech.*, 28, 705-707.

Heisler L and Mortiz J. (1975), PD 16(5): Problems in treatment and production of sour natural gas from ultra deep wells. In: 9th World Petroleum Conference, 11–16 May 1975. Tokyo: Society of Petroleum Engineers.

Hope I. (2008), World perspectives for the oil and gas industry. In: Rio Oil & Gas, 2008 Rio de Janeiro, Brazil. Instituto Brasilero de Petróleo, gás e biocombustíveis.

ISO (2004), Petroleum and natural gas industries – Steel pipes for use as casing or tubing for wells, Geneva: International Organization for Standardization.

Jellison M, Chandler R, Payne M and Shepard J (2008), Ultradeep drilling pushes drillstring technology innovations. *SPE Drilling & Completion*, 23, 190–200.

Johnson B V, Al-Ghafri M J H, Al-Harthy A S M and Kompally S N (2008), SPE114140: 'The use of carbon steel in highly sour oil and gas production systems'. *SPE International Oilfield Scale Conference*. Aberdeen, UK: Society of Petroleum Engineers.

Kane R D, Watkins M, Jacobs D F and Hancock G L (1977a), Factors influencing the embrittlement of cold worked high alloy materials in H₂S environments. *Corrosion*, 33, 309–320.

Kane R D, Watkins M and Greer J (1977b), The improvement of sulfide stress cracking resistance of 12% chromium stainless steels through heat treatment, *Corrosion*, 33, 231–235.

Kane R D, Greer J B, Jacobs D F, Hanson H R, Berkowitz B H and Vaughn G A (1979), Stress corrosion cracking of nickel and cobalt base alloys in chloride containing environments, *CORROSION/79*, paper no. 174, Houston, TX: NACE International.

Kermani B (2000), Materials optimization for oil and gas sour production. *CORROSION/00*. Orlando, FL: NACE International.

Khan M I (2007), Drilling and production operations. *The Petroleum Engineering Handbook: Sustainable Operations*. Houston, TX: Gulf Publishing Co.

Kimura M, Mochizuki R, Yamazaki Y, Sakata K and Tamari T (2005), Development of New 15Cr stainless steel octg with superior corrosion resistance. *CORROSION/05*. Houston, TX: NACE International.

Leffler W L, Pattarozzi R A and Sterling G (2003), *Deepwater Petroleum Exploration and Production: A Nontechnical Guide*, Tulsa, OK: PennWell Co.

Mack R (2008), A review of laboratory evaluation procedures for materials selection in downhole applications. *CORROSION/08*. New Orleans, LA: NACE International.

Mahmund S E and Chung H E (1990), Localized corrosion of CRA in sulfur-chloride containing environments. *CORROSION/90*. Houston, TX: NACE International.

Morsi K (1989), Field experience with production of deep sour gas reservoir. *SPE Gas Technology Symposium*. Dallas, TX: Society of Petroleum Engineers.

NACE (2000), Technical Committee Report 1F192 (Rev. 2000): Use of Corrosion Resistant Alloys in Oilfield Environments. Houston, TX.

NACE (2003), MR0175/ISO 15156: 'Petroleum and natural gas industries – Materials for use in H2S-containing environments in oil and gas production', Houston, TX: NACE International.

NACE (2004a), MR0175/ISO 15156-2: 'Petroleum and natural gas industries – Materials for use in H_2S-containing environments in oil and gas production – Part 2: Cracking-resistant carbon and low alloy steels, and the use of cast iron', Houston, TX: NACE International.

NACE (2004b), MR0175/ISO 15156-3: 'Petroleum and natural gas industries – Materials for use in H_2S-containing environments in oil and gas production – Part 3: Cracking-resistant CRAs (corrosion resistant alloys) and other alloys)', Houston, TX: NACE International.

NACE (2004c), Standard TM0198, 'Slow strain rate test method for screening corrosion-resistant alloys (CRAs) for stress corrosion cracking in sour oilfield service', Houston, TX: NACE International.

NACE (2005), Standard TM0177, 'Laboratory testing of metals for resistance to specific forms of environmental cracking in H_2S environments', Houston, TX: NACE International.

Perrin D (1999), *Well Completion and Servicing*. Paris: Technip.

Rebak R (2010), Environmentally assisted cracking behavior of nickel alloys in oil and gas applications: a review. *EuroCorr 2010*. Paper 4737, Moscow European Federation of Corrosion.

Rhodes P R (1996), Development of corrosion resistant alloys for the oil and gas industry. *Metarials Performance*, 35, 57–61.

Rhodes P R (2001), Environment-assisted cracking of corrosion-resistant alloys in oil and gas production environments: a review. *Corrosion*, 57, 923–966.

Rhodes P R, Skogsberg L A and Tuttle R N (2007), Pushing the limits of metals in corrosive oil and gas well environments. *Corrosion*, 63, 63–100.

Roberge P R (2000), Materials selection: duplex stainless steels. In *Handbook of Corrosion Engineering*. New York: McGraw-Hill.

Rocha S L, Junqueira P and Roque J (2003), Overcoming deep and ultra deepwater drilling challenges. *Offshore Technology Conference*. Houston, TX: Society of Petroleum Engineers.

Schillmoller C (1988), High performance alloys: how they are used offshore. *World Oil*, 207, 1–7.

Schillmoller C (1989), Selection of corrosion resistant alloy tubulars for offshore applications. *Offshore Technology Conference*. Houston, TX: Society of Petroleum Engineers.

Shaughnessy J, Romo L and Soza R (2003), Problems of ultra-deep high-temperature, high-pressure drilling. *SPE Annual Technical Conference and Exhibition*. Denver, CO: Society of Petroleum Engineers.

Smith J, Jensen D and Meyer B (1970), Liquid hydrogen sulfide in contact with sulfur. *J. Chem. Eng. Data*, 15, 144–146.

Smith L and Craig B (2005), Practical corrosion control measures for elemental sulfur containing environments. *CORROSION/05*. NACE International.

SPE (1975), Panel discussion: the drilling of ultra-deep wells. *9th World Petroleum Congress*. Tokyo: Society of Petroleum Engineers.

Suryanarayana P V and Bjorkevoll K (2009), Design considerations for high-pressure/ high-temperature wells. In: Aadnoy B S, Cooper I, Miska S Z, Mitchell R F and Payne M L (eds), *Advanced Drilling and Well Technology*. Richardson, TX: Society of Petroleum Engineers.

Szklarska-Smialowska S (2005), *Pitting and Crevice Corrosion*, Houston, TX: NACE International.

Tuck C D S, Sykes J M and Garfias-Mesias L F (1994), Influence of specific alloying elements in the control of pitting mechanisms of 25% Cr duplex steels. *4th International Conference on Duplex Stainless Steels*. Glasgow, Scottland.

Turnbull A and Griffiths A (2003), Corrosion and cracking of weldable 13 wt-% Cr martensitic stainless steels for application in the oil and gas industry. *Corr. Eng. Sci. Tech.*, 38, 21–50.

Turnbull A and Saenz de Santa Maria M (1989), The effect of H_2S concentration and pH on hydrogen permeation in AISI 410 stainless steel in 5% NaCl. *Corr. Sci.*, 29, 89.

Tuttle R N (1997), Corrosion in oil and gas production. In: Patton C, Craig B, Drake D, Hilliard H and Lewis R (eds), *SPE Reprint Series No 46: Corrosion*. Richardson, TX: Society of Petroleum Engineers.

Ueda M and Kudo T (1991), Evaluation of SCC resistance of CRAs in sour service. *CORROSION/91*. Cincinnati, OH: NACE International.

Ueda M, Kudo T, Kitayama S and Shida Y (1993), Corrosion behavior of titanium alloys in sulfur-containing H_2S-Co_2-Cl- environments. *Corrosion*, 48, 79–88.

Watkins M and Greer J (1976), Corrosion testing of highly alloyed materials for deep, sour gas well environments. *J. Petrol. Tech.*, 28, 698–704.

Wilhelm S and Kane R (1986), Selection of materials for sour service in petroleum production. *J. Petrol. Tech.*, 38, 1051–1061.

16

Stress corrosion cracking (SCC) in aerospace vehicles

R. J. H. WANHILL, National Aerospace Laboratory NLR,
The Netherlands and R. T. BYRNES and C. L. SMITH,
Defence Science and Technology Organisation (DSTO),
Australia

Abstract: Stress corrosion cracking (SCC) is particularly important for
aerospace vehicles, since it occurs, or can occur, in all major alloy systems
used in their construction. The consequences of stress corrosion failures may
be serious, even leading to loss of an aircraft. This chapter surveys the types
of structures and materials used in aerospace vehicles and the environments
encountered by them. Special mention is made of space vehicles and
platforms, which pose unique problems with respect to avoiding stress
corrosion cracking. Selected case histories from a wide variety of aircraft are
included to illustrate the problems caused by SCC in service, and guidelines
are given for preventing and alleviating these problems.

Key words: stress corrosion cracking, aerospace, case histories, prevention,
remedial measures.

16.1   Introduction

The requirement for speed, strength and higher performance drove the
development of air vehicles from the wood and canvas structures of the early
twentieth century to the all-metal structures of post-World War II. Metallic
aircraft components possessed more consistent properties than wood, and
could be fabricated with a higher degree of reliability. The use of metals also
eliminated the problems of moisture stability and fungus attack found with
wooden structures. However, many alloys deemed suitable for aircraft use
were subsequently discovered to suffer their own forms of environmental
degradation, including corrosion, hydrogen embrittlement and stress corrosion
cracking (SCC). SCC has since become a well-recognised phenomenon in
all major alloy systems used in the aerospace industry.

SCC service failures of aircraft components peaked in the late 1960s, in
no small part due to the widespread use of the aluminium alloys 2024-T3,
7075-T6 and 7079-T6, all of which are highly susceptible to SCC in the
short transverse direction. Improvements in alloy chemistry and processing
(for aluminium alloy components) and control of strength and corrosion

608

protection schemes (for high strength alloy steel components) have helped to reduce the number of service failures in modern aerospace vehicles. Service failures of titanium and magnesium alloy components are rare. The former require a pre-existing notch or crack to initiate SCC, while the latter usually suffer from general corrosion before SCC can become established. Stainless steel components, however, remain an unexpected source of service failure to many aircraft designers and operators, and have the potential to result in loss of aircraft.

Some statistics are available on the frequency of failure modes, including SCC, in aircraft and general engineering (see Table 16.1). This table shows that there are significant differences in failure frequencies, but also similarities, notably for SCC failures. The most significant differences are that (a) fatigue failures are much more frequent in aircraft structures and components than in general engineering, and (b) failures due to corrosion occur more often in general engineering. Fatigue is also the main contributor to serious aircraft accidents and incidents (Campbell and Lahey, 1984; Wanhill, 2009a). This predominance of fatigue reflects the dynamic nature of aircraft service loads and the relatively high design stress levels required to achieve lightweight structures.

On the other hand, Table 16.1 also shows that SCC failures have generally low and fairly constant frequencies of occurrence, 6–8%. Hence it might be thought that SCC in aircraft and space platforms is a minor problem, even though it can occur in many alloys. For aircraft this impression is likely to be upheld by forensic engineering archives. As an example, the NLR has investigated hundreds of service failures over a 40-year period. Only 33 of these failures were attributable to SCC, and only one definitely resulted in loss of an aircraft (Wanhill, 2009b).

Table 16.1 Failure mode frequencies

Failure mechanisms	Percentage of failures		
	Aircraft structures		Engineering industry
	Findley and Harrison (2002)	Brooks and Choudhury (2001)	Brooks and Choudhury (2001)
Corrosion	16	3	29
Fatigue	55	61	25
Brittle fracture	–	–	16
Overload	14	18	11
High temperature corrosion	2	2	7
SCC/corrosion fatigue/hydrogen embrittlement	7	8	6
Creep	–	1	3
Wear/abrasion/erosion	6	7	3

Despite these facts and the impressions they can make, SCC in aircraft is a serious actual and potential problem. It has occurred and still occurs, despite the preferred use of more modern alloys and heat treatments that reduce the susceptibility to SCC. And when SCC occurs, it causes much downtime owing to repairs, component replacements, and extra maintenance, which can include regular (and time-consuming) inspections of suspect components. All of these factors incur considerable costs as well as affecting the operational readiness.

For space platforms there is a difference in emphasis. Since their structures and components generally offer no possibilities for repairs, replacements or inspections, there are stringent design requirements intended to effectively rule out SCC. The stringency of these design requirements can cause problems in selecting suitable structural materials, and there is a protocol for this, see point (4) in Section 16.3.3.

16.2 Structures, materials and environments

Aerospace structures may be placed into three general categories: primary structures, fluid systems and mechanical systems (see Table 16.2), although some might argue that landing gears are primary structures. A number of the fluid system sub-categories are more or less exclusive to manned spacecraft and orbital platforms, as are pyrotechnic devices (exploding bolts for decoupling and interstage separations).

Table 16.2 Categories of aerospace structures (after Korb, 1987)

Category	Types of structures
Primary structures	Airframes
	Space platform shell structures and satellite frames*
	Fasteners
	Additional structural applications
Fluid systems	Hydraulic systems
	Main propulsion systems
	Auxiliary power units
	Plumbing lines and pressure vessels*
	Environmental control and life support systems*
	Electrical power systems (fuel cells)*
	Reaction control systems*
	Orbital manoeuvring systems*
Mechanical systems	Landing gear
	Mechanical devices
	Pyrotechnic devices*

*Exclusive or more or less exclusive to space vehicles and platforms

16.2.1   Primary structures

Because weight is so critical for aerospace vehicles, the choice of materials for primary structures is generally based on high specific strength (strength/density) and high specific modulus (E/density). The corrosion and SCC resistances of metallic materials are important *secondary* considerations, and the design approach in these respects is to use (a) corrosion protection measures, e.g. anodising, cladding, and primer and paint systems, (b) SCC-resistant alloys if possible and feasible, and (c) some engineering design changes to improve the distribution of sustained stresses, including assembly stresses.

Airframes and space platforms

Metallic airframes are constructed primarily of AA2000 and AA7000 series aluminium alloys. Most are used in heat-treatment tempers resistant to SCC, namely the T8XX (AA2000 series) and T7XX (AA7000 series) tempers. An important exception is the industry standard airframe alloy 2024-T3XX, and its more modern variants (AA2X24-T3XX), which have excellent fatigue crack growth and fracture properties but only moderate SCC resistance. Space platform shell structures and satellite frames use these alloys as well, and also aluminium-lithium (Al-Li) alloys owing to their lower densities and higher elastic moduli. These property improvements increase payload weights and improve mission performance. A notable example is the Space Shuttle's 'Super Lightweight External Tank', which is made of Al-Li 2090 and 2195 alloys that replaced conventional AA2000 alloys used in earlier versions of the tank.

Fasteners

Aluminium rivets are commonly used in airframe sheet structures, as are higher-grade fasteners made from low alloy steels and titanium alloys. Optimum corrosion protection of riveted structures requires wet installation of the rivets using corrosion-inhibiting sealants, but this is not always done, especially for air force aircraft: all naval aircraft are mandated to have wet installed fasteners. Low alloy steel fasteners require corrosion protection, which is still cadmium plating – despite environmental concerns – except in space applications. Cadmium plate is banned from space platforms because it can sublime *in vacuo* and redeposit on cooler nearby surfaces where it is not wanted. It should be noted that a number of alternative coatings to cadmium are being developed, e.g. tin-zinc and zinc-nickel plating, but so far they have not demonstrated equivalent durability, especially for high strength alloy steels (CTC/NDCEE, 2002).

Additional structural applications

These include a variety of special structural joints and can require the use of many different materials, including low alloy and stainless steels, titanium alloys, nickel-base superalloys, and beryllium and niobium alloys in space vehicles. Strength restrictions may be placed on the low alloy and stainless steels, in order to avoid or reduce the risks of SCC and also hydrogen embrittlement. These strength restrictions are mandatory for space vehicles (Korb, 1987; ESA, 1981).

Environmental considerations

The potential and actual environments that may be encountered by primary structures include the following that are of most significance for the risk of SCC:

* Air of varying humidity, containing environmental pollutants at ground and near-ground levels, and also sea salt aerosols.
* Potable and waste water from spillages and leaks in manned vehicles and platforms.
* Water condensate, contaminated by mineral salts.

As mentioned above, most airframe aluminium alloys are used in heat treatment tempers resistant to SCC. However, since all the commercial alloys are susceptible to pitting corrosion, they are often coated with protective oxide films produced by anodising or chemical conversion treatments. This protection is augmented by inhibitor-containing paint primers and topcoats; and sometimes also by sacrificial aluminium cladding layers that are bonded to sheet and plate products by conjoint rolling operations, vapour phase deposition or electrodeposition.

16.2.2   Fluid systems

Fluid systems are particularly varied and important for space vehicles, which require many 'exotic' fluids for the main and auxiliary propulsion systems, and also the environmental control and life support systems in manned vehicles and platforms. The general approach in designing these systems is to select alloys compatible with the fluids without protective coatings (Korb, 1987). Aircraft also require fluid systems, notably for fuel, hydraulic and plumbing lines, and these also rely on the basic compatibility of the alloys and fluids. However, fuel tanks are sealed to prevent leaks, and fuel cells have epoxy coatings owing to the generally aggressive fluids held within them.

There are two main types of fluid system components: tubing for fuel, hydraulic and plumbing lines, and pressure vessels. These require other components such as pumps, valves, couplings and nozzles.

Tubing

Stainless steels are used for many applications, including fuel, hydraulic and plumbing lines. These steels generally belong to the 'standard' austenitic AISI 300 series, which are essentially Cr-Ni and Cr-Ni-Mo steels to which small amounts of other elements have been added. An important exception is 21-6-9, which is a Cr-Ni-Mn alloy that is generally stronger than the 300 series at ambient and elevated temperatures. This alloy has been much used in the Space Shuttle (Korb, 1987), also because it is less susceptible to embrittlement by gaseous hydrogen (Nelson, 1988).

Other alloys used for tubing are the nickel-base superalloy Inconel 718 (oxygen lines) and the titanium alloy Ti-3Al-2.5V. Both are used in the Space Shuttle (Korb, 1987), and Ti-3Al-2.5V is widely used in aircraft air-conditioning and de-icing systems.

Pressure vessels

In this context, pressure vessels are used mostly, but not exclusively, in space vehicles and platforms. In aircraft they can also be used for high and low pressure oxygen delivery systems, auxiliary power units (APUs) and in hydraulic accumulators. Several classes of high strength alloys are employed. The Space Shuttle has pressure vessels made from aluminium alloys, Inconel 718 and titanium alloy Ti-6Al-4V (Korb, 1987). The choice of materials is dictated mainly by the fluids to be contained, see below. The fluids can be liquid and gaseous hydrogen and oxygen, various types of water, fuels (hydrazine, N_2H_4; nitrogen tetroxide, N_2O_4; and mixed oxides of nitrogen, MON), liquefied inert gases (helium, nitrogen, freons), and hydraulic oils.

Environmental considerations

Although the alloys in fluid systems are selected for compatibility with the fluids, there are a few points worth noting.

* Liquid hydrogen is benign: problems begin at temperatures above 170 K (−100°C) (Nelson, 1988). Aluminium alloys are the best choice for storing hydrogen, since most are unaffected by high pressure hydrogen up to at least room temperature (Nelson, 1988; Williams and Nelson, 1973). On the other hand, titanium alloys are severely embrittled by hydride formation and exposure of titanium to hydrogen is totally avoided (Korb, 1987).
* Aluminium alloys used for pressure vessels vary in susceptibility to corrosion. Even highly resistant alloys like AA6061-T6 may require internal coatings when used to contain waste water (Korb, 1987).
* Extreme care must be taken to contain liquid or gaseous oxygen, owing

to the danger of ignition. This is why Inconel 718 is used in the Space Shuttle instead of titanium alloys for oxygen-containing pressure vessels and fuel and plumbing lines.

• Titanium alloy pressure vessels can be used to store reactive compounds like ammonia, hydrazine, and nitrogen tetroxide. However, precautions must be taken with nitrogen tetroxide. Firstly, the chemistry must be carefully controlled by the addition of 1.5–3% nitric oxide. This combination is called MON, as mentioned above, and does not cause SCC. Secondly, threaded titanium alloy fasteners cannot be inserted into titanium pressure vessels containing MON, since there is a risk of 'impact' ignition and localised melting of the fasteners. Only aluminium alloy fasteners are permitted to be threaded into MON-containing titanium pressure vessels (Korb, 1987).

16.2.3 Mechanical systems

Materials

As in the case of primary structures, the choice of materials for mechanical systems is usually based on high specific strength and modulus. Protective coatings are used where necessary to increase the resistance to corrosion and the onset of SCC.

Landing gear cylinders and main axles are commonly made from high-strength low alloy steels, but the cylinders are also made from high-strength aluminium alloys. A number of ancillary components such as torque links, rod-ends, levers and brackets are also made from these materials. An exception is the landing gear for the AV-8 Harrier jump-jet. The main components of the Harrier landing gear are made from (very) high strength titanium alloys (Best, 1986). These are more structurally efficient than other alloys (and also much more expensive), but weight savings on the Harrier are at a premium because of its required capabilities of vertical and short take-offs.

Wheels are usually made from aluminium alloys, but magnesium alloys have also been used. However, in general the use of magnesium alloys in any type of component for aerospace vehicles is discouraged, owing to concerns about corrosion, particularly galvanic corrosion, owing to contact with, e.g., high strength steel bolts. Notable exceptions are gear boxes, especially in helicopters, since they combine lightness, strength, moderate elevated temperature capability and good castability for intricately shaped components.

Mechanical devices, including gears, are made typically of high strength low alloy steels. Like landing gear they have to fit into confined spaces, and this restriction is best achieved with high strength/modulus materials. Wear resistance is *the* property requirement for gears, and steels are used because they can be case-hardened. Other materials suitable for mechanical devices include all grades of stainless steels.

Pyrotechnical devices are almost exclusive to space vehicles and platforms. They include frangible nuts, explosive bolts and guillotine blades. In the Space Shuttle these devices are mostly made from Inconel 718, but guillotine blades have also been made from A286 stainless steel and corrosion-protected tool steels (Korb, 1987). However, explosive devices are also used in military aircraft, namely for ejection seat operation and cockpit canopy removal.

Environmental considerations

All high-strength low alloy steels must be protected against corrosion and the risk of SCC. As mentioned on page 612, there may be strength restrictions to reduce the risks of SCC and hydrogen embrittlement, but this is usually not the case for aircraft landing gear cylinders and axles. Instead reliance is made in the first instance on high-quality cadmium, chromium and nickel plating, sometimes in combination with each other. Additional protection is provided by paint systems on external surfaces.

Landing gear components made from aluminium alloys are susceptible to pitting corrosion, even if nominally immune to SCC, although there are cases of SCC in older aircraft (e.g. Wanhill, 2009b). The protection against corrosion is similar to that for airframes, i.e. anodising or chemical conversion coatings augmented by inhibitor-containing paint primers and topcoats. (Cladding is not an option, since these components are made from forgings and extrusions rather than sheet and plate, although vapour phase plating has been used to limit exposure to moisture.)

16.3   Material–environment compatibility guidelines

16.3.1   A comprehensive approach

The previous section has broadly classified and discussed the types of aerospace structures, the materials used, and the compatibilities – or incompatibilities – between the materials and their potential or actual service environments. An important part of this discussion was the provision of guidelines for avoiding the problems of corrosion and SCC. These guidelines are summarised in Table 16.3 and complemented by Table 16.4, which lists the material–environment combinations that *could* result in SCC in aerospace vehicles (Wanhill, 1991).

16.3.2   Guideline limitations

The guidelines in Table 16.3 are comprehensive, but not all of them are adhered to in practice. Other requirements and properties, e.g. weight savings

Table 16.3 Guidelines for prevention of corrosion and SCC in aerospace vehicles

Types of structures	General guidelines	Specific guidelines	
		Materials	Coatings
Primary structures and mechanical systems	• Restrict strength levels • Stress relief • Protective coatings where necessary for corrosion and SCC protection of Al alloys and low alloy steels • Minimise assembly stresses	Al alloys for primary structures and landing gear: – AA2000 series in T8XX tempers – AA7000 series in T7XX tempers	Cladding, anodising or chemical conversion + paint systems
		Low alloy steels: UTS < 1400 MPa (higher strengths for landing gear)	Cd + chemical conversion, Cr or Ni plating, paint systems
		Stainless steels: PH grades ≥ H1000 temper	–
		Ti alloys	–
		Mg alloys	Various anodising and chromate treatments

		Potentially aggressive environments	Materials	
			Pressure vessels	Plumbing lines and components
Fluid systems	• Use materials compatible with fluids without protective coatings • Stress relief • Control fluid chemistry	H_2O (aqueous solutions), Cl^-, H^+	Al alloys, possibly coated; Ti alloys	Stainless steels
		NH_3, N_2H_4, N_2O_4	Ti alloys	AISI 300 series stainless steels
		Freons	Al alloys	21-6-9 stainless steel
		Hydraulic fluids	Cr-plated low alloy steels	Al and Ti alloys; stainless steels

and fabricability, may strongly influence the material selections. Well-known examples are the use of ultrahigh strength low alloy steels in landing gears, and differing choices of materials for pressure vessels and plumbing lines.

Another guideline that requires qualification is the restriction of material strength levels to avoid or reduce the risk of SCC. This guideline is appropriate for aluminium alloys, low alloy steels and all types of stainless steels, but not for titanium alloys. For many titanium alloys the principal determinants of SCC resistance or susceptibility are alloy processing and the type of heat

Table 16.4 Material–environment combinations that *could* result in SCC in aerospace vehicles

Aerospace structural materials potentially at risk of SCC	Environment/aggressive species
Al alloys, Ti alloys, low alloy steels, stainless steels, maraging steels	H_2O (aqueous solutions)
AA7000 Al alloys, low alloy steels, AISI 400 stainless steels, Inconel 718	N_2H_4
Low alloy steels, AISI 400 stainless steels; Ti alloys if N_2O_4 chemistry not controlled by addition of NO (mixed oxides of nitrogen, MON)	N_2O_4
Ti alloys	Organic liquids, including freons
Mg alloys	H_2O (aqueous solutions)

treatment. So-called β processing and/or β heat treatment result in increased SCC resistance over a wide range of strength levels (Wanhill, 1978).

16.3.3   Material–environment combinations that *do* result in SCC in aerospace vehicles

Table 16.4 is similarly comprehensive, but again this deserves comment. There are four main points to discuss:

(1) Aqueous environments, which may be encountered before and during service, have the potential to cause SCC in many aerospace structural alloys. This includes titanium alloys provided they contain surface-connected cracks or sharp-notch defects (Wanhill, 1975). However, to the authors' knowledge there have never been SCC failures in titanium alloy airframe and mechanical systems components. Hot salt SCC of titanium alloy engine components remains a possibility (Gray, 1969), but there appear to have been no cases, probably because the conditions for hot salt SCC to occur are very specific (Boyer *et al.*, 1994).

(2) Despite the long list of materials potentially susceptible to SCC when in contact with hydrazine and nitrogen tetroxide, it is possible to avoid SCC by the correct choice of materials and controlling the fluid chemistry. Titanium alloys can be used for pressure vessels and AISI 300 stainless steels for plumbing lines, but – as mentioned above – it is essential to add 1.5–3% nitric oxide to nitrogen tetroxide, to prevent titanium alloy SCC.

(3) Certain organic liquids can, or could, cause SCC in titanium alloys, specifically during cleaning operations and pre-service pressurization tests. In both cases stringent quality control and/or the use of alternative liquids avoids the problems (Williamson, 1969).

(4) Space agencies (NASA, ESA) have a mandatory classification and screening procedure to prevent SCC. Candidate metals and alloys are placed in three categories of SCC resistance, each with its own requirements for service use (ESA, 1981):

- high resistance alloys are preferred;
- moderate resistance alloys may be considered only when a suitable high resistance alloy is not available – a stress corrosion evaluation form must be submitted as a waiver;
- low resistance alloys may be considered only when it can be demonstrated that the probability of SCC is remote, owing to low sustained tensile stress, suitable corrosion protection, or an innocuous environment – as for the previous category, a waiver must be submitted.

This classification and screening procedure is the result of numerous problems encountered in the early 1960s during stringent pre-service testing and verification. For example, some 40 critical fittings in the lunar module were found to be susceptible to SCC, resulting in several changes. These included changing 7075-T6 components to 7075-T73, introducing liquid shimming at structural joints to minimise clamp-up stresses and provide a perfect fit, shot-peening surfaces, and coating exposed surfaces with protective paint (Weiss, 1973).

In the light of these points it is perhaps no surprise that actual service cases of SCC in aerospace vehicles appear to be confined to aircraft materials in aqueous environments. Furthermore, in the authors' experience the majority of these cases concern aluminium alloys, high strength low alloy steels and stainless steels.

16.4   Selected case histories (aircraft)

16.4.1   Introduction

This section presents selected case histories of aircraft SCC drawn from the authors' experiences. The selection reflects the variety and importance of the cases for each alloy class and – to some extent – their preponderance. This can vary markedly, depending on the type or types of aircraft. However, it does appear that most problems have been experienced with aluminium alloys and stainless steels, followed by high strength low alloy steels. This is possibly because low alloy steels must be protected by well-established plating and painting combinations, while stainless steels are often thought to be immune to corrosion and SCC. In any event, all of the cases concern SCC in aqueous or moisture-containing environments.

There are 33 cases concerning aluminium alloys, 11 cases for stainless steels, 6 cases for high strength low alloy steels, and 2 cases for magnesium alloys.

As far as possible, the cases are treated in groups rather than individually, in order to show commonalities but also to point out significant differences. The measures to alleviate or prevent the recurrence of SCC are listed and/ or briefly discussed for each group or case (excepting the magnesium alloy cases, where straightforward replacement occurred). These measures are placed in the most relevant context in Section 16.5.

16.4.2   Aluminium alloys

Background

Earlier in this chapter, especially on page 611 and Section 16.3.1, we noted the importance of aluminium alloys for aerospace vehicles. In the aircraft industry high strength aluminium alloys are commonly used for primary airframe structures (fuselage skins, stringers and frames; wing and empennage skins, spars and ribs), mechanical systems (landing gear legs, cylinders, forks and struts) and fluid systems (pressure vessels and connectors).

The principal alloy groups are the AlCuMgMn and AlZnMgCu alloys, which have the Aluminium Association series designations of AA2000 and AA7000, respectively. These alloys are age-hardenable. The 2000 series alloys are used in both the naturally aged (T3XX and T4) and artificially aged (T6XX and T8XX) tempers. The 7000 series alloys are used in the artificially aged (T6XX and T7XX) tempers.

In the 1960s it became apparent that thick section products of 2000-T3XX/T4 series and 7000-T6XX series alloys were highly susceptible to SCC in moist air and aqueous environments. This led to developments in heat treatments and alloy chemistry, especially for 7000 series alloys. The heat treatments for optimum SCC resistance are as follows:

- 2000 series alloys: Solution treatment, quench, cold-work and artificial ageing to a T8XX temper. Not all alloys respond to this: the well-known 2014 alloy remains susceptible to SCC.
- 7000 series alloys: Solution treatment, quench, two-stage artificial (over) ageing to a T7XX temper. Again, not all alloys respond to this, especially the notoriously SCC-susceptible 7079 alloy.

The overageing treatment for 7000 series alloys results in a significant loss in strength, about 15%, compared to the peak aged T6XX condition. This has led to the development of alloys which in T7XX conditions achieve the same strength as older alloys in T6XX conditions (Jones, 1992).

The aircraft industry now has design guidelines broadly similar to those of the Space Agencies NASA and ESA, whereby the use of SCC-susceptible alloys is discouraged. However, many aircraft still in use have components made from the older alloys and in SCC-susceptible tempers, and newly manufactured aircraft that are old designs still retain these alloys and

SCC-susceptible tempers. The case histories surveyed and discussed in the following subsections illustrate the key features of SCC in these alloys.

Survey of the aluminium alloy case histories

Table 16.5 surveys the aluminium alloy SCC case histories selected from the DSTO and NLR archives. The majority are from a variety of military aircraft operated by the Royal Netherlands Air Force (RNLAF) and the Royal

Table 16.5 Classification of selected aluminium alloy SCC case histories

Main parameters			Number of cases	Remarks
Aircraft types		Combat	14	Most of the aircraft were designed before 1970, pre-dating the late 1960s/ early 1970s developments in heat treatments and alloy chemistry to improve SCC resistance
		Transport	11	
		Maritime patrol	7	
		Light trainer	1	
Structural areas		Landing gear	14	
		Wing	10	
		Engine/pylon	3	
		Other	6	Window shield, three liquid oxygen fittings, fin pivot bearing housing, fuselage frame
Alloy types		AA2000 series	10	See the above remarks on heat treatments and alloy chemistry
		AA7000 series	23	
SCC causes		Residual stresses	23–24	7 cases starting from corrosion pitting
		Assembly stresses	8	4 cases owing to bearing or bushing inserts
		On-ground tensile stress	1	
		Cold-stamping	1	
Remedial measures	Actual	Repairs	4	1 unsuccessful: incomplete removal of SCC
		NDI* only	2	Some cracking allowed (unusual)
	Proposed	Fleet-wide replacements	27	Only a very few case history reports contain information about whether the proposed remedial measures were taken. However, some are known to have been adopted
		Replacements using better alloys and/or heat treatments	9	
		Improved corrosion protection	6	

Australian Air Force (RAAF). Most cases concerned landing gear and wing components made from 7000 series alloys. In fact, these were effectively just two alloys, 7075-T6XX and 7079-T6XX. The same is true of the 2000 series alloys. These were either 2024-T3XX/T4 or 2014-T6.

The source of most of the SCC problems was residual tensile stresses introduced during heat treatment, sometimes aided by prior corrosion pitting. Assembly stresses also contributed, notably the tensile stresses introduced by interference fit bearing or bushing inserts. Two cases were unusual, where the tensile stresses came from (a) component loading while the aircraft was on the ground, and (b) cold-stamping a groove onto a bearing housing to keep the bearing in place.

The remedial measures obviously included replacing cracked components, except where repairs were feasible or – exceptionally – when some cracking was allowed if the components were regularly inspected. The repair and inspection possibilities are more interesting than simple replacements. These possibilities are discussed in more detail on page 624–630.

Characteristics of aluminium alloy SCC

Aluminium alloy SCC can initiate from nominally undamaged smooth or notched surfaces (e.g., fillet radii and holes) as well as surfaces damaged by corrosion and abrasion. The crack path in commercial alloys is entirely intergranular. This is of major importance to the SCC behaviour of aircraft components made from high strength aluminium alloys. The rolling, forging and extrusion processes required to fabricate half-products and finished components cause the material grains to elongate in the direction of working, resulting in a 'pancake' microstructure. The elongated grain boundaries are usually retained during subsequent heat treatments and provide easy fracture paths, especially for SCC. The direction normal to the pancake microstructure is called the short transverse (ST) direction, and SCC is favoured when there are sustained tensile stresses in this direction. It turns out that this is often the case for residual stresses introduced during heat-treatment (see above and Table 16.5).

Macroscopic characteristics

Figure 16.1 shows two die-forged flap track hinges, one intact and the other failed by SCC along the forging flash line (F). Figure 16.2 is a detail of the failure, showing the SCC origin and progression markings. These markings indicate that SCC progressed under varying environmental conditions and with well-defined and contoured crack fronts. This is not always the case: Fig. 16.3 shows a highly irregular SCC shape in an engine mount bracket. The reason for these differences is probably the crack driving force (CDF).

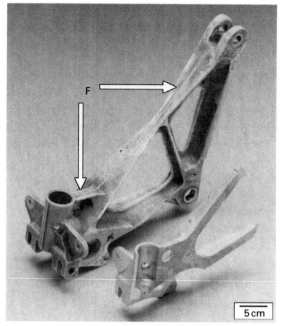

16.1 Intact and SCC-failed flap track hinges made from die-forged 7079-T651 aluminium alloy.

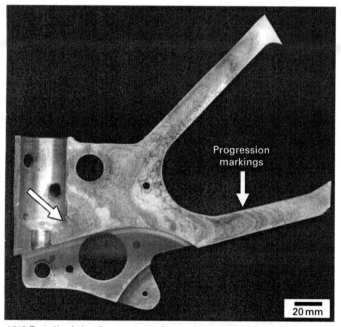

16.2 Detail of the fracture surface of the failed flap track hinge, showing the SCC origin (arrowed) and progression markings: optical fractograph.

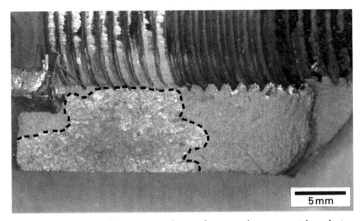

16.3 Broken-open fracture surface of an engine mount bracket made from 2024-T4 aluminium alloy plate, showing a highly irregular SCC shape: optical fractograph.

16.4 Typical 'woody' texture of an SCC fracture surface: optical fractograph.

A lower CDF makes it more difficult for a crack to grow, and the crack 'seeks out' paths of minimum resistance, resulting in an irregular crack front. Similar but less pronounced behaviour has been observed for fatigue cracks approaching the fatigue crack growth threshold (Wanhill *et al.*, 2009). In other words, a highly irregular SCC shape is a possible indication that the crack was growing very slowly or had arrested at local points along the crack front.

Microscopic characteristics

Figures 16.4 and 16.5 show the 'woody' texture and uplifted grains on SCC fracture surfaces from a landing gear linkage arm made from die-forged 7075-T6. The uplifted grains are a diagnostic feature of SCC, also for other materials (Kool *et al.*, 1992), and are the result of ligament deformation

16.5 Uplifted grains on an SCC fracture surface: optical fractograph.

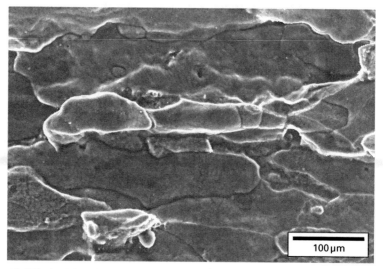

16.6 Typical flat SCC fracture surface: SEM fractograph from a thick-section 2024-T3/T4 aluminium alloy landing gear support strut.

and tearing during the *final* joining of multiple stress corrosion cracks on slightly different levels. This final joining can be the result of in-service final failure of a component or be caused by breaking open a crack for laboratory investigation.

Figure 16.6 shows a typical SCC fracture surface in more detail. Owing to the pancake microstructure, the intergranular characteristics are not obvious. However, sometimes there are recrystallised areas, or a component has undergone bulk recrystallisation. The intergranular nature of SCC is then evident (see e.g. Fig. 16.7).

16.7 Clearly intergranular SCC fracture surface: SEM fractograph from a die-forged 7075-T6 aluminium alloy engine truss mount.

* NDI = non-destructive inspection

Table 16.6 Main aspects of the aluminium alloy SCC case histories involving repair possibilities

Aircraft type	Material and component	Problems	Remedial measures
Transport	DTD 5024 MLG legs (AA7075/9-T6 equivalent)	SCC beyond repair zone	Further repair machining Shot-peening Improved paint scheme Duplex ageing (T7XX temper)
Maritime patrol	AU4SG-T6 MLG legs (AA2014-T6 equivalent)	Corrosion pits and SCC	Repair machining Regular four-monthly inspections
Combat	AA7079-T6 MLG legs	Small (≤4 mm) cracks	Repair machining and/or regular inspections Improved corrosion protection

Repair and inspection possibilities

As noted in Table 16.5, four case histories involved repairs to components damaged by SCC. Three of these cases concerned main landing gear (MLG) legs, which were large forgings. The fourth was a main landing gear linkage arm, also a forging. Table 16.6 summarises the main aspects of these case histories. The remedial measures all included repair machining, but otherwise there were differences. The reason lies in the details of the case histories, which are reviewed briefly in the following paragraphs.

Transport MLG legs

SCC was found in a number of the legs in the 1960s. The manufacturer issued a service bulletin for repair machining. Figure 16.8 illustrates the MLG leg geometry and the repairs, which were pear-shaped holes in the front and rear walls. The operator subsequently found cracks above and below the pear-shaped holes, and the manufacturer again authorised repair machining. However, since this could not go on indefinitely, additional measures were considered. These included shot-peening and an improved paint scheme for existing components, and changing the heat treatment temper for new ones. The changed heat treatment, from T6 to T7XX, was introduced in the mid-1960s.

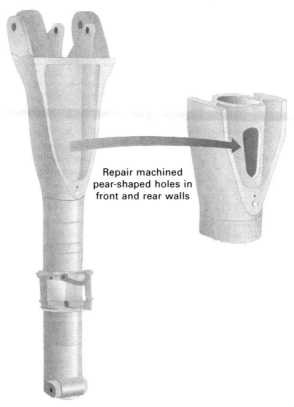

Repair machined
pear-shaped holes in
front and rear walls

16.8 SCC repair machining in outer cylinders of transport MLG legs: DTD 5024 aluminium alloy forgings.

Maritime patrol MLG legs

Several areas of corrosion, sometimes associated with SCC, were found in the late 1960s. The manufacturer issued a service bulletin for repair machining. Figure 16.9 shows the leg geometry and the repair areas, which included the outside and inside surfaces of the legs and the inside surfaces of the two bushed lugs at the top of each leg. The 'clean-up' limits for the repair machining were based on retaining sufficient static strength in the legs. The damage appeared to occur slowly, and so the repair machining was considered adequate in combination with regular inspections. (Chemical conversion coatings were applied after repair machining, and the external surfaces were also re-painted.)

Combat aircraft MLG legs

Small cracks were detected in two legs during routine deep inspections in 1985. The legs were removed from service for destructive investigation.

16.9 Corrosion and SCC repair areas (circled) in maritime patrol MLG legs: AU4SG-T6 aluminium alloy forgings.

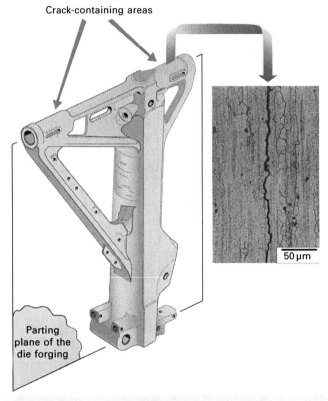

16.10 SCC in combat aircraft MLG legs: 7079-T6 aluminium alloy forgings.

They contained small stress corrosion cracks in the locations shown in Fig. 16.10. Since the legs had seen more than a decade of service, and since the alloy was the notoriously susceptible 7079-T6, it was concluded that the cracks must have arrested owing to complete relaxation of residual stresses: otherwise the legs would have failed. Repair machining and improved corrosion protection were proposed for any other legs with small crack indications. No more cracks were found, and the problem soon disappeared because new legs made from 7075-T73 forgings were introduced in 1986 as part of a major refurbishment.

Maritime patrol MLG linkage arm

Figure 16.11 shows the linkage arm. Four cracks and corrosion pits were detected in the main bore. Two cracks were opened up and found to be SCC, with one crack completely through the bore thickness, see Fig. 16.5. Scratches and abrasion marks were present along the bore, especially at the

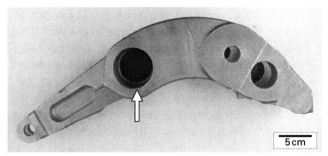

16.11 MLG linkage arm, indicating the main bore where four cracks and corrosion pitting were found: 7079-T6 aluminium alloy forging.

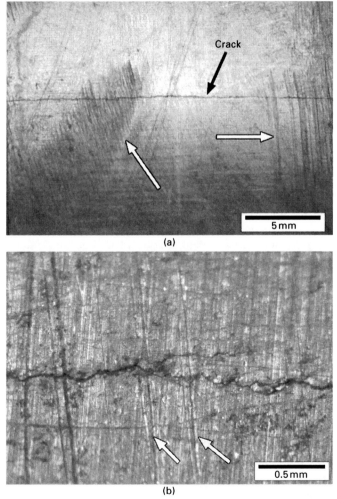

16.12 Scratches and abrasion marks (white arrows) associated with cracks along the main bore of the MLG linkage arm.

crack locations. Examples are given in Fig. 16.12. The scratches, abrasion marks and corrosion pits were all chemical-conversion coated. This indicated that the bore had been reworked (unsuccessfully) to remove the corrosion pits and cracks, followed by re-applying the corrosion protection scheme and returning the linkage arm to service. In view of these results it was recommended to inspect all other linkage arms and replace any cracked ones. It was also suggested to regularly apply water-displacing compounds (WDCs) to the main bores of undamaged linkage arms to provide improved corrosion protection.

16.4.3 Stainless steels

Background

Stainless steels are much used in aircraft engine and fluid systems. Typical applications include fuel and hydraulic tubing, pins, bolts, nuts, clamps, pumps, pistons, bleed air ducts and valves. The alloys include the austenitic grades (e.g. 304, 316, 321, 347, 21-6-9 and Nitronic 60), the martensitic grades (e.g. 410, 431, 440C), and the precipitation-hardenable grades (e.g. PH 13-8 Mo, 17-4 PH, AM350).

All austenitic stainless steels are susceptible to SCC to some degree (Washko and Aggen, 1990). The susceptibility depends on several factors, including the environment, temperature, sensitivity to pitting and crevice corrosion, and metallurgical condition (annealed, cold-worked, welded). SCC initiation can be facilitated by poor design, e.g. sharp corners and crevices, and residual stresses from manufacturing.

A particularly important problem is 'sensitisation' or 'weld decay'. Many austenitic stainless steels are susceptible to sensitisation, which occurs when the steels are exposed to certain temperature ranges, e.g. during welding. Chromium carbides precipitate at the grain boundaries and deplete the adjacent matrix of the chromium content needed to maintain corrosion resistance. The steels become susceptible to intergranular corrosion, which in combination with tensile stresses can be considered a type of SCC.

Sensitisation of the austenitic grades can be avoided by using Ti-containing or (Nb+Ta)-containing stabilised steels like 321 and 347, or low C grades like 304L and 316L, which can take short-term high temperatures during welding (Washko and Aggen, 1990; Vander Voort, 1990). Any sensitisation that does occur can normally be reversed by post-weld annealing. However, if this is not done correctly, then SCC failures can still be expected.

It is also important to note that sensitisation can occur in martensitic and precipitation-hardening stainless steels. Examples are mentioned in the following subsections.

Survey of the stainless steel case histories

Table 16.7 surveys the stainless steel SCC case histories selected from the NLR and DSTO archives. All are from military aircraft operated by the RAAF and RNLAF. Most of the cases concerned hydraulic and fuel systems tubing made from 300 series and 21-6-9 austenitic steels. Many of the SCC problems were due to sensitisation, and one case may have caused the loss of an aircraft owing to a major fuel leak: the wreckage was too damaged to be certain. Another case definitely led to loss of an aircraft (Kool *et al.*, 1992). This case is reviewed below, since it also resulted in world-wide replacement of similar components, using an alloy immune to SCC. Clearly, stainless steel SCC in aircraft can be a major problem, not just an inconvenience.

Table 16.7 Classification of selected stainless steel SCC case histories

Main parameters		Number of cases	Remarks
Aircraft types	Combat	3	2 modern aircraft
	Transport	1	
	Maritime patrol	4	
	Light trainer	1	
	Helicopter	2	2 modern aircraft
Structural areas	Hydraulic system	5	Most of the cases concerned tubing
	Fuel system	2	
	Engine	3	
	Fire extinguisher system	1	
Alloy types	AISI 300 series (austenitic)	5	2 cases of sensitisation close to welds
	AISI 400 series (martensitic)	2	Sensitised close to welds
	21-6-9 (austenitic)	2	Crevice corrosion + SCC
	Nitronic 60 (austenitic)	1	Susceptibility discovered after aircraft loss
	AM350 (precipitation-hardening)	1	Sensitised close to welds
SCC causes	Sensitisation	5	Possible cause of 1 aircraft loss
	Residual stresses/cold-work	2	1 aircraft loss
	Assembly stresses	1	
	Crevices	3	
Remedial measures	World-wide replacements with new alloy	1	
	Individual replacements	9	(2 aircraft losses)
	Recommended NDI of similar components	9?	Number of recommendations uncertain
	Recommended better alloy	2	
	Recommended corrosion protection	3	Water-displacing compounds (WDCs)

Characteristics of stainless steel SCC

Stainless steel SCC can initiate from nominally undamaged smooth and notched surfaces, although some localised pitting or crevice corrosion always precedes SCC. Besides intergranular cracking owing to sensitisation, the SCC can be intergranular in some environments and transgranular in others, notably acidic chloride-containing aqueous solutions (Latanision and Staehle, 1969). SCC in chloride-containing environments occurs at slightly elevated temperatures, typically higher than 50–60°C (Vander Voort, 1990).

Macroscopic characteristics

Figures 16.13 and 16.14 show typical locations for stainless steel SCC problems. Usually all that is visible is an external crack, e.g. Fig. 16.15, but crevice corrosion is sometimes evident. Figure 16.16 shows an exceptional example of crevice-induced corrosion pitting and SCC in a type 304 wire braid from a hose in an MLG hydraulic system. Note that loss of system function in service could have had serious consequences.

Figure 16.17 shows two low-magnification fractographs of intergranular SCC. A brown discoloration due to fracture surface corrosion is typical, and crack front progression markings are sometimes seen. (Greenish-brown fracture surface corrosion and progression markings have been observed for mainly transgranular SCC in a 17-4 PH backstay connector from a yacht; Wanhill, 2003.)

16.13 Location of crevice corrosion and SCC in a 21-6-9 (austenitic stainless steel) hydraulic pressure tube.

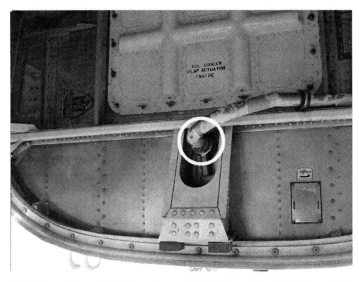

16.14 Location of SCC in a type 431 (martensitic stainless steel) nut from a fire extinguisher system.

16.15 Close-up view of a stress corrosion crack (arrowed) in a type 431 (martensitic stainless steel) nut from a fire extinguisher system.

Microscopic characteristics

Figure 16.18 shows a typical intergranular SCC fracture surface from a type 431 (martensitic) bolt in a fuel tank coupling. Figure 16.19 is a detail from an electrolytically etched metallographic cross-section of the bolt. The grain boundaries are deeply etched in comparison to the matrix, and this deep-etching effect is diagnostic for sensitisation of stainless steels.

16.16 Crevice-induced corrosion pitting and SCC in a type 304 (austenitic stainless steel) wire braid from a hose in an MLG hydraulic system.

Figure 16.20 shows transgranular SCC in a Nitronic 60 (austenitic) pin from a rear compressor variable vane (RCVV) lever arm assembly in an aircraft gas turbine. The pin failure led to loss of the aircraft, reviewed below.

Case history: loss of an F-16

In February 1992 a General Dynamics F-16 crashed between housing blocks in the Dutch city of Hengelo. The crash resulted from engine failure that began with crevice corrosion and SCC fracture of a Nitronic 60 pin in a rear compressor variable vane (RCVV) lever arm assembly (Kool *et al.*, 1992). Figure 16.21 shows the location of the RCVVs in the engine; Fig. 16.22 is a schematic of the RCVV and lever arm assembly and the pin fracture location; and Fig. 16.23 shows the sequence of events leading to engine failure.

At the time of the accident the SCC susceptibility of Nitronic 60 was unknown. Subsequently, many more pins were found to contain cracks, all of which were due to a combination of residual stresses from pin manufacture and salt solutions in crevices between the lever arms and pins. The engine manufacturer took the remedial measures of (a) changing the pin material to the nickel-base superalloy Inconel 625, which is immune to SCC in salt solutions (Kool *et al.*, 1992), and (b) gradual world-wide replacement of all (Inconel 718 + Nitronic 60) lever arm assemblies by (Inconel 718 + Inconel 625) assemblies.

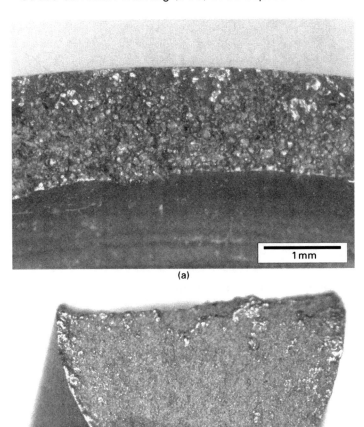

16.17 Intergranular SCC fracture surfaces of (a) a type 431 (martensitic stainless steel) nut from a fire extinguisher system and (b) a type 431 (martensitic) bolt in a fuel tank coupling: optical fractographs.

16.4.4 High strength low alloy steels

Background

High strength low alloy steels have a tempered martensite microstructure. The degree of tempering determines the strength range. These steels are

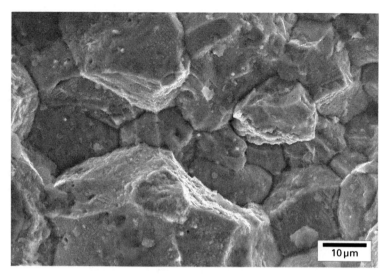

16.18 Intergranular SCC fracture surface: SEM fractograph of a type 431 (martensitic stainless steel) bolt in a fuel tank coupling.

16.19 Deep-etched grain boundaries in a sensitised type 431 (martensitic stainless steel) bolt: optical metallograph.

used mainly in mechanical systems in aircraft, notably for landing gear and gearbox components and high strength bolts and fittings. The alloys include the AISI grades 4330, 4330M and 4340, and 300M, D6ac and H11. All are susceptible to SCC, and also hydrogen embrittlement (Wanhill *et al.*, 2008),

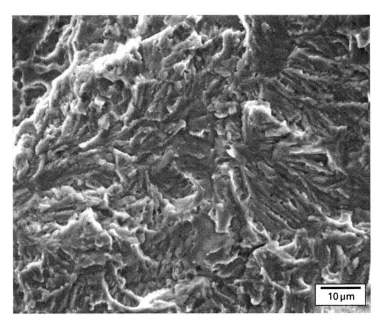

16.20 Transgranular SCC in a Nitronic 60 (austenitic stainless steel) pin from an RCVV lever arm assembly in an aircraft gas turbine.

16.21 F100-PW-220 cutaway with the location (arrowed) of the rear compressor variable vanes (RCVVs).

at yield strengths above 1200 MPa; and they are extremely susceptible at yield strengths above 1400 MPa. This is why the guidelines in Table 16.3 advise restricting the ultimate tensile strength (UTS) to less than 1400 MPa. However, exceptions are made for landing gear, among other items, as is also noted in Table 16.3.

It is important also to note that SCC in high strength steels involves hydrogen embrittlement due to hydrogen generated at the crack tips, and that this can occur in moist air as well as aqueous environments. The SCC fracture characteristics are also similar – if not identical – to those of internal hydrogen embrittlement (IHE), which is due to the presence of solute hydrogen in the steel.

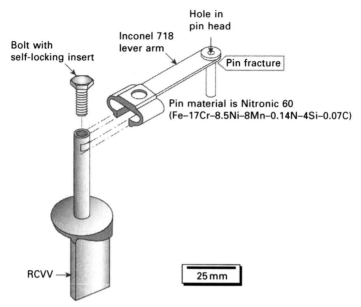

16.22 Schematic of RCVV and lever arm assembly and the pin fracture location.

Salt in by-pass air deposits on lever arms and absorbs moisture during shutdowns

⬇

Concentrated salt solution penetrates crevices between arm and pin

⬇

Crevice corrosion, SCC and fracture of pin

⬇

Mispositioned 5th stage RCVV

⬇

Aerodynamic excitation of 6th stage compressor blades

⬇

Fatigue failure of blade-retaining lugs on 6th stage rotor

⬇

Blade loss and internal disintegration of engine

16.23 Sequence of events leading to the F100-PW-220 engine failure.

Survey of the high strength low alloy steel case histories

Table 16.8 surveys the high strength low alloy steel SCC case histories selected from the NLR and DSTO archives. Five are from military aircraft operated by the RNLAF and RAAF. The sixth was from a commercial transport. Most of the cases concerned landing gear components, and most were due to damage and deterioration (wear) of the corrosion protection systems. One of these cases is reviewed in the next subsection.

There was one unusual case, where SCC started from an in-service fatigue crack. SCC could occur because the aircraft remained on the ground, statically

Table 16.8 Classification of selected high strength low alloy steel SCC case histories

Main parameters		Number of cases
Aircraft types	Combat	4
	Transport (civil)	1
	Helicopter	1
Structural areas	Landing gear	4
	Wing	2
Alloy types	4340, 4340M	3
	300M	1
	D6ac	1
	H11	1
SCC causes	Removed, damaged and/or worn corrosion protection systems (paint, plating)	4
	Shot-peening omitted during manufacture	1
	Fatigue cracking	1
Remedial measures	Individual replacements (1 unnecessary: removal of cracked area sufficient)	5
	NDI of similar components (uncertain)	5?
	Recommended refurbishment of corroded but uncracked components	1

loaded, for about one year during major refurbishment. Also, the cracking was in a very large steel component and could be removed with careful blending or contouring to enable the component to re-enter service.

Characteristics of high strength low alloy steel SCC: a case history

SCC can occur when the corrosion protection systems, paint and/or plating, become damaged or worn or are deliberately removed. If an aqueous environment is present, then SCC can be initiated and accompanied by corrosion pitting. The cracking usually occurs along the prior austenite grain boundaries, resulting in evident intergranular fracture. However, this is also the fracture mode of IHE. The similar fracture characteristics of SCC and IHE can sometimes make it difficult to determine the failure mechanism, especially in the absence of significant corrosion and black or tinted oxide films on the fracture surfaces (Wanhill *et al.*, 2008). Be that as it may, the macroscopic and microscopic SCC characteristics are well illustrated by the following case history.

Helicopter MLG drag beam

In February 2007 a helicopter MLG drag beam failed during a routine landing. Figure 16.24 shows the type of helicopter and the drag beam location. The

16.24 The helicopter type, showing the location of the left-side MLG drag beam.

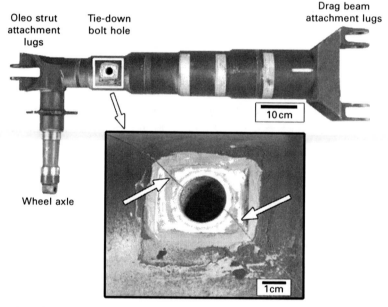

16.25 Cracking (white arrows) at the tie-down bolt hole of the drag beam: 300M steel.

drag beam was ultrahigh strength 300M steel, and investigation showed that it failed from the tie-down bolt hole (see Fig. 16.25), owing to corrosion pitting followed by SCC.

Figures 16.26–16.28 give macroscopic views of the cracking, pointing out the overall fracture characteristics. Despite having been cadmium plated, the tie-down bolt hole was severely corroded and there was rust on much

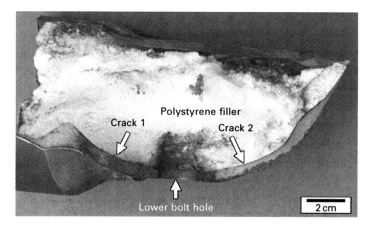

16.26 Macroscopic view of two cracks growing from the tie-down bolt hole: optical fractograph.

16.27 Close-up view of crack 1: optical fractograph.

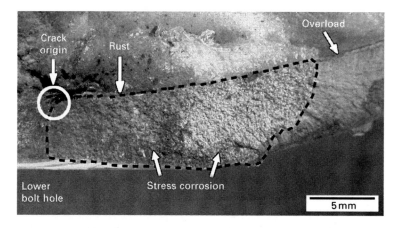

16.28 Close-up view of crack 2: optical fractograph.

of the SCC fracture surfaces. This suggested that the local environmental conditions were severe and that the drag beam could have been cracked for a long time. Figures 16.29 and 16.30 are SEM fractographic details of one of the cracks. These show that SCC started from a corrosion pit and that the non-rusted SCC fracture was classically intergranular.

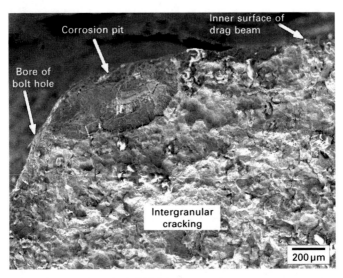

16.29 Origin of crack 2: SEM fractograph.

16.30 Non-rusted intergranular SCC fracture surface of crack 2: SEM fractograph.

Additional investigation showed that the cadmium plating in the bore of the tie-down hole had disappeared and that it was also degrading on the outside of the hole. Besides replacement of the failed drag beam, it was recommended to (a) inspect all other drag beams in the fleet at regular intervals, (b) replace any cracked ones, and (c) check the feasibility of refurbishing uncracked but corroded drag beams.

16.4.5   Magnesium alloys

Background

As stated on page 614, the use of magnesium alloys in aerospace vehicles is generally discouraged owing to concerns about corrosion. Protective coating systems are available, and the basic corrosion resistance has been improved by alloy modifications and additions, but the most serious risk is galvanic corrosion in high conductivity environments, e.g. aqueous salt solutions (Anon., 1979).

Be that as it may, magnesium alloys have a long history of employment in aircraft components, including wheels, flying controls and gearboxes. The use of magnesium alloy gearboxes is more or less standard for helicopters, since (a) the alloys combine lightness, strength, moderate elevated temperature capability (up to about 175°C) and good castability, and (b) the local environment (oils and greases) is innocuous.

Magnesium alloy case histories

A search of the NLR and DSTO archives found only two magnesium alloy failures owing to SCC. These were a nose-gear wheel and a gearbox in an engine system from two combat aircraft. The components were sand castings made from the MgAlZn alloys AZ91C and AZ92A respectively. For the wheel the sustained stresses enabling SCC were attributed to the tyre pressure, and for the gearbox to a misaligned attachment bolt. The remedial actions were straightforward replacements.

Characteristics of magnesium alloy SCC

Magnesium alloy SCC can be initiated by corrosion pitting, as was the case for the nose gear wheel. The cracking can be both intergranular and transgranular (Speidel *et al.*, 1972; Winzer, *et al.*, 2007), but transgranular SCC is stated to be the intrinsic type (Winzer *et al.*, 2007). Transgranular SCC has a complicated cleavage-like and stepped fracture topography that distinguishes it from magnesium alloy fatigue fracture, which is flatter and often shows fatigue striations and well-defined crack front progression markings (Slater *et al.*, 2007). Another distinguishing feature is that SCC crack fronts can be irregular and jagged, e.g. Fig. 16.31, unlike the fatigue crack fronts (Slater *et al.*, 2007; Lynch and Trevena, 1988).

16.5   Preventative and remedial measures

Preventative and remedial measures to avoid SCC in aerospace vehicle alloys have been discussed throughout the previous sections of this chapter. In this section we shall endeavour to place these measures in the most relevant context by discussing and commenting upon the prevention guidelines in Table 16.3, and reviewing the remedial actions from the case histories in Section 16.4.

16.5.1   Preventative measures

As mentioned in Section 16.3.3, actual service cases of SCC in aerospace vehicles appear to be confined to aircraft materials in aqueous environments. This is because the Space Agencies NASA and ESA have instituted stringent material selection criteria. These criteria are based partly on some disastrous

16.31 Irregular and jagged magnesium SCC crack front (Lynch and Trevena, 1988).

early experiences (Williamson, 1969) and also on the virtual impossibility of in-service repairs, replacements or inspections. Furthermore, the majority of aircraft SCC problems concern aluminium alloys, high strength low alloy steels and stainless steels. In the case of aluminium alloys it is primarily the older aircraft, and the older generation aircraft, using 7000 series alloys in peak aged T6XX tempers, that have these problems. A widely used older generation aircraft that still has SCC problems in relatively new models is the Lockheed Martin C-130 Hercules, which retains much of the primary structural design and materials from the early 1950s. These problems are returned to in the discussion on repairs in Section 16.5.2.

Owing to the above considerations, the first part of Table 16.3 is of most relevance to the present discussion. This is presented as Table 16.9 in a modified form, specifically for aircraft and with additions (italicised) that take account of the case histories in Section 16.4. For stainless steels there is some additional information to consider with respect to alloy selection. Higher-chromium and especially higher-molybdenum austenitic stainless steels are more resistant to crevice corrosion (Washko and Aggen, 1990), and a more recent possibility is the use of duplex (austenite + ferrite) stainless

Table 16.9 Guidelines for prevention of corrosion and SCC in aircraft

Types of structures	General guidelines	Specific guidelines	
		Materials	Coatings
Primary structures and mechanical systems	• Restrict strength levels • Protective coatings where necessary for corrosion and SCC protection of Al alloys and low alloy steels	Al alloys for primary structures and landing gear: – AA2000 series in T8XX tempers *(if feasible)* – AA7000 series in T7XX tempers – *Reduce residual and assembly stresses*	Cladding, metallic coatings (IVD), anodising or chemical conversion + paint systems
		Low alloy steels: – UTS < 1400 MPa (higher strengths for landing gear) – *wear-resistant coatings (whenever feasible)*	Cd, Cr or Ni plating + paint systems
		Stainless steels: – PH grades ≥ H1000 temper – *Avoid sensitisation and residual stresses* – *Avoid crevices if possible: use WDCs otherwise* – *Use alloys more resistant to SCC*	–

steels. Duplex stainless steels can substitute for austenitic stainless steels at any strength level and with high resistance to SCC (Ekman and Pettersson, 2009). The duplex steels also have a higher surface hardness, which makes them more resistant to abrasion and wear.

16.5.2   Remedial measures

There are two basic remedial actions: replacement and repair. Table 16.10 gives remedial measure guidelines for both of these actions. There are some similarities in these remedial measures, and also some similarities with the preventative measures listed in Table 16.9. For example, two generally

Table 16.10 Guidelines for remedial measures against corrosion and SCC in aircraft

Corroded and/or cracked components	Alloy classes	Guidelines for remedial measures
Replace	All	Straightforward replacement Improved corrosion protection, e.g. use WDCs Inspect similar components, replace as necessary Regular inspections of replaced and retained components
	Aluminium alloys Stainless steels	Replace with alloys and/or heat treatments more resistant to corrosion and SCC
Repair	Aluminium alloys	Repair machining of corroded and/or cracked components Shot-peening Improved corrosion protection, e.g. paint systems and WDCs Re-ageing to SCC-resistant tempers* Composite patches** Inspect similar components, repair (or replace) as necessary Regular inspections of repaired (and replaced) components
	High strength low alloy steels	Refurbishment of corroded but uncracked components Improved corrosion protection, e.g. use WDCs Inspect similar components, repair (or replace) as necessary Regular inspections of repaired (and replaced) components

*Proposed and **actual repairs for Lockheed Martin C-130H 7075-T6 extrusions.

applicable measures are the use of alloys and/or heat treatments more resistant to SCC and the provision of improved corrosion protection.

Repair options are generally limited, and none have been indicated for stainless steel components, which are generally easily replaced. In fact, all components that are simple and easily removed will most likely be scrapped. This avoids the costly auditing, certification and changed inspection schedules associated with repairs. More complex and difficult to remove components will also most likely be scrapped unless repairs can be done *in situ* or replacements are unavailable.

The need for repairs is particularly relevant to older aircraft that continue in service well beyond their original target service lives. Table 16.10 shows that there are several guidelines and options for aluminium alloys, some of which were discussed with respect to the MLG case histories on pages 624–630. The remaining two guidelines are the use of composite patches and reheat treatment. The reheat treatment is a special one called retrogression and re-ageing (RRA). This was invented in the early 1970s for increasing the SCC resistance of thin sheets of 7000 series alloys in the T6 temper (Cina and Gan, 1974). Subsequently, the National Research Council (NRC) in Canada modified the RRA treatment to make it suitable for thicker sections (Wallace, *et al.*, 1981) and *in-situ* application on airframes (Raizenne *et al.*, 2002). There is also the potential to treat replacement components prior to insertion on an aircraft so as to forestall the problem reoccurring in the future. The DSTO are currently (2010) using data supplied by the NRC, and data it has collected itself, to certify the use of RRA 7075-T6 extrusions on RAAF C-130J aircraft.

16.6   Conclusions

Stress corrosion cracking (SCC) is particularly important for aerospace vehicles since it occurs, or can occur, in all classes of high strength alloys used in their construction. The consequences of SCC failures can be and have been serious, including the destruction of spacecraft and rocket components during the early days of pre-mission testing in the 1960s, and even the loss of an aircraft, though this is rare. The Space Agencies NASA and ESA have mandatory procedures to prevent SCC, and the chances of it occurring in modern space vehicles and platforms are remote.

Actual service cases of SCC appear to be confined to aircraft materials in aqueous environments. Selected case histories from a wide variety of aircraft indicate that the majority of SCC problems concern aluminium alloys and stainless steels, followed by high strength low alloy steels. This may be because low alloy steels must be protected against corrosion and SCC by well-established plating and painting combinations, while stainless steels are often thought to be immune to corrosion and SCC.

With respect to aluminium alloys, it is primarily older aircraft and new aircraft of older designs, using 7000 series alloys in peak aged T6XX tempers, that have SCC problems. However, this restriction does not apply to stainless steels and high strength low alloy steels, which are just as susceptible in modern aircraft. Considered as a material class, stainless steels are actually far from 'stainless'. Many are susceptible to crevice corrosion, leading to SCC; and they can also be sensitised to SCC if fabricated and welded without due regard to material composition and post-weld annealing. Finally, all high strength low alloy steels are susceptible to corrosion and SCC, and rely completely on high quality plating and painting systems.

There are reasonably well-defined guidelines for preventing and alleviating SCC problems in aircraft aluminium alloys, stainless steels and high strength low alloy steels. These guidelines include limited possibilities for repair of components made from aluminium alloys and high strength low alloy steels.

16.7   References

Anon. (1979), 'Corrosion resistance of magnesium and magnesium alloys', in *Metals Handbook Ninth Edition, Vol. 2 Properties and Selection: Nonferrous Alloys and Pure Metals*, Metals Park, OH, ASM International, 596–609.

Best K F (1986), 'High strength materials for aircraft landing gear', *Aircraft Engineering and Aerospace Technology*, 58(7), 14–24.

Boyer R, Welsch G and Collings E W (eds) (1994), *Materials Properties Handbook Titanium Alloys*, Materials Park, OH, ASM International, 294.

Brooks C R and Choudhury A (2001), *Failure Analysis of Engineering Materials*, New York, McGraw-Hill.

Campbell G S and Lahey R (1984), 'A survey of serious aircraft accidents involving fatigue fracture', *International Journal of Fatigue*, 6(1), 25–30.

Cina B and Gan R (1974), 'Reducing the susceptibility of alloys, particularly aluminium alloys, to stress corrosion cracking', US Patent 3856584, December 24, 1974.

CTC/NDCEE (2002), *Validation of alternatives to electrodeposited cadmium for corrosion protection and threaded part lubricity applications*, Joint Test Report BD-R-1-1.

Ekman S and Pettersson R (2009), 'Desirable duplex', *Materials World*, 17(9), 28–30.

ESA (1981), 'Material selection for controlling stress-corrosion cracking', *Procedures, Standards and Specifications ESA PSS-01-736*, Noordwijk, European Space Agency.

Findley S J and Harrison N D (2002), 'Why aircraft fail', *Materials Today*, 5(11), 18–25.

Gray H R (1969), 'Hot-salt stress corrosion of titanium alloys', *Special Publication NASA SP-227*, Cleveland, Ohio, NASA Lewis Research Center.

Jones R H (ed.) (1992), 'Stress-corrosion cracking of aluminium alloys', in *Stress-Corrosion Cracking: Materials Performance and Evaluation*, Materials Park, OH, ASM International, 233–249.

Kool G A, Kolkman H J and Wanhill R J H (1992), 'Aircraft accident F16/J-054: F100-PW-220 6th stage disk and 5th stage RCVV pin investigation', *Contract Report NLR CR 92424 C*, Amsterdam, National Aerospace Laboratory NLR.

Korb L J (1987), 'Corrosion in the aerospace industry', in *Metals Handbook Ninth Edition, Vol. 13 Corrosion*, Metals Park, OH, ASM International, 1058–1100.

Latanision R M and Staehle R W (1969), 'Stress corrosion cracking of iron-nickel-chromium alloys', in *Proceedings of Conference: Fundamental Aspects of Stress Corrosion Cracking*, Houston, TX, National Association of Corrosion Engineers, 214–296.

Lynch S P and Trevena P (1988), 'Stress corrosion cracking and liquid metal embrittlement in pure magnesium', *Corrosion*, 44, 113–124.

Nelson H G (1988), 'Hydrogen and advanced aerospace materials', *SAMPE Quarterly*, 20(1), 20–23.

Raizenne D, Sjoblom P, Rondeau R, Snide J and Peeler D (2002), 'Retrogression and re-aging of new and old aircraft parts', in *Proceedings of the 6th Joint FAA/DoD/NASA Conference on Aging Aircraft* (CD-ROM), San Francisco, CA.

Slater S L, Wood C A and Turk S D (2007), 'Preliminary failure assessment of T56-A-15 reduction gearbox case cracking', *Contract Report DSTO-CR-2007-0256*, Melbourne, Defence Science and Technology Organisation.

Speidel M O, Blackburn M J, Beck T R and Feeney J A (1972), 'Corrosion fatigue and stress corrosion crack growth in high strength aluminium alloys, magnesium alloys, and titanium alloys exposed to aqueous solutions', in *Corrosion Fatigue: Chemistry, Mechanics and Microstructure*, Houston, TX, National Association of Corrosion Engineers, 324–342.

Vander Voort G F (1990), 'Embrittlement of steels', in *Metals Handbook Tenth Edition, Vol. 1 Properties and selection: Irons, Steels, and High-Performance Alloys Corrosion*, Materials Park, OH, ASM International, 707.

Wallace W, Beddoes J C and de Malherbe M C (1981), 'A new approach to the problem of stress corrosion cracking in 7075-T6 aluminum', *Canadian Aeronautics and Space Journal*, 27, 222–232.

Wanhill R J H (1975), 'Aqueous stress corrosion in titanium alloys', *British Corrosion Journal*, 10, 69–78.

Wanhill R J H (1978), 'Microstructural influences on fatigue and fracture resistance in high strength structural materials', *Engineering Fracture Mechanics*, 10, 337–357.

Wanhill R J H (1991), 'Fracture control guidelines for stress corrosion cracking of high strength alloys', *Technical Publication NLR TP 91006*, Amsterdam, National Aerospace Laboratory NLR.

Wanhill R J H (2003), 'Failure of backstay rod connectors on a luxury yacht', *Practical Failure Analysis*, 3(6), 33–39.

Wanhill R J H (2009a), 'Some notable aircraft service failures investigated by the National Aerospace Laboratory (NLR), *Structural Integrity and Life*, 9(2), 71–87.

Wanhill R J H (2009b), 'Aircraft stress corrosion in the Netherlands', *Technical Publication NLR-TP-2009-520*, Amsterdam, National Aerospace Laboratory NLR.

Wanhill R J H, Barter S A, Lynch S P and Gerrard D R (2008), 'Prevention of hydrogen embrittlement in high strength steels with the emphasis on reconditioned aircraft components', *Contract Report NLR-CR-2008-678*, Amsterdam, National Aerospace Laboratory NLR.

Wanhill R J H, Barter S A, Hattenberg T, Huls R A, Ubels L C and Bos M (2009), 'Fractography-assisted observations on fatigue thresholds in an aerospace aluminium alloy', *Technical Publication NLR-TP-2009-596*, Amsterdam, National Aerospace Laboratory NLR.

Washko S D and Aggen G (1990), 'Wrought stainless steels', in *Metals Handbook Tenth Edition, Vol. 1 Properties and selection: Irons, Steels, and High-Performance Alloys Corrosion*, Materials Park, OH, ASM International, 873.

Weiss S P (1973), 'Apollo experience report – lunar module structural subsystem', *Technical Note NASA TN D-7084,* Houston, TX, NASA Manned Spacecraft Center.

Williams D P and Nelson H G (1973), 'Gaseous hydrogen embrittlement of aerospace materials', in *Astronautical Research 1971*, Boston, MA, D. Reidel Publishing, 323–330.

Williamson J G (1969), 'Stress corrosion cracking of Ti-6Al-4V titanium alloy in various fluids', *Technical Memorandum NASA TM X-53971*, Huntsville, AL, NASA – George C. Marshall Space Flight Center.

Winzer N, Atrens A, Dietzel W, Song G and Kainer K U (2007), 'Stress corrosion cracking in magnesium alloys: characterization and prevention', *Journal of Metals*, 59(8), 49–53.

Prediction of stress corrosion cracking (SCC) in nuclear power systems

P. L. ANDRESEN and F. P. FORD (retired),
GE Global Research, USA

Abstract: This chapter addresses the phenomenon of stress corrosion cracking in light water reactors, and specifically the prediction of the crack propagation rate in stainless steel components in boiling water reactors (BWR). Attention is focused on the various approaches that may be used for life prediction. These vary from analyses of plant incidents, to empirical correlations between the propagation rate and various engineering parameters in laboratory experiments, to algorithms based on knowledge of the processes and mechanism of crack propagation. The chapter concludes with a comparison between prediction and observation for cracking of stainless steels in both unirradiated and irradiated BWR components.

Key words: boiling water reactors, stress corrosion cracking, slip-oxidation model, stainless steels, piping, irradiated core components, water chemistry, life prediction.

17.1   Introduction

Environmentally assisted cracking (EAC) is a materials degradation mode that encompasses stress corrosion cracking (SCC), strain induced corrosion cracking (SICC) and corrosion fatigue, where these nomenclatures originate from the loading details; namely, constant stress (or strain), monotonically increasing strain and cyclic loading. There is a spectrum of cracking susceptibility across these three submodes of environmentally assisted cracking, all of which are encountered in light water reactor (LWR) systems. They form an important subset of localized corrosion phenomena since the cracking is often hard to detect, and can have a severe impact on operational economics and, potentially, on plant safety.

The cracking morphology can be intergranular (interdendritic in weld metals), transgranular or mixed mode (granulated), and the morphology and degree of cracking susceptibility may change in any given alloy/environment system with relatively subtle changes in the combinations of material, stress and environment conditions. The interactions between these conditions are conjoint: that is, as illustrated schematically in Fig. 17.1, all of the individual conditions must be present, although a given SCC susceptibility can result from many combinations of conditions.

651

The Venn diagram in Fig. 17.1 was widely used in the 1970s and early 1980s to provide a qualitative description of the effect of the various system parameters on stress corrosion cracking. The limitation of this representation became apparent with the need for a more quantitative analysis of the benefits of various mitigation actions, and the realization that the actual cracking process was more complex than depicted in Fig. 17.1. For instance, system parameters such as cold work could affect two of the circles of influence (material and stress) independently, and irradiation could affect all three circles of influence.

Although environmentally assisted cracking has been observed in many of the structural materials in both pressurized water reactors (PWR) and BWRs, attention in this chapter is focused on the quantitative prediction of stress corrosion cracking of stainless steel components in BWRs. Such cracking has been observed since the late 1950s, and has had a significant impact on the economics of reactor operation. These degradation issues have been documented, for example, in the proceedings of a series of biennial symposia and conferences held since 1983 (Proceedings of Environmental Degradation Conferences 1983–2009), in EPRI reports on water chemistry issues (EPRI, 2000, 2004) and in information notices and reports published on the US Nuclear Regulatory Commission website (http://www.nrc.gov/reading-rm/doc-collections/).

The chronology of events during cracking in many alloy/environment systems involves the following.

- A 'precursor period' during which specific stress, metallurgical or

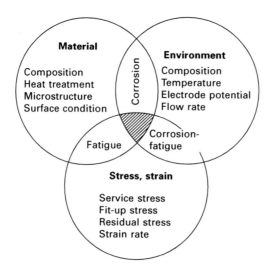

17.1 Conjoint material, stress and environment requirements for stress corrosion cracking (Speidel, 1984).

environmental conditions may develop at the metal/solution interface that are conducive to subsequent crack initiation (Staehle, 2007).

- The initiation of cracks when the local environment, microstructure, stress and crack geometry conditions have reached a critical state. These microstructurally small cracks (i.e., <50 µm dimensions) may then grow and coalesce to form a dominant crack (Fig. 17.2). The precursor, crack initiation, coalescence and 'short crack' growth phenomena dominate the 'engineering crack initiation' period that may be defined practically as the time to develop a crack that is detectable by current commercial non-destructive examination techniques. The criteria for crack initiation (and the subsequent short crack growth) have been extensively reviewed elsewhere (Andresen *et al.*, 1990 and Hickling, 2005).
- Propagation of a single dominant crack at a rate that, as will be discussed below, is very dependent on the material, stress and environmental conditions. The growth rate is rarely constant, as is too often assumed, but may accelerate or decelerate over time depending on the stress and strain profiles, the material, irradiation conditions, plant operating conditions, etc.

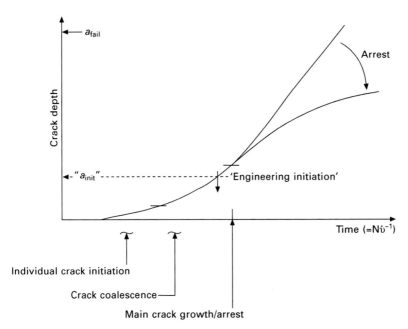

17.2 Sequence of crack initiation, coalescence and growth during sub-critical cracking in aqueous environments. Note the arbitrary definition of 'engineering initiation' which generally coincides with the NDE resolution limit.

It is apparent that the development of a life prediction algorithm of the general form:

$$\text{Time to failure} = f \, (\text{Material, Environment, Stress}) \qquad 17.1$$

that can conceptually take into account all of the factors listed above, poses considerable challenges. For instance, there are complicated interactions among the various system parameters (Fig. 17.3) that control the various submodes (e.g., pitting, intergranular attack, crack coalescence, crack propagation, etc.). Moreover, it will become apparent in the following sections that small changes in these interacting factors can significantly affect the stress corrosion cracking susceptibility.

This immediately leads to the conclusion that there will be a measure of uncertainty (albeit quantifiable) in the life prediction capabilities and this uncertainty can be attributed to two sources. The first are *aleatory*

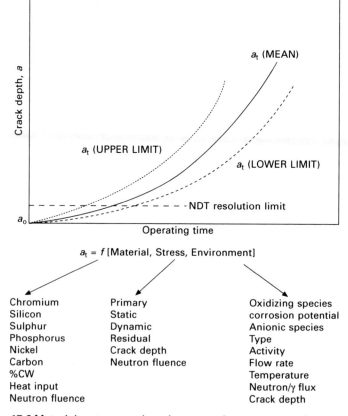

17.3 Materials, stress and environmental parameters relevant to environmentally assisted cracking of stainless steels in BWRs (Ford *et al.*, 1988).

uncertainties that arise out of random, stochastic events (such as pitting, intergranular attack, crack coalescence) that occur on a smooth surface, and these may dominate the onset of engineering initiation and the failure time of thin specimens or components. Such uncertainties may be assessed via, for instance, Weibull statistics of cracking of tubing in PWR steam generators (Staehle *et al.*, 1994; Stavropoulos *et al.*, 1991).

The second are *epistemic* uncertainties that arise out of incompleteness of the degradation model that is adopted, the dispersion in the model inputs, and the inadequacy of the system definition or control. These epistemic uncertainties, unlike the aleatory uncertainties, may be reduced by improvements in the degradation model and the control/definition of the model inputs.

With this as background, attention is now focused on the state of understanding of the phenomenology of stress corrosion cracking in stainless steels in BWRs.

17.2   Life prediction approaches

To predict the incidents of SCC in stainless steels in BWRs, it is necessary to develop a damage vs. time algorithm (Eq. 17.1) that addresses the interactions among all of the relevant material, environmental and stress components (Fig. 17.3). There are three complementary approaches:

- general analyses of plant experience
- empirical correlations of plant and laboratory data
- a quantitative understanding of the mechanism of cracking.

17.2.1   Plant experience

There is no question that past plant experience is of crucial importance, since this acts as a 'calibration point' that *must* be explained by any life prediction methodology. For instance, the methodology must be able to explain quantitatively the relative contributions of the system parameters (such as cold work, sensitization, weld repairs, etc.) to the incidents of stress corrosion cracking that have been observed in a given operating period. To accomplish this it is necessary to have an event database, as exists in, for instance, the STRYK system in Sweden (Gott, 1999), or the 'Generic Aging Lessons Learned' (GALL) report in the US (USNRC, 2006). Although such databases provide useful information for, for instance, defining criteria for a risk-informed inspection strategy (Gott, 2001), they are limited in providing the necessary damage vs time information. This is illustrated in Fig. 17.4 for the case of cracking in stainless steel piping in GE BWRs, where it is apparent that the scatter in data makes it impossible to define the desired algorithm from this information alone.

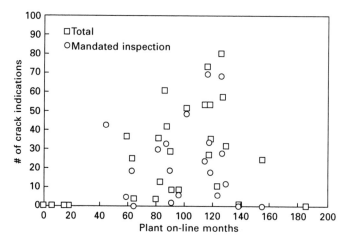

17.4 Relationship between incidences of cracking in BWR piping and operational time (Klepfer, 1975).

There are various reasons for the scatter in results from event databases:

- Some sub-phenomena in the chronology of cracking are random, stochastic events, thereby leading to the conclusion from Weibull analyses that the probability of failure in a particular classification of pipe may span orders of magnitude of time (Fig. 17.5). In addition, the 'incubation periods' before the first cracking incident is detected can be very short; for example, this 'incubation time' may be less than a year for sensitized steels in BWRs, but this incubation period may increase to approximately 7 years for core internals.
- Not all data in the event database are, in fact, comparable. For instance, some of the plants from which the database originates may be operating under slightly different water chemistries over varying time scales and this can, as will be discussed in Section 17.3, have profound effects on the cracking susceptibility.
- Different criteria may be used for 'failure' of components recorded in the same dataset. These criteria may range from a definition that corresponds to a depth sufficient to give leakage in a pressure boundary, to the observation of a shallow surface crack via visual examination.
- The number of cracking events are directly relatable to the nature and number of inspections, and this needs to be taken into account in quantifying the plant experience. For instance, the high frequency of cracking incidents of stainless steel BWR piping recorded in the USA in the early 1980s was driven just as much by the increased number of mandatory inspections in that time period as by the less-than-optimal water chemistry in the years prior to the inspections.

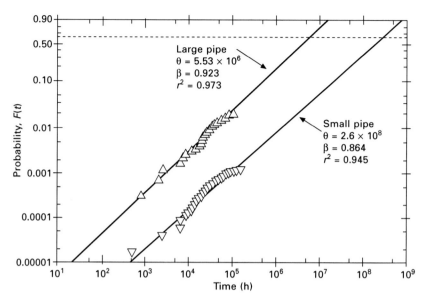

17.5 Probability of failure of Type 304 stainless steels in 2″ and 4″ diameter pipes in operating BWRs (Eason and Shusto, 1986).

Thus, plant experience is of limited value if it is the *sole* input to the algorithm for predicting the development of cracking over time. This is because adequate definitions of the component fabrication and operating conditions are rarely available, and the criteria for 'failure' are not always consistent. Consequently, attention will now be focused on life prediction methodologies that are based on either empirical correlations between the component life and the governing stress, material and environment parameters and/or an understanding of the mechanism of cracking.

17.2.2   Empirical correlations of plant and laboratory data

The empirical correlation approach relies on an identification of all system variables that can impact the kinetics of the cracking process, and an experimental determination of the magnitude of these dependencies via 'separate effects' tests.

The advantage of this empirical correlation approach is that the tests can be conducted, at least in principle, under laboratory conditions where all the system parameters are adequately defined and controlled, and the crack propagation can be monitored with good resolution and accuracy. Based on this assumption there have been numerous developments that have attempted to formulate the stress corrosion susceptibility in terms of the controlling system parameters, albeit with an appreciation of the aleatory uncertainties associated with predicting crack initiation (Akashi and Nakayama, 1995) and the epistemic uncertainties associated with the distribution of inputs to the chosen model (Wei and Harlow, 1995). Some of the developments have been based on an assumed cracking mechanism (Jiang and Staehle, 1997), while others (Yamauchi *et al.*, 1995) have been based on the interactions of several 'engineering' parameters (carbon content, dissolved oxygen content, etc.) and their effect on the cracking susceptibility. These developments have been valuable and some (Yamauchi *et al.*, 1995) have been used to make, for instance, materials selection for reactor service.

The common problem with these empirical approaches is that they are very dependent on the quality and the 'relevance to operational conditions' of the data upon which the prediction formulae are based. Although improved testing techniques and material characterization have been developed in the last 25 years, the fact is that many of these empirical life prediction approaches were based on earlier data that may be questionable, e.g., because of scatter (Ford, 2007). Consequently, in the last 25 years extensive attention has been paid to experimental control and data quality (Andresen 1999; Andresen *et al.*, 1999a) and old databases have been examined and, where necessary, censored with respect to agreed-upon quality criteria. This decreases the data variance at the cost of (often substantially) reducing the number of data points (Andresen, 1999; Hickling, 1994; Kim and Andresen, 2003).

Even after screening the laboratory data, there can be considerable variance in the crack propagation rates as a function of changes in the system conditions. This is illustrated in Fig. 17.6, which compares the empirically derived prediction algorithm (Eq. 17.2) with the data for crack propagation rates in Type 304 stainless steel in high temperature water as a function of the corrosion potential. The prediction algorithm in Fig. 17.6 is a multi-parameter least squares fit to the fracture mechanics database that was available in 1995 for sensitized Type 304 stainless steels (EPRI, 1996):

$$\ln(V) = 2.181 \ln(K) - 0.787 \ \kappa\text{-}0.586$$

$$+ \ 0.00362 \ \varphi_c + 6730/T - 35.567 \qquad\qquad 17.2$$

where: V = crack propagation rate (mm/s), K = stress intensity factor (MPa$\sqrt{m}$), κ = solution conductivity at 25°C (μS/cm), φ_c = corrosion potential (mV$_{SHE}$) and T = absolute temperature (°K).

It is apparent that in this case the correlation factor between prediction and observation is very low ($R^2 < 0.1$), thereby raising questions regarding the quality of the data upon which the correlation was based and/or the completeness of the model (i.e., whether it takes into account the interactions between all the relevant system variables). In spite of this low correlation factor the approach was accepted by the USNRC on the basis that the use of the 95th percentile provides a conservative upper bound to the data

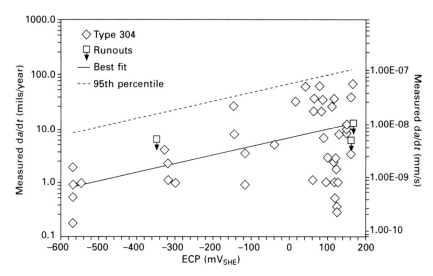

17.6 Comparison between prediction (according to Eq. 17.2) and observation of the crack propagation rate vs corrosion potential relationship for Type 304 stainless steel in BWR water (EPRI, 1996).

(available at that time), and was suitable therefore for evaluation of flaws and the determination of structural integrity. It should be pointed out that there has been a concerted and successful effort in France to apply Weibull statistics to the laboratory and field data for nickel-base alloys that fail by stress corrosion cracking in PWR primary systems (Scott *et al.*, 1995; Scott and Benhamou, 2001).

Similar empirical correlations based on plant and laboratory experience have not been developed for stainless steels in BWR environments, largely because of the deficiencies in the plant database, as previously discussed, and the sensitivity of the cracking susceptibility to (unrecorded) combinations of corrosion potential and water purity.

17.2.3 Life prediction based on an understanding of the mechanism of cracking

The scope of this section is limited to an outline of the development of a mechanisms-based life prediction capability for stainless steels in unirradiated conditions in BWRs. This will be expanded upon in Section 17.3.4 to cover the effects of irradiation.

In these mechanisms-based approaches attention was focused on the crack *propagation* period in the chronology of phenomena discussed earlier in Section 17.1. This was done for two reasons. First, there was an urgent need to develop justifiable propagation rate vs stress intensity factor disposition relationships that provide an estimate of the crack advance that might occur before the next inspection. Second, from the viewpoint of predicting overall life, there was experimental evidence (Fig. 17.7) which indicated that crack propagation of a single dominant crack occurred relatively quickly after loading, with 'deep crack' response being observed when the crack depth was greater than about 50–100 μm. In other words, the initial precursor, and microscopic crack initiation and coalescence periods were less significant than the propagation period in these BWR systems.

Slip-oxidation mechanism for crack propagation

Numerous crack advancement theories have been proposed that relate crack propagation to oxidation and the stress/strain conditions at the crack tip, and these have been supported by a correlation between the average oxidation current density on a straining surface and the crack propagation rate for a number of systems (Parkins, 1979; Ford, 1982a; 1982b).

Experimentally validated elements of these earlier hypotheses have been incorporated into the current slip-oxidation model, which relates crack propagation to the oxidation that occurs when the protective film at the crack tip is damaged (Logan, 1952; Vermilyea, 1972; Beck, 1974). In particular, quantitative predictions of the crack propagation rate are based on the Faradaic relationship between the oxidation charge density on a surface and the amount of metal transformed from the metallic state to the oxidized state (e.g., MO or M^{z+}_{aq}). The change in oxidation charge density with time following the rupture of a protective film at the crack tip is shown schematically in Fig.

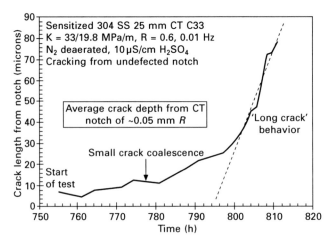

17.7 Crack depth-time relationship for intergranular cracks initiating, coalescing and propagating in notched 1T CT specimen of sensitized stainless steel in 288°C water (Andresen *et al.*, 1990).

17.8(a) and the associated oxidation current density vs time relationship is shown in Fig. 17.8(b).

Once the protective oxide, with a fracture strain, ε_f, is ruptured, the crack will advance rapidly into the metal by oxidation (dissolution or solid state oxide formation) of the bare surface but will, within a matter of milliseconds, begin to slow down as the thermodynamically stable and protective oxide grows at the crack tip. Sustained crack advance depends, therefore, on maintaining a strain rate in the vicinity of the crack tip that will promote further rupture of the oxide film after a time period $[\varepsilon_f/(d\varepsilon/dt)_{ct}]$.

Thus the crack propagation rate, V, is governed (Ford, 1996) by the relationship:

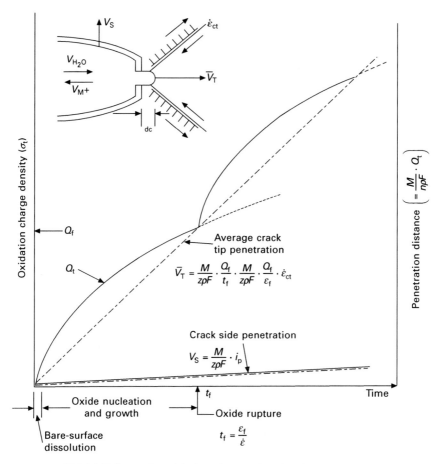

17.8 (a) Schematic oxidation charge density-time relationship for strained crack tip and unstrained crack sides (Ford, 1982a, 1982b). (b) Changes in oxidation current density following the rupture of the oxide at the crack tip (Ford, 1988).

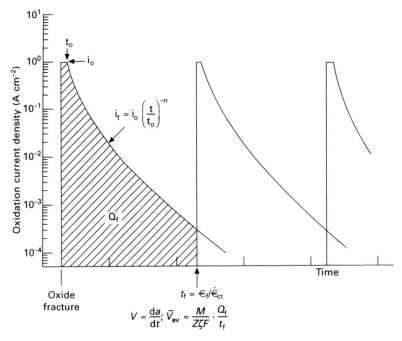

$$i_t = i_o \left(\frac{t}{t_o}\right)^{-n}$$

Q_f

Oxide
fracture

$$t_f = \epsilon_f / \dot{\epsilon}_{ct}$$

$$V = \frac{da}{dt}; \; \bar{V}_{av} = \frac{M}{Z\zeta F} \cdot \frac{Q_f}{t_f}$$

17.8 Continued.

$$V = [MQ_f/z\rho F] \; (1/\varepsilon_f) \; (d\varepsilon/dt)_{ct} \qquad\qquad 17.3$$

and if the passivation relationship in Eq. 17.4 is used,

$$i = i_0 \; [t/t_o]^{-n} \qquad\qquad 17.4$$

then Eq. 17.3 becomes:

$$V = (M/z\rho F) \; (i_o t_o{}^n/(1-n) \; \varepsilon_f{}^n) \; (d\varepsilon/dt)_{ct}{}^n \qquad\qquad 17.5$$

where M, ρ = atomic weight and density of the crack tip metal, F = Faraday's constant, z = Number of electrons involved in the overall oxidation of an atom of metal, i_o, t_o = bare surface oxidation (dissolution) current density and its duration time, ε_f = fracture strain of the oxide at the crack tip, and $(d\varepsilon/dt)_{ct}$ = crack tip strain rate.

It has long been recognized from plant and laboratory observations that there are conjoint material, stress and environment requirements that have to be met to sustain stress corrosion cracking in stainless steels (and other materials) in BWR systems (Fig. 17.1). The reason why the slip-oxidation model was adopted as the working hypothesis was that the basic controlling parameters at the crack tip in that model (creation of a localized environment at the crack tip, the periodic rupture of the protective oxide and the oxidation kinetics on the bared crack tip surface) (Fig. 17.9) could be correlated with these empirical observations (Fig. 17.10).

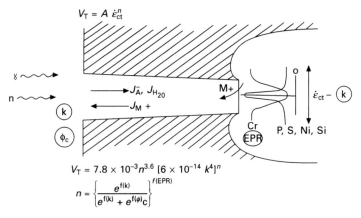

$$V_T = A\,\dot{\varepsilon}_{ct}^{\,n}$$

$$V_T = 7.8 \times 10^{-3} n^{3.6} \left[6 \times 10^{-14}\, k^4 \right]^n$$

$$n = \left\{ \frac{e^{f(k)}}{e^{f(k)} + e^{f(\phi)c}} \right\}^{f(EPR)}$$

17.9 Schematic of the crack enclave and the relevant phenomena associated with the slip-oxidation mechanism of crack advance (Ford, 1988; Ford *et al.*, 1987).

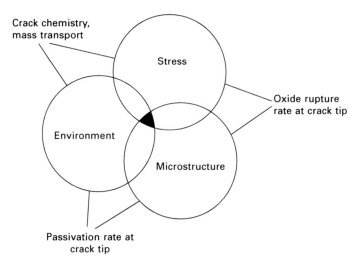

17.10 Relationship between the empirical conjoint factors necessary for EAC and the crack tip mechanistic factors, oxide rupture frequency, passivation rate and solution renewal rate (Ford *et al.*, 1987).

To reduce the prediction algorithm (Eq. 17.5) according to the slip-oxidation model to practical usefulness, it is necessary to redefine that algorithm in terms of measurable engineering or operational parameters. This involves:

- defining the crack tip alloy/environment composition in terms of bulk alloy composition (and fabrication heat treatment), anionic concentration or conductivity, dissolved oxygen content or corrosion potential, etc.

- measuring the reaction rates for the simulated crack tip alloy/environment system that corresponds to the 'engineering' system, and
- defining the crack tip strain rate in terms of continuum parameters such as stress, stress intensity factor, loading frequency, etc.

Extensive work (Ford, 1988; Ford et al., 1987; Andresen, 1987) has been conducted in these areas and, for brevity, it will not reviewed in any detail. The crack propagation rate for stainless steels in 288°C water is given by:

$$da/dt = 7.8 \times 10^{-3} \, n^{3.6} \, (d\varepsilon/dt)_{ct}{}^{n}$$
17.6

where da/dt is in units of cm/s, n varies between 0.33 and 1.0 (Fig. 17.11) depending on the material sensitization (or EPR), anionic impurity concentration (or conductivity, κ) and corrosion potential (Φ_c):

$$n = \left\{ \frac{e^{f(\kappa)}}{e^{f(\kappa)} + e^{g(\phi_c)}} \right\}^{h(\mathrm{EPR})}$$
17.7

The formulation of the crack tip strain rate in Eq. 17.6 has proved particularly difficult given the necessity to describe this parameter in terms of, for instance, a stress intensity factor. However, Shoji and co-workers (Shoji and Suzuki, 1978; Shoji et al., 1995, 2010; Shoji, 2003; Eason et al., 2005) formulated the crack tip strain rate relationship (Eq. 17.8) (a) to take into account these complex strain rate factors in front of an advancing crack tip and (b) to include the expected contributions due to work hardening coefficient, yield strength, degree of plastic constraint, and dynamic applied loads:

$$(d\varepsilon/dt)_{ct} = \beta \cdot (\sigma_y/E) \cdot (m/(m-1)) \cdot ((2 \, (dK/dt)/K)$$

$$+ (da/dt)/r) \cdot (\ln \cdot ((\lambda/r) \cdot (K/\sigma_y)^2))^{1/(m-1)}$$
17.8

where m = strain hardening coefficient in the Gao equation (Gao and Hwang, 1981), σ_y, E = yield strength and elastic modulus, respectively, K = stress intensity factor, and λ, β are constants.

The general validity of the prediction methodology described by Eq. 17.6 is indicated in Fig. 17.12 which covers sensitized Type 304 stainless steel in 8 ppm oxygenated water at 288°C, stressed over a wide range of constant load, monotonically increasing load and cyclic load conditions (Ford et al., 1987; Ford, 1988). The solid line is the theoretical relationship from Eq. 17.6 and the appropriate strain rate formula for the stressing mode. The agreement between prediction and observation illustrates the applicability of the methodology to the whole stress corrosion/corrosion fatigue spectrum for these very specific conditions of corrosion potential, anionic conductivity and degree of grain boundary sensitization.

The discussion of the mechanism of stress corrosion crack propagation for stainless steels in BWRs is now continued from the standpoint of providing a

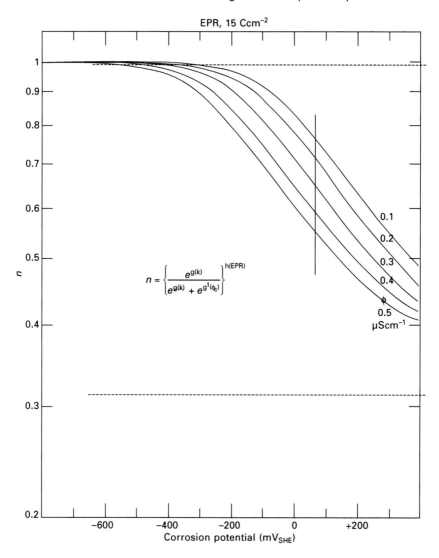

17.11 Relationships between the factor *n* in Eqs 17.4 and 17.5 and the corrosion potential and bulk solution conductivity for a sensitized (EPR = 15 C/cm²) Type 304 stainless steel stainless steel in water at 288°C (Ford *et al.*, 1987). The dashed lines indicate the maximum and minimum values of *n*.

basic understanding of the sensitivities of the cracking susceptibility observed in the laboratory and in the plant to stress, material and environmental parameters, and the effect of *interactions* between these parameters.

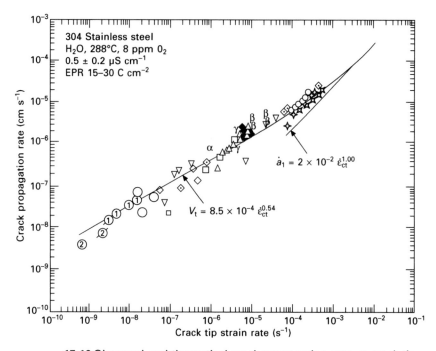

17.12 Observed and theoretical crack propagation rate vs crack tip strain rate relationships for Type 304 sensitized 304 stainless steel in oxygenated water at 288°C (Ford *et al.*, 1987; Ford, 1988). The data points that are numbered were obtained under constant load, the Greek letter points under constant applied strain rate and the geometric symbols under fatigue conditions with varying mean stress, stress amplitude and frequency.

17.3   Parametric dependencies and their prediction

The observed and predicted dependencies of the crack propagation rate of unirradiated and irradiated stainless steels on various BWR system parameters are now addressed.

17.3.1   Environmental parameters

The two prime environmental factors that control the stress corrosion crack propagation rate for austenitic stainless steels in BWR coolant are the corrosion potential and the anionic impurity concentration. Temperature also plays a role, but generally is a small factor unless BWR start-up and shutdown are considered.

The specific corrosion potential of a metal surface is determined by the corrosion rate of the metal surface (e.g., oxidized vs freshly machined), the dissolved O_2, H_2O_2, H_2 concentrations, the water flow rate and the temperature. A representative relationship between corrosion potential for oxidized stainless steel and the dissolved oxygen content is shown in Fig. 17.13; these data were obtained from operating BWR plant and show considerable variability, indicative that the dissolved oxygen content is not the only controlling variable. This factor is emphasized in Fig. 17.14, which shows the significant impact of the dissolved hydrogen and hydrogen peroxide concentrations on the corrosion potential (Lin *et al.*, 1996).

Other variables of importance are the flow rate and hydrodynamics at the metal surface (since the reactant concentrations and reaction rates at that surface can be dominated by liquid diffusion) and the temperature (since the reaction rates are temperature activated).

It is apparent in Fig. 17.15 that the crack propagation rate of sensitized Type 304 stainless steel is very sensitive to the corrosion potential. These data were obtained under constant load conditions where the corrosion potential has been changed via irradiation and/or the concentration of dissolved oxygen and hydrogen, and spans the range of measured and predicted values for core components, unirradiated piping (e.g., recirculation lines) and components operating under hydrogen and noble metal chemistry (Andresen, 1995b). This sensitivity is accurately predicted via the slip-oxidation mechanism, as indicated by the solid lines.

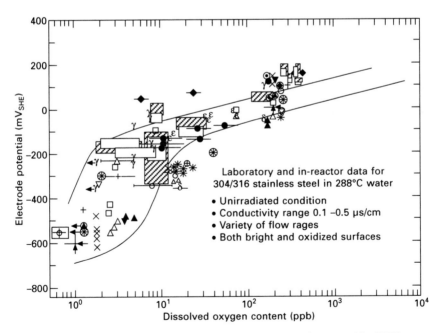

17.13 Variation of corrosion potential for stainless steel in BWR coolant as a function of the dissolved oxygen content (Ford *et al.*, 1987).

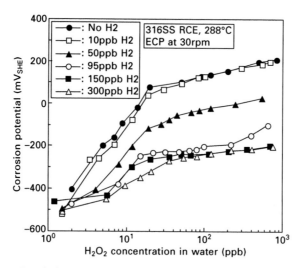

17.14 Laboratory data indicating variation of corrosion potential of stainless steel with dissolved hydrogen peroxide and hydrogen (Lin *et al.*, 1996).

Many corrosion problems may be encountered on metal surfaces where cation or anion concentration increases or metal salt/oxide precipitation occurs; these phenomena may be associated with the presence of crevices in oxidizing environments, or boiling at heat transfer surfaces. To minimize these potential localized corrosion problems, the BWR water coolant purity is maintained at a high level. For instance, in many current operating reactors the coolant purity approaches that of theoretical purity water (whose room temperature conductivity is 0.055 μS/cm), with dissolved chloride and sulfate activities typically below 5 ppb. Early observations (Brown and Gordon, 1987; Gordon and Brown, 1989) indicated that the incidence of plant component cracking correlated well with the plant reactor water conductivity. Subsequent laboratory measurements on the austenitic (and ferritic) alloys confirmed correlations between the crack propagation rate and conductivity under both steady state and transient water purity conditions (Fig. 17.16 and 17.17, respectively).[1]

The observed relationships between the crack propagation rate and the corrosion potential and conductivity shown in Figs 17.15, 17.16 and 17.17 can be predicted via the slip-oxidation model, as indicated in these graphs. As a result, the model forms part of the fundamental rationale in the EPRI Water Chemistry Guidelines (EPRI, 2000, 2004) and has been used, for example,

[1]Note that the range in predicted crack propagation rates shown in Fig. 17.16 (a) and (b) correspond to the range of corrosion potentials expected for an environment defined solely as '200 ppb O_2'.

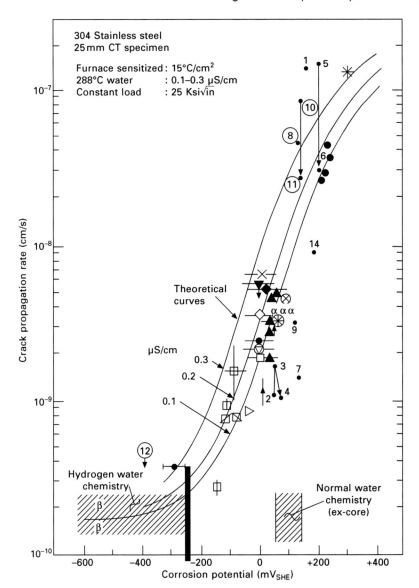

17.15 Observed and predicted relationships between the crack propagation rate and corrosion potential for sensitized Type 304 stainless steel in 288°C water under constant load. Water conductivity in range 0.1–0.3 μS/cm. The three predicted relationships are for 0.1, 0.2 and 0.3 μS/cm. Figure adapted from Ford *et al.* (1987).

to assess the damage in plant due to conductivity transients occasioned by the failure of deep bed demineralizers (White and Cubicciotti, 1989).

The crack tip anionic activity – not the conductivity of the bulk water – is

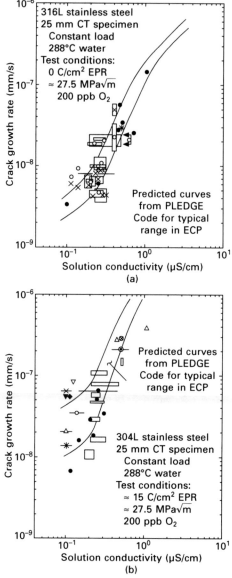

17.16 Observed and predicted relationships between the crack propagation rate and water conductivity for (a) Type 316L stainless steel and (b) sensitized Type 304 stainless steel under constant load (25 ksi√in) in 288°C water containing 200 ppb oxygen (Ford *et al.*, 1987; Ford, 1988). Figure adapted from Ford *et al.* (1987).

of fundamental importance, and the cracking susceptibility is anion specific. For instance, strong acid-forming anions such as sulfate (Davis and Indig, 1983; Ruther and Kassner, 1983; Andresen, 1983, 1991, 1993a; Indig *et*

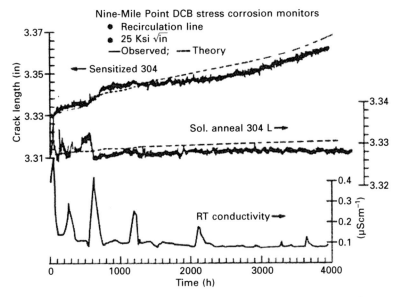

17.17 Observed and predicted crack depth vs time relationships for sensitized Type 304 and solution annealed Type 304L stainless steels during conductivity transients observed on pre-cracked specimens inserted into the recirculation line at Nine Mile Point (Ford *et al.*, 1989).

al., 1983; Kurtz, 1983; EPRI, 1985; Shack *et al.*, 1983, 1985, 1986; 1987; Ruther *et al.*, 1988; Ljungberg *et al.*, 1989, 1995; Hale, 1991), or chloride (Gordon, 1980; Congelton *et al.*, 1990; Andresen, 1991), which can originate from either resin decomposition in the demineralizer beds or leakage of impurities in defective condenser tubes can have a significant effect on the crack initiation and propagation rate of all the ferritic and austenitic alloys in BWR environments. The theoretical predictions of the effect of corrosion potential and anionic activity on the intergranular stress corrosion crack propagation rate of sensitized Type 304 stainless steel is indicated in Fig. 17.18 for chloride anions. Under 'normal' water chemistry conditions corresponding to a corrosion potential of approximately +50 mV$_{SHE}$ for a recirculation line, increasing the anionic activity from a preferred operating value of <5 ppb to an 'actionable' (EPRI, 2000) level of 20 ppb or, in a worst case scenario of 100 ppb (warranting immediate plant shutdown), the crack propagation rate could increase up to two orders of magnitude. Under lower corrosion potential conditions associated with hydrogen water chemistry or noble metal technology, it is predicted and confirmed (Andresen *et al.*, 2003b) that the increases in propagation rate for stainless steels due to such anionic transients are markedly lower. Other anions (such as nitrate, nitrite and chromate) do not have the effect on crack propagation rate that

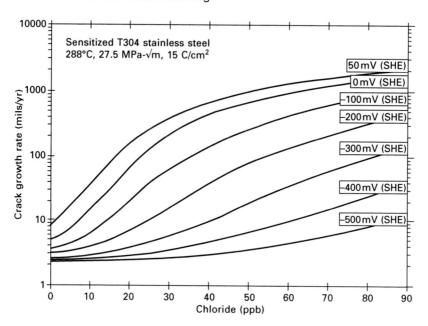

17.18 Crack growth rate predictions as a function of chloride concentration for sensitized stainless steel stressed at 27 MPa√m in 288°C water at various corrosion potentials (EPRI, 2000, 2004).

chloride and sulphate anions do, primarily because they are not stable at the low corrosion potentials at the crack tip. Effects of other dissolved species such as zinc (Andresen, 1992a; Andresen and Angeliu, 1995), and silicon (either as an alloying element or dissolved species) (Andresen and Morra, 2005; Andresen, 1993b; Hettiarachchi *et al.*, 1995a) have been addressed and interpreted in terms of their effects on the cracking mechanism (Angeliu, 1995). These effects and flow rate effects are discussed in detail elsewhere (Andresen 1991, 1993c, 1995a, 1997; EPRI, 1992a, 1992b, 1999; Fong, 2007; Molander *et al.*, 1997).

The stress corrosion susceptibilities for BWR material/environment systems are thermally activated, and this is particularly important during start-up and shutdown. This is observed (Hale, 1986; Hale and Pickett, 1984; Ruther *et al.*, 1988; Andresen, 1980, 1984, 1992b; EPRI, 1997; Ford and Povich, 1979; Indig, 1980; Kawakubo *et al.*, 1980) in both initiation and propagation rate tests for all BWR structural materials. The data in Fig. 17.19 illustrate that the EAC susceptibility/temperature relationship for stainless steels is complicated and does not necessarily follow an expected 'Arrhenius' relationship because in many cases there are several rate controlling processes operative over the range in temperature. From a practical perspective, the strong temperature dependence noted for cracking in the laboratory is mirrored in the plant

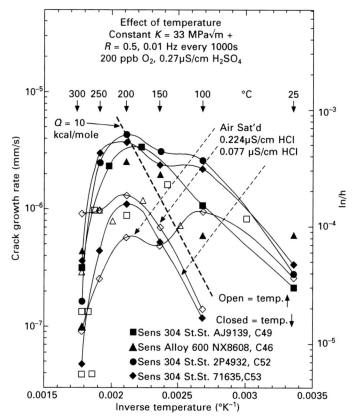

17.19 Crack propagation rate as a function of test temperature for sensitized stainless steel under constant load in water of various purities (Andresen, 1992b).

where cracking is largely confined to components operating at temperatures > 250°C (Kilian *et al.*, 2005).

On the basis of the observations and theoretical predictions of the effects of corrosion potential and coolant purity on the EAC susceptibility, EPRI issued 'BWR Water Chemistry Guidelines' (Fruzzetti and Blok, 2005) in 2000 (EPRI, 2000) and 2004 (EPRI, 2004). In part, these guidelines are designed to minimize stress corrosion cracking in Type 304/316 stainless steels during different operating modes (start-up/hot standby, power operation, etc.). The VGB (VGB, 2006) also issued water chemistry guidelines for German BWRs that were slightly less restrictive, partly because the SCC susceptibility of the stabilized stainless steels use in Germany tends to be lower.

17.3.2   Material condition

The material conditions of stainless steels that are of relevance to stress corrosion cracking in high temperature water include:

- the alloy composition
- the alloy microstructure, especially the extent of grain boundary sensitization
- the yield strength
- the extent of surface damage/cold work.

Historically, the material attribute that has been associated most with stress corrosion cracking susceptibility in austenitic stainless steels has been the extent of chromium depletion at the grain boundary due to the growth of $M_{23}C_6$ carbides, including how this is altered by material composition (e.g., the presence of stabilizing elements), stress (i.e., stress-induced martensite formation) and heat treatment (Fig. 17.20). Grain boundary sensitization can lead to intergranular crack initiation and propagation, with the crack propagation rate depending on the other material attributes (yield strength, etc.), as well as the water chemistry and loading. This grain boundary sensitization effect

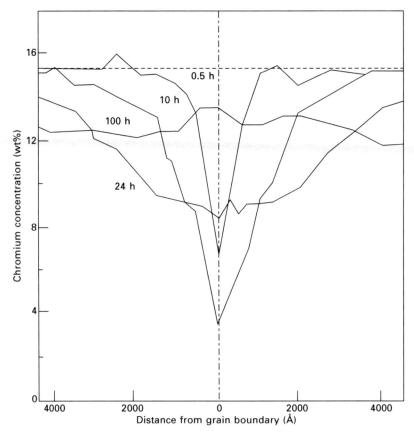

17.20 Chromium depletion associated with $Cr_{23}C_6$ precipitation following various aging times at 700°C.

was illustrated in Fig. 17.16 where the crack propagation rate at a given water purity and corrosion potential for sensitized Type 304 stainless steel is an order of magnitude higher than for unsensitized Type 316L stainless steel. As indicated in Fig. 17.21, cracking is sensitive to small degrees of grain boundary sensitization and, in particular, the minimum chromium content. A detailed discussion of methods of measuring and diminishing the effect of sensitization on the SCC of stainless steels in BWRs is outside the scope of this chapter, but is found in references such as Bruemmer *et al.* (1988a, 1988b), Bose and De (1987), Briant (1982), Clarke (1980), Clarke *et al.* (1978), Kilian *et al.* (1995), Solomon (1980, 1984), Solomon and Lord (1980) and Stickler and Vinckier (1961).

In the light of this understanding, well-proven and predictable mitigation actions based on minimizing the extent of grain boundary sensitization include:

- controlling the heat input to the structure during fabrication
- decreasing the carbon content (as in the L-grade 300-series stainless steels) – this causes a drop in yield and tensile strength, which can be counteracted by the addition of nitrogen (i.e., the LN or NG grade 300-series stainless steels)
- annealing the material prior to welding to remove the cold work that accelerates the sensitization process

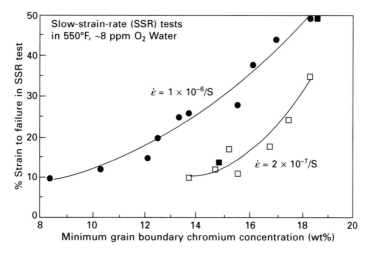

17.21 Influence of grain boundary chromium content on the % strain to failure during a slow strain rate test on sensitized stainless steel in 288°C water. Note that less than a 1% drop in Cr content adjacent to the grain boundary from the 18% in the bulk material can have a significant effect on the IGSCC susceptibility. The two curves are associated with the two applied strain rates used in the slow strain rate SCC test (Bruemmer *et al.*, 1993).

- decreasing the heat input or increasing the heat extraction rates during welding by using low weld heat inputs and low interpass temperatures, and/or by water cooling the inside of pipes after the initial weld root pass
- adding other strong carbide formers which counteract the grain boundary $Cr_{23}C_6$ carbide precipitation, such as niobium (Type 347 stainless steel) and titanium (Type 321 stainless steel).

It has been recognized for a long time that cold work may promote the initiation and propagation of stress corrosion cracks in structural alloys in BWRs and PWRs. This recognition started with observations in the 1960s of the effect of 'abusive' grinding applied to welds in BWR stainless steel piping. More recently, similar deleterious effects have been noted for Type 304 stainless steel BWR core components that have been abusively ground and/or cold worked during fabrication (Medoff, 1996; Horn et al., 1997; Suzuki et al., 2004; Ando and Nakata, 2007). In these more recent cases, the observations have been made on L-grade stainless steels operating with good water purity. It is apparent that the beneficial effects of modifying the alloy composition and microstructure, or controlling the water chemistry, may be negated if there is a significant degree of surface or bulk cold work in the structure.

The interactions between various forms of cold work and the crack propagation rate are illustrated schematically in Fig. 17.22. The primary effects of cold work are on: (a) the residual stress profile, (b) the yield strength (or hardness), and (c) transformations in the material microstructure. These primary effects are impacted by secondary issues such as the local

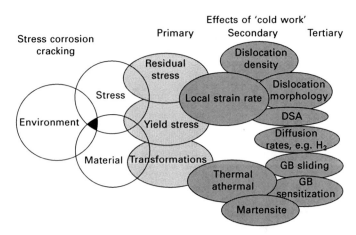

17.22 Interactions between the various parameters associated with 'cold work' and their effect on the conjoint 'material' and 'stress' conditions for stress corrosion crack propagation.

strain rate and the details of the thermally activated and athermal structural transformations, and these in turn, will be impacted by a plethora of tertiary issues such as dislocation density and morphology, diffusion rates of absorbed hydrogen or nitrogen, dynamic strain aging, etc. These secondary and tertiary effects will not be discussed further in this chapter. Attention, however, is now focused on the prediction of the effect of cold work and yield strength effects on SCC. The effect on residual stress and hence SCC will be discussed in the next section.

The fundamental reason for the propagation rate vs yield strength relationship (Fig. 17.23) lies in the formulation of the crack tip strain rate (Eq. 17.8). The absolute accuracy of such a formulation may be questioned in view of the fact that it does not explicitly account for primary and secondary creep phenomena that are likely to dominate the crack tip strain rate under static loading at low crack propagation rates. However, this algorithm has gained wide acceptance and is of practical relevance in that it addresses relevant engineering parameters such as the yield strength, stress intensity (and variations in stress intensity amplitude and loading rate), strain hardening coefficient, etc. The coupling of the crack tip strain rate algorithm (Eq. 17.8) and the slip-oxidation crack propagation rate model provides good prediction of the effect of yield strength on the propagation rate for stainless steels and Ni-base alloys (Figs 17.24 and 17.25).

The effect of cold work/yield strength on the SCC propagation rate is especially noteworthy for the non-sensitized or stabilized stainless steels at low potential, because at low potential the growth rate of sensitized stainless steels is lower (Fig. 17.26). Mitigating SCC by adopting hydrogen water

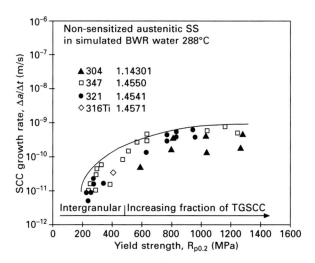

17.23 Crack propagation rate for non-sensitized stainless steels as a function of yield stress (Speidel and Magdowski, 1989).

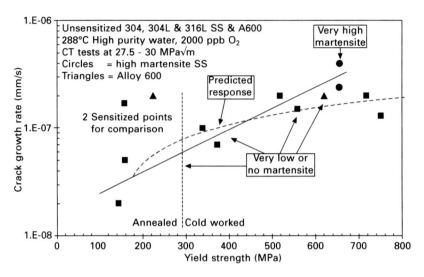

17.24 Effect of yield strength on the crack propagation rate on stainless steel and Alloy 600 in 288°C high purity water at high potential (Andresen *et al.*, 2002).

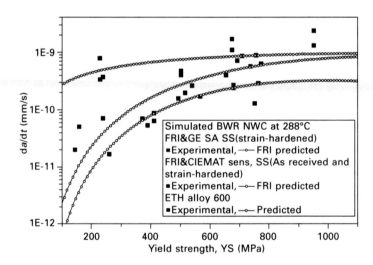

17.25 Observed and predicted crack propagation rate vs yield strength relationships for cold worked Alloy 600, solution annealed 316L SS and sensitized 304 SS in BWR simulated environment at 288°C (Shoji *et al.*, 2010).

chemistry and using unsensitized stainless steels has revealed the sensitivity of SCC to the deleterious effect of bulk and surface cold work, and weld residual strain.

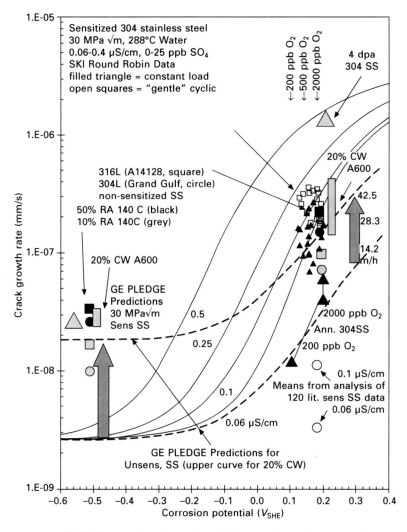

17.26 Observed and predicted changes in crack propagation rate for non-sensitized stainless steels with changes in degree of cold work at reducing and oxidizing corrosion potentials. The arrows indicate the changes in crack propagation rate associated with the transition between annealed and cold worked stainless steel in high purity water. Adapted from Andresen *et al.* (2003b).

17.3.3   Stress and strain localization

Based on experience with many material/environment cracking systems, the stresses must be tensile to initiate and propagate EAC. Shear and compression loading can also restrict access of liquid to the crack tip. Sources of primary and secondary tensile stresses in the BWR reactor components include those originating from fabrication and reactor pressurization/operation. Fabrication stresses are introduced during fit-up and assembly in the shop or in the field, including machining or forming operations (such as surface grinding, welding and cold straightening). The extent of tensile stress due to some of these sources can be surprisingly large even when associated with 'cosmetic' repair; for example, abusive grinding can introduce surface tensile stresses far above the original yield strength. Weld residual stresses will vary with distance from the weld and through the component thickness (Fig. 17.27), and depend on welding parameters such as heat input, welding speed, number of welding passes, etc. (Klepfer *et al.*, 1975; Chrenko, 1980; Shack and Ellingson, 1980; Shack, 1980).

Although these effects are well understood on the basis of data and modeling, the complex parameter interactions give rise to scatter in the measured residual stresses (Fig. 17.27(b)). Furthermore, not all sources of tensile stress are immediately obvious; for instance, sufficient tensile stress for EAC may occur due to growth of corrosion product in creviced geometries where there are no consequential sources of applied or residual stress (e.g., the crevice regions of certain designs of top guides). Neutron fluence relaxes the weld residual stresses in irradiated core structures, but neutron irradiation may also introduce deleterious grain boundary segregation and irradiation-induced hardening (Andresen *et al.*, 1989; Andresen and Ford, 1995; Nelson and Andresen, 1991; Bruemmer *et al.*, 1996, 1999) and such competing effects will give non-monotonic cracking kinetics.

Finally, cyclic stresses from vibrations or from changes in reactor operation can also contribute. Such varying stresses are of great importance in initiating cracks or in restarting stopped cracks because they provide continuing dynamic plastic strain.

An early compilation of crack propagation rate data (da/dt) as a function of stress intensity factor (K) for sensitized stainless steel in oxygenated water at 288°C is shown in Fig. 17.28. Also shown in this figure is the 'USNRC disposition curve' (Hazleton and Koo, 1988), which is used for disposition of cracks in BWR piping:

$$\mathrm{d}a/\mathrm{d}t = 3.590 \times 10^{-8} \cdot K^{2.161} \text{ ins/h} \qquad\qquad 17.9$$

where K = stress intensity factor (ksi√in) and da/dt = crack propagation rate (ins/h). It is apparent that there is a significant variance in the dataset that may be attributed to the range of corrosion potential expected for 0.2 ppm dissolved oxygen and the measured range in solution purity (0.3–0.5 μS/cm) (Ford *et al.*, 1987, 1989).

The crack propagation rate vs stress intensity factor relationships for a non-sensitized or stabilized stainless steel in 200 ppb oxygenated water with a conductivity range 0.3–0.5 μS/cm is indicated in Fig. 17.29(a). The corresponding data for a non-sensitized stainless steel in higher purity water

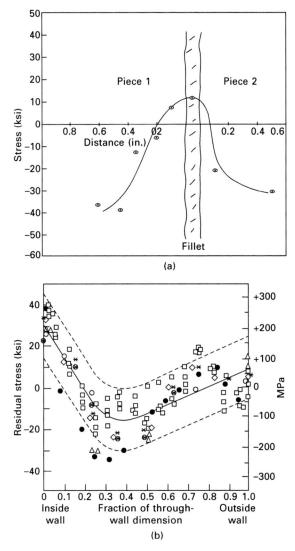

17.27 (a) Variation in longitudinal residual stress adjacent to a weld on the internal pipe surface of a 4′ pipe (Chrenko, 1980). (b) Variation in through-wall residual stress in 28′ dia sch. 80 piping close to the fusion line (Shack, 1980).

is shown in Fig. 17.29(b). In both cases the theoretical ranges in crack propagation rates (corresponding to the range in conductivity) are shown, as is the NRC disposition line (Eq. 17.9). The comparison of prediction vs observation is reasonable (given the ranges of uncontrolled or unmonitored conductivity and corrosion potential) for these combinations of material sensitization and water purity.

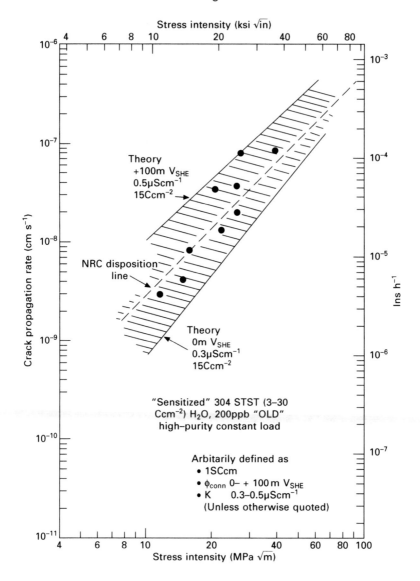

17.28 Crack propagation rate/stress intensity factor dependence for sensitized (EPR 15C/cm^2) stainless steel in water at 288°C with 0.2 ppm oxygen and conductivity 0.3–0.5 μS/cm conductivity. Note the insertion of bounding theoretical relationships based on the defined range in water conductivity range and the range in corrosion potentials associated with '200 ppb O$_2$'. Also shown is the NRC regulatory relationship (Eq. 17.9) used for structural integrity evaluations (Hazelton and Koo, 1988). Figure adapted from Ford *et al.* (1987).

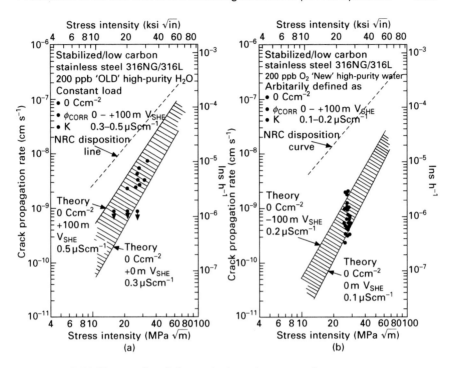

17.29 Observed and theoretical crack propagation rate vs stress intensity factor relationships for non-sensitized, or stabilized stainless steel in 200 ppb oxygenated 288°C water with conductivity in (a) the range 0.3–0.5 μS/cm and (b) 0.1–0.3 μS/cm. Figure adapted from Ford *et al.* (1987).

The corresponding observed and theoretical crack propagation rate vs stress intensity factor relationships for a sensitized (EPR value 15 C/cm^2) stainless steel in hydrogenated water (low corrosion potential) are illustrated in Fig. 17.30. It is apparent from Figs 17.28, 17.29 and 17.30 that the accuracy of the predicted crack propagation rate vs stress intensity factor relationships depend on the accuracy of the definition of the system conditions. It is further apparent that the original NRC disposition relationship based on data produced in relatively impure water is over-conservative given the current specifications for water purity in operating BWRs. The benefit of decreasing the degree of grain boundary sensitization is apparent (Fig. 17.28 vs 17.29(a)) even in 'impure' water, as is the benefit of increasing the purity of the water (Fig. 17.29(a) vs 17.29(b)) and in decreasing the corrosion potential (Fig. 17.30). In all of these examples, the benefit of decreasing the grain boundary sensitization or improving the water purity and decreasing the corrosion potential is reduced as the stress intensity factor increases. This is a direct result of the fact that the *K* dependency is a function of the '*n*' value (i.e.,

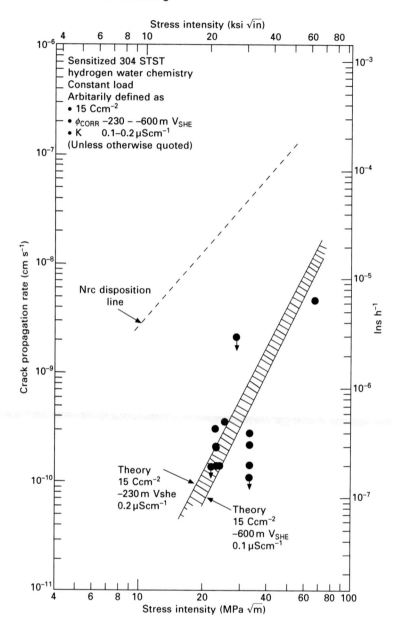

17.30 Observed and theoretical crack propagation rate vs stress intensity factor relationships for a sensitized (EPR 15 C/cm^2) stainless steel in hydrogenated 288°C water with a corrosion potential < –230 mV$_{she}$ in 288°C water with conductivity in the range 0.1–0.2 μS/cm. Figure adapted from Ford *et al.* (1987).

$da/dt = f(K^{4n})$, and the EAC response tends to converge at high K or ΔK (Figs 17.29 and 17.30).

The prediction of the effects of the stress intensity factor on the crack propagation rate is relatively straightforward since the crack tip strain rate may be formulated in terms of that parameter (Eq. 17.8). Prediction of the crack propagation rate is made more difficult, however, when the cold work/yield strength effects occur over very restricted geometries such as, for example, surface machining or plastic constraint immediately adjacent to welds (Fig. 17.31).

17.3.4   Irradiation

The earliest incidents of irradiation assisted stress corrosion cracking (IASCC) in BWRs occurred during the early 1960s and were associated with

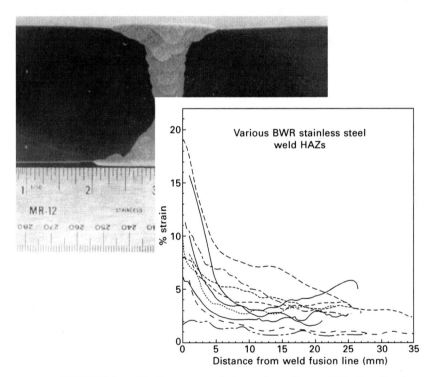

17.31 Weld residual strain vs distance from the weld fusion line for stainless steel welds. If the number of welding passes is limited, the peak residual strain can be below 10% equivalent room temperature tensile strain. However, most pipe welds that have been analyzed show residual strains in the range of 15–20%, with some slightly above 25%. The residual strain is also highest near the root of the weld (Angeliu *et al.*, 1999).

cracking of Type 304 stainless steel fuel cladding, where the driving forces for cracking were the increasing tensile hoop stress in the cladding due to the swelling fuel and the highly oxidizing conditions in the water. Major concerns for IASCC developed in the mid-1970s due to an increasing number of observations of IASCC in highly stressed core components exposed to fast neutron (>1 MeV) irradiation when a 'threshold' fluence level (5×10^{20} n/cm^2) was exceeded (Fig. 17.32). These high stress components included neutron source holders and control rods where the component was being dynamically strained due to swelling Be or B_4C, or in components where there were high fabrication stresses.

Subsequently, IASCC was observed in lower stressed components (instrumentation tubes, control blade handles, control rod sheaves) at fluences $>2 \times 10^{21}$ n/cm^2. Although the presence of crevices in these components exacerbated the problem, it became apparent that other factors were involved. For instance, water purity was observed (Fig. 17.33) to play a major role in the extent of IASCC of IRM/SRM instrumentation tubes (Brown and Gordon, 1987). In addition, cracking of control blade handles was largely confined to those that were cold formed and, in the case of the control blade sheaves, cracking was primarily in the upper portion of the stainless steel sheaving adjacent to the spot welds.

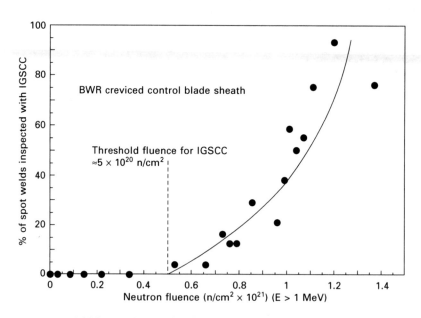

17.32 Dependence of IASCC on fast neutron fluence for creviced control blade sheath in high conductivity BWRs where the tensile stress is high due to the spot welds and dynamic due to the swelling B_4C absorber tubes (Gordon and Brown, 1989).

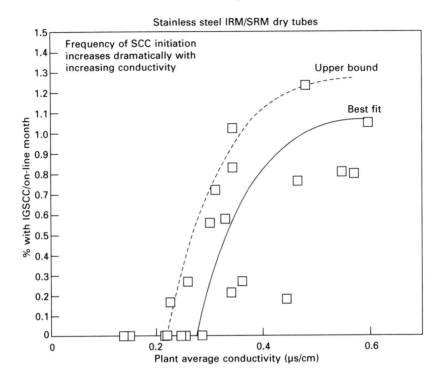

17.33 The effect of average coolant conductivity on the incidence of cracking of IRM/SRM instrumentation tubes (Brown and Gordon, 1987).

IASCC was also observed at *lower* fluence levels than 5×10^{20} n/cm^2 adjacent to the welds in larger stainless steel core structures, such as core shrouds, with the extent of cracking depending on the specific combinations of prior thermal-sensitization due to the fabrication procedure, fast neutron flux and fluence, weld residual stress and coolant purity (Andresen *et al.*, 1989; Nelson and Andresen, 1991; Andresen and Ford, 1995; Bruemmer *et al.*, 1996; Was *et al.*, 2006). Cracking was also observed in top guides which (at least for the GE designs) did not involve welding and where there was no consequential source of applied or residual tensile stresses.

Slow strain rate testing on irradiated stainless steel samples indicated the arbitrary nature of imposing 'threshold fluences' for cracking to occur on reactor components. For instance, the fluence level above which increased amounts of intergranular cracking occurs depends on the dissolved oxygen content and/or the presence of dissolved hydrogen. This is undoubtedly associated with changes in corrosion potential, and this conclusion is supported by the more extensive database in Fig. 17.34 in which it is apparent that, although the extent of cracking increases with increasing fluence, it could

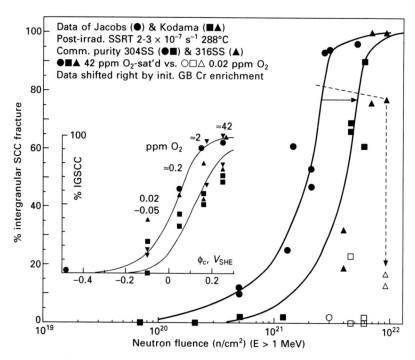

17.34 Dependence of IASCC on fast neutron fluence in slow strain rate tests at $3.7 \times 10^{-7}\,s^{-1}$ on irradiated Type 304 stainless steel in 288°C water (Jacobs *et al.*, 1986).

be minimized under these aggressive loading conditions by lowering the O_2 level to 5 ppb.

Changes in grain boundary chromium content due to both thermally induced fabrication effects and/or irradiation-induced segregation can have a significant effect on the cracking susceptibility of irradiated stainless steels over time. This conclusion is supported by the observation that although the substitution of Type 304 stainless steels with L-grade or stabilized stainless steels in the irradiated components increased the operating time before cracking was observed, IASCC incidents still occurred (Medoff, 1996; Suzuki *et al.*, 2004; Horn *et al.*, 1997) after approximately 8–9 effective full power years (EFPY), especially if the component was severely surface cold worked by grinding, straightening, etc.

It became apparent from plant experience that the operating stress, environment and material parameters that affected SCC in irradiated components were also noted in non-irradiated components. Thus, it was reasonable to propose that IASCC was not a new mechanism of cracking, but was merely mirroring irradiation-induced changes to the rate-controlling parameters in the various phases in crack development described in Section 17.1.

Irradiation effects include changes to the corrosion potential, applied tensile stress, grain boundary composition and yield strength. Those changes that are caused by neutron *flux* (i.e., corrosion potential) have an effect immediately on the cracking susceptibility, whereas *fluence*-driven changes (such as radiation-induced segregation, radiation hardening and stress relaxation) have a cumulative effect over an extended time.

It was hypothesized that the slip-oxidation mechanism for crack propagation was also applicable to irradiated stainless steels, and that Eq. 17.6 applied. Detailed prediction required the identification of the roles of neutron flux and fluence and gamma irradiation on the corrosion potential, stress relaxation, the yield strength and the grain boundary chemistry.

As indicated schematically in Fig. 17.35, this approach was fully consistent with the empirical observations of the conjoint requirements of stress, environment and material condition, and merely expanded on the approach taken for unirradiated stainless steels. The specific effect of irradiation on the fundamental parameters that control crack propagation in stainless steels was quantified by considering:

- the corrosion potential and its change with radiation flux
- irradiation-induced changes in grain boundary composition
- irradiation-induced hardening and
- for displacement loaded structures, radiation creep stress relaxation.

These developments and the background information of, for example,

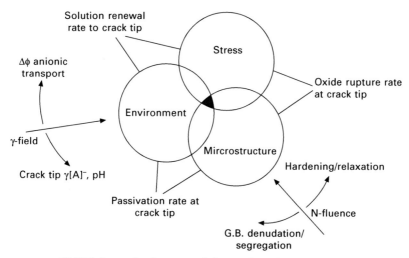

17.35 Schematic diagram of the engineering parameters (stress, environment and microstructure), underlying the fundamental rate controlling processes (mass transport, oxide rupture, and repassivation rates) and effects of radiation (Andresen and Ford, 1995; Andresen *et al.*, 1989).

irradiation-induced grain boundary segregation, hardening, etc., have been extensively reviewed elsewhere (Andresen *et al.*, 1989; Nelson and Andresen, 1991; Andresen and Ford, 1995; Bruemmer *et al.*, 1996; Was *et al.*, 2006), and this chapter will provide highlights.

Radiolysis of H_2O leads to various constituent species, including H_2O_2 and H_2 (the longer-lived species) as well as radicals (e.g., e_{aq}^-, H, OH, HO_2) (Lin, 1985; Dawson and Sowden, 1963; Cohen, 1969) and, while stoichiometric quantities of oxidizing and reducing species are formed, the corrosion potential generally increases, sometimes dramatically. The effect of this corrosion potential increase by itself is predictable as illustrated by the data at the higher corrosion potential ranges in Fig. 17.15, which were obtained in a variable energy cyclotron where the irradiation could be turned off and on.

Further evidence of the predictable effect on the cracking susceptibility due to irradiation-induced changes in corrosion potential is shown in Fig. 17.36, where the observed and predicted crack depth vs time relationships are compared for furnace sensitized stainless steel specimens exposed to water in either the unirradiated recirculation line or in a core instrumentation tube. The observed and predicted effect of irradiation in increasing the crack propagation rate via its effect on the corrosion potential alone is apparent.

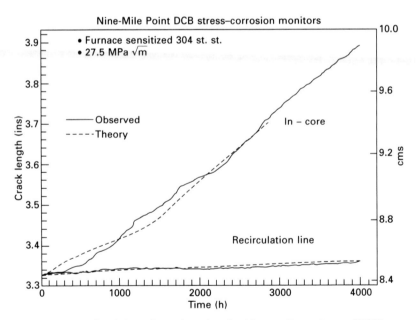

17.36 Crack length vs time for double cantilever beam (DCB) specimens of sensitized Type 304 stainless steel exposed in core (high corrosion potential from radiolysis) and in the recirculation system (Andresen and Ford, 1995).

There is a well-recognized tendency for segregation and depletion of alloying elements at the austenite grain boundaries due to the presence of irradiation-induced increases in interstitials and vacancies, and these compositional changes may be considerable with accumulating fluence. For instance, silicon may increase by a factor of >5–10 after 5 dpa. Initial attention has been placed on the chromium depletion (Fig. 17.37) since this is known to have reproducible effects on the SCC susceptibility. The following empirical relationship was developed to account for, the effect of the initial degree of chromium depletion due to thermal sensitization and the further grain boundary sensitization from irradiation induced segregation.

$$\text{EPR}_{\text{effective}} = \text{EPR}_0 + 3.36 \times 10^{-24} (\text{fluence})^{1.17} \qquad 17.10$$

where EPR_0 is the grain boundary sensitization due to thermally activated chromium depletion and the fluence is in units of n/cm^2. Note here that in the analysis of SCC of welded and irradiated structures, it is assumed that the radiation-induced chromium depletion is additive to thermally induced sensitization (from heat treatment or welding). SCC is controlled by the minimum grain boundary Cr content, even for very narrow Cr depletion profiles, which can be generated during complex, multi-step heat treatments (Andresen, 2002). In the mechanisms-based prediction methodology, the minimum Cr level (or effective EPR value) controls the value of n in Eq. 17.7.

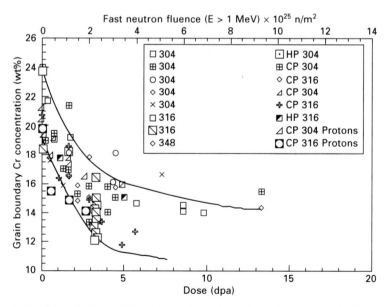

17.37 Dose dependence of grain boundary chromium concentration for 300 series austenitic stainless steels irradiated at temperatures about 300°C (Was et al., 2006).

Irradiation generated defects also give rise to an increase in yield strength (Fig. 17.38) and a localization of deformation to channels in the material. An increase in the yield strength from 150–200 MPa up to 750–1000 MPa is commonly observed, with saturation after several dpa. In the mechanisms-based prediction methodology, this irradiation fluence modified yield strength is inserted into a crack tip strain rate algorithm (Eq. 17.8). The final irradiation modification is associated with irradiation induced creep, which leads to stress relaxation in displacement loaded structures, such as welded components, bolts, etc. (Fig. 17.39).

Fast neutron irradiation can affect all of the controlling parameters for stress corrosion cracking, and these effects take place over different time and geometry scales. For instance, the neutron flux will have an immediate effect on the corrosion potential, whereas radiation-induced segregation at grain boundaries, increases in yield strength, and relaxation of residual stresses will all take effect over time as fluence increases. With regard to the geometrical aspect, the neutron flux will be attenuated through the thickness dimension of a thick component such as a core shroud; thus, not only is the fluence changing over time, but it is changing at the crack tip as it propagates into the component.

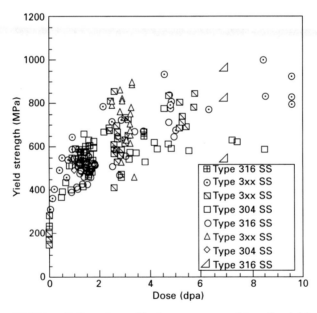

17.38 Irradiation dose effects on measured tensile yield strength for several 300-series stainless steels, irradiated and tested at a temperature of about 300°C (Bruemmer *et al.*, 1999; Was *et al.*, 2006).

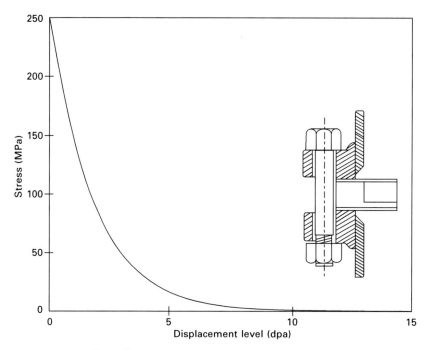

17.39 The effects of radiation-induced creep on load relaxation of stainless steel in a constant displacement (bolt) condition (Van Duysen *et al.*, 1993).

17.4   Prediction of stress corrosion cracking (SCC) in boiling water reactor (BWR) components

The preceding discussion has focused on the dependency of SCC propagation on various system parameters, and the prediction capability for these dependencies is based on an understanding of the mechanism of crack propagation. The comparisons between observation and prediction have been made on data produced in the laboratory. The discussion in this section uses the current mechanistic understanding to predict the SCC behaviour of BWR components (Andresen *et al.*, 1996, 1999b).

It is assumed in the subsequent discussion that 'microscopic' crack initiation corresponds to the creation of a crack of depth 50 μm, and that this is achieved relatively early in the overall life of the component. The factors behind this assumption were discussed in Section 17.2.3, along with the experimental evidence in Fig. 17.7. Thus, the subsequent crack depth vs time relationship may be calculated by integrating Eq. 17.6 for an appropriately defined alloy/environment system. A major uncertainty in this process is the definition of the residual stress profile adjacent to welds which, as illustrated in Fig. 17.40 can exhibit variability. Mean and bounding values

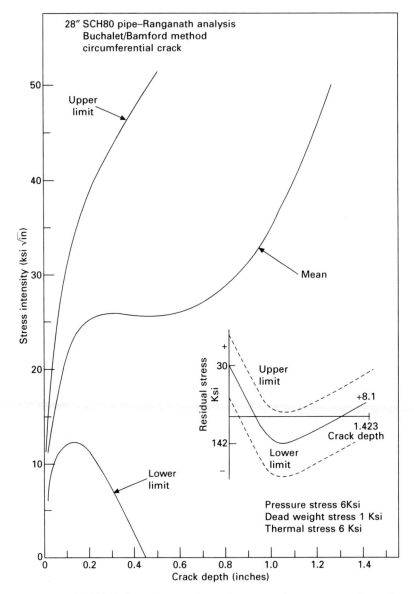

17.40 Variation of stress intensity as crack progresses through wall thickness for 28″ sch. 80 piping associated with the scatter of residual stress shown in Fig. 17.27(b).

for the residual stress profile can be defined for a given weld geometry, and the corresponding stress intensity factor as a function of defect size can be computed (Buchelet and Bamford, 1976).

Figure 17.41 shows the comparison between predicted and observed cracking adjacent to welds in 28″ schedule 80 recirculation piping at two BWR plants with markedly different water purity. The residual stress profile was not known, and thus the prediction can only evaluate the range in crack depth vs time relationships for the range in residual stresses expected for this pipe weld (Fig. 17.40).

Despite uncertainties, it is possible to predict the time before cracks will be observed (i.e., when the crack depth exceeds the NDE resolution limit) and to predict the range of crack depths within a given plant piping. The effect of water purity on the cracking susceptibility of these recirculation lines is apparent in Fig. 17.41 in that cracks are predicted and observed after only 32 operating months if the water purity control is poor (i.e., 0.326 µS/cm), but it is predicted that it would be more than 220 months before cracks would be observed in higher purity water (0.15 µS/cm). The specific effect of water purity control on the predicted and observed cracking at a number of plants for 28″ sch. 80 recirculation lines is further illustrated in Fig. 17.42, where the predicted relationships are based on the mean residual stress profile in Fig. 17.40.

A more expanded database is examined in Fig. 17.43 for a series of plants where the time for the crack to penetrate to a quarter wall thickness is recorded as a function of the geometric mean of the water conductivity at those plants. Against each data point is recorded the number of cracks noted. These observed data exhibit a large scatter that is hard to analyze if the data were the only input to the analysis. It is seen, however, that the data are consistent with the mechanisms-based predictions that indicate the theoretical relationships for three combinations of assumed degree of sensitization and residual stress profile. For instance, the earliest failures, which would occur at the worst combination of sensitization and stress, are

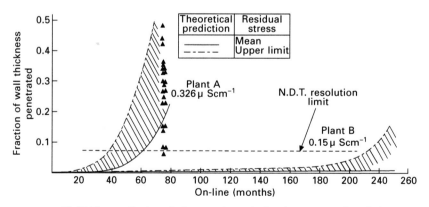

17.41 Theoretical and observed crack depth vs operational time relationships for 28″ diameter sch. 80 304 stainless steel piping for two boiling water reactors operating at different mean coolant conductivities. Figure adapted from Ford et al. (1988).

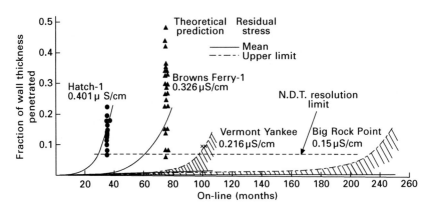

17.42 Theoretical and observed crack depth vs operational time relationships for 28" diameter sch. 80 304 stainless steel piping for various boiling water reactors operating at different mean coolant conductivities. Figure adapted from Ford *et al.* (1988).

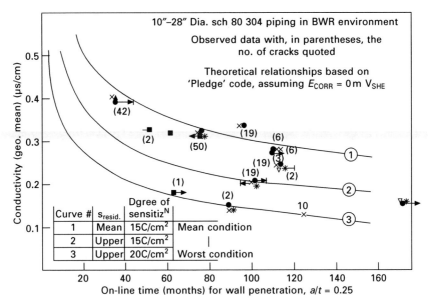

17.43 Predicted vs observed relationships between the time for a crack to penetrate to a quarter wall thickness and the plant's mean coolant conductivity. Figure adapted from Ford *et al.* (1988).

in fact predicted, whereas the larger number of incidences correspond to the mean conditions of sensitization and stress.

Similar predictions may be made for the cracking propensity for different system parameters such as a change in water chemistry (Fig. 17.44) and a change in stainless steel composition (304 vs 316NG vs 321) or the effect

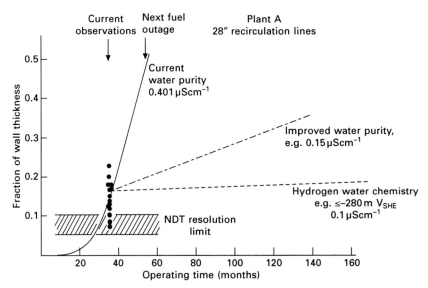

17.44 Predicted response of defected piping for defined changes in water chemistry in BWR plant. Figure adapted from Ford *et al.* (1988).

of surface grinding and its effect on surface residual stresses. Obviously, there will be epistemic uncertainties in the predictions associated with uncertainties in the values of the model parameters (e.g., residual stress, coolant conductivity, corrosion potential, sensitization, etc.), but these can be treated via knowledge of the distributions of these parameters and their input, via a Monte Carlo assessment, into the deterministic algorithm (Eq. 17.6). This was done early on for BWR piping systems (Harris *et al.*, 1986) in the development of the PRAISE-B and PRAISE-CC codes using an empirically derived damage accumulation algorithm rather than the mechanisms-informed algorithm described above.

Mechanisms-informed predictions, together with an assessment of the uncertainty in the prediction associated with epistemic uncertainties in the system definition, can define the cracking susceptibility of unirradiated stainless steel components in a specific BWR plant for proposed material, stress or environmental modifications (Ford *et al.*, 1989).

There are two basic problems in verifying the prediction capabilities for IASCC of irradiated stainless steel core components in BWRs. The first issue is the definition of the inputs to the prediction model, and these have been recognized above in terms of the effect of the irradiation flux on the corrosion potential, and the effect of fluence on the grain boundary sensitization, the yield strength and the stress relaxation in displacement loaded structures. The second issue is the accuracy of NDE measurements on defected components, which act as confirmation (or denial) of the proposed life prediction model. This latter aspect is significant (Pathania *et al.*, 2009).

Prediction of IASCC in BWR components, followed by their validation via comparison against observation, can be made even accepting these issues. For instance, an example of a predicted crack depth vs time relationship is shown in Fig. 17.45 for a welded stainless steel core shroud in which the crack depth is predicted to change with time (fluence) in a manner that reflects the time-dependent relaxation in weld residual stress (which decreases the cracking susceptibility) and the increase in grain boundary sensitization (which increases the cracking susceptibility). In this case the sensitization is attributed (Eq. 17.9) to the initial thermally induced chromium depletion at the grain boundary due to the welding process (i.e., 20 C/cm^2), plus the additional Cr depletion that occurs with fluence at that particular point in the shroud. Note that this diagram does not show the increase in yield strength with fluence, which will increase the cracking susceptibility due to its effect on the crack tip strain rate. The crack growth rate is not constant but will vary, mirroring these competing effects of stress relaxation, increasing yield strength and grain boundary sensitization, etc.

The predicted variation in crack propagation rate with time can become complicated as illustrated in Fig. 17.46, due to the interactions and competing effects of a decrease in susceptibility associated with stress relaxation and

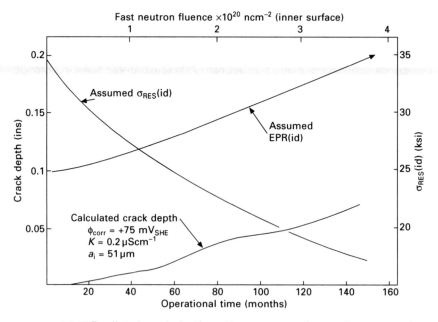

17.45 Predicted crack depth vs time response for cracks propagating in the beltline weld of a core shroud, illustrating the non-monotonic variation in crack propagation due to the competing effects of fast neutron irradiation on grain boundary sensitization and on residual stress relaxation (Andresen *et.al.*, 1996; Andresen and Ford, 1995).

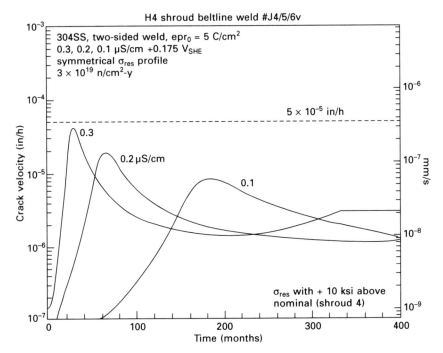

17.46 Predicted variation in crack propagation rate in an irradiated core shroud belt line weld as a function of time (fluence) indicating that the NRC disposition propagation rates cannot be sustained.

an increase in susceptibility associated with grain boundary sensitization and increasing yield stress. As indicated in Fig. 17.46, these crack propagation rate responses will change with water conductivity (purity). They will also change with the initial degree of thermal sensitization (e.g., the use of L-grade steels), extent of surface cold work, position in the reactor core (e.g., flux), etc.

A comparison is given in Fig. 17.47 between the observed and predicted crack depths adjacent to a horizontal beltline weld in a BWR core shroud, where the observations have been made during consecutive reactor outages. For the reasons given above, there is a range in observed crack depths adjacent to the horizontal welds due to the uncertainty in NDE measurements. It is also apparent that there is a range in predicted crack depths due to the uncertainty in defining the residual stress profile both around the circumference of the weld as well as through the 2″ thickness of the shroud.

Given these uncertainties, the only rational way of assessing the overall accuracy of the prediction methodology is to compare the ranges in observed and predicted crack depths for different welds and stainless steels. This comparison is made in Fig. 17.48 for various BWR core shroud welds

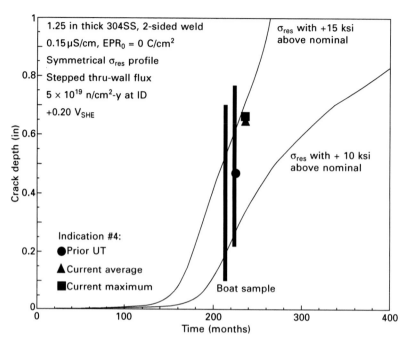

17.47 Crack length vs time predictions and observations for a Type 304 stainless steel core shroud with multiple inspections and multiple cracks (Andresen *et al.*, 1989; Andresen and Ford, 1995).

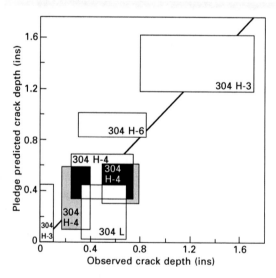

17.48 Comparison of observed and predicted crack depth for a number of BWR core shrouds (Andresen *et al.*, 1989; Andresen and Ford, 1995).

manufactured with different steels (Type 304, 304L). Over this population the agreement between observation and prediction is reasonable, given the uncertainties in observation and definition of the system.

17.5   Conclusions

Prediction of EAC can be made in many ways, ranging from simple estimation based on plant history, to empirical correlation with engineering parameters, to quantitative predictions based on knowledge of the processes and mechanisms. The ideal case is a mechanistic framework that permits interpretation, quantitative description of interdependencies and anticipation/ extrapolation of new variables and their interactions. Such fundamental approaches must be validated with detailed laboratory and plant data. Provided that this is done then the mechanistically informed approach to life prediction acts as a powerful underpinning for the mitigation actions that are robust and fundamentally valid. For stainless steels in BWR, environments in BWRs, such mitigation actions have included material (e.g., control of grain boundary sensitization), reduction in tensile stress, and especially control of certain anionic impurities and lowering the corrosion potential (by, for instance, 'Hydrogen Water Chemistry' and 'Noble Metal Technology'). These latter approaches are widely accepted and used in the BWR fleet worldwide (Cowan, 1994; Andresen *et al.*, 2003a, 2007; Gordon, 1984a, 1984b; Gordon *et al.*, 1983; Hettiarachchi, 2001 2003, 2005; Hettiarachchi *et al.*, 1995a, 1995b, 1997, 2009; EPRI, 2000, 2004; Kim *et al.*, 2007).

17.6   Future trends

Major progress in the prediction of EAC will depend on improvements in the quality and reproducibility of experimental data, especially SCC data (both initiation and growth). While a breakthrough in the conceptual, mechanistic description of EAC cannot be discounted, there are more ideas and concepts than there are definitive data to support or refute a given approach. Isolated, detailed observations, such as of tight SCC tips, raise qualitative questions about mechanisms, but they have not provided a definitive insight into the cracking mechanism.

For the relatively narrow conditions that comprise EAC of austenitic stainless steels (and nickel alloys) in high temperature LWR coolants, it is reasonable to propose that the controlling processes and mechanisms of EAC are common. As these common elements are more clearly defined and quantified, and their interdependencies understood, the ability to predict all forms of EAC, including environmental effects on fracture, will improve. A fundamental element will be definitive and accurate experimental data.

17.7   Sources of further information

There are many excellent proceedings and reviews of EAC in structural materials in LWR water, and the reader is encouraged to examine reviews and individual papers in the *Proceedings of the First to Fourteenth International Conference on Environmental Degradation of Materials in Nuclear Power Systems – Water Reactors*. Additional details of the behavior and prediction of EAC can be found in publications such as Ford and Andresen (1994), Andresen (2010) and Ford *et al.* (2006). These sources are listed among the references.

17.8   References

Akashi M. and Nakayama G. (1995), 'Stress Corrosion Crack Initiation Process Model for BWR Plant Materials' in *Plant Aging and Life Prediction of Corrodible Structures*, Eds T. Shibata and T. Shoji, Sapporo, pp. 99–106.

Ando M. and Nakata K. (2007), 'Crack Growth Rate Behavior of Low Carbon Stainless Steels of Hardened Heat Affected Zone in PLR Piping Weld Joints', *Proc. 13th Int. Symp. on Environmental Degradation of Materials in Nuclear Power Systems – Water Reactors*, Whistler, BC, August, 2007.

Andresen P.L. (1980), 'The Effects of Surface Preparation, Stress and Deaeration on the Stress Corrosion Cracking of Type 304 Stainless Steel in Simulated BWR Start-Up Cycles', *Proc. Seminar on Countermeasures for Pipe Cracking in BWRs*, Vol. 3, January 22–24, EPRI Reports WS 79-174, May.

Andresen P.L. (1983), 'The Effects of Aqueous Impurities on Intergranular Stress Corrosion Cracking of Sensitized 304 Stainless Steel', EPRI Report NP-3384, November.

Andresen P.L. (1984), 'Laboratory Results on Effects of Oxygen Control during a BWR Start up on the IGSCC of 304 Stainless Steel', *Proc. 2nd Seminar on Countermeasures for Pipe Cracking in BWRs*, November 15–18, EPRI Report NP-3684-SR Sept. 1984, Vol. 2, p. 12–1.

Andresen P.L. (1987), 'Modeling of Water and Material Chemistry Effects on Crack Tip Chemistry and Resulting Crack Growth Kinetics', *Proc. 3rd Int. Conf. Environmental Degradation of Materials in Nuclear Power Systems – Water Reactors*, Traverse City, AIME, pp. 301–314.

Andresen P.L. (1991), 'Effects of Specific Anionic Impurities on Environmental Cracking of Austenitic Materials in 288°C Water', *Proc. 5th Int. Symp. on Environmental Degradation of Materials in Nuclear Power Systems – Water Reactors*, Monterey, August 25–29, 1991, Eds D. Cubicciotti, E. Simonen, American Nuclear Society, pp. 209–218.

Andresen P.L. (1992a), 'Effects of Zinc Additions on the Crack Growth Rate of Sensitized Stainless Steel and Alloys 600 and 182 in 288C Water', Paper 72, *Water Chemistry of Nuclear Reactor Systems 6*, BNES, London, 1992.

Andresen P.L. (1992b), 'Effect of Temperature on Crack Growth Rate in Sensitized Type 304 Stainless Steel and Alloy 600', Paper 89, *NACE Corrosion/92*, Nashville, April 1992.

Andresen P.L. (1993a), 'Specific Anion and Corrosion Potential Effects on Environmentally Assisted Cracking in 288C Water', GE-CRD Report 93CRD215, December.

Andresen P.L. (1993b), 'Effects of Dissolved Silica on the Crack Growth Rate of Sensitized Stainless Steel', GE-CRD Report 93CRD212, December.

Andresen P.L. (1993c), 'The Effects of Nitrate on the Stress Corrosion Cracking of Sensitized Stainless Steel at 288C', GE-CRD Report 93CRD213, December.

Andresen P.L. (1995a), 'Effects of Nitrate on the Stress Corrosion Cracking of Sensitized Stainless Steel in High Temperature Water', *Proc. 7th Int. Symp. on Environmental Degradation of Materials in Nuclear Power Systems – Water Reactors*, Breckenridge, August 7–10, 1995. Eds R. Gold, A. McIlree, National Association of Corrosion Engineers, pp. 609–619.

Andresen P.L. (1995b), 'Application of Noble Metal Technology for Mitigation of Stress Corrosion Cracking in BWRs', *Proc. 7th Int. Symp. on Environmental Degradation of Materials in Nuclear Power System – Water Reactors*, NACE, Houston, TX, p. 563.

Andresen P.L. (1997), 'Effects of Flow Rate on SCC Growth Rate Behavior in BWR Water', *Proc. 8th Int. Symp. on Environmental Degradation of Materials in Nuclear Power Systems – Water Reactors*, Amelia Island, Eds A. McIlree, S. Bruemmer, American Nuclear Society, August 10–14, 1997, pp. 603–614.

Andresen P.L. (1999), 'Stress Corrosion Cracking Testing and Quality Considerations', *Proc. 9th Int. Symp. on Environmental Degradation of Materials in Nuclear Power Systems – Water Reactors*, Newport Beach, August 1–5, 1999. Eds S. Bruemmer, F.P. Ford, The Metallurgical Society, pp. 411–421.

Andresen P.L. (2002), 'Similarity of Cold Work and Radiation Hardening in Enhancing Yield Strength and SCC Growth of Stainless Steel in Hot Water', *Corrosion/02*, Paper 02509, NACE.

Andresen P.L. (2010), 'Stress Corrosion Cracking in Austenitic Stainless Steels', in *Understanding and Mitigating Ageing in Nuclear Power Plants*, Ed. P G Tipping, Woodhead Publishing, Cambridge.

Andresen P.L. and Angeliu T.M. (1995), 'Effects of Zinc Additions on the Stress Corrosion Crack Growth Rate of Sensitized Stainless Steel, Alloy 600 and Alloy 182 Weld Metal in 288C Water', Paper 409, *Corrosion/95*, NACE, Orlando, FL, March.

Andresen P.L. and Ford F.P. (1995), 'Modeling and Prediction of Irradiation Assisted Cracking', *Proc. 7th Int. Symp. on Environmental Degradation of Materials in Nuclear Power Systems – Water Reactors*, Breckenridge, August 7–10, 1995. Eds R. Gold, A. McIlree. National Association of Corrosion Engineers, pp. 893–908.

Andresen P.L. and Morra M.M. (2005), 'Effects of Si on SCC of Irradiated and Unirradiated Stainless Steels and Nickel Alloys', *Proc. 12th Int. Symp. on Environmental Degradation of Materials in Nuclear Power Systems – Water Reactors*, TMS, Snowbird, August 2005, pp. 87–108.

Andresen P.L., Ford F.P., Murphy S.M. and Perks J.M. (1989), 'State of Knowledge of Irradiation Effects on EAC in LWR Core Materials', *Proc. 4th Int. Symp. on Environmental Degradation of Materials in Nuclear Power Systems – Water Reactors*, Jekyll Island, August 6–10, 1989, Eds D. Cubicciotti, E. Simonen, National Association of Corrosion Engineers, pp. 1.83–1.121.

Andresen P.L., Vasatis I.P. and Ford F.P. (1990), 'Behavior of Short Cracks in Stainless Steel at 288°C', Paper 495, NACE Conference, Las Vegas, April.

Andresen P.L., Ford F.P., Higgins J.P., Suzuki I., Koyama M., Akiyama M., Okubo Y., Mishima Y., Hattori S., Anzai H., Chujo H. and Kanazawa Y. (1996), 'Life Prediction of Boiling Water Reactor Internals', *Proc. ICONE-4 Conference*, ASME.

Andresen P.L., Gott K. and Nelson J.L. (1999a), 'Stress Corrosion Cracking of Sensitized

Stainless Steel – a Five Lab Round Robin', *Proc. 9th Int. Symp. on Environmental Degradation of Materials in Nuclear Power Systems – Water Reactors*, Newport Beach, CA, August 1–5, 1999, Eds S. Bruemmer, F.P. Ford, The Metallurgical Society, pp. 423–433.

Andresen P.L., Ford F.P., Angeliu T.M., Solomon H.D. and Cowan R.L. (1999b), 'Prediction of Environmentally-Assisted Cracking and its Relevance to Life Management in BWRs', *Proc. 9th Int. Symp. on Environmental Degradation of Materials in Nuclear Power Systems – Water Reactors*, Newport Beach, CA, August 1–5, 1999, Eds S. Bruemmer, F.P. Ford, The Metallurgical Society, pp. 373–381.

Andresen P.L., Emigh P.E. and Young L.M. (2002), 'Mechanistic and Kinetic Role of Yield Strength/Cold Work/Martensite, H_2, Temperature, and Composition on SCC of Stainless Steels', *Proc. 10th Anniversary INSS Symp. on SCC in Nuclear Power Systems*, Osaka, Japan, May.

Andresen P.L., Diaz T.P. and Hettiarachchi S. (2003a), 'Effect on Stress Corrosion Cracking of Electrocatalysis and Its Distribution Within Cracks', *Proc. 11th Int. Conf. on Environmental Degradation of Materials in Nuclear Power Systems – Water Reactors*, Skamania Lodge, August 5–9, 2003, Eds G. Was, L. Nelson, American Nuclear Society.

Andresen P.L., Emigh P.W., Morra M.M. and Horn R.M. (2003b), 'Effects of Yield Strength, Corrosion Potential, Stress Intensity Factor, Silicon and Grain Boundary Character on SCC of Stainless Steels', *Proc. 11th Int. Conf. on Environmental Degradation of Materials in Nuclear Power Systems – Water Reactors*, Skamania Lodge, August 5–9, 2003, Eds G.S. Was, J.L. Nelson, American Nuclear Society, pp. 816–832.

Andresen P.L., Kim Y-J., Diaz T.P. and Hettiarachchi S. (2007), 'Mitigation of SCC by On-line NobleChem', *Proc. 13th Int. Conf. on Environmental Degradation of Materials in Nuclear Power Systems – Water Reactors*, Whistler, August 20–24, 2007, Eds P. King, T. Allen, The Canadian Nuclear Society.

Angeliu T.M. (1995), 'The Effect of Zinc Additions on the Oxide Rupture Strain and Repassivation Kinetics of Fe-Base Alloys in 288C Water', Paper 411, *Corrosion/95*, NACE, Orlando, FL, March.

Angeliu T.M., Andresen P.L., Sutliff J.A. and Horn R.M. (1999), 'Intergranular Stress Corrosion Cracking of Unsensitized Stainless Steels in BWR Environments', *Proc. 9th Int. Symp. on Environmental Degradation of Materials in Nuclear Power Systems – Water Reactors*, Newport Beach, CA, August 1–5, 1999, Eds S. Bruemmer, F.P. Ford, The Metallurgical Society, pp. 311–318.

Beck T.R. (1974), 'Reactions and Kinetics of newly generated titanium surfaces and relevance to stress corrosion cracking', *Corrosion* 30, p. 408.

Bose A. and De P.K. (1987), 'An EPR Study on the Influence of Prior Cold Work on the Degree of Sensitization of AISI 304 Stainless Steel', *Corrosion* 43, pp. 624–631.

Briant C.L. (1982), 'Effects of Nitrogen and Cold Work on the Sensitization of Austenitic Stainless Steels', EPRI Report NP-2457, June.

Brown K.S. and Gordon G.M. (1987), 'Effects of BWR Coolant Chemistry on Propensity for IGSCC Initiation and Growth in Creviced Reactor Internals Components', *Proc. 3rd Int. Symp. on Environmental Degradation of Materials in Nuclear Power Systems – Water Reactors*, Traverse City, August 30–Sept 3, 1987, Eds G.J. Theus, W. Berry, The Metallurgical Society, pp. 243–248.

Bruemmer S.M., Charlot L.A. and Arey B.W. (1988a), 'Sensitization Development in Austenitic Stainless Steel: Correlation Between STEM-EDS and EPR Measurements', *Corrosion* 44, pp. 328–333.

Bruemmer S.M. Charlot L.A. and Atteridge D.G. (1988b), 'Sensitization Development in Austenitic Stainless Steel: Measurement and Prediction of Thermochemical History Effects', *Corrosion* 44, pp. 427–434.

Bruemmer S.M., Arey B.W. and Charlot L.A. (1993), 'Grain Boundary Chromium Concentration Effects on the IGSCC and IASCC of Austenitic Stainless Steels', *Proc. 6th Int. Symp. on Environmental Degradation of Materials in Nuclear Power Systems – Water Reactors,* San Diego, CA, August 1–5, 1993, Eds E. Simonen, R. Gold, National Association of Corrosion Engineers, pp. 227–287.

Bruemmer S.M. Cole J.I., Garner F.A., Greewood L.R., Hamilton M.L., Reid B.D., Simonen E.P., Lucas G.E., Was G.S., Andresen P.L. and Pettersson K. (1996), 'Critical Issues Reviews for the Understanding and Evaluation of Irradiation Assisted Stress Corrosion Cracking', EPRI TR-107159, November.

Bruemmer S.M., Simonen E.P., Scott P.M., Andresen P.L., Was G.S. and Nelson J.L. (1999),'Radiation-induced materials changes and susceptibility to intergranular failure of light-water-reactor core internals', *J. Nucl. Mater.*, 274, pp. 299–314.

Buchelet C.B. and Bamford W.H. (1976), 'Stress Intensity Factor Solutions for Continuous Surface Flaws in Reactor Pressure Vessels', in *Mechanics of Crack Growth*, ASTM STP S90 American Society for Testing and Materials, pp. 385–402.

Chrenko R.M. (1980), 'Residual Stress Measurements on Type 304 Stainless Steel Welded Pipes', in *'EPRI, Proceedings of Seminar on Countermeasures for Pipe Cracking in BWRs.* Volume 2, January 22–24, 1980, EPRI Reports WS 79-174.

Clarke W.L. (1980), 'The ECP Method for Detecting Sensitization in Stainless Steels', NUREG/CR-1095, February 1980.

Clarke W.L., Cowan R.L. and Walker W.L. (1978), 'Comparative Methods for Measuring Degree of Sensitization in Stainless Steels', ASTM STP 656.

Cohen P. (1969), *Water Coolant Technology of Power Reactors*, Gordon and Breach Science Publishers, 1969.

Congleton J., Zheng W. and Hua H. (1990), 'The stress corrosion cracking behaviour of annealed 316 stainless steel in low oxygen 5 ppm chloride content water at 300°C, *Corrosion Science* 27(6/7), p. 555–567.

Cowan R.L. (1994), 'Optimizing Water Chemistry in US Boiling Water Reactors', *VGB Conference Plant Chemistry Essen*, Germany, October.

Davis R.B. and Indig M.E. (1983), 'The Effect of Aqueous Impurities on the Stress Corrosion Cracking of Austenitic Stainless Steel in High Temperature Water', Paper 128, *Corrosion/83*, Anaheim, CA, National Association of Corrosion Engineers, April.

Dawson J.K. and Sowden R.G. (1963), 'Chemical Aspects of Nuclear Reactors', in *Water Cooled Reactors*, Vol. 2, Butterworths.

Eason E. and Shusto L.M. (1986), 'Analysis of Cracking in Small Diameter BWR Piping', EPRI NP-4394, EPRI, Palo Alto.

Eason E.D., Pathania R. and Shoji T. (2005), 'Evaluation of the Fracture Research Institute Theoretical Stress Corrosion Cracking Model', *Proc. 12th Int. Conf. on Environmental Degradation of Materials in Nuclear Power Systems – Water Reactors*, Skamania Lodge, August 5–9, 2005, Eds L. Nelson, P. King, The Metallurgical Society, pp. 145–154.

EPRI (1985), 'BWR Coolant Impurity Identification Study', EPRI Report NP-4156, August.

EPRI (1992a), 'BWR Chromium Chemistry', EPRI Report TR-100792, October.

EPRI (1992b), 'The Effect of Chromates on IGSCC in BWR Environments', EPRI Report TR-100853, July.

EPRI (1996), 'Evaluation of Crack Growth in BWR Stainless Steel RPV Internals (BWRVIP-14)', EPRI report TR-105873, March.

EPRI (1997), 'BWR Vessel and Internals Project 'IGSCC in BWRs during Transient Conditions', EPRI Report BWR VIP-40, TR-107093, August.

EPRI (1999), 'BWR Vessel and Internals Project 'Effects of Flow Rate on Intergranular Stress Corrosion Cracking and Electrochemical Corrosion Potential', EPRI Report VIP-64, TR-112314, March.

EPRI (2000), 'BWR Water Chemistry Guidelines – 2000 Revision', EPRI Report TR-103515-R2, February.

EPRI (2004), 'BWR Water Chemistry Guidelines – 2004 Revision', EPRI Technical Report TR-1008192, October, VIP-130.

Fong C. (2007), 'Flow Rate Effects on the Stress Corrosion Cracking of Sensitized Stainless Steels in BWR Environments', *Proc. 13th Int. Conf. on Environmental Degradation of Materials in Nuclear Power Systems – Water Reactors*, Whistler, August 20–24, 2007, Eds P. King, T. Allen, The Canadian Nuclear Society.

Ford F.P. (1982a), 'Mechanisms of Environmental Cracking Peculiar to the Power Generation Industry', Report NP2589, EPRI, Palo Alto, September.

Ford F.P. (1982b), 'Stress Corrosion Cracking', in *Corrosion Processes*, Ed. R.N. Parkins, Applied Science.

Ford F.P. (1988), 'The Crack Tip System and its Relevance to the Prediction of Cracking in Aqueous Environments', *Proc. 1st Inter. Conf. on Environmentally Assisted Cracking of Metals*, Kohler, WI, October 2–7 1988, Eds R. Gangloff and B. Ives, NACE, pp. 139–165.

Ford F.P. (1996), 'Quantitative Prediction of Environmentally Assisted Cracking', *Corrosion* 52, pp. 375–395.

Ford F.P. (2007), 'Technical and Management Challenges Associated with Structural Materials Degradation in Nuclear Reactors in the Future', *Proc. Int. Conf. Environmental Degradation of Materials in Nuclear Power Systems*, Whistler, British Columbia, August 19–23, 2007, Canadian Nuclear Society.

Ford F.P. and Andresen P.L. (1994), 'Corrosion in Nuclear Systems: Environmentally Assisted Cracking in Light Water Reactors', in *Corrosion Mechanisms*, Eds P. Marcus and J. Ouder, Marcel Dekker, pp. 501–546.

Ford F.P. and Povich M.J. (1979), 'Effect of Oxygen and Temperature Combinations on the Stress Corrosion Cracking of Sensitized Type 304 Stainless Steel', *Corrosion* 35, 569.

Ford F.P., Taylor D.F., Andresen P.L. and Ballinger R.G. (1987), 'Corrosion Assisted Cracking of Stainless Steel and Low Alloy Steels in LWR Environments', Report NP5064S, EPRI, Palo Alto, February.

Ford F.P., Andresen P.L., Benz M.G. and Weinstein D. (1988), 'On-Line BWR Materials Monitoring and Plant Component Lifetime Prediction', *Proc. Nuclear Power Plant Life Extension*, Snowbird, UT, June 1988, American Nuclear Society, Vol. 1, pp. 355–366.

Ford F.P., Andresen P.L., Solomon H.D., Gordon B.M., Ranganath S., Weinstein D. and Pathania R. (1989), 'Application of Water Chemistry Control, On-Line Monitoring and Crack Growth Rate Models for Improved BWR Materials Performance', *Proc. 4th Int. Symp. on Environmental Degradation of Materials in Nuclear Power Systems – Water Reactors*, Jekyll Island August 6–10, 1989, Eds D. Cubicciotti, E. Simonen, National Association of Corrosion Engineers, pp. 4.26-4.51.

Ford F.P., Gordon B.M. and Horn R.M. (2006), 'Corrosion in Boiling Water Reactors', in *ASM Materials Handbook Vol 13C, Corrosion: Environments and Industries*, pp. 341–361.

Fruzzetti K.P. and Blok J. (2005), 'A Review of the EPRI Chemistry Guidelines', *Proc. Int. Conf. on Water Chemistry of Nuclear Reactor Systems*, San Francisco.

Gao Y.C. and Hwang K.C. (1981), 'Elastic Plastic Fields in Steady Crack Growth in a Strain Hardening Material', *Proc. 5th Int. Conf. on Fracture*, pp. 669–682.

Gordon B.M. (1980), 'The Effect of Chloride and Oxygen on Stress Corrosion Cracking of Stainless Steel; Review of Literature', *Materials Performance* 19(4), p. 29.

Gordon B.M. (1984a), 'Laboratory Studies of Materials Performance in Hydrogen Water Chemistry', Paper presented at *EPRI Seminar on BWR Corrosion, Chemistry and Radiation Control*, Palo Alto, CA, Electric Power Research Institute, October.

Gordon B.M. (1984b), 'Corrosion and Corrosion Control in BWRs', GE Class 1 Report NEDO-24819A, December.

Gordon B.M., Cowan R.L., Jewett C.W. and Pickett A.E. (1983), 'Mitigation of Stress Corrosion Cracking through Suppression of Radiolytic Oxygen', *Proc. 1st Int. Symp. on Environmental Degradation of Materials in Nuclear Power Systems – Water Reactors*, Myrtle Beach, August 22–25, 1983, Eds J. Roberts, W Berry, National Association of Corrosion Engineers, pp. 893–932.

Gordon G.M. and Brown K.S. (1989), 'Dependence of Creviced BWR Component IGSCC Behavior on Coolant Chemistry', *Proc. 4th Int. Symp. on Environmental Degradation of Materials in Nuclear Power Systems – Water Reactors*, Jekyll Island, August 6–10, 1989, Eds D. Cubicciotti, E. Simonen, National Association of Corrosion Engineers, pp. 14.46–14.62.

Gott K. (1999), 'The History of Cracking in the Reactor Coolant Pressure Boundary of Swedish BWR Plants', *Proc. 9th Int. Symp. on Environmental Degradation of Materials in Nuclear Power Systems – Water Reactors*, Newport Beach, August 1–5, 1999, Eds S. Bruemmer, F.P. Ford, The Metallurgical Society.

Gott K. (2001), 'Cracking Data Base as a Basis for Risk Informed Inspection', *Proc. 10th Int. Conf. on Environmental Degradation of Materials in Nuclear Power Systems – Water Reactors*, Lake Tahoe, August 5–9, 2001, Eds F.P. Ford, G.S. Was, National Association of Corrosion Engineers.

Hale D.A. (1986), 'The Effect of BWR Startup Environments on Crack Growth in Structural Alloys', *Trans ASME* 108, p. 44.

Hale D.A. (1991), 'BWR Coolant Impurities Program at Peach Bottom Units 2&3', EPRI Report NP-7310-L, May.

Hale D.A. and Pickett A.E. (1984), 'Material Performance in a Startup Environment', *Proc. 2nd Seminar on Countermeasures for Pipe Cracking in BWRs*, November 15–18, 1983, EPRI Report NP-3684-SR, Vol. 2, pp. 13–1, September.

Harris D.O., Dedia D.D., Eason E.D. and Patterson S.D. (1986), 'Probability of Failure in BWR Reactor Coolant Piping; Probabilistic Treatment of Stress Corrosion Cracking in 304 and 316NG BWR Piping Weldments', NUREG/CR-4792, December.

Hazelton W.S. and Koo W.H. (1988), 'Technical Report on Materials Selection and Processing Guidelines for BWR Coolant Pressure Boundary Piping', NUREG-0313 Rev.2, USNRC, January.

Hettiarachchi S. (2001), 'NobleChem from Concept to Operating Commercial Power Plant Application', *Proc. 10th Int. Conf. on Environmental Degradation of Materials in Nuclear Power Systems – Water Reactors*, Lake, Tahoe August 5–9, 2001, Eds F.P. Ford, G.S, Was, National Association of Corrosion Engineers.

Hettiarachchi S. (2003), 'BWR SCC Mitigation Strategies and their Effectiveness',

Proc. 11th Int. Conf. on Environmental Degradation of Materials in Nuclear Power Systems – Water Reactors, Skamania Lodge, August 5–9, 2003, Eds G.S. Was, J.L. Nelson, American Nuclear Society.

Hettiarachchi S. (2005), 'BWR SCC Mitigation Experiences with Hydrogen Water Chemistry', *Proc. 12th Int. Conf. on Environmental Degradation of Materials in Nuclear Power Systems – Water Reactors*, Skamania Lodge, August 5–9, 2005, Eds J.L. Nelson, P. King, The Metallurgical Society, pp. 685.

Hettiarachchi S., Wozadlo G.P. and Diaz T.P. (1995a), 'Influence of Zinc Additions on the Intergranular Stress Corrosion Crack Initiation and Growth of Sensitized Stainless Steel in High Temperature Water', Paper 410, *Corrosion/95*, NACE, Orlando, FL, March.

Hettiarachchi S., Wozadlo G., Andresen P.L., Diaz T. and Cowan R.L. (1995b), 'The Concept of Noble Metal Addition Technology for IGSCC Mitigation of Structural Materials', *Proc. 7th Int. Symp. on Environmental Degradation of Materials in Nuclear Power Systems – Water Reactors*, Breckenridge, August 7–10, 1995, Eds R. Gold, A. McIlree, National Association of Corrosion Engineers, pp. 735–746.

Hettiarachchi S., Law R.J., Diaz T.P., Cowan R.L., Keith W. and Pathania R.S. (1997), 'The First In-Plant Demonstration of Noble Metal Chemical Addition (NMCA) Technology for IGSCC Mitigation of BWR Internals', *Proc. 8th Int. Symp. on Environmental Degradation of Materials in Nuclear Power Systems – Water Reactors*, Amelia Island, August 10–14, 1997, Eds A. McIlree, S. Bruemmer, American Nuclear Society, pp. 535–543.

Hettiarachchi S., Horn R.M., Andresen P.L. and Kim Y.J. (2009), 'Electrochemical Corrosion Potential (ECP) Reduction and Crack Mitigation Experiences with NobleChem™ and On-Line NobleChem™', *Proc. 14th Int. Conf. on Environmental Degradation of Materials in Nuclear Power Systems – Water Reactors*, Virginia Beach, August 24–28, 2009, Eds T. Allen, J. Busby, American Nuclear Society.

Hickling J. (1994), 'Evaluation of Acceptance Criteria for Data on Environmentally Assisted Cracking in Light Water Reactors', SKI Report 94.14, September.

Hickling J. (2005), 'Status Review of Initiation of Environmentally Assisted Cracking and Short Crack Growth', EPRI Report # 1011788, December.

Horn R.M., Gordon G.M., Ford F.P. and Cowan R.L. (1997), 'Experience and Assessment of Stress Corrosion Cracking in L-Grade Stainless Steel BWR Internals', *Nuclear Engineering and Design* 174, pp. 313–325.

Indig M. (1980), 'Controlled Potential Simulation of Deaerated and Non-Dearated Startup', *Proc. Seminar on Countermeasures for Pipe Cracking in BWRs*, Vol. 3, January 22–24, 1980, EPRI Reports WS 79–174, May.

Indig M.E., Gordon G.M. and Davis R.B. (1983), 'The Role of Water Purity on Stress Corrosion Cracking', *Proc. 1st Int. Symp. on Environmental Degradation of Materials in Nuclear Power Systems – Water Reactors*, Myrtle Beach, August 22–25, 1983, Eds J. Roberts, W Berry, National Association of Corrosion Engineers, pp. 506–531.

Jacobs A.J., Hale D.A. and Siegler M. (1986), 'Unpublished GE data.

Jiang X.C. and Staehle R.W. (1997), 'Application of the Chemical–Mechanical Correlation Equation for SCC', *Proc. 8th Int. Symp. on Environmental Degradation of Materials in Nuclear Power Systems – Water Reactors*, Amelia Island, August 10–14, 1997, Eds A. McIlree, S. Bruemmer, American Nuclear Society, pp. 582–596.

Kawakubo T., Hishida M. and Arii M. (1980), 'A Model for Intergranular Stress Corrosion Cracking Initiation in BWR Pipes and the Virtue of Startup Deaeration', *Proc. Seminar on Countermeasures for Pipe Cracking in BWRs*, Vol. 3, January 22–24, 1980, EPRI Reports WS 79–174, May.

Kilian R., Brummer G., Ilg U., Maier V., Teichmann H. and Wachter O. (1995), 'Characterization of Sensitization and Stress Corrosion Behaviour of Stabilized Stainless Steels under BWR Conditions', *Proc. 7th Int Symp. on Environmental Degradation of Materials in Nuclear Power Systems – Water Reactors*, Breckenridge, August 7–10, 1995, Eds R. Gold, A. McIlree, National Association of Corrosion Engineers.

Kilian R., Hoffmann H., Ilg U., Küster K., Nowak E., Wesseling U. and Widera M. (2005), 'German Experience with Intergranular Cracking in Austenitic Piping in BWRs and Assessment of Parameters Affecting the In-Service IGSCC Behaviour Using an Artificial Neural Network', *Proc. 12th Int. Conf. on Environmental Degradation of Materials in Nuclear Power Systems – Water Reactors*, Skamania Lodge, August 5–9, 2005, Eds L. Nelson, P. King, The Metallurgical Society, pp. 803–812.

Kim Y.J. and Andresen P.L. (2003), 'Data Quality, Issues and Guidelines for ECP Measurements in High Temperature Water', *Corrosion* 59, pp. 584–596.

Kim Y.J., Andresen P.L., Hettiarachchi S. and Diaz T.P. (2007), 'Effect of Variations in Noble Metal Chemical Addition Process on Electrochemical Catalytic Response in High Temperature Water', *Proc. 13th Int. Conf. on Environmental Degradation of Materials in Nuclear Power Systems – Water Reactors*, Whistler, August 20–24, 2007, Eds P. King, T. Allen, The Canadian Nuclear Society.

Klepfer H.H. *et al.* (1975), 'Investigation of Cause of Cracking in Austenitic Stainless Steel Piping', NEDO-21000, General Electric, July.

Kurtz R.J. (1983), 'Evaluation of BWR Resin Intrusions on Stress Corrosion Cracking of Reactor Structural Materials', EPRI Report NP-3145, June.

Lin C.C. (1985), 'An Overview of Radiation Chemistry in Reactor Coolants', *Proc. 2nd Int. Symp. on Environmental Degradation of Materials in Nuclear Power Systems – Water Reactors*, Monterey, September 9-12, 1985, Eds J. Roberts, J. Weeks, American Nuclear Society, pp. 160–172.

Lin C.C., Kim Y.J., Niedrach L. and Ramp K.S. (1996), 'Electrochemical Potential Models for BWR Applications', *Corrosion* 52, p. 618.

Ljungberg L.G., Cubicciotti D. and Troll M.E. (1989), 'The Effect of Sulfate on Environmental Cracking in BWRs Under Constant Load or Fatigue', Paper 617, *Corrosion/89*, NACE, New Orleans, LA, April.

Ljungberg L.G., Ortnas A., Stigenberg M., Hofling C.G., Sahlberg A and Moller J. (1995), 'Stress Corrosion Cracking of Alloys 600 and 182 in BWRs', EPRI Report TR-104972, May.

Logan H.L. (1952), 'Film Rupture Mechanism of Stress Corrosion', *J. Natl. Bur. Stand* 48, p. 99.

Medoff J. (1996), 'Status Report; Intergranular Stress Corrosion Cracking of BWR Core Shrouds and other Internals Components', USNRC NUREG-1544, March.

Molander A., Bengtsson B., Jansson C., Takagi J. and Sakamoto H. (1997), 'Influence of Flow Rate on the Critical Potential for IGSCC of Stainless Steel in Simulated BWR Environment – A SSRT Study', *Proc. 8th Int. Symp. on Environmental Degradation of Materials in Nuclear Power Systems – Water Reactors*, Amelia Island, August 10–14, 1997, Eds A. McIlree, S. Bruemmer, American Nuclear Society, pp. 615–621.

Nelson J.L. and Andresen P.L. (1991), 'Review of Current Research and Understanding of Irradiation Assisted Stress Corrosion Cracking', *Proc. 5th Int. Symp. on Environmental Degradation of Materials in Nuclear Power Systems – Water Reactors*, Monterey, August 25–29, 1991, Eds D. Cubicciotti, E. Simonen, American Nuclear Society, pp. 10–26.

Parkins R.N. (1979), 'Environment-Sensitive Fracture – Controlling Parameters', *Proc.*

3rd Int. Conf. on Mechanical Behavior of Materials, Cambridge, 20–24 August, 1979, Pergamon, Eds K.J. Miller and R.F. Smith, Vol. 1, pp. 139–164.

Pathania R. Carter R., Horn R.N. and Andresen P.L. (2009), 'Crack Growth Rates in Irradiated Stainless Steels in BWR Internals', *Proc. 14th Int. Conf. on Environmental Degradation of Materials in Nuclear Power Systems – Water Reactors*, Virginia Beach, August 24–28, 2009, Eds T. Allen, J. Busby, American Nuclear Society.

Proceedings of First International Symposium on Environmental Degradation of Materials in Nuclear Power Systems – Water Reactors, Myrtle Beach, August 22–25, 1983, Eds J. Roberts, W. Berry, National Association of Corrosion Engineers, 1983.

Proceedings of Second International Symposium on Environmental Degradation of Materials in Nuclear Power Systems – Water Reactors, Monterey, September 9–12, 1985, Eds J. Roberts, J. Weeks, American Nuclear Society.

Proceedings of Third International Symposium on Environmental Degradation of Materials in Nuclear Power Systems – Water Reactors, Traverse City, August 30–Sept 3, 1987, Eds G.J. Theus, W. Berry, The Metallurgical Society.

Proceedings of Fourth International Symposium on Environmental Degradation of Materials in Nuclear Power Systems – Water Reactors, Jekyll Island, August 6–10, 1989, Eds D. Cubicciotti, E. Simonen, National Association of Corrosion Engineers.

Proceedings of Fifth International Symposium on Environmental Degradation of Materials in Nuclear Power Systems – Water Reactors, Monterey, August 25–29, 1991, Eds D. Cubicciotti, E. Simonen, American Nuclear Society.

Proceedings of Sixth International Symposium on Environmental Degradation of Materials in Nuclear Power Systems – Water Reactors, San Diego, August 1–5, 1993, Eds E. Simonen, R. Gold, National Association of Corrosion Engineers.

Proceedings of Seventh International Symposium on Environmental Degradation of Materials in Nuclear Power Systems – Water Reactors, Breckenridge, August 7–10, 1995, Eds R. Gold, A. McIlree, National Association of Corrosion Engineers.

Proceedings of Eighth International Symposium on Environmental Degradation of Materials in Nuclear Power Systems – Water Reactors, Amelia Island, August 10–14, 1997, Eds A. McIlree, S.M. Bruemmer, American Nuclear Society.

Proceedings of Ninth International Symposium on Environmental Degradation of Materials in Nuclear Power Systems – Water Reactors, Newport Beach, August 1–5, 1999, Eds S.M. Bruemmer, F.P. Ford, The Metallurgical Society.

Proceedings of Tenth International Conference on Environmental Degradation of Materials in Nuclear Power Systems – Water Reactors, Lake Tahoe, August 5–9 2001, Eds F.P. Ford, G.S. Was, National Association of Corrosion Engineers.

Proceedings of Eleventh International Conference on Environmental Degradation of Materials in Nuclear Power Systems – Water Reactors, Skamania Lodge, August 5–9, 2003, Eds G.S. Was, J.L. Nelson, American Nuclear Society.

Proceedings of Twelfth International Conference on Environmental Degradation of Materials in Nuclear Power Systems – Water Reactors, Skamania Lodge, August 5–9, 2005, Eds J.L. Nelson, P. King, The Metallurgical Society.

Proceedings of Thirteenth International Conference on Environmental Degradation of Materials in Nuclear Power Systems – Water Reactors, Whistler, August 20–24, 2007, Eds P. King, T. Allen, The Canadian Nuclear Society.

Proceedings of Fourteenth International Conference on Environmental Degradation of Materials in Nuclear Power Systems – Water Reactors, Virginia Beach, August 24–28, 2009, Eds T. Allen, J. Busby, American Nuclear Society.

Ruther W.E. and Kassner T.F. (1983), 'Effect of Sulphuric Acid, Oxygen and Hydrogen in High Temperature Water on Stress Corrosion Cracking of AISI 304 Stainless Steel', Paper 125, *Corrosion/83*, Anaheim, April.

Ruther W.E., Soppet W.K. and Kassner T.F. (1988), 'Effect of Temperature and Ionic Impurities at Very Low Concentrations on Stress Corrosion Cracking of Type 304 Stainless Steel', Paper 102, *Corrosion/85*, Boston, MA, March also *Corrosion* 44(11), p. 791.

Scott P.M., Meyzaud Y. and Benhamou C. (1995), 'Prediction of Stress Corrosion Cracking of Alloy 600 Components Exposed to PWR Primary Water', *Proc. Int. Symp. on Plant Aging and Life Prediction of Corrodible Structures*, May 15–18, 1995, Sapporo, Eds T. Shoji and T. Shibata, JSCE and NACE, pp. 285–293.

Scott P.M. and Benhamou C. (2001), 'An Overview of Recent Observations and Interpretations of IGSCC in Nickel-Base Alloys in PWR Primary Water', *Proc. 10th Int. Conf. on Environmental Degradation of Materials in Nuclear Power Systems – Water Reactors*, Lake Tahoe, August 5–9 2001, Eds F.P. Ford, G.S. Was, National Association of Corrosion Engineers.

Shack W.J. (1980), 'Measurement of Through Wall Residual Stresses in Large Diameter Piping Butt Welds Using Strain Gage Techniques', *Proc. Seminar on Countermeasures for Pipe Cracking in BWRs*, Vol. 2, January 22–24, 1980, EPRI Reports WS 79-174, May.

Shack W.J. and Ellingson W.A. (1980), 'Measured Residual Stresses in Type 304 Stainless Steel Piping Weldments', *Proc. Seminar on Countermeasures for Pipe Cracking in BWRs*', Vol. 2, January 22–24, 1980.

Shack W.J. *et al.* (1983), 'Environmentally Assisted Cracking in Light Water Reactors: Annual Report, October 1981–September 1982', NUREG CR3292, NRC, February.

Shack W.J. *et al.* (1985), 'Environmentally Assisted Cracking in Light Water Reactors: Annual Report, October 1983–September 1984', NUREG/CR-4287, ANL-85-33, June.

Shack W.J. *et al.* (1986), 'Environmentally Assisted Cracking in Light Water Reactors: Semiannual Reports', NUREG/CR-4667, Vol. 1, ANL-86-31, Argonne National Laboratory, June.

Shack W.J. *et al.* (1987), 'Environmentally Assisted Cracking in Light Water Reactors: Semi Annual Report April–September 1986', NUREG/CR-4667, Vol. III, ANL-87-37, September.

Shoji T. (2003), 'Progress in Mechanistic Understanding of BWR SCC and Implication to Predictions of SCC Growth Behaviour in Plants', *Proc. 11th Int. Conf. on Environmental Degradation of Materials in Nuclear Power Systems – Water Reactors*,' Skamania Lodge, August 5–9, 2003, Eds G.S. Was, J.L. Nelson, American Nuclear Society.

Shoji T. and Suzuki M. (1978), 'Corrosion Fatigue Aspects in BWR Pipe Cracking', in *Predictive Methods for Assessing Corrosion Damage in BWR Piping and PWR Steam Generators*, Eds H. Okada, R.W. Staehle, Fuji, May, p. 58.

Shoji T., Suzuki S. and Ballinger R.G. (1995), 'Theoretical Prediction of Stress Corrosion Crack Growth – Threshold and Plateau Growth Rate', *Proc. 7th Int. Symp. on Environmental Degradation of Materials in Nuclear Power Systems – Water Reactors*, Breckenridge, August 7–10, 1995, Eds R. Gold, A. McIlree, National Association of Corrosion Engineers, pp. 881–891.

Shoji T., Lu Z. and Murakami H. (2010), 'Formulating Stress Corrosion Crack Growth Rates by Combination of Crack Tip Mechanics and Crack Tip Oxidation Kinetics', *Corrosion Science* 52, pp. 769–779.

Solomon H.D. (1980), 'Weld Sensitization', *Proc. Seminar on Countermeasures for*

Pipe Cracking in BWRs, Vol. 2, January 22–24, 1980, EPRI Reports WS 79–174, May.

Solomon H.D. (1984), 'Influence of Composition on Continuous Cooling Sensitization of Type 304 Stainless Steel', *Corrosion* 40, pp. 51–60.

Solomon H.D. and Lord D.C. (1980), 'Influence of Strain During Cooling on the Sensitization of Type 304 Stainless Steel', *Corrosion* 36, pp. 395–399.

Speidel M.O. (1984), 'Stress Corrosion and Corrosion Fatigue Mechanics', in *Corrosion in Power Generating Equipment*, Eds M.O. Speidel, A. Atrens, Plenum Press, New York.

Speidel M.O. and Magdowski R. (1989), 'Stress Corrosion Cracking of Stabilized Austenitic Stainless Steels in Various Types of Nuclear Power Plants', *Proc. Int. Symp. on Environmental Degradation of Materials in Nuclear Power Systems – Water Reactors*, Jekyll Island, August 6–10, 1989, Eds D. Cubicciotti, E Simonen, National Association of Corrosion Engineers, pp. 325–329.

Staehle R.W. (2007), 'Predicting Failures Which Have Not Yet Been Observed', *Expert Panel Report on Proactive Materials Degradation Assessment*, USNRC NUREG Report, CR-6923, February.

Staehle R.W., Gorman J.A., Stavropoulos K.D. and Welty C.S. (1994), 'Application of Statistical Distributions to Characterizing and Predicting Corrosion of Tubing in Steam Generators of Pressurized Water Reactors', *Proceedings of Life Prediction of Corrodible Structures*, Ed. R.N. Parkins, NACE International, Houston, TX, pp. 1374–1439.

Stavropoulos K.D., Gorman J.A., Staehle R.W. and Welty C.S. (1991), 'Selection of Statistical Distributions for Prediction of Steam Generator Tube Degradation', *Proc. 5th Int. Symp. on Environmental Degradation of Materials in Nuclear Power Systems – Water Reactors*, Monterey, August 25–29, 1991, Eds D. Cubicciotti, E. Simonen, American Nuclear Society, pp. 731–744.

Stickler R. and Vinckier A. (1961), 'Morphology of Grain Boundary Carbides and its Influence on Intergranular Corrosion of 304 Stainless Steel', *Trans. ASM* 54, pp. 362–380.

Suzuki S. Takamori K., Kumagai K., Suguru O., Toshihiko F., Hironobu F. and Tsuneo F. (2004), 'Evaluation of SCC Morphology on L-grade Stainless Steels in BWRs', *Journal of High Pressure Institute of Japan* 42(4), p. 197.

USNRC (2006), 'Generic Aging Lessons Learned (GALL) Report', NUREG-1801, Rev 1.0, Vols. 1 and 2.

Van Duysen J.C., Todeschini P. and Zacharie G. (1993), 'Effects of Neutron Irradiations at Temperatures below 550°C on the Properties of Cold Worked 361 Stainless Steels: A Review', *Effects of Radiation on Materials: 16th Int. Symposium*, STP 1175, ASTM, pp. 747–776.

Vermilyea D.A. (1972), 'A Theory for the Propagation of Stress Corrosion Cracks in Metal', *J. Electrochem. Soc.* 119, p. 405.

VGB (2006), VGB Water Chemistry Guideline (Richtlinie) R401J DWR Chemie.

Was G.S., Busby J. and Andresen P.L. (2006), 'Effect of Irradiation on Stress Corrosion Cracking and Corrosion in Light Water Reactors', in *ASM Handbook on Corrosion; Environments and Industries*, Vol. 13C, pp. 386–414.

Wei R.P. and Harlow D.G. (1995), 'A Mechanistically-Based Probability Approach for Life Prediction', in *Plant Aging and Life Prediction of Corrodible Structures*, Eds T. Shibata and T. Shoji, Sapporo, pp. 47–56.

White B.C. and Cubicciotti D. (1989), 'Evaluation of IGSCC Effects of BWR Chemistry

Excursions', *Proc 4th Int. Symp. on Environmental Degradation of Materials in Nuclear Power Systems – Water Reactors*, Jekyll Island, August 6–10, 1989, Eds D. Cubicciotti, E. Simonen, National Association of Corrosion Engineers, pp. 4.52–4.56.

Yamauchi K., Hattori S., Kuniya J. and Asami T. (1995), 'Methodology for Formulating Predictions of Degradation of Plant Components Induced by Synergistic Effect of Various Influencing Factors', in *Plant Aging and Life Prediction of Corrodible Structures*, Eds T. Shibata and T. Shoji, Sapporo, pp. 69–78.

18

Failures of structures and components by metal-induced embrittlement

S. P. LYNCH, Defence Science and Technology
Organisation (DSTO), Australia

Abstract: The basic aspects of liquid-metal embrittlement (LME) and solid metal-induced embrittlement (SMIE) are concisely reviewed, followed by case histories of failures involving: (i) LME of an aluminium-alloy pipe by mercury in a natural-gas plant, (ii) SMIE of a brass valve in an aircraft-engine oil-cooler by internal lead particles, (iii) LME of a cadmium-plated steel screw from a crashed helicopter, (iv) LME of a cold-formed steel stiffener during galvanising, and (v) LME of a steel gear by a copper alloy from an overheated bearing. These case histories illustrate how failures by LME and SMIE can be diagnosed and distinguished from other failure modes. The underlying causes of the failures and how they might be prevented are also discussed. Several beneficial uses of LME in failure analysis are then outlined.

Key words: liquid-metal embrittlement, solid-metal-induced embrittlement, failure analysis, fractography, metallography, aluminium alloys, steels, copper alloys.

Note: this chapter, which is © The Commonwealth of Australia, was first published in the *Journal of Failure Analysis and Prevention*, Vol. 8 (2008), pp. 259–274, published by ASM International. Some photographs and micrographs which appear in colour in the original paper are reproduced in black and white in this chapter.

'Quicksilver ... so penetrant is this liquor, that there is no vessel in the world but it will eat and break through it ...' Pliny, AD78 [1]

18.1 Introduction

Failures of components by liquid-metal embrittlement (LME) and solid-metal-induced embrittlement (SMIE) are less common than failures by other environmentally assisted cracking phenomena such as hydrogen embrittlement (HE), stress-corrosion cracking (SCC), and corrosion-fatigue (CF), but a significant number of industrial failures by LME (and very occasionally by SMIE) do occur [2–4]. There are remarkably diverse, potential sources of embrittling metal environments [4–7], summarised in Table 18.1, and failures

714

Table 18.1 Some sources of embrittling metal environments in industry

- Coatings, e.g. Cd, Zn on steels, inadvertently exposed to high temperatures.
- Solders, brazes, and galvanising baths (if applied/residual stresses are high).
- Liquid-metal coolants for nuclear reactors, nuclear fission products, and neutron spallation targets (Pb-Bi).
- Overheated bearing materials containing Sn, Cu, Pb.
- Internal sources, e.g. Pb particles in free-machining alloys, Na-K impurity phases in Al-Li alloys.
- Mercury from broken thermometers, etc.
- Mercury as an impurity in natural gas.
- Molten metal from fires, especially Zn from galvanised steel dripping on stainless steel.
- Zn-based paint splashes onto hot equipment.
- Lubricants containing metals such as Pb.
- Pick-up from contaminated grinding wheels.
- Cu from electrical resistance-heating contacts.
- Deposition from aqueous environments after corrosion/erosion-corrosion of embrittling metals, e.g. Cu, Pb.
- Airborne ZnS (near mining facilities) reduced to Zn on hot surfaces.

Table 18.2 Common embrittling liquid-metal environments for some structural alloys

Structural material	Embrittling environments
High strength martensitic steel	Cu, In, Sn, Pb, Cd, Zn (Hg)
Stainless steels	Zn, Cu
Titanium alloys	Hg, Cd, Ag, Au
Aluminium alloys	Hg, Ga, In, Pb, Na
Copper alloys	Hg, Ga, Bi, Zn, Sn, Pb

can occur if engineers who design, operate, and maintain equipment are not aware of these potential sources and the dangers involved.

The most common embrittling metals for some structural alloys are listed in Table 18.2. More comprehensive information on embrittling (and non-embrittling) couples can be found in the literature [8–10]. Generally, embrittling couples have low mutual solubilities and do not form intermetallic compounds, although there are some exceptions, such as the compound-forming Fe-Zn couple. Moreover, some couples do not exhibit embrittlement despite satisfying the above criteria. The reason for this 'specificity' is not well understood at the fundamental level and merits further work, especially using the latest quantum-mechanical modelling techniques.

LME is certainly one of the more dramatic phenomena in materials science and engineering, in that some normally ductile metals, with a reduction of area approaching 100% when tensile tested in air, can exhibit brittle behaviour with essentially 0% reduction of area when tested in specific liquid-metal environments. For specimens held under constant stress at levels well below the yield stress, brittle fracture can occur 'instantaneously' when they are exposed to specific liquid-metal environments. Sometimes, however, there are long incubation times prior to sudden fracture (Fig. 18.1) [11]. For fracture-mechanics specimens, brittle sub-critical crack growth in liquid metal environments can occur at velocities as high as several hundred millimeters per second at stress-intensity factors as low as 10% of those required for fast fracture in air (Fig. 18.2) [12–14].

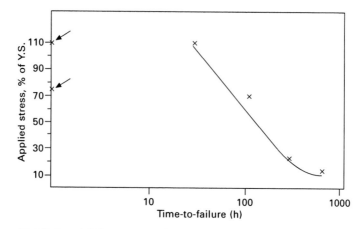

18.1 Delayed-failure curve (time-to-failure vs applied stress as a percentage of the yield stress) for aluminium-alloy weld (5083) specimens exposed to liquid mercury at 20°C. Note that failure can be almost immediate (see arrows) or ~100 h for the same stress level [11].

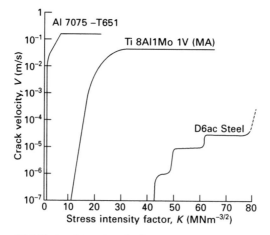

18.2 Plots of crack-velocity vs stress-intensity factor for an aluminium alloy (7075-T651) [12], a titanium alloy (Ti 8%Al 1%Mo 1%V) [13] and a high-strength steel (D6aC) [14], tested in liquid mercury at 20°C.

An essential requirement for the occurrence of LME (and SMIE) is that there must be intimate contact ('wetting') between the embrittling metal and substrate, with no intervening oxide or other films [2–8]. The long incubation times sometimes observed for LME under sustained loads are probably associated with breakdown of oxide films by localised micro-plasticity (creep) or by other processes. In some cases, oxide films need to be removed chemically (by fluxing) or mechanically (by abrasion) before LME can occur. Even with fluxes and abrasion, it can be difficult to achieve wetting in some systems, e.g. liquid mercury and high-strength steels.

18.2   Mechanisms and rate-controlling processes for liquid-metal embrittlement (LME) and solid-metal-induced embrittlement (SMIE)

Since there is neither the time nor the tendency for diffusion of embrittling atoms ahead of crack tips in many systems, it is generally accepted [4–8] that LME results from adsorption-induced weakening of interatomic bonds at surfaces and crack tips, which facilitates dislocation-emission or decohesion. There are, however, a few systems, such as the Al-Ga and Ni-Bi couples [15, 16], where the embrittling atoms diffuse rapidly along grain boundaries (even in unstressed material). This results in liquid films of nanometer-scale thickness along grain boundaries, and the material disintegrates into individual grains when very low stresses are applied – a phenomenon that is equally as dramatic as adsorption-induced LME.

Adsorption-induced LME most commonly occurs along grain boundaries in polycrystals, but cleavage-like cracking occurs preferentially for some couples. Cleavage-like cracking can also occur for most couples when there are no suitably oriented grain boundaries (and in single crystals), with some exceptions such as copper alloys. Intergranular and cleavage-like fracture surfaces sometimes exhibit small shallow dimples (Fig. 18.3) or flutes, and there can be significant localised plasticity around crack tips. These observations are consistent with an adsorption-induced dislocation-emission (AIDE)/microvoid-coalescence (MVC) process, as proposed by the present author [5, 17]. In most cases, fracture surfaces appear flat and largely featureless except for isolated tear ridges in some cases (Fig. 18.4) and there are no obvious signs of slip around crack tips, consistent with an adsorption-induced decohesion process. However, dimples can be so small and shallow that they are not resolved by scanning electron microscopy (SEM), and slip can be extremely localised, so that it is debatable whether LME occurs by adsorption-induced decohesion or by an AIDE/MVC process in such cases.

The degree of LME is usually greatest at, and just above, the melting temperature of the embrittling metal, with embrittlement being less severe

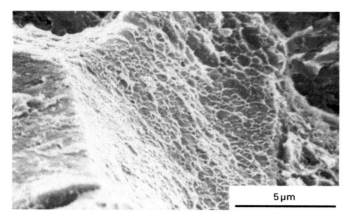

5 µm

18.3 SEM of dimpled intergranular fracture surface produced by sub-critical crack growth for a high strength D6aC steel (tempered at 650°C to 41HRC) in liquid mercury at 20°C [5].

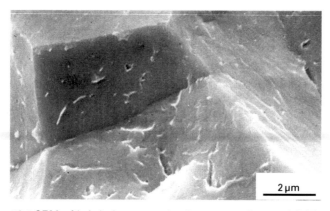

2 µm

18.4 SEM of brittle intergranular fracture surface, exhibiting smooth areas with tear ridges, produced by sub-critical crack growth for a high strength D6aC steel (tempered at 400°C to 51HRC) in liquid mercury at 20°C [5].

at high temperatures (Fig. 18.5), probably because stress-relaxation by slip processes at potential crack-initiation sites occurs more readily at higher temperatures [4, 5, 8, 9]. LME can occur below the melting point of the embrittling metal if a low-melting-point phase such as a eutectic forms between the embrittling metal and the substrate metal. When embrittling metals are solid, embrittlement can still occur (by SMIE) but it is much less severe than LME, in that times to failure are much longer, the crack velocities are much slower, and the extent of crack growth is much less (Fig. 18.6) [5, 18, 19].

SMIE is much less severe than LME because the rate-controlling process

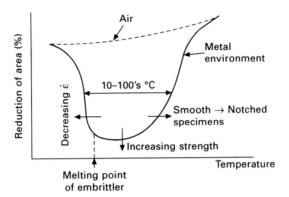

18.5 Schematic plot showing the effect of temperature on the severity of embrittlement (measured by the reduction of area of tensile specimens).

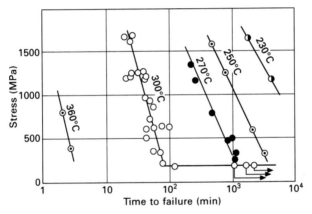

18.6 Delayed-failure plots for a high-strength steel exposed to liquid and solid cadmium at various temperatures (cadmium melts at 321°C) [18].

for SMIE, namely surface self-diffusion of embrittling atoms from their source to the crack tip, is much slower than the rate-controlling step for LME, namely capillary flow of liquid metal to the crack tip (or possibly adsorption at the crack tip) [18–21]. SMIE would be expected to occur to a significant extent at temperatures above $0.5T/T_m$, where T is the test temperature and T_m is the melting temperature in degrees Kelvin. For example, lead ($T_m \sim 600\,K$) and cadmium ($T_m \sim 594\,K$) can produce significant sub-critical crack growth at room temperature ($\sim 300\,K$) in susceptible materials such as steels and aluminium alloys.

The rates of cracking by SMIE increase with increasing temperature, but decrease with increasing crack length as the distance from the source of the embrittling metal to the crack tip increases. Rates of cracking by SMIE can also decrease (or cracks can be arrested) if air gains access to cracks, which becomes more likely as cracks get longer. Oxide films are likely to inhibit surface-diffusion, and preferential adsorption of oxygen or oxide-film formation at crack tips will inhibit adsorption of embrittling metal atoms. Thus, the extent of crack growth by SMIE from external sources of embrittling metals such as coatings is limited (usually less than several millimeters) [19], and failures by SMIE are rare. On the other hand, when closely spaced, *internal* embrittling particles are present, extensive crack growth and failure by SMIE can occur since distances from embrittling metal sources to crack tips are always small, and no air is present within internal cracks [3].

18.3   Evidence for liquid-metal embrittlement (LME) and solid-metal-induced embrittlement (SMIE)

For LME, globules or films of embrittling metal are usually evident (using optical microscopy and SEM) on brittle intergranular or transgranular (cleavage-like) fracture surfaces and within cracks. Liquid-metal globules rather than films are present when 'de-wetting' occurs after fracture owing to diffusion of oxygen through liquid-metal films and repassivation of the fracture surface. The presence and nature of deposits can sometimes be obvious from their colour during visual examination, but in general the composition of any deposits can be determined by energy-dispersive spectroscopy (EDS) or by other techniques.

For SMIE, embrittling metal films on fracture surfaces are also sometimes evident from their colour, e.g. for copper and oxidised cadmium, but are often not obvious even using high-resolution SEM/EDS. The films are also too thin to be apparent within secondary cracks on metallographic sections. Very thin embrittling-metal films can, however, be detected using surface-sensitive techniques such as Auger-electron spectroscopy or secondary-ion mass spectroscopy. For SMIE from internal sources of embrittling metal, particles may be apparent on metallographic sections, and their identity can be determined by EDS if they are sufficiently large. In such cases, EDS of fracture surfaces (regardless of the fracture mode) will also show that embrittling metal is present – but not necessarily on fracture surfaces, since EDS probes significant volumes of material beneath fracture surfaces.

For components that are partially cracked by LME, and then cooled below the melting point of the embrittling metal, subsequent crack extension results in a localised ductile or brittle fracture through the embrittling metal, producing a dimpled or vein pattern on fracture surfaces when the embrittling metal is ductile [5]. The presence of films of embrittling metal along grain boundaries is sometimes mistakenly viewed as evidence that LME involves diffusion of embrittling metal atoms along grain boundaries prior to their separation. Diffusion can occur in some systems (as already mentioned), but it is more common for liquid metal to be drawn into cracks by capillary action during the process of adsorption-induced crack growth.

More details of the foregoing aspects of embrittlement, and on the effects of variables such as strength and microstructure, can be found in the numerous reviews of LME and SMIE [4–10]. The rest of the present chapter describes some case histories of failures involving LME and SMIE, with which the author has been involved, to illustrate some of the typical characteristics associated with failure by these modes. The underlying causes and how such failures might have been prevented are also discussed. Most of the examples have not been previously reported, while one has been published previously but is worth reviewing in order to emphasise important points. A final, brief section summarises some beneficial uses of LME in failure analysis.

18.4   Failure of an aluminium-alloy inlet nozzle in a natural gas plant [22]

A catastrophic rupture of a 5083-0 Al-Mg alloy inlet nozzle carrying pressurised, liquefied gas to a heat-exchanger led to the escape of a large quantity of gas, and a resultant gas-cloud explosion that caused substantial damage to the plant, but fortunately no injuries to personnel. The rupture in the 400 mm diameter pipe, which had a wall thickness of 14 mm, extended longitudinally for about 500 mm upstream from the flange, and also extended around part of the circumference of the flange (Fig. 18.7). Examination of

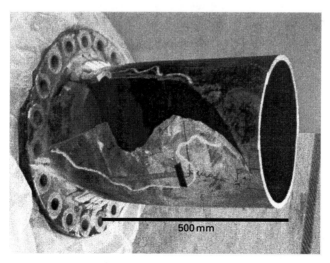

18.7 Macroscopic view of ruptured 5083-0 Al-Mg alloy inlet nozzle to a heat-exchanger in a natural-gas plant. Extensive secondary cracking and delaminations occurred within the area outlined.

the ruptured pipe, using both a stereo-microscope and radiography, showed that there were substantial quantities of liquid mercury on the pipe surfaces, the fracture surfaces, and within the extensive network of secondary cracks (Fig. 18.8). There were also white, flaky, oxide whiskers on fracture surfaces, which are typical of those that form when mercury has wetted aluminium surfaces in the presence of atmospheric moisture [23].

Fracture surfaces exhibited numerous steps and delaminations normal to the fracture surfaces of the pipe except at a circumferential weld, where fracture surfaces were not as rough and were discoloured to a greater extent (Fig. 18.9). SEM showed that the fracture surfaces (cleaned by immersion in concentrated nitric acid) were intergranular, and that grains were fairly equi-axed in the weld, but elongated in the parent plate (from which the pipe was formed), as would be expected (Figs 18.10 and 18.11). Metallographic sections through the fracture showed that micro-crack branching was extensive, both in the welds and in the parent plate (Fig. 18.12). For the latter, intergranular delaminations normal to the main fracture surface extended to depths of several millimeters.

It was concluded from the above observations that failure had occurred by LME (by mercury). Mercury-induced degradation of aluminum-alloy components in natural-gas plants, usually involving 'amalgam corrosion' or LME, has been reported previously [24–29]. Mercury is present in natural-gas basins, typically at 30–250 ppb concentrations, and substantial quantities

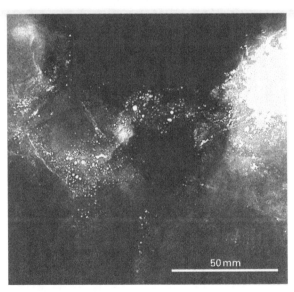

18.8 Radiograph of outlined area in Fig. 18.7 showing white areas (not penetrated by x-rays) owing to the presence of mercury in secondary cracks.

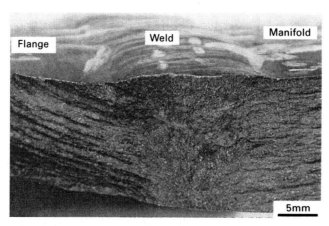

18.9 Macroscopic view of fracture surface in the region of the circumferential weld.

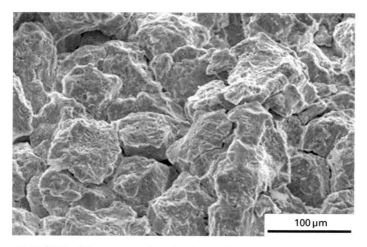

18.10 SEM of fracture surface in the weld region of inlet nozzle showing intergranular cracking through the equi-axed grain structure.

of mercury can accumulate in plants after many years of operation unless preventative measures are taken. The stresses required for LME are generally present in the pipework of natural gas plants, owing to the combined effects of internal pressures and, near flanges, stresses from bolting pipework together. Stresses owing to subsidence of foundations, and residual stresses from welding, can also be significant.

Once it was established that LME was the failure mode, a number of issues required further detailed investigation. These issues included: (i) Why had the plant been operating for almost 20 years with mercury in the

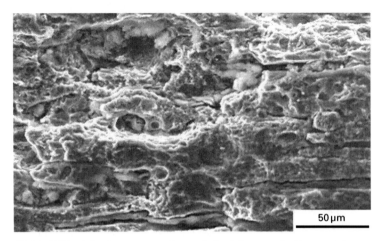

18.11 SEM of the fracture surface of the inlet nozzle (formed from rolled plate) showing intergranular fracture through elongated grain structure, with delaminations normal to the main fracture surface.

system without a failure, and what events initiated the failure? (ii) Where did the failure initiate? (iii) Why was cracking so extensive, resulting in a catastrophic rupture rather than a small leak, as had occurred in other natural gas plants exposed to mercury? and (iv) What steps could be taken to minimise the risk of further failures?

With regard to the first question, the plant generally operated at temperatures below the melting point of mercury (−39°C) so that LME was not possible. Just prior to the failure, however, 'process upsets' resulted in temperature fluctuations from about −42°C to about −30 to −35°C. Nevertheless, the plant had undoubtedly experienced similar temperature excursions previously, and had been at ambient temperature (up to 40°C) during maintenance periods, without incidents. The reason crack initiation had not occurred earlier may have been because oxide films prevented wetting. (Once cracking initiated, rupture would occur almost immediately owing to the very high crack growth rates typical of LME of aluminium alloys in liquid mercury.)

Wetting might have occurred at the time it did, and not on previous occasions, because greater volumes of mercury had accumulated, resulting in greater surface areas, including weaker spots in the oxide, being covered with liquid mercury. Thermal-cycling may also have been greater on this occasion than previously, thereby resulting in greater strains and, hence, oxide-film rupture by slip processes. Statistically rare events, such as impacts of hard particles that ruptured the oxide film, could also have been responsible. Such particles could include ice and solid mercury in the gas stream, with wetting subsequently occurring when mercury became molten. (Repassivation at oxide-rupture sites may have been very slow since little if any oxygen was present in the gas stream.) The wetting process is, however,

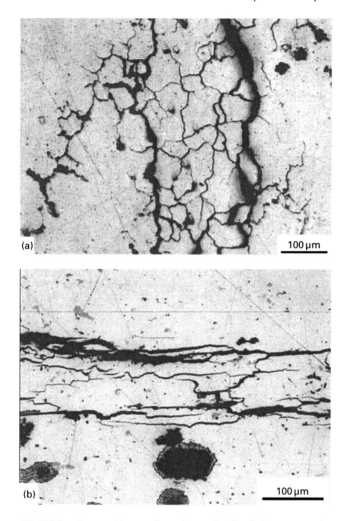

18.12 Metallographic sections (unetched) of the ruptured 5083-0 Al-Mg alloy inlet nozzle showing extensive intergranular cracking (a) in the weld, and (b) in the plate.

not well understood even under simple laboratory conditions (Fig. 18.1), and further work is required to answer the first question.

The second question regarding where failure initiated was relatively easy to answer from 'chevron' markings on fracture surfaces and patterns of crack branching – both of which indicated that fracture initiated from the circumferential weld near the flange (at the 6 o'clock position) on the inside surface of the pipe. Welds, with their equi-axed grain structure and residual-stress fields, are known to be favoured sites for crack initiation. In this case, the proximity of the weld to the flange (where bolting stresses were high) is an additional factor favouring crack initiation at the weld. Geometric factors may have also led to mercury 'pooling' around this location.

Cracking was more extensive for this failure compared with previous failures in natural gas plants possibly because there was a greater volume of mercury in the system (perhaps owing to longer times in operation without mercury traps). During LME, embrittling metal films are left on fracture surfaces behind crack tips and, hence, crack tips can 'run out of' embrittling metal and arrest (providing stress-intensity factors are not sufficiently high as to result in unstable fast fracture). Thus, greater volumes of mercury at the crack-initiation site and along crack paths can lead to more extensive cracking (Fig. 18.13) [30].

Measures that have been taken, or considered, to prevent LME in natural gas plants include the following [24–29]:

- Installing mercury traps, such as sulfur-impregnated activated-charcoal filters upstream of the Al alloy pipework/equipment.
- Coating Al alloy components with thicker, less easily ruptured films than the naturally formed oxide.
- Shot-peening to introduce compressive residual stresses into near-surface regions.
- Avoiding temperatures at which mercury is liquid whenever practical.
- Using materials less susceptible to embrittlement by mercury, such as stainless steels (providing such alternative materials have other requisite properties).
- Eliminating crevices or other 'geometries' where mercury can collect.

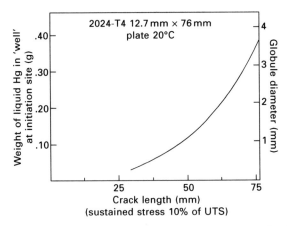

18.13 Plot showing the variation in the extent of sub-critical cracking for a 2024-T4 aluminium alloy plate with the amount of mercury available at the crack-initiation site (from Ref. [30], with modifications).

• Positioning welds away from the most highly stressed areas wherever possible.

18.5   Failure of a brass valve in an aircraft-engine oil-cooler [31]

An investigation into the cause of a cockpit-warning light indicating low oil pressure in an F-111 engine oil-cooler revealed that the crimped end of the housing of a thermostatically controlled valve was cracked around the circumference (Figs 18.14 and 18.15). The valve operated via a spring that was compressed by the thermal expansion of a wax contained in the valve housing when the oil temperature reached ~140°C. Thus, the valve would not have been operational with a completely cracked housing. The valve housing would have been subjected to internal pressures from the expansion of the wax, and probably high tensile residual stresses at the inside radius of the crimped region if a stress-relieving heat treatment had not been carried out after crimping (see later).

The valve housing was manufactured from a free-machining brass (10%Zn+2%Pb), and metallographic sections showed that there were numerous lead particles throughout the microstructure (Fig. 18.16). The sections, and SEM of the fracture surfaces, showed that cracking was intergranular near the inside surface of the housing, gradually transitioning to a dimpled transgranular fracture about one quarter of the way through the wall thickness (Figs 18.17 and 18.18), suggesting that cracking had initiated at the inside

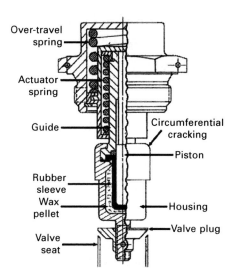

18.14 Simplified engineering drawing of thermostatically controlled valve, showing position of cracking.

Published by Woodhead Publishing Limited, 2011

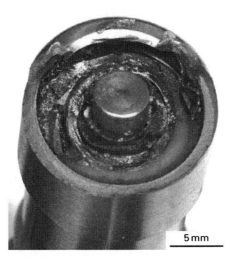

18.15 Macroscopic view of failed valve after dismantling and partial cleaning, showing fracture through the outside wall of the housing near the crimped end.

18.16 Metallographic section near inside surface of failed valve examined by SEM (back-scattered mode) showing crack-path topography typical of intergranular cracking, and distributed lead particles (white spots).

radius of the crimped region. At high magnifications, shallow, flat-bottomed dimples centred on small holes, and numerous slip steps, were evident on intergranular facets (Fig. 18.19).

The unusual fracture surface appearance was considered most likely to have

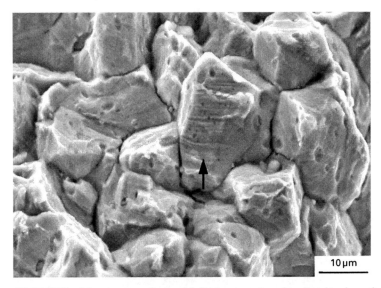

18.17 SEM of fracture surface of failed valve showing predominantly intergranular fracture, with (possibly) a small area of transgranular fracture exhibiting fatigue striations (arrowed), near the inside surface of the valve.

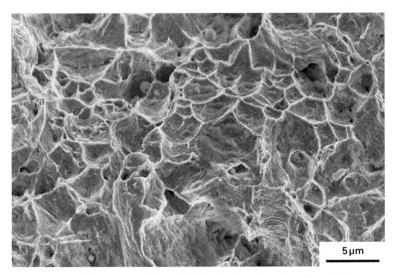

18.18 SEM of fracture surface of failed valve showing dimpled transgranular fracture near the mid-thickness position.

been produced by internal SMIE owing to lead particles in the microstructure. For SMIE by internal phases, voids are initiated at the particles, and grow by an adsorption-induced decohesion or dislocation-emission process until they

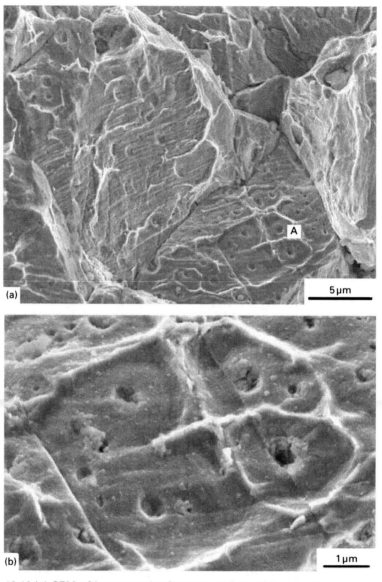

18.19 (a) SEM of intergranular fracture surface of failed valve showing shallow dimples, centered on holes, and slip markings, and (b) area A at a higher magnification.

coalesce with each other. The rapid coalescence process involves localised necking, leaving small cusps on the fracture surface. Holes are left on the fracture surface where the particles had been since the embrittling metal diffuses over the fracture surfaces to crack tips and leaves the embrittling metal as a thin film on the fracture surface. Deformation during the process, especially during void coalescence, occurs to some extent behind the crack tip and produces slip steps on the fracture surface.

LME by internal phases can produce a similar appearance to that observed [32], but was not considered possible in this case since the operating temperature of the valve (up to 140°C) was well below the melting point of lead (327°C). Surface-self-diffusion of lead can occur at significant rates even at room temperature, and would be relatively rapid at 140°C – but not so rapid that embrittlement would occur at high crack growth rates. Rapid crack growth appears to have occurred when crack depths reached about one-quarter of the way through the wall thickness. At this stage, the stress-intensity factor presumably reached the level required for fast fracture, so that normal transgranular, dimple rupture occurred.

Stress-corrosion cracking (SCC) can produce intergranular fracture in copper alloys, and this possibility was initially considered by one investigator. However, SCC can be discounted since intergranular SCC fracture surfaces are generally relatively featureless compared with the appearance in the present case [33]. More basically, environments known to produce SCC in brass were not present in the valve during its operation. The wax in the housing (a hydrocarbon containing fluoride, plus fine copper particles to produce the right 'consistency'), was thought to be inert, and would exclude any other environment. Testing to conclusively prove that the wax would not produce SCC was not considered to be worthwhile given the evidence in support of SMIE.

Fatigue in inert environments at low ΔK can also produce intergranular fracture in copper (and other) alloys [34], but was not considered a likely primary fracture mode in view of the evidence for SMIE. However, the valve housing would have been subject to cyclic stresses owing to temperature variations that caused changes in internal wax pressure. The valve had probably been in service for many years (a detailed history was not available) and, hence, could have experienced significant numbers of cycles. In some areas of the fracture surface, 'ductile' striations were apparent (Fig. 18.17), and it is possible that these striations resulted from fatigue crack growth through ligaments of material between internal cracks produced by SMIE.

Striation-like markings can be produced by processes other than fatigue, and it is sometimes difficult to distinguish fatigue striations from other markings, such as slip lines and crack-arrest markings produced by discontinuous sustained-load cracking [35]. In the present case, some markings appeared to be slip lines while others appeared more like fatigue striations, and it would have been helpful if mating halves of opposite fracture surfaces could have been examined. Slip lines generally do not match on opposite fracture surfaces whereas fatigue striations generally do match. However, both halves of the fracture surface were not available for examination, since one half had been sectioned for metallographic examination (arguably prematurely) by a previous investigator.

In conclusion, there is no doubt that the primary failure mode was SMIE owing to the presence of internal lead particles, and that the underlying cause of the failure was the incorrect selection of a free-machining brass for a component that operated at elevated temperatures in the presence of high stresses. Whether residual tensile stresses at the inside radius of the crimped area were present in addition to stresses from internal pressures is not known since intact valves were not available for testing, e.g. using the mercurous-nitrate test described in the final part of this chapter. If high residual stresses were present, stress-relieving after crimping would eliminate them, and might prevent failures or extend the life of the valves, providing cracking did not occur by SMIE or LME during a stress-relief heat treatment.

For any future manufacture of valves, the use of a 'normal' brass (without potentially embrittling phases in the microstructure) would be a solution. In the present case, it was considered unnecessary to replace all the valves in service, even though further failures would be expected, since this failure was only the second one to occur and the consequences of failure were considered not to be serious. Nevertheless, due consideration should generally be given to scenarios where the simultaneous failures of several 'minor' components might lead to serious consequences.

18.6   Failure of a screw in a helicopter fuel-control unit [36]

Investigation of the wreckage of a helicopter that crashed and caught fire while attempting an emergency landing revealed that one of two cadmium-plated steel screws securing a cover-plate on an engine fuel-control unit had fractured at the head-to-shank region. The surfaces of the fractured screw, and intact screws (including those on other cover plates), were covered in cadmium globules, indicating that the screws had experienced temperatures above the melting point of cadmium during the post-crash fire (Fig. 18.20).

The fracture surface of the failed screw was covered by a cadmium film, as was evident from macroscopic observations showing blue/yellow tints typical of oxidised cadmium (Fig. 18.21) and from EDS during SEM examination. The SEM observations also showed that fracture surfaces were intergranular (Fig. 18.22) except for very small areas of dimpled (overload) fracture. Optical and SEM/EDS examination of unetched metallographic sections showed that there were networks of secondary intergranular cracks completely filled with cadmium right up to the crack tips (Fig. 18.23). Metallographic sections through the failed screw and intact screws also showed (after etching) that they had a tempered-martensite structure, and hardness measurements gave

18.20 Macroscopic view of one of the intact cadmium-plated steel screws showing cadmium globules on the surface.

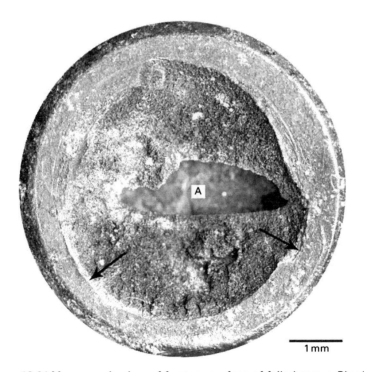

18.21 Macroscopic view of fracture surface of failed screw. Blue/yellow colours typical of oxidised cadmium were evident. Area A is where the fracture intersected a hole in the head of the screw. Arrows indicate the location of small areas of overload fracture.

18.22 SEM of fracture surface of failed screw in region where the topography was not obscured by a layer of cadmium.

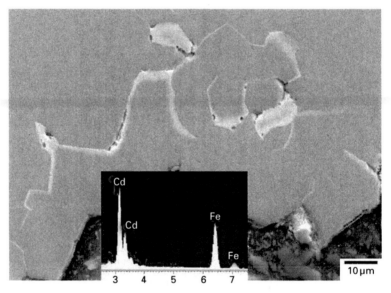

18.23 SEM of metallographic section showing secondary intergranular cracks filled with cadmium. Inset shows EDS spectrum for the lighter areas.

readings of 42HRC (Rockwell C scale), well above the specified hardness of 26-32HRC.

The foregoing observations left little doubt that the screw had failed by LME during the post-crash fire. However, two other (independent) investigators had previously concluded that the screw had failed during flight owing to cracking by SCC or HE, which also produce intergranular cracking. The previous investigations had concluded that the failure of the screw led to the crash by causing a fuel leak, a loss in fuel pressure, and a consequent loss in engine power. The fact that the screw was much harder than specified was then a major issue since high-strength steels are much more susceptible to HE and SCC than low-strength steels. For LME, the steel strength is not as important, with both high and low strength steels being embrittled by cadmium.

Evidence cited in favour of a HE or SCC failure mode rather than LME by these previous investigators included: (i) the apparent presence of cadmium (from EDS) on the small areas of dimpled overload areas (as well as intergranular areas of the fracture surface) suggesting that cadmium had flowed over the whole pre-existing fracture surface during the post-crash fire, (ii) the apparent absence of secondary cracks filled with cadmium (typical of LME failures) during the initial investigations, and (iii) the fact that other screws subject to the same conditions during the post-crash fire had not failed by LME.

The above arguments against failure by LME (and by default supporting a HE or SCC failure mode) were discounted by subsequent investigators (including the present author) [36] on the grounds that: (i) the apparent EDS detection of cadmium on the dimpled overload areas was an artifact resulting from back-scattered electrons from the overload areas impinging on intergranular areas covered by cadmium, thereby generating secondary fluorescent X-rays with energies characteristic of cadmium, (ii) the apparent absence of cadmium-filled secondary cracks reported by the initial investigators probably arose because *etched* sections had been examined, with the etching process dissolving the cadmium, and (iii) other screws did not fail during the post-crash fire because there were very thin oxide films between the electroplated cadmium and the underlying steel that precluded wetting.

Tests on new cadmium-plated screws did, in fact, show that LME usually did not eventuate when they were torqued to a high stress and heated to above the melting point of cadmium. On the other hand, some cadmium-plated notched bars tested in three-point bending at 350°C did fail by LME, presumably because the electroplating conditions in this case were such that wetting did occur when the cadmium melted. The failed screw was therefore probably plated under conditions enabling wetting during the post-crash fire, whereas the other screws were plated under (probably subtly) different conditions that precluded wetting.

Finally, it is worth noting that pre-existing fracture surfaces produced by HE or SCC would be covered by oxide fims, and the difficulties of wetting in the presence of oxide films would probably prevent cadmium from flowing over a pre-existing fracture surface. Oxide films would certainly preclude cadmium from penetrating to the tips of fine, secondary intergranular cracks, as was observed.

18.7   Collapse of a grain-storage silo [37]

A large grain-storage silo (~150 000 tons capacity) collapsed many years ago after only three months in use as it was gradually filled to capacity (Fig. 18.24). The collapse occurred at lunchtime when workers were either off-site or indoors, and there were no injuries. The silo was constructed from corrugated, galvanised steel sheet supported by galvanised carbon-steel stiffeners (~5 mm thick) that had been formed by bending them through 90° at 20°C into an L-profile.

Visual examination of the stiffeners after the collapse showed that some had large cracks extending from the outside radii of the bends to depths of up to 4 mm. Others were completely fractured, with fracture surfaces covered by a thick layer of zinc from the outside of the bend up to at least the mid-thickness position, and generally beyond. The remaining (uncoated) fracture surfaces had a rusty appearance, presumably because they were exposed to the weather for some time after the collapse (Fig. 18.25).

Close examination of metallographic sections through partially cracked stiffeners showed that, not only were there large cracks extending from the outside radii, but there were also small cracks (up to 0.5 mm long) near the inside radii. These small, narrow cracks were intergranular and filled with zinc, while those extending from the outside radii were much wider though still filled with zinc (Fig. 18.26). Areas of the steel near the galvanised surfaces and adjacent to cracks were more lightly etched than other areas,

18.24 General view of the scene after the grain-storage silo collapse. Workers were, fortunately, having lunch in the hut (arrowed) at the time of the collapse.

Published by Woodhead Publishing Limited, 2011

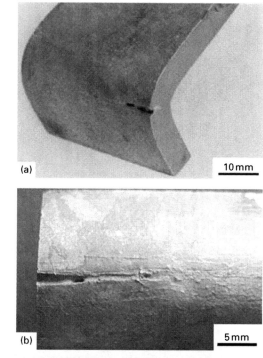

18.25 Macroscopic views of the galvanised carbon-steel stiffeners showing (a, b) cracks at the outside of the bend, and (c) fracture surface, which was partially covered by a film of zinc.

and Fe-Zn intermetallic compounds were present at the interfaces between the zinc and the steel. The lighter etched regions were attributed to the greater galvanic protection of the steel closer to the zinc during etching rather than any differences in the composition of the steel (such as decarburisation) near surfaces and adjacent to cracks.

It was concluded from the above observations that the gaping cracks extending from the outside radii of the bends occurred during the forming operation prior to galvanising, while the narrow intergranular cracks extending

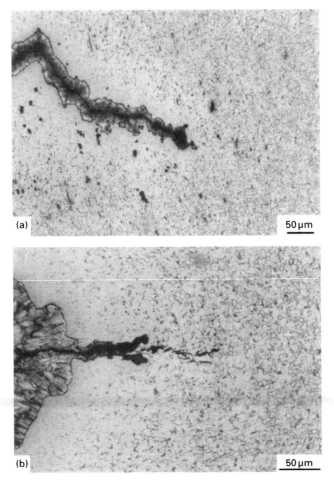

18.26 Metallographic sections after etching showing (a) the crack tip region for a large crack that initiated from the outside radius of the bend, and (b) the area around the inside radius of the bend showing a small intergranular crack. Both cracks were filled with zinc and intermetallic Fe-Zn compounds.

from the inside radii had occurred by LME during galvanising. Since the LME cracks were relatively small, their contribution to the weakening of the stiffeners (and their subsequent overload failure as the silo was filled) would have been negligible. However, the observations do show that LME of steel can occur during galvanising, although it appears to be uncommon [6, 38]. It occurred in this case probably because tensile residual stresses at the inside radii of the bends were very high, such that crack-initiation and growth occurred rapidly before there was time for formation of Fe-Zn intermetallic compounds, which normally prevent LME from occurring.

The observations also illustrate that the presence of adherent metal deposits on fracture surfaces does not necessarily indicate that LME has occurred. In this case, zinc flowed into gaping pre-existing cracks at the outside radii during galvanising, aided by the presence of a flux so that wetting occurred. Other circumstances where metal deposits (or metal salts) can be produced on fracture surfaces by modes other than LME and SMIE include:

- Infiltration of cracks by fretting products from coatings during fatigue.
- Rubbing of fracture surfaces against soft metal surfaces, as could occur, for example, if two pieces of a fractured, cadmium-plated component were carelessly put in the same container for transportation.
- Electroplating partially cracked components, as can occur when used components are stripped for inspection, and then re-plated (and re-used) without sufficiently careful inspection for cracks.
- Exposure of fracture surfaces (or cracks) to metal-ion-rich aqueous solutions, e.g. produced by corrosion of coatings, followed by displacement reactions or evaporation of the solutions.

With respect to the underlying cause of the silo collapse, the (presumably) non-existent inspection of the stiffeners after forming is obviously crucial. Even after galvanising, when cracks would have been partially obscured by zinc deposits, the cracks at the outside radii should have been obvious. Consideration at the design stage should also have been given to the fact that bending 5 mm thick steel through 90° at 20°C was a severe forming operation that could have produced cracking. Analysis of the steel showed that it had a higher-than-normal phosphorus concentration, which would have reduced its formability, so that there were also deficiencies in material selection or in checking material specifications.

Using steel with adequate formability, and bending at elevated temperatures, would be a solution, providing the elevated-temperature bending reduced residual stresses sufficiently that LME did not occur during galvanising. (For the cold-formed stiffeners, if cracking had not occurred from the outside surface during forming, the extent of the tensile residual stress fields at the inside radii would have been greater, and probably would have led to greater extents of cracking by LME during galvanising.)

18.8 Failure of planetary gears from centrifugal gearboxes [39]

Several large case-hardened steel planetary gears (~200 mm diameter) from centrifugal gearboxes fractured at diametrically opposite locations into two halves. It was obvious from the blackened and scored bearing surface, and from the 'temper colours' on the ends of the gears, that severe overheating had occurred (Fig. 18.27). The fracture surfaces exhibited black and copper-

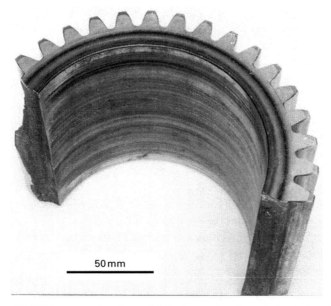

50 mm

18.27 Macroscopic view of failed case-hardened steel planetary gear from a centrifugal gearbox. Note the blackened inside diameter and temper colours at the end surface (move obvious in colour).

Copper coloured deposits

20 mm

18.28 Macroscopic view of fracture surface of planetary gear shown in Fig. 18.27, which exhibited copper-coloured deposits, along with markings indicating that crack growth occurred from the inside bore (in contact with the copper-alloy bearing material) towards the outside.

coloured deposits (Fig. 18.28), and secondary intergranular cracks filled with copper-coloured material were evident on metallographic sections. It was therefore concluded that failure had occurred primarily by LME owing to the melting of the bearing material – analysed as Cu 12%Sn 0.24%Zn, 0.18%Ni, 0.38%Pb. For some gears, copper-coloured deposits did not extend over the entire fracture surfaces, and the remaining section had fractured by fatigue, with characteristic progression markings on fracture surfaces.

The gears had operated satisfactorily for many years, and the underlying cause of the failure was attributed to a sudden loss of oil lubrication (for reasons that are not known to the author). A contributory cause would have been that the bearing material did not conform to specification, having a lead content of 0.38% compared with the specified level of 1–2%. The higher lead content provides some 'dry' lubrication if oil supplies become marginal. Presumably, LME or SMIE of the steel by lead (which can occur in some circumstances) does not eventuate because there are compressive residual stresses present at the surface of the steel resulting from case-hardening. During severe overheating, to temperatures ~900°C required to cause melting of the copper-based bearing material in this case, the steel would become austentised, compressive residual stresses would be lost, and copper-induced LME would not be inhibited by compressive residual stresses.

Initiation and growth of cracks by LME (or SMIE), with continued crack growth by fatigue has been observed previously [40], although it is much more common for components to separate into two (or more) pieces by LME (along with small areas of overload in some cases). Crack growth by LME can arrest, and then possibly progress by other modes, for a number of reasons, such as when:

- There are only very limited supplies of liquid metal so that cracks 'run out of' embrittling metal, although this is unusual because very small volumes of liquid metal can produce extensive crack growth.
- There are decreasing temperature gradients across the component so that the liquid metal solidifies after some crack growth, as may have occurred in one of the gears, since frictional heating was involved.
- There are decreasing stress gradients so that the stress-intensity factor at LME crack tips falls below threshold values, although since such values are usually lower than threshold values for other fracture modes in most systems, further crack growth by other modes is unlikely.

18.9   Beneficial uses of liquid-metal embrittlement (LME) in failure analysis

18.9.1   Residual stress tests

The low threshold stress levels and high crack growth rates characteristic of LME enable a rapid, simple (albeit qualitative) test to be carried out to determine whether residual tensile stresses have played a role in a failure, especially for copper alloys [41, 42]. The test involves immersing an intact, cleaned component (manufactured by the same method as the failed component) in a mercurous-nitrate solution at 20°C for 30 minutes, washing, and then visual inspection for cracks (Fig. 18.29) [43]. Liquid mercury 'plates out' onto the surfaces by a displacement reaction, and cracking occurs by LME if there are surface residual tensile stresses. Very fine cracks can be obscured

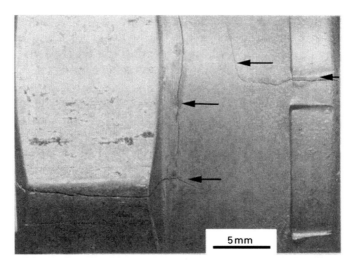

18.29 Macroscopic views of crimped nickel-aluminium-bronze hydraulic couplings after the ASTM mercurous-nitrate residual-stress test showing longitudinal and circumferential cracks (arrowed). Note silvery colouration owing to the presence of liquid mercury [43].

by mercury films, but evaporating the mercury under vacuum at ~150°C (using appropriate precautions to prevent mercury vapour escaping into the atmosphere) allows fine cracks to be detected. Major cracks can then also be broken open in air to reveal the extent of LME cracking and thereby get an indication of the depth of the tensile residual stresses.

18.9.2   Examining grain-boundary microstructures

LME that causes either adsorption-induced or diffusion-induced intergranular cracking can be used to determine the extent of grain-boundary precipitation by examining resultant brittle intergranular fracture surfaces (cleaned if necessary) by high-resolution SEM – a much quicker and, in terms of areas examined, a more representative technique than transmission electron microscopy of thin foils. Grain-boundary precipitation can be of interest in failure analyses because excessive precipitation can contribute to failure by facilitating intergranular corrosion or SCC, and by decreasing fracture toughness of some materials.

For the failure of the 5083 Al-Mg alloy inlet nozzle by mercury-induced LME discussed earlier, a question arose regarding whether the weld material was 'sensitised', i.e. whether the grain boundaries exhibited excessive amounts of β-phase (Al_3Mg_2), thereby perhaps facilitating LME crack initiation. Unstressed specimens (1–2 mm thick) from the weld were therefore lightly abraded in the presence of liquid gallium (melting point ~29°C) to facilitate

Published by Woodhead Publishing Limited, 2011

wetting and left at 30–40°C for 30–60 minutes.* Excess gallium was then carefully removed by wiping with paper tissues, and the specimens were then fractured by bending. These exposure conditions resulted in gallium diffusing along grain boundaries to almost (but not quite) the mid-thickness position of the specimen so that fracture was intergranular except for small regions of overload near the centre. The intergranular facets nearest the overload region were covered with the thinnest layers of gallium (several nanometres), and grain-boundary precipitates were most readily visible in this region using high-resolution SEM.

The observations for the 5083-0 Al-Mg inlet-nozzle welds showed that extensive grain-boundary precipitation was present in some areas of the failed pipe, but that welds from other pipes (presumably welded under the same conditions and in service for the same time) exhibited little grain-boundary precipitation (Fig. 18.30). From these observations, and other considerations, it was concluded that the weld material from the failed pipe had been sensitised during a post-failure fire rather than during welding or subsequent use and, hence, sensitisation had not contributed to failure.

18.9.3   Understanding other failure modes

Comparisons of the characteristics of adsorption-induced LME with HE and SCC have helped understand these relatively more complex phenomena [17, 44]. The remarkably similar characteristics produced by LME, HE, and SCC in some materials (Figs 18.31 and 18.32), along with other considerations, suggest that the atomistic mechanisms are the same for some fracture modes and circumstances. Since LME involves only adsorption-induced weakening of interatomic bonds at crack tips for the systems studied, it has therefore been proposed [17, 44] that HE and SCC in these systems also occurs by an adsorption mechanism (usually involving hydrogen), as opposed to mechanisms based on solute hydrogen or dissolution. Understanding the fundamental mechanisms of various fracture modes does help in failure analysis, partly because 'What one sees or observes depends on what one knows and understands' [45]. A better fundamental understanding of fracture modes should also help in materials selection, and should lead to the development of improved materials and a better ability to predict crack-growth rates.

---

*The optimum exposure conditions depend sensitively on the alloy and grain structure and, hence, trials are required.

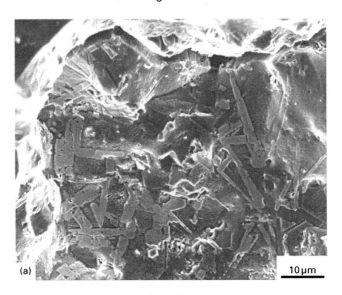

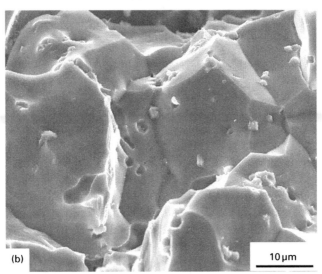

18.30 SEM of brittle intergranular fracture produced by pre-exposure of 5083-0 Al-Mg weld material to liquid gallium showing (a) extensive plate-like β-phase particles on grain-boundary facets for a weld that had experienced elevated temperatures during a fire, and (b) relatively clean grain-boundary facets for a weld away from the fire.

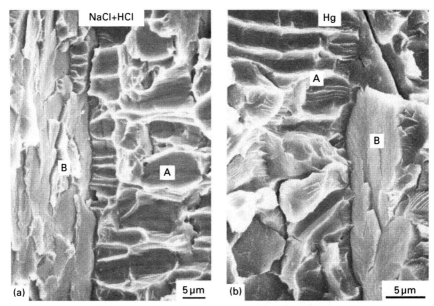

18.31 SEMs of fracture surfaces produced by (a) SCC in aqueous environment, and (b) LME in liquid mercury, for a Ti6%Al4%V alloy, showing fluted areas (A) and cleavage-like areas (B) [17].

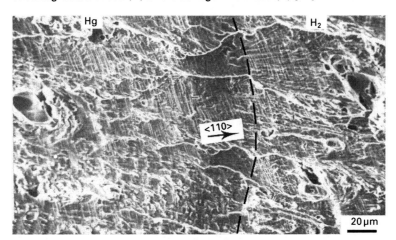

18.32 SEM of fracture surface of notched, fatigue-pre-cracked nickel single crystal cracked first in liquid mercury and then (after completely evaporating the mercury) in hydrogen gas (101 kPa) at 20°C showing essentially identical appearance (tear ridges, slip lines, and isolated dimples) and crystallography for both environments. Specimens were unloaded slightly after evaporating the mercury to produce a marking on the fracture surface to determine the transition position. The fracture-surface appearance depended on the crystal orientation, but was the same in both environments for each orientation. (The appearance was also the same for specimens tested separately in mercury and hydrogen.) [17]

18.10   References

1. C. Plinius Secundus (or Pliny): *The History of the World commonly called The Natural History*, translated by P. Holland, McGraw-Hill, pp. 364–365, 1964.
2. S. P. Lynch: 'Failures of Structures and Components by Environmentally Assisted Cracking', *Engineering Failure Analysis*, 1994, 1, pp. 77–90.
3. S. P. Lynch: 'Failure of Engineering Components due to Environmentally Assisted Cracking', *Practical Failure Analysis*, 2003, 3, pp. 33–42.
4. M. H. Kamdar: 'Liquid Metal Embrittlement', *Failure Analysis and Prevention, ASM Handbook*, American Society for Metals, Metals Park, OH, 1986, 11, pp. 225–238.
5. S. P. Lynch: 'Metal Induced Embrittlement of Materials', *Mater. Charact.*, 1992, 28, pp. 279–289.
6. P. J. L. Fernandes, R. E. Clegg and D. R. H. Jones: 'Failure by Liquid Metal Induced Embrittlement', *Engineering Failure Analysis*, 1994, 1, pp. 51–63.
7. C. F. Old: 'Liquid Metal Embrittlement of Nuclear Materials', *J. Nuclear Materials*, 1980, 92, pp. 2–25.
8. M. H. Kamdar: 'Embrittlement by Liquid Metals', *Progress in Materials Science*, 1973, 15, pp. 289–374.
9. F. A. Shunk and W. R. Warke: 'Specificity of Liquid Metal Embrittlement', *Scripta Metall.*, 1974, 8, pp. 519–526.
10. M. G. Nicholas: 'A Survey of Literature on Liquid Metal Embrittlement of Metals and Alloys', in *Embrittlement by Liquid and Solid Metals*, Ed. M. H. Kamdar, The Metallurgical Society of AIME, 1984, pp. 27–50.
11. D. R. McIntyre and J. W. Oldfield: 'Environmental Attack of Ethylene Plant Alloys by Mercury', in *Proc. Conf. 'Corrosion Prevention in the Process Industries … What European Industry is Doing'*, Amsterdam, The Netherlands, NACE, 1989, pp. 239–252.
12. M. O. Speidel: 'Current Understanding of Stress Corrosion Crack Growth in Aluminum Alloys', *The Theory of Stress Corrosion Cracking in Alloys*, Ed. J. C. Scully, NATO, Brussels, 1971, pp. 289–344.
13. J. A. Feeney and M. J. Blackburn: 'The Status of Stress Corrosion Cracking of Titanium Alloys in Aqueous Solutions', *The Theory of Stress Corrosion Cracking in Alloys*, Ed. J. C. Scully, NATO, Brussels, 1971, pp. 355–398.
14. S. P. Lynch: 'Metallographic and Fractographic Aspects of Liquid-Metal Embrittlement', in *Environmental Degradation of Materials in Aggressive Environments*, Eds M. R. Louthan, Jr., R. P. McNitt and R. D. Sisson, Jr., Virginia Polytechnic Inst., 1981, pp. 229–244.
15. E. Pereiro-López, W. Ludwig and D. Bellet: 'Discontinuous Penetration of Liquid Gallium into Grain Boundaries of Al Polycrystals', *Acta Mater.*, 2004, 52, pp. 321–332.
16. N. Marié, K. Wolski and M. Biscondi: 'Grain Boundary Penetration of Nickel by Liquid Bismuth as a Film of Nanometric Thickness', *Scripta Mater.*, 2000, 43, pp. 943–949.
17. S. P. Lynch: 'Environmentally Assisted Cracking: Overview of Evidence for an Adsorption-Induced Localised Slip Process', *Acta Metall.*, 1988, 20, Overview No. 74, pp. 2639–2661, and references therein.
18. A. P. Drushitz and P. Gordon: 'Solid Metal Induced Embrittlement of Materials', in *Embrittlement by Liquid and Solid Metals*, Ed. M. H. Kamdar, The Metallurgical Society of AIME, 1984, pp. 285–316 and other papers in this conference proceedings.

19. S. P. Lynch: 'Solid Metal-Induced Embrittlement of Aluminium Alloys and Other Materials', *Mat. Sci. and Engng*, 1989, A108, pp. 203–212.

20. P. Gordon: 'Metal-Induced Embrittlement of Metals – An Evaluation of Embrittler Transport Mechanisms', *Metallurgical Transactions A*, 1978, 9A, pp. 267–273.

21. R. E. Clegg: 'A Fluid Flow Based Model to Predict Liquid Metal Induced Embrittlement Crack Propagation Rates', *Engng. Frac. Mech.*, 2001, 68, pp. 1777–1790.

22. R. Coade, D. Coldham, H. Moss and S. P. Lynch: Unpublished work, 2004.

23. M. R. Pinnel and J. E. Bennet: 'Voluminous Oxidation of Aluminium by Continuous Dissolution in a Wetting Mercury Film', *Journal of Materials Science*, 1972, 7, pp. 1016–1026.

24. J. E. Leeper: 'Mercury – LNG's Problem', *Hydrocarbon Processing*, 1980, 59, no. 11, pp. 237–240.

25. S. M. Wilhelm, A. McArthur and R. D. Kane: 'Methods to Combat Liquid Metal Embrittlement in Cryogenic Aluminum Heat Exchangers', in *Proc. 73rd GPA Annual Convention*, March 1994, pp. 62–71.

26. D. L. Lund, 'Causes and Remedies for Mercury Exposure to Aluminum Cold Boxes', in *Proc. 75th GPA Annual Convention*, 1996, pp. 282–287.

27. D. R. Nelson, 'Mercury Attack of Brazed Aluminum Heat Exchangers in Cryogenic Gas Service', in *Proc. 73rd GPA Annual Convention*, March 1994, pp. 178–183.

28. J. J. English and D. J. Duquette, 'Mercury Liquid Embrittlement Failure of 5083–0 Aluminum Alloy Piping', in *Handbook of Case Histories in Failure Analysis*, vol. 2, ASM, 1993, pp. 207–213.

29. R. Coade and D. Coldham: 'The Interaction of Mercury and Aluminium in Heat Exchangers in a Natural Gas Plant', in *Proc. 8th Int. Conference & Exhibition, Operating Pressure Equipment*, Melbourne, 2005, pp. 183–190.

30. W. Rostoker, J. M. McCaughey and H. Markus, *Embrittlement by Liquid Metals*, Reinhold Publishing, New York, 1960.

31. S. P. Lynch, R. T. Byrnes and G. Redmond: Australian Department of Defence, unpublished work, 2007.

32. E. D. Sweet, S. P. Lynch, C. G. Bennett, R. B. Nethercott and I. Musulin: 'Effects of Alkali-Metal Impurities on Fracture Toughness of 2090 Al-Li-Cu Extrusions', *Metall. and Mater. Trans. A*, 1996, 27A, pp. 3530–3541.

33. S. P. Lynch: unpublished work.

34. S. E. Stanzl and H. M. Ebenberger: 'Concepts of Fatigue Crack Growth Thresholds Gained by the Ultrasound Method', in *Fatigue Crack Growth Threshold Concepts*, Ed. D. L. Davidson and S. Suresh, Met. Soc. AIME, 1984, pp. 399–416.

35. S. P. Lynch: 'Progression Markings, Striations and Crack-Arrest Markings on Fracture Surfaces', *Mater. Sci. and Engng A*, 2007, 468–470, pp. 74–80.

36. S. P. Lynch, D. P. Edwards and A. Crosky: 'Failure of a Screw in a Helicopter Fuel-Control Unit: Was it the Cause of a Fatal Crash?', *J. Fail. Anal. Prev.*, 2004, 4, pp. 39–49.

37. S. P. Lynch: DSTO, Australia, unpublished work, with information and specimens from M. Broadhurst, CAPSIS, UK.

38. *ASM Metals Handbook, vol. 5*, pp. 323–332, 'Hot Dip Galvanised Coatings', American Society for Metals, Metals Park, OH, 1986.

39. S. P. Lynch: unpublished work, with information and specimens from I. Wills, Fesca Qualos Gears, 1992.

40. I. Le May, A. K. Koul and R. V. Dainty: 'Fracture Mechanisms in a Series of Locomotive Axle Failures', *Mater. Charact.*, 1991, 26, pp. 235–251.

41. ASTM, B154–89: Standard Test Method for Mercurous Nitrate Test for Copper and Copper Alloys, 1990.

42. ISO 196–1978 (E) Wrought Copper and Copper Alloys – Detection of Residual Stress – Mercury (I) Nitrate Test.

43. S. P. Lynch, D. P. Edwards, R. B. Nethercott and J. L. Davidson: 'Failures of Nickel-Aluminium Bronze Hydraulic Couplings, with Comments on General Procedures for Failure Analysis', *Practical Failure Analysis*, 2002, 2, pp. 50–61.

44. S. P. Lynch: 'Progress Towards Understanding Mechanisms of Hydrogen Embrittlement and Stress-Corrosion Cracking', Paper 07493, *NACE Corrosion Conference Proceedings*, 2007.

45. D. Hull: *Fractography – Observing, Measuring and Interpreting Fracture Surface Topography*, Cambridge University Press, Cambridge, 1999.

19
Stress corrosion cracking in pipelines

W. ZHENG, M. ELBOUJDAINI, and R. W. REVIE,
CANMET Materials Technology Laboratory, Canada

Abstract: Welded steel pipelines that are used to transport oil and gas under high pressure have been known to fail by stress corrosion cracking (SCC), and the cracking can be intergranular, transgranular, or a combination of both, depending on the environmental and operational conditions. In this chapter, the practical aspects of SCC that occurs on the external surfaces of these underground pipelines are discussed, along with preventive measures that can be considered for mitigating failures by SCC.

Key words: stress corrosion cracking, pipelines, high pH stress corrosion cracking, near-neutral pH stress corrosion cracking, coating, cathodic protection, pipeline integrity.

19.1   Introduction

Intergranular stress corrosion cracking (SCC) of pipelines has been known to cause pipeline failure since the 1960s and is associated with a high pH, above 9, in the electrolyte in direct contact with the external pipe surface. This form of SCC is found on pipelines with cathodic protection that is less than optimum, i.e., more noble to the usually accepted cathodic protection potential of -850 mV vs $Cu/CuSO_4$ reference electrode.

Since the mid-1980s, numerous colonies of transgranular cracks have been found on natural gas and oil pipelines across Canada, and subsequently, in many other countries. Some of these stress corrosion cracks have caused major ruptures in gas and oil pipelines. In contrast to intergranular cracking, transgranular SCC is associated with a near-neutral pH, around pH 7, typically in areas of the pipeline that cathodic protection has not reached, such as under disbonded coatings that shield the underlying pipe from cathodic protection.

19.1.1   High pH SCC

Crack initiation and growth in the high pH environment occur by selective dissolution of the grain boundaries, while a passive film forms on the remainder of the surface and on the crack sides that prevents general corrosion at those locations. When an unstressed, fresh metal surface of line-pipe steel is

749

exposed to a high pH carbonate/bicarbonate environment at the appropriate potential for SCC, etching of the grain boundaries occurs with no noticeable corrosion of the grain faces. A strong correlation has been found between the maximum rate of crack growth in highly stressed metal and the maximum corrosion rate that can be sustained in that environment.

19.1.2   Near-neutral pH SCC

Although the mechanism of near-neutral pH SCC is not as well understood as the mechanism of high pH SCC, evidence from field failures suggests that corrosion pits are a common site for crack initiation. Crack initiation from seemingly unpitted surfaces was sometimes also observed. In experiments with full-scale pipes, the growth of pre-made cracks produced identical quasi-cleavage fracture surface morphologies to those of the actual field failures. On polished, smooth specimens, researchers at CANMET observed clusters of transgranular cracks with appearances similar to near-neutral pH stress corrosion cracks that have been found in the field, although the experiments were carried out under accelerated conditions.

Characteristics of the two forms of SCC are summarized in Table 19.1.

19.2   Mechanisms of stress corrosion cracking (SCC) in pipelines

19.2.1   General characterization

In order for SCC to occur in a metal or alloy in a specific environment, three conditions must be satisfied simultaneously:

1.   a tensile stress higher than the threshold stress, frequently including some dynamic or cyclic component of the stress
2.   a material that is susceptible to SCC and
3.   a cracking-inducing chemical environment.

When SCC occurs on pipelines, it first appears as colonies of parallel cracks on the external surface of the pipe. Over several years, cracks may coalesce to form larger cracks with a united crack front that can lead to pipe leak or rupture. Most commonly, cracks are in the longitudinal direction of the pipe, perpendicular to the hoop stress, although transverse cracks can also develop as a result of longitudinal stresses caused by pipe bending or ground movement. Residual stresses from the thermomechanical processing of the steel, from pipe forming operations, or from welding can also contribute to cracking.

If a crack grows to sufficient length, albeit without penetrating the pipe wall, the mechanical loading may be such that fracture toughness of the steel

Table 19.1 Characteristics of high pH and near-neutral pH SCC in pipelines as observed in field studies. Adapted from [1]

Factor	Near-neutral pH SCC	High pH SCC
Location	65% occurred between the compressor station and the 1st downstream block valve (distances between valves are typically 16 to 30 km) 12% occurred between the 1st and 2nd valves 5% occurred between the 2nd and 3rd valves 18% occurred downstream of the 3rd valve SCC associated with soils that tend to disbond or damage coatings or soils that have high resistance for CP current.	Typically within 20 km downstream of compressor stations Number of failures falls markedly with increased distance from compressor and lower pipe temperature SCC associated with specific terrain conditions, alternate wet-dry soils, and soils that tend to disbond or damage coating
Temperature	No apparent correlation with temperature of pipe Appear to occur in the colder climates where CO_2 concentration in groundwater is higher	Growth rate decreases exponentially with temperature decrease
Associated electrolyte	Dilute ground water solution with a neutral pH in the range of 5.5 to 7.5	Concentrated carbonate-bicarbonate solution with an alkaline pH greater than 9.3
Electrochemical potential	At free corrosion potential: from about –700 to about –800 mV (Cu/ $CuSO_4$) Cathodic protection does not reach pipe surface at SCC sites	–600 to –750 mV (Cu/$CuSO_4$) Cathodic protection is effective to achieve these potentials. Cathodic protection current at the pipe surface causes the increase of pH [2–4]
Crack path and morphology	Primarily transgranular Wide cracks with evidence of corrosion of crack side wall	Primarily intergranular Narrow, tight cracks with no evidence of secondary corrosion of the crack wall

is approached. With steels of low fracture toughness, the resulting fast fracture can propagate for considerable distances beyond the length of the initiating SCC before arresting. With higher toughness pipe, failure may involve local bulging and overloading, whereas if a slowly growing crack penetrates the wall before the critical crack size for brittle fracture is reached, then a leak occurs.

The sides of *intergranular* stress corrosion cracks show little evidence of lateral corrosion, whereas *transgranular* cracks are characterized by appreciable lateral corrosion of the crack walls, as is apparent from Fig.

Published by Woodhead Publishing Limited, 2011

19.1(a) and (b), respectively. These differences reflect the different passivity of the steel in the environments involved with the two forms of cracking.

When the coating fails, allowing the underlying steel to be in contact with groundwater, the steel is vulnerable to corrosion. Cathodic protection (CP) is

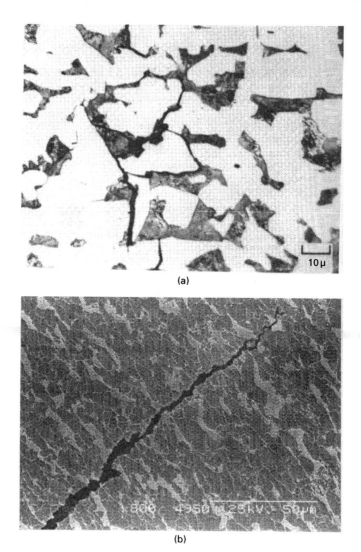

(a)

(b)

19.1 Examples of (a) intergranular SCC of pipeline steel (reproduced from R.N. Parkins, 'Overview of Intergranular Stress Corrosion Cracking Research Activities', Report No. PR-232-9401, published by American Gas Association, Arlington, VA, 1994) and (b) transgranular SCC of pipeline steel (reproduced with permission from R.N. Parkins, W.K. Blanchard, Jr., and B.S. Delanty, *Corrosion*, 50(5), 394 (1994). © NACE International 1994).

used to mitigate corrosion by passing an electrical current through the soil, giving the pipeline a cathodic potential. Concentrated carbonate–bicarbonate (CO_3–HCO_3) solution, the most probable environment responsible for high pH SCC, may develop as a result of the interaction between hydroxyl ions produced by the cathode reaction and carbon dioxide (CO_2) in the soil generated by the decay of organic matter. CP current causes the pH of the electrolyte to increase, and carbon dioxide readily dissolves in the elevated pH electrolyte, resulting in the generation of the concentrated carbonate–bicarbonate electrolyte. The pH of this electrolyte depends on the relative concentration of carbonate and bicarbonate.

19.2.2 Crack characteristics and stages of crack growth

There are many similarities between the two forms of SCC. Both occur as colonies of multiple parallel cracks that are generally perpendicular to the direction of the highest stress on the external pipe surface. These cracks can vary in depth and length and grow in two directions. The stages of SCC crack growth can be modeled as a four-stage process as shown in Fig. 19.2. The first stage is the development of conditions conducive to SCC and is followed by the crack 'initiation' stage, in which small cracks develop into sizable individual cracks. These cracks then continue to grow and coalesce, while additional crack initiation also occurs during stage 3. Finally, in stage 4, large cracks coalesce, deepen further into the pipe wall, and finally failure occurs [3].

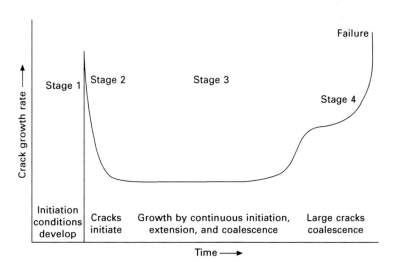

19.2 Four-stage process of SCC growth [2].

19.3   Factors contributing to stress corrosion cracking (SCC) in pipelines

Factors that may contribute to SCC in pipelines include:

1. Environment
2. Metallurgy
3. Manufacturing
4. Mechanical properties
5. Pipeline operating conditions
6. Coatings
7. Soil conditions.

These factors will be reviewed in the following sections.

19.3.1   Environment

The incubation period before groundwater reaches the pipe surface may involve:

- coating deterioration leading to the formation of holidays and/or the diffusion of water, carbon dioxide and other species through the coating;
- disbonding of the coating from the pipe surface;
- the type of coating, since different coatings have different properties that affect degradation;
- the moisture content of the soil, with its seasonal variations and the type of terrain, soil chemistry, and resistance of the soil to electrical currents associated with the cathodic protection system;
- stresses imparted to the coating by movements in the soil;
- temperature at the pipe surface; and
- condition of the pipe surface when the coating was applied.

While these factors, many of which are interlinked, are known to influence the deterioration of coatings, they are not yet sufficiently quantifiable to allow prediction of failure leading to the ingress of groundwater onto the pipe surface [5]. The integrity of the coating or coating systems is a key factor.

Defects in a coating, such as holidays and disbondments, allow groundwater to contact the steel surface. If no significant amount of cathodic current reaches the exposed surface, the groundwater between the coating and pipe can cause transgranular SCC. However, if cathodic current does flow to the steel surface, hydroxyl ions are generated and the *pH of the solution increases*, as shown in Fig. 19.3 [4]. As interpreted by Charles and Parkins [4], the data in Fig. 19.3 show resistance to the pH increasing above a value of about 6.5, but once that pH is achieved, it is followed by a relatively rapid rise

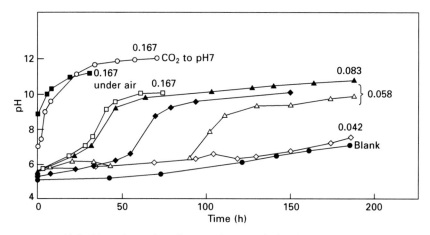

19.3 pH vs time plots for specimens of pipeline steel polarized at various cathodic current densities (in mA/cm^2) immersed in simulated groundwater, saturated with CO_2 before the start of test [4]. Reproduced with permission, © NACE International 1995.

until resistance to change at about pH 9.5. Parkins found that the time to the relatively rapid change between 6.5 and 9.5 is reduced with higher applied current density for the CO_2-saturated solutions, whereas for those solutions that had higher initial pH values, reflecting their lower CO_2 contents, the pH increased very rapidly to 11 or more. The holding of pH at about 6.5 or 9.5 is based on the conversion of free CO_2 to bicarbonate ions at about pH 6.5 and the conversion of bicarbonate to carbonate ions at the higher pH. The calculated pH for equilibrium between these two pairs of species is in good agreement with the observed pH values for the two forms of SCC [5].

For pipelines with an effective CP system, cathodic current reaches the pipe surface, hydroxyl ions are generated, and the pH increases to a value that depends on the partial pressure of CO_2. In the absence of CO_2, the pH rises to 11–12 and neither high pH SCC nor near-neutral pH SCC conditions develop. When the partial pressure of CO_2 is in a certain range, the pH remains around 9.5, buffered due to equilibrium between bicarbonate and carbonate ions [5].

As suggested by Charles and Parkins [4], starting with a dilute, relatively low pH, i.e., a CO_2-containing solution that promotes transgranular SCC, it is possible with appropriate hydroxyl ion generation and ion transport from cathodic protection, together with slow evaporation, to generate the high pH intergranular cracking environment. Despite the experimental evidence in support of this suggestion, field inspection results have yet to show the occurrence of such transitions in practice. Cathodic protection plays a critical role, and the implication is that the near-neutral pH solution will persist if inadequate cathodic current flows at the pipe surface. Such circumstances are expected to occur with poorly conducting coatings, conforming to the field experience of transgranular failures being associated with tape coatings of poor conductivity, or with very high resistivity soils, e.g., asphalt coatings in high-resistivity regions [1].

19.3.2   Metallurgy

Over the past few decades, the yield strength of line pipe has been gradually increased, in part by the addition of microalloying elements, such as vanadium, niobium, and/or titanium. The addition of these elements and the thermomechanical processing of the steel reduce the grain size, increasing both strength and toughness of the steel. Controlled rolling of the steel plate used to manufacture line pipe results in fine grained ferrite and bainitic microstructures and the presence of V, Nb and Ti in the steel helps limit the amount of grain growth in the heat-affected zone (HAZ) of the pipe welds.

Crack initiation and growth studies for near-neutral pH conditions have led to mechanistic discussion of the cracking mechanism [6]. However, there has not been extensive work on the effect of metallurgy. In one laboratory study of high and near-neutral pH SCC, a correlation of susceptibility to SCC and steel composition or microstructure was found to be limited. However, when results of the compact-tension test specimens are compared, the modern steels had a significantly lower threshold stress intensity (K_{ISCC}) values compared to the older X52 steels [7]. In the long-term static loading tests carried out in CANMET, discussed in Section 19.4, X-80 steel developed deeper and more numerous cracks than an X-65 steel under the same test conditions. Threshold stresses, above which high pH SCC occurs, have been measured using tapered specimens. By analyzing data on threshold stresses, SCC susceptibility of specific steels for service under specified conditions of maximum operating stress and R value, as shown in Fig. 19.4, can be assessed [2].

19.3.3   Manufacturing

From reported cases of SCC, no definitive generalizations can be made regarding the effect of manufacturing process on susceptibility to SCC [8]. However, spirally welded pipes seem to be associated with fewer field SCC failures than long straight-seam pipes.

19.3.4   Mechanical properties

For gas transmission pipelines, the specified minimum yield strength (SMYS) and the toughness are the mechanical properties of greatest interest. Higher

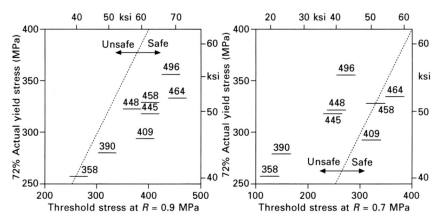

19.4 Threshold stresses for intergranular SCC of various steels at $R =$ 0.9 and 0.7 against 72% of the actual yield strengths of the steels [2].

strength steels make possible the design and construction of pipelines that operate at higher pressure than would otherwise be possible. The higher the actual strength over and above the SMYS, the greater the margin of allowance for SCC. This is because the pipelines are operated to a maximum pressure level that is a specific percentage of the SMYS (e.g., for Class I locations, the maximum allowable pressure is 80% of SMYS [1]), whereas the initiation of SCC cracks is determined by the ratio of the stress level to the actual yield point. Increase in toughness increases the critical size of cracks that a pipeline can sustain before the final rupture.

19.3.5 Pipeline operating conditions

The stress or stress intensity factor must be above a threshold level for high pH SCC to occur. The threshold stress is reduced by small-amplitude, low-frequency stress fluctuations superimposed on the mean stress. As shown in Fig. 19.5, the threshold stress decreased to 40% of the yield strength when stress fluctuations as low as 1.5% of the mean stress were applied twice a month [9].

The longitudinal stress caused by pressure can be up to half of the hoop stress. Pipeline axial loading, as produced by ground movement or slope instability, can result in additional longitudinal stress, so that the total longitudinal stress can exceed the hoop stress. Circumferential SCC cases reported by the Canadian Energy Pipeline Association (CEPA) have been related to undulating terrain where pipeline loading resulted from soil movement or localized bending [10]. Localized bending may also occur at dents.

For near-neutral pH SCC, as discussed in Section 19.4, pressure fluctuation

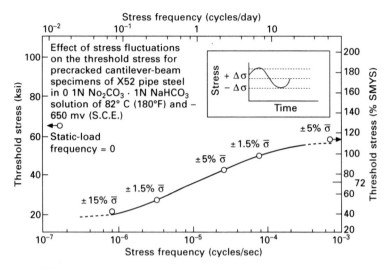

19.5 Effect of low-amplitude (high *R*) stress cycles on the threshold stress of an X-52 steel exposed to a 1N solution carbonate + 1N sodium bicarbonate solution at 75°C and – 650 MV (SCE) [9].

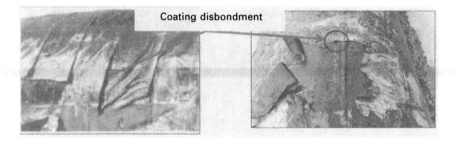

19.6 Disbonded coating on line pipe.

is the most significant factor in the crack growth produced on full-scale pipes buried in test soil, once the applied stress level is above a threshold value.

19.3.6 Coatings

SCC will not occur under a coating that prevents groundwater from contacting the pipe surface. Some coatings, such as tape coatings, tend to be susceptible to disbondment (Fig. 19.6). In addition, coatings may disbond when excessive cathodic protection current is applied. When a coating fails by disbonding from the pipe surface, the coating may shield the pipe from the cathodic protection current designed to protect the steel from corrosion.

19.3.7   Soil conditions

Delanty and O'Beirne [11, 12] developed an SCC severity rating model for near-neutral pH SCC for the tape-coated portions of TCPL's system in eastern Canada. The predictors in that model were soil type, drainage, and topography. The soil classifications were based on method of deposition. The cracking severity seems to be related to soil drainage characteristics. Very poorly or poorly drained soils were the most aggressive.

The soil effects seem to depend on the coating type. In the asphalt-coated pipelines, SCC was found mostly (over 80%) in very dry terrains such as sandy soils or a mixture of sand and bedrock. CP was very difficult to be applied in these soils. For tape-coated pipe, the most aggressive soil was the lacustrine type, followed by organics over glaciofluvial, and organics over lacustrine types [13].

19.4   CANMET studies of near-neutral pH stress corrosion cracking (SCC)

By using the full-scale pipes, actual loading conditions and the soil environment can be reproduced in the lab setting to simulate those of the actual operating pipeline. Soil conditions [1] and operating conditions, in particular the level of pressure fluctuations [11, 14, 15], were identified in the early CANMET full-scale tests as key factors affecting the growth rate of cracks, and this was confirmed by a number of independent studies [16–18].

The renewed intention to build new high-pressure pipelines in the northern parts of Canada and the United States has triggered a series of research projects on SCC of high-strength steels, i.e. X-80 and even X-100 grades. SCC studies of these new materials have been reported in recent years [19].

19.4.1   Crack growth behavior and models for crack growth

In view of the important role of cyclic loading in crack growth, a corrosion fatigue approach had been adopted by some workers in analyzing their laboratory results [20–22], in which the total crack growth was modeled to consist of two parts: a fatigue growth as if the steel were fatigued in an inert environment and an SCC growth at a constant rate as if there were no cyclic loading. The total crack growth during a load cycle is therefore expressed by:

$$\frac{da}{dN}(\text{total}) = \frac{da}{dN}(\text{fatigue}) + \frac{1}{f}\frac{da}{dt}(\text{scc}) \qquad 19.1$$

When conducting full-scale testing, a direct current potential drop technique was used for detection of crack growth. The full-scale SCC tests in the first few years of the program at CANMET [14, 15] showed no detectable crack growth when the pressure fluctuation component was removed from the load spectrum, with the resolution of the crack detection tool being about 30 microns. In the subsequent work, the crack growth rates during a re-loading part of a load cycle (i.e., when the loading was increased from the minimum level of the load cycle) were correlated with the increase in mechanical driving force that was applied to a given crack [23]. The growth rates were found, on a log-log scale, to increase linearly with the time rate of J, expressed as $\Delta J/\Delta t$ [24]:

$$\log(\Delta a/\Delta t) = k \log(\Delta J/\Delta t) + c \qquad 19.2$$

where $\Delta a/\Delta t$ is the effective crack growth rate under dynamic loading, k and c are the constants that reflect the intrinsic sensitivity of a particular steel microstructure to near-neutral pH SCC and the potency of the environment, respectively.

In another full-scale test on X-52 steels, the crack growth rates from a full-scale pipe were analyzed using parameters of linear elastic fracture mechanics (LEFM), i.e., using K and K range as the crack tip stress descriptor [25]. In a given load cycle, the rate of increase in J would be related to K range by: $\Delta J/\Delta t = (\Delta K)2/(E'\Delta t)$ [25]. The test data are presented in Fig. 19.7. For large plastic deformation, at very high K levels (K greater than 40 MPam$^{1/2}$), the data points shifted away from the low-K data trend, whereas the data at lower loading levels had a linear relationship with the loading rate.

The mechanistic significance of Fig. 19.7 is apparent because the dynamic crack growth rate seems to depend on the rate of opening of the crack tip in this near-neutral pH condition. It was hence hoped [26, 27] that a model such as Eq. 19.2 can be used to predict growth rates under either hoop loading conditions or an axial loading situation. However, the findings of crack initiation and growth in statically loaded X-65 and X-80 steels suggest that the total crack growth needs to take into account the portion of crack advance under static load as well [28]. This point is further discussed in Section 19.4.3.

Nevertheless, the sensitivity of growth rate to the loading rate, as shown in Fig. 19.7, can be seen as an indication of the sensitivity of the cracking to strain rate. Beavers and Jaske [29] carried out SCC growth studies using pre-cracked CT specimens at various stress intensity ranges at different loading frequencies, ranging from 10^{-5} to 10^{-2} Hz. The ΔK values were 20, 10 and 7 MPam$^{1/2}$. For $\Delta K = 20$ MPam$^{1/2}$, the crack growth rate showed gradual increase with time at all frequencies. For $\Delta K = 10$ MPam$^{1/2}$, the crack subject to fast loading (10^{-2} Hz) showed faster growth with time while the sample loaded to 10^{-5} Hz showed considerable decrease in the growth rate with test time. For a given ΔK value, the higher the frequency, the higher

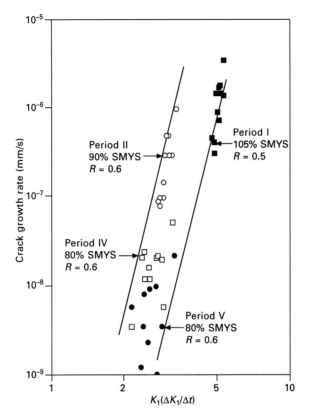

19.7 Dependence of crack growth rates on $K_1(\Delta K_1/\Delta t)$ for an X-52 steel [25]. The actual yield strength of the X-52 steel was about 460 MPa, about 100 MPa higher than SMYS. Period III at 72% SMYS and $R =$ 0.5 resulted in no crack growth.

the rate of crack opening during re-loading. This work supports the growth rate versus the loading rate trend demonstrated in Fig. 19.7.

19.4.2   Crack dormancy

It had been hoped that the loading rate (or, the related strain rate effect) in SCC growth could be utilized to explain the occurrence of crack dormancy observed in the field [26, 30]. However, even under static loading, i.e., when $\Delta J/\Delta t = 0$, cracking does take place, although it occurs at a very slow rate. In tests that lasted for 110 days, typical crack depth at the end of the testing ranged from 10 to 15 microns [19]. Cracks continued to deepen and new cracks continued to initiate as the test time increased to 220 days. Thus it is proved that dynamic loading is effective in accelerating crack growth, but is not essential in the formation and growth of cracks.

Published by Woodhead Publishing Limited, 2011

19.4.3   Crack initiation and growth

Because of the slow rate at which SCC initiates and grows in linepipe steels, high-quality SCC data require long-term testing using realistic conditions. In the static tests involving stressing the steels to 95% of the yield point, SCC incubation time in an NS4 solution is longer than 50 days. Under less aggressive conditions, this incubation period should be even longer.

Acceleration of the SCC process is usually involved in laboratory studies of crack initiation because the real cracks on operating pipelines take many years to develop. For initiation tests, a number of research groups have used various forms of test specimens, such as round or flat tensile bars [31, 32], compact toughness (CT) pre-cracked specimens [33, 34] and curved pipe strip cut from the hoop direction [21]. Other types of test specimens have also been used, but, in general, initiating near-neutral pH SCC in the laboratory under stressing conditions that are in the 'nominal' operating pressure range has proven to be very difficult, raising the question as to what the true stress level is on operating pipelines.

There are several factors that can change the net total stress level at a specific site on a pipeline, including the effect of stress concentration by surface profiles such as the long seam weld. In one finite element calculation where the applied stress is 340 MPa (77% of the specified minimum yield stress (SMYS) for the pipe in question), there is a zone of a few millimeters width from the weld toe in which the actual local stress is close to the SMYS of the base steel [23]. In a CANMET experimental study of residual stress in various pipes retrieved from service, tensile residual stresses in the range of 20% SMYS were often found to exist within the pipe wall up to a depth of about 1 mm [35].

These factors were considered in designing the static-loading SCC tests of X-80 and X-65 steels, which were loaded to 95% of their respective yield points. Despite the relatively high stress level, SCC took over 50 days to incubate in a NS4 solution, and this observation applies to both the X-65 and the X-80 steels tested. Data for the X-80 steel from this series are shown in Fig. 19.8 [19]. After 26 days, only one 'crack-like' feature of depth less than 2 micrometers (microns) was observed in the cross-section. Five small cracks of depth ranging from 3 to 8 micrometers (microns) were seen after 54 days of testing; after 110 days of testing, the number of cracks increased to 27 and, more significantly, the crack depth increased significantly. It is important to note that only one center-line cross-sectioning was made for each SCC specimen so the actual number of cracks on the whole specimen would be many times that seen on the cross-section. The time-averaged crack growth rate in the 110-day test was in the order of 10^{-9} mm/s [19].

This result from the static tests suggests that cracking does indeed occur in stressed steel without pressure fluctuation and therefore an update of the

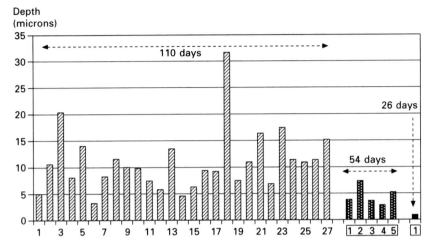

19.8 Histograms of cracks found, by cross-sectioning, on an X-80 test sample after various lengths of test time [19].

growth model shown in Eq. 19.2 is needed. Taking into account the natural corrosion process inside a crack, which is unavoidable even if there is no stress applied, a more comprehensive description of the penetration of a crack should be:

Δa (total growth in a loading period, Δt)

$$= \int f_1(da/dt) \times \Delta t + \int f_2(da/dt) \times \Delta t + \int \beta \times \Delta t \qquad 19.3$$

where $f_1(da/dt)$ is a term that describes the contribution of dynamic crack growth due to increasing load, as defined in Eq. 19.2; $f_2(da/dt)$ is the term that reflects the crack growth under static loading, i.e., when the term $\Delta J/\Delta t$ in Eq. 19.2 is zero. The rate of growth under static load should be a function of the stress level applied and of the microstructural sensitivity of the tested material; the term β represents the penetration of a crack by a pure corrosion process alone, i.e., without any significant stress applied. The crack under this condition would advance in all directions by the same amount and logically cracking blunting would occur if this form of 'general corrosion', instead of cracking, is maintained for sufficiently long time. This condition could exist, for example, when a pipeline is shut down when the stress level in a pipe is below the threshold for cracking. For a generally near-neutral pH soil without excessive microbial activities, this value can be assumed to be in the range of 5–50 µm per year or $0.15–1.7 \times 10^{-9}$ mm/s. This term is determined by the corrosivity of the soil, which reflects the ionic species present in the soil water, the amount of dissolved gases and the pH. The data from the dozen or so full-scale tests completed in CANMET so far suggest that the growth rate due to dynamic loading, $f_1(da/dt)$, is the most significant part and it is generally between 5×10^{-9} mm/s and 1×10^{-5} mm/s; the second term, $f_2(da/dt)$, which is the cracking under static loading, should be in the range 20–150 μm per year or $0.6–5 \times 10^{-9}$ mm/s, based on the data produced so far in NS4 and NS4 plus clay solutions.

19.4.4   Effect of the soil environment

Field SCC failures in the near-neutral pH condition are generally believed to be associated with anaerobic soils at the pipe level [1, 36]. Of course we know from the Pourbaix diagram that corrosion potential of iron in aqueous media is affected by the presence of oxygen. The lower the oxygen content in the soil water, the lower (more cathodic) will be the corrosion potential, and therefore, a higher level of hydrogen production, which is part of the overall corrosion process of steel.

Another major influence on near-neutral pH SCC is ground CO_2 level, which controls the concentration of the CO_3^{2-} and HCO_3^- and pH value in the soil electrolyte. An increase in CO_2 content in the soil electrolyte generally enhances corrosion. However, soils that are prone to cause rapid uniform corrosion at open-circuit condition (OCP) do not promote the formation of sharp surface pits, which are often associated with the initiation of cracks. Weight loss corrosion measurements indicated that the rate of dissolution of various synthetic soil electrolytes is not very different from one another at the beginning of immersion testing, but the difference becomes significant after an extended testing time (typically after 10 days) [37]. This observation suggests that corrosion products formed in the initial stages of corrosion affect the subsequent long-term corrosion rate.

In the 110-day static-loading tests carried out at CANMET, the effect of adding clay soil to the NS4 solution, with the ratio of the clay soil volume to the solution volume set at 50%, was quite pronounced. Figure 19.9 shows the comparison of the SCC histograms for an X-65 steel in de-aerated solution of very similar pH level. When a clay soil was added to the test cell, the number of cracks developed was significantly reduced [19]. The natural soil water is generally comprised of various dissolved and un-dissolved soil minerals, so that the mechanisms of the soil chemistry influence on SCC can be quite complicated. Cross-sections of the test samples showed a surface corrosion product layer that is generally thicker and more porous in the clay-NS4 mixture than in the NS4 solution [19].

19.4.5   Effects of steel microstructure and surface conditions

In early CANMET studies, clusters of transgranular cracks were produced using cyclic loading [38]. These clusters of cracks initiated mainly at non-

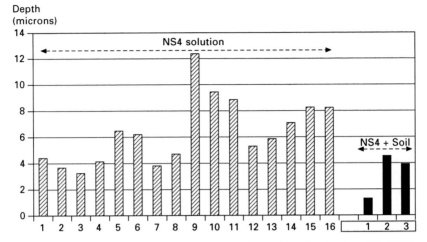

19.9 Effects of clay soil on the number of cracks produced in X-65 steel sample tested for 110 days [19].

metallic inclusions, and then more randomly on the polished surface. However, crack initiation in these studies generally took hundreds of thousands of load cycles at relatively low R ratios. Such frequent pressure variations at low R values could potentially be encountered in some of the oil pipelines, but unlikely in gas lines which tend to have more stable pressure profiles.

19.5   Prevention of stress corrosion cracking (SCC) failures

In this section, current approaches available to control SCC failures in existing pipelines and prevent SCC in new pipelines are reviewed.

19.5.1   Hydrostatic testing

Hydrostatic testing has been effective in removing stress corrosion cracks greater than the critical size corresponding to the high level of pressure applied during hydrotesting. As the critical crack size for normal operating pressure is much greater than the failure flaw size at hydrotest pressure, a margin of safety is achieved following a hydrotest. In the United States, pipelines operating at a maximum allowable operating pressure (MAOP) of 72% of SMYS are required to be hydrotested at 90% of the SMYS or higher, and the minimum high-pressure hold time is 8 hours [8]. In Canada, pipelines known to be affected by SCC are required to be hydrotested at a pressure that is up to 110% of SMYS, and the hold time is at least several hours.

In a full-scale hydrotest study [39], cracks previously growing in a simulated soil environment were subject to a full hydrotest load spectrum, and then these cracks were tested again under lower pressure with some degree of pressure fluctuation. It was found that all cracks grew at much reduced rates after the hydrotest and this was attributed to the creation of compressive residual stress at and near the crack tip during the hydrotest over-loading.

19.5.2 In-line inspection (ILI)

Once a pipeline is known to have developed stress corrosion cracks, ILI is usually conducted by the pipeline operator to assess the flaw size. Repeated ILI on the same pipe segment can allow the operator of the pipeline to estimate the rate at which the cracking is progressing. Magnetic flux leakage (MFL) and shear-wave ultrasonic thickness detection (UT) are two widely used pipeline inspection tools. Shear-wave ultrasonic thickness (UT) measurement has recently seen significant improvement in both detection probability and sizing accuracy [40]. MFL inspection tools can provide a thickness mapping of the pipe wall, and as cracking tends to be associated with pitting, this tool can sometimes identify cracked areas.

Electromagnetic acoustic transduction (EMAT) is the newest technology being explored for ILI application; at this time there is no wide-scale commercial use.

19.5.3 Coatings

Coatings are the first line of defence against corrosion. It is essential to follow best practices and current standards for coating selection and application [41–49]. The CEPA members have recommended that the following coatings be considered for new construction based on SCC performance [41]:

- fusion bonded epoxy (FBE)
- liquid epoxy
- urethane
- extruded polyethylene
- multilayer or composite coatings.

FBE, liquid epoxies, and urethane coatings meet all three requirements of an effective coating; they have high adhesive strength and are resistant to disbondment, and they are typically applied over a white or near white grit-blasted surface. Multilayer or composite coatings typically consist of an FBE inner layer and a polyolefin outer layer with an adhesive between the two layers. These new coatings are promising from the standpoint of resistance to disbondment, mechanical damage, and soil stresses, but the polyolefin outer layer will shield CP current should disbondment occur. Regardless of the coating selected, the line pipe surface should be prepared to a white

(NACE No. 1/SSPC-SP 5) or near white (NACE No. 2/SSPC-SP 10) finish to remove mill scale and to impart sufficient residual compressive stresses to retard SCC initiation.

19.5.4   Improvement in pipeline steels

There are several studies on the effects of steel microstructure on its SCC properties. The results by Chen *et al.* [50] and Chu *et al.* [51] showed preferential corrosion at metallurgical discontinuities, such as grain boundaries, pearlitic colonies, and banded structures in the steel. Kushida *et al.* [52] noted an increased susceptibility of line pipe steels with non-uniform microstructures such as ferrite-pearlite.

Bulger *et al.* [53] reported a 30% higher crack growth rate for the heat-affected zone next to the pipe weld; they also found that steels with a microstructure of fine-grained bainite plus ferrite have a better combination of strength and SCC resistance than ferrite/pearlite structures. However, when the bainite/ferrite steel is welded, decomposition of the bainitic/ferrite structure caused reduction in the SCC resistance as a result of the heating effect of the welding.

The presence of mill-scales on the steel surface would increase the risk of crack initiation. Beavers and Harper [54] have shown some correlation between residual stresses and locations of SCC on a given joint of pipe. In a laboratory study, Van Boven [55] has confirmed that the formation of micro-pits and microcracks (less than 50 μm in depth) occurred preferentially in areas where the tensile residual stresses were about 200 MPa, whereas little pitting or general corrosion or microcracks occurred on surfaces with compressive residual stresses.

The knowledge gained from the study of older steels is bound to benefit the development and use of new pipeline steels. The yield strengths of the newest pipeline steels are approaching 100 ksi and beyond. Using stronger pipeline steels, thinner pipelines can be made, leading to lower pipeline construction costs. In 2007, X90, X100, and X120 were accepted as standardized materials as ISO3183 and API 5L. Along with the increase in strength level, the new strain-based design approach for pipeline construction is being developed. Although segments of pipeline made of X-100 steels have been installed in Canada on a trial basis, comprehensive field operating experience of these high-strength steel is yet to be accumulated.

19.6   Conclusions

Research on pipeline SCC has been reviewed, including characteristics of high pH and near-neutral pH SCC. Initiation, growth modeling, and other aspects have been discussed. Much of the work reviewed was carried out since the 1990s, indicating continuing active advances of research in this field. The recent results of SCC under static loading conditions prove that the cracking in the near-neutral pH condition is indeed an 'authentic' SCC process, but the rate is very sensitive to dynamic loading. Studies into the influence of soil chemistry and steel microstructure have also produced many insightful results.

In order to prevent SCC initiation, the single most important recommendation is the emphasis on coatings that remain bonded to the pipe. The quality of the surface preparation and field application is also critical, and this applies to both new pipe installations as well as to coating replacements.

For pipelines already affected by stress corrosion cracking, high-pressure hydrostatic testing is an effective way of removing flaws reaching the critical size. Full-scale testing results suggest there is likely some crack growth retardation effect introduced by hydrotesting. In addition to hydrotesting, in-line inspection (ILI) is another way to detect and size the flaws in operating liquid and gas lines. The advances in ILI detection speed and accuracy are bound to impact in a positive way the management of pipeline integrity of both old and new pipelines, the latter to be constructed of thin-wall high-strength steels with significantly modified microstructure.

19.7   References

1. National Energy Board, Report of the Inquiry, Public Inquiry Concerning Stress Corrosion Cracking on Canadian Oil and Gas Pipelines, National Energy Board, Calgary, Alberta, November 1996, p. 16.
2. Parkins, R.N., 'Overview of Intergranular Stress Corrosion Cracking Research Activities', Report PR-232-9401 to PRCI at the American Gas Association, November 1994.
3. Parkins, R.N., 'A Review of Stress Corrosion Cracking of Pipelines in Contact with Near-Neutral (Low) pH Solutions', Report PR-232-9701 to PRCI at the American Gas Association, April 1999.
4. Charles, E.A. and R.N. Parkins, 'Generation of stress corrosion cracking environments at pipeline surfaces'. *Corrosion*, 51, 518, 1995.
5. Parkins, R.N., 'A Review of Stress Corrosion Cracking of High-Pressure Gas Pipelines', *CORROSION/2000*, Paper No. 363, NACE International, Houston, TX, 2000.
6. Beavers, J.A. and B.A. Harle, 'Mechanisms of High pH and Near Neutral pH SCC of Underground Pipelines'. *Journal of Offshore Mechanics and Arctic Engineering*, August 2001.
7. Danielson, M.J. and R.H. Jones, 'Effect of Microstructure and Microchemistry on the SCC Behavior of Archival and Modern Pipeline Steels in a High pH Environment'. *Corrosion 2001*, Paper No. 01211, NACE International, Houston, TX, 2001.
8. Baker, M., Jr. Inc., 'Stress Corrosion Cracking Study', US Department of Transportation, Research and Special Programs Administration, Office of Pipeline Safety, January, 2005.
9. Fessler, R.R., 'Combination of Conditions Causes Stress-Corrosion Cracking', *The Oil and Gas Journal*, 74 (7), 81–83, 1976.

Published by Woodhead Publishing Limited, 2011

10. CEPA, *The CEPA Report on Circumferential Stress Corrosion Cracking. Submitted to the National Energy Board.* Canadian Energy Pipeline Association, December 1997.

11. Delanty, B.S. and J. O'Beirne, 'Major Field Study Compares Pipeline SCC with Coatings'. *Oil and Gas Journal*, 90, 24, 1992.

12. Delanty, B.S. and J. O'Beirne, 'Low-pH Stress Corrosion Cracking', in *Proceedings from the 6th Symposium on Line Pipe Research*, PRCI, L30175, 1991.

13. Beavers, J.A. and W.V. Harper, 'Stress Corrosion Cracking Prediction Model', *CORROSION 2004*, Paper 04189, NACE International, Houston, TX, 2004.

14. Zheng, W., R.W. Revie, W.R. Tyson, G. Shen and F.A. Macleod, 'Effects of Pressure Fluctuation on the Growth of Stress Corrosion Cracks in an X-52 Linepipe Steel', *8th International Conf. on Pressure Vessel Technology*, ICPVT-8, ASME, New York, 1996, pp. 321–326.

15. Zheng, W., F.A. Macleod, R.W. Revie, W.R. Tyson, G. Shen, D. Kiff, M. Skaff and E.-W. Wong, 'Recent Progress in the Study of Transgranular SCC in Linepipe Steels', *Corrosion Deformation Interactions CDI '96*, Ed. T. Magnin (Nice, France: Published for the European Federation of Corrosion by the Institute of Metals), 1996, p. 282.

16. Wilmott, M.J. and R.L. Sutherby, 'The Role of Pressure and Pressure Fluctuations in the Growth of SCC in Line Pipe Steels', *International Pipeline Conf.*, 1, ASME, New York, 1998, p. 409.

17. Zhang, X.-Y., S.B. Lambert, R. Sutherby and A. Plumtree, 'Transgranular stress corrosion cracking of X-60 pipeline steel in simulated ground water.' *Corrosion*, 55, 297, 1999.

18. Wang, Y.Z., R.W. Revie and M.T. Shehata, 'Early Stage of Stress Corrosion Crack Development of X-65 Pipeline Steel in Near-Neutral-pH Solution', *Materials for Resource Recovery and Transport*, Ed. L. Collins (Montreal, Quebec: The Metallurgical Society of the Canadian Institute of Mining, Metallurgy and Petroleum), 1998, p. 71.

19. Zheng, W., D. Bibby and J. Li, 'Near-Neutral pH SCC of Two Line Pipe Steels under Quasi-Static Stressing Conditions', *Proc. of International Pipeline Conf.*, Calgary, Alberta, Canada, Sept., Paper # IPC2006-10084, ASME, New York, 2006.

20. Plumtree, A., S.B. Lambert and R. Sutherby, 'Stress Corrosion Crack Growth of Pipeline Steels in Simulated Ground Water', in *Corrosion-Deformation Interaction CDI'96*, Nice, France, 1996.

21. Zhang, X.Y., S.B. Lambert, R. Sutherby and A. Plumtree, 'Transgranular Stress Corrosion Cracking of X-60 Pipeline Steel in Simulated Ground Water', *Corrosion*, 55, 297–305, 1999.

22. Lambert, S.B., J.A. Beavers, B. Delanty, R. Sutherby and A. Plumtree, 'Mechanical Factors Affecting Stress Corrosion Crack Growth Rates in Buried Pipelines', in *Proceedings of International Pipeline Conference*, ASME, New York, 2000, pp. 961–966.

23. Zheng, W., F.A. Macleod, R.W. Revie, W.R. Tyson, G. Shen, M. Shehata, G. Roy and J. McKinnon, 'Growth of Stress Corrosion Cracks in Pipelines in Near-Neutral pH Environment: The CANMET Full-Scale Tests – Final Report to the CANMET/Industry Consortium', CANMET/MTL, Ottawa, MTL97-48 (CF), 1997.

24. Zheng, W., 'Stress Corrosion Cracking of Pipelines: An Overview of Key Factors', *Second International Conference on Environment Sensitive Cracking and Corrosion Damage*, Hiroshima Japan, Oct. 29–Nov. 2, 2001.

25. Zheng, W., R.W. Revie, F.A. MacLeod, O. Dinardo, D. Kiff and J. McKinnon, 'Low pH SCC: Environmental Effects on Crack Propagation', PRCI Contract No. PR-230-9413, Final Report, July 1998.
26. Zheng, W., B. Tyson, G. Shen, R.W. Revie, G. Williams and D. Bibby, 'Effects of Operating Practices on Crack Dormancy and Growth', GRI-04/0121, January 2005. GRI, Des Plaines, IL.
27. Chen, W. and W. Zheng, 'Progress in Understanding of Pipeline SCC: Initiation and Propagation', *Symposium of Pipelines for the 21st Century, Conference of Metallurgists*, Calgary, Canada, Aug. 21–24, 2005.
28. Zheng, W., 'Stress Corrosion Cracking of Oil and Gas Pipelines In Near Neutral pH Environment: Review of Recent Research', *J. Energy Materials: Materials Science and Engineering for Energy Systems*, 3(4), 220–226, 2008.
29. Beavers, J.A. and C. Jaske, 'Effects of Pressure Fluctuations on SCC Propagation, Final Report to PRCI', Project PR 186-9706. Catalogue No. L51872, 2002.
30. Beavers, J.A., 'Near-Neutral pH SCC: Dormancy and Re-Initiation of Stress Corrosion Cracks', GRI-05/0009, GRI, August 2004.
31. Parkins, R.N. and B.S. Delanty, 'The Initiation and Early Stages of Growth of Stress Corrosion Cracks in Pipeline Steel Exposed to a Dilute Near-Neutral pH Solution', in *Proceedings from the Ninth Symposium on Pipeline Research*, PRCI, Catalogue No. L51746, 1996, pp. 19-1–19-14.
32. Gu, B., J. Luo and X. Mao, 'Hydrogen-Facilitated Anodic Dissolution-Type Stress Corrosion Cracking of Pipeline Steels in Near-Neutral pH Solution', *Corrosion*, 55, 96–106, 1999.
33. Chen, W. and R.L. Sutherby, 'Environmental Effect of Crack Growth Rate of Pipeline Steel in Near-Neutral pH Soil Environments', *Proceedings of IPC 2004, International Pipeline Conference*, October 4–8, 2004, Calgary, Alberta, Canada, Paper No. IPC04-0449.
34. Eadie, R.L., K.E. Szklarz and R.L. Sutherby, 'Corrosion Fatigue and Near-Neutral pH Stress Corrsion Cracking of Pipeline Steel and the Effect of Hydrogen Sulfide', *Corrosion*, 16(2), 167, 2005.
35. Roy, G., D. Kiff, R.W. Revie, E.J.-C. Cousineau, G. Williams and E. Sinigaglia, 'Residual Stresses in Linepipe Specimens', MTL Report 94-17(TR), CANMET Materials Technology Laboratory, 1994.
36. Wilmott, M.J. and D.A. Diakow, 'Factors Influencing Stress Corrosion Cracking of Gas Transmission Pipelines: Detailed Studies Following a Pipeline Failure'. *International Pipeline Conference*, ASME, Calgary, Canada, 1996, p. 573.
37. Chen, W., M.L. Wilmott and T.R. Jack, 'Hydrogen Permeation Behaviour of X-70 Pipeline Steel in NS4 Neutral pH Environment', *International Pipeline Conference*, ASME, New York, Vol. 2, 2000, pp. 953–960.
38. Wang, Y-Z., R.W. Revie and R.N. Parkins, *Corrosion 99*, Paper No. 143, NACE International, Houston, TX, 1999.
39. Zheng, W., W.R. Tyson, R.W. Revie, G. Shen and J.E.M. Brade, *International Pipeline Conference*, Vol. 1, 1998, pp. 459–472.
40. Anderson, T., 'Advanced Assessment of Pipeline Integrity Using ILI Data', *J. of Pipeline Eng.*, 9(3), 29–38, 2010.
41. *Stress Corrosion Cracking – Recommended Practices*, 2nd edn. An Industry leading document detailing the management of transgranular SCC – CEPA, December 2007.
42. Berry, W.E., 'Stress Corrosion Cracking Laboratory Experiments', in *Proceedings from the Fifth Symposium on Line Pipe Research*. PRCI, L30174, 1974.

43. Beavers, J.A., *Assessment of the Effects of Surface Preparation and Coating on the Susceptibility of Line Pipe to Stress Corrosion Cracking*. PRCI, L51666, 1992.

44. Barlo, T.J. and R.R. Fessler, 'Shot Peening, Grit Blasting Make Pipe Steels More Resistant to Stress-Corrosion Cracking', *The Oil and Gas Journal*, 79, 68, 1981.

45. Beavers, J.A., N.G. Thompson and K.E.W. Coulson, 'Effects of Surface Preparation and Coatings on SCC Susceptibility of Line Pipe: Phase 1 – Laboratory Studies', *CORROSION/1993*, Paper No. 93597, NACE International, Houston, TX, 1993.

46. NACE No. 1/SSPC-SP 5 (Reaffirmed 1999). *White Metal Blast Cleaning*. NACE International.

47. NACE No. 2/SSPC-SP 10 (Reaffirmed 1999). *Near-White Metal Blast Cleaning*. NACE International.

48. NACE Standard RP 0204. 2004. *Stress Corrosion Cracking (SCC) Direct Assessment Methodology*. NACE International.

49. CSA Z662-03 and Z662.1-03 – *Oil and Gas Pipeline Systems and Commentary on CSA Standard Z662-03*, Oil and Gas Pipeline Systems.

50. Chen, W., S.-H. Wang, R. Chu, F. King, T.R. Jack and R.R. Fessler, 'Effect of Pre-Cyclic Loading on SCC Initiation in an X-65 Pipeline Steel Exposed To Near-Neutral pH Soil Environment', *Metall. and Mater. Trans.*, 34A, 2601–2608, 2003.

51. Chu, R., W. Chen, S.-H. Wang, F. King, T.R. Jack and R.R. Fessler, 'Microstructure Dependence of SCC Initiation in X-65 Pipeline Steel Exposed to a Near-Neutral pH Soil Environment', *Corrosion*, 60(3), 275–283, 2004.

52. Kushida, T., K. Nose, H. Asahi, M. Kimura, Y. Yamane, S. Endo and H. Kawano, 'Effects of Metallurgical Factors and Test Conditions on Near-Neutral pH SCC of Pipeline Steels', *Corrosion 2001*, Paper No. 01213, NACE International, Houston, TX, 2001.

53. Bulger, J.T., B.T. Lu and J.L. Luo, 'Microstructural Effect on Near-Neutral pH Stress Corrosion Cracking Resistance of Pipeline Steels', *Journal of Materials Science*, 41(15), 5001–5005, 2006.

54. Beavers, J.A. and W.V. Harper, 'Stress Corrosion Cracking Prediction Model', *Corrosion 2004*, Paper 04189, NACE International, Houston, TX, 2004.

55. Van Boven, G.J., 'The Influence of Residual Stress on the Initiation and Propagation of Transgranular Stress Corrosion Cracking in High Pressure Natural Gas Transmission Pipelines', MSc Thesis, University of Alberta, 2003, 105.

Published by Woodhead Publishing Limited, 2011

Index